D1053353

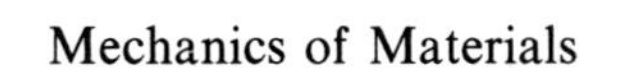

Mechanics of Materials

THIRD EDITION

ARCHIE HIGDON

Engineering and Technology Education Consultant

EDWARD H. OHLSEN

Department of Engineering Science and Mechanics
Iowa State University

WILLIAM B. STILES

Department of Aerospace Engineering, Mechanical
Engineering and Engineering Mechanics
The University of Alabama

JOHN A. WEESE

School of Engineering
Old Dominion University

WILLIAM F. RILEY

Department of Engineering Science and Mechanics
Iowa State University

Mechanics of Materials

JOHN WILEY & SONS, INC. New York · London · Sydney · Toronto

Library of Congress Cataloging in Publication Data:

Higdon, Archie, 1905–
  Mechanics of materials.

  Includes index.
  1. Strength of materials. I. Title.
TA405.H5 1976 620.1'12 75–28453
ISBN 0–471–38812–2

Printed in the United States of America

10 9 8 7 6 5 4 3 2 1

Book Design by Angie Lee

Preface

The primary objectives of a course in mechanics of materials are: (1) to develop a working knowledge of the relations between the loads applied to a nonrigid body made of a given material and the resulting deformations of the body, (2) to develop a thorough understanding of the relations between the loads applied to a nonrigid body and the stresses produced in the body, (3) to develop a clear insight into the relations between stress and strain for a wide variety of conditions and materials, and (4) to develop adequate procedures for finding the required dimensions of a member of a specified material to carry a given load subject to stated specifications of stress and deflection. These objectives involve the concepts and skills that form the foundation of all structural and machine design.

The principles and methods used to meet the general objectives are drawn largely from prerequisite courses in mechanics, physics, and mathematics together with the basic concepts of the theory of elasticity and the properties of engineering materials. This book is designed to emphasize the required fundamental principles, with numerous applications to demonstrate and develop logical orderly methods of procedure. Instead of deriving numerous formulas for all types of problems, we have stressed the use of free-body diagrams and the equations of equilibrium, together with the geometry of the deformed body and the observed relations between stress and strain, for the analysis of the force system acting on a body.

This book is designed for a first course in mechanics of deformable bodies. Because of the extensive subdivision into different topics, the book will provide flexibility in the choice of assignments to cover courses of different length and content. The developments of structural applications include the inelastic as well as the elastic range of stress; however, the material is so organized that the book will be found satisfactory for elastic coverage only.

Extensive use has been made of prerequisite course material in statics and calculus. A working knowledge of these subjects is considered an essential prerequisite to a successful study of mechanics of materials as presented in this book.

New topics for this edition include thick-walled cylindrical pressure vessels, elastic flexural stresses in beams with unsymmetrical cross sections and in beams curved in the plane of loading, introduction of the metric (SI) system of units, and complete coverage of second moments of areas.

The illustrative examples and problems have been selected with special attention devoted to problems that require an understanding of the principles of mechanics of materials without demanding excessive time for computational work. A large number of problems are included so that problem assignments may be varied from term to term. Approximately half of the problems are given in SI units.

The answers to about half the problems are included at the end of this book. We feel that the first assignment on a given topic should include some problems for which the answers are given; and, since the simpler problems are usually reserved for this first assignment, the answers are provided for the first four or five problems of each article and thereafter are given in general for alternate problems. We feel that answers for certain problems should not be given; hence we have not followed the flat rule of answers for all even-numbered problems. Since the convenient designation of problems for which answers are provided is of great value to those who make up assignment sheets, the problems for which answers are provided are indicated by means of an asterisk after the number.

In general, all given data are assumed to be composed of three significant figures regardless of the number of figures shown. Answers are, therefore, given to three significant figures, unless the number lies between 1 and 2 or any decimal multiple thereof, in which case four significant figures are reported. In Chapter 10, the problems involve material properties, which are quite variable, and possibly stress concentration factors, which are difficult to obtain to more than two significant figures from the graphs; therefore we hesitate to give results to more than two significant figures. However, in order to avoid confusion, the answers are usually given to three significant figures.

We are grateful for comments and suggestions received from colleagues and from users of the earlier editions of this book. We appreciate also the careful appraisal made by the publisher's consultant and have incorporated those recommended changes that we considered important.

Archie Higdon
Edward H. Ohlsen
William B. Stiles
John A. Weese
William F. Riley

Contents

Symbols and Abbreviations*

A	area
avg.	average
b	breadth, width
C	compression, Celsius
c	distance from neutral axis or from center of twist to extreme fiber
E	modulus of elasticity in tension or compression
e	eccentricity
Eq.	equation or equations
F	force, or body force, Fahrenheit
f	normal force per unit of length
ft	foot or feet
ft-lb	foot-pound
G	modulus of rigidity (modulus of elasticity in shear)
g	acceleration of gravity
h	height, depth of beam
hp	horsepower
I	moment of inertia or second moment of area
in.	inch or inches
in-lb	inch-pound
J	polar moment of inertia of area, joule (N-m)
K	stress concentration factor
k	modulus of spring, factor of safety
kg	kilogram
kip or k	kilopound (1000 lb)
ksi	kilopounds (or kips) per square inch
L	length
lb	pound
M	bending moment
m	metre
max	maximum

*The symbols and abbreviations used in this book conform essentially with those approved by the International Standards Organization.

min	minimum
N	normal force, newton
n	ratio of elastic moduli, number of cycles
P	concentrated load, force
p	fluid pressure
Pa	pascal (N/m^2)
psf	pounds per square foot
psi	pounds per square inch
Q	first moment of area
q	shearing force per unit of length
R	radius, modulus of rupture, resultant force
r	radius of gyration
rad	radian
rpm	revolutions per minute
rps	revolutions per second
S	surface force, resultant stress, stress vector
sec	second
sq	square
T	torque, temperature, tension
t	thickness, tangential deviation
U	strain energy
u	strain energy per unit volume
u, v, w	components of displacement
V	shearing force
W	weight, watt (N-m/s)
W_k	work
w	load per unit of length
y	deflection
Z	section modulus ($Z = I/c$)
α (alpha)	coefficient of thermal expansion
γ (gamma)	shearing strain (unit), specific weight
δ (delta)	total deformation
Δ (delta)	maximum deflection under energy load
ϵ (epsilon)	normal strain (unit)
η (eta)	efficiency, portion of energy load which is effective
θ (theta)	total angle of twist in radians, slope of deflected beam
μ (mu)	micro-inches per inch (10^{-6} in. per in.)
ν (nu)	Poisson's ratio
ρ (rho)	radius
σ (sigma)	normal stress (unit)
τ (tau)	shearing stress (unit)

ω (omega)	angular velocity
W16×88	a 16-in. wide flange rolled section with a weight of 88 lb per foot.
S10×35	a 10-in. American standard rolled section (I-beam) with a weight of 35 lb per foot.
C12×30	a 12-in. rolled channel section with a weight of 30 lb per foot.
L6×4×$\frac{1}{2}$	a rolled angle section with 6-in. and 4-in. legs $\frac{1}{2}$-in. thick
UB610×113	a 610-mm rolled universal beam with a mass of 113 kg/m
°	degree

1 Analysis of Stress–Concepts and Definitions

1–1 INTRODUCTION

The primary objective of a course in mechanics of materials is the development of the relations between the loads applied to a nonrigid body and the resulting *internal forces* and *deformations* induced in the body. Ever since the time of Galileo Galilei (1564–1642), men of scientific bent have studied assiduously the problem of the load-carrying capacity of structural and machine elements and the mathematical analysis of the internal forces and deformations induced by applied loads. The experience and observations of these scientists and engineers of the last three centuries are the heritage of the engineer of today, providing him with fundamental knowledge from which he may progress to the development of theories and techniques that permit him to design, with competence and assurance, structures and machines of unprecedented size and complexity.

The subject matter of this book forms the basis for the solution of three general types of problems, as follows:

1. Given a certain function to perform (the transporting of traffic over a river by means of a bridge, conveying scientific instruments to Mars in a space vehicle, the conversion of water power into electric power), of what materials should the machine or structure be constructed, and what should be the sizes and proportions of the various elements? This is the designer's task, and obviously there is no single solution to any given problem.

2. Given the completed design, is it adequate? That is, does it perform the function economically and without excessive deformation? This is the checker's problem.

3. Given a completed structure or machine, what is its actual load-carrying capacity? The structure may have been designed for some purpose other than the one for which it is now to be used. Is it adequate for the proposed use? For example, a building may have been designed for an office building, but is later found to be desirable for use as a warehouse. In such a case, what maximum loading may the floor safely support? This is the rating problem.

Since the complete scope of these problems is obviously too comprehensive for mastery in a single course, this book is restricted to a study of individual members and very simple structures or machines. The design courses that follow will consider the entire structure or machine, and will provide essential background for the complete analysis of the three problems listed.

The principles and methods used to meet the objective stated at the beginning of this chapter depend to a great extent on prerequisite courses in mathematics and mechanics, supplemented by additional concepts from the theory of elasticity and the properties of engineering materials. The equations of equilibrium from *statics* are used extensively, with one major change in the free-body diagrams; namely, most free bodies are isolated by *cutting through* a member instead of removing a pin or some other connection. The forces transmitted by the cut sections are *internal* forces. The intensities of these internal forces (force per unit area) are called *stresses*.

It will frequently be found that the equations of equilibrium (or motion) are not sufficient to determine all the unknown loads or reactions acting on a body. In such cases it is necessary to consider the geometry (the change in size or shape) of the body after the loads are applied. The *deformation* per unit length in any direction or dimension is called *strain*. In some instances, the specified maximum deformation and not the specified maximum stress will govern the maximum load that a member may carry.

Some knowledge of the physical and mechanical properties of materials is required in order to create a design, to evaluate a given design properly, or even to write the correct relation between an applied load and the resulting deformation of a loaded member. Essential information will be introduced as required, and more complete information can be obtained from textbooks and handbooks on properties of materials.

1–2 LOAD CLASSIFICATION

Certain terms are commonly used to describe applied loads; their definitions are given here so that the terminology will be clearly understood.

Loads may be classified with respect to time.

---

1. A *static load* is a gradually applied load for which equilibrium is reached in a relatively short time.

2. A *sustained load* is a load that is constant over a long period of time, such as the weight of a structure (called *dead load*). This type of load is treated in the same manner as a static load;. however, for some materials and conditions of temperature and stress, the resistance to failure may be different under short-time loading and under sustained loading.

3. An *impact load* is a rapidly applied load (an energy load). Vibration normally results from an impact load, and equilibrium is not established until the vibration is eliminated, usually by natural damping forces.

4. A *repeated load* is a load that is applied and removed many thousands of times. The helical springs that close the valves on automobile engines are subjected to repeated loading.

Loads may also be classified with respect to the area over which the load is applied.

---

1. A *concentrated load* is a load or force applied at a point. Any load applied to a relatively small area compared with the size of the loaded member is assumed to be a concentrated load; for example, a truck wheel load on the longitudinal members of a bridge.

2. A *distributed load* is a load distributed along a length or over an area. The distribution may be uniform or nonuniform. The weight of a concrete bridge floor of uniform thickness is an example of a uniformly distributed load.

Loads may be classified with respect to the location and method of application.

---

1. A *centric load* is one in which the resultant force passes through the centroid of the resisting section. If the resultant passes through the centroids of all resisting sections, the loading is termed *axial*. In Fig. 1–1*a*, the line of action of the load P passes through the centroid of section a–a; therefore, the load is centric for section a–a. The force P in Fig. 1–1*b* is applied through the plate to the shank of the rivet and is resisted by the force V on the shaded section b–b. If the thickness t is not too large (the usual case for riveted joints), the force P can be assumed to be collinear with V, thus satisfying the definition of a centric load. The loading illustrated in this figure is called *single shear*, and the force V is a shearing force. If the plate A were sandwiched between two plates B with the rivet extending through all three plates, the rivet would be subjected to two shearing forces. Such loading is called *double shear*. Figure 1–1*c* represents an eyebar used for trusses, tie rods, and linkages. This is an example of axial loading.

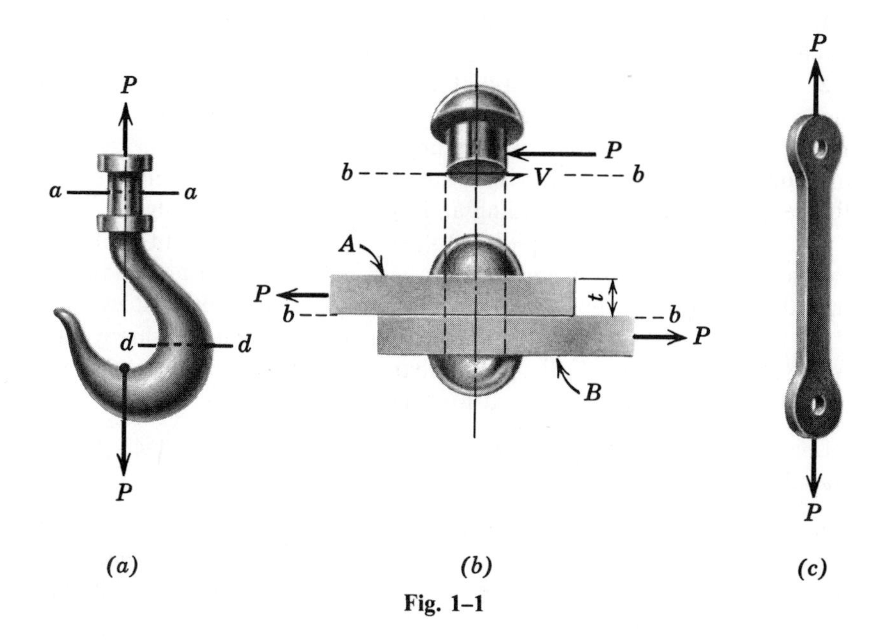

Fig. 1–1

2. A *torsional load* is one that subjects a shaft or some other member to couples that twist the member. If the couples lie in planes transverse to the axis of the member, as in Fig. 1–2, the member is subjected to pure torsion.

Fig. 1–2

3. A *bending* or *flexural load* is one in which the loads are applied transversely to the longitudinal axis of the member. The applied load may include couples that lie in planes parallel to the axis of the member. A member subjected to bending loads bends or bows along its length. Figure 1–3 illustrates a beam subjected to flexural loading consisting of a concentrated load, a uniformly distributed load, and a couple.

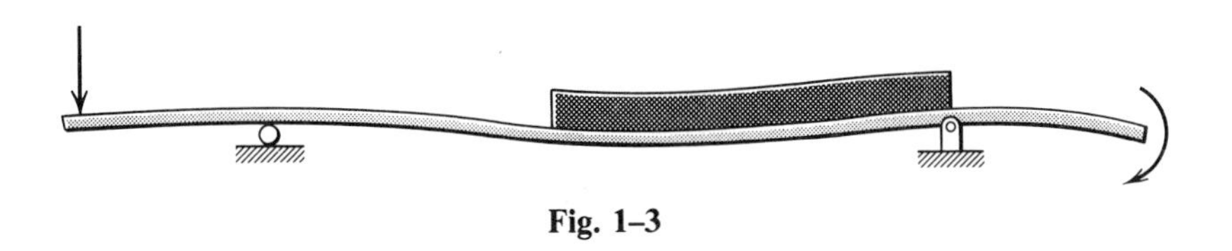

Fig. 1–3

4. A *combined loading* is a combination of two or more of the previously defined types of loading.

1–3 CONCEPT OF STRESS

In every subject area there are certain fundamental concepts of paramount importance to a satisfactory comprehension of the subject matter. For the subject mechanics of materials, a thorough mastery of the physical significance of stress and strain is paramount. The discussion of stress will be undertaken first, and the study of strain will be taken up in later articles.

The eyebar AB from the pile extractor of Fig. 1–4 (used to reclaim steel sheet piles from cofferdams and similar installations) will serve as an example for the introduction of the concept of stress. Although the bar is subjected to dynamic loading in use, an equivalent static load (a static load that will produce the same stresses and strains as the dynamic load) may be used in the design. Since the eyebar is a two-force member, the pin reactions at A and B are collinear with the axis of the bar. A review of two-force members and their application in trusses, as presented in a course in statics, is strongly recommended. When the bar is cut by any plane, such as plane c-c, and the free-body diagram of Fig. 1–5 is drawn, equilibrium may be established by placing on the cut section a series of parallel forces (stresses), each of which acts on a unit area. The term *stress* is sometimes used to denote the resultant force on a section; in other instances it is used to indicate the force per unit area. *In this book stress will always refer to the force per unit area, unless otherwise stated.* The equilibrium equation $\Sigma F_y = 0$ gives the expression

$$\text{applied force} = (\text{stress})(\text{area})$$

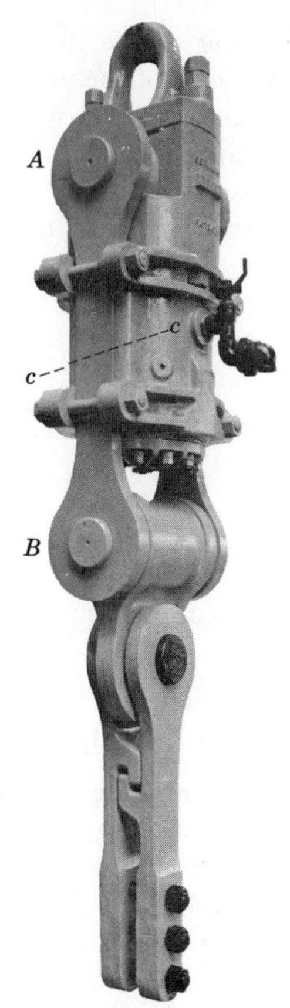

Fig. 1–4 Courtesy: Vulcan Iron Works Inc., Chicago, Ill.

It is immediately apparent that the magnitude of the stress on plane *c–c* is a function of the angle θ, and this fact distinguishes a stress from an ordinary force vector. The aspect of the plane on which the stress acts is an additional characteristic, which means that a stress is a tensor of the second order,[1] not a vector, and that the laws of vector addition do not apply to stresses that act on different planes. This need not be cause for concern if, in the application of the equations of equilibrium (or motion), one always replaces a stress with a total force (stress multiplied by the appropriate area), thus reducing the problem to one involving ordinary force vectors. However, stresses that act on a particular plane can be

[1]A scalar is a tensor of zero order that can be defined by one component. A vector is a tensor of the first order and, in general, has three components. A stress involves a force and the area on which it acts, and is a tensor of the second order requiring six components to specify it completely. See *Tensor Analysis*, I. S. Sokolnikoff, Second Edition, Wiley, New York, 1964.

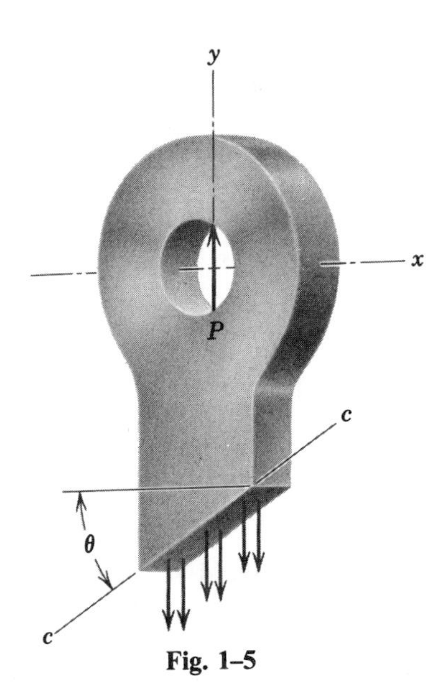

Fig. 1–5

treated as vectors (see Art. 1–5), as acknowledged in the following paragraph.

Because the stress indicated on plane c–c of Fig. 1–5, if evaluated, would convey little information regarding the adequacy of design, it is customary to replace this stress with two components: one perpendicular to the plane, called the *normal stress* (which may be tension T or compression C) and designated by the Greek letter σ; and the other parallel to the plane, called the *shearing stress* and designated by the letter τ. These stresses, when evaluated, may be compared with experimentally determined values for the strength of the material in tension, compression, and shear, thus yielding significant information. Such stresses are indicated in Fig. 1–6*a*. For this simple case of axial loading, the sense of the normal and shearing stresses has been determined by inspection. Again, the aspect of the plane as an additional characteristic possessed by a stress must be emphasized. In Fig. 1–6*b* the stresses are replaced by their resultants, the force vectors N and V; therefore, the equations of equilibrium may be applied to this free-body diagram. In the preceding discussion, the assumption is made that the stress is uniformly distributed over the cut section. A necessary (but not sufficient) condition for the distribution to be uniform is that the loading be *centric*. Non uniform stress distribution under centric loading will be discussed in Art. 1–10. In this book, unless otherwise indicated, a uniform stress distribution is to be assumed for all centric (or assumed centric) loading.

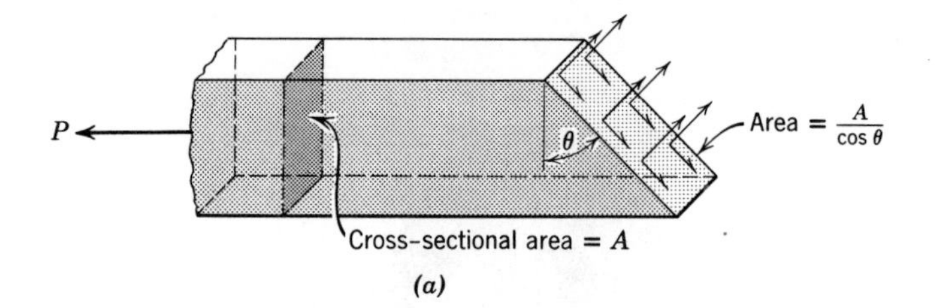

(a)

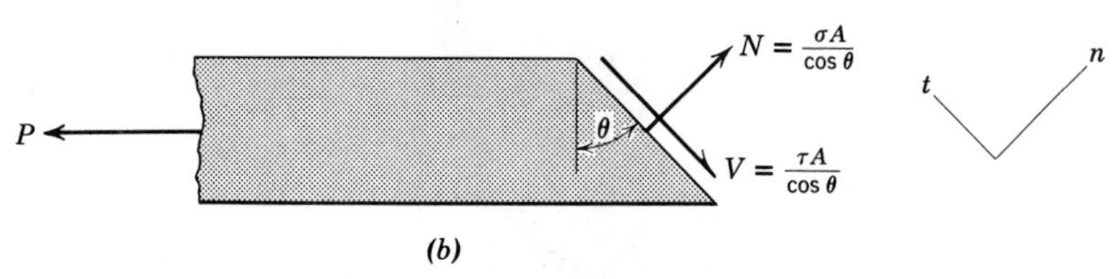

(b)

Fig. 1–6

1–4 STRESSES UNDER CENTRIC LOADING

If the equations of equilibrium are applied to the free-body diagram of Fig. 1–6*b*, the following expressions are obtained:

For $\Sigma F_n = 0$ $$P\cos\theta = N = \frac{\sigma A}{\cos\theta}$$

from which $$\sigma = \frac{P}{A}\cos^2\theta = \frac{P}{2A}(1+\cos 2\theta) \tag{a}$$

and for $\Sigma F_t = 0$ $$P\sin\theta = V = \frac{\tau A}{\cos\theta}$$

from which $$\tau = \frac{P}{A}\sin\theta\cos\theta = \frac{P}{2A}(\sin 2\theta) \tag{b}$$

The graph of these expressions in Fig. 1–7 indicates that σ is maximum when θ is 0° or 180°, that τ is maximum when θ is 45° or 135°, and also that $\tau_{max} = \frac{1}{2}\sigma_{max}$. Therefore the maximum stresses are given by the expressions

$$\sigma_{max} = P/A \tag{1–1}$$

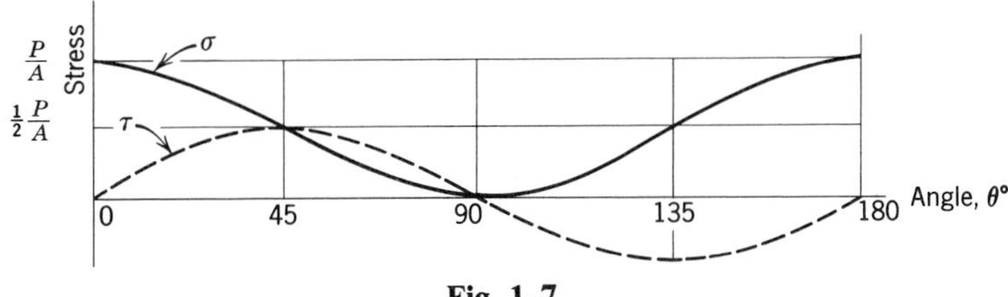

Fig. 1–7

and

$$\tau_{max} = P/(2A) \tag{1–2}$$

It must be emphasized that Eq. 1–2 is valid only for centric tensile or compressive loading in one direction. For convenience, the maximum normal stress under axial loading is called the *axial stress*. Note that the normal stress is either maximum or minimum on planes for which the shearing stress is zero. In Art. 1–7 it is shown that the shearing stress is always zero on the planes of maximum or minimum normal stress.

Laboratory experiments indicate that both shearing and normal stresses under axial loading are important since a brittle material loaded in tension will fail in tension on the transverse plane, whereas a ductile material loaded in tension will fail in shear on the 45° plane.

One observes that when θ of Fig. 1–7 is greater than 90°, the sign of the shearing stress in Eq. (b) changes. The magnitude of the shearing stress for any angle θ, however, is the same as that for $90° + \theta$. The sign change merely indicates that the shear force vector changes sense, being directed toward the top of the element instead of toward the bottom as in Fig. 1–6. Normal and shearing stresses on planes having aspects θ_1 and $90° + \theta_1$ are shown in Fig. 1–8.

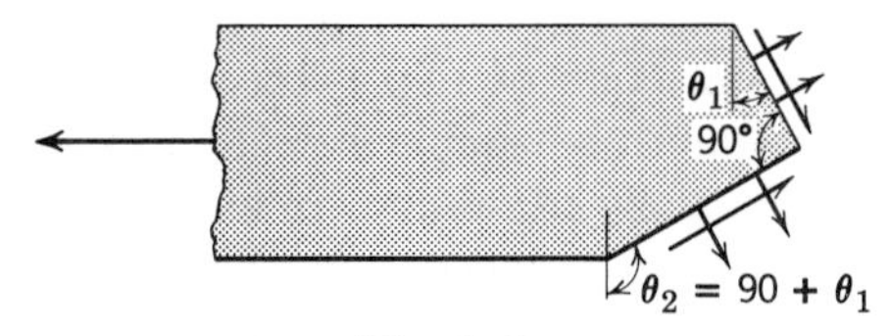

Fig. 1–8

The equality of shearing stresses on orthogonal planes can be demonstrated by applying the equations of equilibrium to the free-body diagram of a small rectangular block of thickness dz, shown in Fig. 1-9. If

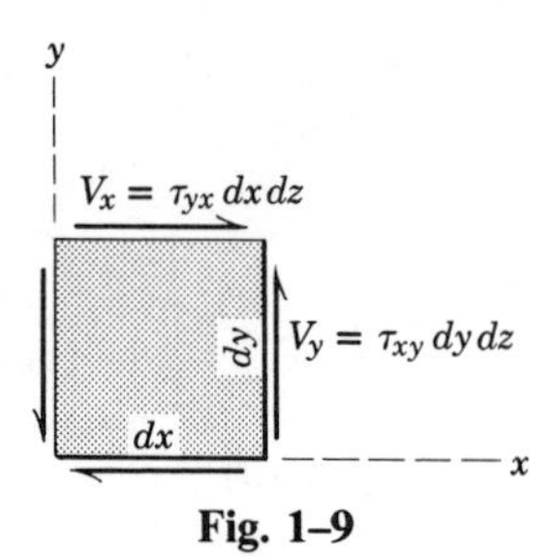

Fig. 1–9

a shearing force[2] $V_x = \tau_{yx}\,dx\,dz$ is applied to the top surface of the block, the equation $\Sigma F_x = 0$ will dictate the application of an oppositely directed force V_x to the bottom of the block, thus leaving the block subjected to a clockwise couple; this couple must be balanced by a counterclockwise couple composed of the oppositely directed forces V_y applied to the vertical faces of the block. Finally, application of the equation $\Sigma M_z = 0$ yields the following:

$$\tau_{yx}(dx\,dz)dy = \tau_{xy}(dy\,dz)dx$$

from which

$$\tau_{yx} = \tau_{xy} \tag{1–3}$$

Therefore, *if a shearing stress exists at a point on any plane, there must also exist at this point a shearing stress of the same magnitude on an orthogonal plane*. This statement is also valid when normal stresses are acting on the planes, since the normal stresses occur in collinear but oppositely directed pairs, and thus have zero moment with respect to any axis.

It will be observed that loads on a structure or machine are transmitted to the individual members through some sort of connection, as, for example, the riveted connection of Fig. 1–1*b*, the pins at the ends of the eyebar of Fig. 1–4, and even so simple a detail as the head (or nut) of a bolt (Fig. 1–10) loaded axially in tension. In all these connections, one of the most significant stresses induced is a shearing stress; hence, these members are often said to be subjected to centric shear loading. If the loading is as indicated in Fig. 1–1*b*, the average shearing stress on plane *b–b* is the significant shearing stress (although it is not the maximum stress

[2]The double subscript on the shearing stress is used to designate both the plane on which the stress acts and the direction of the stress. The first subscript indicates the plane (or rather the normal to the plane), and the second subscript indicates the direction of the stress.

since the shearing stress cannot be uniformly distributed over the area) and is referred to as the *cross shearing stress*. Another type of shear loading is termed *punching shear*, as, for example, the action of a punch in forming rivet holes in a metal plate, the tendency of building columns to punch through footings, and the tendency of a tensile axial load on a bolt to pull the shank through the head. Under a punching shear load, the significant stress is the average shearing stress on the surface described by the periphery of the punching member and the thickness of the punched member;[3] for example, the shaded cylindrical area of Fig. 1–10.

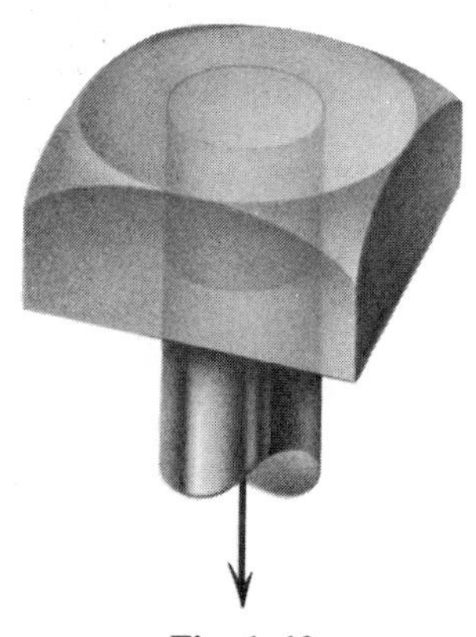

Fig. 1–10

The following example will serve to illustrate the concepts discussed in Art. 1–3 and 1–4. *Note. In all problems in this textbook the weights of the members will be neglected, unless otherwise specified.*

Example 1–1 A hoist is made from a 4×4-in. wood post BC and a 1-in. diameter steel eyebar AB as shown in Fig. 1–11a. The hoist supports a load W of 7000 lb. Determine the axial stresses in members AB and BC and the cross shearing stress in the 1-in. diameter bolt at A which is in double shear.[4]

Solution The members AB and BC are two-force members subjected to axial loading. When the equations of equilibrium are applied to the

[3]This is modified in the case of building columns on concrete footings. Reference should be made to reinforced-concrete design specifications.

[4]Structural timbers specified as rough sawn will be approximately full size. Cross-sectional dimensions of dressed timbers will be between $\frac{3}{8}$ and $\frac{1}{2}$ in. smaller than nominal. For convenience, all timbers in this book will be considered as full size.

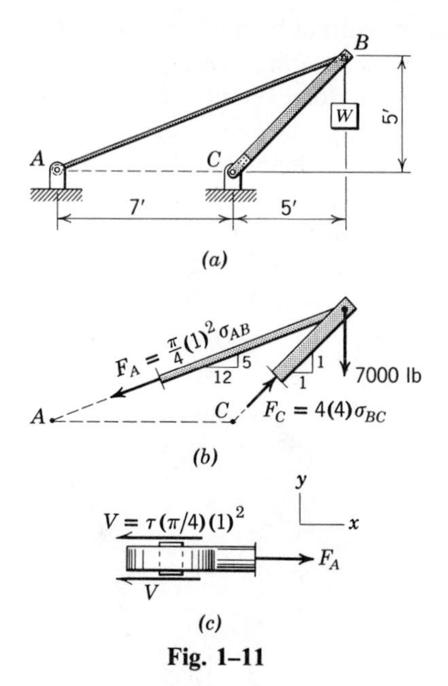

Fig. 1–11

free-body diagram of Fig. 1–11*b* the following results are obtained:

$$\circlearrowleft\!+ \quad \Sigma M_C = 0$$

$$7000(5) - (5/13)(F_A)(7) = 0, \qquad F_A = 13{,}000 \text{ lb } T$$

$$\sigma_{AB} = \frac{F_A}{A} = \frac{13{,}000}{(\pi/4)(1)^2} = \underline{16{,}550 \text{ psi } T} \qquad \text{Ans.}$$

Observe that it is not necessary to solve the equation explicitly for the force F_A in the member if only the stress is required. Hence, for the next equation this step will be omitted; thus,

$$\circlearrowleft\!+ \quad \Sigma M_A = 0$$

$$7000(12) - \frac{1}{\sqrt{2}}(16)(\sigma_{BC})(7) = 0$$

$$\sigma_{BC} = \underline{1061 \text{ psi } C} \qquad \text{Ans.}$$

From the free-body diagram of Fig. 1–11*c*, which is a free-body diagram of the connection at A, and using the value of force F_A as determined above, the cross shearing stress is obtained as follows:

$$\overset{+}{\rightarrow} \quad \Sigma F_x = 0$$

$$F_A - 2V = 0$$

$$V = \tau\pi(1)^2/4 = 6500$$

$$\tau = \underline{8280 \text{ psi}} \qquad \text{Ans.}$$

Stress, being the intensity of internal force, has the dimensions of force per unit area (FL^{-2}). Until recently the commonly used unit for stress in the United States was the pound per square inch, abbreviated as psi. Since metals can sustain stresses of several thousand pounds per square inch, the unit ksi (kips per square inch) is also frequently used (1 ksi = 1000 psi). With the advent of the International System of Units (usually called SI units), units of stress based on the metric system are beginning to be used in the United States and will undoubtedly come into wider use in the next few years.

During the transition, both systems will be encountered, so some problems in this book are given using the English system (pounds and inches) and others with SI units (newtons and metres). For problems with SI units, forces will be given in newtons (N) or kilonewtons (kN), dimensions in metres (m) or millimetres (mm), and masses in kilograms (kg). The SI unit for stress or pressure is a newton per square metre (N/m^2), also known as a pascal (Pa). Stress magnitudes normally encountered in engineering applications are expressed in meganewtons per square metre (MN/m^2), or megapascals (MPa).

Data will usually be given in either the English or the SI system; a combination of the two systems will not be used for a given problem. When it is necessary to convert units from one system to the other, the following conversion factors will be useful:

1 lb = 4.448 N	1 N = 0.2248 lb
1 in. = 25.40 mm	1 mm = 0.03937 in.
1 ft = 0.3048 m	1 m = 3.281 ft
1 ksi = 6.895 MN/m^2	1 MN/m^2 = 0.1450 ksi

Note that one MN/m^2 is approximately one-seventh of one ksi. This one-seventh "rule of thumb" is convenient for interpreting the physical significance of answers in SI units for those more accustomed to English units.

Acceleration due to gravity is 9.81 metres per second per second. From Newton's second law, the force due to gravitational attraction on a mass of 1 kg is 9.81 N. Consequently, when a 1-kg mass is supported by a cord in the earth's gravitational field, the tension in the cord is 9.81 N. Similarly the tension in a cable sustaining a mass of m kg is 9.81 m N. A more detailed discussion of SI units is given in App. E.

Example 1–2 The crane structure shown in Fig. 1–12*a* supports an automobile of mass 1800 kg. If member AB is a steel bar 15 mm in diameter:

(a) Determine the force transmitted by member AB.
(b) Determine the axial stress in member AB.
(c) Convert the answer to part (b) to stress units in the English system.

Solution The tension in the cable at D is (9.81)(1800) or 17,660 N. From the free-body diagram of Fig. 1–12*b*, the force in member AB is obtained by summing moments about pin F. Thus,

$$\text{(a)} \qquad \Sigma M_F = (17{,}660)(8) - 6F_{AB} = 0$$

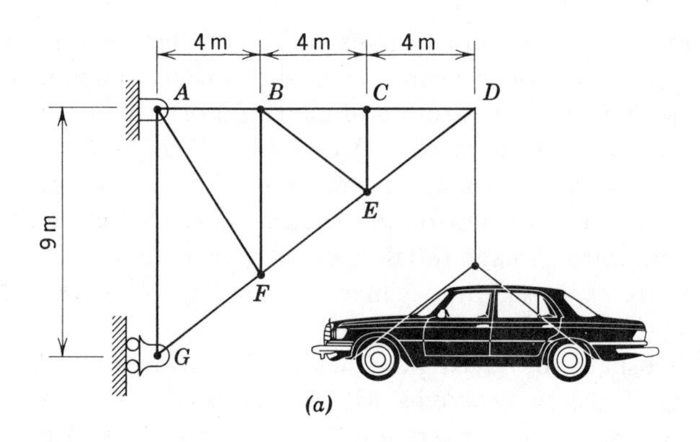

(a)

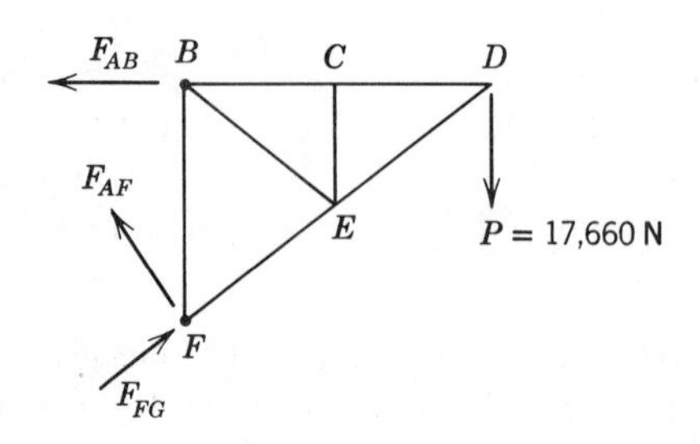

(b)

Fig. 1–12

Solving for F_{AB} yields

$$F_{AB} = (17{,}660)(8/6) = 23{,}547 \text{ N or } \underline{23{,}500 \text{ N}} \qquad \text{Ans.}$$

(b) $\sigma = F_{AB}/A = 23{,}547/(\pi/4)(0.015)^2 = \underline{133.2 \text{ MN/m}^2\ T}$ Ans.

(c) $\sigma = 133.2 \text{ MN/m}^2 = 133.2/6.895 = \underline{19.30 \text{ ksi } T}$ Ans.

PROBLEMS

Note. Unless otherwise specified, all members are assumed to have negligible weight, and all pins used for connections are assumed to be smooth. An asterisk on the problem number indicates that the answer is given in the back of the book.

1–1* Consider the loads of Fig. P1–1 to be axially applied at sections A, B, C, and D to the steel bar, which has a cross-sectional area of 3 sq in. Determine the axial stress in the bar:

(a) On a section 10 in. to the right of A.
(b) On a section 60 in. to the right of A.

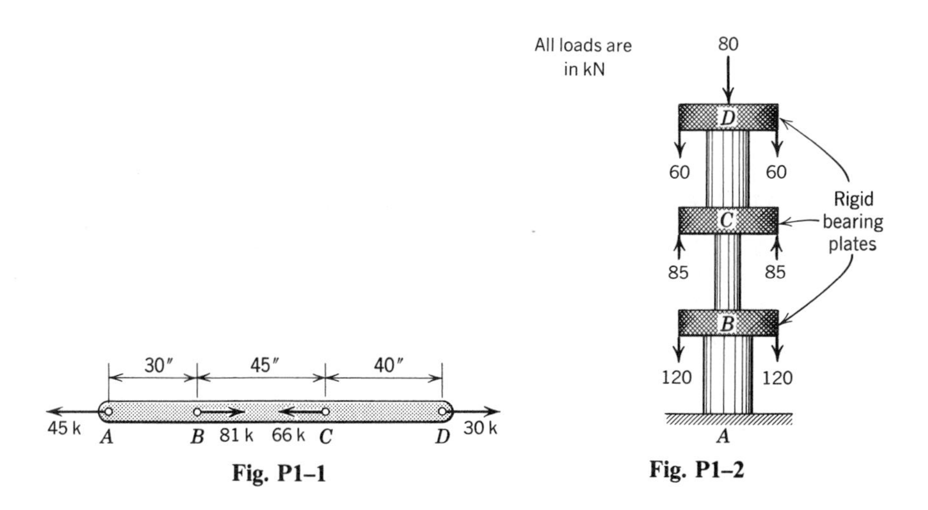

Fig. P1–1

Fig. P1–2

1–2* Consider the loads of Fig. P1–2 to be axially applied at sections B, C, and D to the concrete structure supported at A. The cross-sectional areas are: for CD 8000 mm², for BC 2000 mm², and for AB 12,000 mm². Determine the axial stress in the concrete:

(a) On a section in the CD interval.
(b) On a section in the BC interval.
(c) On a section in the AB interval.

1–3* In Fig. P1–3, the axial stresses are 1500 psi C in the wood post B, and 20,000 psi T in the steel bar A. Determine the load P.

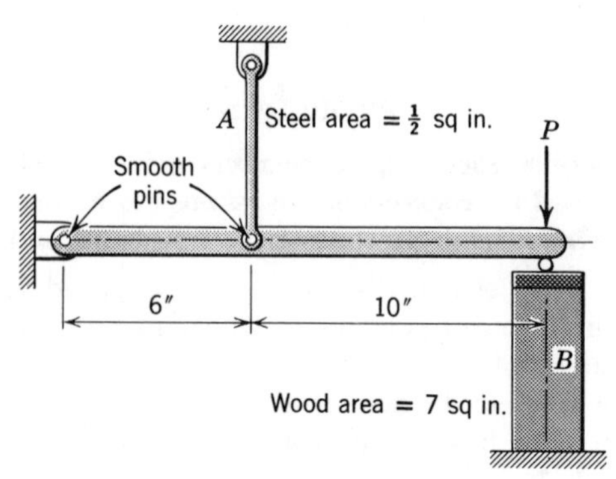

Fig. P1–3

1–4* Member AC of the pin-connected frame ABC of Fig. P1–4 is a 50-mm diameter rod with an axial stress of 120 MN/m^2 T. Determine the force P.

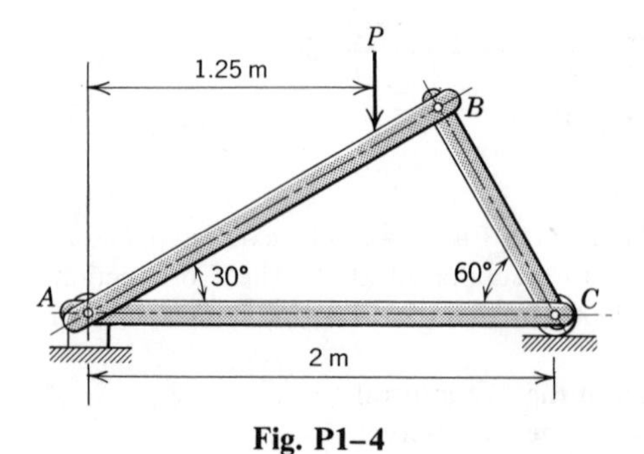

Fig. P1–4

1–5 The timber post D of Fig. P1–5 has a cross-sectional area of 20 sq. in. Due to the loading shown, a pressure cell indicates an average pressure between D and the foundation of 600 psi; the axial stresses in bars A, B, and C are 20, 10, and 10 ksi T, respectively. Determine the cross-sectional areas of bars A, B, and C.

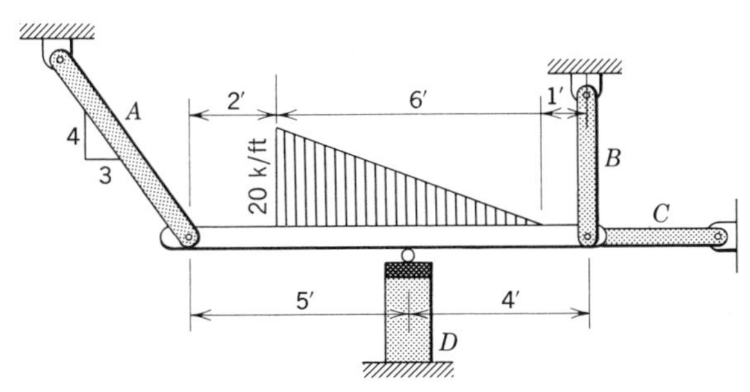

Fig. P1–5

1–6* The concrete cylinder C of Fig. P1–6 has a cross-sectional area of 10,000 mm^2. A pressure-measuring device indicates an average pressure between C and the foundation of 2.5 MN/m^2 excluding the weight of the members. Contact with the horizontal member E is made through a ball and steel cap on C. The axial stresses in rods A and B are 90 and 120 MN/m^2 T, respectively. Determine the cross-sectional areas of A and B.

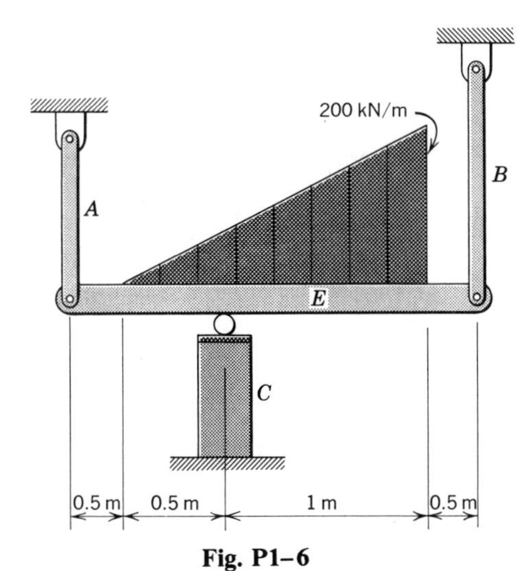

Fig. P1–6

1–7* Two strips of a plastic material are bonded together, as shown in Fig. P1–7. The average shearing stress in the glue must be limited to 50 psi. What length of splice plate is needed if the axial load carried by the joint is 5000 lb?

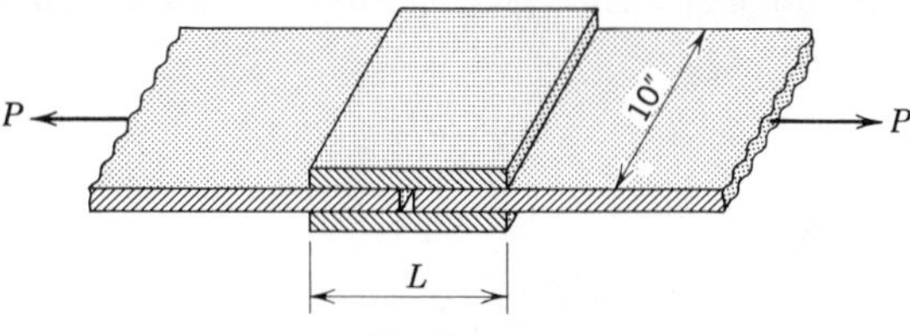

Fig. P1–7

1–8 A punch press is used to punch a 25-mm-diameter hole in a 6.5-mm-thick steel plate. Determine the force exerted by the press on the plate when the average punching shear resistance of the steel plate is 350 MN/m^2.

1–9 The inclined member AB of a timber truss is framed into a 4×6-in. douglas fir bottom chord, as shown in Fig. P1–9. Determine the axial compressive force in member AB when the average shearing stress parallel to the grain in the end of the bottom chord is 120 psi.

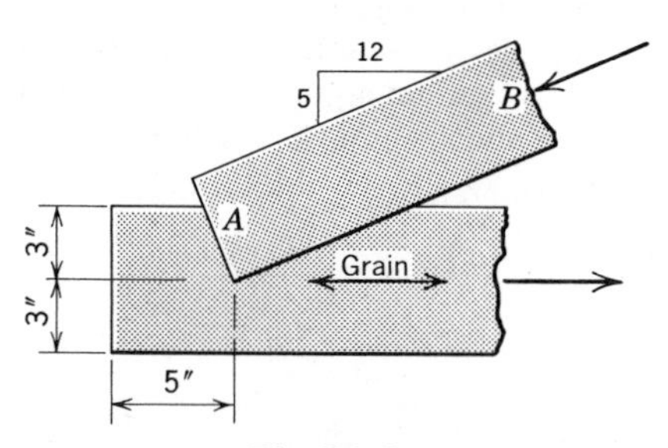

Fig. P1–9

1–10* Member AD of the timber truss of Fig. P1–10 is framed into the 100×150-mm white oak chord ABC as shown in the insert. Determine the dimension a necessary if the average shearing stress parallel to the grain in the end of AB is not to exceed 1.2 MN/m^2.

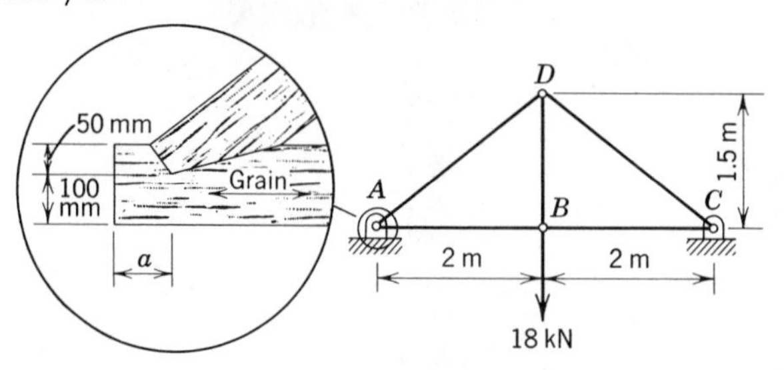

Fig. P1–10

1-11 Member GF of the pin-connected truss of Fig. P1-11 is composed of two $4 \times 3 \times 1/4$-in. steel angles having a total cross-sectional area of 3.38 sq in. Determine the axial stress in GF. The lengths shown are center to center of the pins.

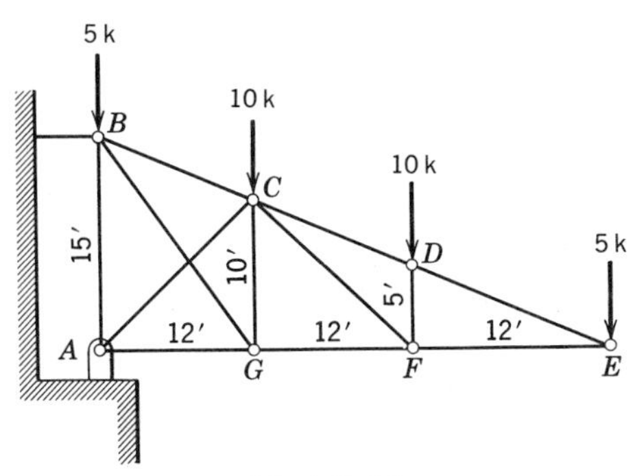

Fig. P1-11

1-12* Member CF of the pin-connected truss of Fig. P1-11 is composed of two $2 \times 1.5 \times 3/16$-in. steel angles having a total cross-sectional area of 1.24 sq in. Determine the axial stress in CF. The lengths shown are center to center of the pins.

1-13 Determine the axial stress in member GF of the pin-connected truss of Fig. P1-13. The cross-sectional area of GF is 3900 mm^2.

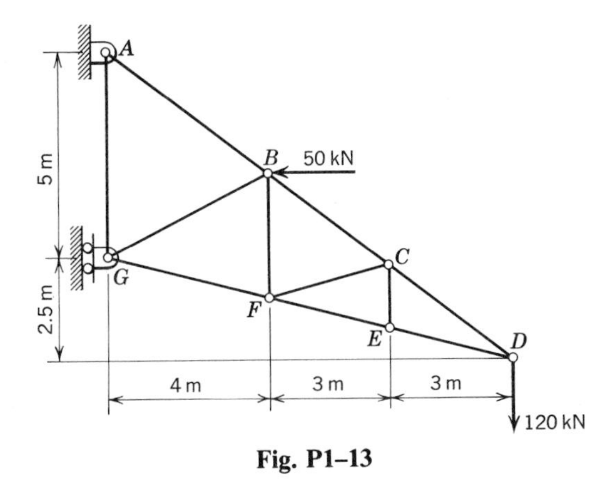

Fig. P1-13

1-14* Determine the axial stress in member BG of the pin-connected truss of Fig. P1-13. The cross-sectional area of BG is 2200 mm^2.

1–15 The tie rod AB of Fig. P1–15 is subjected to an axial stress of 20,000 psi T. Determine:

(a) The cross-sectional area of AB.

(b) The cross-sectional area of pin C if the cross shearing stress in pin C is 15,000 psi.

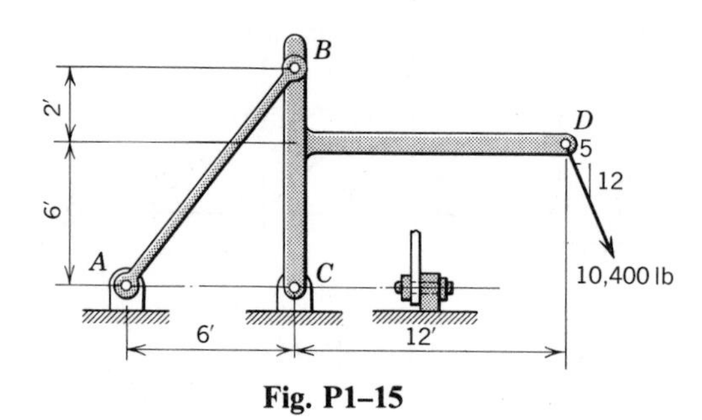

Fig. P1–15

1–16 For the pin-connected frame of Fig. P1–16, determine:

(a) The axial stress in the 20-mm-diameter rod CD.

(b) The cross shearing stress in the 20-mm-diameter pin at B. Pin B is in single shear.

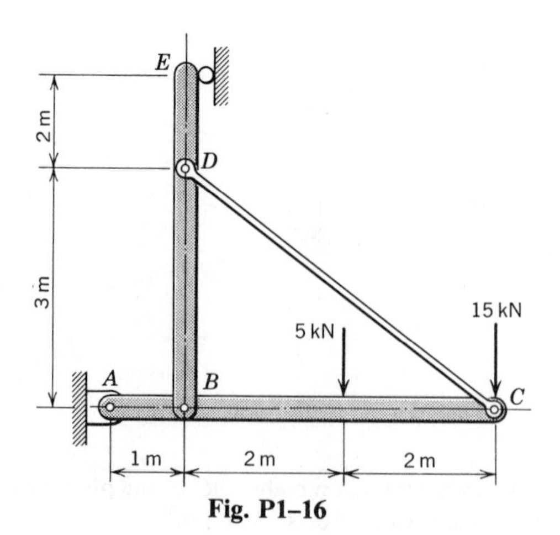

Fig. P1–16

1–17* A lever is attached to the steel gate valve operating shaft of Fig. P1–17 by means of a 1/2×1/2×1-in. key. Determine the average cross shearing stress in the key when a load of 200 lb is applied to the lever.

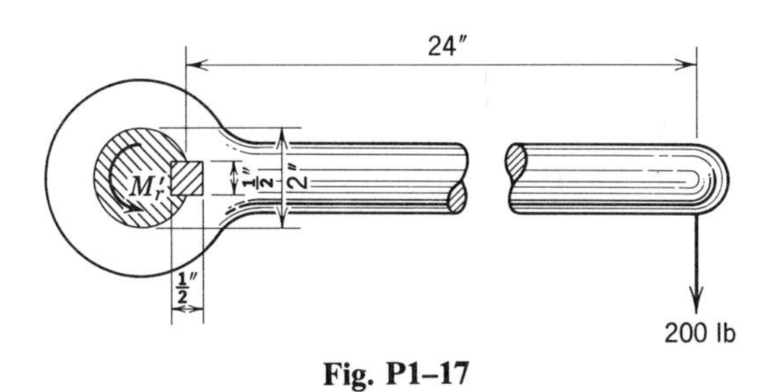

Fig. P1–17

1–18* A vertical shaft is supported by a thrust collar and bearing plate, as shown in Fig. P1–18. Determine the maximum axial load that can be applied to the shaft if the average punching shear stress in the collar and the average bearing stress between the collar and the plate are limited to 40 and 80 MN/m^2, respectively.

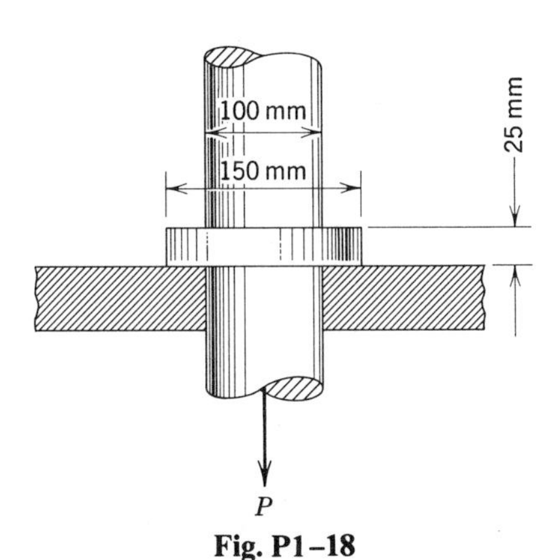

Fig. P1–18

1–19 Member BE of the pin-connected frame of Fig. P1–19 is a $1 \times 3/8$-in. rectangular bar, and the pin at E has a diameter of 1.25 in. The force P is applied to the pin outside members BE and DE. Determine:

(a) The axial stress in member BE.
(b) The maximum cross shearing stress in the pin at E.

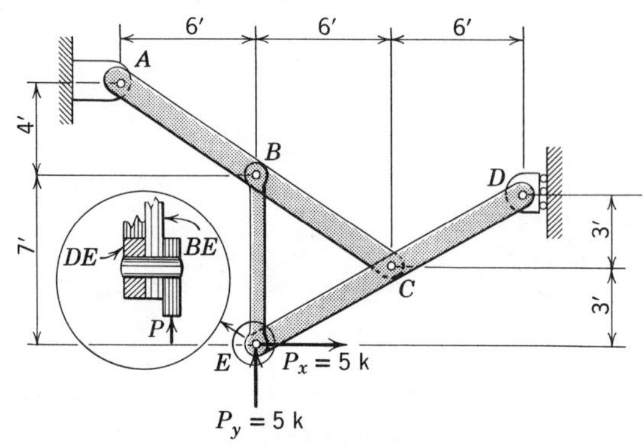

Fig. P1–19

1–20* Member CE of Fig. P1–20 is an aluminum alloy tube pinned at C and E. The cable is attached to the pins and passes over a smooth pulley at D. Determine:

(a) The cross-sectional area of the tube if the axial stress is 25 MN/m^2 C.
(b) The diameter of the pin at C if the maximum cross shearing stress is 70 MN/m^2.

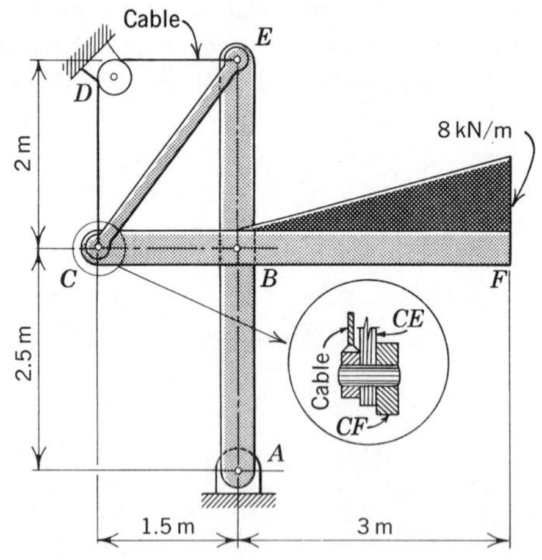

Fig. P1–20

1–5 CONCEPT OF STRESS AT A GENERAL POINT IN AN ARBITRARILY LOADED MEMBER

In Art. 1–3 and 1–4 the concept of stress was introduced by considering the internal force distribution required to satisfy equilibrium in a portion of a bar under centric load. The nature of the force distribution led to uniformly distributed normal and shearing stresses on transverse planes through the bar. In more complicated structural members or machine components the stress distributions will not be uniform on arbitrary internal planes; therefore, a more general concept of the state of stress at a point is needed.

Consider a body of arbitrary shape that is in equilibrium under the action of a system of applied forces. The nature of the internal force distribution at an arbitrary interior point O can be studied by exposing an interior plane through O as shown in Fig. 1–13*a*. The force distribution required on such an interior plane to maintain equilibrium of the isolated part of the body, in general, will not be uniform; however, any distributed force acting on the small area ΔA surrounding the point of interest O can be replaced by a statically equivalent resultant force ΔF_n through O and a couple ΔM_n. The subscript n indicates that the resultant force and couple are associated with a particular plane through O—namely, the one having an outward normal in the n direction at O. For any other plane through O the values of ΔF and ΔM could be different. Note that the line of action of ΔF_n or ΔM_n may not coincide with the direction of n. If the resultant force ΔF_n is divided by the area ΔA, an average force per unit area (average resultant stress) is obtained. As the area ΔA is made smaller and smaller, the couple ΔM_n vanishes as the force distribution becomes more and more uniform. In the limit a quantity known as the *stress vector*[5] or resultant stress is obtained. Thus,

$$S_n = \lim_{\Delta A \to 0} \frac{\Delta F_n}{\Delta A}$$

In Art. 1–3 it was pointed out that materials respond to components of the stress vector rather than the stress vector itself. In particular, the components normal and tangent to the internal plane were important. As shown in Fig. 1–13*b* the resultant force ΔF_n can be resolved into the components ΔF_{nn} and ΔF_{nt}. A normal stress σ_n and a shearing stress τ_n are then defined as

$$\sigma_n = \lim_{\Delta A \to 0} \frac{\Delta F_{nn}}{\Delta A}$$

and

$$\tau_n = \lim_{\Delta A \to 0} \frac{\Delta F_{nt}}{\Delta A}$$

For purposes of analysis it is convenient to reference stresses to some coordinate system. For example, in a Cartesian coordinate system the

[5]For a particular plane, the stresses can be treated as vectors.

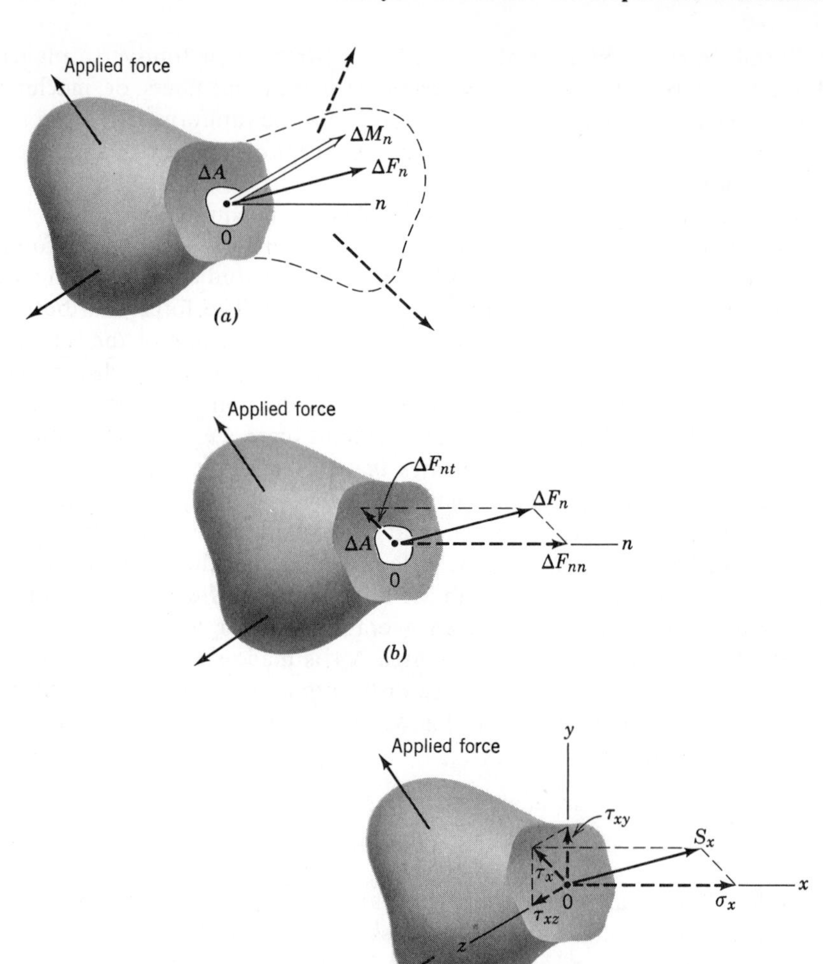

Fig. 1–13

stresses on planes having outward normals in the x, y, and z directions are usually chosen. Consider the plane having an outward normal in the x direction. In this case the normal and shear stresses on the plane will be σ_x and τ_x, respectively. Since τ_x, in general, will not coincide with the y or z axes, it must be resolved into the components τ_{xy} and τ_{xz}, as shown in Fig. 1–13c.

Unfortunately the state of stress at a point in a material is not completely defined by these three components of the stress vector since the stress vector itself depends on the orientation of the plane with which it is associated. An infinite number of planes can be passed through the point, resulting in an infinite number of stress vectors being associated with the point. Fortunately it can be shown (see Art. 1–9) that the specification of stresses on three mutually perpendicular planes is sufficient to describe completely the state of stress at the point. The rectangular components of stress vectors on planes having outward normals in the coordinate directions are shown in Fig. 1–14. The six faces of the small element are denoted by the directions of their outward normals so that the positive x face is the one whose outward normal is in the direction of the positive x axis. The coordinate axes x, y, and z are arranged as a right-hand system.

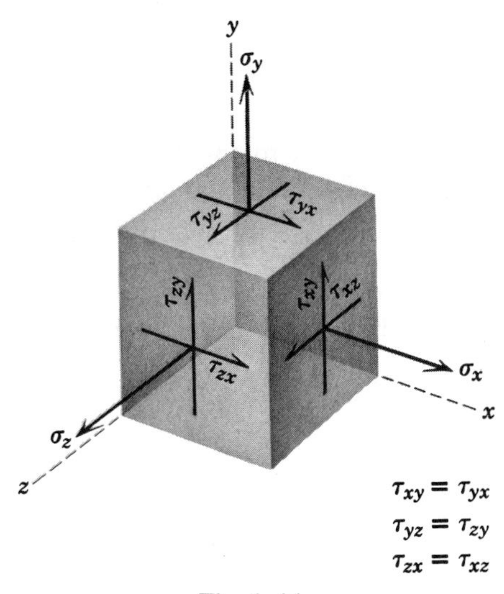

Fig. 1–14

The sign convention for stresses is as follows. Normal stresses (indicated by the symbol σ and a single subscript to indicate the plane on which the stress acts) are positive if they point in the direction of the outward normal. Thus, normal stresses are positive if tensile. Shearing stresses are denoted by the symbol τ followed by two subscripts; the first subscript designates the plane on which the shearing stress acts and the second the coordinate axis to which it is parallel. Thus τ_{xz} is the shearing stress on an x plane parallel to the z axis. A positive shearing stress points in the positive direction of the coordinate axis of the second subscript if it acts on a surface with an outward normal in the positive direction. Conversely, if the outward normal of the surface is in the negative direction, then the positive shearing stress points in the negative direction of the coordinate axis of the second subscript. The stresses shown on the element in Fig. 1–14 are all positive.

1–6 TWO-DIMENSIONAL OR PLANE STRESS

Before proceeding to consideration of the general three-dimensional state of stress, considerable insight into the nature of stress distributions can be gained by considering a less complicated state known as two-dimensional or *plane stress*. For this case two parallel faces of the small element of Fig. 1–14 are assumed to be free of stress. For purposes of analysis let these faces be perpendicular to the z axis. Thus,

$$\sigma_z = \tau_{zx} = \tau_{zy} = 0$$

From Eq. 1–3, however, this also implies that

$$\tau_{xz} = \tau_{yz} = 0$$

The components of stress present for plane stress analysis will be σ_x, σ_y, and $\tau_{xy} = \tau_{yx}$. For convenience this state of stress is usually represented by the two-dimensional sketch shown in Fig. 1–15. The three-dimensional element of which the sketch is a plane projection should be kept in mind at all times. Normal and shearing stresses on an arbitrary plane such as plane A-A in Fig. 1–15 can be obtained by using the free-body diagram method discussed for centric loading in Art. 1–4. The solution to the following example illustrates this method of approach under plane stress conditions.

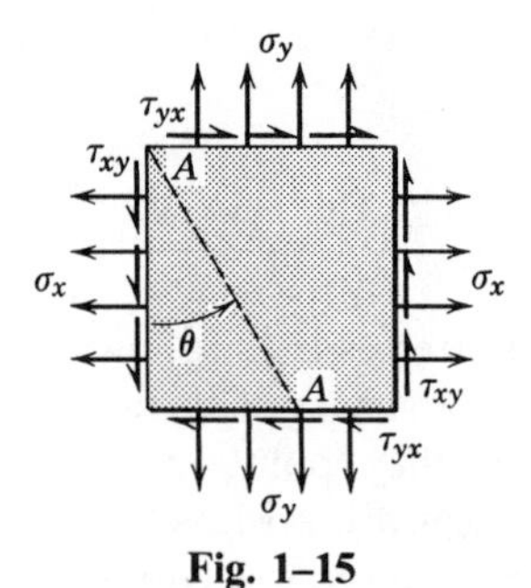

Fig. 1–15

Example 1–3 At a given point in a machine element, the following stresses were evaluated: 8000 psi T and zero shear on a horizontal plane and 4000 psi C on a vertical plane. Determine the stresses at this point on a plane having a slope of 3 vertical to 4 horizontal.

Solution As an aid to visualization of the data, it is suggested that the differential block of Fig. 1–16a be drawn (a stress picture, not a free-body diagram). The next step is to draw a free-body diagram subjected to ordinary force vectors. Various satisfactory free-body diagrams can be used, such as the wedge-shaped element defined by the three given planes in Fig. 1–16*b*, in which the shaded area indicates the plane on which the stresses to be evaluated are acting. To this area is assigned arbitrarily the magnitude dA, and the corresponding areas of the horizontal and vertical faces of the element are 0.8 dA and 0.6 dA, respectively. The forces acting on these areas will be as indicated in the diagram. It must be emphasized that *force vectors* act on these areas. For practical purposes, the two-dimensional free-body diagram of Fig. 1–16*c* will be adequate. Summing forces in the *n* direction yields the following:

$$(+\searrow)\quad \Sigma F_n = 0$$

$$\sigma\, dA + 0.6\, dA\,(4000)(0.6) - 0.8\, dA\,(8000)(0.8) = 0$$

from which

$$\sigma = \underline{3680 \text{ psi } T} \qquad \text{Ans.}$$

When forces are summed in the t direction, the shearing stress is found to be

$$\tau = \underline{5760 \text{ psi}} \qquad \text{Ans.}$$

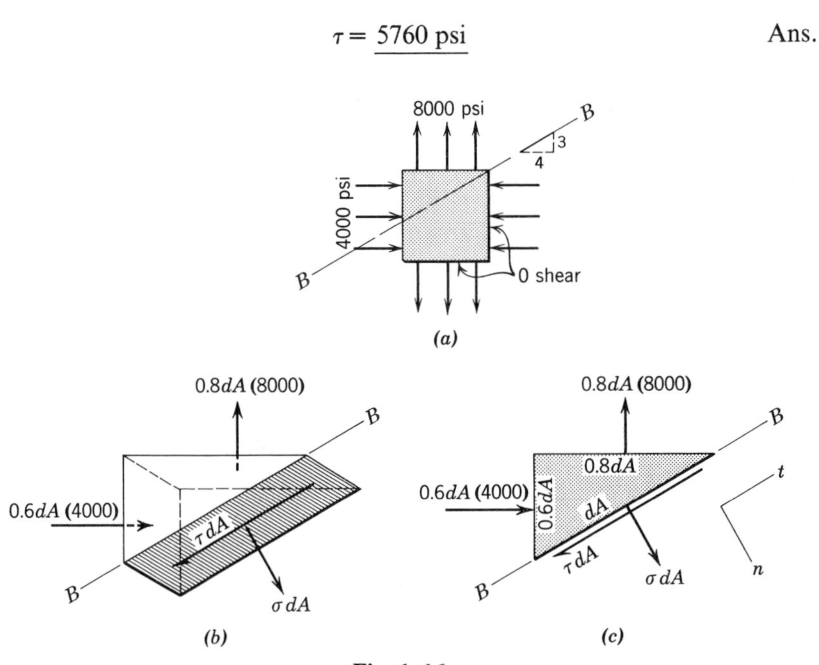

Fig. 1–16

Note that the normal stress should be designated as *tension* or *compression*. The presence of shearing stresses on the horizontal and vertical planes, had there been any, would merely have required two more forces on the free-body diagram: one parallel to the vertical face and one parallel to the horizontal face. Note, however, that the magnitude of the shearing stresses (not the forces) must be the same on any two orthogonal planes (see Art. 1–4) and that the vectors should be directed as in Fig. 1–9 or both vectors reversed, depending on the given data.

PROBLEMS

1–21* A structural steel eyebar with a 2×6-in. rectangular cross section is subjected to an axial tensile load of 240 kips. Determine:

(a) The normal and shearing stresses on a plane through the bar that makes an angle of 30° with the direction of the load.
(b) The maximum normal and shearing stresses in the bar.

1–22* A structural steel bar 30 mm in diameter is subjected to an axial tensile load of 27π kN. Determine:

(a) The normal and shearing stresses on a plane through the bar that makes an angle of 30° with the direction of the load.
(b) The maximum normal and shearing stresses in the bar.

1–23* The normal stress on plane *AB* of the rectangular block of Fig. P1–23 is 600 psi *C* when θ is 25°. Determine:

(a) The load *P*.
(b) The shearing stress on plane *AB*.
(c) The maximum shearing stress in the block.

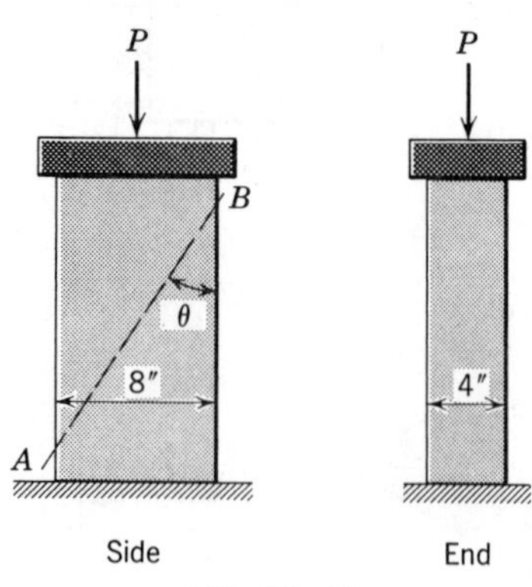

Fig. P1–23

1–24* An axially loaded post similar to that of Fig. P1–23 has a 100×150-mm rectangular cross section. The shearing stress on plane *AB* is 10 MN/m^2 when θ is

20°. Determine:

(a) The load P.

(b) The normal stress on plane AB.

(c) The maximum shearing stress in the post.

1–25 Determine the maximum axial load P that may be applied as shown to the $4 \times 4 \times 12$-in. birch post of Fig. P1–25 without exceeding a shearing stress of 185 psi parallel to the grain or a normal stress of 365 psi C perpendicular to the grain.

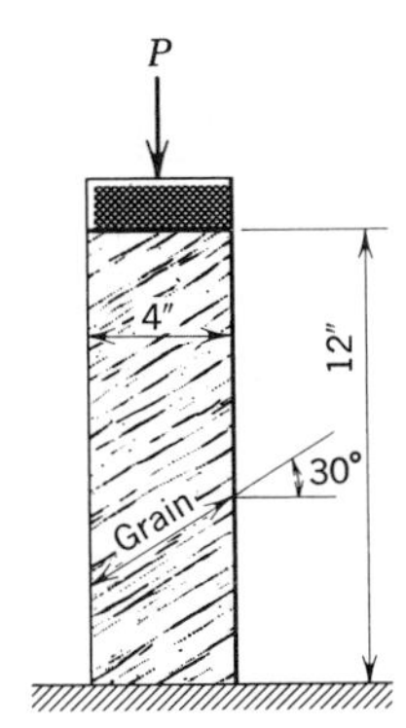

Fig. P1–25

1–26* Specifications for the $75 \times 75 \times 250$-mm block of Fig. P1–26 require that the normal stress and shearing stress on plane A-A not exceed 2.40 and 0.84 MN/m^2, respectively. Determine the maximum value of the load P that can be applied without exceeding the given requirements.

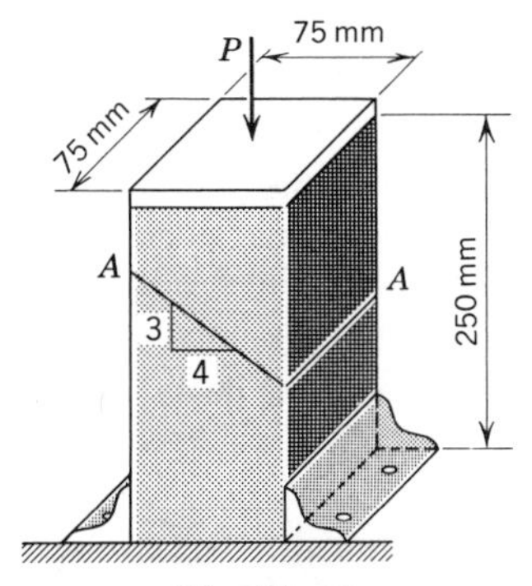

Fig. P1–26

1–27* Specifications for the timber block of Fig. P1–27 require that the stresses not exceed the following: shear parallel to the grain 130 psi, compression perpendicular to the grain 250 psi. Determine the maximum value of the axial load P that can be applied without exceeding the given requirements.

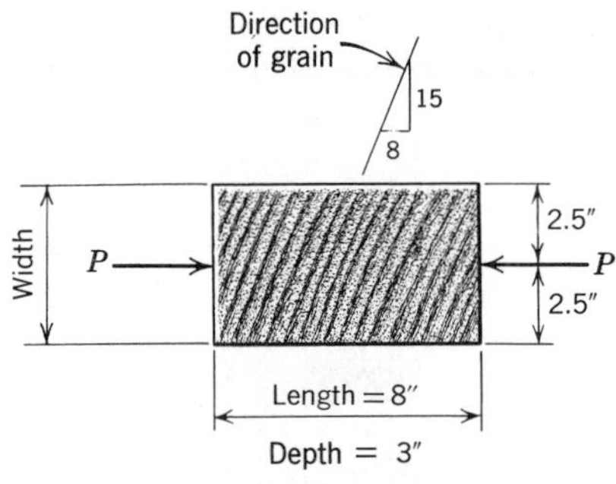

Fig. P1–27

1–28 Solve Prob. 1–27 with the following data changes: dimensions of block 80 mm deep × 130 mm wide × 250 mm long; slope of grain 5 horizontal to 12 vertical; specified stresses 0.80 MN/m^2 shear and 1.40 MN/m^2 compression.

1–29* Because of internal pressure in a boiler, the stresses at a particular point in the boiler plate were found to be as shown in the stress picture of Fig. P1–29. Determine the normal stress at this point on the inclined plane shown.

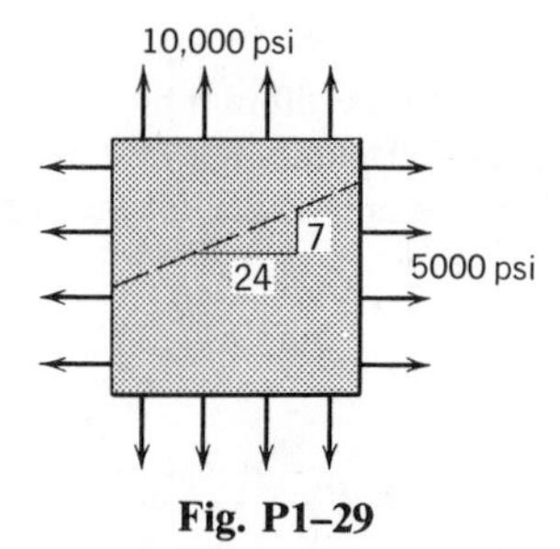

Fig. P1–29

1–30 At a point in a stressed body, there are normal stresses of 16 ksi C on a vertical plane and 8 ksi T on a horizontal plane. The shearing stresses on these planes are zero. Determine the normal and shearing stresses on a plane making an angle of 40° with the horizontal plane and an angle of 50° with the vertical plane.

1–31 Solve Prob. 1–29 with the following data changes: the stresses on horizontal and vertical planes are 80 and 40 MN/m^2, respectively; the inclined plane has a slope of 15 horizontal to 8 vertical.

1–32* At a point in a stressed region, there are stresses of 120 MN/m^2 T and zero shear on a vertical plane and 80 MN/m^2 C on a horizontal plane. Determine the normal and shearing stresses on a plane making an angle of 30° with the horizontal plane and 60° with the vertical plane.

1–33 Figure P1–33 is an incomplete free-body diagram for the determination of stresses at a point in a machine element. Draw the complete free-body diagram and determine the normal stress on plane AC. The stresses indicated are in psi.

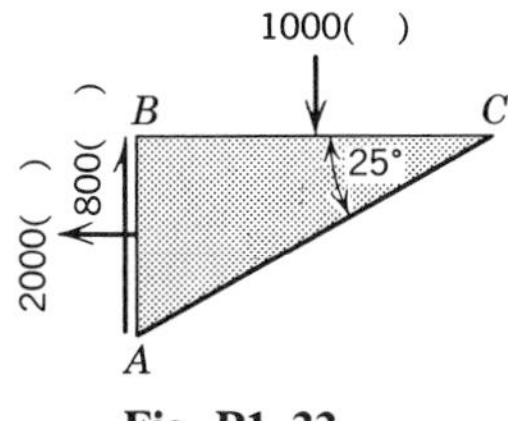

Fig. P1–33

1–34* On a vertical plane through a point in a structural member, there is a shearing stress of 15 MN/m^2, as shown in Fig. P1–34. There are no normal stresses on horizontal or vertical planes through the point. Determine the normal stress on plane B-B.

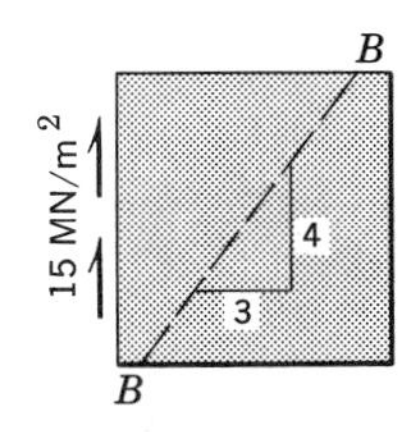

Fig. P1–34

1–35* At point A in the structural element in Fig. P1–35, the stresses on the inclined plane are $\sigma = 3200$ psi T and $\tau = 2000$ psi in the directions indicated, and the normal stress on a vertical plane is zero. Determine the normal stress on a horizontal plane through this point.

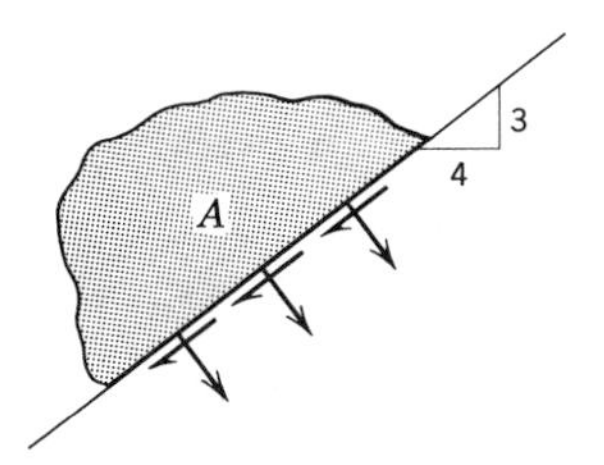

Fig. P1–35

1–36 At point A of Prob. 1–35, the normal stress on a horizontal plane is 4400 psi T. Determine the normal stress on the plane that is perpendicular to the inclined plane shown. All planes mentioned are perpendicular to the plane of the page.

1–37* At point A in the structural element of Fig. P1–35, the known stresses are 20 MN/m^2 T on the inclined plane and 15 MN/m^2 T and zero shear on the horizontal plane. Determine the normal stress on a vertical plane through the point.

1–38 At a point in a body stresses are as shown in Fig. P1–38. The normal stress on plane AB is 32 MN/m^2 T. Determine the stress σ_x.

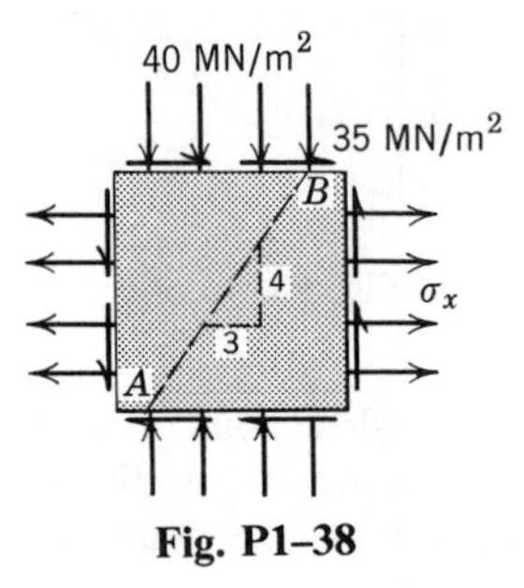

Fig. P1–38

1–39* Known stresses at point A in a structural member (see Fig. P1–39) are 20 ksi T and zero shear on plane B-B and 45 ksi C on plane C-C. Determine the stresses on the vertical plane at the point.

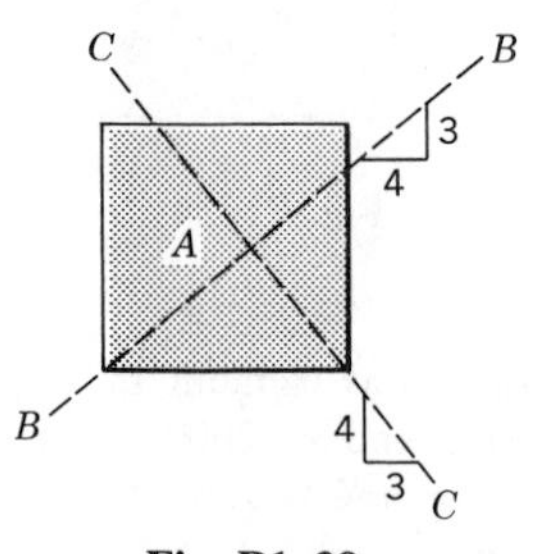

Fig. P1–39

1–40 Solve Prob. 1–39 with the given stresses changed from 20 ksi T and 45 ksi C to 150 MN/m^2 T and 250 MN/m^2 C, respectively.

1–7 PRINCIPAL STRESSES—PLANE STRESS

As mentioned in Art. 1–3, a thorough understanding of the physical significance of stress is fundamental in mechanics of materials. The concept of stress (internal force per unit area) at a point in a body on a plane through the point must be kept constantly in mind. The fact that the stresses are, in general, different for different planes through any one point has been emphasized previously.

Three important facts about stresses were developed in Art. 1–4: (1) The maximum normal stress for a centric tensile or compressive load in one direction is in the direction of the load on a plane normal to the load, and its magnitude is $\sigma = P/A$. (2) The maximum shearing stress for a centric tensile or compressive load in one direction is on a plane at an angle of 45° to the load, and its magnitude is $\tau = \sigma/2 = P/(2A)$. (3) For all types of loads, a shearing stress at any point on any plane is accompanied by a shearing stress of equal magnitude on an orthogonal plane through the point.

Once the normal and shearing stresses on orthogonal planes through a point in a body are known, the normal and shearing stresses on any specific oblique plane through the same point can be determined by the free-body diagram method of Art. 1–6. Usually the critical stresses are the maximum tensile and compressive stresses and the maximum shearing stresses. Therefore, the method of Art. 1–6 is inadequate, unless the slopes of the planes on which the maximum stresses act are known. For a centric load in one direction, the planes on which maximum and minimum normal stresses and maximum shearing stresses act are readily available from the developments of Art. 1–4. As more complicated loads are studied, more general methods for finding the critical stresses become necessary. Two such methods will be presented.

Equations relating the desired normal and shearing stresses σ_n and τ_{nt} on an arbitrary plane through a point oriented at an angle θ with respect to a reference x axis and the known stresses σ_x, σ_y, and $\tau_{xy} = \tau_{yx}$ on the reference planes can be developed using the free-body diagram method of Art. 1–6. Consider the plane stress situation indicated in Fig. 1–17*a*, where the dotted line *A*-*A* represents the trace of any plane through the point (all planes are perpendicular to the plane of zero stress—the plane of the paper). In the following derivation, a counterclockwise angle θ is positive.

Figure 1–17*b* is a free-body diagram of a wedge-shaped element in which the areas of the faces are dA for the inclined face (plane *A*–*A*), $dA\cos\theta$ for the vertical face, and $dA\sin\theta$ for the horizontal face. The force equation of equilibrium in the n direction gives

$$\sigma_n\,dA - \tau_{yx}(dA\sin\theta)\cos\theta - \tau_{xy}(dA\cos\theta)\sin\theta - \sigma_x(dA\cos\theta)\cos\theta - \sigma_y(dA\sin\theta)\sin\theta = 0$$

from which, since $\tau_{yx} = \tau_{xy}$,

$$\sigma_n = \sigma_x\cos^2\theta + \sigma_y\sin^2\theta + 2\tau_{xy}\sin\theta\cos\theta \qquad (1\text{–}4a)$$

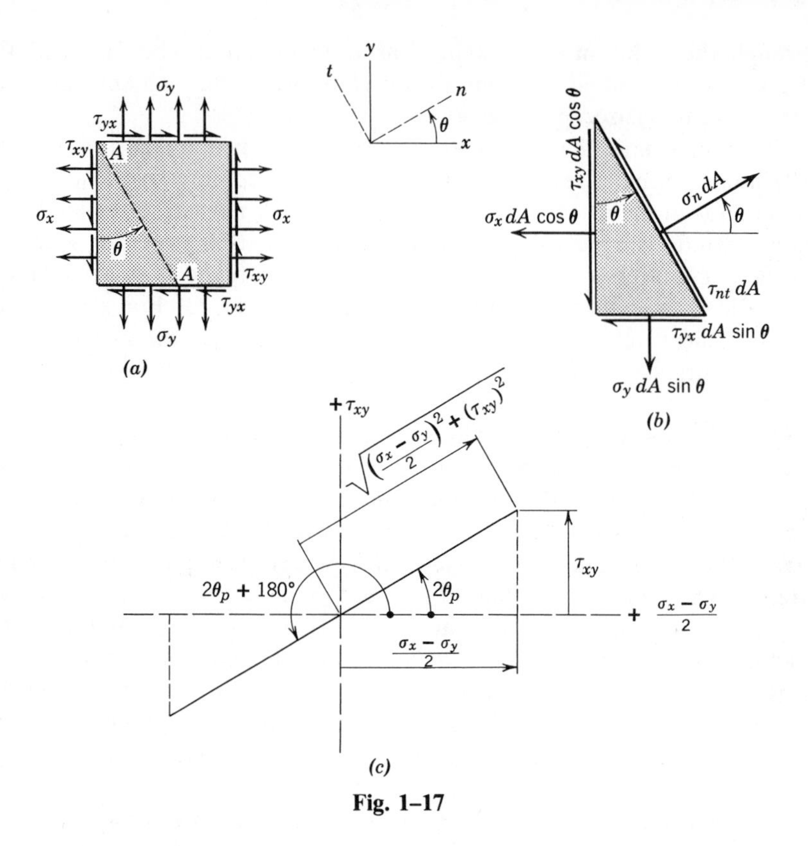

Fig. 1–17

or, in terms of the double angle,

$$\sigma_n = \frac{\sigma_x(1+\cos 2\theta)}{2} + \frac{\sigma_y(1-\cos 2\theta)}{2} + \frac{2\tau_{xy}(\sin 2\theta)}{2}$$

$$= \frac{\sigma_x+\sigma_y}{2} + \frac{\sigma_x-\sigma_y}{2}\cos 2\theta + \tau_{xy}\sin 2\theta \tag{1–4b}$$

For the free-body diagram of Fig. 1–17*b*, the force equation of equilibrium in the t direction gives

$$\tau_{nt}\,dA - \tau_{xy}(dA\cos\theta)\cos\theta + \tau_{yx}(dA\sin\theta)\sin\theta + \sigma_x(dA\cos\theta)\sin\theta$$

$$- \sigma_y(dA\sin\theta)\cos\theta = 0$$

from which

$$\tau_{nt} = -(\sigma_x - \sigma_y)\sin\theta\cos\theta + \tau_{xy}(\cos^2\theta - \sin^2\theta) \tag{1–5a}$$

or, in terms of the double angle,

$$\tau_{nt} = -\frac{\sigma_x - \sigma_y}{2}\sin 2\theta + \tau_{xy}\cos 2\theta \tag{1–5b}$$

Equations 1–4 and 1–5 provide a means for determining normal and shearing stresses on any plane whose outward normal is oriented at an angle θ with respect to the reference x axis. Maximum and minimum values of σ_n will occur at the values of θ for which $d\sigma_n/d\theta$ is equal to zero. Differentiation of σ_n with respect to θ gives

$$\frac{d\sigma_n}{d\theta} = -(\sigma_x - \sigma_y)\sin 2\theta + 2\tau_{xy}\cos 2\theta \tag{a}$$

Setting Eq. (a) equal to zero and solving gives

$$\tan 2\theta_p = \frac{\tau_{xy}}{(\sigma_x - \sigma_y)/2} \tag{1–6}$$

Note that the expression for $d\sigma_n/d\theta$ from Eq. (a) is numerically twice the value of the expression for τ_{nt} from Eq. 1–5*b*. Consequently, the shearing stress is zero on planes experiencing maximum and minimum values of normal stress. Planes free of shear stress are known as *principal planes*. The normal stresses occurring on principal planes are known as *principal stresses*. The values of θ_p from Eq. 1–6 give the orientation of two principal planes. A third principal plane for the plane stress state has an outward normal in the z direction. For a given set of values of σ_x, σ_y, and τ_{xy}, there are two values of $2\theta_p$ differing by 180° and, consequently, two values of θ_p 90° apart. This proves that the principal planes are normal to each other.

When τ_{xy} and $(\sigma_x - \sigma_y)$ have the same sign, $\tan 2\theta_p$ is positive and one value of $2\theta_p$ is between 0° and 90°, with the other value 180° greater, as shown in Fig. 1–17*c*. Consequently, one value of θ_p is between 0° and 45°, and the other one is 90° greater. In the first case, both $\sin 2\theta_p$ and $\cos 2\theta_p$ are positive, and in the second case both are negative. When these functions of $2\theta_p$ are substituted into Eq. 1–4*b*, two in-plane principal stresses σ_{p1} and σ_{p2} are found to be

$$\sigma_{p1,p2} = \frac{\sigma_x + \sigma_y}{2} + \frac{\sigma_x - \sigma_y}{2}\left[\pm \frac{(\sigma_x - \sigma_y)/2}{\sqrt{\left(\frac{\sigma_x - \sigma_y}{2}\right)^2 + \tau_{xy}^2}}\right]$$

$$+ \tau_{xy}\left[\pm \frac{\tau_{xy}}{\sqrt{\left(\frac{\sigma_x - \sigma_y}{2}\right)^2 + \tau_{xy}^2}}\right]$$

which reduces to

$$\sigma_{p1,p2} = \frac{\sigma_x + \sigma_y}{2} \pm \sqrt{\left(\frac{\sigma_x - \sigma_y}{2}\right)^2 + \tau_{xy}^2} \tag{1–7}$$

Equation 1–7 gives the two principal stresses in the xy plane, and the third one is $\sigma_{p3} = \sigma_z = 0$. Equation 1–6 gives the angles θ_p and $\theta_p + 90°$ between the x (or y) plane and the mutually perpendicular planes on which the principal stresses act. When $\tan 2\theta_p$ is positive, θ_p is positive, and the rotation is counterclockwise from the x and y planes to the planes on which the two principal stresses act. When $\tan 2\theta_p$ is negative, the rotation is clockwise. Note that one value of θ_p will always be between positive and negative 45° (inclusive) with the other value 90° greater. The numerically greater principal stress will act on the plane that makes an angle of 45° or less with the plane of the numerically larger of the two given normal stresses. This statement can be confirmed in any case by substitution of the value of θ_p from Eq. 1–6 into Eq. 1–4b to obtain the corresponding principal stress.

Note that if one or both of the principal stresses from Eq. 1–7 is negative, the algebraic maximum stress can have a smaller absolute value than the "minimum" stress.

The *maximum in-plane shearing stress* τ_p occurs on planes located by values of θ where $d\tau_{nt}/d\theta$ is equal to zero. Differentiation of Eq. 1–5b gives

$$\frac{d\tau_{nt}}{d\theta} = -(\sigma_x - \sigma_y)\cos 2\theta - 2\tau_{xy}\sin 2\theta$$

When $d\tau_{nt}/d\theta$ is equated to zero, the value of θ_τ is given by the expression

$$\tan 2\theta_\tau = -\frac{(\sigma_x - \sigma_y)/2}{\tau_{xy}} \tag{b}$$

Comparison of Eqs. (b) and 1–6 reveals that the two tangents are negative reciprocals. Therefore, the two angles $2\theta_p$ and $2\theta_\tau$ differ by 90°, and θ_p and θ_τ are 45° apart. This means that the planes on which the maximum in-plane shearing stresses occur are 45° from the principal planes. The maximum in-plane shearing stresses are found by substituting values of angle functions obtained from Eq. (b) in Eq. 1–5*b*. The results are

$$\tau_p = -\frac{\sigma_x - \sigma_y}{2}\left[\mp\frac{(\sigma_x - \sigma_y)/2}{\sqrt{\left(\frac{\sigma_x - \sigma_y}{2}\right)^2 + \tau_{xy}^2}}\right]$$

$$+\tau_{xy}\left[\pm\frac{\tau_{xy}}{\sqrt{\left(\frac{\sigma_x - \sigma_y}{2}\right)^2 + \tau_{xy}^2}}\right]$$

$$\tau_p = \pm\sqrt{\left(\frac{\sigma_x - \sigma_y}{2}\right)^2 + \tau_{xy}^2} \tag{1–8}$$

which is observed to have the same magnitude as the second term of Eq. 1–7.

A useful relation between the principal stresses and the maximum in-plane shearing stress is obtained from Eqs. 1–7 and 1–8 by subtracting the values for the two in-plane principal stresses and substituting the value of the radical from Eq. 1–8. The result is

$$\tau_p = (\sigma_{p1} - \sigma_{p2})/2 \tag{1–9}$$

or, in words, the maximum value of τ_{nt} is equal in magnitude to one-half the difference between the two in-plane principal stresses.

In general, when stresses act in three directions it can be shown (see Art. 1–9) that there are three orthogonal planes on which the shearing stress is zero. These planes are known as the principal planes, and the stresses acting on them (the principal stresses) will have three values: one maximum, one minimum, and a third stress between the other two. The maximum shearing stress, $\tau_{\max}$ on any plane that could be passed through the point, is one-half the difference between the maximum and minimum principal stresses and acts on planes that bisect the angles between the planes of the maximum and minimum normal stresses.

When a state of plane stress exists, one of the principal stresses is zero. If the values of σ_{p1} and σ_{p2} from Eq. 1–7 have the same sign, then the third principal stress, σ_{p3} equals zero, will be either the maximum or the minimum normal stress. Thus, the maximum shearing stress may be $(\sigma_{p1}-\sigma_{p2})/2$, $(\sigma_{p1}-0)/2$, or $(\sigma_{p2}-0)/2$, depending on the relative magnitudes and signs of the principal stresses. These three possibilities are illustrated in Fig. 1–18, in which one of the two orthogonal planes on which the maximum shearing stress acts is hatched for each example.

The direction of the maximum shearing stress can be determined by drawing a wedge-shaped block with two sides parallel to the planes having the maximum and minimum principal stresses, and with the third side at an angle of 45° with the other two sides. The direction of the maximum shearing stress must oppose the larger of the two principal stresses.

Another useful relation between the principal stresses and the normal stresses on the orthogonal planes, shown in Fig. 1–17, is obtained by adding the values for the two principal stresses as given by Eq. 1–7. The result is

$$\sigma_{p1}+\sigma_{p2}=\sigma_x+\sigma_y \tag{c}$$

or, in words, *for plane stress, the sum of the normal stresses on any two orthogonal planes through a point in a body is a constant or invariant.*

In the preceding discussion, "maximum" and "minimum" stresses were considered as algebraic quantities, and it has already been pointed out that

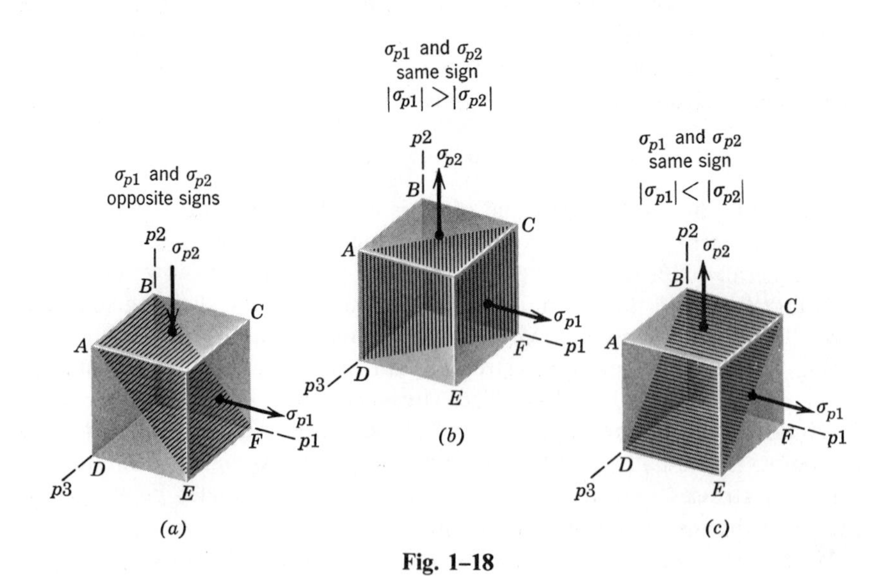

Fig. 1–18

the minimum algebraic stress may have a larger magnitude than the maximum stress. However, in the application to engineering problems (which includes the problems in this book), the term *maximum* will always refer to the largest absolute value (largest magnitude).

The application of the formulas and procedures developed in this article are illustrated by the following examples.

Example 1–4 At a point in a structural member subjected to plane stress there are stresses on horizontal and vertical planes through the point, as shown in Fig. 1–19*a*.

(*a*) Determine the principal stresses and the maximum shearing stress at the point.

(*b*) Locate the planes on which these stresses act and show the stresses on a complete sketch.

Solution (a) On the basis of the established sign convention, σ_x is positive, whereas σ_y and τ_{xy} are negative. For use in Eqs. 1–6 and 1–7, the given

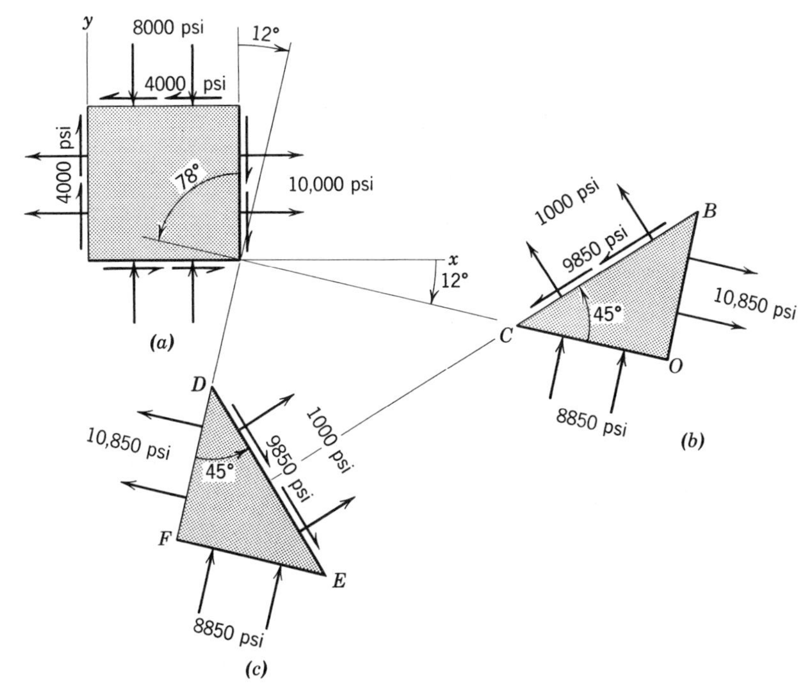

Fig. 1–19

values are

$$\sigma_x = +10{,}000 \text{ psi}, \qquad \sigma_y = -8000 \text{ psi}, \qquad \text{and} \quad \tau_{xy} = -4000 \text{ psi}$$

When these values are substituted in Eq. 1–7, the principal stresses are found to be

$$\sigma_p = \frac{10{,}000 - 8000}{2} \pm \sqrt{\left(\frac{10{,}000 + 8000}{2}\right)^2 + (-4000)^2}$$

$$= 1000 \pm 9850$$

$$= \underline{10{,}850 \text{ psi } T \text{ and } 8850 \text{ psi } C} \qquad \text{Ans.}$$

$$\sigma_z = \underline{0} \qquad \text{Ans.}$$

Since σ_{p1} and σ_{p2} are of opposite sign, the maximum shearing stress is

$$\tau_{max} = \frac{10{,}850 - (-8850)}{2} = \underline{9850 \text{ psi}} \qquad \text{Ans.}$$

(b) When the given data are substituted in Eq. 1–6, the results are

$$\tan 2\theta_p = -4000/9000 = -0.4444$$

from which

$$2\theta_p = -23.96°$$

and

$$\theta_p = -11.98° \text{ or } 12° \circlearrowright$$

The required sketch is shown in Fig. 1–19*b* or *c*.

From Eq. (c), the sum of the normal stresses on any two orthogonal planes is a constant for plane stress. Planes *CB* and *DE* of Fig. 1–19 are orthogonal, and the normal stresses on them are obviously equal. Therefore,

$$2\sigma_n = \sigma_x + \sigma_y = 10{,}000 - 8000 = 2000$$

and

$$\sigma_n = 1000 \text{ psi on the planes of maximum shear.}$$

Example 1–5 Solve Exam. 1–4 when σ_y is 8000 psi T with σ_x and τ_{xy} as shown (see Fig. 1–20a).

Solution (a) The principal stresses, expressed in ksi, are

$$\sigma_{p1,p2} = \frac{10+8}{2} \pm \sqrt{\left(\frac{10-8}{2}\right)^2 + 4^2}$$

$$= 9 \pm 4.123$$

$$= \underline{13.12 \text{ ksi } T \text{ and } 4.88 \text{ ksi } T} \qquad \text{Ans.}$$

Also

$$\sigma_{p3} = \sigma_z = \underline{0} \qquad \text{Ans.}$$

Since σ_{p1} and σ_{p2} have the same sign, the maximum shearing stress is

$$\tau_{\max} = (13.12 - 0)/2 = \underline{6.56 \text{ ksi}} \qquad \text{Ans.}$$

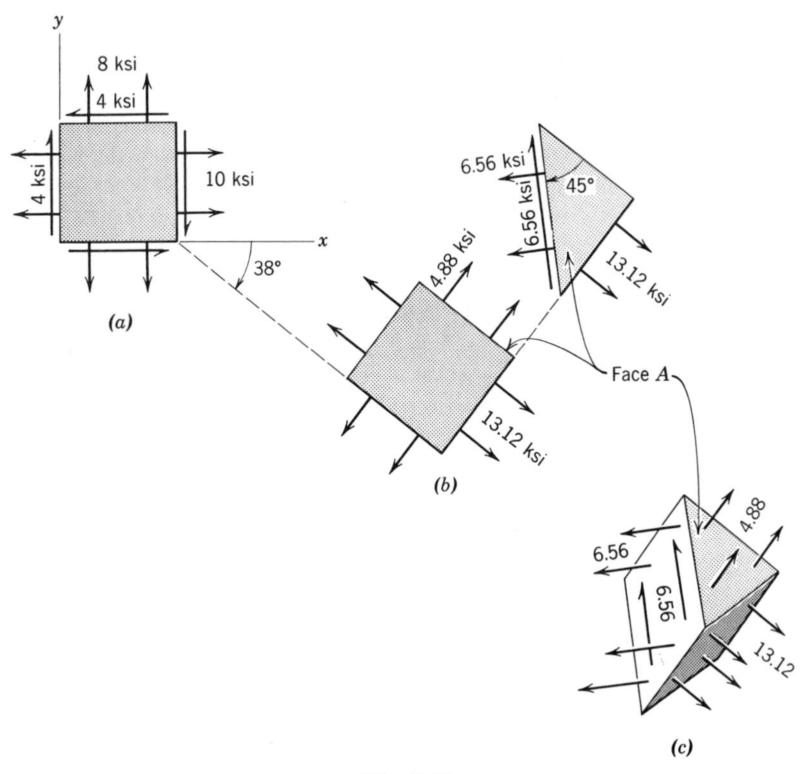

Fig. 1–20

(b) When the given data are substituted in Eq. 1–6, the results are

$$\tan 2\theta_p = -4/1 = -4$$

from which

$$2\theta_p = -75.97°$$

and

$$\theta_p = -37.985° \text{ or } 38° \circlearrowright$$

The maximum shearing stress occurs on the plane making an angle of 45° with the planes of maximum and minimum normal stresses—in this case, 13.12 ksi and zero. The complete sketch is given in Fig. 1–20*b* where the upper (wedge-shaped) block is the orthographic projection of the lower block. The three-dimensional Fig. 1–20*c* is presented as an aid to the visualization of Fig. 1–20*b*.

PROBLEMS

Note. The following problems are assumed to be plane stress problems unless otherwise noted. The required sketches are to be complete, indicating all pertinent angles and stresses. For interpretation of signs on shearing stresses, the horizontal (*x*) and vertical (*y*) directions are positive to the right and upward, respectively.

1–41* At a point in a stressed body, the normal stresses are 12,000 psi *T* on a vertical plane and 4000 psi *C* on a horizontal plane. A negative shearing stress of 6000 psi acts on a vertical plane at the point. Determine, and show on a sketch, the principal stresses and the maximum shearing stress at the point.

1–42* At a point in a stressed machine part, the known stresses are 160 MN/m^2 *T* and −50 MN/m^2 shear on a vertical plane, and 80 MN/m^2 *C* on a horizontal plane. Determine, and show on a sketch, the principal stresses and the maximum shearing stress at the point.

1–43* At a point in a stressed body, the normal and shearing stresses acting on a horizontal plane are 30 ksi *T* and +7 ksi shear, respectively. The normal stress acting on a vertical plane through the point is 18 ksi *C*. Determine, and show on a sketch, the principal and maximum shearing stresses at the point.

1–44* At a point in a loaded member, stresses are 90 MN/m^2 *C* and +40 MN/m^2 shear on a vertical plane, and 60 MN/m^2 *T* on a horizontal plane. Determine, and show on a sketch, the principal and maximum shearing stresses at the point.

1–45 The vertical plane through a point in a member has a normal stress of 6 ksi *T* and a shearing stress of −6 ksi. The horizontal plane through the point has a normal stress of 20 ksi *C*. Determine, and show on a sketch, the principal and maximum shearing stresses at the point.

1–46 The known stresses at a point in a loaded member are 50 MN/m^2 T and +70 MN/m^2 shear on a vertical plane, and 150 MN/m^2 C on a horizontal plane. Determine, and show on a sketch, the principal and maximum shearing stresses at the point.

1–47* The stresses shown in Fig. P1–47 act at a point in a structural member. The tensile principal stress is known to be 1200 psi. Determine:

(a) The maximum shearing stress at the point.
(b) The orientation of the planes on which the stress of part *a* acts.
(c) The shearing stress on the horizontal plane.

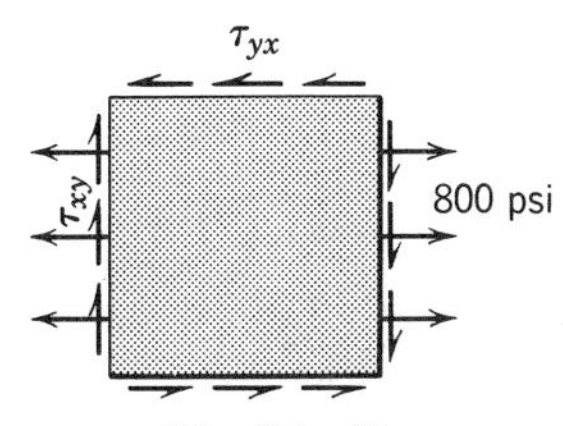

Fig. P1–47

1–48* At a point in a stressed region, on a vertical plane there is a normal stress of 130 MN/m^2 T and an unknown negative shearing stress. The maximum principal stress at the point is 150 MN/m^2 T, and the maximum shearing stress has a magnitude of 100 MN/m^2. Determine the unknown stresses on the horizontal and vertical planes and show these stresses as well as the principal stresses on a sketch.

1–49 A point within a structural member is subjected to the stresses in Fig. P1–49.

(a) Determine, and show on a sketch, the principal and maximum shearing stresses at the point.
(b) Determine, by the free-body diagram method, the shearing and normal stresses on plane *AB*.
(c) Determine the shearing and normal stresses on plane *AB* by using Eqs. 1–4*a* and 1–5*a*.

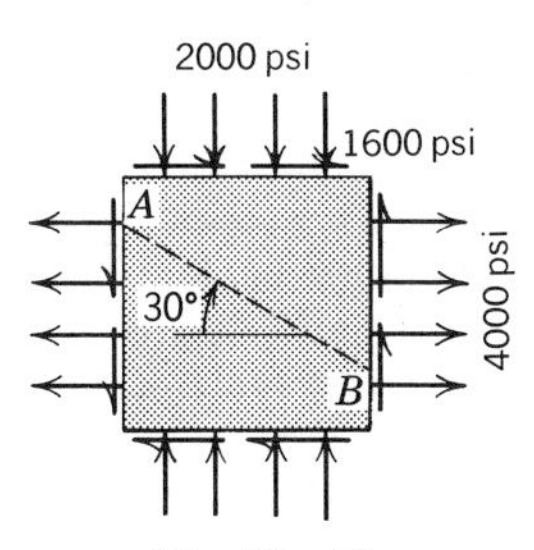

Fig. P1–49

Problems 1–50 through 1–55 are problems in plane stress in which certain stresses or angles are given. Determine the unknown quantities for each problem, and draw a sketch indicating the stresses and planes on which they act. Note that in some problems there might be more than one possible value of θ_p depending on the sign of τ_{xy}.

	Units	σ_x	σ_y	τ_{xy}	σ_{p1}	σ_{p2}	τ_{max}	θ_p
1–50*	ksi	2	−8		10			
1–51	ksi	7.5			13		8	
1–52*	ksi			5	30			15°
1–53	MN/m^2	28	−42		76			
1–54*	MN/m^2	42			62		38	
1–55	MN/m^2			21	55			30°

1–8 MOHR'S CIRCLE FOR PRINCIPAL STRESSES

The equations for finding the principal stresses and the maximum shearing stresses at a point in a stressed member were developed in Art. 1–7. The German engineer Otto Mohr (1835–1918) developed a useful pictorial or graphic interpretation of these equations.[6] This method, commonly called Mohr's circle,[7] involves the construction of a circle in such a manner that the coordinates of each point on the circle represent the normal and shearing stresses on one plane through the stressed point, and the angular position of the radius to the point gives the orientation of the plane. The proof that normal and shearing components of stress on an arbitrary plane through a point can be represented as points on a circle follows from the general expressions for normal and shearing stresses

[6]The particular case for plane stress was presented by the German engineer K. Culmann (1829–1881) about sixteen years before Mohr's paper.

[7]Mohr's circle is also used as a pictorial interpretation of the formulas for obtaining maximum and minimum second moments of areas and maximum products of inertia. See *Engineering Mechanics*, Higdon and Stiles, Prentice-Hall, Englewood Cliffs, N.J., 1962, 1968. Mohr's circle is also used for plane strain, three-dimensional stress, and maximum and minimum moments of inertia of mass.

derived in Art. 1–7. Recall Eqs. 1–4*b* and 1–5*b* and rewrite in the following form:

$$\sigma_n - \frac{\sigma_x + \sigma_y}{2} = \frac{\sigma_x - \sigma_y}{2}\cos 2\theta + \tau_{xy}\sin 2\theta$$

$$\tau_{nt} = -\frac{\sigma_x - \sigma_y}{2}\sin 2\theta + \tau_{xy}\cos 2\theta$$

Squaring both equations, adding, and simplifying yields

$$\left(\sigma_n - \frac{\sigma_x + \sigma_y}{2}\right)^2 + \tau_{nt}^2 = \left(\frac{\sigma_x - \sigma_y}{2}\right)^2 + \tau_{xy}^2$$

This is the equation of a circle in terms of the variables σ_n and τ_{nt}. The circle is centered on the σ axis at a distance $(\sigma_x + \sigma_y)/2$ from the τ axis, and the radius of the circle is given by

$$r = \sqrt{\left(\frac{\sigma_x - \sigma_y}{2}\right)^2 + \tau_{xy}^2}$$

Normal stresses are plotted as horizontal coordinates, with tensile stresses (considered positive) plotted to the right of the origin and compressive stresses (considered negative) plotted to the left. Shearing stresses are plotted as vertical coordinates; since, for a given state of stress, the shearing stresses ($\tau_{xy} = \tau_{yx}$) have only one sign (both are positive or both are negative), it becomes necessary to introduce a new convention in order to have Mohr's circle give not only the magnitudes of the stresses but also the correct directions. The convention used here is that shearing forces that apply a clockwise couple (considered positive) to the small elemental stress cube are plotted above the σ axis, and forces that produce a counterclockwise couple (considered negative) are plotted below the σ axis. The method for interpreting the sign to be associated with a particular shear stress value obtained from a Mohr's circle analysis will be illustrated in Exam. 1–6.

The Mohr's circle for any point subjected to plane stress can be drawn when the stresses on two mutually perpendicular planes through the point are known. Consider, for example, the stress situation of Fig. 1–17*a* with σ_x greater than σ_y, and plot in Fig. 1–21 the points representing the given stresses. The coordinates of point *V* of Fig. 1–21 are the stresses on the vertical plane through the stressed point of Fig. 1–17*a*, and point *H* is determined by the stresses on the horizontal plane through the point. Line

VH is the diameter of Mohr's circle for the stresses at the point. Because τ_{yx} is equal to τ_{xy}, point C, the center of the circle, is on the σ axis. Line CV on Mohr's circle represents the plane (the vertical plane of Fig. 1–17*a*) through the stressed point from which the angle θ is measured. The coordinates of each point on the circle represent σ_n and τ_{nt} for one particular plane through the stressed point, the abscissa representing σ_n and the ordinate representing τ_{nt}. To demonstrate this statement, draw any radius CF in Fig. 1–21 at an angle 2θ counterclockwise from radius CV. From the figure, it is apparent that

$$OF' = OC + CF\cos(2\theta_p - 2\theta)$$

and since CF equals CV, the above equation reduces to

$$OF' = OC + CV\cos 2\theta_p \cos 2\theta + CV\sin 2\theta_p \sin 2\theta$$

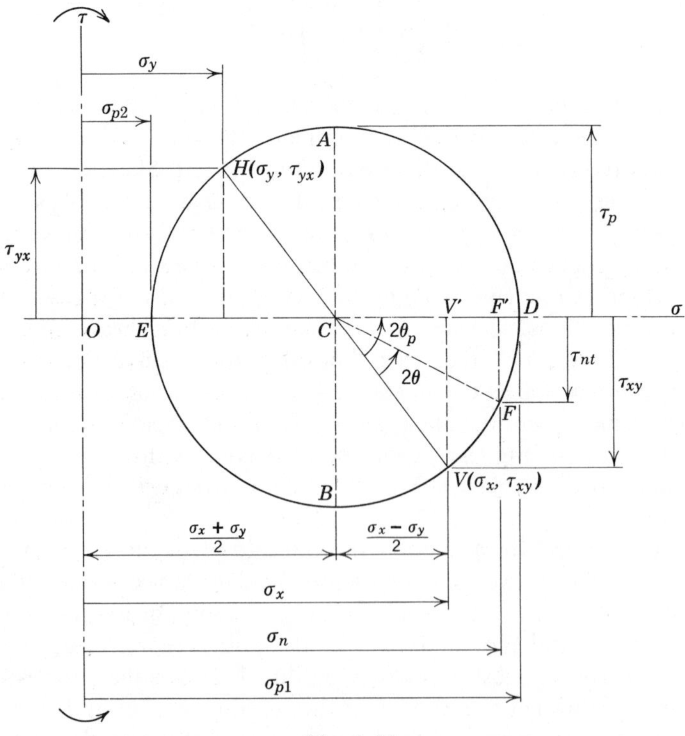

Fig. 1–21

Referring to Fig. 1–21, note that

$$CV\cos 2\theta_p = CV' = (\sigma_x - \sigma_y)/2$$

$$CV\sin 2\theta_p = VV' = \tau_{xy}$$

and

$$OC = (\sigma_x + \sigma_y)/2$$

therefore

$$\begin{aligned} OF' &= OC + CV'\cos 2\theta + VV'\sin 2\theta \\ &= \frac{\sigma_x + \sigma_y}{2} + \frac{\sigma_x - \sigma_y}{2}\cos 2\theta + \tau_{xy}\sin 2\theta \end{aligned}$$

This expression is identical to Eq. 1–4*b* of Art. 1–7. Therefore, OF' is equal to σ_n. In a similar manner,

$$\begin{aligned} F'F &= CF\sin(2\theta_p - 2\theta) \\ &= CV\sin 2\theta_p \cos 2\theta - CV\cos 2\theta_p \sin 2\theta \\ &= V'V\cos 2\theta - CV'\sin 2\theta \\ &= \tau_{xy}\cos 2\theta - \frac{\sigma_x - \sigma_y}{2}\sin 2\theta = \tau_{nt} \end{aligned}$$

which is Eq. 1–5*b* of Art. 1–7.

Since the horizontal coordinate of each point on the circle represents a particular value of σ_n, the maximum normal stress is represented by OD, and its value is

$$\begin{aligned} \sigma_{p1} &= OD = OC + CD = OC + CV \\ &= \frac{\sigma_x + \sigma_y}{2} + \sqrt{\left(\frac{\sigma_x - \sigma_y}{2}\right)^2 + \tau_{xy}^2} \end{aligned}$$

which agrees with Eq. 1–7.

Likewise, the vertical coordinate of each point on the circle represents a particular value of τ_{nt} (called the in-plane shearing stress), which means that the maximum in-plane shearing stresses are represented by CA and

CB, and their value is

$$\tau_p = CA = CV = \sqrt{\left(\frac{\sigma_x - \sigma_y}{2}\right)^2 + \tau_{xy}^2}$$

which agrees with Eq. 1–8. As noted in Art. 1–7, if the two nonzero principal stresses have the same sign, the maximum shearing stress at the point will not be in the plane of the applied stresses.

The angle $2\theta_p$ from CV to CD is counterclockwise or positive, and its tangent is

$$\tan 2\theta_p = \frac{\tau_{xy}}{(\sigma_x - \sigma_y)/2}$$

which is Eq. 1–6. From the derivation of Eq. 1–7, the angle between the vertical plane and one of the principal planes was θ_p. In obtaining the same equation from Mohr's circle, the angle between the radii representing these same two planes is $2\theta_p$. In other words, all angles in Mohr's circle are twice the corresponding angles for the actual stressed body. The angle from the vertical plane to the horizontal plane in Fig. 1–17*a* is 90°, but in Fig. 1–21 the angle between line CV (which represents the vertical plane) and line CH (which represents the horizontal plane) in Mohr's circle is 180°.

The results obtained from Mohr's circle have been shown to be identical with the equations derived from the free-body diagram of Art. 1–7. Consequently, Mohr's circle provides an extremely useful aid in the solution for, and visualization of, the stresses on various planes through a point in a stressed body in terms of the stresses on two mutually perpendicular planes through the point. Although Mohr's circle can be drawn to scale and used to obtain values of stresses and angles by direct measurements on the figure, it is probably more useful as a pictorial aid to the analytic determination of the desired quantities. The relationships between the stresses and the planes on which they act are readily apparent from the circle. Problems of the type presented in Art. 1–6 as well as those of Art. 1–7 can readily be solved by this semigraphic method, as will be illustrated in the following example.

Example 1–6 At a point in a structural member subjected to plane stress there are stresses on horizontal and vertical planes through the point, as shown in Fig. 1–22*a*. Determine, and show on a sketch:

(a) The principal stresses and maximum shearing stress at the point.

(b) The normal and shearing stresses on plane A–A through the point.

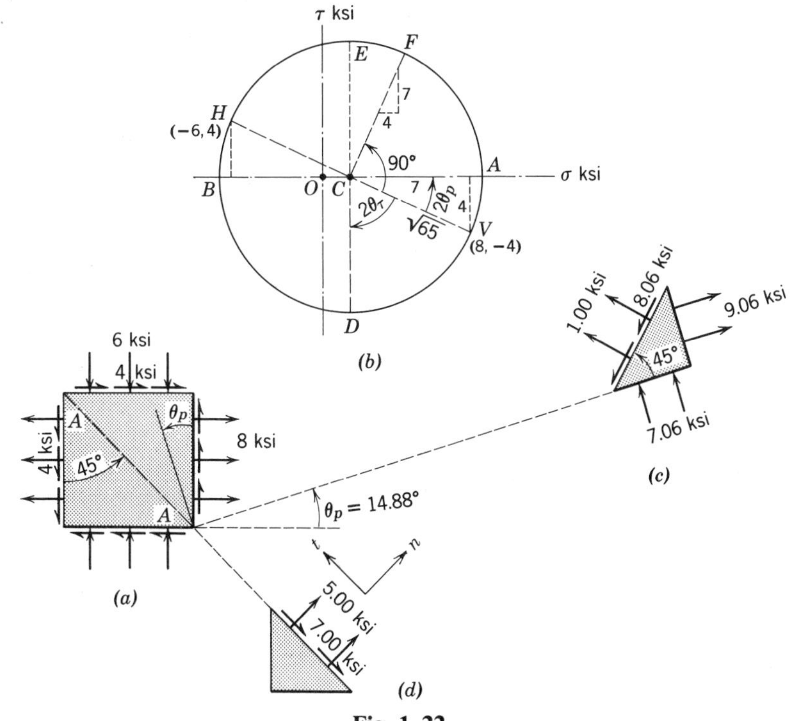

Fig. 1–22

Solution Mohr's circle is constructed from the given data as follows: on a set of coordinate axes (Fig. 1–22*b*) plot point V (representing the stresses on the vertical plane) at $(8, -4)$ because the stresses on the vertical plane are 8 ksi T and -4 ksi (counterclockwise) shear. Likewise, point H (representing the stresses on the horizontal plane) has the coordinates $(-6, 4)$. Draw line HV, which is a diameter of Mohr's circle, and note that the center of the circle is at $(1, 0)$. The radius of the circle is

$$CV = \sqrt{7^2 + 4^2} = 8.06 \text{ ksi}$$

(a) The principal stresses and the maximum shearing stress at the point are

$$\sigma_{p1} = OA = 1 + 8.06 = \underline{9.06 \text{ ksi } T} \qquad \text{Ans.}$$

$$\sigma_{p2} = OB = 1 - 8.06 = -7.06 = \underline{7.06 \text{ ksi } C} \qquad \text{Ans.}$$

$$\sigma_{p3} = \sigma_z = \underline{0} \qquad \text{Ans.}$$

Since σ_{p1} and σ_{p2} have opposite signs, the maximum shearing stress is

$$\tau_{\max} = CE = CD = \underline{8.06 \text{ ksi}} \qquad \text{Ans.}$$

The principal planes are represented by lines CA and CB, where

$$\tan 2\theta_p = 4/7 = 0.5714$$

which gives

$$2\theta_p = 29.75° \circlearrowleft$$

Since the angle $2\theta_p$ is counterclockwise, the principal planes are counterclockwise from the vertical and horizontal planes of the stress block, as shown in Fig. 1–22*c*. To determine which principal stress acts on which plane, note that as the radius of the circle rotates counterclockwise, the end of the radius CV moves from V to A, indicating that as the initially vertical plane rotates through 14.88°, the stresses change to 9.06 ksi T and zero. Note also that the end H of radius CH moves to B, indicating that as the initially horizontal plane rotates through 14.88°, the stresses change to 7.06 ksi C and zero. The required sketch is Fig. 1–22*c*.

(b) Plane A–A is 45° counterclockwise from the vertical plane; therefore, the corresponding radius of Mohr's circle is 90°, 2θ, counterclockwise from the line CV, and is shown as CF on Fig. 1–22*b*. The coordinates of the point F are seen to be (5,7), which means that the stresses on plane A–A are 5.00 ksi T and 7.00 ksi shear in a direction to produce a clockwise (previously defined as positive) couple on a stress element having plane A–A as one of its faces (see Fig. 1–22*d*). This shear stress would be defined as a negative shear stress since to produce a clockwise couple it must be directed in the negative t direction associated with plane A–A (see Fig. 1–22*d*). Note that Fig. 1–22*d* does not show all of the stresses acting on the element.

PROBLEMS

Note. The bodies in the following problems are assumed to be in a state of plane stress. For interpretation of signs on shearing stresses, the horizontal (x) and vertical (y) directions are positive to the right and upward, respectively.

1–56* Solve Prob. 1–41 by means of Mohr's circle.

1–57* Solve Prob. 1–42 by means of Mohr's circle.

1–58* Solve Prob. 1–43 by means of Mohr's circle.

1–59* Solve Prob. 1–44 by means of Mohr's circle.

1–60 Solve Prob. 1–45 by means of Mohr's circle.

1–61 Solve Prob. 1–46 by means of Mohr's circle.

1–62* Solve Prob. 1–47 by means of Mohr's circle.

1–63* Solve Prob. 1–48 by means of Mohr's circle.

1–64* Solve Prob. 1–50 by means of Mohr's circle.

1–65 Solve Prob. 1–51 by means of Mohr's circle.

1–66 Solve Prob. 1–53 by means of Mohr's circle.

1–67 Solve Prob. 1–54 by means of Mohr's circle.

1–68 At a point in a machine element, the normal stresses are 16,000 psi *T* on a vertical plane and 4000 psi *C* on a horizontal plane. A negative shearing stress of 5000 psi acts on the vertical plane at the point. Determine, and show on a sketch:

(a) The stresses on a plane 30° clockwise from the vertical plane.
(b) The principal and maximum shearing stresses at the point.

1–69* At a point in a stressed body, the stresses on a vertical plane are 55 MN/m^2 *C* and +14 MN/m^2 shear. The normal stress on the horizontal plane is 15 MN/m^2 *T*. Determine and show on a sketch:

(a) The stresses on a plane 60° counterclockwise from the vertical plane.
(b) The principal and maximum shearing stresses at the point.

1–70 At a point in a stressed body, there are stresses of 16 ksi *T* on a vertical plane and 24 ksi *C* on a horizontal plane. A negative unknown shearing stress acts on the vertical plane, and the maximum shearing stress at the point has a magnitude of 25 ksi. Determine, and show on a sketch, the principal stresses and the shearing stress on the vertical plane.

1–71* At a point in a stressed member, on the horizontal plane there is a normal stress of 64 MN/m^2 *C* and an unknown positive shearing stress. One principal stress at the point is 8 MN/m^2 *C*, and the maximum shearing stress has a magnitude of 95 MN/m^2. Determine the unknown stresses on the horizontal and vertical planes and show these stresses as well as the principal stresses on a sketch.

1–72 At a point in a plane-stressed body, the nonzero principal stresses are oriented as indicated in Fig. P1–72. Determine, and show on a sketch, the stresses acting on the horizontal and vertical planes and on the plane *a–a*. Also evaluate the maximum shearing stress at the point.

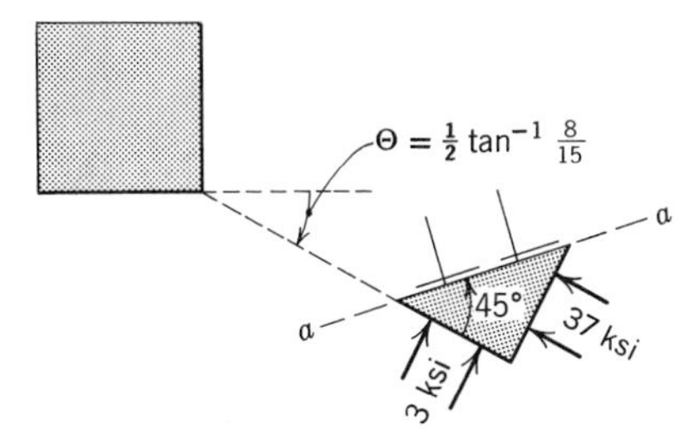

Fig. P1–72

1–73* At a point in a stressed body, there are stresses on horizontal and vertical planes as shown in Fig. P1–73. The principal stresses at the point are 100 MN/m^2 *C* and 30 MN/m^2 *T*. Determine σ_x and σ_y and show on a complete sketch the principal and maximum shearing stresses at the point.

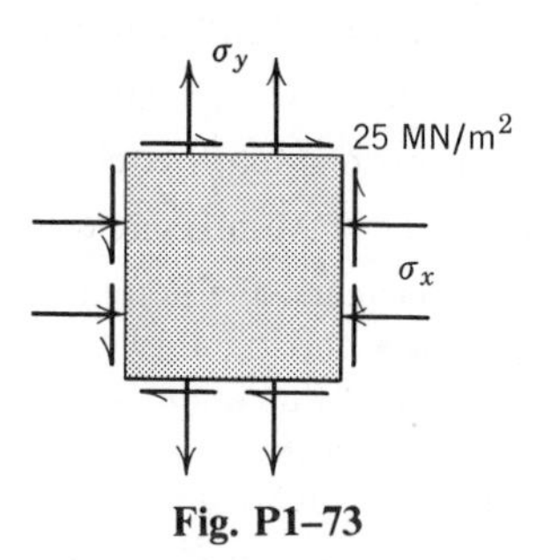

Fig. P1–73

1–74 The principal compressive stress on a vertical plane through a point in a wooden block is equal to four times the principal compressive stress on a horizontal plane. The plane of the grain is 30° clockwise from the vertical plane, and the normal and shearing stresses on this plane shall not exceed 250 psi *C* and 120 psi shear. Using Mohr's circle, determine the maximum permissible stress on the horizontal plane.

1–75 The principal stresses at a point in a region of plane stress are as shown on the stress block *B* of Fig. P1–75. Determine, and show on a sketch:

(a) The unknown stresses on plane *a–a* of block *B*.
(b) The unknown stresses on the horizontal and vertical planes of block *A*.
(c) The maximum shearing stress at the point.

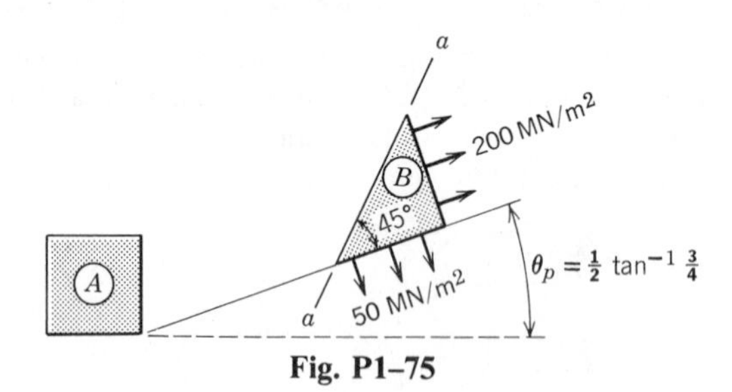

Fig. P1–75

1–76* At a point in a stressed member, on the horizontal plane there is a normal stress of 7500 psi *T* and an unknown negative shearing stress. One principal stress at the point is 20,000 psi *T*, and the maximum in-plane shearing stress has a magnitude of 8500 psi. Determine the unknown stresses on the horizontal and vertical planes and show these stresses as well as the principal stresses on a sketch.

1–77 Solve Prob. 1–71 with the same data except that the 95 MN/m^2 is the magnitude of the maximum in-plane shearing stress. Also determine the maximum shearing stress at the point.

1–9 GENERAL STATE OF STRESS AT A POINT

The general state of stress at a point was previously illustrated in Fig. 1–14. Expressions for the stresses on any oblique plane through the point in terms of stresses on the reference planes can be developed with the aid of the free-body diagram of Fig. 1–23, where the n axis is normal to the oblique (shaded) face. The areas of the faces of the element are dA for the oblique face and $dA\cos\theta_x$, $dA\cos\theta_y$, and $dA\cos\theta_z$ for the x, y, and z faces, respectively.[8] The resultant force F on the oblique face is $S\,dA$, where S is the *resultant stress* (stress vector—see Art. 1–5) on the area and is equal to $\sqrt{\sigma_n^2+\tau_{nt}^2}$. The forces on the x, y, and z faces are shown as three components, the magnitude of each being the product of the area by the appropriate stress. Using l, m, and n for $\cos\theta_x$, $\cos\theta_y$, and $\cos\theta_z$, respectively, the three force equations of equilibrium in the x, y, and z directions are

$$F_x = S_x\,dA = \sigma_x\,dA\,l + \tau_{yx}\,dA\,m + \tau_{zx}\,dA\,n$$

$$F_y = S_y\,dA = \sigma_y\,dA\,m + \tau_{zy}\,dA\,n + \tau_{xy}\,dA\,l$$

$$F_z = S_z\,dA = \sigma_z\,dA\,n + \tau_{xz}\,dA\,l + \tau_{yz}\,dA\,m$$

from which the three orthogonal components of the resultant stress are

$$\begin{aligned} S_x &= \sigma_x\,l + \tau_{yx}\,m + \tau_{zx}\,n \\ S_y &= \tau_{xy}\,l + \sigma_y\,m + \tau_{zy}\,n \\ S_z &= \tau_{xz}\,l + \tau_{yz}\,m + \sigma_z\,n \end{aligned} \tag{a}$$

The normal component σ_n of the resultant stress S equals $S_x\,l + S_y\,m + S_z\,n$; therefore, from Eq. (a), the following equation for the normal stress on any oblique plane through the point is obtained:

$$\sigma_n = \sigma_x\,l^2 + \sigma_y\,m^2 + \sigma_z\,n^2 + 2\tau_{xy}\,lm + 2\tau_{yz}\,mn + 2\tau_{zx}\,nl \tag{1–10}$$

[8]These relationships can be established by considering the volume of the tetrahedron in Fig. 1–23. Thus, $V=(1/3)\ dn\,dA=(1/3)\ dx\,dA_x=(1/3)\ dy\,dA_y=(1/3)\,dz\,dA_z$. But $dn = dx\cos\theta_x = dy\cos\theta_y = dz\cos\theta_z$; therefore, $dA_x = dA\cos\theta_x$, $dA_y = dA\cos\theta_y$, $dA_z = dA\cos\theta_z$.

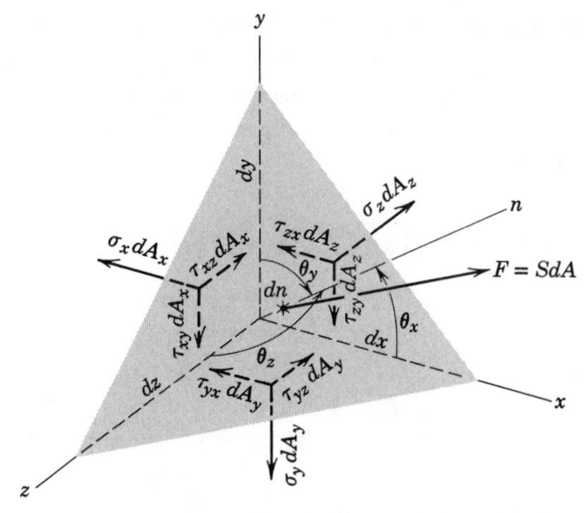

Fig. 1–23

The shearing stress τ_{nt} on the oblique plane can be obtained from the relation $S^2 = \sigma_n^2 + \tau_{nt}^2$. For a given problem, the values of S and σ_n will be obtained from Eqs. (a) and 1–10.

A principal plane was previously defined as a plane on which the shearing stress τ_{nt} is zero. The normal stress σ_n on such a plane was defined as a principal stress σ_p. If the oblique plane of Fig. 1–23 is a *principal plane*, then $S = \sigma_p$ and $S_x = \sigma_p l$, $S_y = \sigma_p m$, $S_z = \sigma_p n$. When these components are substituted into Eq. (a), the equations can be rewritten to produce the following homogeneous linear equations in l, m, and n:

$$\begin{aligned}(\sigma_p - \sigma_x)l - \tau_{yx}m - \tau_{zx}n &= 0\\ (\sigma_p - \sigma_y)m - \tau_{zy}n - \tau_{xy}l &= 0\\ (\sigma_p - \sigma_z)n - \tau_{xz}l - \tau_{yz}m &= 0\end{aligned} \qquad \text{(b)}$$

This set of equations has a nontrivial solution only if the determinant of the coefficients of l, m, and n is equal to zero. Thus,

$$\begin{vmatrix} (\sigma_p - \sigma_x) & -\tau_{yx} & -\tau_{zx} \\ -\tau_{xy} & (\sigma_p - \sigma_y) & -\tau_{zy} \\ -\tau_{xz} & -\tau_{yz} & (\sigma_p - \sigma_z) \end{vmatrix} = 0$$

Expansion of the determinant yields the following cubic equation for determining the principal stresses:

$$\sigma_p^3-(\sigma_x+\sigma_y+\sigma_z)\sigma_p^2+(\sigma_x\sigma_y+\sigma_y\sigma_z+\sigma_z\sigma_x-\tau_{xy}^2-\tau_{yz}^2-\tau_{zx}^2)\sigma_p$$
$$-(\sigma_x\sigma_y\sigma_z-\sigma_x\tau_{yz}^2-\sigma_y\tau_{zx}^2-\sigma_z\tau_{xy}^2+2\tau_{xy}\tau_{yz}\tau_{zx})=0 \quad (1\text{–}11)$$

For given values of $\sigma_x, \sigma_y, \ldots \tau_{zx}$, Eq. 1–11 gives three values for the principal stresses $\sigma_{p1}, \sigma_{p2}, \sigma_{p3}$. By substituting these values for σ_p, in turn, into Eq. (b) and using the relation

$$l^2+m^2+n^2=1 \quad \text{(c)}$$

three sets of direction cosines may be determined for the normals to the three principal planes. The foregoing discussion verifies the existence of three mutually perpendicular principal planes for the most general state of stress.

In developing equations for maximum and minimum normal stresses, the special case will be considered in which $\tau_{xy}=\tau_{yz}=\tau_{zx}=0$. No loss in generality is introduced by considering this special case since it involves only a reorientation of the reference x,y,z axes to coincide with the principal directions. Since the x,y,z planes are now principal planes, the stresses $\sigma_x, \sigma_y, \sigma_z$ become σ_{p1}, σ_{p2}, and σ_{p3}. Solving Eq. (a) for the direction cosines yield

$$l=S_x/\sigma_{p1} \qquad m=S_y/\sigma_{p2} \qquad n=S_z/\sigma_{p3}$$

By substituting these values into Eq. (c), the following equation is obtained:

$$\frac{S_x^2}{\sigma_{p1}^2}+\frac{S_y^2}{\sigma_{p2}^2}+\frac{S_z^2}{\sigma_{p3}^2}=1 \quad \text{(d)}$$

The plot of Eq. (d) is the ellipsoid shown in Fig. 1–24. It can be observed that the magnitude of σ_n is everywhere less than that of S (since $S^2=\sigma_n^2+\tau_{nt}^2$) except at the intercepts, where S is σ_{p1}, σ_{p2}, or σ_{p3}. Therefore, it can be concluded that two of the principal stresses (σ_{p1} and σ_{p3} of Fig. 1–24) are the maximum and minimum normal stresses at the point. The third principal stress is intermediate in value and has no particular significance. The discussion above demonstrates that the set of principal stresses includes the maximum and minimum normal stresses at the point.

Continuing with the special case where the given stresses σ_x, σ_y, and σ_z are principal stresses, equations can be developed for the maximum shearing stress at the point. The resultant stress on the oblique plane is

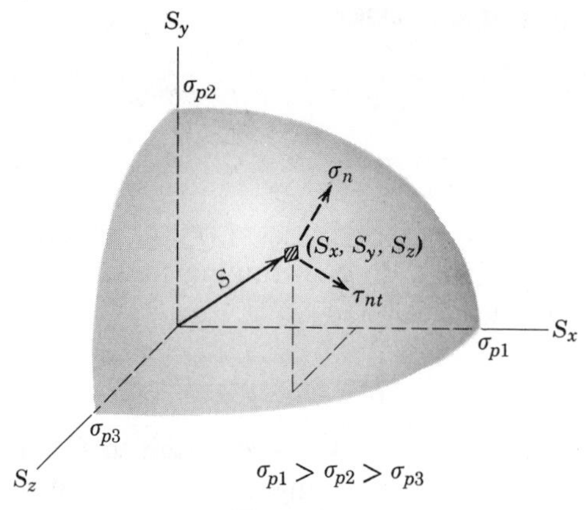

Fig. 1–24

given by the expression

$$S^2 = S_x^2 + S_y^2 + S_z^2$$

Substitution of values for S_x, S_y, and S_z from Eq. (a), with zero shearing stresses, yields the expression

$$S^2 = \sigma_x^2 l^2 + \sigma_y^2 m^2 + \sigma_z^2 n^2 \tag{e}$$

Also, from Eq. 1–10,

$$\sigma_n^2 = \left(\sigma_x l^2 + \sigma_y m^2 + \sigma_z n^2\right)^2 \tag{f}$$

Since $S^2 = \sigma_n^2 + \tau_{nt}^2$, an expression for the shearing stress τ_{nt} on the oblique plane is obtained from Eqs. (e) and (f) as

$$\tau_{nt} = \sqrt{\sigma_x^2 l^2 + \sigma_y^2 m^2 + \sigma_z^2 n^2 - \left(\sigma_x l^2 + \sigma_y m^2 + \sigma_z n^2\right)^2} \tag{1–12}$$

The planes on which maximum and minimum shearing stresses occur can be obtained from Eq. 1–12 by differentiating with respect to the direction cosines l, m, and n. One of the direction cosines in Eq. 1–12 (n for example) can be eliminated by solving Eq. (c) for n^2 and substituting into Eq. 1–12. Thus

$$\tau_{nt} = \sqrt{(\sigma_x^2 - \sigma_z^2) l^2 + \left(\sigma_y^2 - \sigma_z^2\right) m^2 + \sigma_z^2 - \left[(\sigma_x - \sigma_z) l^2 + (\sigma_y - \sigma_z) m^2 + \sigma_z\right]^2} \tag{g}$$

By taking the partial derivatives of Eq. (g)—first with respect to l and then with respect to m and equating to zero—the following equations are obtained for determining the direction cosines associated with planes having maximum and minimum shearing stresses:

$$l\left[\tfrac{1}{2}(\sigma_x-\sigma_z)-(\sigma_x-\sigma_z)l^2-(\sigma_y-\sigma_z)m^2\right]=0 \tag{h}$$

$$m\left[\tfrac{1}{2}(\sigma_y-\sigma_z)-(\sigma_x-\sigma_z)l^2-(\sigma_y-\sigma_z)m^2\right]=0 \tag{i}$$

One solution of these equations is obviously $l=m=0$. Then from Eq. (c) $n=\pm 1$. Solutions different from zero are also possible for this set of equations. Consider first that $m=0$; then from Eq. (h) $l=\pm\sqrt{1/2}$, and from Eq. (c) $n=\pm\sqrt{1/2}$. Also if $l=0$, then from Eq. (i) $m=\pm\sqrt{1/2}$, and from Eq. (c) $n=\pm\sqrt{1/2}$. Repeating the above procedure by eliminating l and m in turn from Eq. (g) yields other values for the direction cosines that make the shearing stresses maximum or minimum. All the possible solutions are listed in Table 1–1. In the last line of the table the planes corresponding to the direction cosines in the column above are shown shaded. Note that in each case only one of the two possible planes is shown.

TABLE 1–1 Direction Cosines for Shearing Stresses

	Minimum			Maximum		
	1	2	3	4	5	6
l	± 1	0	0	$\pm\sqrt{1/2}$	$\pm\sqrt{1/2}$	0
m	0	± 1	0	$\pm\sqrt{1/2}$	0	$\pm\sqrt{1/2}$
n	0	0	± 1	0	$\pm\sqrt{1/2}$	$\pm\sqrt{1/2}$

Substituting values for the direction cosines from column 4 of Table 1–1 into Eq. 1–12, with σ_x, σ_y, σ_z replaced with σ_{p1}, σ_{p2}, σ_{p3} yields

$$\tau_{\max}=\sqrt{\tfrac{1}{2}\sigma_{p1}^2+\tfrac{1}{2}\sigma_{p2}^2+0-\left(\tfrac{1}{2}\sigma_{p1}+\tfrac{1}{2}\sigma_{p2}\right)^2}=\frac{\sigma_{p1}-\sigma_{p2}}{2}$$

Similarly, using the values of the cosines from columns 5 and 6 gives

$$\tau_{max} = \frac{\sigma_{p1} - \sigma_{p3}}{2} \qquad \text{and} \qquad \tau_{max} = \frac{\sigma_{p2} - \sigma_{p3}}{2}$$

Of these three possible results, the largest magnitude will be the maximum stress; hence, the expression for the maximum shearing stress is

$$\tau_{max} = \frac{\sigma_{max} - \sigma_{min}}{2} \tag{1–13}$$

which verifies the statement in Art. 1–7 regarding maximum shearing stress.

PROBLEMS

1–78 Demonstrate that Eq. 1–11 reduces to Eq. 1–7 for the state of plane stress.

1–79* At a point in a stressed body, the known stresses are: $\sigma_x = 4$ ksi T, $\sigma_y = 2$ ksi C, $\sigma_z = 2$ ksi T, $\tau_{xy} = +4$ ksi, $\tau_{yz} = \tau_{zx} = 0$. Determine the normal and shearing stresses on a plane whose outward normal is oriented at angles of 57°, 57°, and 50.4° with the x, y, and z axes, respectively.

1–80* At a point in a stressed body, the known stresses are: $\sigma_x = 70\ \text{MN/m}^2\ T$, $\sigma_y = 60\ \text{MN/m}^2\ T$, $\sigma_z = 50\ \text{MN/m}^2\ T$, $\tau_{xy} = +20\ \text{MN/m}^2$, $\tau_{yz} = -20\ \text{MN/m}^2$, $\tau_{zx} = 0$. Determine the normal and shearing stresses on a plane whose outward normal is oriented at angles of 40°, 60°, and 66.2° with the x, y, and z axes, respectively.

1–81 At a point in a stressed body, the known stresses are: $\sigma_x = 6000$ psi T, $\sigma_y = 9000$ psi T, $\sigma_z = 6000$ psi T, $\tau_{xy} = +12{,}000$ psi, $\tau_{yz} = +7500$ psi, $\tau_{zx} = +9000$ psi. Determine the normal and shearing stresses on a plane whose outward normal is oriented at angles of 70.5°, 48.2°, and 48.2° with the x, y, and z axes, respectively.

1–82 At a point in a stressed body, the known stresses are: $\sigma_x = \sigma_y = \sigma_z = 0$, $\tau_{xy} = +30\ \text{MN/m}^2$, $\tau_{yz} = 0$, $\tau_{zx} = +40\ \text{MN/m}^2$. Determine the normal and shearing stresses on a plane whose outward normal makes equal angles with the x, y, and z axes.

1–83* At a point in a stressed body, the known stresses are: $\sigma_x = 6\ \text{MN/m}^2\ T$, $\sigma_y = 4\ \text{MN/m}^2\ C$, $\sigma_z = 2\ \text{MN/m}^2\ T$, $\tau_{xy} = +2\ \text{MN/m}^2$, $\tau_{yz} = \tau_{zx} = 0$. Determine the principal stresses and the maximum shearing stress at the point.

1–84* At a point in a stressed body, the known stresses are: $\sigma_x = 6$ ksi T, $\sigma_y = 4$ ksi C, $\sigma_z = 0$, $\tau_{xy} = +\sqrt{10.5}$ ksi, $\tau_{yz} = 0$, $\tau_{zx} = +\sqrt{10.5}$ ksi. Determine the principal stresses and the maximum shearing stress at the point.

1–85* At a point in a stressed body, the known stresses are: $\sigma_x = 60\ \text{MN/m}^2\ T$, $\sigma_y = 40\ \text{MN/m}^2\ T$, $\sigma_z = 20\ \text{MN/m}^2\ T$, $\tau_{xy} = +40\ \text{MN/m}^2$, $\tau_{yz} = +20\ \text{MN/m}^2$, $\tau_{zx} = +30\ \text{MN/m}^2$. Determine the principal stresses and the maximum shearing stress at the point.

1–86* For the stresses given in Prob. 1–79, determine the principal stresses and the maximum shearing stress at the point.

1–87* For the stresses given in Prob. 1–80, determine the principal stresses and the maximum shearing stress at the point.

1–88 For the stresses given in Prob. 1–81, determine the principal stresses and the maximum shearing stress at the point.

1–89 For the stresses given in Prob. 1–82, determine the principal stresses and the maximum shearing stress at the point.

1–90* At a point in a stressed body, the known stresses are: $\sigma_x = 10$ ksi T, $\sigma_y = 10$ ksi T, $\sigma_z = 10$ ksi T, $\tau_{xy} = +20$ ksi, $\tau_{yz} = 0$, $\tau_{zx} = +10$ ksi. Determine:

(a) The normal and shearing stresses on a plane whose outward normal is oriented at angles of 69.8°, 59.6°, and 37.8° with the x, y, and z axes, respectively.
(b) The principal stresses and the maximum shearing stress at the point.
(c) The orientation of the plane on which the maximum tensile stress acts.

1–91* At a point in a stressed body, the known stresses are: $\sigma_x = 90$ MN/m^2 T, $\sigma_y = 60$ MN/m^2 T, $\sigma_z = 30$ MN/m^2 T, $\tau_{xy} = +30$ MN/m^2, $\tau_{yz} = +30$ MN/m^2, $\tau_{zx} = +60$ MN/m^2. Determine:

(a) The principal stresses and the maximum shearing stress at the point.
(b) The orientation of the plane on which the maximum tensile stress acts.

1–10 STRESS CONCENTRATION

In the foregoing articles, it was assumed that the average stress, as determined by the expression σ or τ equals P/A, is the significant or critical stress. For many problems, this is true; for other problems, however, the maximum normal stress on a given section may be considerably greater than the average; for certain combinations of loading and material, the maximum rather than the average is the important stress. If there exists in the structural or machine element a discontinuity that interrupts the stress path (called a stress trajectory),[9] the stress at the discontinuity may be considerably greater than the nominal (average, in the case of centric loading) stress on the section; thus, there is a *stress concentration* at the discontinuity. This is illustrated in Fig. 1–25 in which a type of discontinuity is shown in the upper figure and the approximate distribution of normal stress on a transverse plane is shown in the accompanying lower figure. The ratio of the maximum stress to the nominal stress on the section is known as the *stress concentration factor*. Thus, the expression for the maximum normal stress in a centrically loaded member becomes

$$\sigma = K(P/A)$$

[9]A stress trajectory is a line everywhere parallel to the maximum normal stress.

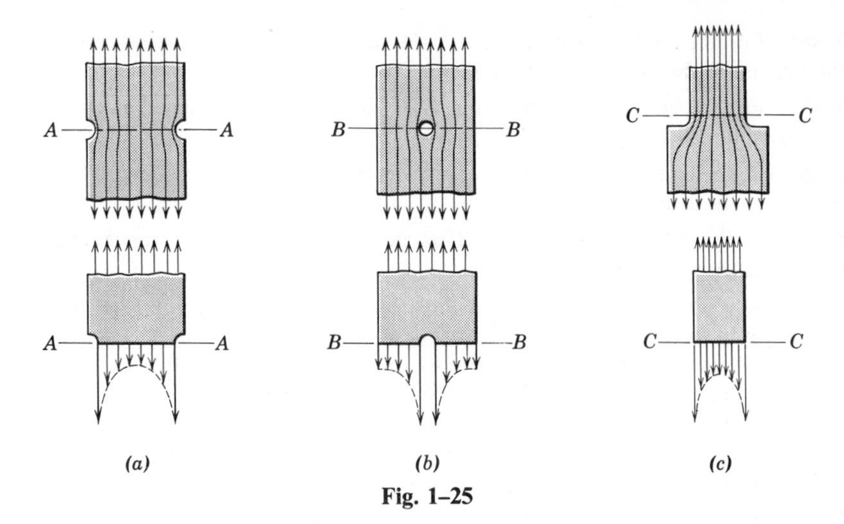

Fig. 1–25

where A is either the gross or the net area (area at the reduced section) depending on the value used for K, the stress concentration factor. Curves for stress concentration factors will be found in Chap. 10. It is important that the user of such curves (or tables of factors) ascertain whether the factors are based on the gross or net section.

A classic example of the solution of a problem involving a localized redistribution of stress occurs in the case of a small circular hole in a wide plate under uniform unidirectional tension.[10] The theory of elasticity solution is expressed in terms of a radial stress σ_r, a tangential stress σ_θ, and a shearing stress $\tau_{r\theta}$, as shown in Fig. 1–26. The equations are

$$\sigma_r = \frac{\sigma}{2}\left(1 - \frac{a^2}{r^2}\right) - \frac{\sigma}{2}\left(1 - \frac{4a^2}{r^2} + \frac{3a^4}{r^4}\right)\cos 2\theta$$

$$\sigma_\theta = \frac{\sigma}{2}\left(1 + \frac{a^2}{r^2}\right) + \frac{\sigma}{2}\left(1 + \frac{3a^4}{r^4}\right)\cos 2\theta$$

$$\tau_{r\theta} = \frac{\sigma}{2}\left(1 + \frac{2a^2}{r^2} - \frac{3a^4}{r^4}\right)\sin 2\theta$$

[10]This solution was obtained by G. Kirsch; see *Z. Ver. deut. Ing.*, Vol. 42, 1898. See also A. P. Boresi and P. P. Lynn, *Elasticity in Engineering Mechanics*, Prentice-Hall, Englewood Cliffs, N.J., 1974, pp. 304–309.

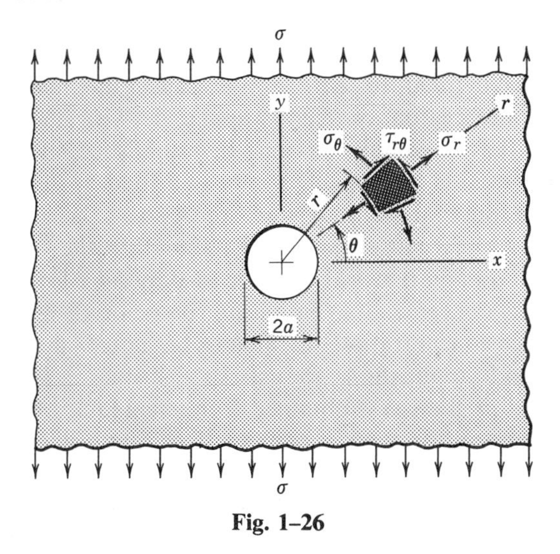

Fig. 1–26

On the boundary of the hole (at $r = a$) these equations reduce to

$$\sigma_r = 0$$

$$\sigma_\theta = \sigma(1 + 2\cos 2\theta)$$

$$\tau_{r\theta} = 0$$

At $\theta = 0°$, the tangential stress σ_θ equals 3σ where σ is the uniform tensile stress in the plate in regions far removed from the hole. Thus, the stress concentration factor associated with this type of discontinuity is 3.

The localized nature of a stress concentration can be evaluated by considering the distribution of the tangential stress σ_θ along the x axis ($\theta = 0°$). Here

$$\sigma_\theta = \frac{\sigma}{2}\left(2 + \frac{a^2}{r^2} + \frac{3a^4}{r^4}\right)$$

At a distance $r = 3a$ (one hole diameter from the hole boundary) this equation yields $\sigma_\theta = 1.074\sigma$. Thus, the stress that began as three times the nominal at the boundary of the hole has decayed to a value only 7 percent greater than the nominal at a distance of one diameter from the hole. This rapid decay is typical of the redistribution of stress in the neighborhood of a discontinuity.

Stress concentration is not significant in the case of static loading of a ductile material (defined in Art. 2–7) because the material will yield inelastically in the region of high stress and, with the accompanying redistribution of stress, equilibrium may be established and no harm done. However, if the load is an impact or repeated load, instead of the action described above, the material may fracture. Also, if the material is brittle, even a static load may cause fracture. Therefore, in the case of impact or repeated loading on any material or static loading on a brittle material, the presence of stress concentration should not be ignored. Before leaving the subject of stress concentration, it should be noted that in regions of support and load application, the stress distribution varies from the nominal (defined as the stress obtained from elementary theories of stress distribution—uniform for centric loading). This fact was discussed in 1864 by Barre de Saint-Venant (1797–1886), a French mathematician. Saint-Venant observed that although localized distortions in such regions produced stress distributions different from the theoretical, these localized effects disappeared at some distance (the implication being that the distance is not of great magnitude[11]) from such locations. This statement is known as *Saint-Venant's principle*, and is constantly used in engineering design. A more complete discussion of stress concentration is included in Chap. 10.

[11]For example, it can be shown mathematically that the localized effect of a concentrated load on a beam may disappear at a section slightly greater than the depth of the beam away from the load. See *Theory of Elasticity*, S. Timoshenko and J. N. Goodier, Third Edition, McGraw-Hill, New York, 1970.

2 Analysis of Strain and the Stress-Strain Relationship

2–1 INTRODUCTION

The state of stress at an arbitrary point in a body resulting from a system of loads was discussed in Chap. 1. The relationships between stresses on planes having different aspects at a point were developed using equilibrium considerations. No assumptions involving deformations or materials used in fabricating the body were made; therefore, the results are valid for an idealized rigid body or for a real deformable body. In the design of structural elements or machine components, the deformations experienced by the body as a result of the applied loads often represent as important a design consideration as the stresses previously discussed. For this reason the nature of the deformations experienced by a real deformable body as a result of internal force or stress distributions will be studied, and methods to measure or compute deformations will be established.

2–2 DISPLACEMENT, DEFORMATION, AND THE CONCEPT OF STRAIN

When a system of loads is applied to a machine component or structural element, individual points of the body generally move. This movement of a point with respect to some convenient reference system of axes is a vector quantity known as a *displacement*. In some instances displacements are associated with a translation and/or rotation of the body as a whole. Since the size or shape of the body is not changed by this type of displacement (referred to as a rigid-body displacement), they have no significance in the analysis of deformations. When displacements are induced by an applied load or a temperature change, individual points of the body move relative to each other; therefore, the size and/or shape of the body is altered. The change in any dimension associated with these load- or temperature-induced displacements is known as a *deformation* and will be designated by the Greek letter δ.

Under general conditions of loading, deformations will not be uniform throughout the body. Some line segments will experience extensions while others will experience contractions. Different segments (of the same length) along the same line may experience different amounts of extension or contraction. Similarly, angle changes between line segments may vary with position and orientation in the body. This nonuniform nature of load-induced deformations requires an analysis of deformations similar to the analysis of internal forces developed in Chap. 1.

Strain is a quantity used to provide a measure of the intensity of a deformation (deformation per unit length) just as stress is used to provide a measure of the intensity of an internal force (force per unit area). In Chap. 1 two types of stresses were defined: normal stresses and shearing stresses. This same classification is used for strains. *Normal strain*, designated by the letter ϵ, is used to provide a measure of the elongation or contraction of an arbitrary line segment in a body during deformation. *Shearing strain*, designated by the letter γ, is used to provide a measure of angular distortion (change in angle between two lines that are orthogonal in the undeformed state). The deformation or strain may be the result of a change in temperature, of a stress, or of other physical phenomena such as grain growth or shrinkage. In this book only strains resulting from changes in temperature or stress are considered. The term *strain* is sometimes used to denote total deformation δ; however, *in this book, strain will always refer to deformation per unit length for normal strains or changes in angle for shearing strains.*

2–3 NORMAL AND SHEARING STRAINS

The deformation (change in length and width) of a simple bar under an axial load (see Fig. 2–1*a*) can be used to illustrate the idea of a normal strain. The average axial strain (a normal strain, hereafter called *axial* strain) ϵ_{avg} over the length of the bar is obtained by dividing the axial

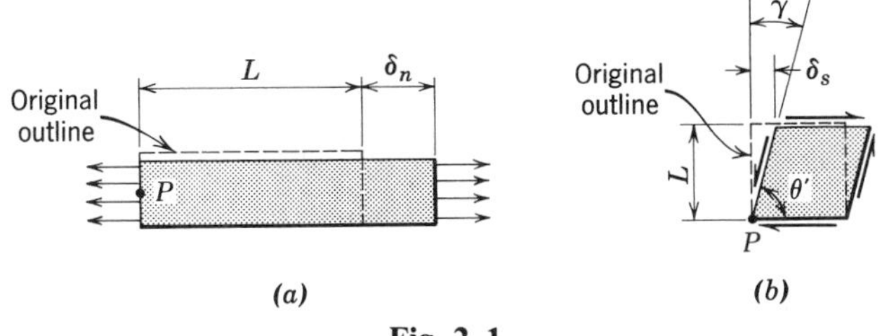

(a) (b)

Fig. 2–1

deformation δ_n by the length L of the bar. Thus

$$\epsilon_{\text{avg}} = \delta_n / L \tag{a}$$

In those cases where the deformation is nonuniform along the length of the bar (a long bar hanging under its own weight, for example), the average axial strain given by Eq. (a) may be significantly different from the true axial strain at an arbitrary point P along the bar. The true axial strain at a point can be determined by making the length over which the axial deformation is measured smaller and smaller. In the limit a quantity defined as the axial strain at the point, $\epsilon(P)$, is obtained. This limit process is indicated by the expression

$$\epsilon(P) = \lim_{L \to 0} \frac{\delta_n}{L} = \frac{d\delta_n}{dL} \tag{b}$$

In a similar manner a deformation involving a change in shape (distortion) can be used to illustrate a shearing strain. An average shearing strain γ_{avg} associated with two reference lines that are orthogonal in the undeformed state (two edges of the element shown in Fig. 2–1*b*) can be obtained by dividing the shearing deformation δ_s (displacement of the top edge of the element with respect to the bottom edge) by the distance L between these two edges. The shearing strain is defined here for small deformations only (i.e., $\tan\gamma = \gamma$). Thus

$$\gamma_{\text{avg}} = \frac{\delta_s}{L} \tag{c}$$

Again, for those cases where the deformation is nonuniform, the shearing strain at a point, $\gamma(P)$, associated with two orthogonal reference lines is obtained by measuring the shearing deformation as the size of the element is made smaller and smaller. In the limit

$$\gamma(P) = \lim_{L \to 0} \frac{\delta_s}{L} = \frac{d\delta_s}{dL} \tag{d}$$

Since shearing strain is the tangent of the angle of distortion, which for small angles is equal to the angle in radians, an equivalent expression for shearing strain that is sometimes useful for calculations is

$$\gamma(P) = \frac{\pi}{2} - \theta' \tag{e}$$

In this expression θ' is the angle in the deformed state between the two initially orthogonal reference lines.

Equations (a) through (e) indicate that both normal and shearing strains are dimensionless quantities; however, normal strains are frequently expressed in units of in./in. or microin./in. while shearing strains are expressed in radians or microradians. The symbol μ is frequently used to indicate micro (10^{-6}).

From the definition of normal strain given by Eq. (a) or (b) it is evident that normal strain is positive when the line elongates and negative when the line contracts. In general, if the axial stress is tensile the axial deformation will be an elongation. Therefore, positive normal strains are referred to as *tensile strains*. The reverse will be true for compressive axial stresses; therefore, negative normal strains are referred to as *compressive strains*. From Eq. (e) it is evident that shearing strains will be positive if the angle between the reference lines decreases. If the angle increases the shearing strain is negative. Positive and negative shearing strains are not given special names. Normal and shearing strains for most engineering materials in the elastic range (see Art. 2–8) seldom exceed values of 0.2 percent, which is equivalent to 0.002 in./in. or 0.002 rad.

PROBLEMS

2–1* A structural steel bar was loaded in tension to fracture. The 8.00-in. gage length of the rod was marked off in 1.00-in. lengths before loading. After the rod broke, the 1.00-in. segments were found to have lengthened to 1.20, 1.22, 1.25, 1.35, 1.78, 1.27, 1.23, and 1.20 in., consecutively. Compare the average strain over the 8.00-in. gage length with the maximum average strain over any 2.00-in. length.

2–2* A 13-mm round structural steel bar was loaded to fracture in tension. The 200-mm gage length was marked off in 25-mm lengths before loading. After the rod broke, the strain in the 50-mm length containing the fracture was found to be 0.480, and the total elongation of the other 150 mm was found to be 40 mm. Determine the average strain for the 200-mm gage length.

2–3* A rigid (does not bend) steel plate is supported as shown in Fig. P2–3. Posts *B* and *D* are hard rubber 4×5×8 in. Post *C* is a 2×2×7.8-in. timber. Given that all three posts are compressed when the load (plate) is applied, determine the relation between the axial strains in the two materials.

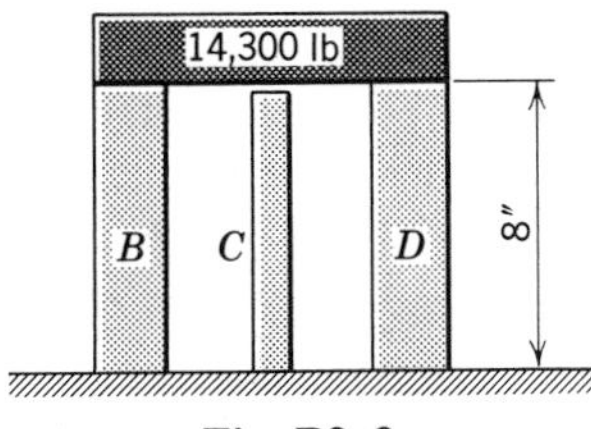

Fig. P2–3

2–4* A rigid steel plate is supported by three rods as shown in Fig. P2–4. There is no strain in the rods before the load *P* is applied. After load *P* is applied, the axial strain in rod *C* is found to be 0.0024. Determine the axial strain produced in rod *B*. The plate remains horizontal as the load is applied.

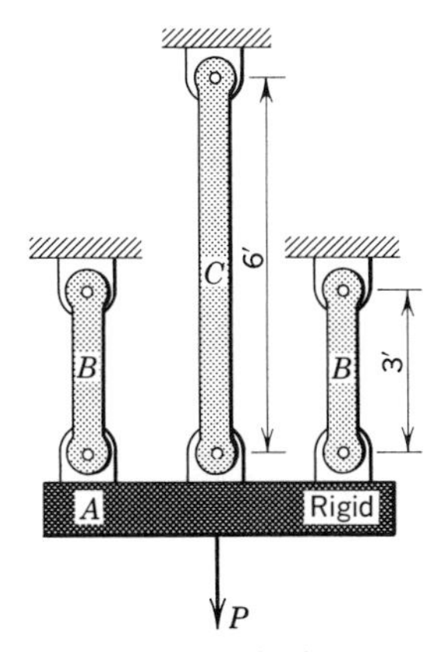

Fig. P2–4

2–5 Solve Prob. 2–4 when there is a 0.006-in. slack in the connection between *A* and *C* before the load is applied.

2–6* Solve Prob. 2–4 with the following changes in data: (1) lengths of rods *B* and *C* are 2 m and 3 m, respectively, (2) the axial strain in rod *C* is 0.0020.

2–7 Solve Prob. 2–6 when there is a slack of 0.6 mm in the connection between *A* and *C* before the load is applied.

2–8 A steel cable is used to support an elevator cage at the bottom of a 2000-ft deep mineshaft. A uniform axial strain of 250μ is produced in the cable by the weight of the cage. At each point the weight of the cable produces an additional axial strain that is proportional to the length of the cable below the point. If the total axial strain at a point at the upper end of the cable is 500μ, determine the total elongation of the cable.

2–9 Solve Prob. 2–8 with the following changes in data: (1) the mineshaft is 500 m deep, (2) the uniform axial strain produced by the weight of the cage is $500\ \mu$, (3) the axial strain produced by the weight of the cable at the midlength of the cable is 100μ.

2–10 The axial strain in a suspended bar of varying cross section (due to its weight) is given by the expression $\epsilon = (\gamma y)/(3E)$ where γ is the specific weight of the material, y is the distance from the free (bottom) end of the bar, and E is a material constant. Determine the average axial strain over the length L of the bar in terms of γ, L, and E.

2–11* A thin rectangular plate is uniformly deformed as shown in Fig. P2–11. Compute the shearing strain at P associated with the two edges (PQ and PR) that were orthogonal in the undeformed plate.

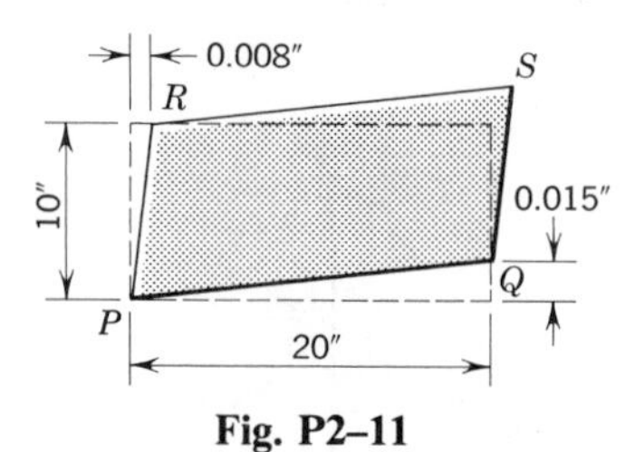

Fig. P2–11

2–12* A thin triangular plate is uniformly deformed as shown in Fig. P2–12. Compute the shearing strain at P associated with the two edges (PQ and PR) that were orthogonal in the undeformed plate.

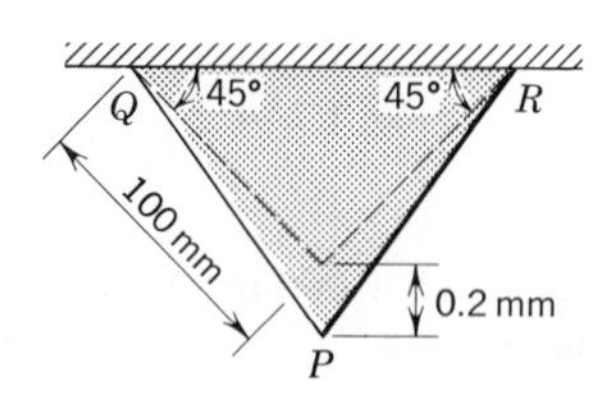

Fig. P2–12

2–13* The thin triangular plate shown in Fig. P2–13 is uniformly deformed such that $\epsilon_{PQ} = +0.0015$, $\epsilon_{QR} = 0$, and $\epsilon_{PR} = +0.0025$. Compute the shearing strain at P

associated with the two edges (PQ and PR) that were orthogonal in the undeformed plate.

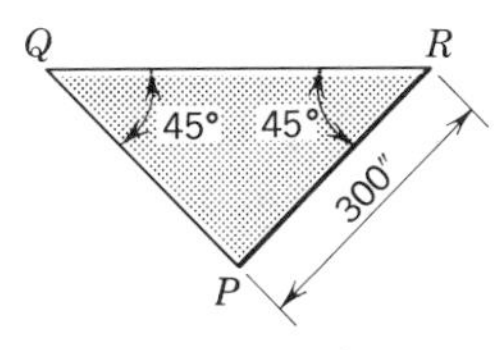

Fig. P2–13

2–14* The thin rectangular plate shown in Fig. P2–14 is uniformly deformed such that $\epsilon_{PQ} = +0.0020$, $\epsilon_{PR} = +0.0010$, and $\epsilon_{PS} = +0.0030$. Compute the shearing strain at P associated with the two edges (PQ and PR) that were orthogonal in the undeformed plate.

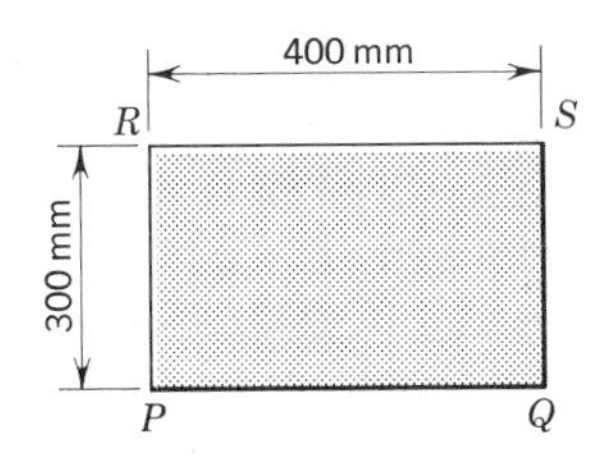

Fig. P2–14

2–15 A brake block has the shape of a circular ring segment. During application of the brake, the outer surface of the block rotates with respect to the inner surface, as shown in Fig. P2–15b. Determine, in terms of ϕ, R_1, and R_2, the shearing strains at points A and B in the block (associated with the radial and circumferential directions). Note that radial lines before deformation are assumed to be straight lines after deformation. Solve first by assuming that the angle ϕ is not small. Simplify the results for the case of a small angle.

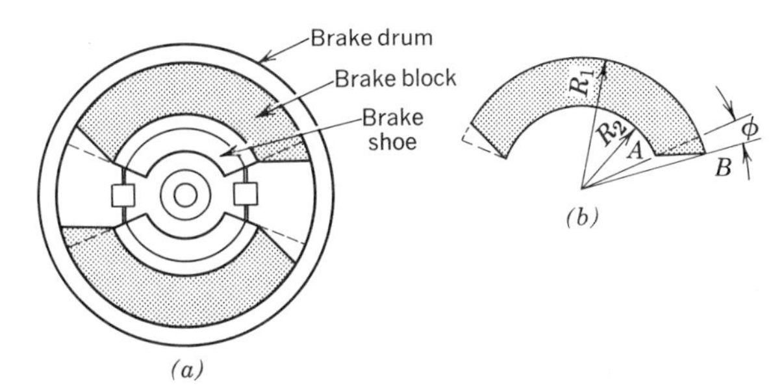

Fig. P2–15

2–4 THE STATE OF STRAIN AT A POINT

The material of Art. 2–3 serves to convey the concept of strain as a unit deformation, but it is inadequate for other than one-directional loading. The extension of the concept to biaxial loading is essential because of the important role played by the strain in experimental methods of stress evaluation. In many practical problems involving the design of structural or machine elements, the configuration and loading are too complicated to permit stress determination solely by mathematical analysis; hence, this technique is supplemented by laboratory measurements.

Strains can be measured by several methods but, except for the simplest cases, stresses cannot be obtained directly. Consequently, the usual procedure in experimental stress analysis is to measure strains and calculate the state of stress from the stress-strain equations presented in Art. 2–8.

The complete state of strain at an arbitrary point P in a body under load can be determined by considering the deformation associated with a small volume of material surrounding the point. For convenience the volume is normally assumed to have the shape of a rectangular parallelepiped with its faces oriented perpendicular to the reference axes x, y, and z in the undeformed state, as shown in Fig. 2–2*a*. Since the element of volume is very small, deformations are assumed to be uniform; therefore, parallel planes remain plane and parallel and straight lines remain straight in the deformed element, as shown in Fig. 2–2*b*. The final size of the deformed element is determined by the lengths of the three edges dx', dy', and dz'. The distorted shape of the element is determined by the angles θ'_{xy}, θ'_{yz}, and θ'_{zx} between faces.

The Cartesian components of strain at the point can be expressed in terms of the deformations by using the definitions of normal and shearing strain presented in Art. 2–3. These are the strain components associated with the Cartesian components of stress discussed in Art. 1–5 and shown in Fig. 1–13. Thus,

$$\begin{aligned}
\epsilon_x &= \frac{dx' - dx}{dx} = \frac{d\delta_x}{dx} & \gamma_{xy} &= \frac{\pi}{2} - \theta'_{xy} \\
\epsilon_y &= \frac{dy' - dy}{dy} = \frac{d\delta_y}{dy} & \gamma_{yz} &= \frac{\pi}{2} - \theta'_{yz} \\
\epsilon_z &= \frac{dz' - dz}{dz} = \frac{d\delta_z}{dz} & \gamma_{zx} &= \frac{\pi}{2} - \theta'_{zx}
\end{aligned} \tag{2–1a}$$

In a similar manner the normal strain component associated with a line oriented in an arbitrary n direction and the shearing strain component associated with two arbitrary orthogonal lines oriented in the n and t

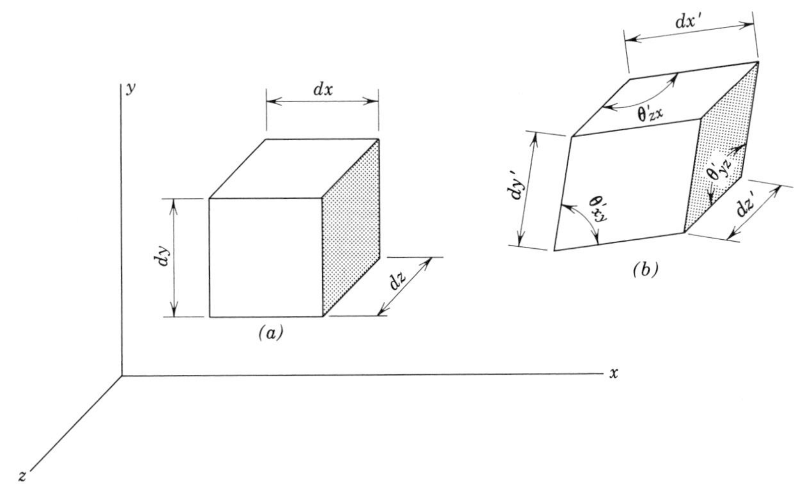

Fig. 2–2

directions in the undeformed element are given by

$$\epsilon_n = \frac{dn' - dn}{dn} = \frac{d\delta_n}{dn} \qquad \gamma_{nt} = \frac{\pi}{2} - \theta'_{nt} \tag{2–1b}$$

Alternative forms of Eq. 2–1, which will be useful in later developments, are

$$\begin{aligned} dx' &= (1+\epsilon_x)dx & \theta'_{xy} &= \frac{\pi}{2} - \gamma_{xy} \\ dy' &= (1+\epsilon_y)dy & \theta'_{yz} &= \frac{\pi}{2} - \gamma_{yz} \\ dz' &= (1+\epsilon_z)dz & \theta'_{zx} &= \frac{\pi}{2} - \gamma_{zx} \\ dn' &= (1+\epsilon_n)dn & \theta'_{nt} &= \frac{\pi}{2} - \gamma_{nt} \end{aligned} \tag{2–2}$$

2–5 TWO-DIMENSIONAL OR PLANE STRAIN

The method of relating the components of strain associated with a Cartesian coordinate system to the normal and shearing strains associated with other orthogonal directions will be illustrated by considering the *two-dimensional* or *plane strain* case. If the *xy* plane is taken as the reference plane, then for conditions of plane strain $\epsilon_z = \gamma_{zx} = \gamma_{zy} = 0$. Consider Fig. 2–3 in which the shaded rectangle represents the small unstrained element of material having the configuration of a rectangular

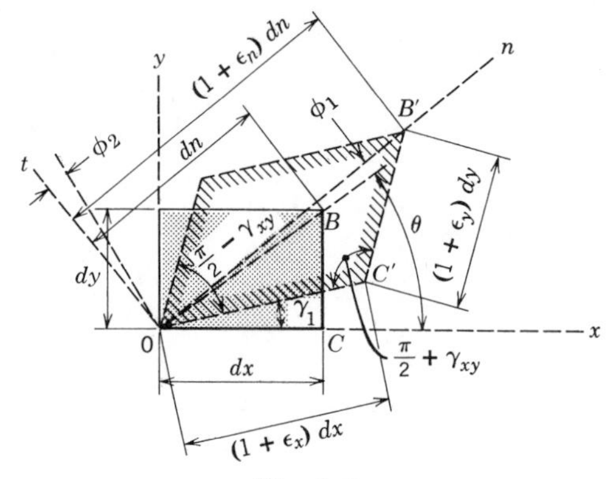

Fig. 2–3

parallelepiped. When the body is subjected to a system of loads, the element assumes the shape indicated by the dashed lines in Fig. 2–3. The dimensions of the deformed element are given in terms of strains obtained from Eq. 2–2.

An expression for normal strain in the n direction, ϵ_n, can be obtained by applying the law of cosines to the triangle $OC'B'$ shown in Fig. 2–3. Thus,

$$(OB')^2=(OC')^2+(C'B')^2-2(OC')(C'B')\cos\left(\frac{\pi}{2}+\gamma_{xy}\right)$$

or in terms of strains

$$[(1+\epsilon_n)dn]^2=[(1+\epsilon_x)dx]^2+[(1+\epsilon_y)dy]^2 \\ -2[(1+\epsilon_x)dx][(1+\epsilon_y)dy][-\sin\gamma_{xy}] \qquad \text{(a)}$$

Substituting $dx=dn\cos\theta$ and $dy=dn\sin\theta$ into Eq. (a) yields,

$$(1+\epsilon_n)^2(dn)^2=(1+\epsilon_x)^2(dn)^2(\cos^2\theta)+(1+\epsilon_y)^2(dn)^2(\sin^2\theta) \\ +2(dn)^2(\sin\theta)(\cos\theta)(1+\epsilon_x)(1+\epsilon_y)(\sin\gamma_{xy}) \qquad \text{(b)}$$

Since the strains are small, it follows that $\epsilon^2\ll\epsilon$, $\sin\gamma\approx\gamma$, etc.; hence, all second-degree terms such as ϵ^2, $\gamma\epsilon$, etc., can be neglected as Eq. (b) is expanded to become

$$1+2\epsilon_n=(1+2\epsilon_x)\cos^2\theta+(1+2\epsilon_y)\sin^2\theta+2\gamma_{xy}\sin\theta\cos\theta,$$

which reduces to

$$\epsilon_n = \epsilon_x \cos^2\theta + \epsilon_y \sin^2\theta + \gamma_{xy} \sin\theta \cos\theta \tag{2–3a}$$

or in terms of the double angle

$$\epsilon_n = \frac{\epsilon_x + \epsilon_y}{2} + \frac{\epsilon_x - \epsilon_y}{2} \cos 2\theta + \frac{\gamma_{xy}}{2} \sin 2\theta \tag{2–3b}$$

An expression for the shearing strain γ_{nt} can be obtained by applying the law of sines to the triangle $OC'B'$ (shown in Fig. 2–3) to obtain the rotations ϕ_1 and ϕ_2 associated with the n and t directions. Thus

$$\frac{(1+\epsilon_n)dn}{\sin\left(\frac{\pi}{2} + \gamma_{xy}\right)} = \frac{(1+\epsilon_y)dy}{\sin(\theta + \phi_1 - \gamma_1)} \tag{c}$$

Since the strains are small

$$\frac{(1+\epsilon_y)dy}{(1+\epsilon_n)dn} \approx (1 + \epsilon_y - \epsilon_n)\sin\theta$$

and

$$\frac{\sin(\theta + \phi_1 - \gamma_1)}{\sin\left(\frac{\pi}{2} + \gamma_{xy}\right)} \approx \sin\theta + (\phi_1 - \gamma_1)\cos\theta$$

Hence, Eq. (c) can be reduced to

$$(\epsilon_y - \epsilon_n)\sin\theta = (\phi_1 - \gamma_1)\cos\theta \tag{d}$$

Substituting Eq. 2–3a into Eq. (d) and solving for ϕ_1 yields

$$\phi_1 = -(\epsilon_x - \epsilon_y)\sin\theta\cos\theta - \gamma_{xy}\sin^2\theta + \gamma_1 \tag{e}$$

The rotation ϕ_2 associated with the t direction is obtained by substituting $\pi/2 + \theta$ for θ in Eq. (e). Thus

$$\phi_2 = (\epsilon_x - \epsilon_y)\sin\theta\cos\theta - \gamma_{xy}\cos^2\theta + \gamma_1$$

The shearing strain γ_{nt} is equal to the difference between the rotations ϕ_1 and ϕ_2. Therefore,

$$\gamma_{nt} = -2(\epsilon_x - \epsilon_y)\sin\theta\cos\theta + \gamma_{xy}(\cos^2\theta - \sin^2\theta) \tag{2–4a}$$

or in terms of the double angle

$$\gamma_{nt} = -(\epsilon_x - \epsilon_y)\sin 2\theta + \gamma_{xy}\cos 2\theta \tag{2–4b}$$

The similarity between Eqs. 2–3 and 2–4 for plane strain and Eqs. 1–4 and 1–5 for plane stress in Art. 1–7 indicates that all of the relationships developed for plane stress can be applied to plane strain by substituting ϵ_x for σ_x, ϵ_y for σ_y, and $\gamma_{xy}/2$ for τ_{xy}. Thus, from Eqs. 1–6, 1–7, and 1–8, expressions are obtained for determining the in-plane principal directions, the in-plane principal strains, and the maximum in-plane shearing strain

$$\tan 2\theta_p = \frac{\gamma_{xy}}{\epsilon_x - \epsilon_y} \tag{2–5}$$

$$\epsilon_{p1}, \epsilon_{p2} = \frac{\epsilon_x + \epsilon_y}{2} \pm \sqrt{\left(\frac{\epsilon_x - \epsilon_y}{2}\right)^2 + \left(\frac{\gamma_{xy}}{2}\right)^2} \tag{2–6}$$

$$\gamma_p = 2\sqrt{\left(\frac{\epsilon_x - \epsilon_y}{2}\right)^2 + \left(\frac{\gamma_{xy}}{2}\right)^2} \tag{2–7}$$

In the previous equations, normal strains that are tensile and shearing strains that decrease the angle between the faces of the element at the origin of coordinates (see Fig. 2–3) are positive.

When a state of plane strain exists, Eq. 2–6 gives the two in-plane principal strains while the third principal strain, ϵ_{p3}, is $\epsilon_z = 0$. An examination of Eqs. 2–6 and 2–7 indicates that the maximum in-plane shearing strain is the difference between the in-plane principal strains, but this may not be the maximum shearing strain at the point. The maximum shearing strain at the point may be $(\epsilon_{p1} - \epsilon_{p2})$, $(\epsilon_{p1} - 0)$, or $(0 - \epsilon_{p2})$, depending on the relative magnitudes and signs of the principal strains. The lines associated with the maximum shearing strain bisect the angles between lines experiencing maximum and minimum normal strains. The three possibilities are illustrated in Fig. 2–4.

The pictorial or graphic representation of Eqs. 1–4 and 1–5, known as Mohr's circle for stress, can be used with Eqs. 2–3 and 2–4 to yield a Mohr's circle for strain. The equation for the strain circle obtained from the equation for the stress circle by using a change in variables is

$$\left(\epsilon_n - \frac{\epsilon_x + \epsilon_y}{2}\right)^2 + \left(\frac{\gamma_{nt}}{2}\right)^2 = \left(\frac{\epsilon_x - \epsilon_y}{2}\right)^2 + \left(\frac{\gamma_{xy}}{2}\right)^2$$

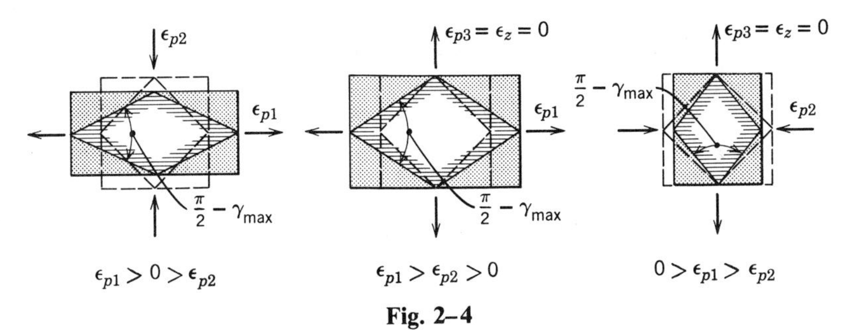

Fig. 2–4

The variables in this equation are ϵ_n and $\gamma_{nt}/2$. The circle is centered on the ϵ axis at a distance $(\epsilon_x + \epsilon_y)/2$ from the origin and has a radius

$$r = \sqrt{\left(\frac{\epsilon_x - \epsilon_y}{2}\right)^2 + \left(\frac{\gamma_{xy}}{2}\right)^2}$$

Mohr's circle for the strains of Fig. 2–3 (with $\epsilon_x > \epsilon_y$) is given in Fig. 2–5. It is apparent that the sign convention for shearing strain needs to be extended to cover the construction of Mohr's circle. Observe that for positive shearing strain (indicated in Fig. 2–3), the edge of the element

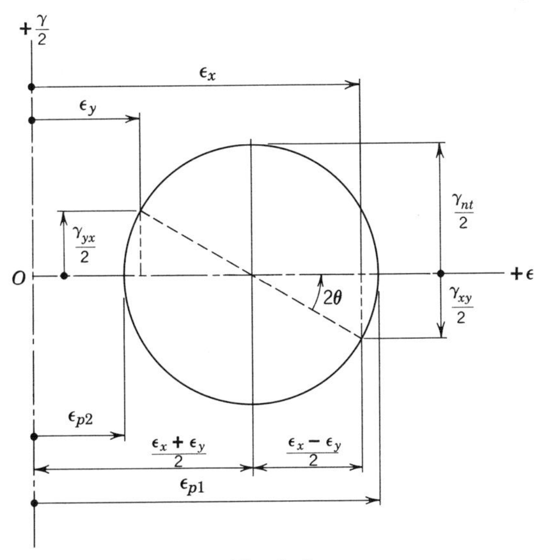

Fig. 2–5

parallel to the x axis tends to rotate counterclockwise while the edge parallel to the y axis tends to rotate clockwise. For Mohr's circle construction, the clockwise rotation will be designated positive and the counterclockwise designated negative. This is consistent with the sign convention for shearing stresses given in Art. 1–8.

Example 2–1 The strain components at a point in a body under a state of plane strain are $\epsilon_x = +1200\mu$, $\epsilon_y = -600\mu$, and $\gamma_{xy} = +900\mu$. Determine the principal strains and the maximum shearing strain at the point. Show the principal strain deformations and the maximum shearing strain distortion on a sketch.

Solution The given data are substituted in Eqs. 2–5, 2–6, and 2–7 to yield the in-plane principal strains at the point, their orientations, and the maximum in-plane shearing strain.

$$\tan 2\theta_p = \frac{900}{1200+600}$$

$$(\epsilon_{p1}, \epsilon_{p2})(10^6) = \frac{1200-600}{2} \pm \sqrt{\left(\frac{1200+600}{2}\right)^2 + \left(\frac{900}{2}\right)^2}$$

$$\gamma_p(10^6) = 2\sqrt{\left(\frac{1200+600}{2}\right)^2 + \left(\frac{900}{2}\right)^2}$$

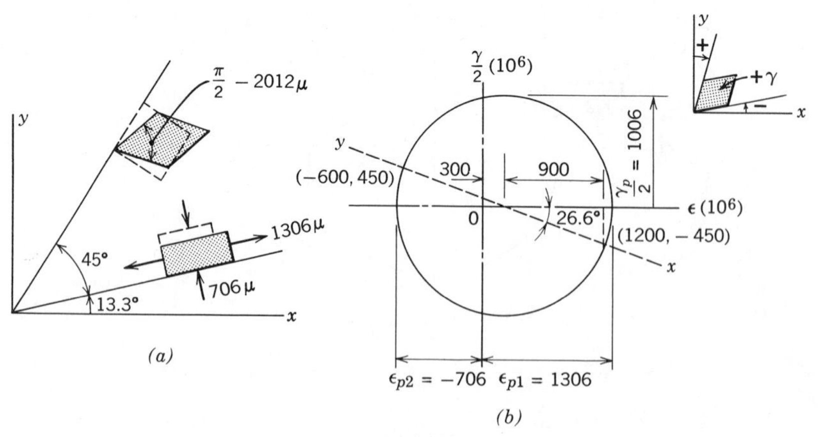

Fig. 2–6

from which

$$\epsilon_{p1} = \underline{+1306\mu}$$

$$\epsilon_{p2} = \underline{-706\mu}$$

$$\epsilon_{p3} = \underline{0} \qquad \text{Ans.}$$

$$\theta_p = \underline{13.3°}$$

$$\gamma_p = \underline{+2012\mu = \gamma_{max}}$$

The required sketch is given in Fig. 2–6*a*. The given data can also be used to construct a Mohr's circle for the state of strain at the point. The Mohr's circle solution to the problem is shown in Fig. 2–6*b*.

PROBLEMS

Note. The bodies in the following problems are assumed to be in a state of plane strain. The symbol μ is used to represent micro (10^{-6}).

2–16* The strain components at a point are $\epsilon_x = +1200\mu$, $\epsilon_y = -400\mu$, and $\gamma_{xy} = -1200\mu$. Determine the principal strains and the maximum shearing strain at the point. Show the principal strain deformations and the maximum shearing strain distortion on a sketch.

2–17* Solve Prob. 2–16 if the strain components at the point are $\epsilon_x = +1600\mu$, $\epsilon_y = -800\mu$, and $\gamma_{xy} = -1000\mu$.

2–18 Solve Prob. 2–16 if the strain components at the point are $\epsilon_x = -900\mu$, $\epsilon_y = +600\mu$, and $\gamma_{xy} = +800\mu$.

2–19* Solve Prob. 2–16 if the strain components at the point are $\epsilon_x = +1000\mu$, $\epsilon_y = +800\mu$, and $\gamma_{xy} = +1000\mu$.

In Probs. 2–20 through 2–27 certain strains or angles are given. Determine the unknown quantities for each problem and draw a sketch indicating the in-plane principal strain directions. Note that in some problems there might be more than one possible value of θ_p depending on the sign γ_{xy}.

	ϵ_x	ϵ_y	γ_{xy}	ϵ_{p1}	ϵ_{p2}	γ_{max}	θ_p
2–20	$+1300\mu$	$+1500\mu$	$+1200\mu$				
2–21	-1000μ	-600μ	-800μ				
2–22*	$+200\mu$	-800μ		$+1000\mu$			
2–23*	$+750\mu$			$+1300\mu$		$+1600\mu$	
2–24*			$+1000\mu$	$+3000\mu$			15°
2–25	$+560\mu$	-840μ		$+1520\mu$			
2–26	$+420\mu$			$+620\mu$		$+760\mu$	
2–27			$+840\mu$	$+1100\mu$			30°

2–6 PRINCIPAL STRAINS—PLANE STRESS

In most experimental work involving strain measurement, the strains are measured on a free (unstressed) surface of a member where a state of plane stress exists. If the outward normal to the surface is taken as the z-direction, then $\sigma_z = \tau_{zx} = \tau_{zy} = 0$. Since this state of stress offers no restraint to out-of-plane deformations, a normal strain ϵ_z develops in addition to the in-plane strains ϵ_x, ϵ_y, and γ_{xy}. The shearing strains γ_{zx} and γ_{zy} remain zero; therefore, the normal strain ϵ_z is a principal strain. In Art. 2–5, expressions were developed for the plane strain case relating the in-plane principal strains (ϵ_{p1} and ϵ_{p2}) and their orientations to the in-plane strains ϵ_x, ϵ_y, and γ_{xy}. For the plane stress case, which involves the normal strain ϵ_z in addition to the in-plane strains, similar expressions are needed.

As an illustration of the effects of an out-of-plane displacement on the deformation (change in length) of a line segment originally located in the xy plane, consider line AB of Fig. 2–7. As a result of the loads imposed on the member, the line AB is displaced and extended into the line $A'B'$. The displacements associated with point A' are u in the x-direction, v in the y-direction, and w in the z-direction. Point B' displaces $u + du$ in the x-direction, $v + dv$ in the y-direction, and $w + dw$ in the z-direction. The deformation δ_{AB} is obtained from the original length of the line and the displacements du, dv, and dw of point B' with respect to A'. Thus

$$(L + \delta_{AB})^2 = (L + du)^2 + (dv)^2 + (dw)^2$$

or after squaring both sides

$$L^2 + 2L\delta_{AB} + \delta_{AB}^2 = L^2 + 2L(du) + (du)^2 + (dv)^2 + (dw)^2$$

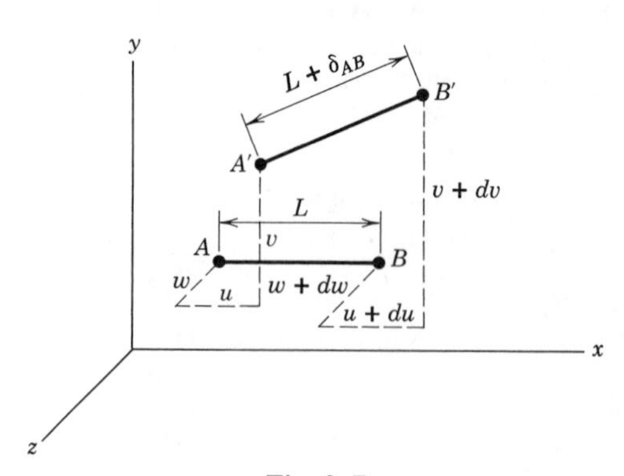

Fig. 2–7

If the deformations are small, the second degree terms can be neglected; hence

$$\delta_{AB} = du$$

This indicates that the normal strain along AB (δ_{AB} divided by L) is not affected by the presence of the out-of-plane displacements. In fact, none of the in-plane strains are affected; therefore, Eqs. 2–3 and 2–4 and their Mohr's circle representation are valid not only for the plane strain case but also for the plane stress case present when strain measurements are made on a free surface.

Electrical resistance strain gages have been developed to provide accurate measurements of normal strain. The resistance gage may be a wire gage consisting of a length of 0.001-in. diameter wire arranged as shown in Fig. 2–8*a* and cemented between two pieces of paper, or it may be an etched foil conductor mounted on epoxy or Bakelite backing (see

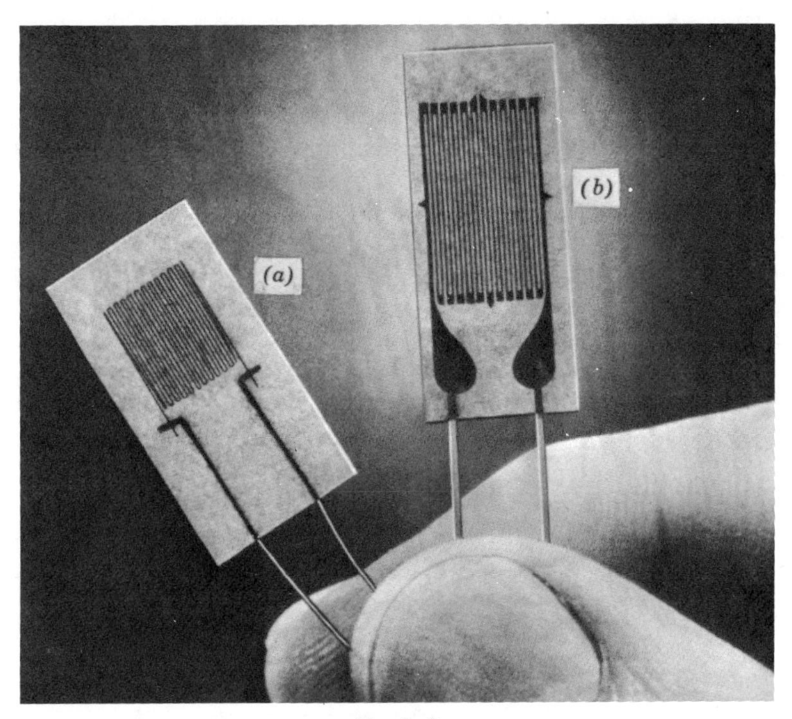

Fig. 2–8

Fig. 2–8*b*). The wire or foil gage is cemented to the material for which the strain is to be determined. As the material is strained, the wires are lengthened or shortened; this changes the electrical resistance of the gage. The change in resistance is measured by means of a Wheatstone bridge, which may be calibrated to read strain directly. Figure 5–1 shows SR-4 wire strain gages mounted on an aluminum beam. Shearing strains are more difficult to measure directly than normal strains and are often obtained by measuring normal strains in two or three different directions. The shearing strain γ_{xy} can be computed from the normal strain data by using Eq. 2–3*a*. For example, consider the most general case of three arbitrary normal strain measurements as shown in Fig. 2–9. From Eq. 2–3*a*

$$\epsilon_a = \epsilon_x \cos^2\theta_a + \epsilon_y \sin^2\theta_a + \gamma_{xy} \sin\theta_a \cos\theta_a$$

$$\epsilon_b = \epsilon_x \cos^2\theta_b + \epsilon_y \sin^2\theta_b + \gamma_{xy} \sin\theta_b \cos\theta_b$$

$$\epsilon_c = \epsilon_x \cos^2\theta_c + \epsilon_y \sin^2\theta_c + \gamma_{xy} \sin\theta_c \cos\theta_c$$

From the measured values of ϵ_a, ϵ_b, and ϵ_c and a knowledge of the gage orientations θ_a, θ_b, and θ_c with respect to the reference x axis, the values of ϵ_x, ϵ_y, and γ_{xy} can be determined by simultaneous solution of the three equations. In practice the angles θ_a, θ_b, and θ_c are selected to simplify the calculations. Multiple-element strain gages used for this type of measurement are known as strain rosettes. Several rosette configurations marketed commercially are shown in Fig. 2–10.

In this book the angles used to identify the normal strain directions of the various elements of a rosette will always be measured counterclockwise from the reference x axis. Once ϵ_x, ϵ_y, and γ_{xy} have been determined, Eqs. 2–5, 2–6, and 2–7 or the corresponding Mohr's circle can be used to determine the in-plane principal strains, their orientations, and the maximum in-plane shearing strain at the point. If the measured strains are at angles of 0°, 45°, and 90°, the determination of ϵ_x, ϵ_y, and γ_{xy} from Eq.

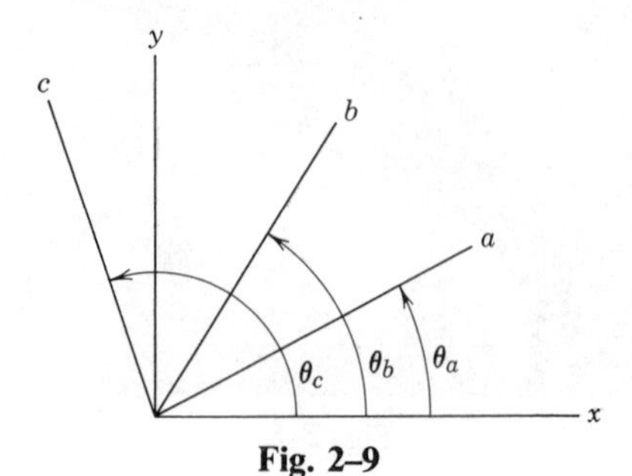

Fig. 2–9

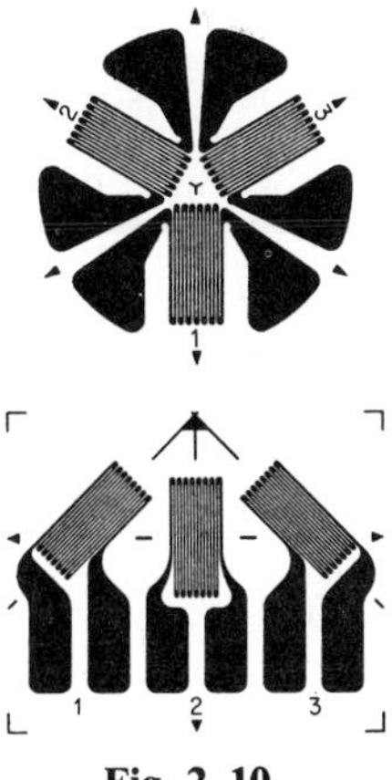

Fig. 2–10

2–3a can be eliminated, and the principal strains can be conveniently determined by direct application of Mohr's circle, as illustrated in Exam. 2–3.

In Art. 2–8 it will be shown that for plane stress

$$\epsilon_z = \epsilon_{p3} = -\frac{\nu}{1-\nu}(\epsilon_x + \epsilon_y) \tag{2–8}$$

where ν is Poisson's ratio (a property of the material used in fabricating the member), which is defined in Art. 2–7. For the case of plane stress this out-of-plane principal strain is important, since the maximum shearing strain at the point may be $(\epsilon_{p1} - \epsilon_{p2})$, $(\epsilon_{p1} - \epsilon_{p3})$, or $(\epsilon_{p3} - \epsilon_{p2})$, depending on the relative magnitudes and signs of the principal strains at the point.

Strain-measuring transducers such as electrical resistance strain gages measure the average normal strain along some gage length and not strain at a point. So long as the gage length is kept small, errors associated with such measurements can be kept within acceptable limits. The following examples illustrate the application of Eq. 2–3a and Mohr's circle to principal strain and maximum shearing strain determinations under conditions of plane stress.

Example 2–2 A strain rosette, composed of three resistance gages making angles of 0°, 60°, and 120° with the x axis, was mounted on the free surface of a material for which Poisson's ratio is 1/3. Under load the following strains were measured:

$$\epsilon_0 = \epsilon_x = +1000\mu, \qquad \epsilon_{60} = -650\mu, \qquad \epsilon_{120} = +750\mu$$

Determine the principal strains and the maximum shearing strain. Show the directions of the in-plane principal strains on a sketch.

Solution The given data are substituted in Eq. 2–3*b* to yield the following (in terms of μ):

$$\epsilon_x = +1000 \text{ (measured)}$$

$$-650 = \frac{1000+\epsilon_y}{2} + \frac{1000-\epsilon_y}{2}(\cos 120°) + \frac{\gamma_{xy}}{2}(\sin 120°)$$

$$= \frac{1000+\epsilon_y}{2} + \frac{1000-\epsilon_y}{2}\left(-\frac{1}{2}\right) + \frac{\gamma_{xy}}{2}\left(\frac{\sqrt{3}}{2}\right)$$

$$750 = \frac{1000+\epsilon_y}{2} + \frac{1000-\epsilon_y}{2}(\cos 240°) + \frac{\gamma_{xy}}{2}(\sin 240°)$$

$$= \frac{1000+\epsilon_y}{2} + \frac{1000-\epsilon_y}{2}\left(-\frac{1}{2}\right) + \frac{\gamma_{xy}}{2}\left(-\frac{\sqrt{3}}{2}\right)$$

from which $\epsilon_y = -266.7$ and $\gamma_{xy} = -1616.6$.These values are then used to construct the Mohr's circle of Fig. 2–11. The extensions of the radii to the

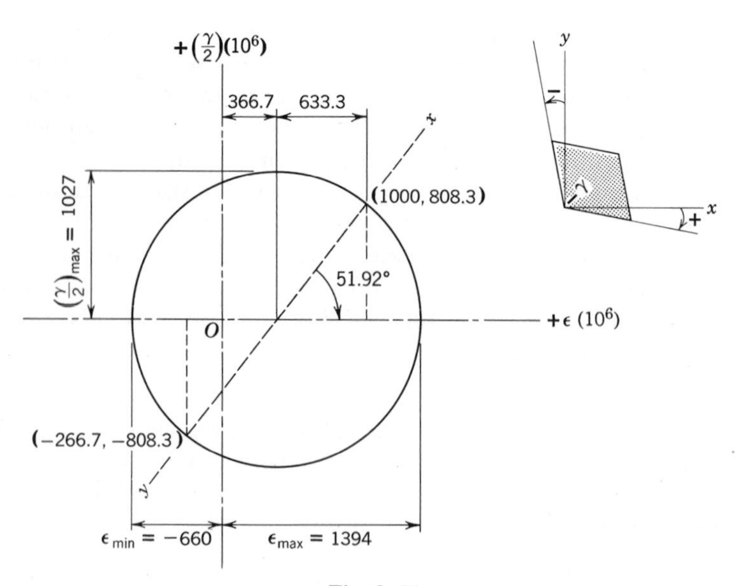

Fig. 2–11

plotted points ϵ_x, ϵ_y are designated x and y as an aid in determining the principal directions. Since γ_{xy} is negative, the angle between the edges of the element at the origin increases; hence, the edge parallel to the x axis rotates clockwise ($\gamma_{xy}/2$ is plotted positive), and the edge parallel to the y axis rotates counterclockwise ($\gamma_{xy}/2$ is plotted negative). The in-plane principal strains obtained from the circle are $\epsilon_{p1} = 1394\mu$ and $\epsilon_{p2} = -660\mu$. The principal strain perpendicular to the surface, $\epsilon_z = \epsilon_{p3}$, is found from Eq. 2–8 as follows:

$$\epsilon_z = \epsilon_{p3} = -\frac{1/3}{1-(1/3)}(1000\mu - 267\mu) = -367\mu$$

Since ϵ_{p3} is less compressive than the in-plane principal strain ϵ_{p2}, the maximum shearing strain at the point is the maximum in-plane shearing strain γ_p, obtained from the circle, as

$$\gamma_{max} = \gamma_p = \pm(1394\mu + 660\mu) = \pm 2054\mu = \underline{\pm 2050\mu} \qquad \text{Ans.}$$

The principal strains are

$$\epsilon_{p1} = \epsilon_{max} = \underline{+1394\mu}$$

$$\epsilon_{p3} = \epsilon_{int} = \underline{-367\mu} \qquad \text{Ans.}$$

$$\epsilon_{p2} = \epsilon_{min} = \underline{-660\mu}$$

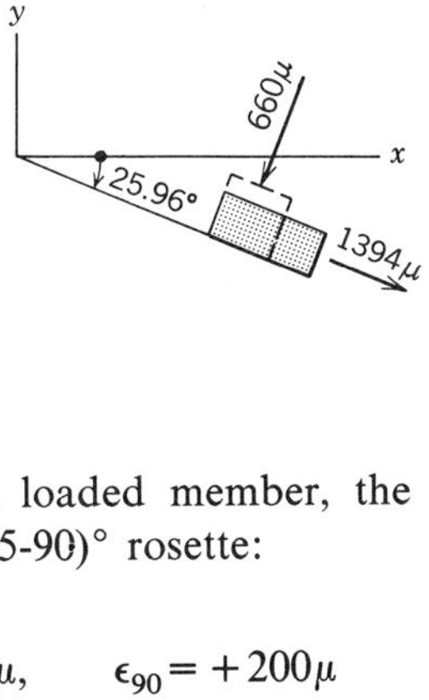

as shown in the accompanying sketch.

Example 2–3 On the free surface of a loaded member, the following strains were measured by means of a (0-45-90)° rosette:

$$\epsilon_0 = -300\mu, \qquad \epsilon_{45} = -200\mu, \qquad \epsilon_{90} = +200\mu$$

Determine the in-plane principal and maximum shearing strains by direct application of Mohr's circle. Show the directions of the principal strains on a sketch.

Solution Let the direction of ϵ_0 be the x axis. Note that the two strains ϵ_x and ϵ_y (ϵ_0 and ϵ_{90}) will be at opposite ends of a diameter of Mohr's circle with the larger (algebraic) value at the right; points A and B in Fig. 2–12a. Then, since the gage angles are measured counterclockwise from the x axis, the radius to the ϵ_{45} strain CD must be at 90° counterclockwise from the radius CA. Thus, the correct relative positions of the points representing the three measured strains are as shown in Fig. 2–12*a*. Having sketched (or visualized) the relative positions of the three points, and assuming that ϵ_x and ϵ_y are not principal strains, the triad of radii is rotated clockwise or counterclockwise to such a position that the values of the measured strains represented by the three end points increase in correct sequence from left to right. In this problem, the rotation is clockwise, as shown in Fig. 2–12*b*, where $\epsilon_{90} > \epsilon_{45} > \epsilon_0$. Referring back to Fig. 2–5, the center of the circle is located at

$$\frac{\epsilon_x + \epsilon_y}{2} = \frac{-300+200}{2} = -50$$

as shown in Fig. 2–12*b*. Also,

$$\frac{\epsilon_x - \epsilon_y}{2} = \frac{-300-200}{2} = -250$$

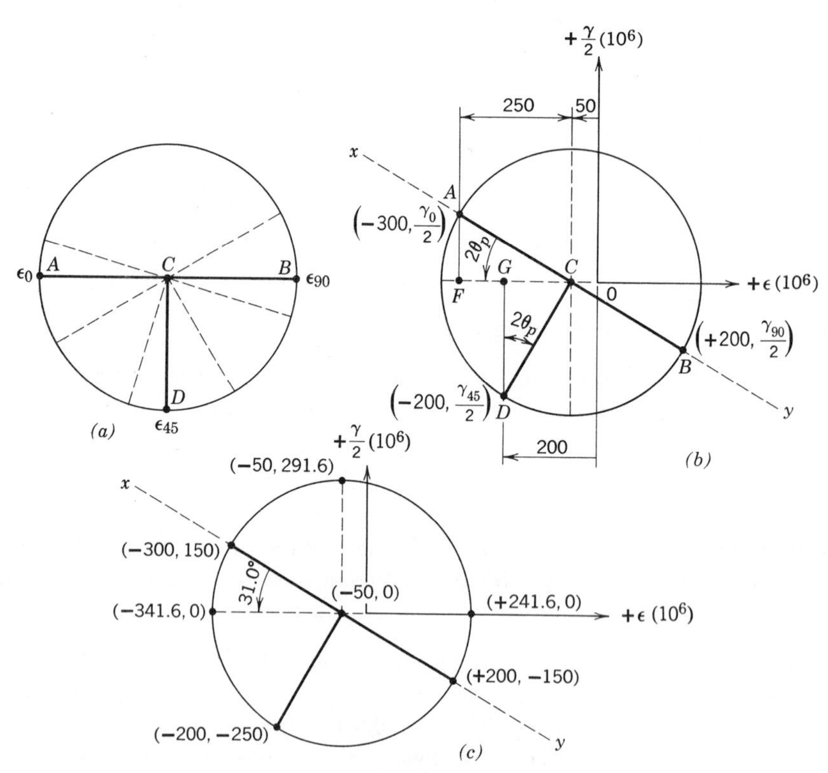

Fig. 2–12

which is the base CF of the triangle ACF. Since the triangle DCG is similar to ACF, line DG also has a magnitude of 250. From similar triangles, the magnitude of line AF is the same as that of GC, which is $200-50=150$. Therefore, the radius of the circle is $\sqrt{250^2+150^2}=291.6$. The complete circle is shown in Fig. 2–12c. From the circle the results are

$$\epsilon_{p1}=\epsilon_{\max}=\underline{+242\mu}$$

$$\epsilon_{p2}=\epsilon_{\min}=\underline{-342\mu}$$

$$\gamma_p=\gamma_{\max}=\underline{\pm 583\mu}$$

Ans.

PROBLEMS

Note. In the the following plane stress problems, the symbol μ is used to represent micro (10^{-6}). For the determination of ϵ_z, assume Poisson's ratio is 0.3.

2–28* At a point on the free surface of a machine part, the known strains are $\epsilon_x=+200\mu$, $\epsilon_y=-120\mu$, and $\gamma_{xy}=+240\mu$. Determine the principal strains and the maximum shearing strain at the point. Show the directions of the in-plane principal strains on a sketch.

2–29 If the strains at the point of Prob. 2–28 had been measured with a (0-60-120)° rosette, what should the strain readings have been?

2–30* An investigator measured the following average normal strains at a point on the free surface of a stressed body. The strains measured at 30°, 60°, and 135° counterclockwise from the horizontal x axis are $+600\mu$, -200μ, and $+100\mu$, respectively. Determine the principal strains and the maximum shearing strain at the point. Show the directions of the in-plane principal strains on a sketch.

2–31 Strains measured on a free surface with a strain rosette composed of three gages at angles of 0°, 60°, and 120° with the x axis are -400μ, $+800\mu$, and -600μ, respectively. Determine the principal strains and the maximum shearing strain at the point. Show the directions of the in-plane principal strains on a sketch.

2–32* Solve Prob. 2–31 if the measured strains are $+500\mu$, $+100\mu$, and $+300\mu$, respectively.

2–33 Strains measured on a free surface with a strain rosette composed of three gages at angles of 0°, 45°, and 90° with the x axis are -400μ, -200μ, and $+200\mu$, respectively. Determine the principal strains and the maximum shearing strain at the point. Show the directions of the in-plane principal strains on a sketch.

2–34* Solve Prob. 2–33 if the measured strains are $+400\mu$, $+260\mu$, and -80μ, respectively.

2–35 If the strains in the region of Prob. 2–34 had been measured with a (0-60-120)° rosette, what should the strain readings have been?

2–36* At a point on the free surface of a stressed region, the strains measured at angles of 0°, 45°, and 90° with the x axis are $+700\mu$, $+350\mu$, and -500μ, respectively. Determine the principal strains and the maximum shearing strain at the point. Show the directions of the in-plane principal strains on a sketch.

2–37* At a point on the free surface of a machine part, the strains are $\epsilon_x = +900\mu$, $\epsilon_y = +400\mu$, and $\gamma_{xy} = +1000\mu$. Use Mohr's circle to determine the strain that would be indicated by the strain gage shown in Fig. P2–37.

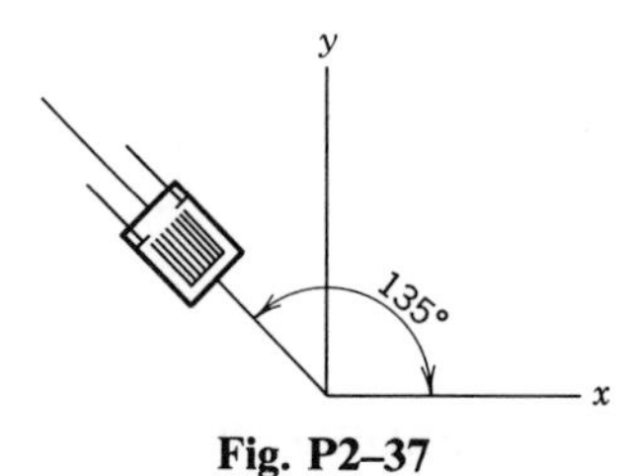

Fig. P2–37

2–38 At a point on the free surface of a machine part, the strain rosette shown in Fig. P2–38 was used to obtain the following normal strain data: $\epsilon_a = +1000\mu$, $\epsilon_b = +200\mu$, and $\epsilon_c = +500\mu$. Use Mohr's circle to determine the principal strains and the maximum shearing strain at the point. Show the principal strain deformations and the maximum shearing strain distortion on a sketch.

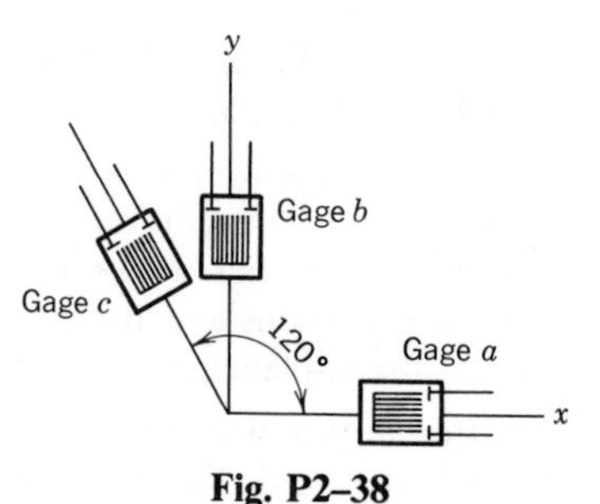

Fig. P2–38

2–39 The strain rosette shown in Fig. P2–39 was used to obtain the following normal strain data at a point on the free surface of a machine part: $\epsilon_a = +900\mu$, $\epsilon_b = +300\mu$, and $\epsilon_c = -300\mu$. Use Mohr's circle to determine the strain components ϵ_x, ϵ_y, and γ_{xy}.

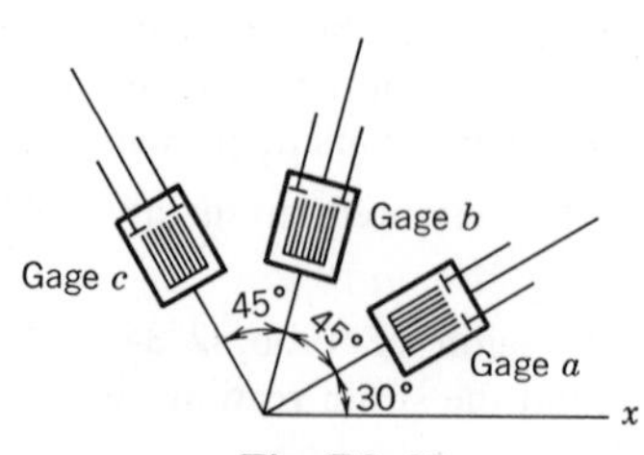

Fig. P2–39

2–40 For the strains given in Prob. 2–39, determine the principal strains and the maximum shearing strain at the point. Show the principal strain deformations and the maximum shearing strain distortion on a sketch.

2–7 PROPERTIES OF MATERIALS—STATIC

The satisfactory performance of a structure frequently is determined by the amount of deformation or distortion which can be permitted. A deflection of a few thousandths of an inch might make a boring machine useless, whereas the boom on a dragline might deflect several inches without impairing its usefulness. It is often necessary to relate the loads on a structure, or on a member in a structure, to the deflection the loads will produce. Such information can be obtained by plotting diagrams showing loads and deflections for each member and type of loading in a structure, but such diagrams will vary with the dimensions of the members, and it would be necessary to draw new diagrams each time the dimensions were varied. A more useful diagram is one showing the relation between the stress and strain. Such diagrams are called stress-strain diagrams.

Data for stress-strain diagrams are usually obtained by applying an axial load to a test specimen and measuring the load and deformation simultaneously. A testing machine (Fig. 2–13) is used to strain the specimen and to measure the load required to produce the strain. The stress is obtained by dividing the load by the initial cross-sectional area of the specimen. The area will change somewhat during the loading, and the stress obtained using the initial area is obviously not the exact stress occurring at higher loads. It is the stress most commonly used, however, in designing structures. The stress obtained by dividing the load by the actual area is frequently called the *true stress* and is useful in explaining the fundamental behavior of materials. Strains are usually relatively small in materials used in engineering structures, often less than 0.001, and their accurate determination requires special measuring equipment. Normal strain can be obtained by measuring the deformation δ in a length L and dividing δ by L. Instruments for measuring the deformation δ are called strain gages, or sometimes extensometers or compressometers, and obtain the desired accuracy by multiplying levers, dial indicators, beams of light, or other means. The electrical resistance strain gage described in Art. 2–6 is widely used for this type of measurement.

True strain, like true stress, is computed on the basis of the actual length of the test specimen during the test and is used primarily to study the fundamental properties of materials. The difference between nominal stress and strain, computed from initial dimensions of the specimen, and true stress and strain is negligible for stresses usually encountered in engineering structures, but sometimes the difference becomes important with larger stresses and strains.

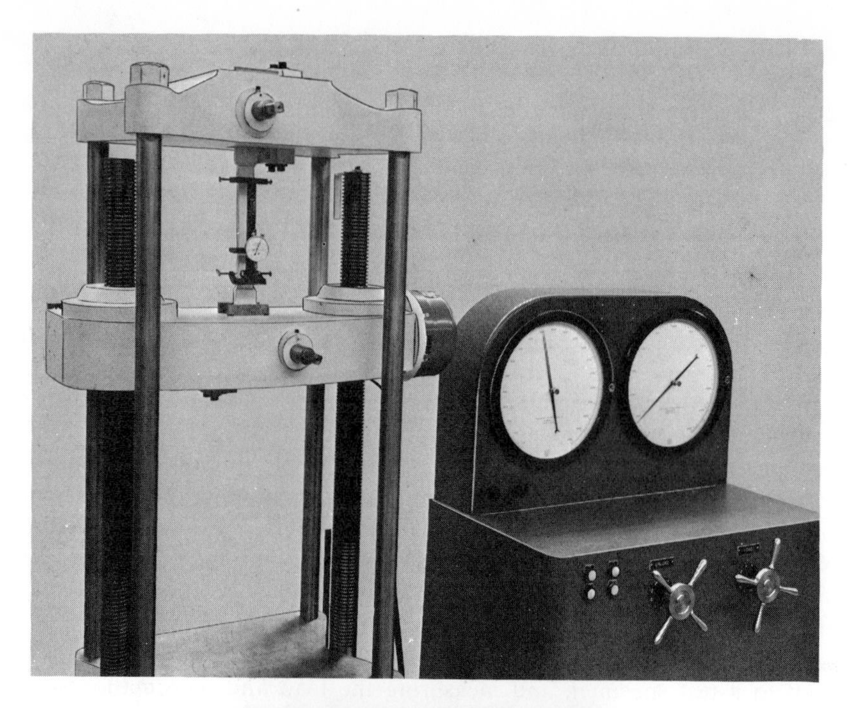

Fig. 2–13 Hydraulic testing machine set up for a tension test. The load on the specimen is indicated on one of the large dials at upper right. Inset shows an extensometer (for measuring strains) attached to the specimen.

A more complete discussion of the experimental determination of stress and strain will be found in various books on experimental stress analysis.[1,2]

Figures 2–14*a* and *b* show tensile stress-strain diagrams for structural steel (a low-carbon steel) and for a magnesium alloy. These diagrams will be used to explain a number of properties useful in the study of mechanics of materials.

The initial portion of the stress-strain diagram for most materials used in engineering structures is a straight line. The stress-strain diagrams for some materials, such as gray cast iron and concrete, show a slight curve even at very small stresses, but it is common practice to draw a straight line to average the data for the first part of the diagram and neglect the curvature. The proportionality of load to deflection was first recorded by Robert Hooke who observed in 1678, "*Ut tensio sic vis*" (as the stretch so the force); this is frequently referred to as Hooke's law. Thomas Young, in 1807, suggested what amounts to using the ratio of stress to strain to measure the stiffness of a material. This ratio is called *Young's modulus* or the *modulus of elasticity* and is the slope of the straight line portion of the stress-strain diagram. Young's modulus is written as

$$E=\sigma/\epsilon \quad \text{or} \quad G=\tau/\gamma$$

where E is used for normal stress and strain and G (sometimes called the modulus of rigidity) is used for shearing stress and strain. The maximum stress for which stress and strain are proportional is called the *proportional limit* and is indicated by the ordinates at points A on Fig. 2–14*a* or *b*.

For points on the stress-strain curve beyond the proportional limit (such as point C on Fig. 2–14*c*), other quantities such as the *tangent modulus* and the *secant modulus* are used as measures of the stiffness of a material. The tangent modulus E_t is defined as the slope of the stress-strain diagram at a particular stress level. Thus, the tangent modulus is a function of the stress (or strain) for stresses greater than the proportional limit. For stresses less than the proportional limit, the tangent modulus is the same as Young's modulus. The secant modulus E_s is the ratio of the stress to the strain at any point on the diagram. Young's modulus E, the tangent modulus E_t, and the secant modulus E_s are all illustrated in Fig. 2–14*c*.

The action is said to be *elastic* if the strain resulting from loading disappears when the load is removed. The *elastic limit* is the maximum stress for which the material acts elastically. For most materials it is found that the stress-strain diagram for unloading is approximately parallel to the

[1]*Handbook of Experimental Stress Analysis*, M. Hetenyi, Wiley, New York, 1950.

[2]*Experimental Stress Analysis*, J. W. Dally and W. F. Riley, McGraw Hill, New York, 1965.

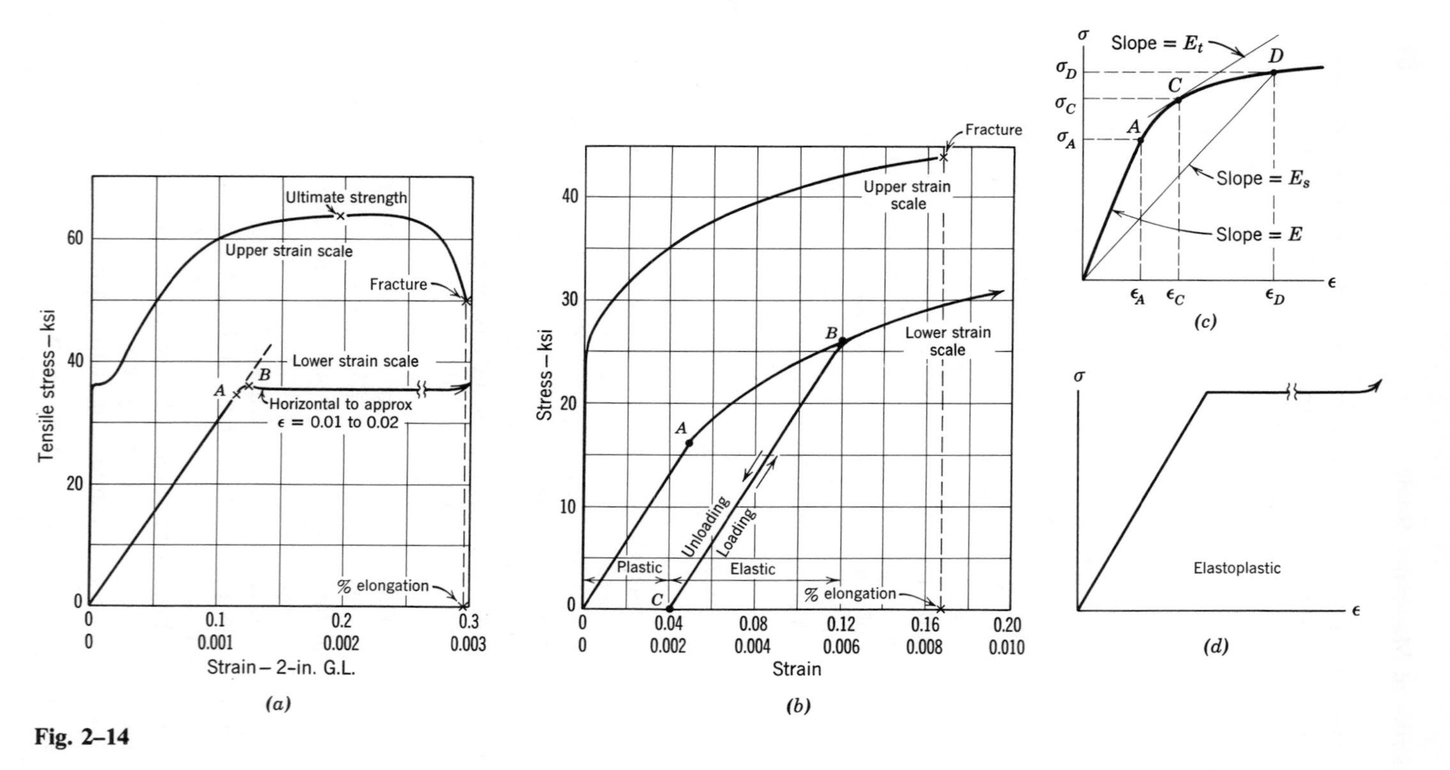
Ultimate strength
Upper strain scale
Fracture
Lower strain scale
Horizontal to approx
ε = 0.01 to 0.02
% elongation
Tensile stress – ksi
Strain – 2-in. G.L.
(a)
Fracture
Upper strain scale
Lower strain scale
Unloading
Loading
Plastic
Elastic
% elongation
Stress – ksi
Strain
(b)
Slope = E_t
Slope = E_s
Slope = E
(c)
Elastoplastic
(d)

Fig. 2–14

loading portion (see line BC in Fig. 2–14*b*). If the specimen is again loaded, the stress-strain diagram will usually follow the unloading curve until it reaches stress a little less than the maximum stress attained during the initial loading, at which time it will start to curve in the direction of the initial loading curve. As indicated in Fig. 2–14*b* the proportional limit for the second loading is greater than that for the initial loading. This phenomenon is called *strain hardening* or *work hardening*.

When the stress exceeds the elastic limit (or proportional limit for practical purposes), it is found that a portion of the deformation remains after the load is removed. The deformation remaining after an applied load is removed is called *plastic deformation*. Plastic deformation independent of the time duration of the applied load is known as *slip*. *Creep* is plastic deformation that continues to increase under a constant stress. In many instances creep continues until fracture occurs; however, in other instances the rate of creep decreases and approaches zero as a limit. Some materials are much more susceptible to creep than are others, but most materials used in engineering exhibit creep at elevated temperatures. The total strain is thus made up of elastic strain, possibly combined with plastic strain which results from slip, creep, or both. When the load is removed, the elastic portion of the strain is recovered, but the plastic part (slip and creep) remains as permanent set.[3]

A precise value for the proportional limit is difficult to obtain, particularly when the transition of the stress-strain diagram from a straight line to a curve is gradual. For this reason, other measures of stress which can be used as a practical elastic limit are required. The yield point and the yield strength for a specified offset are frequently used for this purpose.

The *yield point* is the stress at which there is an appreciable increase in strain with no increase in stress, with the limitation that, if straining is continued, the stress will again increase. This latter specification indicates that there is a kink or "knee" in the stress-strain diagram, as indicated in Fig. 2–14*a*. The yield point is easily determined without the aid of strain-measuring equipment because the beam of the testing machine drops (or if a dial indicator is used, the needle halts) at the yield point. Unfortunately, few materials possess this property, the most common examples being the low-carbon steels.

The *yield strength* is defined as the stress that will induce a specified permanent set, usually 0.05 to 0.3 percent, which is equivalent to a strain of 0.0005 to 0.003. The yield strength is particularly useful for materials with no yield point. The yield strength can be conveniently determined from a

[3]In some instances a portion of the strain which remains immediately after the stress is removed may disappear after a period of time. This reduction of strain is sometimes called *recovery*.

stress-strain diagram by laying off the specified offset (permanent set) on the strain axis as OC in Fig. 2–14*b* and drawing a line CB parallel to OA. The stress indicated by the intersection of CB and the stress-strain diagram is the yield strength.

The maximum stress, based on the original area, developed in a material before rupture is called the *ultimate strength* of the material, and the term may be modified as the ultimate tensile, compressive, or shearing strength of the material. Ductile materials undergo considerable plastic tensile or shearing deformation before rupture. When the ultimate strength of a ductile material is reached, the cross-sectional area of the test specimen starts to decrease or neck down, and the resultant load which can be carried by the specimen decreases. Thus, the stress based on the original area decreases beyond the ultimate strength of the material, although the true stress continues to increase until rupture.

Many engineering structures are designed so that the stresses are less than the proportional limit, and Young's modulus provides a simple and convenient relationship between the stress and strain. When the stress exceeds the proportional limit, the deformation becomes partially plastic, and no simple relation exists between the stress and strain. Various empirical equations have been proposed relating the stress and strain beyond the proportional limit. The following equation, known as the Ramberg-Osgood[4] formula, is one that has been used in plastic analysis:

$$\epsilon = \epsilon_0 \left[\frac{\sigma}{\sigma_0} + \frac{3}{7} \left(\frac{\sigma}{\sigma_0} \right)^n \right]$$

in which ϵ and σ are the strain and stress to be related, and ϵ_0, σ_0, and n are experimentally determined constants. Corresponding values of stress and strain can also be read from stress-strain diagrams. A stress-strain diagram similar to the one in Fig. 2–14*d* is frequently assumed for mild steel or other material with similar properties in order to simplify the calculations. In this book, material with such a stress-strain diagram will be designated *elastoplastic*, and the proportional limit and yield points will be assumed to be the same. For mild steel the plastic strain that occurs at the yield point with no increase in stress is 15–20 times the elastic strain at the proportional limit.

Strength and stiffness are not the only properties of interest to a design

[4]"Description of Stress-Strain Curves by Three Parameters," W. Ramberg and W. R. Osgood, *NACA* TN 902, 1943.

engineer. Another important property is *ductility*, defined as the capacity for plastic deformation in tension or shear. This property controls the amount of cold forming to which a material may be subjected. The forming of automobile bodies, the bending of concrete-reinforcing bars, and the manufacture of fencing and other wire products all require ductile materials. Ductility is also an important property of materials used for fabricated structures. Under static loading the presence of stress concentration in the region of rivet holes or welds may be ignored, since the ductility permits considerable plastic action to take place in the region of high stresses, with a resulting redistribution of stress and the establishment of equilibrium. Two commonly used quantitative indices of ductility are the *ultimate elongation* (expressed as a percentage elongation of the gage length at rupture) and the *reduction of cross-sectional area* at the section where rupture occurs (expressed as a percentage of the original area).

The property indicating the resistance of a material to failure by creep is known as the *creep limit* and is defined as the maximum stress for which the plastic strain will not exceed a specified amount during a specified time interval at a specified temperature. This property is important when designing metal parts subjected to high temperatures and sustained loads—for example, the turbine blades in a turbojet engine.

A material loaded in one direction will undergo strains perpendicular to the direction of the load as well as parallel to it. The ratio of the lateral or perpendicular strain to the longitudinal or axial strain is called *Poisson's ratio*, after Simeon D. Poisson, who identified this constant in 1811, and it is a constant for stresses below the proportional limit. Poisson's ratio varies from $\frac{1}{4}$ to $\frac{1}{3}$ for most metals. The symbol ν is used for Poisson's ratio, which is given by the equation

$$\nu = \epsilon_{\text{lateral}} / \epsilon_{\text{longitudinal}}$$

It will be shown in Art. 2–8 that Poisson's ratio is related to E and G by the formula

$$E = 2(1+\nu)G$$

The properties discussed in this article are primarily concerned with static or continuous loading. Properties that are important when considering energy or impact loads and repeated loads are discussed in Art. 2–9 and

Chap. 10. A more complete discussion of properties of materials is contained in several excellent properties books.[5,6]

PROBLEMS

Note. For simplicity of calculations in the solution of problems in this book, use the approximate values of $30(10^6)$ psi (or 200 GN/m^2) and $12(10^6)$ psi (or 80 GN/m^2) for the moduli of elasticity and rigidity, respectively, for steel without identification as to kind. If the steel is identified (structural steel, etc.) the values given in App. A will be used. Unless the proportional limit is given or referred to in the problem, it is assumed that the action is elastic. If the proportional limit is given without a stress-strain diagram, it is assumed that the diagram is elastoplastic (see Fig. 2–14*d*). When the stress-strain diagram is given or referred to, it is to be used to check the possibility of plastic action.

2–41* At the proportional limit, an 8-in. gage length of a 0.505-in. diameter alloy bar has elongated 0.036 in., and the diameter has been reduced 0.00074 in. The total axial load carried was 9600 lb. Determine the following properties of this material:

(a) Modulus of elasticity.
(b) Poisson's ratio.
(c) Proportional limit.

2–42* An axial load of 430 kN is slowly applied to a $25\times100\times2500$-mm rectangular bar. When loaded (the action is elastic) the 100-mm side measures 99.965 mm, and the length has increased 2.500 mm. Determine Poisson's ratio and Young's modulus for the material.

2–43 A $\frac{1}{4}\times2$-in. flat alloy bar elongates 0.048 in. in a length of 5 ft under a total axial load of 8400 lb. The proportional limit of the material is 44,000 psi.

(a) Determine the axial stress in the bar.
(b) What is the modulus of elasticity of this material?
(c) If Poisson's ratio for the material is 0.32, what will be the total change in each lateral dimension?

2–44* A 40-mm diameter rod 6 m long elongates 12 mm under a load of 240 kN. The diameter decreased 0.028 mm during the loading. Determine the following properties of the material:

(a) Poisson's ratio.
(b) Young's modulus.
(c) The modulus of rigidity.

2–45* A tensile test specimen having a diameter of 0.505-in. and a gage length of 2.000-in. was tested to fracture. Load and deformation data obtained during the

[5] *The Structure and Properties of Materials, Vol. III*, "Mechanical Behavior," H. W. Hayden, W. G. Moffatt, and John Wulff, Wiley, New York, 1965.

[6] *Elements of Material Science*, L. H. VanVlack, second edition, Addison-Wesley, Reading, Mass., 1964.

test were as follows:

Load (lb)	Change in length (in.)	Load (lb)	Change in length (in.)
0	0	12,600	0.0600
2,200	0.0008	13,200	0.0800
4,300	0.0016	13,900	0.1200
6,400	0.0024	14,300	0.1600
8,200	0.0032	14,500	0.2000
8,600	0.0040	14,600	0.2400
8,800	0.0048	14,500	0.2800
9,200	0.0064	14,400	0.3200
9,500	0.0080	14,300	0.3600
9,600	0.0096	13,800	0.4000
10,600	0.0200	13,000	Fracture
11,800	0.0400		

Determine:
(a) The modulus of elasticity.
(b) The proportional limit.
(c) The ultimate strength.

2-46 For the data given in Prob. 2-45, determine:
(a) The yield strength (0.05 percent offset).
(b) The yield strength (0.20 percent offset).

2-47 If the final diameter of the specimen at the location of the fracture in Prob. 2-45 was 0.425-in., determine:
(a) The fracture stress.
(b) The true fracture stress.

2-48 Determine the tangent modulus and the secant modulus at a stress level of 46,000 psi for the specimen of Prob. 2-45.

2-8 GENERALIZED HOOKE'S LAW FOR ISOTROPIC MATERIALS

Hooke's law (see Art. 2-7) may be extended to include biaxial and triaxial states of stress frequently encountered in engineering applications. Consider, for example, the differential cube of Fig. 2-15 subjected to *biaxial* principal stresses in the elastic range (obeying Hooke's law). The shaded square indicates the unstrained configuration of the cube. Under the action of the stress σ_x, the cube extends in the x direction and contracts in the y direction to the configuration indicated by the dotted lines (deformations greatly exaggerated). Then, under the action of the stress σ_y superimposed on σ_x, the cube assumes the outline shown by the hatched dashed lines. If the material is isotropic, Young's modulus has the same value for all directions and the final deformation in the x direction is

$$\delta_x = \epsilon_x\,dx = \frac{\sigma_x}{E}\,dx - \nu\frac{\sigma_y}{E}\,dx$$

and the three principal strains are

$$\begin{aligned} \epsilon_x &= (\sigma_x - \nu\sigma_y)/E \\ \epsilon_y &= (\sigma_y - \nu\sigma_x)/E \\ \epsilon_z &= -\nu(\sigma_x + \sigma_y)/E \end{aligned} \tag{2–9}$$

The analysis above is readily extended to triaxial principal stresses, and the expressions for strain become

$$\begin{aligned} \epsilon_x &= [\sigma_x - \nu(\sigma_y + \sigma_z)]/E \\ \epsilon_y &= [\sigma_y - \nu(\sigma_x + \sigma_z)]/E \\ \epsilon_z &= [\sigma_z - \nu(\sigma_x + \sigma_y)]/E \end{aligned} \tag{2–10}$$

In these expressions, tensile stresses and strains are considered positive, and compressive stresses and strains are considered negative.

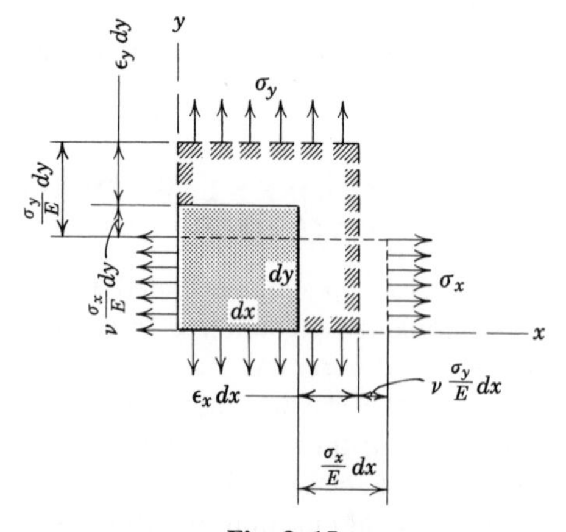

Fig. 2–15

When Eq. 2–9 are solved for the stresses in terms of the strains, they give

$$\sigma_x = \frac{E}{1-\nu^2}(\epsilon_x + \nu\epsilon_y)$$
$$\sigma_y = \frac{E}{1-\nu^2}(\epsilon_y + \nu\epsilon_x) \tag{2–11}$$

Equations 2–11 can be used to calculate normal stresses from measured or computed normal strains (see Art. 2–6). When Eq. 2–10 are solved for stresses in terms of strains, they give

$$\sigma_x = \frac{E}{(1+\nu)(1-2\nu)}\left[(1-\nu)\epsilon_x + \nu(\epsilon_y + \epsilon_z)\right]$$

$$\sigma_y = \frac{E}{(1+\nu)(1-2\nu)}\left[(1-\nu)\epsilon_y + \nu(\epsilon_z + \epsilon_x)\right] \tag{2–12}$$

$$\sigma_z = \frac{E}{(1+\nu)(1-2\nu)}\left[(1-\nu)\epsilon_z + \nu(\epsilon_x + \epsilon_y)\right]$$

In order to simplify the physical concept, the derivation of the previous equations was based on a cube of material subjected to principal stresses (see Fig. 2–15). However, as demonstrated in Art. 2–6, normal strains are unaffected by the presence of displacements perpendicular to the normal strain direction (such as those produced by shearing strains); therefore, Eqs. 2–9 through 2–12 are valid when shearing stresses exist on the faces of the cube.

Torsion test specimens are used to study material behavior under pure shear, and it is observed that a shearing stress produces only a single corresponding shearing strain. Thus, Hooke's law extended to shearing stresses is simply

$$\tau = G\gamma \tag{a}$$

Since $\tau_{max} = G\gamma_{max}$, and since the maximum shearing strain for the case of plane stress may be dependent on the out-of-plane strain $\epsilon_z = \epsilon_{p3}$, Eq. 2–8 of Art. 2–6 is frequently used to determine the third principal strain from measured in-plane strain data. This equation is derived by substituting Eq.

2–11 into Eq. 2–9 as follows:

$$\epsilon_z = -\frac{\nu(\sigma_x+\sigma_y)}{E} = -\frac{\nu}{E}\frac{E}{1-\nu^2}\left[(\epsilon_x+\nu\epsilon_y)+(\epsilon_y+\nu\epsilon_x)\right]$$

$$= -\frac{\nu}{(1-\nu)(1+\nu)}\left[(1+\nu)\epsilon_x+(1+\nu)\epsilon_y\right]$$

$$\epsilon_z = -\frac{\nu}{1-\nu}(\epsilon_x+\epsilon_y) \qquad (2\text{–}8)$$

It is important to note that the three elastic constants E, G, and ν are not independent. A useful relationship between these constants can be developed as follows: when only shearing stresses (no normal stresses) act on a pair of orthogonal (x,y) faces of an element (see Fig. 2–16), it is said to be loaded in *pure shear*, and Mohr's circles for the stresses and strains at the point have their centers at the origins of coordinates, as shown in Figs. 2–16*c* and *d*. Applying Eq. 2–11 to Fig. 2–16*b* and Eq. (a) to Fig. 2–16*a* yields

$$\sigma_p = \frac{E}{1-\nu^2}(\epsilon_p - \nu\epsilon_p) = \tau_{max}$$

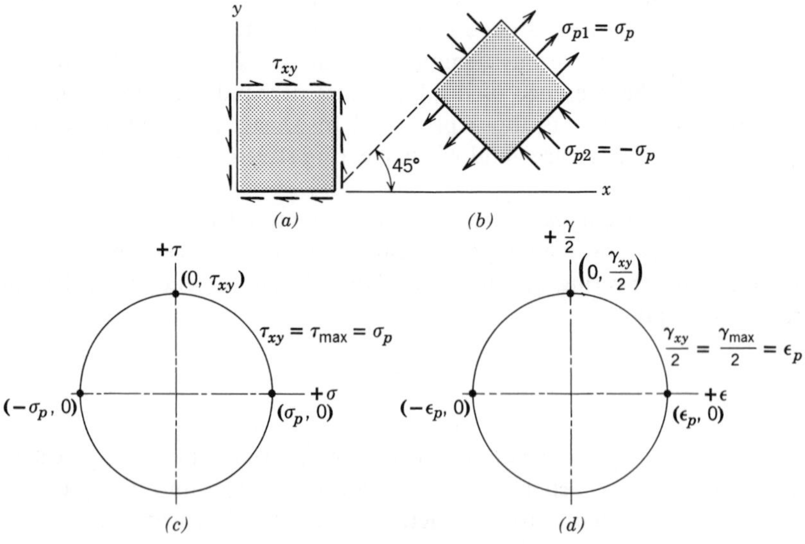

Fig. 2–16

However,

$$\tau_{max} = G\gamma_{max} = G(2\epsilon_p)$$

Therefore,

$$\frac{E}{1+\nu}\epsilon_p = 2G\epsilon_p$$

or

$$E = 2(1+\nu)G \tag{2–13}$$

Equation 2–13 can also be derived by using the concepts of elastic strain energy discussed in Art. 2–9, (refer to Prob. 2–60).

Equation 2–13 substituted in Eq. (a) yields an alternate form of generalized Hooke's law for shearing stress and strain (isotropic material); thus

$$\begin{aligned} \tau_{xy} &= G\gamma_{xy} = \frac{E}{2(1+\nu)}\gamma_{xy} \\ \tau_{yz} &= G\gamma_{yz} = \frac{E}{2(1+\nu)}\gamma_{yz} \\ \tau_{zx} &= G\gamma_{zx} = \frac{E}{2(1+\nu)}\gamma_{zx} \end{aligned} \tag{2–14}$$

Since Eq. 2–11 is so widely used in experimental stress determination, the following example is offered to illustrate two methods of application.

Example 2–4 Assume the strains of Exam. 2–2 were measured on a model constructed of steel for which the modulus of elasticity and Poisson's ratio may be taken as $30(10^6)$ psi and 0.30, respectively. Determine the principal and maximum shearing stresses, and show them on a complete sketch.

Solution The principal strains from Exam. 2–2 are

$$\epsilon_{max} = +1394(10^{-6}) \qquad \epsilon_{min} = -660(10^{-6}) \qquad \gamma_{max} = 2054(10^{-6})$$

From Eq. 2–11

$$\sigma_{max} = \frac{30(10^6)}{1-0.09}\left[1394 + 0.3(-660)\right](10^{-6}) = \frac{30}{0.91}(1196)$$

$$= \underline{39{,}400 \text{ psi } T} \qquad \text{Ans.}$$

$$\sigma_{min} = \frac{30}{0.91}\left[-660 + 0.3(1394)\right] = \underline{7980 \text{ psi } C} \qquad \text{Ans.}$$

From Eq. 2–14

$$\tau_{max} = \frac{30(10^6)}{2(1+0.3)}(2054)(10^{-6}) = \underline{23{,}700 \text{ psi}} \qquad \text{Ans.}$$

Note that the value for τ_{max} checks that given by Eq. 1–9; thus,

$$\tau_{max} = \frac{\sigma_{p1} - \sigma_{p2}}{2} = \frac{39{,}400 + 7{,}980}{2} = 23{,}690 \text{ psi}$$

The directions of the principal stresses will be the same as those for the principal strains; hence, from Fig. 2–11, the directions are obtained for the sketch of Fig. 2–17.

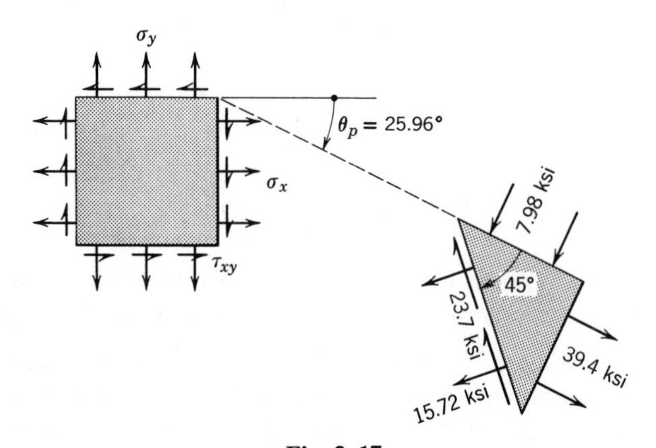

Fig. 2–17

Alternate solution Additional computed strains from Exam. 2–2 are

$$\epsilon_y = -266.7(10^{-6}) \qquad \gamma_{xy} = -1616.6(10^{-6})$$

From Eq. 2–11

$$\sigma_x = \frac{30(10^6)}{1-0.09}[1000 + 0.3(-266.7)](10^{-6}) = \frac{30}{0.91}(920) = 30{,}330 \text{ psi}$$

$$\sigma_y = \frac{30}{0.91}[-266.7 + 0.3(1000)] = 1098 \text{ psi}$$

and from Eq. 2–14

$$\tau_{xy} = \frac{30(10^6)}{2(1+0.3)}(-1616.6)(10^{-6}) = -18{,}650 \text{ psi}$$

The Mohr's circle of Fig. 2-18 is constructed from these data and the following results are obtained:

$$\sigma_{max} = \underline{39.4 \text{ ksi } T} \qquad \text{Ans.}$$

$$\sigma_{min} = \underline{7.98 \text{ ksi } C} \qquad \text{Ans.}$$

$$\tau_{max} = \underline{23.7 \text{ ksi}} \qquad \text{Ans.}$$

$$\theta_p = \underline{25.96°} \circlearrowright \qquad \text{Ans.}$$

The complete answer is given in Fig. 2-17.

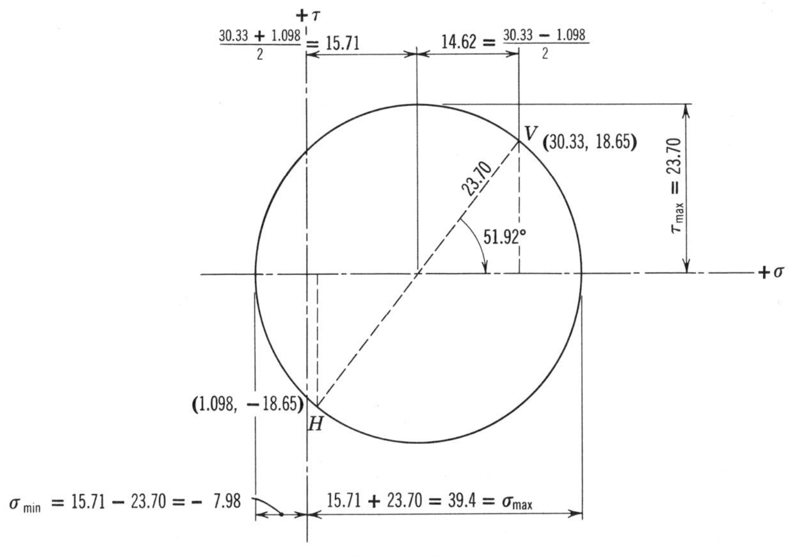

Fig. 2-18

PROBLEMS

Note. In the following problems, when the material is named, refer to Table A-1 in App. A for properties. Use Eq. 2-13 to determine Poisson's ratio.

2-49* At a point on the free surface of a member subjected to plane stress, the known strains are $\epsilon_x = +200\mu$, $\epsilon_y = -120\mu$, and $\gamma_{xy} = +240\mu$. The material is aluminum alloy 2024-T4. Determine the principal stresses and the maximum shearing stress at the point by first computing σ_x, σ_y, and τ_{xy} and using these stresses in Mohr's circle for stress. Show all of the stresses on a sketch.

2–50* Determine, and show on a sketch, the principal stresses and the maximum shearing stress at the point of Prob. 2–49 by first computing the principal strains at the point and using these strains to compute the stresses. The modulus of elasticity for the material is 73 GN/m^2, and Poisson's ratio is 1/3.

2–51 Develop Eq. 2–11 from Eq. 2–9.

2–52* At a point on the free surface (x-y plane) of a structural steel machine part, measured strains yield the following strain data: $\epsilon_{p1} = +750\mu$, $\epsilon_{p2} = -550\mu$, $\epsilon_{p3} = -93.3\mu$, and $\gamma_{xy} = +500\mu$. Determine, and show on a sketch, the principal stresses, their directions, and the maximum shearing stress.

2–53* At a point on the free surface (x-y plane) of a machine part, the known strains are $\epsilon_x = +1000\mu$, $\epsilon_y = -300\mu$, and $\gamma_{xy} = -1300\mu$. The material has a modulus of elasticity of 200 GN/m^2, a proportional limit of 275 MN/m^2, and a Poisson's ratio of 0.30. Determine, and show on a sketch:

(a) The stresses σ_x, σ_y, and τ_{xy}.
(b) The principal stresses and the maximum shearing stress at the point.

2–54 For the data given in Prob. 2–53, determine:

(a) The principal strains and the maximum shearing strain at the point. Show the directions of the in-plane principal strains on a sketch.
(b) The principal stresses and the maximum shearing stress at the point. Show these stresses on a sketch.

2–55* At a point on the free surface of a plane-stressed region, measured strains reduce to the following: $\epsilon_x = +400\mu$, $\epsilon_y = -80\mu$, and $\gamma_{xy} = +200\mu$. The material on which the strains were measured is cold-rolled red brass. Determine, and show on a sketch:

(a) The stresses σ_x, σ_y, and τ_{xy}.
(b) The principal stresses and the maximum shearing stress.

2–56 The strains of Prob. 2–55 were evaluated on the free surface of a material for which the modulus of elasticity is 70 GN/m^2 and Poisson's ratio is 1/3.

(a) Determine the principal strains, the in-plane principal strain directions, and the maximum shearing strain.
(b) Using the results of part (a), determine, and show on a sketch, the principal stresses and the maximum shearing stress.

2–57 At a point on the free surface of a structural element, the measured strains reduce to the following: $\epsilon_x = -500\mu$, $\epsilon_y = +100\mu$, and $\gamma_{xy} = -250\mu$. The material is structural steel.

(a) Determine the principal strains, the directions of the in-plane principal strains, and the maximum shearing strain.
(b) Using the results of part (a), determine, and show on a sketch, the principal stresses and the maximum shearing stress.

2–58* Measured strains at a point on the free surface of a machine element yield the following strain data: $\epsilon_x = +100\mu$, $\epsilon_y = +180\mu$, and $\gamma_{xy} = +60\mu$. The material is magnesium alloy with a modulus of elasticity of 45 GN/m^2 and Poisson's ratio of 0.35.

(a) Determine the principal strains and the maximum shearing strain.

(b) Using the results of part (a), determine, and show on a sketch, the principal stresses and the maximum shearing stress at the point.

2–59 At a point on the free surface of a plane-stressed region, strains measured at angles of 15°, 60°, and 105° with the x axis are $+150\mu$, $+260\mu$, and -150μ, respectively. The material is steel with a modulus of elasticity of 30,000 ksi and a modulus of rigidity of 12,000 ksi.

(a) Determine the principal strains, the in-plane principal strain directions, and the maximum shearing strain.

(b) Using the results of part (a), determine, and show on a sketch, the principal stresses and the maximum shearing stress at the point.

2–9 STRAIN ENERGY

A significant topic involving stress and strain is the strain energy concept, which is not only important to an understanding of the behavior of materials under dynamic loading, but also has useful applications in the solution of certain types of problems involving static loading.

The concept of strain energy is illustrated by Fig. 2–19, in which (*a*) represents a bar of uniform cross section subjected to a *slowly applied* axial load P and held at the upper end by a support assumed to be rigid. From

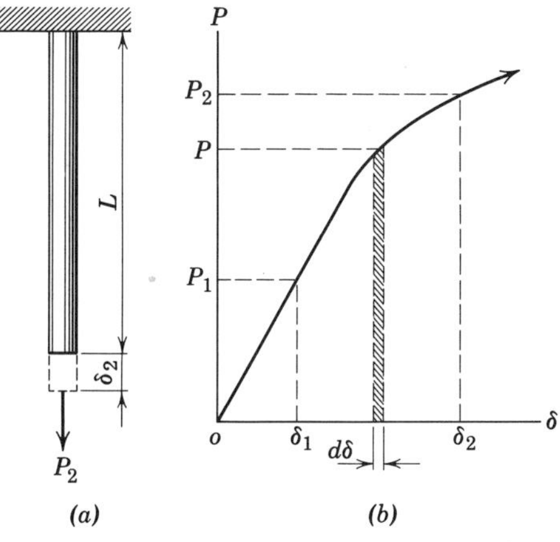

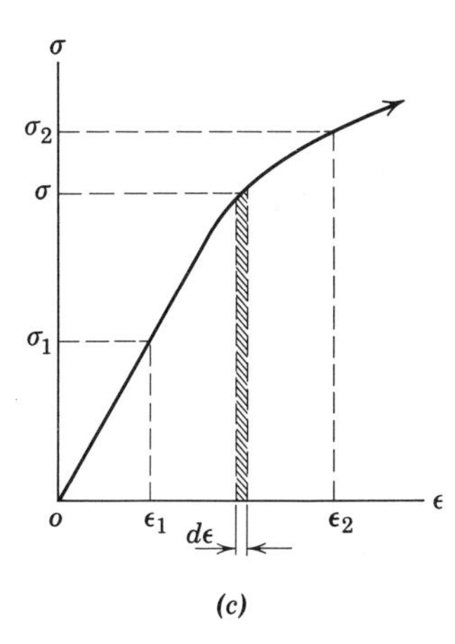

Fig. 2–19

the load-deformation diagram (Fig. 2–19*b*) the work done in elongating the bar an amount δ_2 is

$$W_k = \int_0^{\delta_2} P\, d\delta \tag{a}$$

where P is some function of δ. The work done on the bar must equal the change in energy of the material,[7] and this energy change, because it involves the strained configuration of the material, is termed *strain energy* U. If δ is expressed in terms of axial strain ($\delta = L\epsilon$) and P in terms of axial stress ($P = A\sigma$), Eq. (a) becomes

$$W_k = U = \int_0^{\epsilon_2} (\sigma)(A)(L)\, d\epsilon = AL \int_0^{\epsilon_2} \sigma\, d\epsilon \tag{b}$$

where σ is a function of ϵ (see Fig. 2–19*c*). If Hooke's law applies,

$$\epsilon = \sigma / E \qquad d\epsilon = d\sigma / E$$

and Eq. (b) becomes

$$U = \left(\frac{AL}{E} \right) \int_0^{\sigma_1} \sigma\, d\sigma$$

or

$$U = AL \left(\frac{\sigma_1^2}{2E} \right) \tag{c}$$

Equation (c) gives the *elastic strain energy* (which is, in general, recoverable[8]) for axial loading of a material obeying Hooke's law. The quantity in parentheses, $\sigma_1^2/(2E)$, is the elastic strain energy u in tension or compression per unit volume, or strain energy intensity, for a particular value of σ. For shear loading, the expression would be identical except that σ would be replaced by τ and E by G.

The integral $\int \sigma\, d\epsilon$ of Eq. (b) represents the area under the stress-strain curve and, if evaluated from zero to the elastic limit (for practical purposes, the proportional limit), yields a property known as the *modulus of resilience*. The modulus is defined as the maximum strain energy per unit volume that a material will absorb without inelastic deformation. Customary units are inch-pounds per cubic inch or newton-metres per cubic

[7]Known as *Clapeyron's theorem*, after the French engineer B. P. E. Clapeyron (1799–1864).
[8]Elastic hysteresis is neglected here as an unnecessary complication.

millimetre. The area under the entire stress-strain curve from zero to rupture gives the property known as the *modulus of toughness* and denotes the energy per unit volume necessary to rupture the material. The values of these moduli for the steel of Fig. 2–14a are

$$u_R = (35{,}000)(0.001{,}17)/2 = 22 \text{ in-lb/cu in.}$$

and

$$u_T = 57{,}000(0.3) = 17{,}000 \text{ in-lb/cu in.,}$$

where the 57,000 is an estimated average value of stress for a rectangle having approximately the same area as that under the upper curve of Fig. 2–14a.

The concept of elastic strain energy can be extended to include biaxial and triaxial loadings by writing the expression for strain energy intensity u as $\sigma\epsilon/2$ and adding the energies due to each of the stresses. Since energy is a positive scalar quantity, the addition is the arithmetic sum of the energies. For a system of triaxial principal stresses σ_x, σ_y, σ_z, the total elastic strain energy intensity is

$$u = (1/2)\sigma_x\epsilon_x + (1/2)\sigma_y\epsilon_y + (1/2)\sigma_z\epsilon_z \tag{d}$$

When the expressions for strains in terms of stresses from Eq. 2–10 of Art. 2–8 are substituted into Eq. (d), the result is

$$u = \frac{1}{2E}\left\{\sigma_x\left[\sigma_x - \nu(\sigma_y + \sigma_z)\right] + \sigma_y\left[\sigma_y - \nu(\sigma_z + \sigma_x)\right] + \sigma_z\left[\sigma_z - \nu(\sigma_x + \sigma_y)\right]\right\}$$

from which

$$u = \frac{1}{2E}\left[\sigma_x^2 + \sigma_y^2 + \sigma_z^2 - 2\nu(\sigma_x\sigma_y + \sigma_y\sigma_z + \sigma_z\sigma_x)\right] \tag{e}$$

If $\sigma_x, \sigma_y, \sigma_z$ are not principal stresses, the strain energy due to the shearing stresses on the x, y, z planes must be added to Eq. (d) or (e) to obtain the total elastic strain energy intensity; thus,

$$u = \frac{1}{2}(\sigma_x\epsilon_x + \sigma_y\epsilon_y + \sigma_z\epsilon_z + \tau_{xy}\gamma_{xy} + \tau_{yz}\gamma_{yz} + \tau_{zx}\gamma_{zx}) \tag{2–15a}$$

$$u = \frac{1}{2E}\left[\sigma_x^2 + \sigma_y^2 + \sigma_z^2 - 2\nu(\sigma_x\sigma_y + \sigma_y\sigma_z + \sigma_z\sigma_x)\right] + \frac{1}{2G}(\tau_{xy}^2 + \tau_{yz}^2 + \tau_{zx}^2) \tag{2–15b}$$

The concepts discussed and expressions developed in this article will be found useful in the development of certain beam deflection expressions in Chap. 6 and a theory of failure in Chap. 8. There are other practical applications in structural engineering.

PROBLEMS

2–60 Develop Eq. 2–13 by writing the elastic strain energy intensity expressions for the element of Fig. 2–16*a* in terms of the shearing stress τ_{xy} and for the element of Fig. 2–16*b* in terms of the principal stresses σ_p and noting that the energy can have only one value for a particular state of stress at a point.

2–61* Determine the elastic strain energy intensity at the point of Exam. 2–4 by:
(a) Using the values for the principal stresses and strains.
(b) Using the values for the stresses and strains associated with the x,y planes (note ϵ_x from Exam. 2–2 is $+1000\mu$).

2–62* Determine the elastic strain energy intensity at the point of Prob. 2–58 by:
(a) Using Eq. 2–15*a*.
(b) Using Eq. 2–15*b*.
Refer to the answer section for stress and strain values.

2–63 Determine the elastic strain energy intensity at the point of Prob. 2–53 by:
(a) Using the stresses and strains associated with the x,y planes.
(b) Using the principal stresses.
Refer to the answer section for stress values.

2–64* Determine the elastic strain energy intensity at the point of Prob. 2–49 by:
(a) Using the stresses $\sigma_x = 1908$ psi T, $\sigma_y = 652$ psi C, $\tau_{xy} = +960$ psi and the associated strains.
(b) Using the principal stresses obtained from the answer section.

2–65 Determine the elastic strain energy intensity at the point of Prob. 2–55 by:
(a) Using the stresses and strains associated with the x,y planes.
(b) Using the principal stresses.
Refer to the answer section for stress values.

2–66* Determine the elastic strain energy intensity at the point of Prob. 1–85 in Chap. 1 by:
(a) Using the stresses given in the problem statement.
(b) Using the principal stresses given in the answer section.
The moduli of elasticity and rigidity for the material are 200 and 80 GN/m^2, respectively.

2–67 Determine the elastic strain energy intensity at the point of Prob. 1–79 in Chap. 1 by:
(a) Using the stresses given in the problem statement.
(b) Using the principal stresses as given in the answer section for Prob. 1–86.
The material is aluminum alloy 2024-T4.

3 Axial Loading Applications and Pressure Vessels

3–1 INTRODUCTION

The problem of determining internal forces and deformations at all points within a body subjected to external forces is extremely difficult when the loading or geometry of the body is complicated. A refined analytical method of analysis that attempts to obtain general solutions to such problems is known as the *theory of elasticity*. The number of problems solved by such methods has been limited; therefore practical solutions to most design problems are obtained by what has become known as the *mechanics of materials approach*. With this approach, real structural elements are analyzed as idealized models subjected to simplified loadings and restraints. The resulting solutions are approximate, since they consider only the effects that significantly affect the magnitudes of stresses, strains, and deformations.

In Chaps 1 and 2, the concepts of stress and strain were developed and a discussion of material behavior led to the development of equations relating stress to strain. In the remaining chapters of the book, the stresses and deformations produced in a wide variety of structural members by axial, torsional, and flexural loadings will be considered. The "mechanics of material analyses", as presented here, are somewhat less rigorous than the "theory of elasticity approach", but experience indicates that the results obtained are quite satisfactory for most engineering problems.

3–2 DESIGN LOADS, WORKING STRESSES, AND FACTOR OF SAFETY

As mentioned in Art. 1–1, the designer is required to select a material and properly proportion a member therefrom to perform a specified function without failure. Failure is defined as the state or condition in which a member or structure no longer functions as intended. Several types of failure warrant discussion at this point. *Elastic failure* is characterized by excessive elastic deformation—a bridge may deflect elastically under traffic to an extent that may result in discomfort to the vehicle passengers; a close-fitting machine part may deform sufficiently under load to prevent proper operation of the machine. When a structure is designed to avoid elastic failure, the stiffness of the material, indicated by Young's modulus, is the significant property. *Slip failure* is characterized by excessive plastic deformation due to slip. Yield strength, yield point, and proportional limit are used as indices of strength with respect to failure by slip for members subjected to static loads. *Creep failure* is characterized by excessive plastic deformation over a long period of time under constant stress. For machines or structures that are to be subjected to relatively high stress, high temperature, or both over a long period of time, creep is a design consideration, and the creep limit is the strength index to be used. The creep limit normally decreases as the temperature increases. *Failure by fracture* is a complete separation of the material. The ultimate strength of a material is the index of resistance to failure by fracture under static loads where creep is not involved.

Most design problems involve many unknown variables. The load that the structure or machine must carry is usually estimated; the actual load may vary considerably from the estimate, especially when loads at some future time must be considered. Since testing usually damages the material, the properties of the material used in the structure cannot be evaluated directly but are normally determined by testing specimens of a similar material. Furthermore, the actual stresses that will exist in a structure are unknown because the calculations are based on assumptions as to the distribution of stresses in the material. Because of these and other unknown variables, it is customary to design for the load required to produce failure, which is larger than the estimated actual load, or to use a working (or design) stress below the stress required to produce failure.

Figure 3–1 is a photograph showing the tail section of a Boeing 707 transport plane being prepared for a vibration test. Such tests can be used to measure the ultimate load-carrying capacity of the structure.

A *working stress* or *allowable stress* is defined as the maximum stress permitted in the design computation. The *factor of safety* may be defined as the ratio of a failure-producing load to the estimated actual load. Factor of safety may also be defined as the ratio of the strength of a material to the maximum computed stress in the material. The latter factor of safety may be used to determine an allowable (or working) stress by dividing the strength of a material by the factor of safety; however, because actual sections of structural and machine elements, in general, are not the same as the minimum computed sections, the actual (computed) factor of safety is the strength divided by the maximum computed stress based on actual sections of members in the design.

Factor of safety is also defined as the ratio of the strain energy per unit volume required to cause failure (slip or fracture) to the maximum computed strain energy per unit volume in a material under dynamic loading.

It will be shown in Chap. 5 that the factor of safety based on load may be greater than that based on stress. In other words, the term is meaningless without a statement of the criterion on which it is based. In this book,

Fig. 3–1 Vibration test of tail section of airplane. Courtesy Boeing Airplane Co., Seattle, Washington.

unless otherwise indicated, the term *factor of safety* will denote the *ratio of the strength of the material to the maximum computed stress (unit strain energy for energy loading)*. For a given design, the factor of safety based on the ultimate strength is, in general, quite different from that based on the elastic strength (proportional limit, yield point, or yield strength); hence, the term *strength* is ambiguous without a qualifying statement regarding the pertinent mode of failure. The following example illustrates the use of a factor of safety.

Example 3–1 The improvised system of cables in Fig. 3–2*a* is used to remove the pile Q. A tractor with a maximum drawbar pull of 16 kips is available to apply the force P. Cable A is a 1-in. diameter iron hoisting cable with a breaking load of 29.0 kips. Cables B, C, and D are parts of a continuous $\frac{3}{4}$-in. diameter carbon steel aircraft cable with a breaking strength of 49.6 kips. Cable E is a $\frac{7}{8}$-in. diameter steel hoisting cable with a nominal area of 0.60 sq in. and an ultimate tensile strength, based on the nominal area, of 71.6 ksi. The minimum factor of safety for each cable, based on fracture, is to be 2.0. Determine:

(a) The maximum allowable force that can be exerted on the pile.

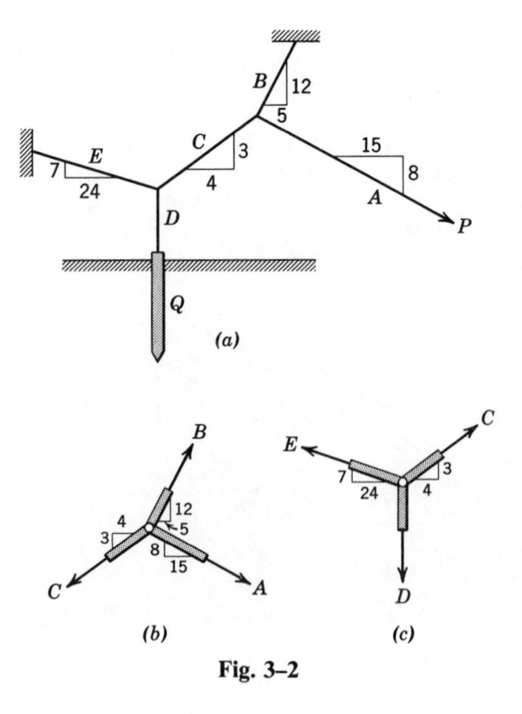

Fig. 3–2

(b) The factor of safetv for each of the five cables when the load in part (a) is applied.

Solution (*a*) The two free-body diagrams needed to solve the problem are shown in Fig. 3–2*b* and *c*. The maximum allowable tension in each cable can be obtained from the strength of the cable and the factor of safety because for axial loading the factor of safety based on load is the same as that based on stress. The allowable tensions are

$$A = 29.0/2 = 14.5 \text{ kips}$$

$$B = C = D = 49.6/2 = 24.8 \text{ kips}$$

$$E = 0.60(71.6)/2 = 21.5 \text{ kips}$$

The maximum tractor pull (16.0 kips) is greater than the allowable force in cable A; therefore, the maximum tractor pull cannot be utilized without exceeding the specifications. Assuming that the force in A is 14.5 kips, the equations of equilibrium applied to Fig. 3–2*b* give

$$\frac{5}{13}B + \frac{15}{17}14.50 - \frac{4}{5}C = 0$$

and

$$\frac{12}{13}B - \frac{8}{17}14.50 - \frac{3}{5}C = 0$$

from which $B = 25.9$ kips and $C = 28.4$ kips. Since the maximum allowable load on cables B and C is only 24.8 kips, it is apparent that the force in A must be less than the allowable load. The force in B is less than that in C; therefore, cable C is the controlling member for Fig. 3–2*b*. The equations of equilibrium for this free-body diagram with $C = 24.8$ kips give the results

$$A = 12.65 \text{ kips} \qquad \text{and} \qquad B = 22.6 \text{ kips}$$

The equations of equilibrium for Fig. 3–2*c* are

$$\frac{4}{5}C - \frac{24}{25}E = 0$$

and

$$\frac{3}{5}C + \frac{7}{25}E - D = 0$$

When $C = 24.8$ kips is substituted in these equations, the resūlts are

$$D = E = 20.7 \text{ kips}$$

Both these forces are less than the maximum allowable loads in the corresponding cables and are therefore safe loads. The strength of cable C is thus the limiting factor. The maximum pull on the pile is the force in cable D and is

$$\underline{20.7 \text{ kips}} \qquad \text{Ans.}$$

(b) The factor of safety in each of the cables is

$$k_A = 29.0/12.65 = \underline{2.29} \qquad \text{Ans.}$$

$$k_B = 49.6/22.6 = \underline{2.20} \qquad \text{Ans.}$$

$$k_C = 49.6/24.8 = \underline{2.00} \qquad \text{Ans.}$$

$$k_D = 49.6/20.7 = \underline{2.40} \qquad \text{Ans.}$$

$$k_E = (0.60)(71.6)/20.7 = \underline{2.08} \qquad \text{Ans.}$$

As noted previously, failure may be due to excessive elastic deformation. To guard against such failures, maximum deformations may be specified, and for many designs the limiting specification is the maximum deformation rather than the maximum stress. A brief discussion of the computation of axial deformation follows.

When a bar of uniform cross section is axially loaded by forces applied at the ends (two-force member), the axial strain along the length of the bar is assumed to have a constant value, and the elongation (or contraction) of the bar resulting from the axial load P may be expressed as $\delta = \epsilon L$ by definition. If Hooke's law applies, Eq. 2–9 of Art. 2–8 reduces to $\epsilon = \sigma/E$, where σ is the axial stress P/A induced by the load. Thus, the axial deformation may be expressed in terms of stress, or load, as follows:

$$\delta = \epsilon L = \frac{\sigma L}{E} \qquad (3\text{–}1a)$$

or

$$\delta = \frac{PL}{AE} \qquad (3\text{–}1b)$$

The first form will frequently be found convenient in elastic problems where limiting axial stress and axial deformation are both specified. The stress corresponding to the specified deformation can be obtained from Eq. 3–1a and compared to the specified working stress, the smaller of the two values being used to compute the unknown load or cross-sectional area. In general, Eq. 3–1*a* is the preferred form when the problem involves the determination or comparison of stresses. If a bar is subjected to a number of axial loads applied at different points along the bar, or if the bar consists of parts having different cross-sectional areas or parts composed of different materials, then the change in length of each part can be computed by using Eq. 3–1*b*. These individual changes in length of the various parts of the bar can then be added algebraically to give the total change in length of the complete bar.

For those cases where the axial force or the cross-sectional area varies continuously along the length of the bar, Eq. 3–1 is not valid. In Art. 2–3 the axial strain at a point for the case of nonuniform deformation was defined as $\epsilon = d\delta / dL$. Thus, the increment of deformation associated with a differential element of length $dL = dx$ may be expressed as $d\delta = \epsilon\, dx$. If Hooke's law applies, the strain may again be expressed as $\epsilon = \sigma / E$, where $\sigma = P_x / A_x$. The subscripts indicate that both the applied load P and the cross-sectional area A may be functions of position x along the bar. Thus,

$$d\delta = \frac{P_x\, dx}{A_x E} \tag{a}$$

Integrating Eq. (a) yields the following expression for the total elongation (or contraction) of the bar:

$$\delta = \int_0^L d\delta = \int_0^L \frac{P_x\, dx}{A_x E} \tag{3–2}$$

Equation 3–2 gives acceptable results for tapered bars, provided the angle between the sides of the bar does not exceed 20°.

PROBLEMS

Note. For simplicity of calculations in the solution of problems in this book, use the approximate values of $30(10^6)$ psi (or 200 GN/m^2) and $12(10^6)$ psi (or 80 GN/m^2) for the moduli of elasticity and rigidity, respectively, for steel without identification as to kind. If the steel is identified (structural steel, etc.), the values given in App. A will be used. Unless the proportional limit is given or referred to in the problem, it is assumed that the action is elastic. For steel, if the proportional limit is given without a stress-strain diagram, it is assumed that the diagram is elastoplastic (see Fig. 2–14*d*). When the stress-strain diagram is given or referred to, it is to be used to check the possibility of plastic action. In the following problems, unless otherwise specified, all members are assumed to have negligible weight, and all pins used for connections are assumed to be smooth.

3–1* A structural tension member of aluminum alloy 6061-T6 (see App. A) has a rectangular cross section 3×1 in. and is 10 ft long. Determine the maximum permissible axial load that may be applied if the total elongation of the member is not to exceed 0.15 in. and the minimum factor of safety with respect to failure by fracture is to be 3.

3–2* A 2-m long steel bar with a rectangular cross section 100×20 mm is to carry an axial tensile load with allowable normal and shearing stresses of 150 and 60 MN/m^2, respectively and an allowable total elongation of 1.40 mm.

(a) Determine the maximum permissible load.
(b) If the tensile elastic and ultimate strengths of the steel are 250 and 450 MN/m^2, respectively, what is the factor of safety with respect to failure by slip when the load of part (a) is applied?

3–3 A structural steel (see App. A) tension member is to be composed of two angles fastened together back to back. The member is to carry an axial load of 65 kips with a minimum factor of safety of 2 with respect to failure by slip. Compute the cross-sectional area needed and select from App. C the lightest pair of angles that will be acceptable. What is the actual factor of safety for the angles chosen?

3–4* The rigid yokes *B* and *C* of Fig. P3–4 are securely fastened to the 50-mm square steel bar *AD*. The elastic and ultimate strengths of the steel are 360 and 580 MN/m^2, respectively. Under the loading shown, determine:

(a) The factor of safety with respect to failure by slip.
(b) The total elongation of the 5.5-m length.

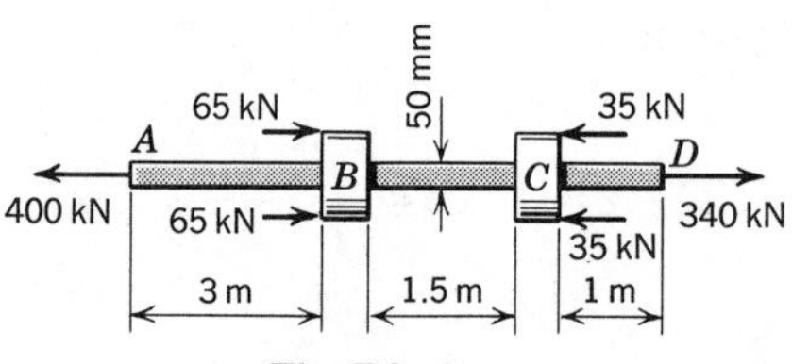

Fig. P3–4

3–5* The load *W* of Fig. P3–5 is supported by two tie rods *A* and *B* as shown. *A* is a 1/2-in. diameter red brass (annealed) rod and *B* is a 3/8-in. diameter bronze (annealed) rod. Determine the maximum load *W* that can be supported with a factor of safety of 4 with respect to failure by fracture (see App. A for properties).

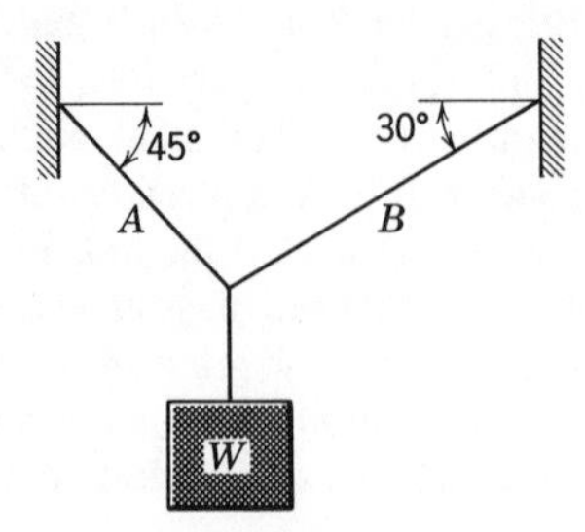

Fig. P3–5

3–6 The 20-mm diameter rod CD and the 20-mm diameter rivet B (which is in single shear) of Fig. P3–6 are made of aluminum alloy, for which the elastic and ultimate strengths in MN/m^2 are 275 tension and 180 shear, and 310 tension and 210 shear, respectively. Determine the factor of safety with respect to failure by fracture for (a) rod CD and (b) rivet B.

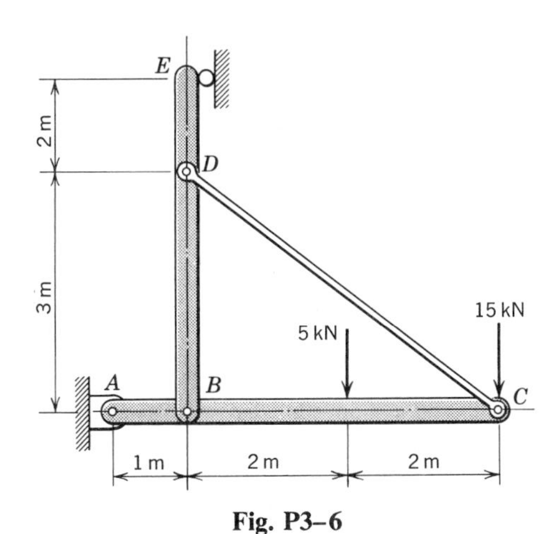

Fig. P3–6

3–7* Member BE of the pin-connected frame of Fig. P3–7 is a $1 \times 3/8$-in. rectangular bar and the pin at E has a diameter of 1.25 in. The force P is applied to the pin outside members BE and DE. All members and pins are made of aluminum alloy 2014-T4 (see App. A). Determine the factor of safety with respect to failure by slip for (a) rod BE and (b) pin E.

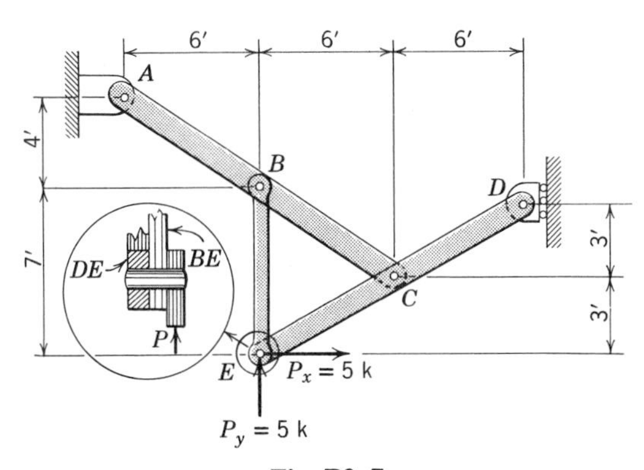

Fig. P3–7

3–8 The load W in Fig. P3–8 is suspended from a 1/4-in. carbon steel aircraft cable with a length L and a breaking strength of 8200 lb. Determine the location of the load (distance a in terms of L) and the maximum value for W with a factor of safety of 2 with respect to failure by fracture.

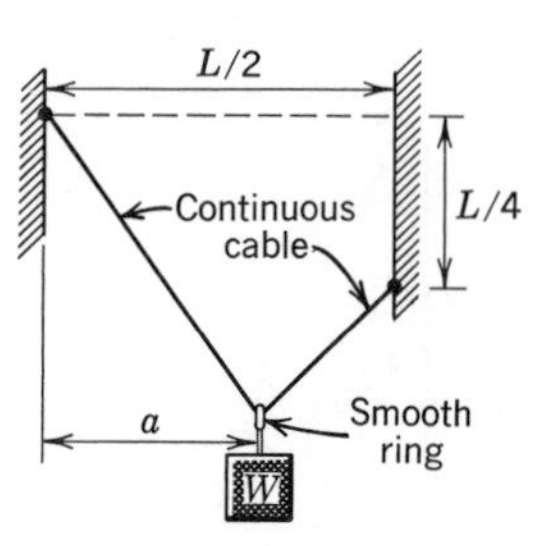

Fig. P3–8

3–9* A bar of uniform cross section and homogeneous material hangs vertically while suspended from one end. Determine, in terms of W, L, A, and E, the extension of the bar due to its own weight.

3–10 Determine the extension, due to its own weight, of the homogeneous conical bar of Fig. P3–10. Give the result in terms of L, E, and the specific weight γ. Assume that the taper of the bar is slight enough for the assumption of a uniform axial stress distribution over a cross section to be valid.

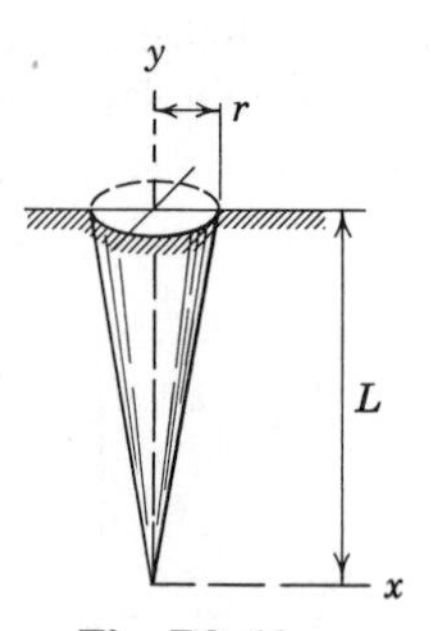

Fig. P3–10

3–11 Determine the extension, due to its own weight, of the homogeneous bar of Fig. P3–11. Give the result in terms of L, E, and the specific weight γ. Assume that the taper of the bar is slight enough that the assumption of a uniform axial stress distribution over a cross section is valid.

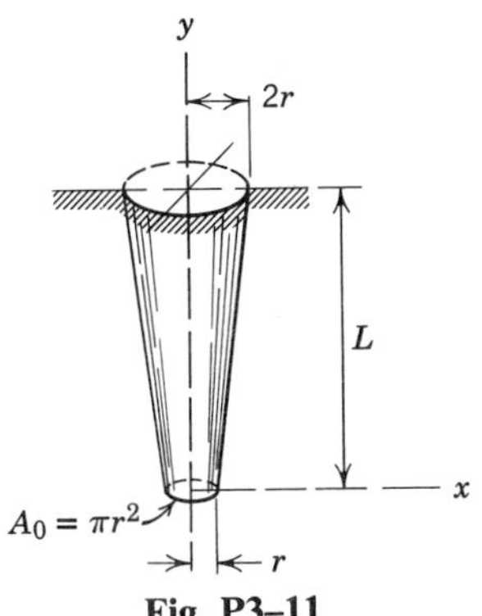

Fig. P3-11

3-12* Solve Prob. 3-11 with an additional axial tensile load of $\gamma A_0 L$ applied at the lower end.

3-3 TEMPERATURE EFFECTS

Most engineering materials when unrestrained expand when heated and contract when cooled. The strain due to a 1° temperature change is designated by α and is known as the coefficient of thermal expansion. The strain due to a temperature change of $\Delta T°$ is

$$\epsilon_T = \alpha(\Delta T)$$

The coefficient of thermal expansion is approximately constant for a considerable range of temperatures (in general, the coefficient increases with an increase of temperature). For a homogeneous, isotropic material, the coefficient applies to all dimensions (all directions). Values of the coefficient of expansion for several materials are included in App. A.

When a member is restrained (free movement prevented) while a temperature change takes place, stresses (referred to as *thermal stresses*) are induced in the member. If the action is elastic, thermal stresses are easily computed by (1) assuming that the restraining influence has been removed and the member permitted to expand or contract freely, (2) applying forces that cause the member to assume the configuration dictated by the restraining influence. For example, the bar AB of Fig. 3-3a is securely fastened to rigid supports at the ends. If the end B is assumed to be cut loose from the wall and the temperature drops, the end will move to B' a distance δ_T, as indicated in Fig. 3-3b. Next a force P, shown in Fig. 3-3c, is applied to B of sufficient magnitude to move end B through a distance

δ_P so that the length of the bar is again L, the distance between the walls. Since the walls do not move, $\delta_T = \delta_P$, where $\delta_T = \alpha L \Delta T$, and $\delta_P = L\epsilon$; if Hooke's law applies, $\epsilon = \sigma / E$, where σ is the induced axial stress.

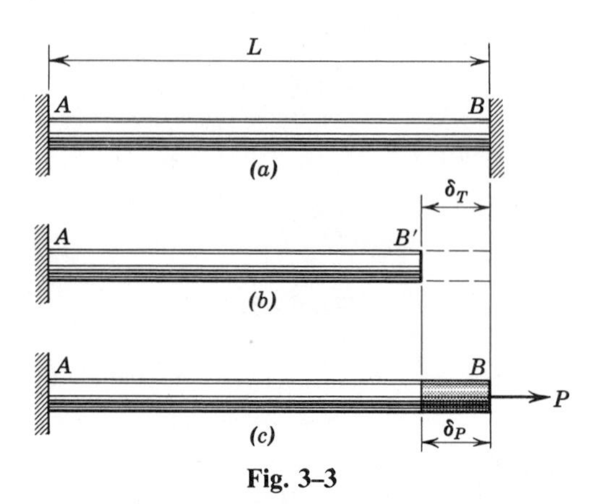

Fig. 3–3

In case a stress exists before the temperature change takes place (the rod of Fig. 3–3*a* might have been tightened by means of a turnbuckle or nuts at the ends), the stress due to the temperature change may be added algebraically to the original stress by using the *principle of superposition*. The principle states that stresses (at a point on a given plane) due to different loads may be computed separately and added algebraically, provided that the sum of the stresses does not exceed the proportional limit of the material and that the structure remains stable. In explanation of this last limitation, consider a member with a length considerably greater than the lateral dimension subjected to axial compression loads. This member may successfully carry either of two applied loads, and the sum of the axial stresses for both loads may be within the proportional limit; yet, upon application of both loads, the member may collapse (this action is discussed further in Chap. 9).

PROBLEMS

Note. Unless otherwise specified, all members are assumed to have negligible weight, and all pins used for connections are assumed to be smooth.

3–13* Determine the horizontal movement of point A in Fig. P3–13 due to a temperature drop of 80° F. Assume member AE to have an insignificant coefficient of thermal expansion. The steel bar is structural steel and the aluminum alloy is 6061-T6 (see App. A).

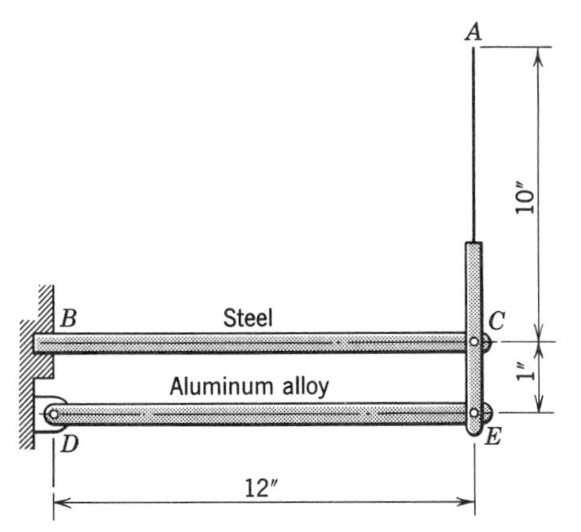

Fig. P3–13

3–14* Determine the movement of the pointer of Fig. P3–14 with respect to the scale zero, when the temperature increases 50°C. The coefficients of thermal expansion are $12(10^{-6})$ per degree C for the steel and $23(10^{-6})$ per degree C for the aluminum.

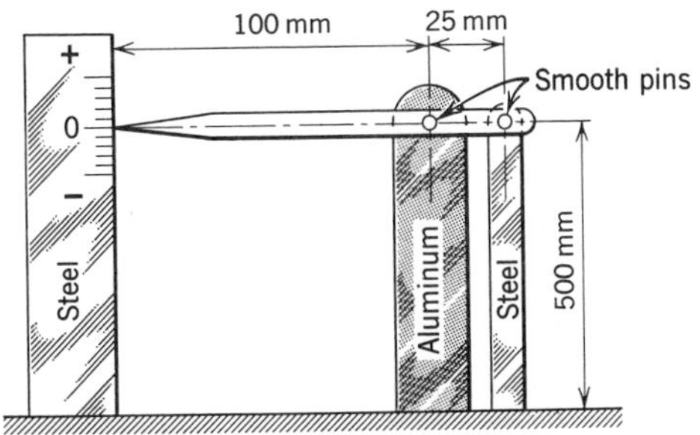

Fig. P3–14

3–15* A 7-ft stainless steel (18-8 annealed) rod and a 5-ft nodular cast-iron rod (see App. A) are welded together end to end and loaded axially at the extreme ends with a tensile load of 5π kips. Both rods have a diameter of 1 in. Determine:

(a) The total change in the 12-ft length due to the load plus a temperature drop of 50°F.

(b) The factor of safety with respect to failure by slip.

3–16 The ends of a 4-m long steel rod are securely attached to rigid supports. At 30°C there is no stress in the rod and the diameter is 50 mm. The steel has the following properties: yield point 260 MN/m^2, ultimate strength 450 MN/m^2, Poisson's ratio 0.25, and coefficient of thermal expansion $12(10^{-6})$ per degree C. For a temperature of $-10°$ C, determine:

(a) The factor of safety with respect to failure by slip.

(b) The change in diameter of the bar.

3–17* A 20-ft long by 2-in. diameter rod of aluminum alloy 6061-T6 (see App. A) is secured at the ends to supports that yield to permit a change in length of 0.04 in. in the rod when stressed. When the temperature is 90° F there is no stress in the rod. For the rod at a temperature of $-10°$ F, determine:

(a) The maximum normal stress.
(b) The maximum shearing stress.
(c) The change in diameter.

3–18 A 6-m rod is composed of a 4-m length of steel securely connected to a 2-m length of aluminum alloy. The diameter of both rods is 25 mm, and the two ends of the composite rod are securely attached to rigid supports. At 40° C there is no stress in the rod. For a temperature of 0° C, determine the factor of safety with respect to failure by slip for each material. Refer to Prob. 3–16 for properties of the steel. The properties of the aluminum are: modulus of elasticity 75 GN/m^2, yield strength 330 MN/m^2, ultimate strength 470 MN/m^2, coefficient of thermal expansion $22.5(10^{-6})$ per degree C.

3–19* An annealed stainless steel (18–8) bolt through a 3-ft long, gray cast-iron tube (see App. A) has the nut tightened until the axial tensile stress in the bolt is 6000 psi at 90° F. Determine the change in axial stress in the bolt when the temperature is $-10°$ F. The change in length of the tube due to the change in stress in the bolt has been found to be -0.0068 in.

3–20 A steel tire is to be heated and placed on a 2.000-m diameter locomotive drive wheel. At 40° C the tire has an inside diameter of 1.998 m. The coefficient of thermal expansion for the steel is $12(10^{-6})$ per degree C, and its yield strength is 400 MN/m^2. Determine:

(a) The temperature increase required for the inside diameter of the tire to be 2.001 m to facilitate placing it on the wheel.
(b) The maximum tensile stress in the tire when it has cooled to $-10°$ C, assuming that the diameter of the wheel does not change.
(c) The factor of safety with respect to failure by slip for the stress obtained in part (b).

3–21* A prismatic bar is fastened at its ends to rigid walls and is free of stress at room temperature. One end of the bar is heated to 100° F above room temperature and the change in temperature ΔT along the bar is proportional to the square of the distance from the unheated end, which is at room temperature. Determine:

(a) The axial stress induced in the bar in terms of α and E.
(b) The axial stress if the bar is made of aluminum alloy 6061-T6 (see App. A).

3–22 A prismatic bar is fastened at its ends to rigid walls and is free of stress at room temperature. The bar is heated in such a manner that the temperature at one end is 45° C above room temperature, and the change in temperature ΔT is proportional to the distance from the other end, which remains at room temperature. Determine:

(a) The axial stress induced in the bar in terms of α and E.
(b) The axial stress if the bar is made of steel having a coefficient of thermal expansion of $12(10^{-6})$ per degree C.

3–4 DEFORMATIONS IN A SYSTEM OF AXIALLY LOADED BARS

It is sometimes necessary to determine axial deformations and strains in a loaded system of pin-connected deformable bars (two-force members). The problem is approached through a study of the geometry of the deformed system, from which the axial deformations δ of the various bars in the system are obtained. Suppose, for example, one is interested in the axial deformations of bars AB, BC, and BD of Fig. 3–4 in which the solid lines represent the unstrained (unloaded) configuration of the system and the dashed lines represent the configuration due to a force applied at B. From the Pythagorean theorem, the axial deformation in bar AB is

$$\delta_{AB} = \sqrt{(L+y)^2 + x^2} - L$$

Transposing the last term and squaring both sides gives

$$\delta_{AB}^2 + 2L\delta_{AB} + L^2 = L^2 + 2Ly + y^2 + x^2$$

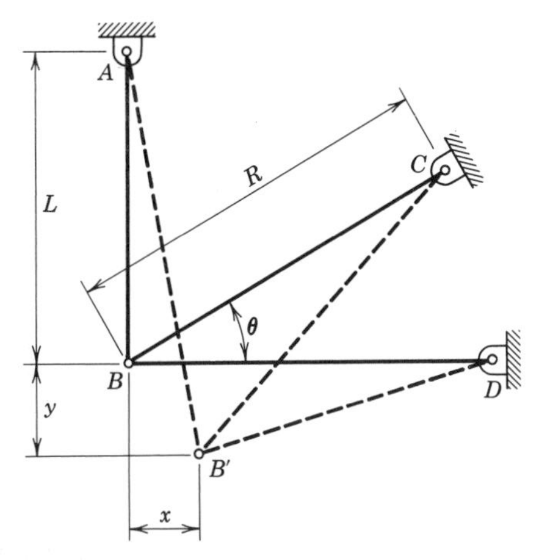

Fig. 3–4

If the displacements are small (the usual case for stiff materials and elastic action), the terms involving the squares of the displacements may be neglected; hence,

$$\delta_{AB} \approx y$$

In a similar manner,

$$\delta_{BD} \approx x$$

The axial deformation in bar BC is

$$\delta_{BC} = \sqrt{(R\cos\theta - x)^2 + (R\sin\theta + y)^2} - R$$

Transposing the last term and squaring both sides gives

$$\delta_{BC}^2 + 2R\delta_{BC} + R^2 = R^2\cos^2\theta - 2Rx\cos\theta + x^2 + R^2\sin^2\theta + 2Ry\sin\theta + y^2$$

Neglecting small second-degree terms and noting that $\sin^2\theta + \cos^2\theta = 1$,

$$\delta_{BC} \approx y\sin\theta - x\cos\theta$$

or in terms of the deformations of the other two bars,

$$\delta_{BC} \approx \delta_{AB}\sin\theta - \delta_{BD}\cos\theta$$

The geometric interpretation of this equation is indicated by the shaded right triangles of Fig. 3–5.

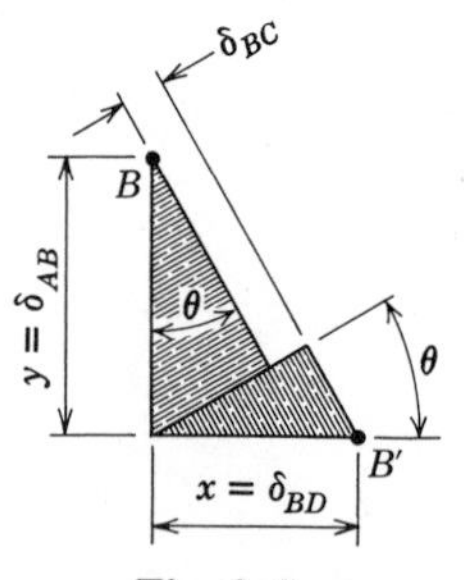

Fig. 3–5

The general conclusion that may be drawn from the above discussion is that, *for small displacements*, the axial deformation in any bar may be assumed equal to the component of the displacement of one end of the bar (relative to the other end) taken in the direction of the unstrained orientation of the bar. Any rigid members of the system will change orientation or position but will not be deformed in any manner. For example, if bar BD of Fig. 3–4 were rigid and subjected to a small downward rotation, point B

could be assumed to be displaced vertically through a distance y, and δ_{BC} would be equal to $y \sin \theta$.

PROBLEMS

Note. For the problems in this article, assume small deformations unless otherwise noted.

3–23* Neglect the bending of bar C in Fig. P3–23. There is no strain in the vertical bars before the load P is applied. After load P is applied, the axial strain in bar A is found to be 0.0024. Determine the axial strain produced in bar B.

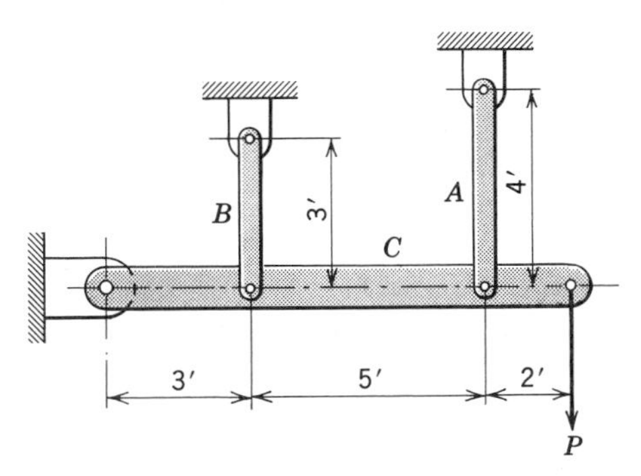

Fig. P3–23

3–24* Solve Prob. 3–23 with the following changes in data: (1) lengths of bars B and A are 2 and 3 m, respectively, (2) dimensions on bar C are, from left to right, 1, 1.5, and 0.5 m, respectively, (3) the axial strain in bar A is 0.0020, (4) there is a slack of 0.6 mm in the connection between B and C before the load is applied.

3–25 The load P of Fig. P3–25 produces an axial strain in the brass post B of 0.0014. Determine the accompanying axial strain in the aluminum-alloy rod A.

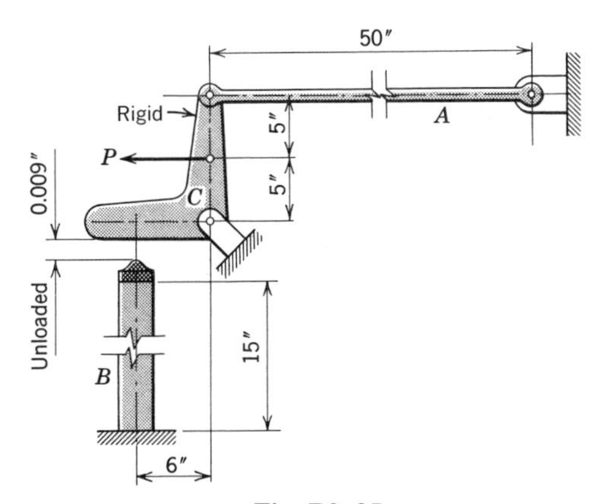

Fig. P3–25

3–26* Solve Prob. 3–25 with the following changes in data: (1) lengths of A and B are 1500 and 400 mm, respectively, (2) the dimensions of the bell crank C are 200 mm from the pin to B, and 150 mm each from the pin to P and from P to A, (3) when unloaded the bell crank C just makes contact with the bearing plate on top of post B, (4) bar A has a slope of 5 vertical on 12 horizontal, and (5) the axial strain in post B is 0.0013 when P is applied.

3–27

(a) Solve Prob. 3–23 when there is a 0.006-in. slack in the connection between B and C before the load is applied.

(b) Determine the factor of safety with respect to failure by slip for each bar (A and B). The materials are: aluminum alloy 2024-T4 for A and hot-rolled 0.4 percent carbon steel for B (see App. A for properties).

3–28* The rigid bar DC of Fig. P3–28 is horizontal under no load, and bars A and B are unstressed. When the load P is applied, the axial strain in bar B is found to be 0.0015. Determine the factor of safety with respect to failure by slip for each bar. The materials are: steel (yield strength = 500 MN/m^2, ultimate strength = 800 MN/m^2) for bar A and cold-rolled brass (yield strength = 400 MN/m^2, ultimate strength = 500 MN/m^2, modulus of elasticity = 100 GN/m^2) for bar B.

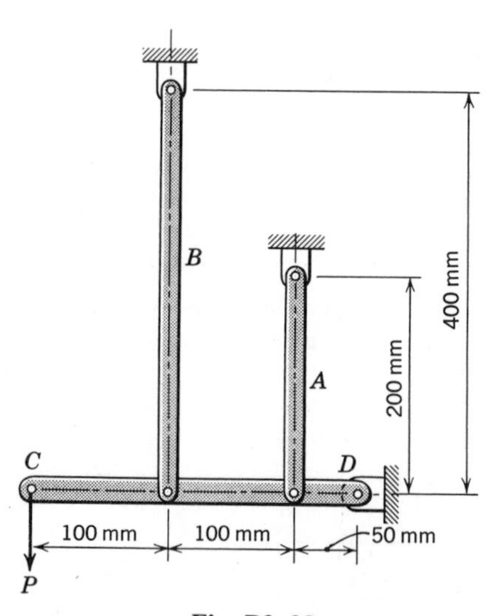

Fig. P3–28

3–29* A mechanism is constructed and loaded exactly like that shown in Fig. P3–25 except that the 50-in. rod A has a slope of 4 right to 3 upward. Rod A is aluminum alloy 6061-T6 and post B is cold-rolled red brass (see App. A). The load P produces an axial strain of 0.0018 in post B. Determine the factors of safety with respect to failure by slip for rod A and post B.

3–30

(a) Solve Prob. 3–26 when there is a gap of 0.65 mm between the bell crank C and the bearing plate on top of post B before the load P is applied.

(b) Determine the factors of safety with respect to failure by slip for rod A and post B. The moduli of elasticity, yield strengths, and ultimate strengths are 70 GN/m^2, 270 MN/m^2, and 300 MN/m^2, respectively, for the aluminum rod A, and 100 GN/m^2, 400 MN/m^2, and 500 MN/m^2, respectively, for the brass post B.

3–5 STATICALLY INDETERMINATE AXIALLY LOADED MEMBERS

In the work so far, it has always been possible to find the internal forces in any member of a structure by means of the equations of equilibrium; that is, the structures were statically determinate. In any structure, if the number of unknown forces and distances exceeds the number of independent equations of equilibrium that are applicable, the structure is said to be *statically indeterminate*, and it is necessary to write additional equations involving the geometry of the deformations in the members of the structure. The following outline of procedure will be helpful in the analysis of problems involving statically indeterminate structures.

---

1. Draw a free-body diagram.

2. Note the number of unknowns involved (magnitudes and positions).

3. Recognize the type of force system on the free-body diagram, and note the number of independent equations of equilibrium available for this system.

4. If the number of unknowns exceeds the number of equilibrium equations, a deformation equation must be written for each extra unknown.

5. When the number of independent equilibrium equations and deformation equations equals the number of unknowns, the equations can be solved simultaneously. Deformations and forces must be related in order to solve the equations simultaneously. Hooke's law and the definitions of stress and strain can be used when all stresses are less than the corresponding proportional limits. If some of the stresses exceed the proportional limits of the materials, the stress-strain diagrams can be used to relate the loads and deformations or simplifying assumptions (such as perfect plasticity, see Fig. 2–14d) can be made to obtain satisfactory results.

It is recommended that a displacement diagram be drawn showing deformations to assist in obtaining the correct deformation equation. The displacement diagram should be as simple as possible (a line diagram), with the deformations indicated with exaggerated magnitudes and clearly dimensioned. Loading members (members assumed to be rigid), especially, should be indicated by single lines. Note that an equilibrium equation and the corresponding deformation equation *must be compatible*; that is, where a tensile force is assumed for a member in the free-body diagram, a tensile deformation must be indicated for the same member in the deformation diagram. If the diagrams are compatible, a negative result will indicate that the assumption was wrong; however, the magnitude of the result will be correct.

In all the following problems and examples, the loading members, pins, and supports are assumed to be rigid, and the mechanism is so constructed that the force system is coplanar. The procedure outlined above is illustrated in the following examples.

Example 3–2 The rigid plate C in Fig. 3–6*a* is fastened to the 0.50-in. diameter steel rod A and to the aluminum pipe B. The other ends of A and B are fastened to rigid supports. When the force P is zero, there are no stresses in A and B. Rod A is made of low-carbon steel assumed to be elastoplastic (see Fig. 2–14*d*) with a proportional limit and yield point of 40 ksi ($E = 30{,}000$ ksi). Pipe B has a cross-sectional area of 2.0 sq in. and is made of an aluminum alloy with the stress-strain diagram shown in Fig. 3–6*b*. A load P of 30 kips is applied to C as shown. Determine the axial stresses in A and B and the displacement of plate C.

Solution Figure 3–6*c* is a free-body diagram of plate C and portions of the members A and B. The free-body diagram contains two unknown forces, and only one equation of equilibrium is available; therefore, the problem is statically indeterminate. As the load is applied to the plate C, it moves downward an amount δ, which represents the total deformation in members A and B; see Fig. 3–6*d*. This observation provides a displacement equation that can be used with the equilibrium equation to solve the problem. The two equations are

$$(\uparrow +) \qquad \Sigma F_y = 0 = P_A + P_B - 30$$

or

$$(\pi/4)(0.50)^2\sigma_A + 2\sigma_B = 30 \tag{a}$$

and

$$\delta_A = \delta_B$$

or

$$10\epsilon_A = 20\epsilon_B$$

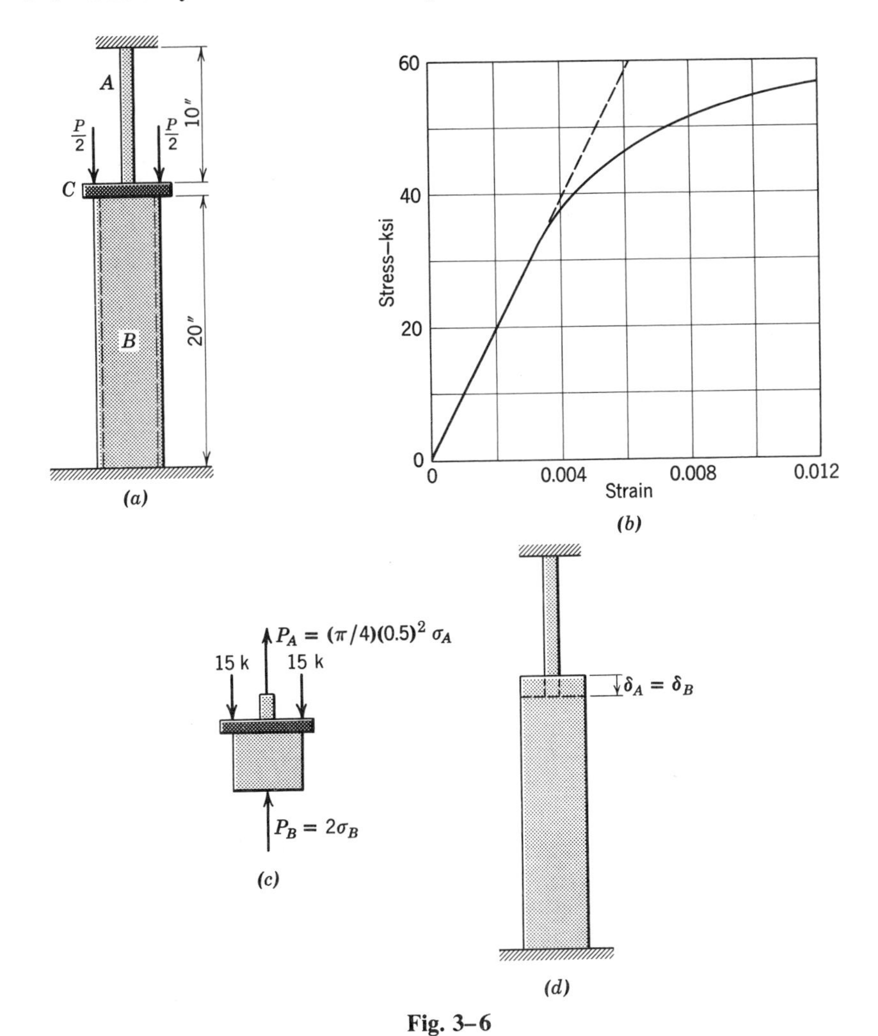

Fig. 3–6

In order to solve these equations, the stresses and strains must be related. If the stresses are less than the corresponding proportional limits, Hooke's law can be used. If Hooke's law is assumed to be valid, the strain equation becomes

$$\frac{10\sigma_A}{30(10^3)} = \frac{20\sigma_B}{10(10^3)}$$

from which

$$\sigma_A = 6\sigma_B \tag{b}$$

The value of $E_B = 10(10^3)$ ksi is obtained as the slope of the stress-strain diagram in Fig. 3–6*b*. The simultaneous solution of Eqs. (a) and (b) gives $\sigma_A = 56.6$ ksi T and $\sigma_B = 9.45$ ksi C.

These results indicate that bar A is stressed beyond the proportional limit of the steel and that Hooke's law does not apply. Since the material is assumed to be perfectly plastic beyond the proportional limit, the stress in bar A must be 40 ksi. When this value is substituted in equilibrium Eq. (a), the equation becomes

$$(\pi/16)(40) + 2\sigma_B = 30$$

from which

$$\sigma_B = \underline{11.07 \text{ ksi } C} \qquad \text{and} \qquad \sigma_A = \underline{40 \text{ ksi } T} \qquad \text{Ans.}$$

The stress in B is less than the proportional limit for aluminum; therefore, Hooke's law can be used to determine the movement of the plate, which is

$$\delta = \delta_B = 20(11.07)/(10^4) = \underline{0.0221 \text{ in. downward}} \qquad \text{Ans.}$$

If the stress in B had been more than the proportional limit of the aluminum, it would have been necessary to determine the strain from the curve in Fig. 3–6*b*. If the material in bar A were not perfectly plastic, a trial-and-error solution would be required, since the stresses in both A and B depend on the movement of C.

Example 3–3 The steel post B of Fig. 3–7*a* is 2 by 2 in. square, and the aluminum alloy bar C is 1.5 in. wide by 1 in. thick. Bar A and the bearing block on B are to be considered rigid; all connections are treated as smooth pins; the temperature is assumed to remain constant; the modulus of elasticity of the aluminum alloy is taken as (10^7) psi. Under no load, the clearance between A and B is 0.002 in. Determine the maximum permissible load P that may be applied as shown if the axial stresses are not to exceed 15,000 psi for the steel and 18,000 psi for the aluminum alloy.

Solution The free-body diagram of body A in Fig. 3–7*b* shows five unknown forces; however, one of the axial stresses, either σ_B or σ_C, will equal the allowable value (which is given), and the other one will be less than the allowable stress (or equal to it in an exceptional case). The force system is coplanar, for which three independent equations of equilibrium are available. With four unknowns and three equations, the system is statically indeterminate, and an additional equation must be obtained from the deformations of B and C. The displacement diagram of Fig. 3–7*c* shows the positions of the members, after the load P is applied, as dashed

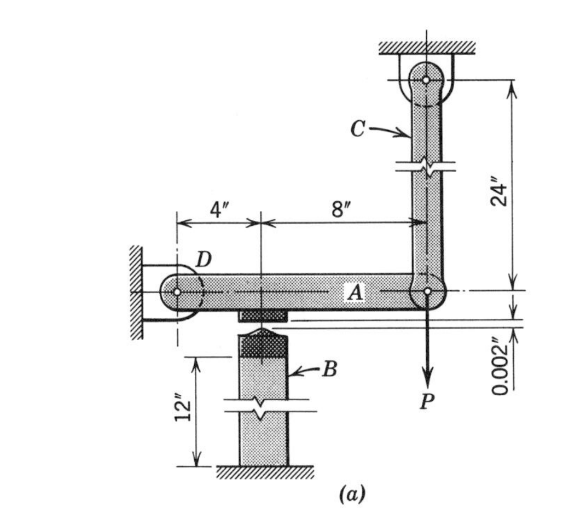

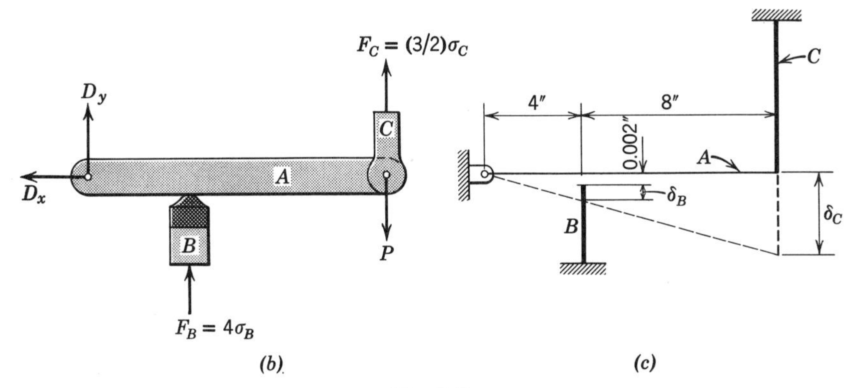

Fig. 3–7

lines. The deformations of members B and C are related by the equation (from similar triangles)

$$\frac{\delta_B + 0.002}{4} = \frac{\delta_C}{12}$$

or

$$3\delta_B + 0.006 = \delta_C$$

In terms of strain ($\delta = L\epsilon$), the relation is

$$3(12)\epsilon_B + 0.006 = 24\epsilon_C$$

Since the stresses are limited to values less than the proportional limit (see App. A), Hooke's law ($\sigma = E\epsilon$) applies, and the equation becomes

$$\frac{36\sigma_B}{30(10^6)} + 0.006 = \frac{24\sigma_C}{10(10^6)}$$

or

$$\sigma_B + 5000 = 2\sigma_C$$

When σ_B equals its allowable value of 15,000 psi, σ_C becomes 10,000 psi as determined from this strain relation. It is apparent that higher values for σ_C would yield a stress in the steel above the specified limit. Therefore, the axial stresses are 15,000 psi in the steel and 10,000 psi in the aluminum. From Fig. 3–7*b* and the equilibrium equation $\Sigma M_D = 0$,

$$4(15{,}000)4 + 12(10{,}000)(3/2) - 12P = 0$$

from which

$$P = \underline{35{,}000 \text{ lb as shown}} \qquad \text{Ans.}$$

Example 3–4 If the mechanism of Exam. 3–3 is loaded with a force P of 10,000 lb as shown and the temperature increases 20° F, what are the axial stresses in members B and C? For the coefficients of thermal expansion per degree F, use $13(10^{-6})$ for the aluminum and $6.5(10^{-6})$ for the steel.

Solution The free-body diagram in Fig. 3–7*b* is valid for this problem. Following the suggestion of Art. 3–3, the thermal deformation is first permitted to take place by removing all restraints (bar A is imagined to be removed). The deformations of B and C due to the change of temperature are indicated as δ_T values in Fig. 3–8*a*. Loads are then applied to B and C, which change their lengths so that the rigid bar A can be connected to them, as indicated in Fig. 3–8*b*. The deformations due to the forces are indicated as δ_P values. The deformation equation is

$$\frac{\delta_{BP} - \delta_{BT} + 0.002}{4} = \frac{\delta_{CP} + \delta_{CT}}{12}$$

or

$$3\delta_{BP} - 3\delta_{BT} + 0.006 = \delta_{CP} + \delta_{CT}$$

When deformations are replaced by their values in terms of strains, the

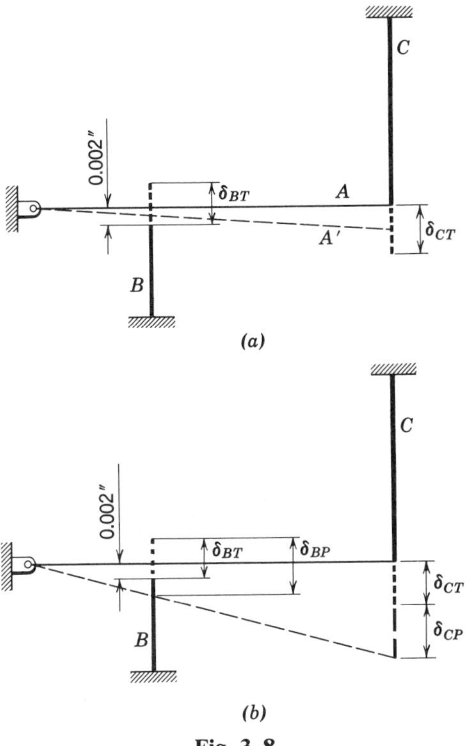

Fig. 3–8

equation becomes

$$3(12)\epsilon_{BP} - 3(12)\epsilon_{BT} + 0.006 = 24\epsilon_{CP} + 24\epsilon_{CT}$$

When the strains due to stress are replaced by their values in terms of stress (assuming Hooke's law is valid), and the strains due to the temperature change by their numerical values, the displacement equation becomes

$$\frac{36\sigma_B}{30(10^6)} - \frac{36(6.5)20}{(10^6)} + \frac{6}{(10^3)} = \frac{24\sigma_C}{10(10^6)} + \frac{24(13.0)20}{(10^6)}$$

or

$$\sigma_B = 2\sigma_C + 4100 \tag{c}$$

From Fig. 3–7*b* and the equilibrium equation $\Sigma M_D = 0$,

$$4\sigma_B(4) + (3/2)\sigma_C(12) - 10{,}000(12) = 0 \tag{d}$$

The simultaneous solution of Eq. (c) and (d) gives

$$\sigma_C = \underline{1088 \text{ psi } T} \qquad \text{Ans.}$$

and

$$\sigma_B = \underline{6280 \text{ psi } C} \qquad \text{Ans.}$$

These stresses are well within the proportional limits of the materials; therefore, Hooke's law is valid.

The use of colored pencils may be found helpful in the construction of the deformation diagram by using one color for temperature deformations and another for load deformations.

From a study of Fig. 3–8*a* it will be observed that with no applied load ($P = 0$), the bar A might have occupied some position such as that shown dashed (marked A') and would therefore have induced compression in both bars B and C. In the solution given, the assumption was made that the force P was of sufficient magnitude to induce tension in bar C. The positive sign on the value for σ_C indicates that the assumption was correct *if the deformation diagram and the free-body diagram are compatible*; that is, where a tensile deformation is assumed in the deformation diagram, a tensile stress must be indicated on the free-body diagram. This very important consideration must not be overlooked.

PROBLEMS

Note. Unless otherwise specified, all members are assumed to have negligible weight, and all pins used for connections are assumed to be smooth. The temperature remains constant unless otherwise specified.

3–31* A steel plate $6 \times 1 \times 20$ in. is bolted to a red oak block $6 \times 6 \times 20$ in. The structure is shortened 0.03 in. by the unknown load P applied as shown in Fig. P3–31. The steel is 0.4 percent carbon (hot rolled) and the oak is dry (see App. A for properties).

(a) Determine the axial stress in each material.
(b) Determine the magnitude of the load P.
(c) Locate the position of P with respect to the outer edge of the steel plate.

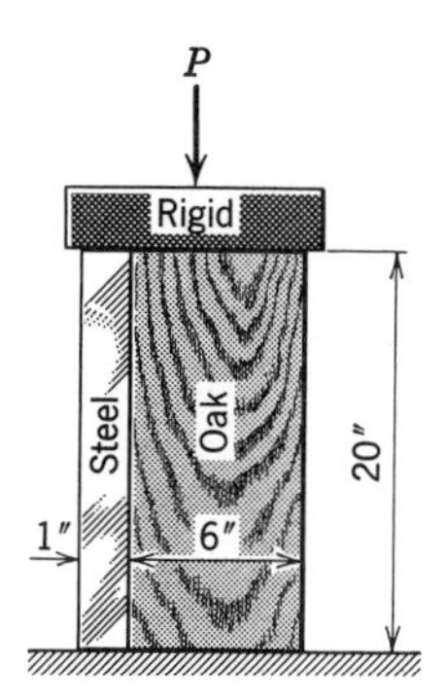

Fig. P3–31

3–32* In Fig. P3–32, A is a steel bar 10×20 mm and B is a brass bar 10×30 mm. The bar C may be regarded as rigid. A load of 18 kN is hung from C in such a place that the bar remains horizontal. The modulus of elasticity of the brass is 100 GN/m^2. Determine:
(a) The axial stress in each material.
(b) The vertical displacement of bar C.
(c) The position of the line of action of the 18-kN force.

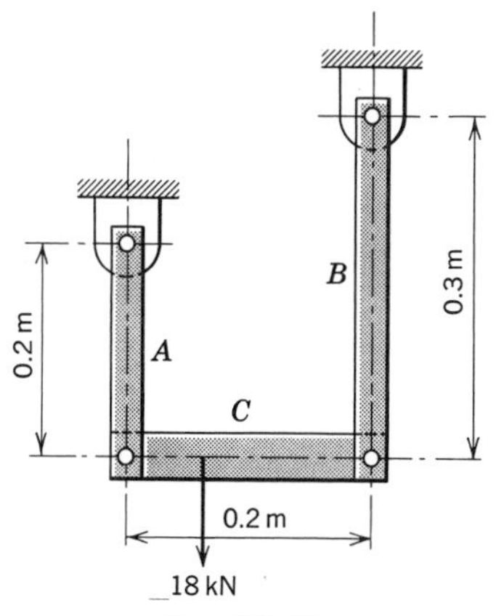

Fig. P3–32

3–33* The load P of Fig. P3–33 is applied to the rigid member B in such a position that B remains horizontal. Bar A is red brass (annealed) and post C is concrete (fairly high strength). The cross-sectional areas of A and C are 3 and 20 sq in., respectively (see App. A for material properties).

(a) Determine the maximum permissible load P if the allowable stresses are 7000 psi T for the brass and 1500 psi C for the concrete.

(b) What is the distance from the center of bar A to the load?

(c) If the restriction that B remain horizontal were removed, what would be the value of the load P, and what would be its location?

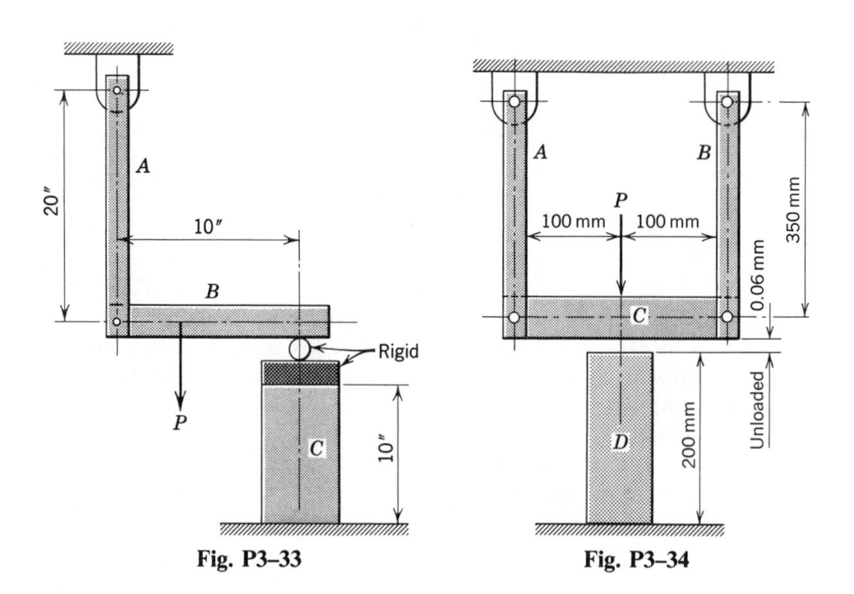

Fig. P3–33 Fig. P3–34

3–34* The system in Fig. P3–34 is initially unstressed. Bar C is rigid and is 0.06 mm above the member D. Determine the magnitude of the load P that will induce an axial stress of 30 MN/m^2 C in post D. The following data apply:

Member	Material	Cross-sectional area (mm^2)	E (GN/m^2)
A	Aluminum alloy	600	70
B	Aluminum alloy	600	70
D	Cast iron	2000	100

3–35* Solve Exam. 3–2 with $P = 70$ kips and the cross-sectional area of the pipe B equal to [illegible].35 sq in.

3–36 Bar CD in the pin-connected structure of Fig. P3–36 is to be considered rigid and is horizontal under no load. The aluminum alloy ($E = 70$ GN/m^2) bars A and B have cross-sectional areas of 400 mm^2 each. Determine the vertical displacement of point C and the axial stress in bar A when the 75-kN load is applied as shown.

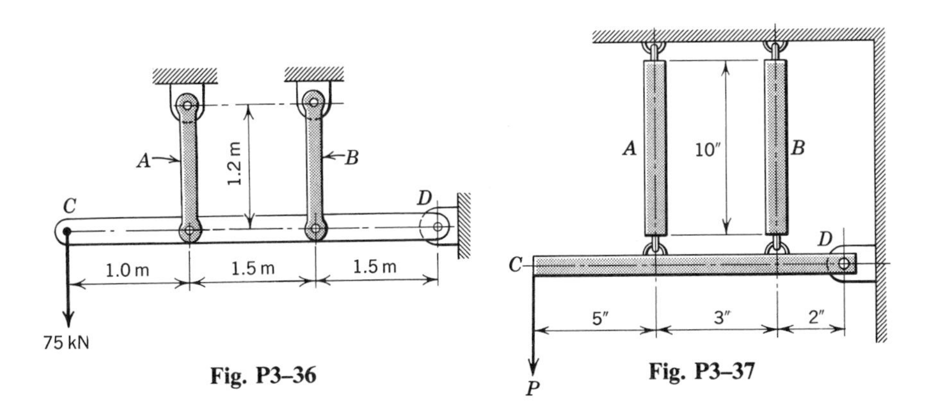

Fig. P3–36

Fig. P3–37

3–37* In Fig. P3–37 bars A and B are connected by rigid links to a fixed support at the top and to a rigid bar CD at the bottom. Determine the axial stresses in bars A and B when the load P is 40 kips applied as shown. The following data apply:

Bar	Material	Cross-sectional area (sq in.)	Proportional limit (ksi)	E (ksi)
A	Aluminum alloy (see Fig. 3–6b)	2	33	$10(10^3)$
B	Elastoplastic steel	2.5	36	$30(10^3)$

3–38 Solve Prob. 3–37 when the load P is increased to 60 kips.

3–39* Bar AB in the system of Fig. P3–39 is to be considered rigid. Bar C is steel (proportional limit = yield point = 240 MN/m^2). Post D is cold-rolled brass (E = 100 GN/m^2 and elastic strength = 410 MN/m^2). The cross-sectional areas are 600 mm^2 for bar C and 2000 mm^2 for post D. Determine the load P that will produce an axial stress of 30 MN/m^2 C in post D.

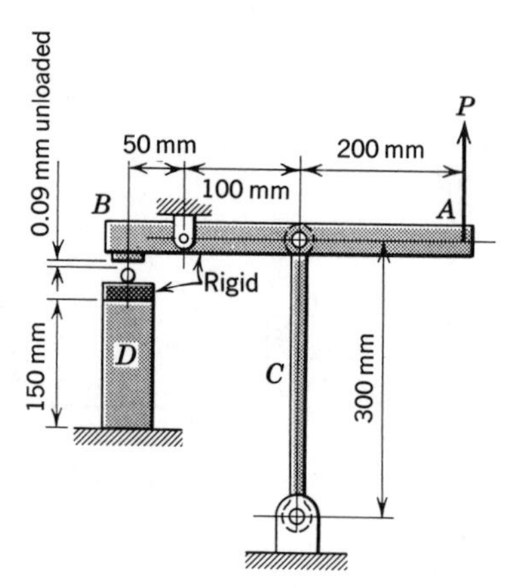

Fig. P3–39

3–40 Determine the vertical displacement of point A of Fig. P3–39 when the load P is 90 kN applied as shown. All data for Prob. 3–39 are unchanged except the stress in post D.

3–41* Bar B of the pin-connected system of Fig. P3–41 is made of aluminum alloy 6061-T6, and bar A is made of 0.2 percent carbon steel (hardened). The cross-sectional areas are 2 sq in. for A and 0.5 sq in. for B. Bar CDE is to be considered rigid. When the system is unloaded, there is a 0.005-in. clearance in the connection at D, which must be taken up before bar A will be stressed. Determine the maximum permissible load that may be applied downward at a point 6 in. to the right of D with a factor of safety of 2 with respect to failure by slip (see App. A for properties).

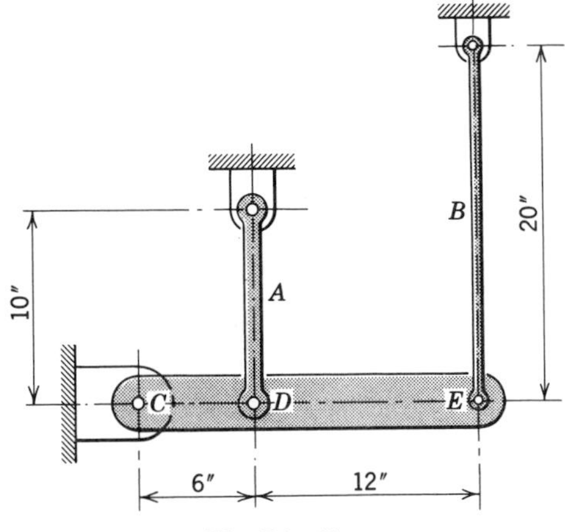

Fig. P3–41

3–42 Bar *BF* of Fig. P3–42 is steel and bar *CE* is aluminum alloy ($E=70$ GN/m^2). The cross-sectional areas are 1000 mm^2 for *BF* and 2000 mm^2 for *CE*. Due to a misalignment of the pinholes at *A*, *B*, and *C*, after the pins *A* and *B* are in place, a force *P* of 30 kN applied upward at *D* is necessary to permit insertion of the pin at *C*. Determine the axial stress in bar *CE* when the force *P* is removed with all pins in place.

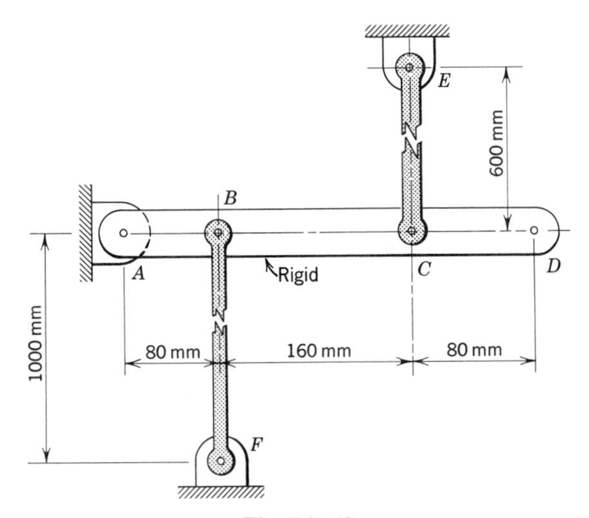

Fig. P3–42

3–43* The pin-connected structure of Fig. P3–43 occupies the position shown under no load. When the loads D (=8 kips) and E are applied, the rigid bar C becomes horizontal. Bar A is aluminum alloy 6061-T6 and bar B is red brass. The cross-sectional areas are 2 sq in. for A and 4 sq in. for B (see App. A for properties). Determine:

(a) The maximum axial stress in the system.
(b) The weight of E.

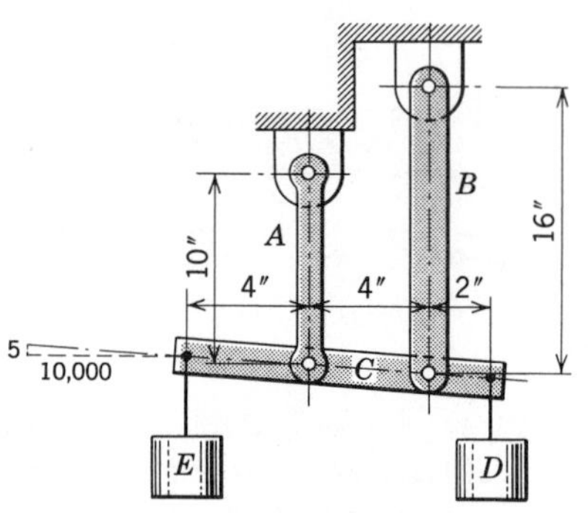

Fig. P3–43

3–44 Bar A of Fig. P3–44 is cold-rolled brass ($E = 100$ GN/m^2) and bar B is steel. Bar C is to be considered rigid. The cross-sectional areas are 200 mm^2 for A and 400 mm^2 for B. Determine the force P that will cause D to move downward 0.3 mm more than point E will move.

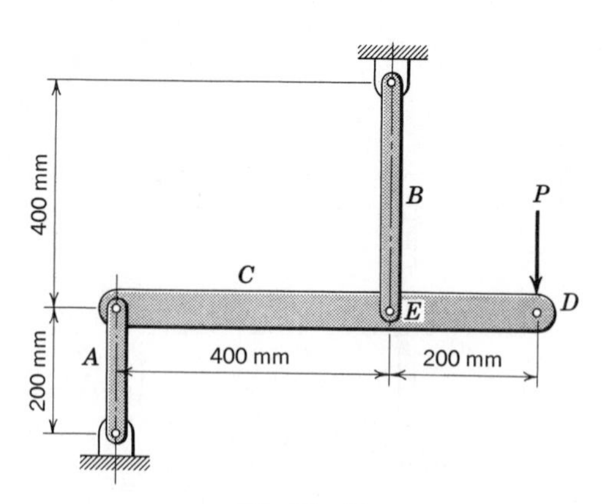

Fig. P3–44

3–45* For the mechanism of Fig. P3–45, assume bar C to be rigid. Rod A is steel (proportional limit = yield point = 36 ksi). Rod B is cold-rolled brass (proportional limit = 50 ksi, $E = 15{,}000$ ksi). The gross cross-sectional areas are 0.50 and 1.20 sq in. for A and B respectively.

(a) Determine the axial stress induced in rod A by advancing the nut 0.10 in. at the top of rod B.

(b) How far must the nut on B be advanced (from the initial position) in order to induce an axial stress of 36 ksi T in rod A, and what would be the axial deformations in rods A and B?

(c) What would be the axial stresses in rods A and B if the nut on B were advanced 0.25 in. from the initial position?

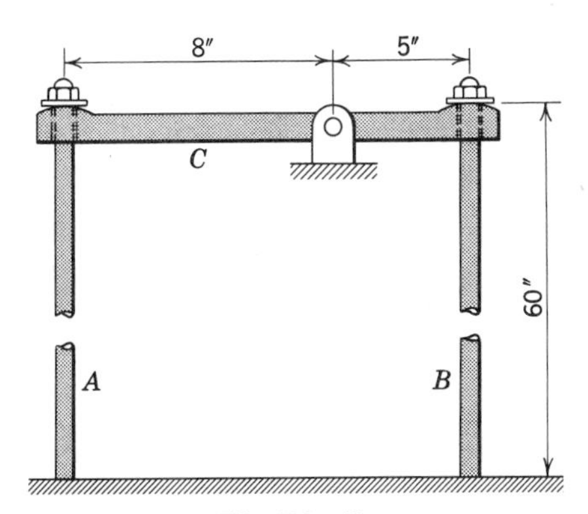

Fig. P3–45

3–46 The rigid body W of Fig. P3–46 is supported by a 6-m steel wire. The body hangs in the position shown when the temperatue is 50° C. Determine the mass of W if the axial stress in the steel wire is 120 MN/m^2 T when the temperature is 5° C. The coefficient of thermal expansion is $12(10^{-6})$ per degree C. Note that the gravitational force on 1 kg of mass is 9.807 N.

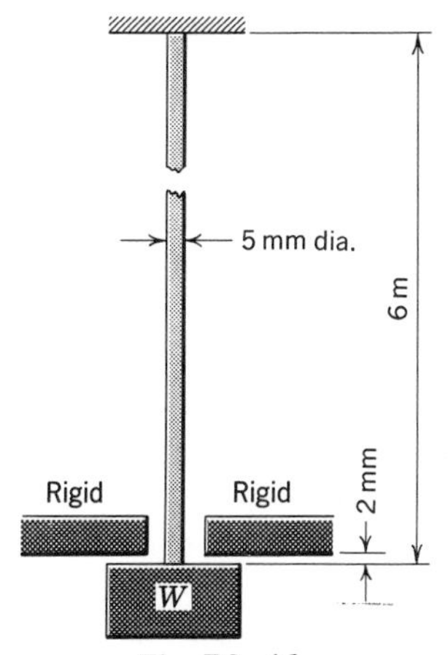

Fig. P3–46

3–47* Bar A of the pin-connected system of Fig. P3–47 is made of aluminum alloy 6061-T6, and bar B is made of low-carbon steel (proportional limit = yield point = 36 ksi, $E = 30{,}000$ ksi) (refer to App. A for properties not given). The cross-sectional areas are 1.5 sq in. for A and 0.25 sq in. for B. Bar CDE is to be considered rigid. Initially the bars are unstressed. Under a temperature increase of 200° F, what vertical force applied to bar CDE 4 in. to the right of D is needed:

(a) To make the axial tensile stress in bar B have the same magnitude as the axial compressive stress in bar A?

(b) To produce an axial stress of 6 ksi C in bar A?

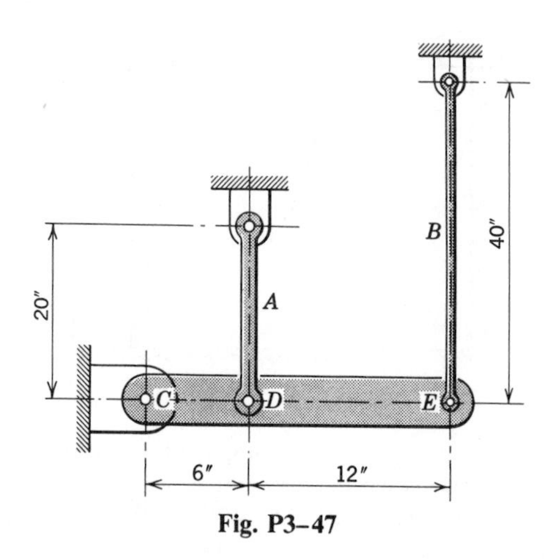

Fig. P3–47

3–48 The assembly in Fig. P3–48 consists of a steel bar A, a rigid block C, and a bronze bar B securely fastened together and fastened to rigid supports at the ends. Initially there is no stress in the members. The temperature drops 25° C, and the load P of 200 kN is applied. Determine the maximum normal stresses in A and B for this condition. The following data apply:

Bar	Cross-sectional area (mm^2)	Young's modulus (GN/m^2)	Coefficient of thermal expansion per degree C
A	500	200	$12(10^{-6})$
B	2000	100	$18(10^{-6})$

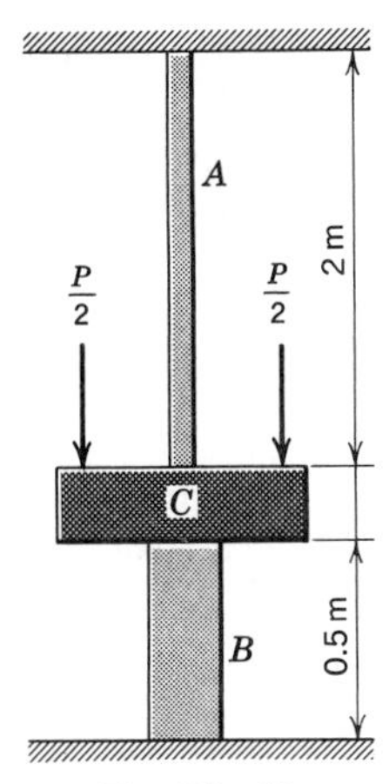

Fig. P3–48

3–49 In Fig. P3–49, bars A and B, having circular cross sections, are securely connected to the rigid plate C and to the rigid supports. The bars are unstressed when the load P is zero, and the temperature remains constant. Bar A has a uniform taper from 1 in. diameter at the bottom to 2 in. diameter at the top (assume a slight taper). The modulus of elasticity of the material of bar B is one-third that of the material of bar A. For elastic action:

(a) Demonstrate that the total axial deformation of bar A is given by the expression $\delta_A = 2P_A L/(\pi E)$, where P_A is the axial load (force) on bar A.

(b) The force P_A is what fraction of the total applied load P?

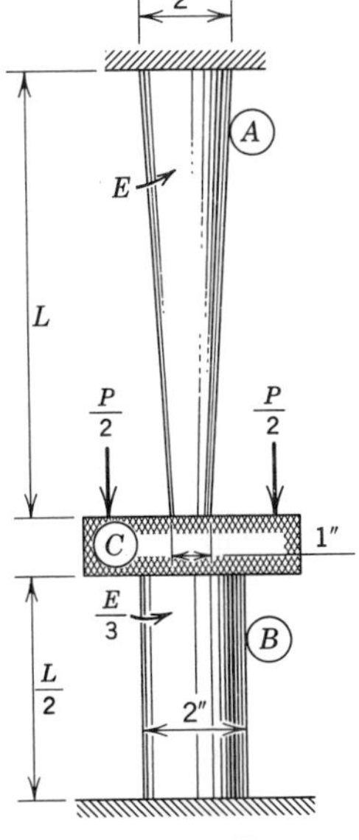

Fig. P3–49

3–50* When the structure of Fig. P3–50 is unloaded, there is a 0.03-mm gap between the rigid bearing plate *D* and the rigid bearing cap on bar *B*. Determine the magnitude of the force *P* that will cause the axial stresses in bars *A*, *B*, and *C* to have the same magnitude. The following data apply:

Bar	Material	Cross-sectional area (mm^2)	Modulus of elasticity (GN/m^2)
A	Aluminum alloy	3000	70
B	Steel	1500	200
C	Brass	1000	100

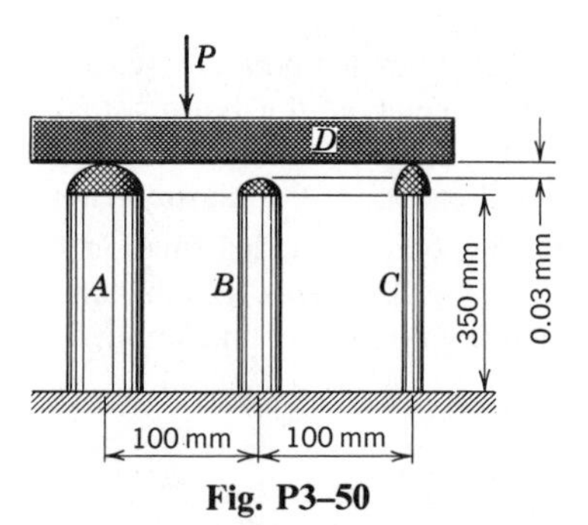

Fig. P3–50

3–51 The data for rods *A*, *B*, and *C* of Fig. P3–51 are tabulated below. Bar *C* is of low-carbon steel (proportional limit = yield point = 36 ksi). Determine the magnitude of the load *P*, applied as shown, that will induce an axial stress of 9 ksi *T* in rod *B*.

Bar	Material	Length (in.)	Cross-sectional area (sq in.)	Modulus of elasticity (ksi)
A	Aluminum alloy	20	0.20	$10(10^3)$
B	Brass	60	0.20	$15(10^3)$
C	Steel	30	0.05	$30(10^3)$

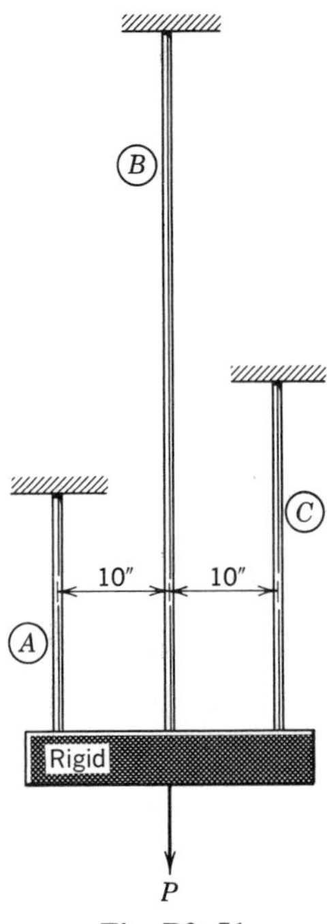

Fig. P3–51

3–52* In Fig. P3–52, bar B is high-strength steel and bar A is aluminum alloy ($E = 70$ GN/m^2). Each bar has a cross-sectional area of 900 mm^2. Determine the maximum permissible value for the load P applied as shown if the allowable axial stresses are 300 MN/m^2 for the steel and 120 MN/m^2 for the aluminum alloy.

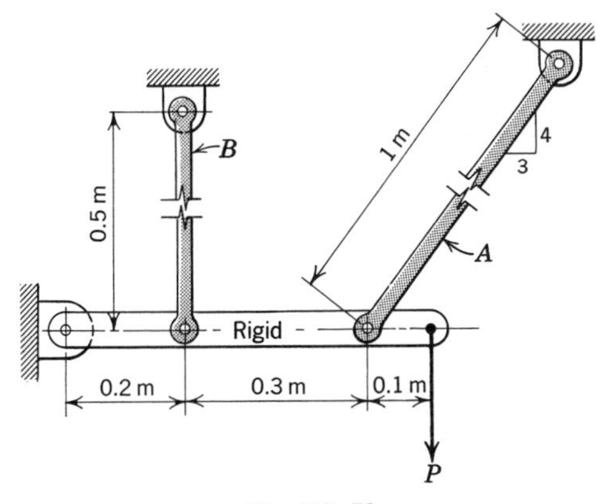

Fig. P3–52

3–53 Bar A of Fig. P3–53 is made of cold-rolled bronze (proportional limit = 60 ksi, $E = 15{,}000$ ksi) with a 3×1-in. rectangular cross section. Bars B are low-carbon steel (proportional limit = yield point = 36 ksi, $E = 30{,}000$ ksi) with 2.5×1-in. cross sections. For the indicated load, determine the axial stresses in the bars and the displacement of pin C.

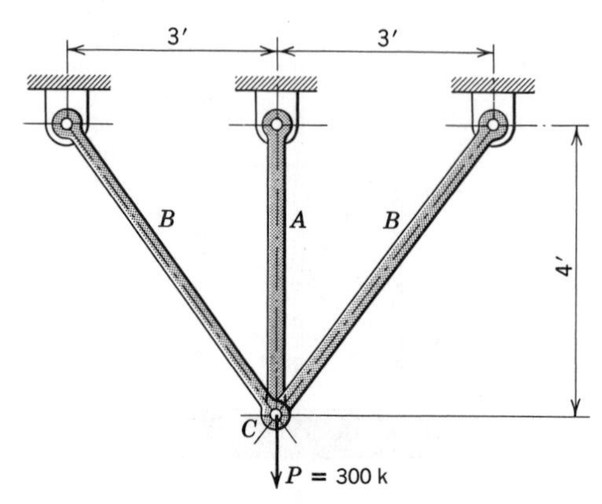

Fig. P3–53

3–54* Solve Prob. 3–53 when there is a clearance of 0.05 in. between the pin at C and the bar A that must be closed before bar A carries any load.

3–55 Determine the axial stresses in the bars of Prob. 3–53 when the load P is 6 kips and the temperature increases 100° F (refer to App. A for coefficients of thermal expansion).

3–56* Three bars, each 50 mm wide by 25 mm thick by 4 m long, are connected and loaded as shown in Fig. P3–56. Determine the maximum permissible load P that may be applied to the crosshead D with a factor of safety of 2 with respect to failure by slip. The temperature is constant, the stresses in the bars are zero when P is zero, and the following data apply:

Member	Material	Modulus of elasticity (GN/m^2)	Proportional limit (MN/m^2)
A	Monel (cold rolled)	180	400
B	Magnesium alloy	40	100
C	Steel (structural)	200	240

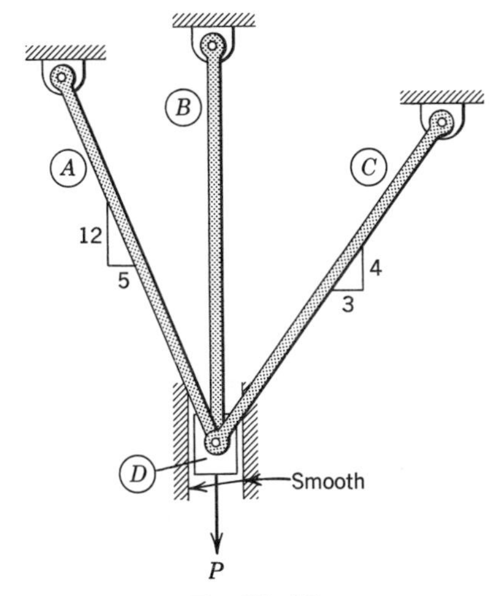

Fig. P3–56

3–57 For the structure of Prob. 3–56, determine:

(a) The load P that will induce an axial stress of 91 MN/m^2 T in bar B.

(b) The movement of the crosshead D under the load evaluated in part (a).

3–58 All members of the structure of Fig. P3–58 are elastic. The deflection of the center of the steel beam C with respect to the pins at the ends is given by the expression $\delta = F/160$, where δ is in inches and F is the resultant force at the center of the beam in kips. Before loading, beam C just touches the rigid bearing block on top of member A. Member A is aluminum alloy 6061-T6 and B is annealed red brass. The cross-sectional areas are 4 sq in. for post A and 0.64 sq in. each for bars B (see App. A for properties). Determine the maximum permissible load P that may be applied with a factor of safety of 2.5 with respect to failure by slip.

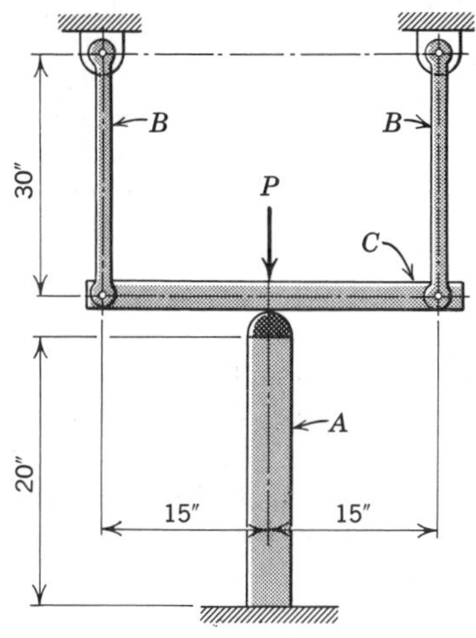

Fig. P3–58

3–59* Members A, B, and C of Fig. P3–59 are elastic, and the bell cranks D and E are to be considered rigid. Determine the maximum permissible value of the load P if the axial stresses are not to exceed 140 MN/m^2 for A, 20 MN/m^2 for B, and 60 MN/m^2 for C. The following data apply:

Bar	Material	Cross-sectional area (mm^2)	Modulus of elasticity (GN/m^2)
A	Steel	300	200
B	Concrete	4000	30
C	Brass	1000	100

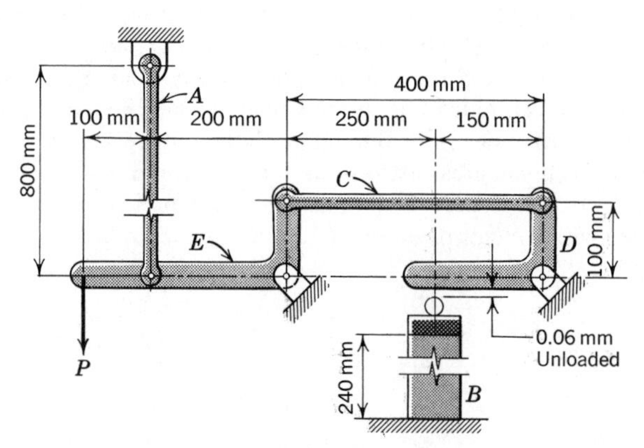

Fig. P3–59

3–60 In Fig. P3–60, the steel bars A and C are pinned to the rigid member D, and the rigid yoke E is securely attached to bar C and the cold-rolled brass bar B (see App. A). The cross-sectional areas are 4 sq in. for A, 2 sq in. for C, and 4 sq in. for B. The axial stresses are not to exceed 20 ksi for the steel and 30 ksi for the brass, and the deflection at E is not to exceed 0.05 in. downward. Determine the maximum allowable value of P.

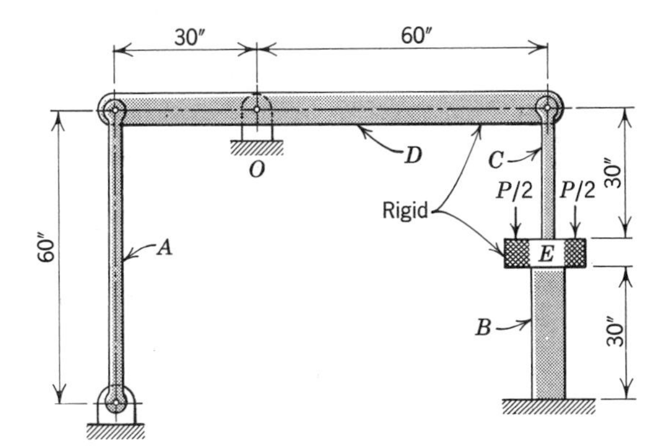

Fig. P3–60

3–6 THIN-WALLED PRESSURE VESSELS

A pressure vessel is described as *thin walled* when the ratio of the wall thickness to the radius of the vessel (an approximately cylindrical or spherical shape is implied) is so small that the distribution of normal stress on a plane perpendicular to the surface of the shell is essentially uniform throughout the thickness of the shell. Actually, this stress varies from a maximum value at the inside surface to a minimum value at the outside surface of the shell, but it can be shown by application of Eq. 3–4 in Art. 3–7 (see Prob. 3–74) that if the ratio of the wall thickness to the inner radius of the vessel is less than 0.1, the maximum normal stress is less than 5 percent greater than the average. Boilers, gas storage tanks, pipelines, metal tires, and hoops are normally analyzed as thin-walled elements. Gun barrels, certain high-pressure vessels in the chemical processing industry, and cylinders and piping for heavy hydraulic presses need to be treated as thick-walled vessels.

Problems involving thin-walled vessels subjected to internal fluid (or gas) pressure p are readily solved with the aid of free-body diagrams of sections of the vessels and the fluid contained therein. Figure 3–9 shows a section of a cylindrical pressure vessel including the fluid. The weights of the fluid and vessel are assumed to be negligible. The force P is the resultant fluid force acting on the fluid remaining within the portion of the container shown, and the forces Q are the resultant internal forces on the sections of the cylinder indicated. For simplicity, the pressures and stresses in the z direction have been omitted. The free-body diagram of Fig. 3–9 can be used to evaluate the normal stress on a longitudinal plane. This

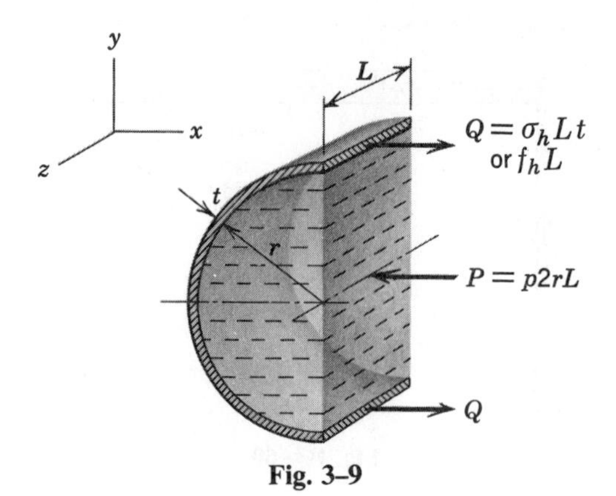

Fig. 3–9

stress component is frequently referred to as the circumferential, tangential, or hoop stress and may be denoted either as σ_t or σ_h.

The normal stress on a transverse plane can be evaluated from a free-body diagram showing stresses and pressures in the z direction. This stress component is known as the longitudinal, meridional, or axial stress and is commonly denoted either as σ_m or σ_a. The free-body diagram will be similar to Fig. 3–10b, where σ_a denotes the stress in the axial direction (the z direction for Fig. 3–9). Analysis of the stresses in a cylindrical vessel, subjected to uniform internal pressure only, indicates that the stress on a longitudinal plane is twice the stress on a transverse plane. Consequently, a longitudinal joint needs to be twice as strong as a transverse (or girth) joint. The internal force per unit length of joint, weld, or section ($f = \sigma t$) is frequently used in pressure vessel calculations.

Note that there are no shearing stresses on the planes shown because there are no loads to induce such shearing stresses. Therefore, the maximum normal stress occurs on one of the planes selected. The stress on the other of the two planes shown in Fig. 3–9 is the intermediate principal stress. The third principal stress is zero for a point at the outside surface of the shell and is equal to $-p$ for a point at the inside surface. Therefore, the maximum shearing stress will not occur on a plane perpendicular to the surface of the vessel. The maximum shearing stress occurs at the inside surface of the vessel on planes inclined 45° to both the radial and tangential (hoop) directions and parallel to the longitudinal (axial) axis of the cylinder.

Example 3–5 The 60-ft diameter Hortonsphere of Fig. 3–10a is constructed of 1.13-in. steel plates with welded joints and contains gas at a

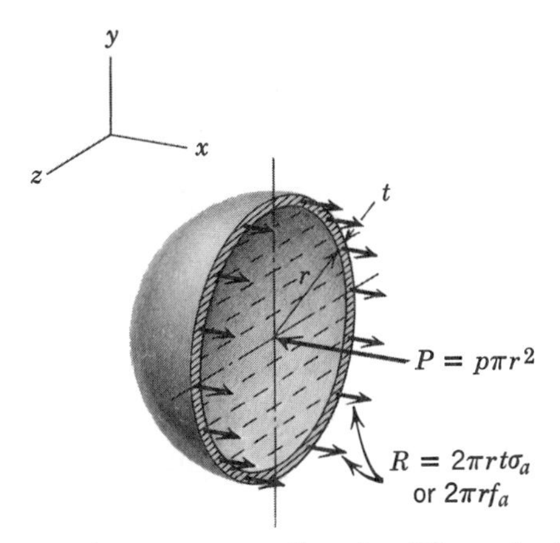

Fig. 3–10 Hortonsphere for gas storage, Superior, Wisconsin. Courtesy: Chicago Bridge and Iron Co., Chicago, Illinois.

pressure of 75 psi gage (meaning above atmospheric pressure). Determine:

(a) The force transmitted across a 10-in. length of a vertical meridian joint.

(b) The maximum (average) normal stress in the shell.

Solution

(a) Using the free-body diagram[1] of Fig. 3–10*b* and summing forces in the *x* direction gives

$$p\pi r^2 = 2\pi r f_a$$

$$75\pi(30^2)(12^2) = 2\pi(30)(12)(f_a)$$

from which

$$f_a = 13{,}500 \text{ lb per in.}$$

and the force on a 10-in. length is

$$F = 13{,}500(10) = \underline{135{,}000 \text{ lb}} \qquad \text{Ans.}$$

(b) The normal stress can be obtained from the force per inch as

$$\sigma_a = \frac{f_a}{t} = \frac{13{,}500}{1.13} = \underline{11{,}950 \text{ psi } T} \qquad \text{Ans.}$$

If only the stress were wanted, it could have been obtained directly from the first equation written by substituting $t\sigma_a$ for f_a.

The previous discussion is limited to cylindrical and spherical vessels. The theory will now be extended to include a more general solution applicable to other shapes. Figure 3–11 represents a small element of a thin shell (or membrane) of uniform thickness and with different curvatures in two orthogonal directions, subjected to an internal pressure *p*. The resultant forces on the various surfaces are shown on the diagram. Summing forces in the *n* direction gives

$$P - 2F_1 \sin d\theta_1 - 2F_2 \sin d\theta_2 = 0$$

from which

$$p(2r_1\,d\theta_1)(2r_2\,d\theta_2) = 2\sigma_1(2tr_2\,d\theta_2)\sin d\theta_1 + 2\sigma_2(2tr_1\,d\theta_1)\sin d\theta_2$$

[1]As a permissible approximation, the inside circumference rather than the mean is used in the expression for *R*.

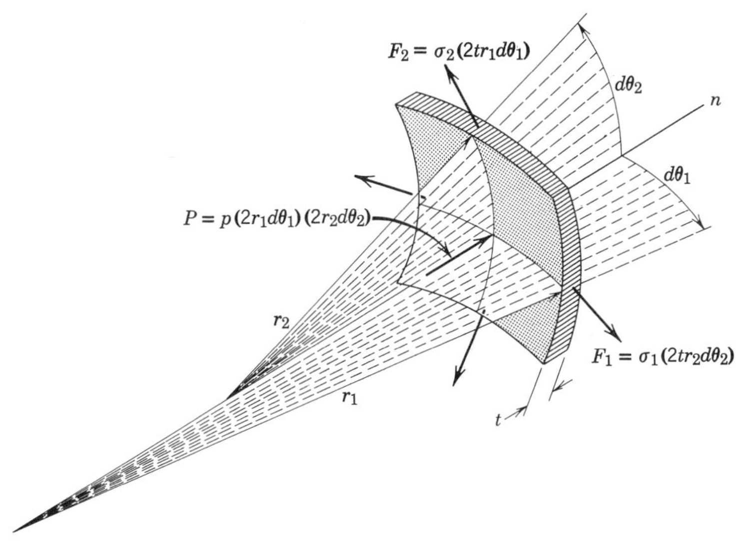

Fig. 3–11

and since, for small angles, $\sin d\theta \approx d\theta$, the above equation becomes

$$\frac{p}{t} = \frac{\sigma_1}{r_1} + \frac{\sigma_2}{r_2} \qquad (3\text{–}3)$$

Since Eq. 3–3 contains two unknown stresses, an additional independent equation is necessary for a solution; this will be an equilibrium equation similar to that in Exam. 3–5. The following example illustrates the application of Eq. 3–3.

Example 3–6 A pressure vessel of 1/4-in. steel plate has the shape of a paraboloid closed by a thick flat plate, as shown in Fig. 3–12*a*. The equation of the generating parabola is $y = x^2/4$, where x and y are in inches. Determine the axial and hoop stresses in the shell at a point 16 in. above the bottom due to an internal gas pressure of 250 psi gage.

Solution The stress σ_a, which must be tangent to the shell, will be determined first with the aid of the free-body diagram of Fig. 3–12*b*, which represents a thin slice of the vessel and gas, with the forces perpendicular to the slice omitted. From the equation of the parabola, the radius x and the slope of the shell dy/dx at $y = 16$ in. are determined to be 8 in. and

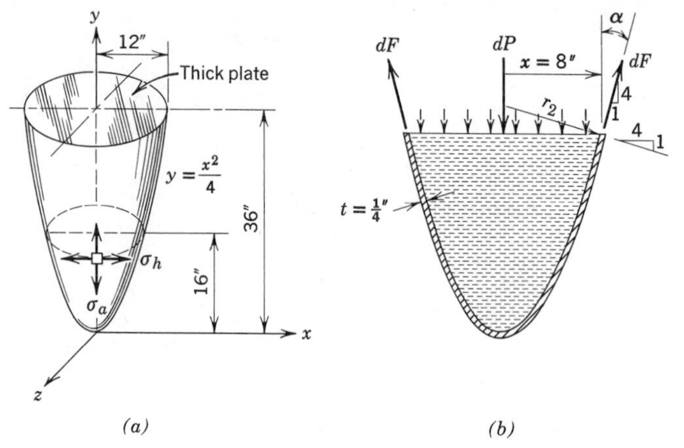

Fig. 3–12

4/1, respectively. Summing forces in the y direction gives

$$-\int_A dP + \int dF \cos\alpha = 0$$

$$-p\pi x^2 + \sigma_a 2\pi x t \cos\alpha = 0$$

Substituting the given data yields

$$-250(\pi)(8^2) + \sigma_a(2\pi)(8)(1/4)(4/\sqrt{17}\,) = 0$$

from which

$$\sigma_a = 4123 = \underline{4120 \text{ psi } T} \qquad \text{Ans.}$$

In order to find σ_h from Eq. 3–3, the radii r_1 and r_2 at the point must be determined. The radius of curvature of the shell in the xy plane r_1 is determined from the expression

$$r_1 = \frac{\left[1 + (dy/dx)^2\right]^{1.5}}{d^2y/dx^2} = \frac{(1+4^2)^{1.5}}{1/2} = 140.19 \text{ in.}$$

and the perpendicular radius r_2 is found from the geometry of Fig. 3–12 as

$$r_2 = 8(\sqrt{17}/4) = 8.246 \text{ in.}$$

Then, from Eq. 3–3,

$$p/t = (\sigma_a/r_1) + (\sigma_h/r_2)$$

$$250(4) = (4123/140.19) + (\sigma_h/8.246)$$

from which

$$\sigma_h = 8003 = \underline{8000 \text{ psi } T} \qquad \text{Ans.}$$

PROBLEMS

3–61* The cylindrical propane tanks of Fig. P3–61 are $124\frac{3}{4}$ in. in diameter and are constructed of 1-in. steel plates with all joints welded. The working pressure for these tanks is 250 psi. Determine:

(a) The maximum (average) normal stress in the shell.
(b) The factor of safety with respect to failure by fracture. The ultimate tensile strength of the steel is 70 ksi.

Fig. P3–61 All-welded horizontal pressure tanks for propane storage. Northlake, Illinois. Courtesy: Chicago Bridge and Iron Co., Chicago, Illinois.

3–62* At Hoover Dam water may be discharged under a maximum head of almost 183 m (pressure approximately 1.8 MN/m^2). What thickness of plate would be required for a 1.5-m diameter pressure pipe at a working tensile stress of 90 MN/m^2 if the joints are welded? What is the factor of safety with respect to failure by fracture if the yield strength and ultimate strength of the steel are 250 and 450 MN/m^2, respectively?

3–63 A spherical gas holder 40 ft in diameter contains gas at a pressure of 50 psi. Determine the force transmitted across an 8-in. length of the joint uniting the two halves of the sphere.

3–64* A 2-m diameter spherical vessel is subjected to an internal pressure of 1.50 MN/m^2. Determine the force transmitted across 200 mm of the joint uniting the two halves of the sphere.

3–65

(a) Prove that the normal stress on a transverse plane of a thin-walled cylindrical pressure vessel is $\sigma_a = pD/(4t)$.

(b) Prove that the normal stress on a longitudinal plane of symmetry of a thin-walled cylindrical pressure vessel is $\sigma_h = pD/(2t)$.

3–66 The cylindrical pressure tank of Fig. P3–66 is made of 8-mm steel plate. The normal stress at A on plane B–B (which is perpendicular to the surface of the plate at A) is 60 MN/m^2 T. Determine the air pressure in the tank.

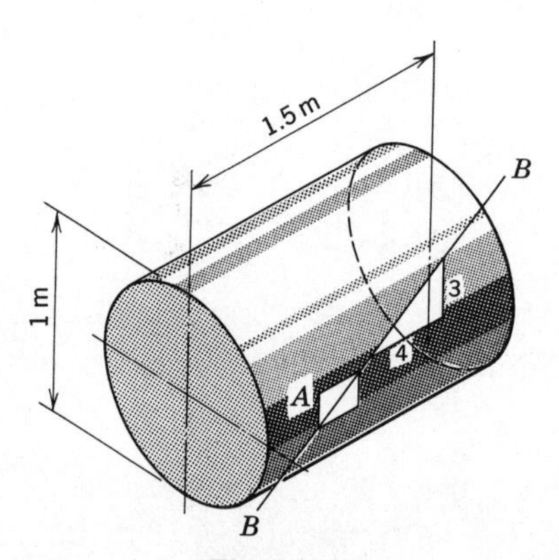

Fig. P3–66

3–67* A cylindrical pressure tank is fabricated by butt welding 5/8 -in. plate with a spiral seam, as shown in Fig. P3–67. The pressure in the tank is 250 psi, and an axial load of 9000π lb is applied to the end of the tank through a rigid bearing plate. Determine the normal and shearing stresses in the weld on the plane of the weld.

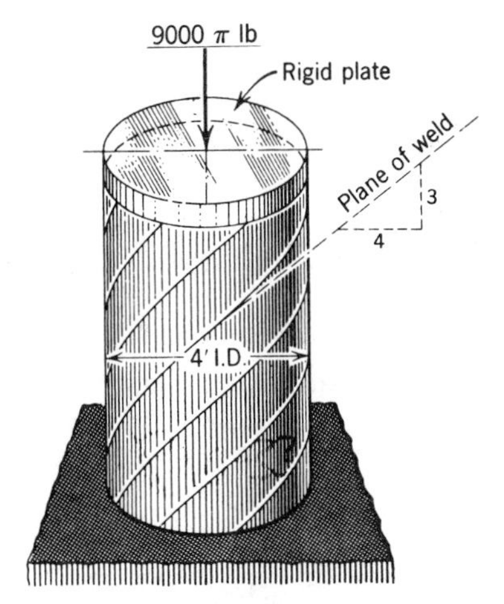

Fig. P3–67

3–68 The strains measured on the outside surface of the cylindrical pressure vessel of Fig. P3–68 are: $\epsilon_1 = +480\mu$ and $\epsilon_2 = +220\mu$, where the angle θ is unknown. The diameter of the vessel is 18 in., thickness of the plate is 1/8 in., and the material is 0.4 percent carbon hot-rolled steel (see App. A; use 0.30 for Poisson's ratio). Determine:

(a) The stresses σ_1 and σ_2 in the shell.
(b) The principal stresses at the outside surface.
(c) The pressure in the vessel.
(d) The factor of safety with respect to failure by fracture.

Note. This problem demonstrates that when the above strains are measured by two orthogonal strain gages, the angle θ is not important.

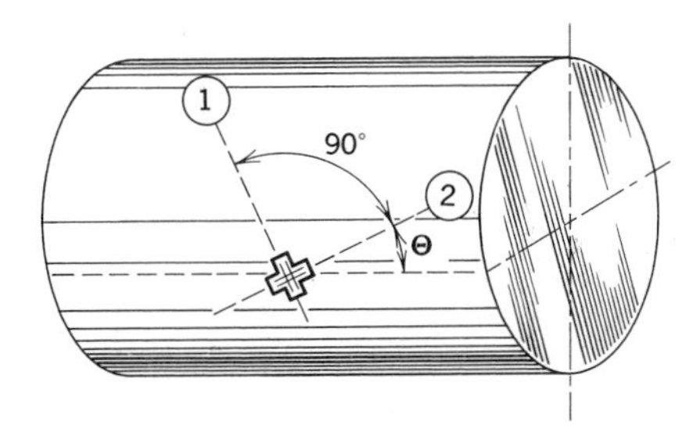

Fig. P3–68

3–69* A vessel similar to that shown in Fig. P3–68 has a diameter of 400 mm and a plate thickness of 4 mm. The material is steel having a yield strength of 250 MN/m^2 and an ultimate strength of 450 MN/m^2 (use 0.30 for Poisson's ratio). The strains measured in the direction 2 at $\theta = 30°$ and 1 at $\theta + 45°$ (not 90° as in the figure) are $+145\mu$ and $+350\mu$. Determine:

(a) The maximum surface strain.
(b) The maximum (average) normal stress in the plate.
(c) The pressure in the vessel.
(d) The factor of safety with respect to failure by fracture.

3–70 A hemispherical tank of radius r is supported by a flange, as shown in the cross section of Fig. P3–70, and is filled with fluid having a specific weight of γ lb per unit volume. Determine, in terms of γ, r, and t, the stresses σ_a and σ_h at a depth $y = r/2$.

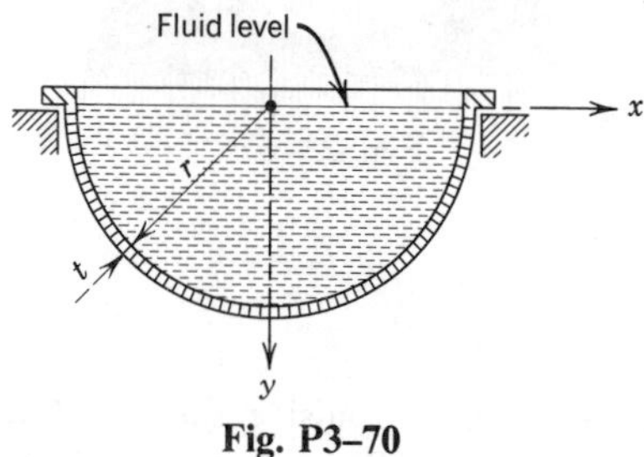

Fig. P3–70

3–71* A reducer in a pipeline (see Fig. P3–71) has a wall thickness of 0.15 in. Assume that the flange bolts at the 6-in. end take all the end thrust (no compression in the flange at the 3-in. end). The pipeline is under a static water pressure of 60 psi (gage). Determine the axial and hoop stresses σ_a and σ_h at a point midway between the flanges.

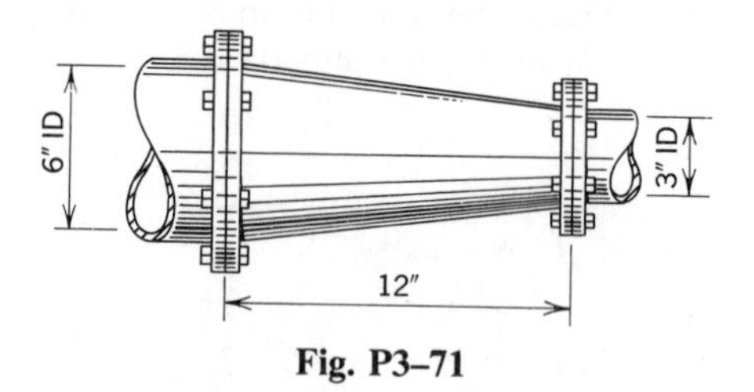

Fig. P3–71

3–72 The conical water tank shown in Fig. P3–72 was fabricated from 1/8-in. steel plate. When the tank is completely full of water (specific weight $\gamma = 62.4$ lb per cu ft), determine the axial and hoop stresses σ_a and σ_h at a point in the wall 8 ft below the apex of the cone.

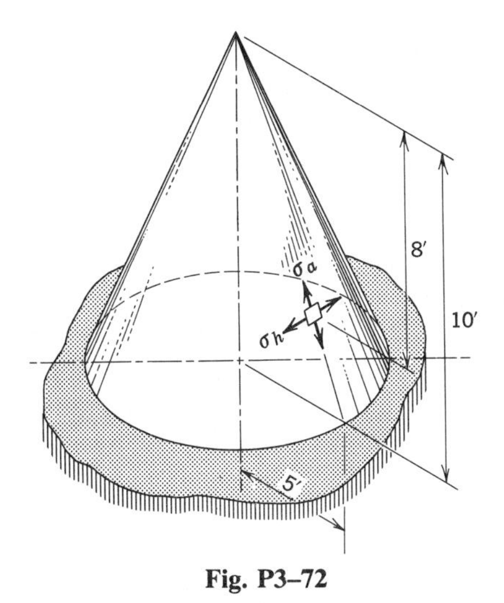

Fig. P3–72

3–73* An internal pressure of 100 psi is applied to the toroidal shell (pressurized doughnut) shown in Fig. P3–73. Determine the axial and hoop stresses σ_a and σ_h at points A and B on the horizontal plane of symmetry of the shell. Hint. The hoop stress σ_h at point A can be determined by using a free body consisting of one quarter of the shell, such as the part shown cross-hatched in Fig. P3–73.

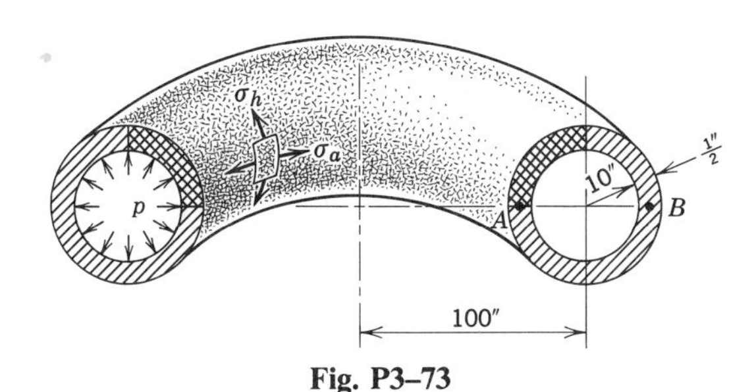

Fig. P3–73

3–7 THICK-WALLED CYLINDRICAL PRESSURE VESSELS

The problem of determining the tangential stress σ_t and the radial stress σ_r at any point in the wall of a thick-walled cylinder, in terms of the pressures applied to the cylinder, was solved by the French elastician G. Lame in 1833. The results can be applied to a wide variety of design situations involving cylindrical pressure vessels, hydraulic cylinders, piping systems, shrink- and press-fit applications, etc.

Consider a thick-walled cylinder having inner radius a and outer radius b, as shown in Fig. 3–13a. The cylinder is subjected to an internal pressure p_i and an external pressure p_o. For purposes of analysis, the thick-walled cylinder can be considered to consist of a series of thin rings. A typical ring located at a radial distance ρ from the axis of the cylinder and having a thickness $d\rho$ is shown by the dashed lines in Fig. 3–13a. As a result of the internal and external pressure loadings, a radial stress σ_r would develop at the interface between rings located at radial position ρ, while a slightly different radial stress $(\sigma_r + d\sigma_r)$ would develop at radial position $(\rho + d\rho)$. These stresses would be uniformly distributed over the inner and outer

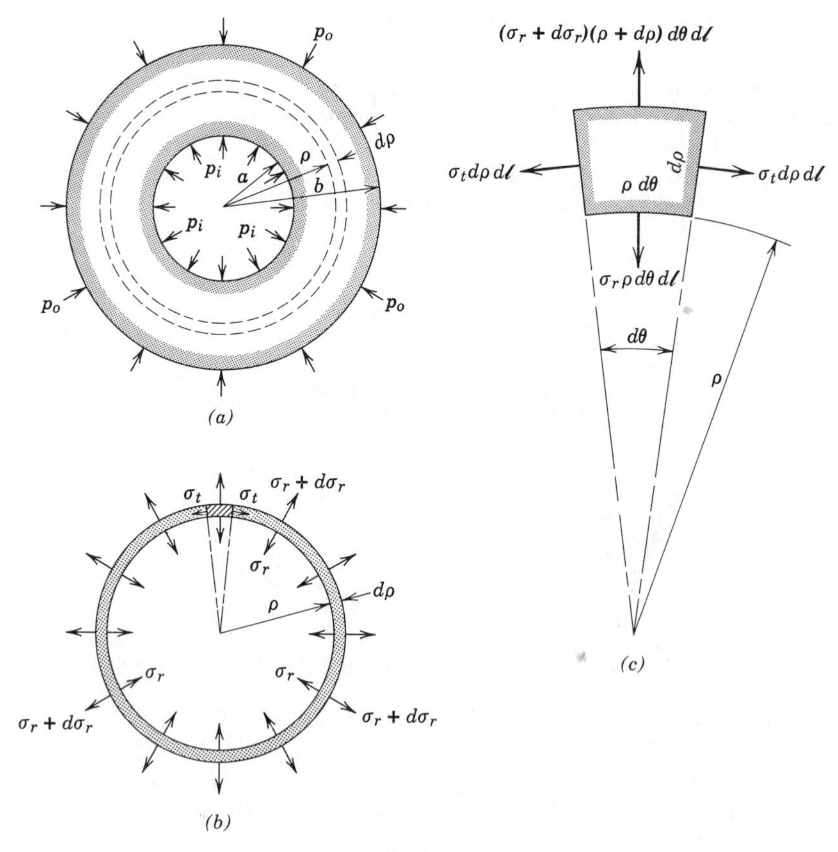

Fig. 3–13

surfaces of the ring, as shown in Fig. 3–13*b*. Shearing stresses would not develop on the inner and outer surfaces of the ring, since the pressure loadings do not tend to force the rings to rotate with respect to one another.

In Art. 3–6, which dealt with thin-walled pressure vessels, it was shown that a tangential or hoop component of stress develops when a pressure difference exists between the inner and outer surfaces of a thin shell or ring. The planes on which these tangential stresses act can be exposed by considering only a small part of a ring, as shown shaded in Fig. 3–13*b*. Since the ring is assumed to be thin, the tangential stress σ_t can be considered to be uniformly distributed through the thickness of the ring. A relationship between radial stress σ_r and tangential stress σ_t can be obtained from equilibrium considerations. A free-body diagram of a small part of a ring, such as the one shown in Fig. 3–13*c* for the shaded part of Fig. 3–13*b*, is useful for this determination. The axial stress σ_a, which may be present in the cylinder, has been omitted from this diagram since it does not contribute to equilibrium in the radial or tangential directions. The free body is assumed to have a length dl along the axis of the cylinder.

From a summation of forces in the radial direction

$$(\sigma_r + d\sigma_r)(\rho + d\rho)d\theta\, dl - \sigma_r \rho\, d\theta\, dl - 2\sigma_t\, d\rho\, dl \sin\frac{d\theta}{2} = 0 \tag{a}$$

By neglecting higher-order terms and noting that for small angles $\sin d\theta/2 \approx d\theta/2$, Eq. (a) can be reduced to

$$\rho\frac{d\sigma_r}{d\rho} + \sigma_r - \sigma_t = 0 \tag{b}$$

Equation (b) cannot be integrated directly, since both σ_r and σ_t are functions of radial position ρ. In previous instances, when such statically indeterminate situations were encountered, the problem was solved by considering deformations of the structure.

For the case of a thick-walled cylinder, the axial strain ϵ_a at any point in the wall of the cylinder can be expressed in terms of σ_a, σ_r, and σ_t by using generalized Hooke's Law (see Eq. 2–14). Thus,

$$\epsilon_a = [\sigma_a - \nu(\sigma_r + \sigma_t)]/E \tag{c}$$

The assumption normally made concerning strains in the thick-walled cylinder, which has been verified by careful measurement, is that axial strain is uniform. This means that plane transverse cross sections before loading remain plane and parallel after the internal and external pressures are applied. So far as the axial stress σ_a is concerned, two cases are of interest in a wide variety of design applications. In the first case, the axial loads induced by the pressure are not carried by the walls of the cylinder ($\sigma_a = 0$). This situation arises in gun barrels and in many types of hydraulic cylinders where pistons carry the axial loads. In the second case, the walls of the cylinder carry the loads. This situation occurs in pressure vessels with various types of end closures or heads. In this second case, in regions of the cylinder away from the ends, it has been found that the axial stress is uniformly distributed over the cross section. Hence ϵ_a, σ_a, E, and ν are constant for the two cases being considered; therefore, it follows from Eq. (c) that

$$\sigma_r + \sigma_t = (\sigma_a - E\epsilon_a)/\nu = 2C_1 \tag{d}$$

The constant is taken as $2C_1$ for convenience in the following derivation.

When the value for σ_t from Eq. (d) is substituted into Eq. (b), this latter equation may be written as

$$\rho \frac{d\sigma_r}{d\rho} + 2\sigma_r = 2C_1 \tag{e}$$

If Eq. (e) is multiplied by ρ, the terms before the equal sign can be expressed as

$$\frac{d}{d\rho}(\rho^2\sigma_r) = 2C_1\rho$$

Integrating yields

$$\rho^2\sigma_r = C_1\rho^2 + C_2$$

where C_2 is a constant of integration. Thus,

$$\sigma_r = C_1 + \frac{C_2}{\rho^2} \tag{f}$$

The tangential stress σ_t is then obtained from Eq. (d) as

$$\sigma_t = C_1 - \frac{C_2}{\rho^2} \tag{g}$$

Values for the constants C_1 and C_2 in Eqs. (f) and (g) can be determined by using the known values for the pressures at the inside and outside surfaces of the cylinder. These values, commonly referred to as boundary conditions, are

$$\sigma_r = -p_i \quad \text{at } \rho = a$$

$$\sigma_r = -p_o \quad \text{at } \rho = b$$

The minus signs indicate that the pressures (normally considered as positive quantities) produce compressive normal stresses at the surfaces on which they are applied. Substituting the boundary conditions into Eq. (f) yields

$$C_1 = \frac{a^2 p_i - b^2 p_o}{b^2 - a^2}$$

$$C_2 = -\frac{a^2 b^2 (p_i - p_o)}{b^2 - a^2}$$

The desired expressions for σ_r and σ_t are obtained by substituting these values for C_1 and C_2 into Eqs. (f) and (g). Thus,

$$\begin{aligned} \sigma_r &= \frac{a^2 p_i - b^2 p_o}{b^2 - a^2} - \frac{a^2 b^2 (p_i - p_o)}{(b^2 - a^2)\rho^2} \\ \sigma_t &= \frac{a^2 p_i - b^2 p_o}{b^2 - a^2} + \frac{a^2 b^2 (p_i - p_o)}{(b^2 - a^2)\rho^2} \end{aligned} \tag{3–4}$$

Radial and circumferential deformations δ_ρ and δ_c play important roles in all press-fit or shrink-fit problems. The change in circumference δ_c of the thin ring shown in Fig. 3–13a, when the pressures p_i and p_o are applied to the cylinder, may be expressed in terms of the radial displacement δ_ρ of a point on the ring as

$$\delta_c = 2\pi\delta_\rho$$

The circumferential deformation δ_c may also be expressed in terms of the tangential strain ϵ_t as

$$\delta_c = \epsilon_t c$$

where $c = 2\pi\rho$ is the circumference of the ring. Thus,

$$\delta_\rho = \epsilon_t \rho$$

For most applications involving shrink fits, the axial stress $\sigma_a = 0$. The tangential strain ϵ_t can then be expressed in terms of the radial stress σ_r and the tangential stress σ_t by using the generalized Hooke's law. Thus,

$$\epsilon_t = (\sigma_t - \nu\sigma_r)/E$$

The radial displacement of a point in the wall is then obtained in terms of the radial and tangential stresses present at the point as

$$\delta_\rho = (\sigma_t - \nu\sigma_r)\rho/E \tag{3–5}$$

As mentioned previously, Eqs. 3–4 and 3–5 can be used to compute stresses and deformations in a wide variety of design situations involving pressure vessels, hydraulic cylinders, etc. Reduced forms of these equations are used with sufficient frequency to warrant consideration of the following special cases.

CASE 1 (INTERNAL PRESSURE ONLY). If the loading is limited to an internal pressure $p_i (p_o = 0)$, Eq. 3–4 reduce to

$$\sigma_r = \frac{a^2 p_i}{b^2 - a^2}\left(1 - \frac{b^2}{\rho^2}\right)$$

$$\sigma_t = \frac{a^2 p_i}{b^2 - a^2}\left(1 + \frac{b^2}{\rho^2}\right)$$

Examination of these equations indicates that σ_r is always a compressive stress, while σ_t is always a tensile stress. In addition, σ_t is always larger than σ_r and is maximum at the inside surface of the cylinder. A typical distribution of the stress σ_t is shown in Exam. 3–7. Substituting the values of σ_r and σ_t from the previous two equations into Eq. 3–5 yields the deformation equation applicable for this special case ($p_o = 0$ and $\sigma_a = 0$). Thus,

$$\delta_\rho = \frac{a^2 p_i}{(b^2 - a^2)E\rho}\left[(1 - \nu)\rho^2 + (1 + \nu)b^2\right]$$

CASE 2 (EXTERNAL PRESSURE ONLY). If the loading is limited to an external pressure $p_o (p_i = 0)$, Eq. 3–4 reduce to

$$\sigma_r = -\frac{b^2 p_o}{b^2 - a^2}\left(1 - \frac{a^2}{\rho^2}\right)$$

$$\sigma_t = -\frac{b^2 p_o}{b^2 - a^2}\left(1 + \frac{a^2}{\rho^2}\right)$$

In this case both σ_r and σ_t are always compressive. The tangential stress is always larger than the radial stress and assumes its maximum value at the inner surface of the cylinder. Substituting the values of σ_r and σ_t from the previous two equations into Eq. 3–5 yields the deformation equation applicable for this special case ($p_i = 0$ and $\sigma_a = 0$). Thus,

$$\delta_\rho = -\frac{b^2 p_o}{(b^2 - a^2) E\rho}\left[(1-\nu)\rho^2 + (1+\nu)a^2\right]$$

CASE 3 (EXTERNAL PRESSURE ON A SOLID CIRCULAR CYLINDER). Radial and tangential stresses σ_r and σ_t and the radial displacement δ_ρ for this special case can be obtained from the expression developed for case 2 by letting the hole in the cylinder vanish ($a = 0$). Thus,

$$\sigma_r = -p_o$$

$$\sigma_t = -p_o$$

and

$$\delta_\rho = -(1-\nu) p_o \rho / E$$

The minus sign in these equations indicates that the stresses are both compressive and the radius of the cylinder is reduced when the external pressure p_o is applied. For this case the stresses are independent of radial position ρ and have a constant magnitude equal to the applied pressure.

In the fabrication of gun barrels and in numerous other applications, a laminated thick-walled cylinder is frequently formed by shrinking an outer cylinder (called a hoop, jacket, or shell) over an inner cylinder (frequently called the cylinder or tube). The purpose of this fabricating procedure is to increase the load-carrying capabilities of the assembly by developing an initial state of compressive tangential stress in the inner cylinder and tensile tangential stress in the outer cylinder. When the stresses induced by an internal pressure are superimposed on this initial state of stress, the resultant stress distribution makes more efficient use of the material than the distribution that would exist without the shrink-fit-induced initial stresses (see Exam. 3–7).

The determination of stresses in such a laminated cylinder requires the solution of a statically indeterminate problem that makes use of deformations. Consider the laminated thick-walled cylinder shown in Fig. 3–14 that has been formed by heating the jacket until it could be slipped over the tube. The heating of the jacket is required for assembly, since the jacket has an inside diameter that is slightly smaller than the outside diameter of the tube. The dotted sections in Fig. 3–14 illustrate jacket and tube configurations (greatly exaggerated) before they were fitted together. As the jacket cools after assembly, an interfacial contact pressure p_s develops and produces stresses and deformations in both the jacket and the tube. The initial difference in diameters between the jacket and the tube at the interface is known as the interference I. The interference is related to radial deformations of the jacket and tube at the interface by the expression

$$I = 2\delta = 2|\delta_1| + 2|\delta_2| \tag{h}$$

The magnitudes of the radial deformations $|\delta_1|$ and $|\delta_2|$ can be determined by using the deformation expressions developed previously for special cases 1 and 2. Thus,

$$|\delta_1| = \frac{b^2 p_s}{(c^2 - b^2)Eb}\left[(1-\nu)b^2 + (1+\nu)c^2\right]$$

$$|\delta_2| = \frac{b^2 p_s}{(b^2 - a^2)Eb}\left[(1-\nu)b^2 + (1+\nu)a^2\right]$$

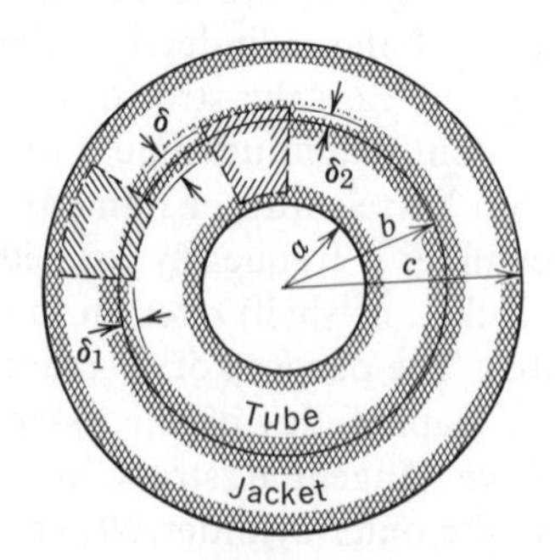

Fig. 3–14

Substituting these values into Eq. (h) yields

$$I = \frac{4b^3 p_s (c^2 - a^2)}{E\,(c^2 - b^2)(b^2 - a^2)} \tag{3–6}$$

or

$$p_s = \frac{EI\,(c^2 - b^2)(b^2 - a^2)}{4b^3(c^2 - a^2)} \tag{3–7}$$

Equation 3–6 provides a means for establishing the interference that can be tolerated without exceeding maximum allowable stress levels in the jacket or the tube. This type of information is required, for example, when bearing races are being designed for press-fit or shrink-fit applications. Equation 3–7 provides a means for establishing the interfacial pressure resulting from a known amount of interference. Once this pressure is known, shrink-fit stress levels can be established in both the jacket and the tube. Load-induced stresses in the laminated cylinder will add to these fabrication-induced values.

Example 3–7 A laminated thick-walled hydraulic cylinder was fabricated by shrink-fitting a steel jacket having an outside diameter of 12 in. onto a steel tube having an inside diameter of 4 in. and an outside diameter of 8 in. The interference was 0.008 in.

(a) Determine the interfacial contact pressure.
(b) Determine the maximum tensile stress in the laminated cylinder resulting from the shrink fit.
(c) Determine the maximum tensile stress in the laminated cylinder after an internal pressure of 40,000 psi is applied.
(d) Compare the stress distribution in the laminated cylinder with the stress distribution in a uniform cylinder for an internal pressure of 40,000 psi.

Solution

(a) The interfacial contact pressure (shrink pressure) can be determined by using Eq. 3–7. Thus,

$$p_s = \frac{(30 \times 10^6)(0.008)(36 - 16)(16 - 4)}{4(64)(36 - 4)} = \underline{7030 \text{ psi}} \qquad \text{Ans.}$$

(b) The maximum tensile stress resulting from this shrink pressure will be the tangential stress σ_t that occurs at the inside surface of the jacket.

This stress can be determined by using the reduced form of Eq. 3–4 for the case of internal pressure only. Thus,

$$\sigma_t = \frac{16(7030)}{36-16}\left(1+\frac{36}{16}\right) = \underline{18{,}280\text{ psi}} \qquad \text{Ans.}$$

(c) Stresses in the laminated cylinder after the internal pressure is applied will result from both the shrink pressure and the applied pressure. Again, the maximum tensile stress will be a tangential stress; however, it may occur either at the inside surface of the jacket or at the inside surface of the tube, depending on the magnitudes of the stresses associated with the two pressure loadings. Reduced forms of Eq. 3–4 can be used for each of the loadings. For the tube,

$$\sigma_t = \sigma_t(\text{tube interference}) + \sigma_t(\text{laminated cylinder due to } p_i)$$

$$= -\frac{16(7030)}{16-4}\left(1+\frac{4}{4}\right) + \frac{4(40000)}{36-4}\left(1+\frac{36}{4}\right)$$

$$= -18{,}750 + 50{,}000 = 31{,}250\text{ psi}$$

For the jacket,

$$\sigma_t = \sigma_t(\text{jacket interference}) + \sigma_t(\text{laminated cylinder due to } p_i)$$

$$= \frac{16(7030)}{36-16}\left(1+\frac{36}{16}\right) + \frac{4(40000)}{36-4}\left(1+\frac{36}{16}\right)$$

$$= 18{,}280 + 16{,}250 = 34{,}530 = \underline{34{,}500\text{ psi}} = \sigma_{max} \qquad \text{Ans.}$$

(d) Radial and tangential stress distributions in the laminated cylinder and in a uniform cylinder for an internal pressure of 40,000 psi are shown in Fig. 3–15.

PROBLEMS

3–74 Demonstrate that for a cylindrical vessel with a shell thickness $(b-a)$ of one-tenth the radius a, subjected only to internal pressure, the error involved in computing the hoop tension σ_t by the method of Art. 3–6 instead of Eq. 3–4 is approximately 5 percent.

3–75* A steel cylinder with an inside diameter of 8 in. and an outside diameter of 12 in. is subjected to an internal pressure of 10,000 psi. Determine:

(a) The maximum tensile stress in the cylinder.

(b) The radial and tangential stresses at a point midway in the cylinder wall.

3–76 A steel cylinder with an inside diameter of 2 in. and an outside diameter of 10 in. is subjected to an internal pressure of 12,000 psi and an external pressure of 4000 psi. Determine the circumferential force per unit length carried by a circular

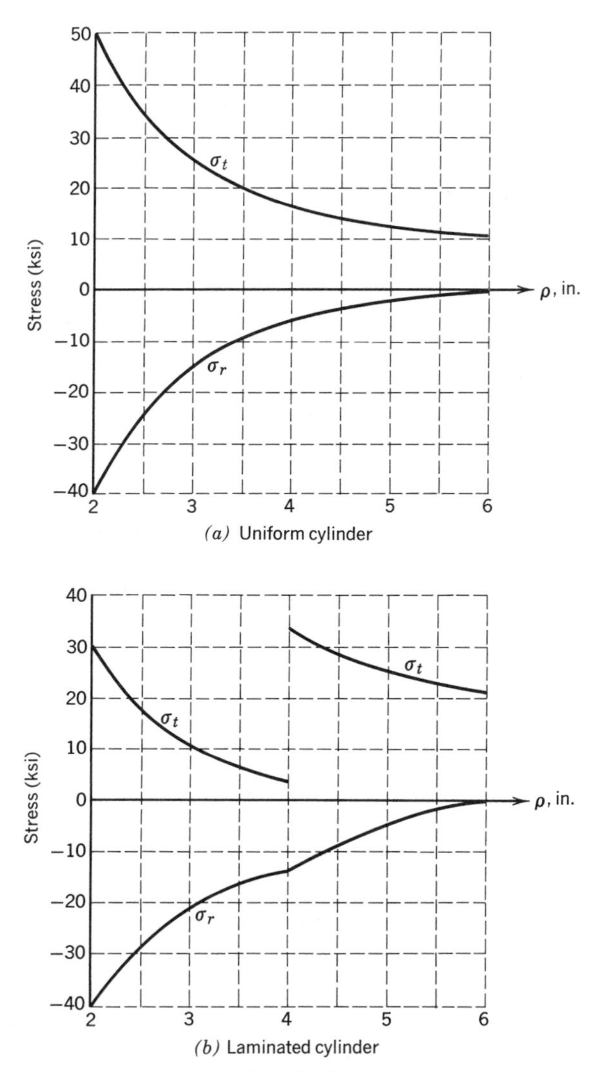

Fig. 3–15

element having an inside diameter of 4 in. and a thickness of 1 in.

3–77* A steel cylinder with an inside diameter of 150 mm and an outside diameter of 250 mm is subjected to an internal pressure of 75 MN/m^2 and an external pressure of 150 MN/m^2. Determine the radial and tangential stresses at a point midway in the cylinder wall.

3–78 A steel cylinder with an inside diameter of 200 mm and an outside diameter of 250 mm is subjected to an internal pressure of 35 MN/m^2. Determine:

(a) The maximum tensile and shearing stresses in the cylinder and their location.
(b) Compare the maximum tensile stress with the average tensile stress assuming a uniform distribution across the thickness.

3–79* A laminated cylinder consists of a jacket with an inside diameter of 10 in. and an outside diameter of 12 in. shrunk onto a tube with an inside diameter of 6 in. The initial shrinkage pressure is 4000 psi. The cylinder is then subjected to an internal pressure of 30,000 psi. Determine:
(a) The initial interference if the cylinder is made of steel ($E = 30{,}000$ ksi).
(b) The maximum tensile stresses in the jacket and in the tube.

3–80* A jacket is to be shrunk onto a tube having an inside diameter of 100 mm and an outside diameter of 200 mm. Determine the minimum thickness for the jacket if the maximum tensile stresses in the jacket and the tube are not to exceed 200 MN/m^2 when the tube is subjected to an internal pressure of 180 MN/m^2.

3–81 A steel cylinder having an internal diameter of 6 in. and an external diameter of 8 in. is shrunk over a tube having an internal diameter of 4 in. If the laminated cylinder is designed so that the maximum tensile stress in the tube will equal the maximum tensile stress in the jacket after an internal pressure of 24,000 psi is applied to the assembly, determine:
(a) The initial shrinkage pressure between the tube and the jacket.
(b) The interference required to produce this shrinkage pressure.

3–82 A laminated cylinder consists of a jacket with an internal diameter of 200 mm and an external diameter of 300 mm shrunk over a tube with an internal diameter of 100 mm. When the laminated cylinder is subjected to an internal pressure of 135 MN/m^2, the maximum tensile stress in the tube is 135 MN/m^2. Determine:
(a) The maximum tensile stress in the jacket.
(b) The initial shrinkage pressure.

3–83* A thick-walled cylinder with an inside diameter of 6 in. was designed with a minimum wall thickness to withstand an internal pressure of 15,000 psi and to develop a maximum tensile stress of 25,000 psi. If the internal pressure must be doubled, what is the minimum thickness for a shrink-fitted jacket that will prevent the maximum tensile stress in the assembly from exceeding 25,000 psi?

3–84* A laminated thick-walled cylinder consists of a jacket with an external diameter of 350 mm and an internal diameter of 250 mm shrunk onto a tube with an internal diameter of 100 mm. If the maximum tensile stresses in the tube and the jacket are not to exceed 140 MN/m^2, determine:
(a) The maximum allowable internal pressure.
(b) The initial shrinkage pressure required.

3–85 A steel cylinder with an internal diameter of 4 in. was designed for a maximum tensile stress of 20,000 psi under an internal pressure of 12,000 psi. It has become necessary to increase the internal pressure to 16,000 psi without increasing the maximum allowable tensile stress in the cylinder. If the cylinder is to be strengthened by winding with square wire 0.10 in. on a side, determine the necessary tensile stress in the wire as it is wound on the cylinder.

4 Torsional Loading

4–1 TORSIONAL SHEARING STRESS

In the design of machinery (and some structures) the problem of transmitting a torque (a couple) from one plane to a parallel plane is frequently encountered. The simplest device for accomplishing this function is a circular shaft such as that connecting an electric motor with a pump, compressor, or other mechanism. For example, the shaft AB of Fig. 4–1 transmits a torque from the driving motor A to the coupling B. A modified free-body diagram (the weight and bearing reactions are not shown because they contribute no useful information to the torsion problem) of a typical installation is shown in Fig. 4–2. The resultant of the electromagnetic forces applied to the armature A of the motor is a couple resisted by the resultant of the bolt forces (another couple) acting on the flange coupling B, and the circular shaft transmits the torque from armature to coupling. The problem is the determination of significant stresses in and the deformation of the shaft.

As a point of departure, a segment of the shaft between transverse planes a–a and b–b of Fig. 4–2 will be studied. In so doing, the complicated stress patterns at the locations of torque-applying devices may be

Fig. 4–1 A 1000-Hp. 180-rpm motor coupled to ball and tube mills. Photo courtesy of Electric Machinery Mfg. Co., Minneapolis, Minn.

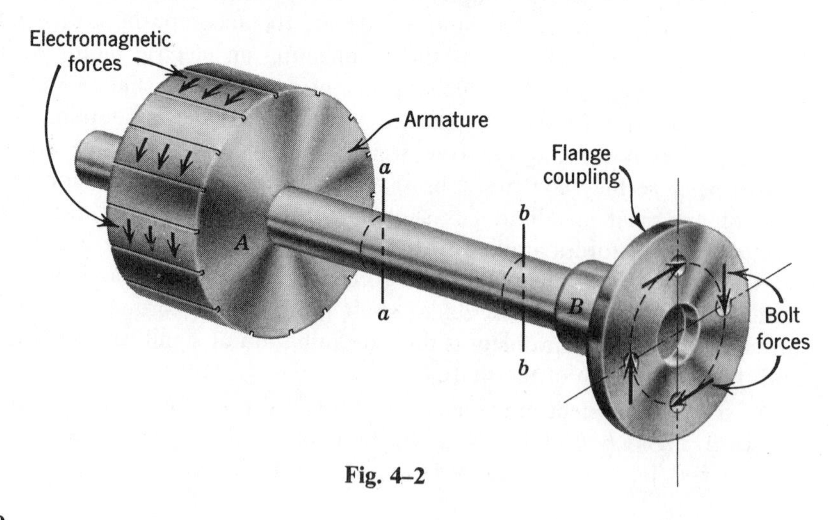

Fig. 4–2

avoided, and Saint-Venant's principle applies. A free-body diagram of the segment of the shaft is shown in Fig. 4–3, with the torque applied by the armature indicated on the left end as T. The resisting torque T_r, indicated at the right end, is the resultant of the differential forces dF acting on the transverse plane b–b. The force dF is equal to $\tau\, dA$ where τ is the shearing stress *on the transverse plane* at a distance ρ from the shaft center, and dA is a differential area. *In keeping with the notation outlined in Art. 1–5, the shearing stress should be designated* $\tau_{z\theta}$ *to indicate that it acts on the z face in the direction of increasing* θ. For the elementary theory of torsion of circular sections, however, the shearing stress on any transverse plane is always perpendicular to the radius to the point, and the stress will be referred to as τ or τ_c for convenience. If the shaft is in equilibrium, the summation of moments about the axis of the shaft indicates that

$$T = T_r = \int_{\text{area}} \rho\tau\, dA \tag{a}$$

The law of variation of the shearing stress on the transverse plane must be known before the integral of Eq. (a) can be evaluated. The approach to the problem of stress distribution will be prefaced by a brief reference to the past.

C. A. Coulomb, a French engineer, in 1784 developed experimentally the relationship between the applied torque and the angle of twist for circular

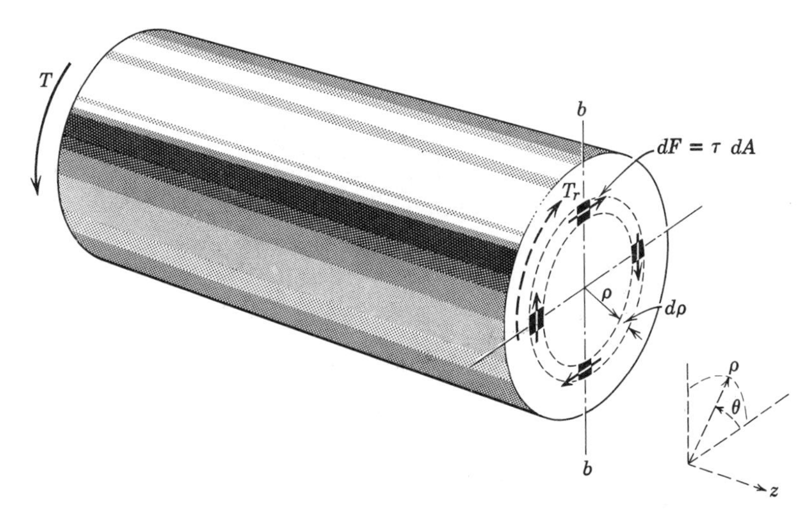

Fig. 4–3

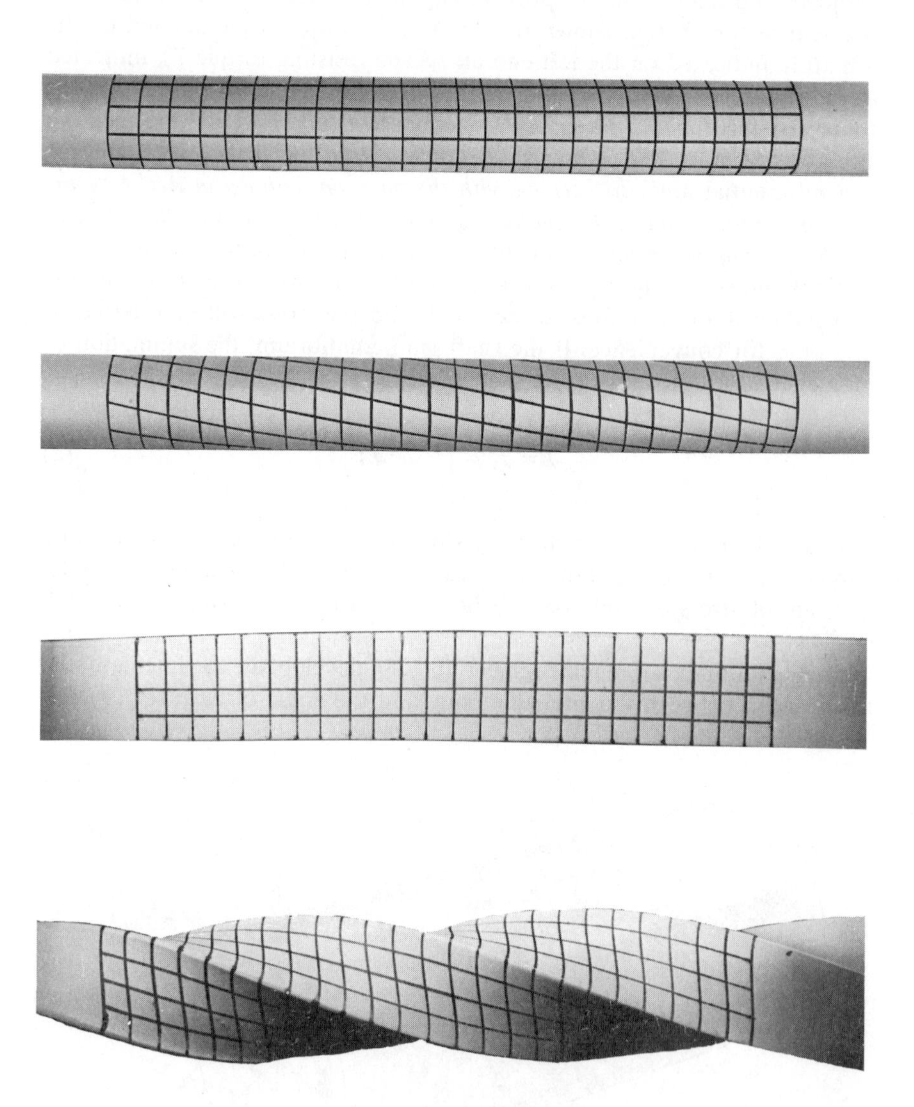

Fig. 4–4

bars.[1] A. Duleau, another French engineer, in a paper published in 1820[1] derived analytically the same relationship by making the assumption that *a plane section before twisting remains plane after twisting and a diameter remains a straight line*. Visual examination of twisted models indicates that these assumptions are apparently correct for circular sections either solid or hollow (provided the hollow section is circular and symmetrical with respect to the axis of the shaft), but incorrect for any other shape. Compare, for example, the distortions of the rubber models with circular and rectangular cross sections shown in Fig. 4-4.

If the assumptions noted above are made, the distortion of the shaft will be as indicated in Fig. 4-5*a*, wherein points B and D on a common radius in a plane section move to B' and D' in the same plane and still on the same radius. The angle θ is called the *angle of twist*. The surface ABB' of Fig. 4-5*a* is shown developed in Fig. 4-5*b*, and a differential cube of the material at B assumes at B' the familiar distortion due to shearing stress. At this point the assumption will be made that *all longitudinal elements have the same length* (which limits the results to straight shafts of constant

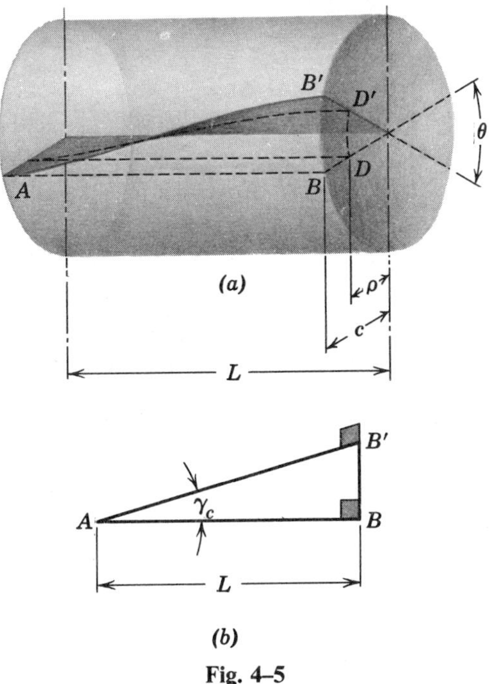

Fig. 4-5

[1]From *History of Strength of Materials*, S. P. Timoshenko, McGraw-Hill, New York, 1953.

diameter). From Fig. 4–5, the following strain relations may be written:

$$\tan\gamma_c = \frac{BB'}{L} = \frac{c\theta}{L}$$

and

$$\tan\gamma = \frac{DD'}{L} = \frac{\rho\theta}{L}$$

or, *if the strain is small*,

$$\gamma_c = \frac{c\theta}{L} \text{ and } \gamma = \frac{\rho\theta}{L} \tag{b}$$

From Eq. (b)

$$\frac{\gamma_c L}{c} = \frac{\gamma L}{\rho}$$

and as the longitudinal elements have the same length, the expression reduces to

$$\gamma = \frac{\gamma_c}{c}\rho \tag{c}$$

which indicates that the shearing strain ($\gamma_{z\theta}$) is proportional to the distance from the axis of the shaft. Note that Eq. (c) is valid for elastic or inelastic action and for homogeneous or heterogeneous materials, provided the strains are not too large ($\tan\gamma \approx \gamma$). Problems in this book will be assumed to satisfy this requirement.

Knowing that the shearing strain $\gamma_{z\theta}$, varies linearly with the radius ρ, the law of variation of the shearing stress τ (meaning $\tau_{z\theta}$) with ρ can be determined by reference to a shear stress-strain diagram and making the appropriate substitutions into Eq. (c) to obtain the τ versus ρ expression. As an aid to the thought sequence, it is suggested that sketches be drawn of γ vs ρ, then τ vs γ, and finally τ vs ρ, as illustrated for two typical examples in Fig. 4–6.

Having arrived at expressions for τ as a function of ρ, such expressions can be substituted into Eq. (a), which is then integrated to evaluate the

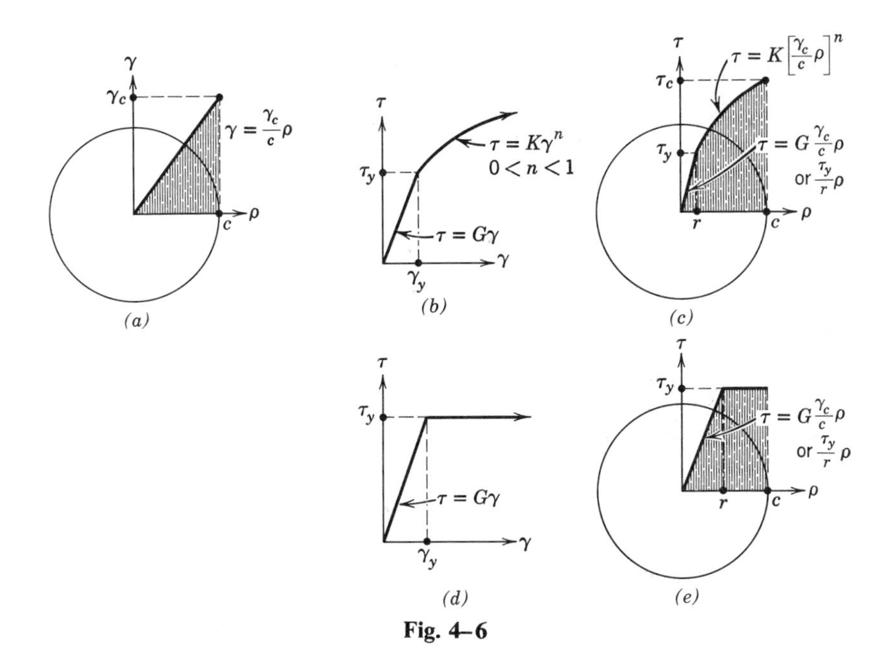

Fig. 4–6

resisting torque (or other unknown if the torque is known).

$$T_r = \int_A \rho\tau \, dA \tag{a}$$

The following examples will illustrate the method. Note that the stress τ is the shearing stress on a *transverse plane*. In Art. 4–4 it will be shown that the transverse plane is a plane of maximum shearing stress for *pure torsion*.

Example 4–1 A solid circular steel shaft 4 in. in diameter is subjected to a pure torque of 7π ft-kips. Assume the steel is elastoplastic having a yield point in shear of 18 ksi and a modulus of rigidity of 12,000 ksi. Determine the maximum shearing stress in the shaft and the magnitude of the angle of twist in a 10-ft length.

Solution Assume first that the maximum stress is less than the proportional limit. In this case, assuming a homogeneous, isotropic material, the expression $\tau = G\gamma$ applies over the entire cross section (see Fig. 4–6*d*), and the τ vs ρ diagram would appear like the γ vs ρ diagram (Fig. 4–6*a*), with

γ_c replaced by τ_c, making the equation of the curve $\tau = (\tau_c/c)\rho$. Substituting this expression for τ into Eq. (a) yields

$$7\pi(12) = \int_0^2 \int_0^{2\pi} \rho\left(\frac{\tau_c}{c}\rho\right)(\rho\, d\rho\, d\theta) = \int_0^2 \rho\left(\frac{\tau_c}{2}\rho\right)(2\pi\rho\, d\rho) = \pi\tau_c\left(\frac{2^4}{4}\right)$$

from which τ_c is 21 ksi, which is greater than 18 ksi. Therefore, the assumption is not valid, and the stress distribution curve should be as in Fig. 4–6*e* with the maximum shearing stress

$$\tau_{max} = \underline{18 \text{ ksi}} \qquad \text{Ans.}$$

Observe that the first integration above replaces the original dA with an lar area of width $d\rho$. In the future dA will be written in this form.
obtain the angle of twist, Eq. (b) will be rewritten thus,

$$\theta = \frac{\gamma L}{\rho} = \frac{\gamma_y L}{r} \tag{d}$$

where the shearing strain at the yield point γ_y equals $\tau_y/G = 18/12{,}000 = 1.5(10)^{-3}$ rad and r is the radius of that part of the cross section that is deforming elastically (Fig. 4–6*e*). The value of r will be found by means of Eq. (a). Thus,

$$T_r = \int_0^r \rho\left(\frac{18}{r}\rho\right)(2\pi\rho\, d\rho) + \int_r^2 \rho(18)(2\pi\rho\, d\rho)$$

$$7\pi(12) = 2\pi\left(\frac{18}{r}\right)\left(\frac{r^4}{4}\right) + 2\pi(18)(1/3)(2^3 - r^3)$$

from which $r^3 = 4$ and $r = 1.587$ in. Now, from Eq. (d), the magnitude of the angle of twist is

$$\theta = \frac{1.5(10)^{-3}(10)(12)}{1.587} = \underline{0.1134 \text{ rad}} \qquad \text{Ans.}$$

Example 4–2 A straight shaft having a hollow circular cross section with outside and inside diameters of 3 and 2 in., respectively, is made of a magnesium alloy having the shear stress-strain diagram of Fig. 4–7. Determine the torque required to twist a 5-ft length of the shaft through 0.240 rad.

Solution From Eq. (b) the shear strains at the outside and inside surfaces are obtained. Thus,

$$\gamma_{1.5} = \frac{\rho\theta}{L} = \frac{(3/2)(0.240)}{5(12)} = 6(10)^{-3}\,\text{rad}$$

$$\gamma_{1.0} = (2/3)\gamma_{1.5} = 4(10)^{-3}\,\text{rad}$$

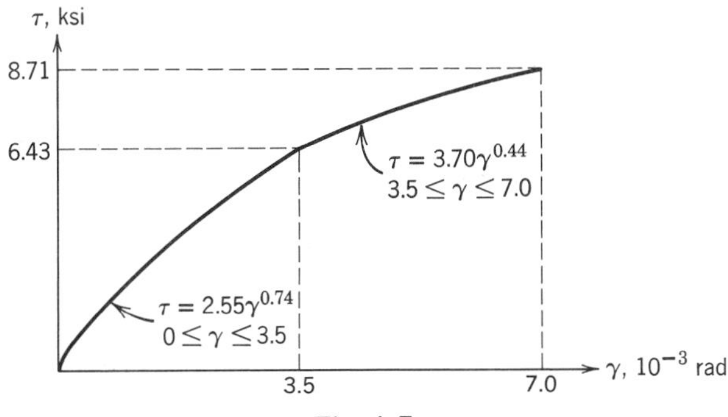

Fig. 4–7

Since both the minimum and maximum shear strains lie within the right-hand portion of the curve of Fig. 4–7, the shear stress-strain function given for this curve applies for the entire cross section. Thus, the γ vs ρ function is

$$\gamma = (\gamma_c/c)\rho = \frac{6}{(3/2)}\rho = 4\rho$$

where γ is in 10^{-3} rad. The τ vs ρ function then becomes

$$\tau = 3.70\gamma^{0.44} = 3.70(4)^{0.44}\rho^{0.44} = 3.70(1.840)\rho^{0.44}$$

Substituting this expression into Eq. (a) yields

$$T_r = \int_{1.0}^{1.5} \rho\left[3.70(1.84)\rho^{0.44}\right](2\pi\rho\, d\rho) = \left|\frac{2\pi(3.70)(1.84)\rho^{3.44}}{3.44}\right|_{1.0}^{1.5}$$

from which

$$T = T_r = \underline{37.7\text{ in-kips}} \qquad \text{Ans.}$$

PROBLEMS

4–1* The maximum shearing stress in the solid circular steel shaft of Fig. P4–1 is 12 ksi. The steel has a proportional limit in shear of 18 ksi. Determine the magnitude of the resisting torque offered by the shaded area. Assume a uniform stress distribution over the shaded area.

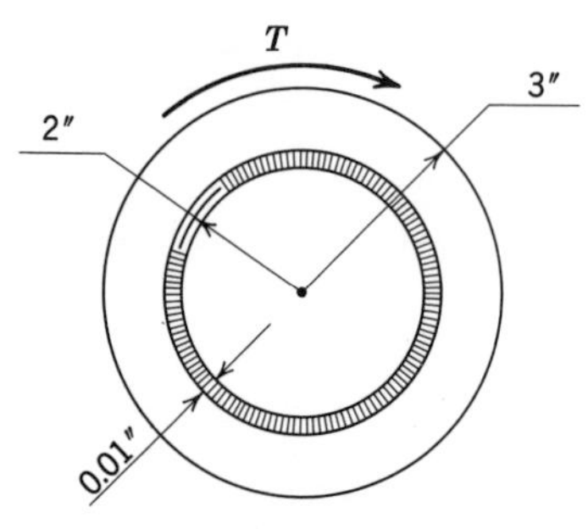

Fig. P4–1

4–2* A pure torque acting on a 400-mm diameter solid circular shaft produces a maximum shearing stress of 140 MN/m^2. Determine by integration the magnitude of the resisting torque offered by the stress on the shaded band in Fig. P4–2. Assume linearly elastic action throughout the cross section.

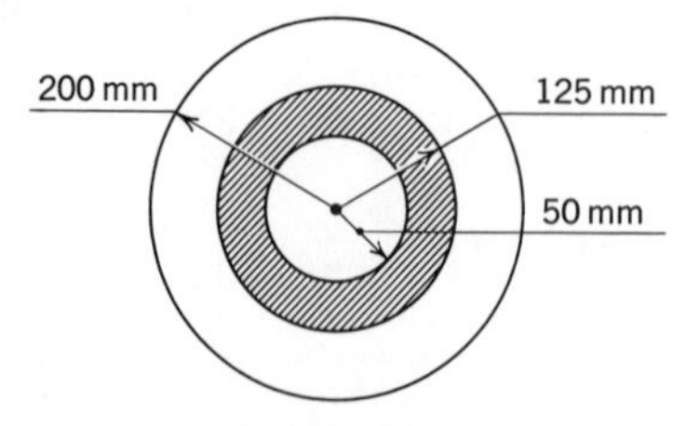

Fig. P4–2

4–3* Given that the torsional shearing stress in a circular shaft is proportional to the distance from the center of the cross section of the shaft, derive an expression for the resisting torque developed by the band of material lying between $c/2$ and $3c/4$ of a solid circular shaft having a radius c. Express the results in terms of c and τ_{max}.

4–4* For the shaft of Exam. 4–1, what would be the magnitude of the torque necessary to produce inelastic action throughout the entire cross section? Note that there must always be some elastic action at the center of the shaft, but this effect is so small that it may be neglected.

4–5* An elastoplastic solid circular steel shaft has a shearing yield point of 24 ksi. Determine:

(a) The shearing strain at the surface of the shaft when the shearing stress at one half of the radius from the center of the shaft reaches the yield point.
(b) The ratio of the torque carried by the shaft of part (a) to the torque carried by an identical shaft when the shearing stress at the outer surface reaches the yield point.

4–6 An elastoplastic hollow steel shaft has an outside diameter of 100 mm and an inside diameter of 50 mm. The shearing yield point is 140 MN/m^2. Determine:

(a) The maximum allowable torque that the shaft can transmit with an allowable shearing stress of 125 MN/m^2.
(b) The percentage increase in the torque of part (a) if the shearing stress at the inside surface is allowed to reach the yield point.

4–7* A 4-in. diameter solid circular shaft of steel having the shearing stress-strain diagram of Fig. P4–7 is twisted through an angle of 0.09 rad in a length of 5 ft by torques applied at the ends. Determine:

(a) The maximum shearing strain in the shaft.
(b) The magnitude of the applied torque.

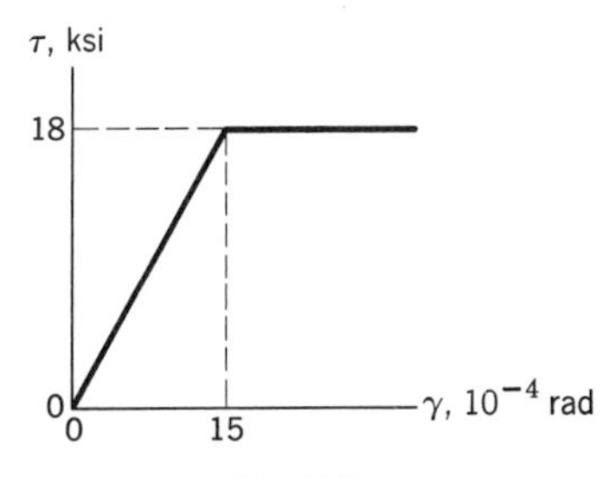

Fig. P4–7

4–8 A hollow shaft of 2014-T6 aluminum alloy (see Fig. P4–8) has inside and outside diameters of 2 and 3 in., respectively. Determine the torque necessary to develop a shearing stress of 30 ksi at the inner radius. Also, evaluate the accompanying angle of twist per foot of length of the shaft.

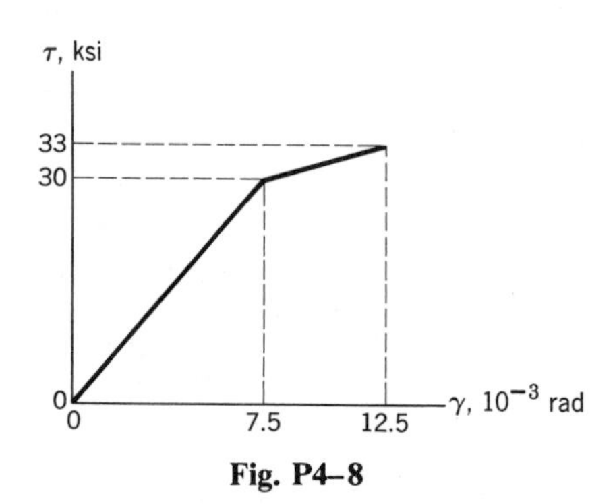

Fig. P4–8

4–9* The shearing stress-strain diagram for an aluminum alloy may be approximated by the two straight lines of Fig. P4–9. Determine the torque necessary to develop a maximum shearing stress of 230 MN/m^2 in a 50-mm diameter solid circular shaft of this material. Determine also the resulting angle of twist per metre of length of this shaft.

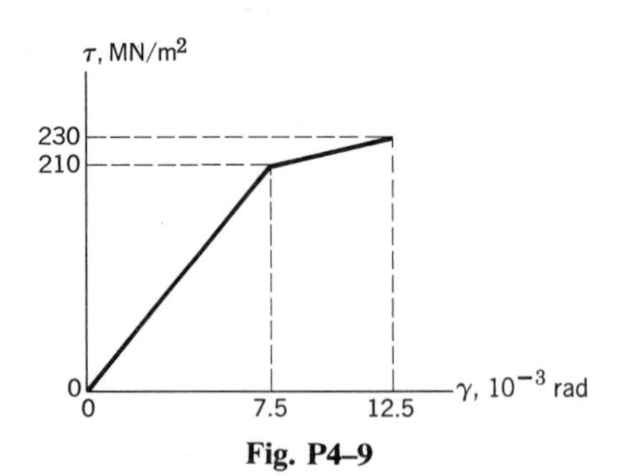

Fig. P4–9

4–10 A solid circular shaft of an aluminum alloy (see Fig. P4–9) is to be subjected to a torque of 3300π N-m. Determine the diameter of the shaft for a shearing stress of 210 MN/m^2 at $\rho = 3c/4$, where c is the radius of the shaft.

4–11 A hollow shaft with an inside diameter of 2 in. and an outside diameter of 4 in. is made of a material with the shearing stress-strain diagram shown in Fig. P4–11. It is proposed that the diagram be approximated by the two straight lines *OA* and *AB* for stresses less than 36 ksi. Determine:

(a) The torque that will be developed in the shaft when the maximum shearing stress is 32 ksi.

(b) The magnitude of the angle of twist per foot of length of the shaft with this stress.

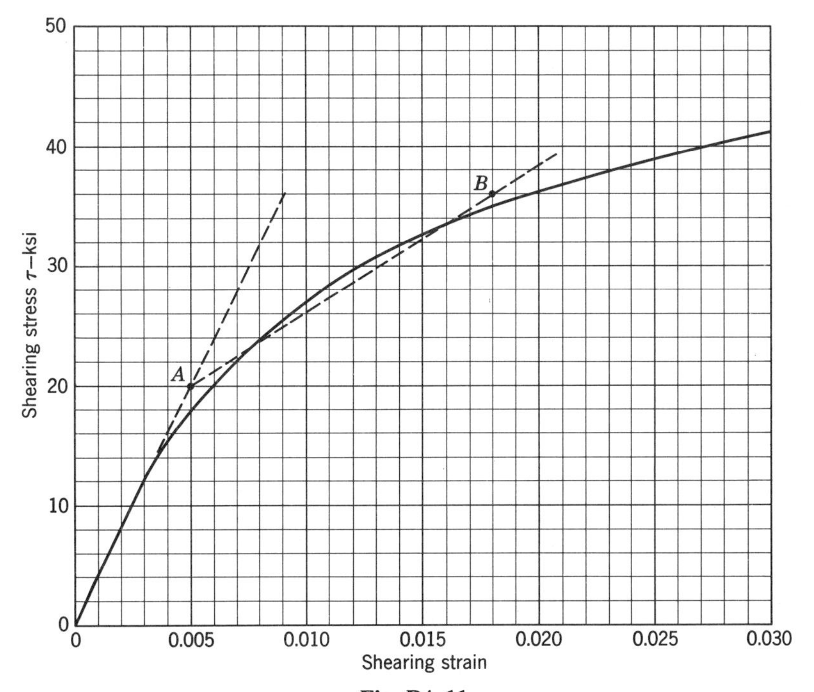

Fig. P4–11

4–12* Solve Prob. 4–11 for a 4-in. diameter solid circular shaft.

4–13 If the shaft of Exam. 4–2 were solid instead of hollow (still 3 in. diameter), determine:

(a) The torque required to twist the shaft through the specified angle.

(b) The percentage increase in the torque.

(c) The percentage increase in the weight of the shaft.

4–2 THE ELASTIC TORSION FORMULA

As noted in Exam. 4–1, when the action is linearly elastic throughout the cross section and the material is assumed to be homogeneous and isotropic, the variation of stress with radius is obtained by substituting the expression $\tau = G\gamma$ into Eq. (c) of Art. 4–1, thus giving the linear relation

$$\tau = \frac{\tau_c}{c}\rho \tag{a}$$

When Eq. (a) is substituted into Eq. (a) of Art. 4–1, the result is

$$T_r = \frac{\tau_c}{c}\int_{\text{area}} \rho^2\, dA = \frac{\tau}{\rho}\int_{\text{area}} \rho^2\, dA \tag{b}$$

The integral in Eq. (b) is the polar second moment of area J (called the polar moment of inertia), with respect to the longitudinal axis of the shaft (see App. B). Equation (b) can then be written as

$$T_r = \frac{\tau J}{\rho}$$

or in terms of the unknown shearing stress as

$$\tau = \frac{T\rho}{J} \tag{4–1}$$

Equation 4–1 is known as the *elastic torsion formula*, in which τ is *the shearing stress on a transverse plane at a distance* ρ *from the axis of the shaft*, and T is the resisting torque, which, in general, is obtained from a free-body diagram and an equilibrium equation. Note that the transverse plane is a plane of maximum shearing stress. Note also that Eq. 4–1 applies only for completely elastic (linearly) action.

4–3 TORSIONAL DISPLACEMENTS

Frequently the amount of twist in a shaft (or structural element) is of paramount importance, making the computation of the angle of twist a necessary accomplishment of the machine or structural designer. The fundamental approach to the problem, involving members satisfying the limitations of Art. 4–1, is provided by the following equations:

$$\gamma = \frac{\rho\theta}{L} \text{ or } \gamma = \frac{\rho\, d\theta}{dL} \tag{4–2}$$

$$\tau = \frac{T\rho}{J} \tag{4–1}$$

$$G = \frac{\tau}{\gamma}$$

Equation 4–2 is Eq. (b) of Art. 4–1, and the second form is necessary where the torque or the cross section is some function of the length (see Exam. 4–4). The equation applies for elastic or inelastic action as shown in Exams. 4–1 and 4–2. Equation 4–1 was derived in Art. 4–2. The third expression is Hooke's law for shearing stresses. The last two expressions are limited to stresses below the proportional limit of the material (elastic action). These three equations can be combined to give several different relationships; for example,

$$\theta = \frac{\gamma L}{\rho} = \frac{\tau L}{\rho G} \tag{4–3a}$$

or

$$\theta = \frac{TL}{JG} \tag{4–3b}$$

The first form of Eq. 4–3 will often be most useful in dual-specification problems (limiting θ and τ are specified).

The angle of twist determined from the above expressions is for a length of shaft of constant diameter sufficiently removed from sections where pulleys, couplings, or other mechanical devices are attached to make Saint-Venant's principle applicable. However, for practical purposes, it is customary to neglect local distortion at all connections and compute angles as though there were no discontinuities, and as though the torques applied were distributed over the cross sections in exactly the same manner in which the stresses given by the torsion formula are distributed. The following examples illustrate the application of the principles developed.

Example 4–3 The steel shaft of Fig. 4–8*a* is in equilibrium under the torques shown. Determine:

(a) The maximum shearing stress in the shaft.

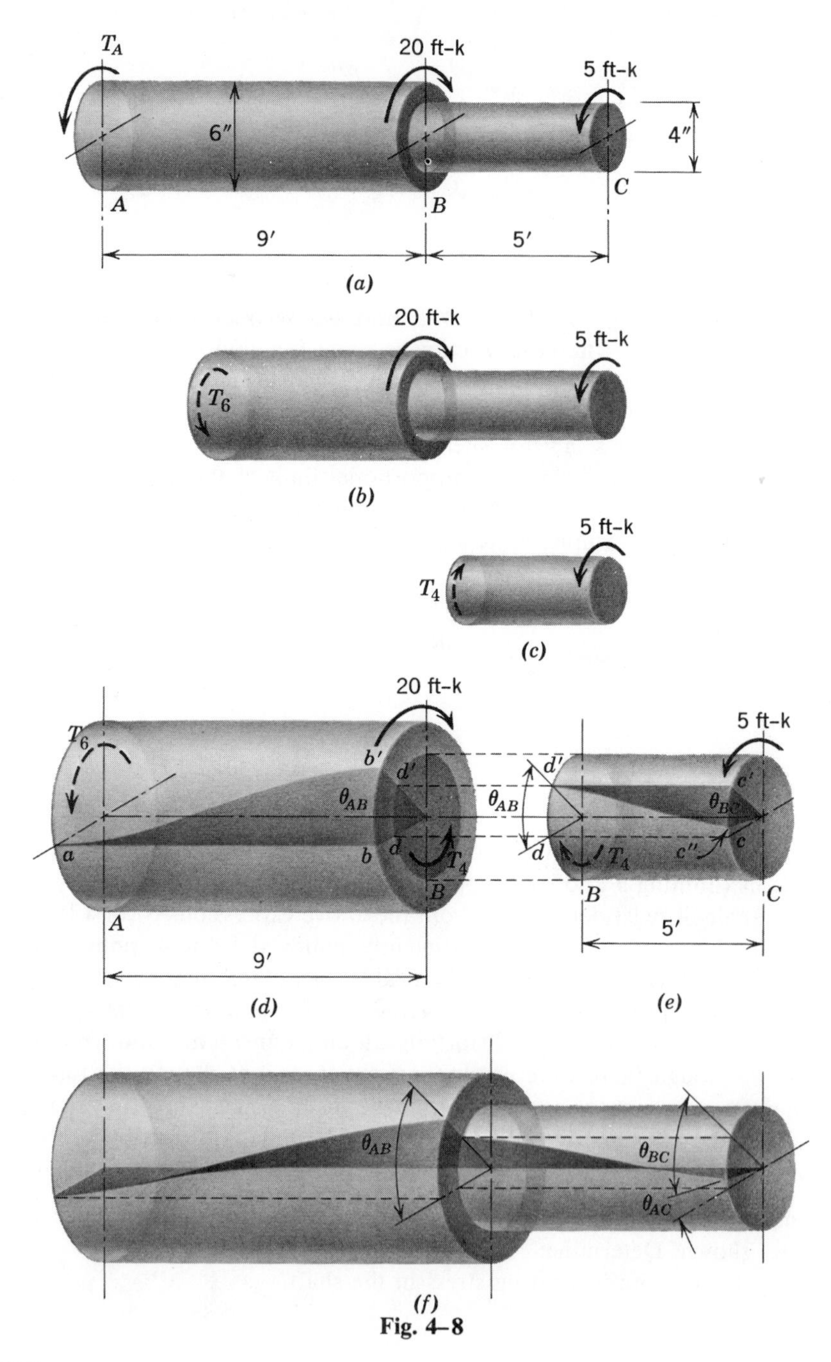

T_A
20 ft-k
5 ft-k
6″
4″
A
B
C
9′
5′
(a)
20 ft-k
5 ft-k
T_6
(b)
5 ft-k
T_4
(c)
20 ft-k
5 ft-k
T_6
b′
d′
θ_{AB}
a
b
d
T_4
B
A
9′
(d)
θ_{AB}
d′
d
T_4
c′
θ_{BC}
c″
c
B
C
5′
(e)
θ_{AB}
θ_{BC}
θ_{AC}
(f)

Fig. 4–8

(b) The angle of twist of end B of the 6-in. segment with respect to end A.

(c) The angle of twist of the end C with respect to end A.

Solution (a) In general, free-body diagrams should be drawn in order to evaluate correctly the resisting torque. Such diagrams are shown in Fig. 4–8*b* and *c*, where in part *b* the shaft is cut by any transverse plane through the 6-in. segment, and T_6 is the resisting torque on this section. Similarly, in Fig. 4–8*c*, the plane is passed through the 4-in. section, and T_4 is the resisting torque on this section. The location of the maximum shearing stress is not apparent; hence, the stress must be checked at both sections. Thus, from Fig. 4–8*b*,

$$\Sigma M = 0, \quad T_6 = 20 - 5 = 15 \text{ ft-kips} = 15{,}000(12) \text{ in-lb}$$

and

$$\tau_{\max} = \frac{Tc}{J} = \frac{15{,}000(12)(3)}{(\pi/2)(3^4)} = \frac{40{,}000}{3\pi}$$

From Fig. 4–8*c*,

$$\Sigma M = 0, \quad T_4 = 5 \text{ ft-kips} = 5000(12) \text{ in-lb}$$

and

$$\tau_{\max} = \frac{5000(12)(2)}{(\pi/2)(2^4)} = \frac{15{,}000}{\pi} > \frac{40{,}000}{3\pi}$$

therefore, the maximum shearing stress is

$$15{,}000/\pi = \underline{4770 \text{ psi}} \qquad \text{Ans.}$$

This stress is less than the shearing proportional limit of any steel; hence, the torsion formula applies. Note that had the larger torque been carried by the smaller section, the maximum stress would obviously occur in the small section, and only one stress determination would have been required.

(b, c) As an aid to visualizing the distortion of the shaft, the segments AB and BC and the torques acting on them are drawn separately in Fig. 4–8*d* and *e* with the distortions greatly exaggerated. As the resultant torque of 15 ft-kips twists segment AB through the angle θ_{AB}, points *b* and *d* move to b' and d' respectively, and segment BC may be considered as a rigid body rotating through the same angle, point c moving to c', after which the torque of 5 ft-kips acting on BC twists this part of the shaft back through the angle θ_{BC}, point c' moving back to c''. The resultant distortion for the entire shaft is shown in Fig. 4–8f.

When computing distortions, it is recommended that the two fundamental relationships of Eqs. 4–1 and 4–2 plus Hooke's law be used; thus,

$$\gamma = \frac{c\theta}{L} = \frac{\tau}{G} = \frac{Tc}{JG}$$

When this expression is applied to Fig. 4–8d, the twist in segment AB is obtained as

$$\theta_{AB} = \frac{(20-5)(12)(9)(12)}{(\pi/2)(3^4)(12)(10^3)} = \frac{40}{\pi(10^3)} = \underline{0.012{,}73 \text{ rad}} \qquad \text{Ans.(b)}$$

From Fig. 4–8e and f,

$$\theta_{AC} = \theta_{AB} - \theta_{BC} = \frac{40}{\pi(10^3)} - \frac{5(12)(5)(12)}{(\pi/2)(2^4)(12)(10^3)}$$

$$= \frac{40-37.5}{\pi(10^3)} = \underline{0.000{,}796 \text{ rad.}} \qquad \text{Ans.(c)}$$

Example 4–4 The solid circular tapered shaft of Fig. 4–9 is subjected to end torques applied in transverse planes. Assuming elastic action and a slight taper, determine the magnitude of the angle of twist in terms of T, L, G, and r.

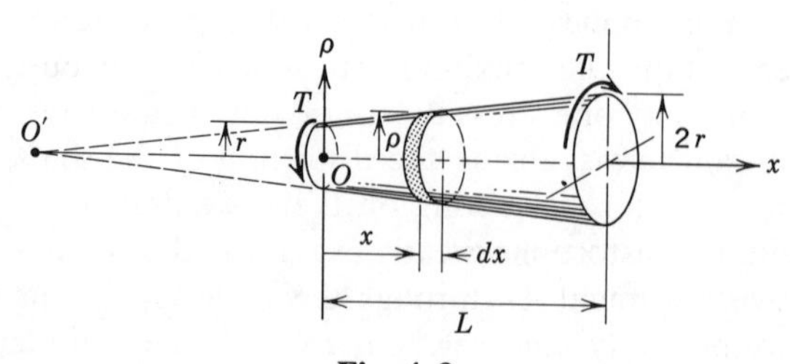

Fig. 4–9

Solution Note that Eq. 4–2 was developed on the assumption that plane cross sections remain plane and all longitudinal elements have the same length. Neither of these assumptions is strictly valid for the tapered shaft, but if the taper is slight, the error involved is negligible; hence, Eq. 4–2 will be used as follows:

$$d\theta = \frac{\gamma}{\rho}\, dx$$

also

$$\gamma = \frac{\tau}{G}$$

and

$$\tau = \frac{T\rho}{J} = \frac{2T\rho}{\pi\rho^4} = \frac{2T}{\pi\rho^3}$$

Therefore,

$$d\theta = \frac{\tau}{G\rho}\, dx = \frac{2T}{G\pi\rho^4}\, dx$$

The radius can be expressed as a function of x; thus,

$$\rho = r + \frac{2r - r}{L}x = \frac{r}{L}(L + x)$$

which, substituted into the expression for $d\theta$, gives

$$d\theta = \frac{2TL^4}{G\pi r^4(L+x)^4}\, dx$$

Integration gives

$$\theta = \frac{2TL^4}{G\pi r^4}\int_0^L \frac{dx}{(L+x)^4} = -\frac{2TL^4}{3G\pi r^4}\left(\frac{1}{8L^3} - \frac{1}{L^3}\right)$$

$$\theta = \frac{7TL}{12G\pi r^4} \qquad \text{Ans.}$$

An alternate setup is to place the origin of coordinates at a distance L to the left of 0 in Fig. 4–9 (point 0′). The function for ρ then becomes

$$\rho = \frac{r}{L}x$$

and

$$\theta = \frac{2TL^4}{G\pi r^4}\int_L^{2L}\frac{dx}{x^4} = \frac{7TL}{12G\pi r^4}$$

PROBLEMS

Note. In the following problems, assume linearly elastic action unless data are given that permit a check on assumptions. Also neglect localized effects at discontinuities, and assume a slight taper for tapered shafts.

4–14* Determine the maximum allowable torque to which a solid circular steel bar 9 ft long and 3 in. in diameter can be subjected when it is specified that the shearing stress must not exceed 12,000 psi and the angle of twist must not exceed 0.048 rad.

4–15 A solid circular steel shaft 3 m long is to resist a torque of 3200 N-m. The allowable shearing stress is 80 MN/m^2, and the maximum allowable angle of twist is 9°. Determine the minimum permissible diameter.

4–16 Specifications for a solid circular aluminum alloy 2024-T4 (see App. A) rod 60 in. long require that it shall be adequate to resist a torque of 22,000 in-lb without twisting more than 15° or exceeding a shearing stress of 14,000 psi. What minimum diameter is required?

4–17* In the factory driveshaft of Fig. P4–17*a* and *b*, a torque of 20 N-m is

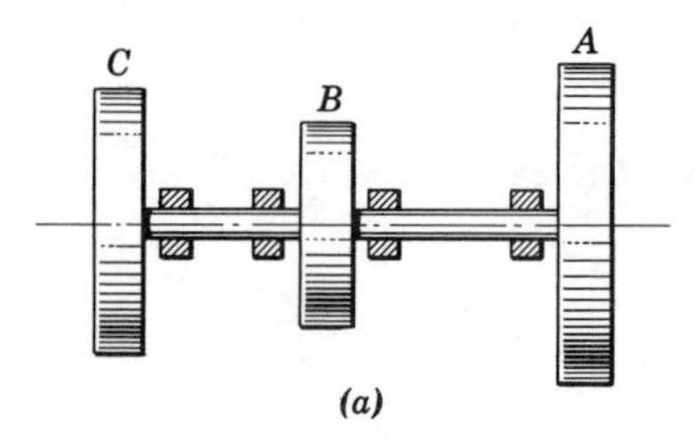

(a)

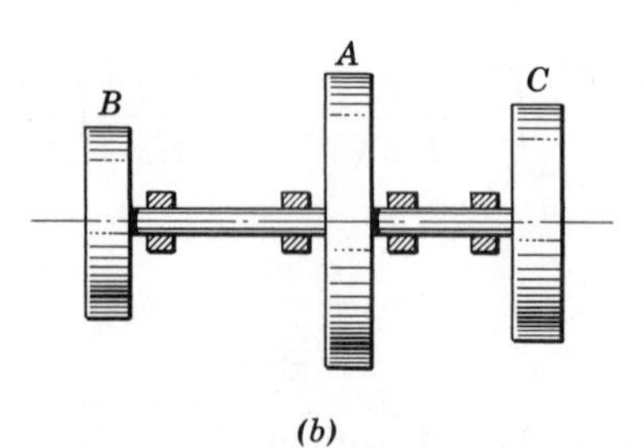

(b)

Fig. P4–17

supplied to the shaft by a belt driving pulley A. A torque of 12 N-m is taken off by pulley B and the remainder by pulley C.

(a) Determine the maximum torque in each portion of each driveshaft.
(b) Which arrangement (a or b) produces the smaller maximum stress in the driveshaft?

4–18 A torque of 100,000 in-lb is supplied to the driving pulley B of Fig. P4–18 by a motor. Pulley A takes off 36,000 in-lb of torque, and the remainder is taken off by pulley C. For an allowable shearing stress of 12 ksi, determine:

(a) The minimum permissible diameters of the two shafts.
(b) The magnitude of the angle of twist between pulleys A and C. Both shafts are made of steel and are rigidly connected to pulley B.

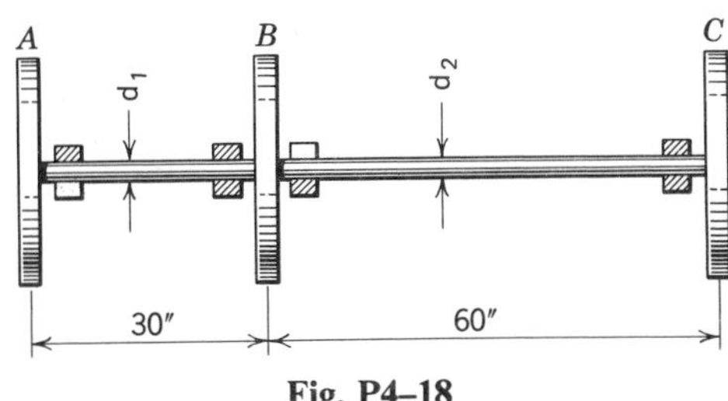

Fig. P4–18

4–19* The member shown in Fig. P4–19 is a solid circular steel shaft 100 mm in diameter. Rod AB is a rigid pointer fastened to the end of the shaft. Determine:

(a) The maximum shearing stress in the shaft.
(b) The distance point A moves, measured from its no-load position.

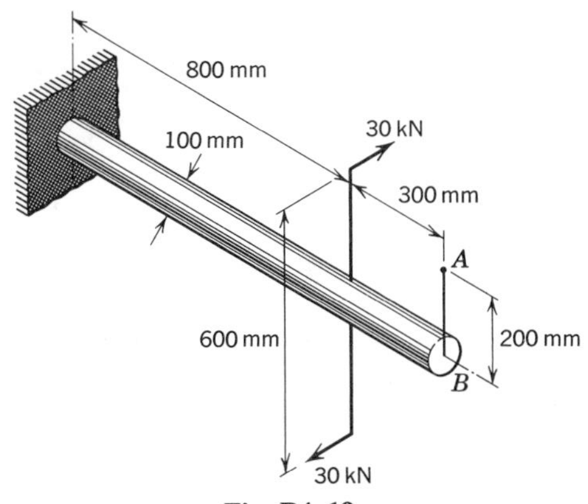

Fig. P4–19

4–20 The 4-in. diameter shaft of Fig. P4–20 is composed of a cold-rolled red brass section securely connected to a structural steel section (see App. A for properties and assume the strengths in shear are 0.6 of the strengths in tension). Determine the maximum allowable torque, applied as shown, if a minimum factor of safety of 2 with respect to failure by slip is specified and the distance AC through which the end of the 10-in. pointer AB moves is not to exceed 1.50 in.

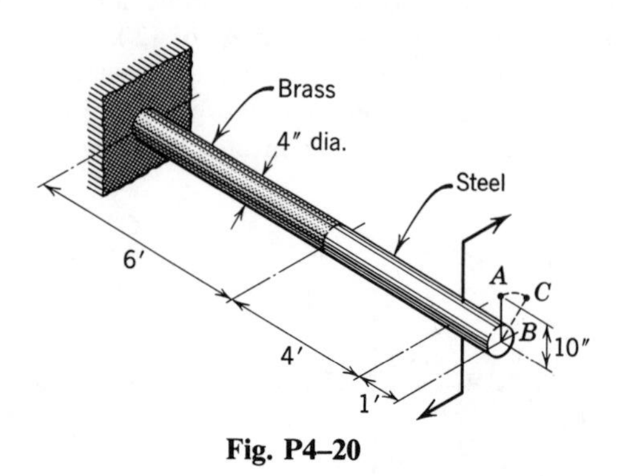

Fig. P4–20

4–21* A solid circular steel shaft with diameters as shown in Fig. P4–21 is subjected to a torque T. The allowable shearing stress is 100 MN/m^2, and the maximum allowable angle of twist in the 2.5-m length is 0.06 rad. Determine the maximum allowable value of T.

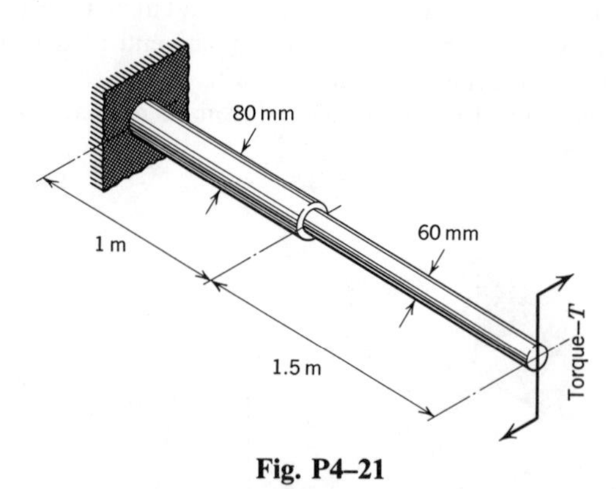

Fig. P4–21

4–22 A solid circular steel shaft is fastened securely to a solid circular bronze shaft, as indicated in Fig. P4–22. Determine the value of the couple T that will produce an angle of twist of 0.02 rad in the 10-ft length. The shearing modulus of elasticity of the bronze is assumed to be 6000 ksi.

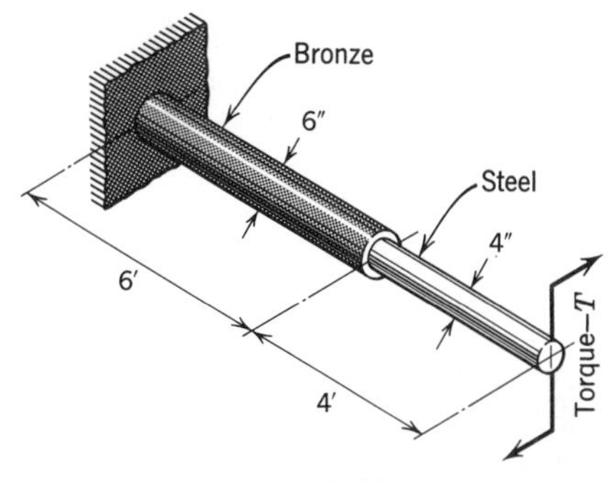

Fig. P4–22

4–23* Solve Prob. 4–22 with an additional torque of $3T$ applied at the right end of the bronze portion, both torques having the same sense.

4–24 The hollow steel shaft of Fig. P4–24 is in equilibrium under the torques indicated. Determine:

(a) The maximum shearing stress in the shaft.

(b) The relative rotation of end D of the shaft with respect to the section at B.

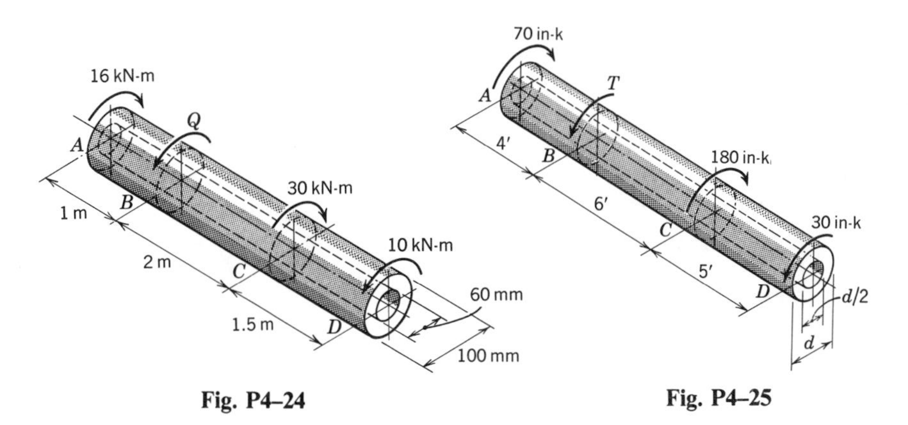

Fig. P4–24

Fig. P4–25

4–25* The hollow circular steel shaft $ABCD$ of Fig. P4–25 is in equilibrium under the torques indicated. The inside diameter is one-half the outside diameter.

(a) Set up, in terms of d, the expression for the rotation of section D with respect to section B. Do not solve, but indicate units and sense for the answer.

(b) Determine the minimum permissible value for the outside diameter for a maximum shearing stress of 16 ksi.

4–26* A steel shaft has the dimensions shown in Fig. P4–26. Determine:

(a) The maximum shearing stress on a section 3 m from the left end.

(b) The angle of twist at the section 2 m from the left end with respect to its no-load position.

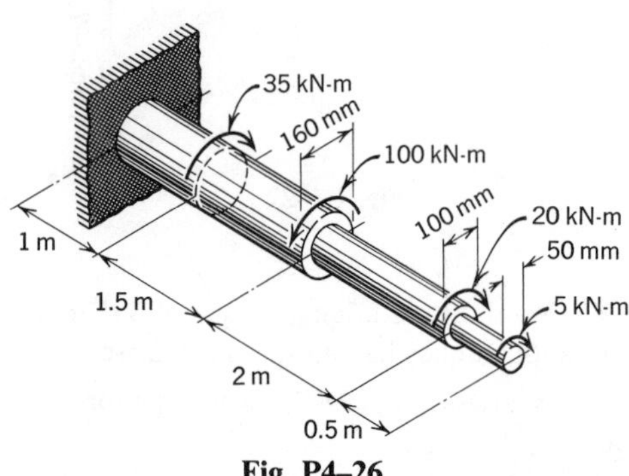

Fig. P4–26

4–27 The steel shaft of Fig. P4–27 is in equilibrium under the torques applied as shown. The maximum shearing stress in the 6-in. segment is 16 ksi, and the rotation of the end C with respect to the section at A is 0.018 rad counterclockwise looking left. Determine the torques T_1 and T_2.

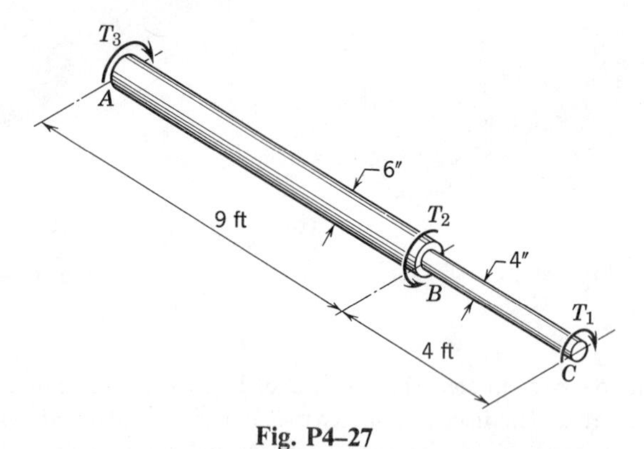

Fig. P4–27

4–28* The torsional loads applied as shown to the steel shaft of Fig. P4–28 produce a maximum shearing stress of 80 MN/m^2 and twist the free end through 0.014 rad. Determine the torques T_1 and T_2.

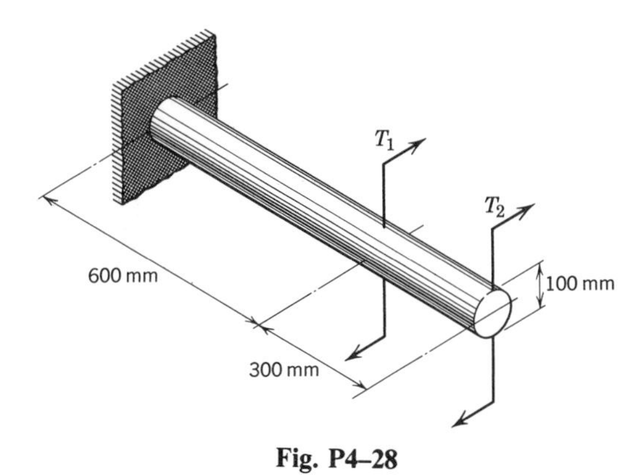

Fig. P4–28

4–29* Under the torques shown in Fig. P4–29, the maximum shearing stress in segment BC of the aluminum alloy ($G=4000$ ksi) shaft is 18 ksi.

(a) Determine by integration the magnitude of the resisting torque carried by the portion of the cross section between $\rho=3/2$ and $\rho=5/2$ in. in segment BC.

(b) Determine the rotation of the section at B when the torques T_1 and T_2 are applied as shown.

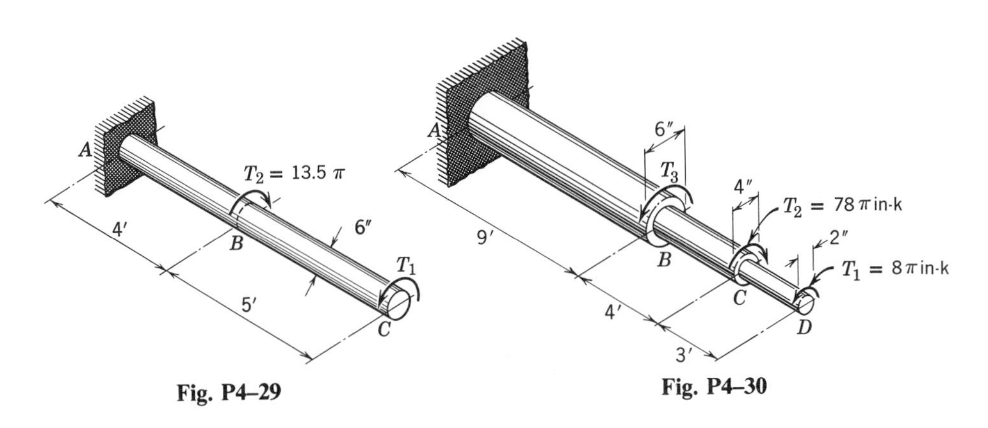

Fig. P4–29 Fig. P4–30

4–30 The shaft of Fig. P4–30 is made of elastoplastic steel ($G=12{,}000$ ksi) having a proportional limit and yield point in shear of 21 ksi. Under the loading shown, the torsional shearing stress in segment AB is 21 ksi at a radius of 2 in. Determine:

(a) The magnitude of torque T_3 (in terms of π).

(b) The rotation of a diameter at D with respect to the no-load position.

4–31 The shaft of Fig. P4–31 is hollow from A to B and solid from B to C. The material is elastoplastic steel ($G = 12{,}000$ ksi) for which the yield point in shear is 24 ksi. Determine:

(a) The maximum value (in terms of π) for T_2 applied as shown if the action is completely elastic.

(b) The maximum value (in terms of π) for T_2 applied as shown if the action in segment AB is completely plastic, with no strain hardening.

(c) The rotation of the section at B under the torques T_1 and T_2 when the stress at the inner surface of the segment AB just reaches 24 ksi.

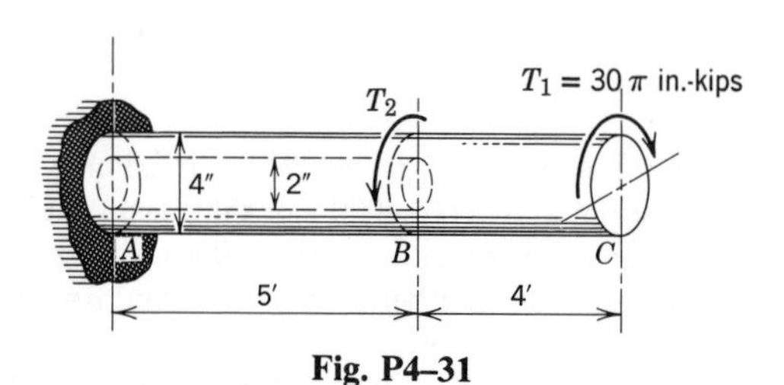

Fig. P4–31

4–32* The hollow steel shaft of Fig. P4–32 is made of steel having the shear stress-strain diagram of Fig. P4–7. Under the two torques, applied as shown, the maximum shearing strain in segment AB is 2000μ. Determine:

(a) The maximum shearing stress in segment BC (is the action elastic?).

(b) The magnitude of the torque T_2.

(c) The rotation of the section at B.

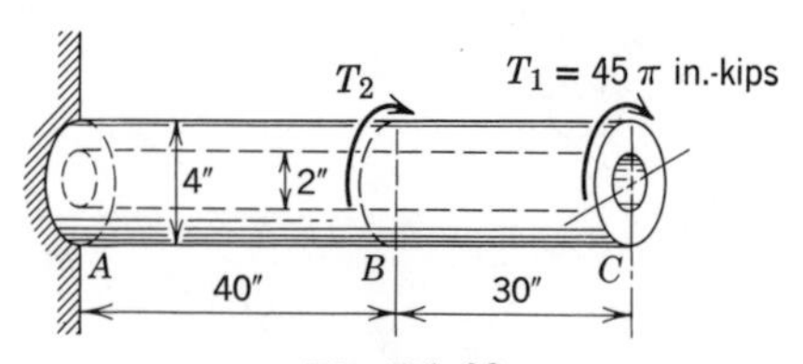

Fig. P4–32

4–33* The shaft of Fig. P4–33 is made of elastoplastic steel ($G = 12{,}000$ ksi) having a proportional limit and yield point in shear of 18 ksi. Determine:

(a) The maximum shearing stress.

(b) The angle of twist of the right end with respect to the unloaded position.

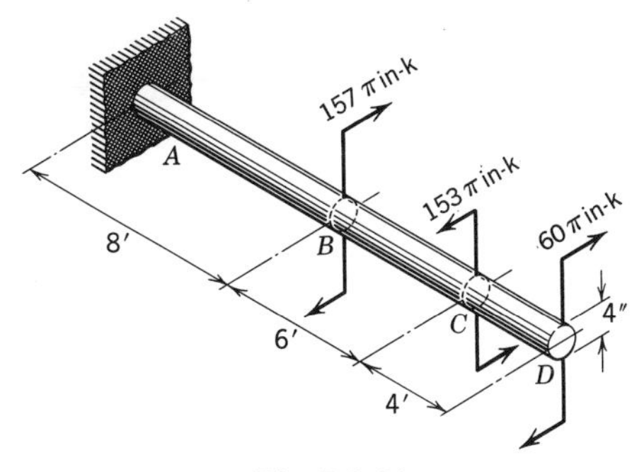

Fig. P4–33

4–34 The solid circular tapered shaft of Fig. P4–34 is subjected to a constant torque T.

(a) Determine the angle of twist in terms of T, L, G, r, and m.

(b) Using the results of part (a), verify the answer to Exam. 4–4 for $m = r/L$.

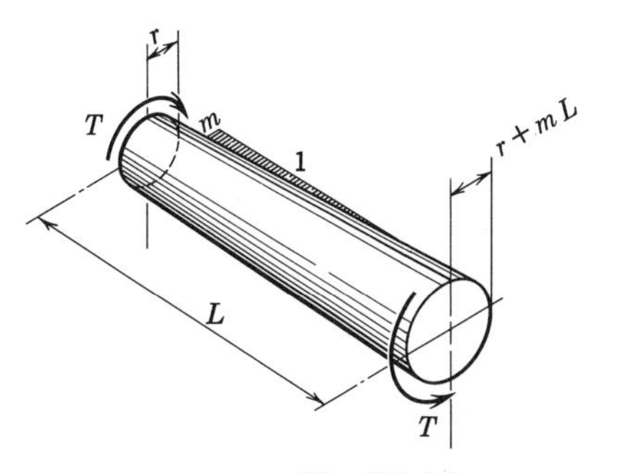

Fig. P4–34

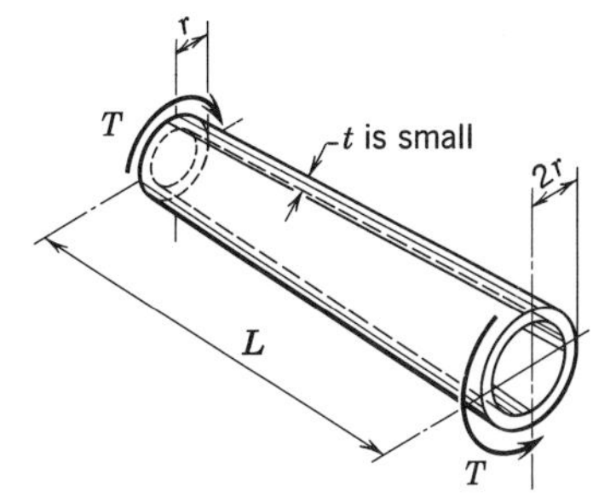

Fig. P4–35

4–35* The hollow tapered shaft of Fig. P4–35 has a constant wall thickness t. Determine the angle of twist for a constant torque T in terms of T, L, G, t, and r. Note that when t is small, the approximate expression for the polar moment of inertia $J = \rho^2 A$ may be used.

4–36 The hollow circular tapered shaft of Fig. P4–36 is subjected to a constant torque T. Determine the angle of twist in terms of T, L, G, and r.

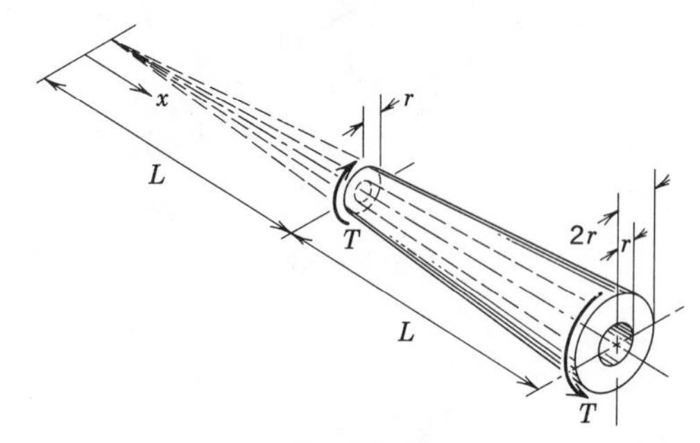

Fig. P4–36

4–37 The tapered circular shaft of Fig. P4–37 has an axial hole of constant diameter throughout its length. Determine the angle of twist due to a constant torque T in terms of T, L, G, R, and r.

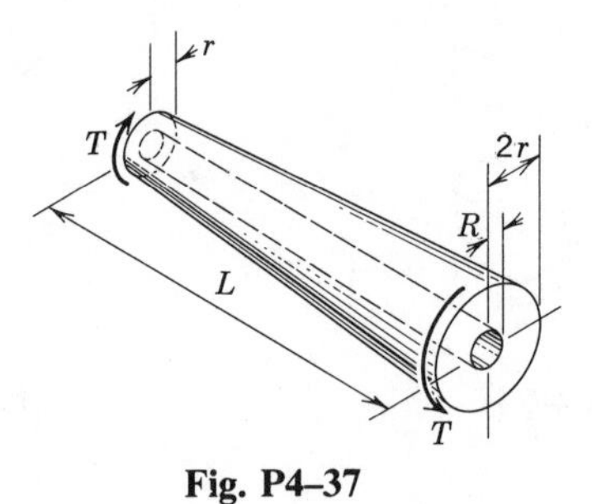

Fig. P4–37

4–38 The solid cylindrical shaft of Fig. P4-38 is subjected to a uniformly distributed torque of q in-lb per in. of length. Determine, in terms of q, L, G, and c, the rotation of the left end under the applied torque.

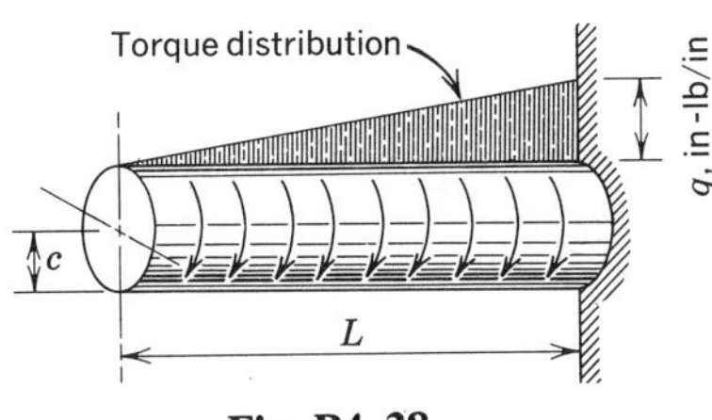

Fig. P4–38

4–39* The solid cylindrical shaft of Fig. P4–39 is subjected to a distributed torque that varies linearly from zero at the left end to q in-lb per in. of length at the right end. Determine, in terms of q, L, G, and c, the rotation of the left end under the applied torque.

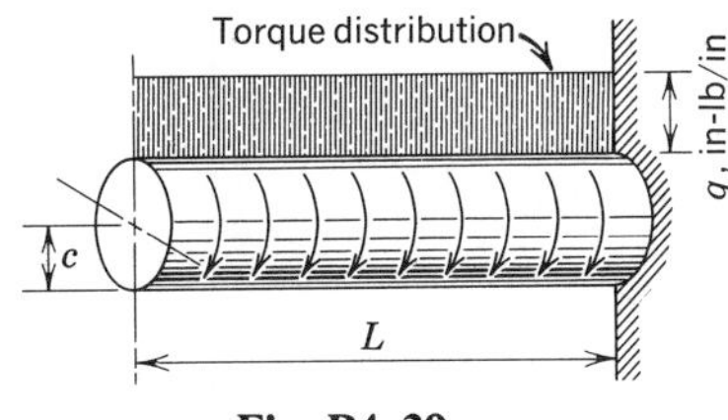

Fig. P4–39

4–4 STRESSES ON OBLIQUE PLANES

As stated in Art. 4–2, the torsion formula can be used to evaluate the shearing stress on a transverse plane. It is necessary to ascertain if the transverse plane is a plane of maximum shearing stress and if there are other significant stresses induced by torsion. For this study, the stresses at point A in the shaft of Fig. 4–10a will be analyzed. Figure 4–10b shows a differential cube taken from the shaft at A and the stresses acting on transverse and longitudinal planes. The stress τ_{xy} may be determined by means of the torsion formula, and $\tau_{yx} = \tau_{xy}$ (see Art. 1–4). Applying the equations of equilibrium to the free-body diagram of Fig. 4–10c, the following results are obtained:

$$(\nwarrow +) \qquad \Sigma F_t = 0$$

$$\tau_{nt}\, dA - \tau_{xy}(dA \cos\alpha)\cos\alpha + \tau_{yx}(dA \sin\alpha)\sin\alpha = 0$$

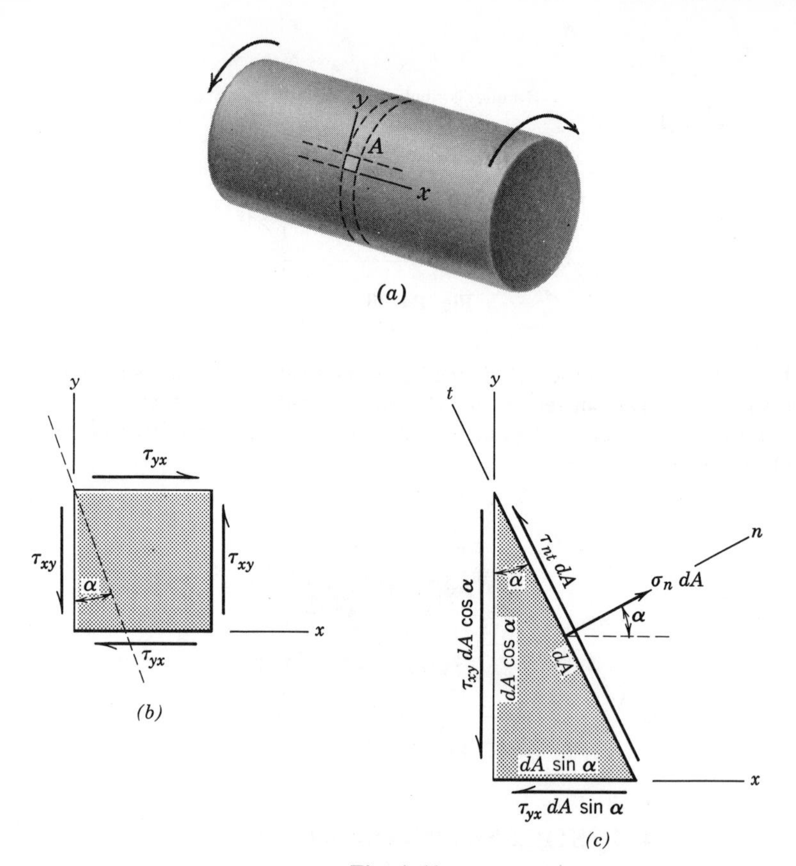

Fig. 4–10

from which

$$\tau_{nt} = \tau_{xy}(\cos^2\alpha - \sin^2\alpha) = \tau_{xy} \cos 2\alpha$$

and

$$\left(+\nearrow\right) \qquad \Sigma F_n = 0$$

$$\sigma_n\, dA - \tau_{xy}(dA \cos\alpha)\sin\alpha - \tau_{yx}(dA \sin\alpha)\cos\alpha = 0$$

from which

$$\sigma_n = 2\tau_{xy}(\sin\alpha)\cos\alpha = \tau_{xy} \sin 2\alpha$$

These results are shown in the graph of Fig. 4–11, from which it is apparent that the maximum shearing stress occurs on transverse and longitudinal (diametral) planes. The graph also shows that the maximum normal stresses occur on planes oriented at 45° with the axis of the bar and perpendicular to the surface of the bar. On one of these planes ($\alpha = 45°$ for Fig. 4–10) the stress is tension, and on the other ($\alpha = 135°$) the stress is compression. Furthermore, *all these stresses have the same magnitude*; hence, the torsion formula will give the magnitude of the maximum normal and shearing stresses at a point in a circular shaft subjected to pure torsion.

Any of the stresses discussed in the preceding paragraph may be significant in a particular problem. Compare, for example, the failures shown in Fig. 4–12. In Fig. 4–12*a*, the steel truck rear axle split longitudinally. One would also expect this type of failure to occur in a shaft of wood with the grain running longitudinally. In Fig. 4–12*b*, the compressive stress caused the thin-walled aluminum alloy tube to buckle along one 45° plane, while the tensile stress caused tearing on the other 45° plane. Buckling of thin-walled tubes (and other shapes) subjected to torsional loading is a matter of paramount concern to the designer. In Fig. 4–12*c*, the tensile stresses caused the gray cast iron to fail in tension—typical of any brittle material subjected to torsion. In Fig. 4–12*d*, the low-carbon steel failed in shear on a plane that is almost transverse—a typical failure for ductile material. The reason the fracture in Fig. 4–12*d* did not occur on a transverse plane is that under the large plastic twisting deformation before rupture (note the spiral lines indicating elements originally parallel to the axis of the bar), longitudinal elements were subjected to axial tensile loading because the grips of the testing machine would not permit the bar

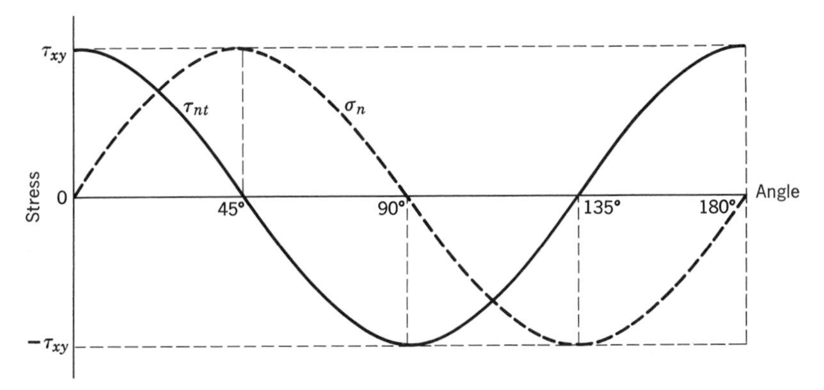

Fig. 4–11

to shorten as the elements were twisted into spirals. This axial tensile stress (not shown in Fig. 4–10) changes the plane of maximum shearing stress from a transverse to an oblique plane (resulting in a warped surface of rupture).[2]

For problems involving stresses on a given oblique plane, the methods of Arts. 1–6, 1–7, and 1–8 are all available, and there is no particular advantage for one method over the others. None of these methods is required for the principal stresses because they are equal in magnitude to the maximum shearing stress as given by the torsion formula.

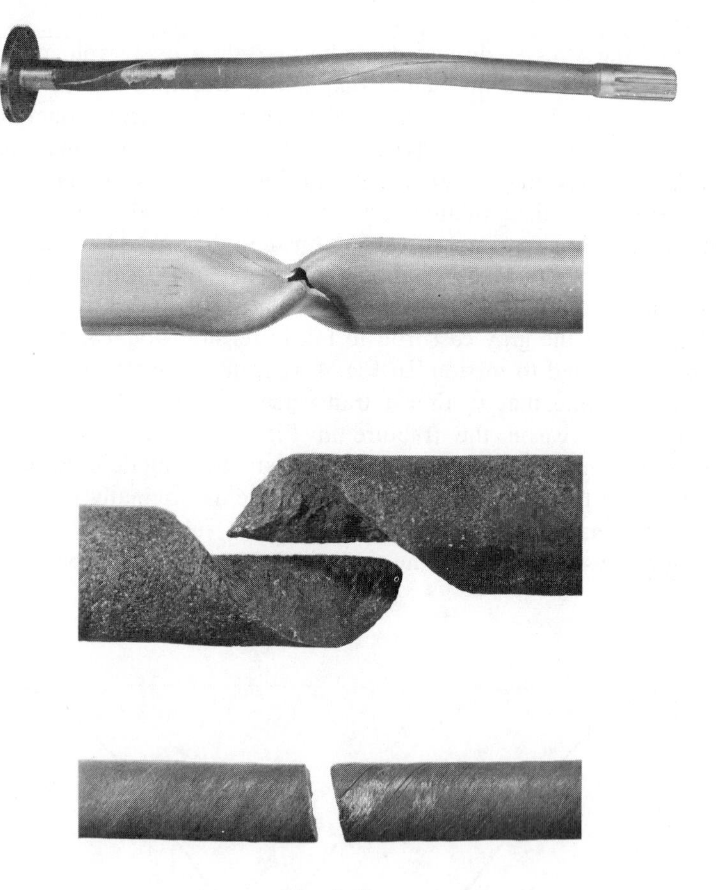

Fig. 4–12

[2]The tensile stress is not entirely due to the grips because the plastic deformation of the outer elements of the bar is considerably greater than that of the inner elements. This results in a spiral tensile stress in the outer elements and a similar compressive stress in the inner elements.

PROBLEMS

Note. Assume linearly elastic action for all of the following problems.

4–40* An aluminum alloy tube is to be used to transmit a torque in a control mechanism. The tube has an outside diameter of 2 in. and a wall thickness of 0.100 in. Because of the tendency of thin sections to buckle, the maximum compressive stress is specified as 8000 psi. Determine the maximum allowable torque that may be applied. The cross section may be treated as a thin annular area for computation of J/c.

4–41* Determine the maximum torque that can be resisted by a hollow circular shaft having an inside diameter of 25 mm and an outside diameter of 50 mm without exceeding a normal stress of 70 MN/m^2 T or a shearing stress of 75 MN/m^2.

4–42* For the shaft shown in Fig. P4–42, determine:

(a) The stresses occurring at point A (at the surface of the shaft) on the plane

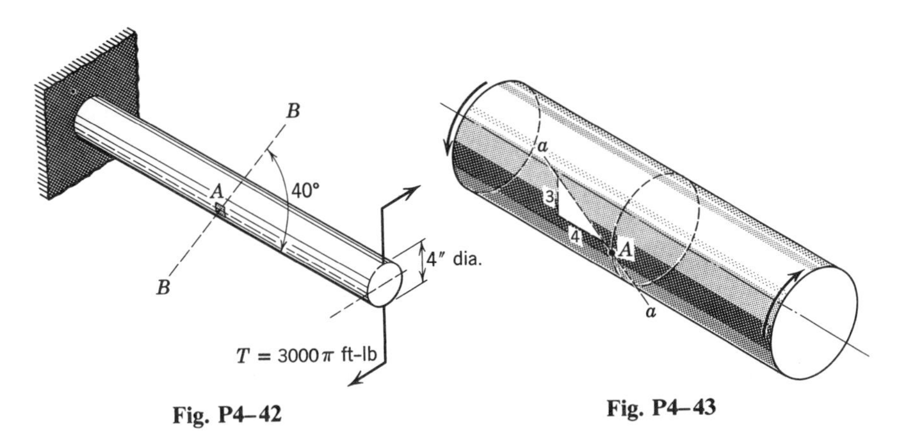

Fig. P4–42

Fig. P4–43

$B-B$ (which is normal to the surface of the shaft at point A and makes an angle of 40° with the axis of the shaft). Show these stresses on an enlarged sketch of the differential area representing A.

(b) The maximum normal stresses occurring at point A. Show these stresses on a sketch representing a differential area around A.

4–43* A solid circular steel shaft 1 m long is required to twist through 0.03 rad. If the shearing stress is not to exceed 60 MN/m^2, determine:

(a) The maximum permissible diameter for the shaft.

(b) The normal stress on plane $a-a$, which is normal to the surface of the shaft at point A and has a slope of 3 to 4 with the axis of the shaft (see Fig. P4–43) when the maximum shearing stress in the shaft is 60 MN/m^2.

4–44 A hollow circular steel shaft, 4 in. outside diameter, 2 in. inside diameter, and 3 ft long, is held at one end and subjected to a torque at the other end, as shown in Fig. P4–44.

(a) Determine the shearing stress at the inner surface on a transverse plane.

(b) Determine the distance point B moves as the torque is applied.

(c) With the outside diameter unchanged, calculate the maximum inside diameter for an allowable compressive stress of 12,000 psi in the shaft.

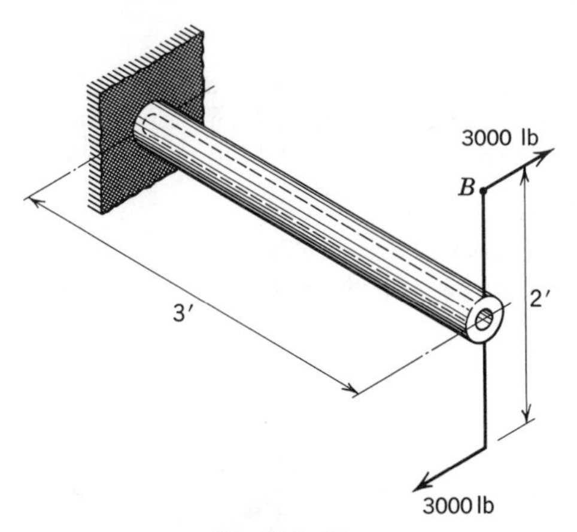

Fig. P4–44

4–45 A cylindrical tube is fabricated by butt welding 6-mm steel plate along a spiral seam, as shown in Fig. P4–45. For an allowable compressive stress in the shell of 80 MN/m^2, determine:

(a) The maximum permissible torque T to which the tube may be subjected (note that the cross section may be treated as a thin annular area for computation of J/c).

(b) The factor of safety with respect to fracture for the weld if the ultimate strengths of the weld metal are 250 MN/m^2 shear and 400 MN/m^2 tension. The thickness of a butt weld is the same as that of the plate.

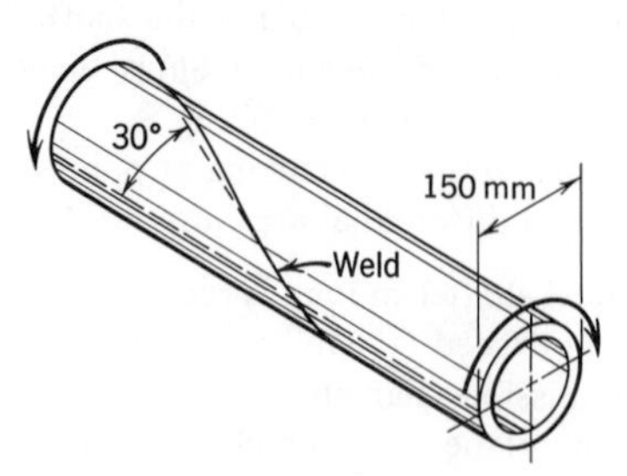

Fig. P4–45

4–46 A hollow circular shaft having an inside diameter of 2 in. and an outside diameter of 4 in. is subjected to equal and opposite torques at the ends. An electrical resistance strain gage bonded to the outside surface of the shaft at an angle of 45° with the axis of the shaft indicates a strain of 2600μ. The material is aluminum alloy 6061-T6. Determine:

(a) The maximum normal and shearing stresses in the shaft.
(b) The magnitude of the applied torques.
(c) The magnitude of the angle of twist per unit length of the shaft.

4–47 When the two torques of Fig. P4–47 are applied to the steel shaft as shown, point A moves 5 mm in the direction indicated by torque T_1. An electrical resistance strain gage bonded to the surface of the 50-mm shaft at an angle of 45° with the axis of the shaft indicates a strain of 1000μ. Determine:

(a) The maximum tensile stress in the 50-mm shaft.
(b) The magnitudes of the two torques.

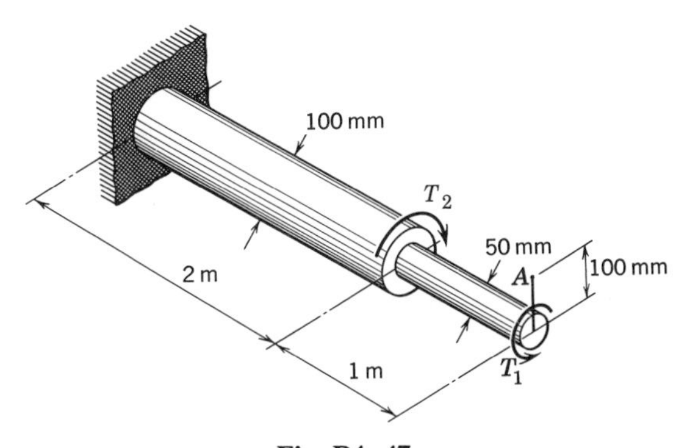

Fig. P4–47

4–5 COMBINED AXIAL AND TORSIONAL LOADS

There are numerous situations where a shaft or other machine member is subjected to both an axial and a torsional load. Examples are the drill rod on a well or other drilling machine and the propeller shaft on a ship (bending stresses, to be studied later, will also probably be present in this case).

For stresses below the proportional limit of the material, the method of superposition (see Art. 3–3) applies. Thus, each stress can be calculated separately and all stresses on a given plane superimposed. Once the stresses on a pair of mutually perpendicular planes through a point have been determined, the procedures of Arts. 1–6, 1–7, and 1–8 are available to calculate the principal stresses and the maximum shearing stress at a point. Since radial and circumferential normal stresses are zero, the combination of axial and torsional loads results in a state-of-plane stress at any point in the body. The axial stress is the same for all interior and exterior points, while the torsional stress is maximum at the outer surface; therefore, the investigation for critical stresses would normally be made for points on the surface of the shaft.

The following example illustrates the procedure for combined torsional and axial loads.

Example 4–5 A hollow circular shaft of outer diameter 4 in. and inner diameter 2 in. is loaded as shown in Fig. 4–13*a*.

(a) Determine the principal stresses and the maximum shearing stress at the point (or points) where the stress situation is most severe.

(b) On a sketch show the approximate directions of these stresses.

Solution The axial stress is constant throughout the shaft and is equal to

$$\sigma = \frac{P}{A} = \frac{24{,}000\pi}{\pi(2^2 - 1^2)} = 8000 \text{ psi } C$$

The torsional load is highest to the right of the torque T_1, and the torsional stress is maximum on the outer surface of the shaft; therefore, the magnitude of the maximum torsional shearing stress is

$$\tau = \frac{Tc}{J} = \frac{22{,}500\pi(2)}{\pi(2^4 - 1^4)/2} = 6000 \text{ psi}$$

These stresses are shown acting on mutually perpendicular planes through a point on the surface of the right portion of the shaft in Fig. 4–13*b*.

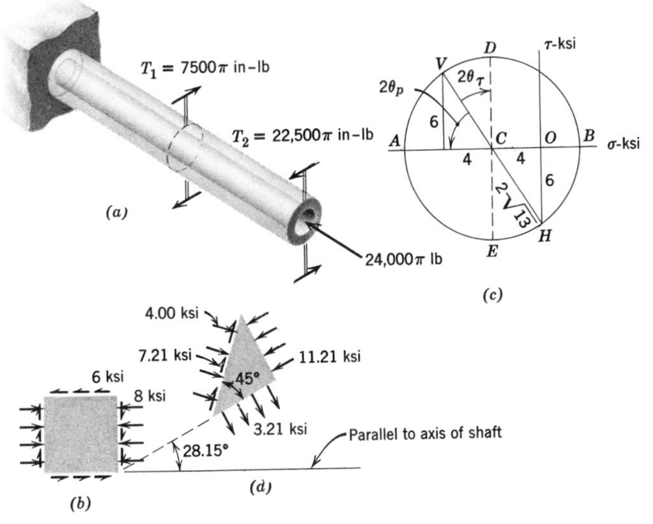

Fig. 4-13

Mohr's circle for the stresses at any point on the surface of the right portion of the shaft is shown in Fig. 4-13*c*.

(a) The principal stresses OA and OB from Mohr's circle are

$$\sigma_{p1} = OA = -4.00 - 2\sqrt{13} = -11.21 = \underline{11.21 \text{ ksi } C} \qquad \text{Ans.}$$

and

$$\sigma_{p2} = OB = -4.00 + 2\sqrt{13} = \underline{3.21 \text{ ksi } T} \qquad \text{Ans.}$$

The maximum shearing stress CD or CE from Mohr's circle is

$$\tau_{max} = CD = 2\sqrt{13} = \underline{7.21 \text{ ksi}} \qquad \text{Ans.}$$

(b) From the Mohr's circle,

$$\tan 2\theta_p = 6/4 = 1.5$$

which gives

$$2\theta_p = 56.30° \circlearrowright$$

and

$$\theta_p = 28.15° ↺$$

The stresses are shown in their proper directions in Fig. 4–13*d*.

PROBLEMS

Note. Assume linearly elastic action for all of the following problems.

4–48* The solid circular shaft of Fig. P4–48*a* has a diameter of 2 in. It is subjected to an axial tensile load of $12{,}000\pi$ lb and a torque of 4000π in-lb, as shown. Determine:

(a) The principal stresses, the maximum shearing stress, and the planes on which these stresses act at point *A* on the surface of the shaft.

(b) The stresses on plane *B–B*, oriented as indicated in Fig. P4–48*b* at point *A*. Draw a complete stress block with answers shown on it.

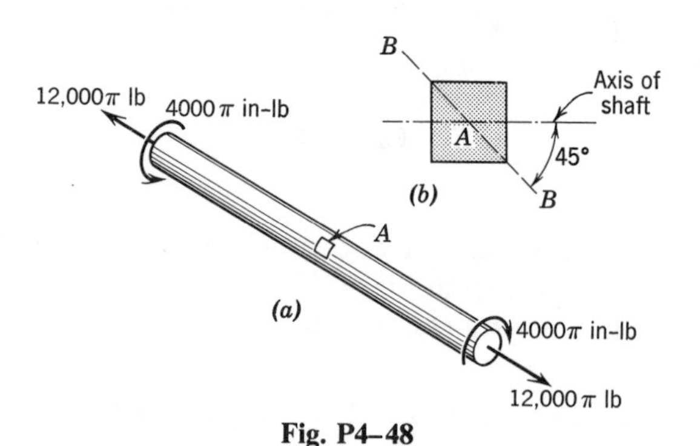

Fig. P4–48

4–49* A shaft 50 mm in diameter is subjected to a torque of 450π N-m and an axial tension of 40π kN, as shown in Fig. P4–49. Determine the principal stresses and the maximum shearing stress at point *A*, and show their magnitudes and approximate directions on a sketch.

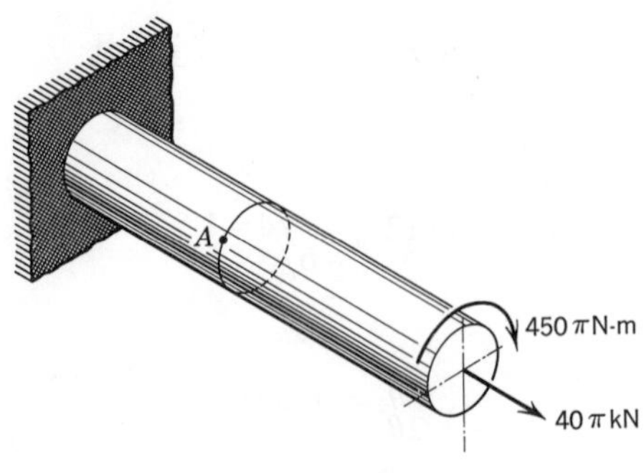

Fig. P4–49

4–50* A solid circular shaft 4 in. in diameter is made of brass and steel segments and is loaded as shown in Fig. P4–50. Determine:

(a) The principal stresses and the maximum shearing stress at a point where the stresses are greatest; show on a sketch.

(b) The angle of twist of section C with respect to section A. Use $6(10^6)$ psi for the modulus of rigidity of the brass.

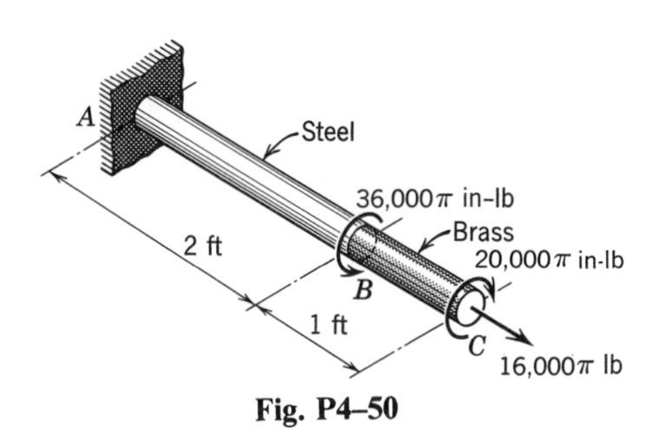

Fig. P4–50

4–51* A 150-mm diameter shaft is loaded as shown in Fig. P4–51.

(a) At what point (or points) are the stresses the highest?

(b) Determine and show on a sketch the principal stresses and the maximum shearing stress at point A.

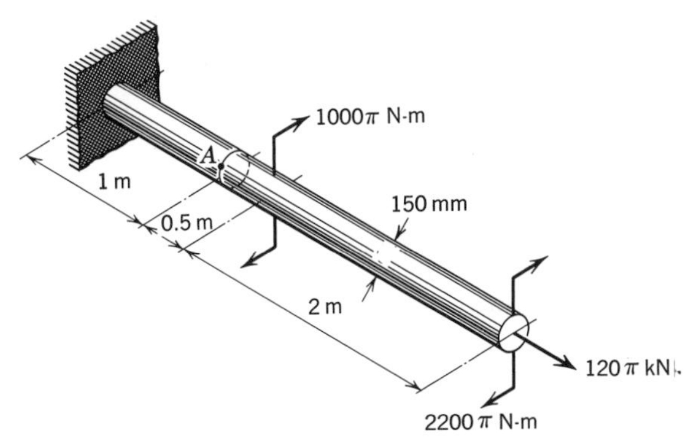

Fig. P4–51

4–52 A short length of 6-in. diameter steel shaft is subjected to a torque of 9000π ft-lb in addition to an axial tensile load P, as indicated in Fig. P4–52. Determine the maximum allowable value of the axial tensile load if the shearing stress is not to exceed 12,000 psi and the tensile stress is not to exceed 18,000 psi.

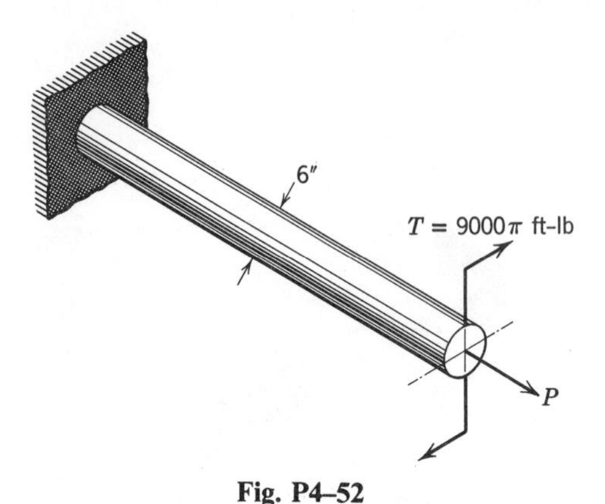

Fig. P4–52

4–53 A solid circular bronze shaft with diameters as shown in Fig. P4–53 is loaded and supported as indicated. The allowable shearing stress is 56 MN/m^2 and the allowable tensile stress is 84 MN/m^2. With T_2 equal to $50P$ mN-m and T_1 equal to $25P$ mN-m, determine the maximum allowable value of P.

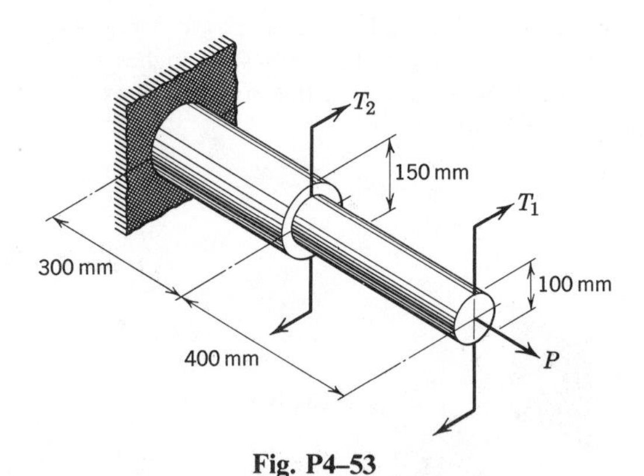

Fig. P4–53

4–54* For the shaft and allowable stresses of Prob. 4–53, with P equal to 400 kN and T_1 equal to 9000 N-m, determine the maximum allowable value of T_2. Note: the stresses in the right portion of the shaft are less than the allowable values.

4–55* A circular steel shaft with the left portion solid and the right portion hollow is loaded as shown in Fig. P4–55. The allowable shearing and normal stresses are 12 and 20 ksi, respectively. Determine the maximum allowable axial load P.

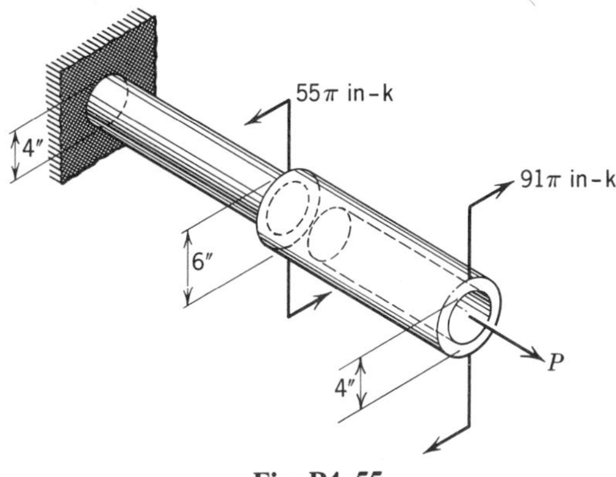

Fig. P4–55

4–6 POWER TRANSMISSION

Almost anyone will recognize that one of the most common uses for the circular shaft is the transmission of power; therefore, no discussion of torsion would be adequate without including this topic. Power is the time rate of doing work, and the basic relationship for work done by a constant torque is $W_k = T\phi$ where W_k is work and ϕ is the angular displacement of the shaft in radians. The time derivative of this expression gives

$$\frac{dW_k}{dt} = T\frac{d\phi}{dt} = T\omega$$

where dW_k/dt is power in foot-pounds per minute (or similar units), T is a constant torque in foot-pounds, and ω is the angular velocity (of the shaft) in radians per minute. All units, of course, may be changed to any other consistent set of units. Since the angular velocity is usually given in revolutions per minute (rpm), the conversion of revolutions to radians will often be necessary. Also, in the English system, power is usually given in units of horsepower, and the relation 1hp = 33,000 ft-lb per min will be found useful. In the metric (SI) system, power is given in watts (N-m/sec). An example of shaft design follows.

Example 4–6 A Diesel engine for a small commercial boat is to operate at 200 rpm and deliver 800 hp through a gearbox with a ratio of 1 to 4 (increase) to the propeller shaft. Both the shaft from engine to gearbox and the propeller shaft are to be solid and made of heat-treated alloy steel. Determine the minimum permissible diameters for the two shafts if the allowable stress is 20 ksi and the angle of twist in a 10-ft length of the propeller shaft is not to exceed 4°. Neglect power loss in the gearbox and assume (incorrectly because of thrust stresses) that the propeller shaft is subjected to pure torsion.

Solution The first step is the determination of the torques to which the shafts are to be subjected. By means of the expression, power $= T\omega$, the torques are obtained as follows:

$$800(33{,}000) = T_1(200)(2\pi)$$

from which

$$T_1 = 66{,}000/\pi \text{ ft-lb}$$

which is the torque at the crank shaft of the engine. Because the propeller shaft speed is four times that of the crankshaft and power loss in the gearbox is to be neglected, the torque on the propeller shaft is one-fourth that on the crankshaft and is equal to $66{,}000/(4\pi)$ ft-lb. Using the torsion formula, the shaft sizes necessary to satisfy the stress specification are determined. For the main shaft,

$$\frac{J}{c} = \frac{T}{\tau} = \frac{66{,}000(12)}{\pi(20)(10^3)} = \frac{(\pi/2)c_1^4}{c_1}$$

$$c_1^3 = 8.02 \quad \text{and} \quad c_1 = 2.002 \text{ in.}$$

or the shaft from engine to gearbox should be

<u>4.00 in. in diameter</u> Ans.

The torque on the propeller shaft is one-fourth that on the main shaft, and this is the only change in the expression for c_1^3; therefore,

$$c_2^3 = 8.02/4 \quad \text{and} \quad c_2 = 1.261 \text{ in.}$$

The size of propeller shaft necessary to satisfy the distortion specification will be determined as follows:

$$\gamma = \frac{c\theta}{L} = \frac{\tau}{G} = \frac{Tc}{JG}$$

from which

$$\frac{4(\pi/180)}{10(12)} = \frac{66{,}000(12)/(4\pi)}{(\pi c_3{}^4/2)(12)(10^6)}$$

$$c_3^4 = 5.75 \quad \text{and} \quad c_3 = 1.548 > 1.261$$

Therefore, the propeller shaft must be

$$\underline{\text{3.10 in. in diameter}} \qquad \text{Ans.}$$

PROBLEMS

Note. In the following problems, the specified allowable stresses are all below the shearing proportional limits of the materials.

4–56* An 8-ft long hollow steel shaft has an outside diameter of 4 in. and an inside diameter of 2 in. The maximum shearing stress in the shaft is 12 ksi and the angular velocity is 200 rpm. Determine:

(a) The horsepower being transmitted by the shaft.
(b) The magnitude of the angle of twist in the shaft.

4–57* A solid circular steel shaft is 80 mm in diameter and 1.5 m long. The maximum permissible angle of twist is 0.036 rad. The working shearing stress is 70 MN/m^2. Determine the maximum power that this shaft can deliver:

(a) When rotating at 225 rpm.
(b) When rotating at 180 rpm.

4–58* A solid circular steel shaft 2 ft long transmits 300 hp at a speed of 400 rpm. If the allowable shearing stress is 10,000 psi and the allowable angle of twist is 0.014 rad, determine:

(a) The minimum permissible diameter for the shaft.
(b) The speed at which this horsepower can be delivered if the stress is not to exceed 8000 psi in a shaft having the diameter determined in part (a).

4–59 A solid circular steel shaft 3 m long transmits 220 kW at 180 rpm. The allowable shearing stress in the shaft is 60 MN/m^2 and the maximum allowable angle of twist is 0.040 rad. Determine the minimum allowable diameter of the shaft.

4–60 A hollow shaft of aluminum alloy 2024-T4 (see App. A) is to transmit 1500 hp at 1800 rpm. The shearing stress is not to exceed 15 ksi, the angle of twist is not to exceed 0.20 rad in a 10-ft length, and the inside diameter shall be three-fourths of the outside diameter. Determine the minimum permissible outside diameter.

4–61* A solid circular steel shaft 5 m long is to transmit 2.25 MW at 1500 rpm. The allowable shearing stress is 85 MN/m^2, and the angle of twist is not to exceed 3° in 3 m. Determine the minimum permissible diameter for the shaft.

4–62* An 8-ft long solid circular shaft is made of brass ($G = 5000$ ksi) and steel sections, each 4 ft long, rigidly connected. The shaft is to be used to transmit 300 hp at 200 rpm. The allowable shearing stress in the shaft is 12 ksi, and the maximum allowable total angle of twist is 0.080 rad. Determine the minimum allowable diameter of the shaft.

4–7 MODULUS OF RUPTURE

Certain routine design problems involving circular torsion members must be designed against failure by fracture (or buckling in the case of tubing) rather than slip, and the method of Art. 4–1 for inelastic action may be too involved for the type of problem considered. In such cases, the torsion formula may possibly be used with modification.

Consider a circular shaft of a material for which the stress-radius distribution under the torque necessary to produce fracture is as shown by line *OB* in Fig. 4–14. A linear stress distribution, represented by line *OC*, may be assumed such that $\int_A \tau dA(\rho)$ for this distribution will have the same value as that for the stress distribution of line *OB*. Naturally, if the distribution of *OC* is assumed, the torsion formula applies thus,

$$T_r = \frac{RJ}{c}$$

where *R* is a fictitious stress known as the *modulus of rupture in torsion*. The magnitude of *R* is obtained by loading a member to fracture (for a brittle material) or buckling (for thin-walled tubing), substituting the magnitude of the torque at fracture (buckling) and J/c for the specimen into the torsion formula and solving for *R*. The magnitude of the modulus of rupture is dependent on the kind of material and, in general, on some size parameter (such as ratio of diameter to wall thickness for tubing); hence, the method is useful only if applicable values of the modulus are available.

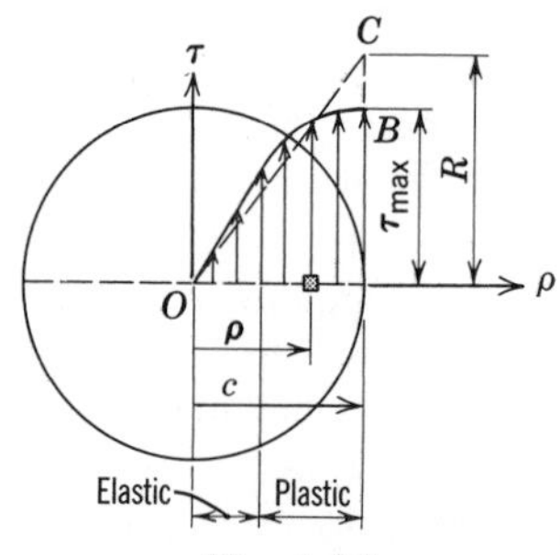

Fig. 4–14

4–8 STATICALLY INDETERMINATE MEMBERS

In all problems discussed in the preceding articles, the equations of equilibrium sufficed to determine the resisting torque at any section. Occasionally, torsionally loaded members are so constructed and loaded that the number of independent equilibrium equations is less than the number of unknowns, and the member is statically indeterminate. When this occurs, it becomes necessary to write the requisite number of distortion equations so that the total number of equations agrees with the number of unknowns to be determined.

The distortion equations will involve the angles of twist, and a simplified twist diagram will often be of assistance in obtaining the correct equations. Where there is a possibility of inelastic action but such action is not specifically indicated, the recommended procedure is to assume elastic action and verify that the resulting stresses are or are not below the proportional limit of the material. The following examples will serve to illustrate the application of the principles.

Example 4–7 The circular shaft AC in Fig. 4–15a is fixed to rigid walls at A and C. The section AB is solid annealed bronze and the hollow section BC is made of aluminum alloy 2024-T4. There is no stress in the shaft before the 20,000-ft-lb torque T is applied. Determine the maximum shearing stress in each shaft after T is applied.

Solution Figure 4–15b is a free-body diagram of the shaft. The torques T_A and T_C, produced by the supports, are unknown, and the summation of moments about the axis of the shaft gives

$$T_A + T_C = 20{,}000(12) \qquad \text{(a)}$$

This is the only independent equation of equilibrium; thus, the problem is statically indeterminate. A second equation can be obtained from the

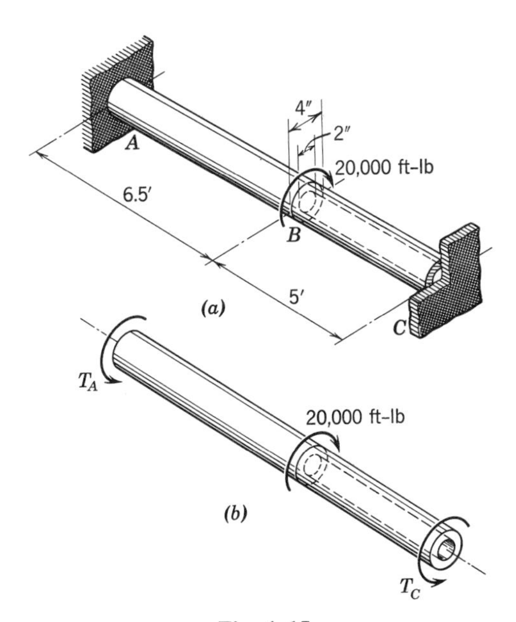

Fig. 4–15

deformation of the shaft, since each section must undergo the same angle of twist; thus,

$$\theta_{AB} = \theta_{BC} \tag{b}$$

Equations (a) and (b) can both be expressed in terms of the shearing stresses in the two sections. When the torsion formula is applied to Eq. (a) it becomes

$$\frac{\tau_{AB}\pi(2^4)/2}{2} + \frac{\tau_{BC}\pi(2^4 - 1^4)/2}{2} = 240{,}000 \tag{c}$$

Since the angle of twist is

$$\theta = \frac{\gamma L}{c} = \frac{\tau L}{Gc}$$

Eq. (b) becomes

$$\frac{\tau_{AB}(6.5)12}{6.5(10^6)2} = \frac{\tau_{BC}(5)12}{4(10^6)2}$$

from which

$$\tau_{AB} = 5\tau_{BC}/4 \tag{d}$$

Equations (c) and (d) can be solved simultaneously to give

$$\tau_{AB} = \underline{10{,}900 \text{ psi}} \qquad \text{Ans.}$$

$$\tau_{BC} = \underline{8{,}730 \text{ psi}} \qquad \text{Ans.}$$

Note: these maximum stresses, which occur at the surfaces of the shafts, do not exceed the proportional limits of the materials; therefore, only elastic behavior is involved and the results are valid.

Example 4–8 In Exam. 4–7 change the length of segment AB to 6 ft and the material to elastoplastic steel with a proportional limit of 18 ksi and modulus of rigidity of 12,000 ksi. Change the applied torque to 30 ft-kips. Make no other changes.

Solution First assume elastic action; then Eqs. (a) and (c) become

$$4\pi\tau_{AB} + \frac{15\pi}{4}\tau_{BC} = 360 \tag{e}$$

and Eq. (b) becomes

$$\frac{\tau_{AB}(6)(12)}{12(10^3)(2)} = \frac{\tau_{BC}(5)(12)}{4(10^3)(2)} \tag{f}$$

from which

$$\tau_{AB} = 5\tau_{BC}/2 \tag{g}$$

Solving Eqs. (e) and (g) gives

$$\tau_{AB} = 20.9 \text{ ksi} > 18 \text{ ksi}$$

Therefore, the steel undergoes some inelastic action, and since for the elastic solution the corresponding aluminum stress is 8.36 ksi, which is much less than the elastic strength of 28 ksi (see App. A), it would be reasonable to assume that the aluminum will undergo elastic action only. Letting r be the radius of the elastic-plastic boundary in the steel segment, the equilibrium equation Eq. (e) becomes

$$\int_0^r \frac{18}{r}\rho(2\pi\rho\,d\rho)\rho + 18\int_r^2 (2\pi\rho\,d\rho)\rho + \frac{15\pi}{4}\tau_{BC} = 360 \tag{h}$$

Also, since the maximum elastic shear strain in the steel is

$$\gamma = \frac{18}{12{,}000} = \frac{r\theta_{AB}}{6(12)}$$

the angle of twist in segment AB is

$$\theta_{AB} = \frac{108(10^{-3})}{r} \tag{i}$$

Equation (f) now becomes

$$\frac{108(10^{-3})}{r} = \frac{\tau_{BC}(5)(12)}{4(10^3)(2)}$$

from which

$$\tau_{BC} = \frac{72}{5r} \tag{j}$$

Now, substituting Eq. (j) into Eq. (h) and integrating gives

$$9\pi r^3 + 12\pi(8 - r^3) + \frac{15\pi}{4}\left(\frac{72}{5r}\right) = 360$$

which after collecting terms and simplifying becomes

$$r^4 + 6.2r = 18$$

A trial-and-error solution of this equation gives a value of 1.665 in. for r, which when substituted into Eq. (j) gives

$$\tau_{BC} = \underline{8.65 \text{ ksi}} \qquad \text{Ans.}$$

which is well below the elastic strength of the aluminum. Also,

$$\tau_{AB} = \underline{18.0 \text{ ksi}} \qquad \text{Ans.}$$

Example 4–9 Segment CD of Fig. 4–16a is a solid shaft of cold-rolled bronze, and segment EF is a hollow shaft of aluminum alloy 2024-T4 with a core of low-carbon steel. The ends C and F are to be considered fixed to rigid walls and the steel core of EF is connected to the flange at E so that the aluminum and steel act together as a unit. The two flanges D and E are bolted together, and the bolt clearance permits flange D to rotate through 0.03 rad before EF will carry any of the load. Determine the maximum shearing stress in any shaft material when the torque of 20 ft-kips is applied (see App. A for mechanical properties).

Solution Figure 4–16b is a free-body diagram of the assembly. It is necessary to place two unknown torques T_A and T_S on the right end because the torques and stresses will need to be related by the torsion formula, which is limited to cross sections of homogeneous material. Summing moments with respect to the axis of the shaft gives the following equation:

$$20(12) = T_B + T_A + T_S \qquad \text{(k)}$$

This is the only independent equilibrium equation for this problem and there are three unknown torques; hence, two distortion equations need to be written. A three-dimensional distortion diagram is not easy to draw, and the cross sectional drawing of Fig. 4–16c is adequate. The angles indicated represent the twist in segment CD at D and in segment EF at E (the coupling is considered rigid). The fact that segment EF is nonhomogeneous does not invalidate the assumption of the plane section remaining

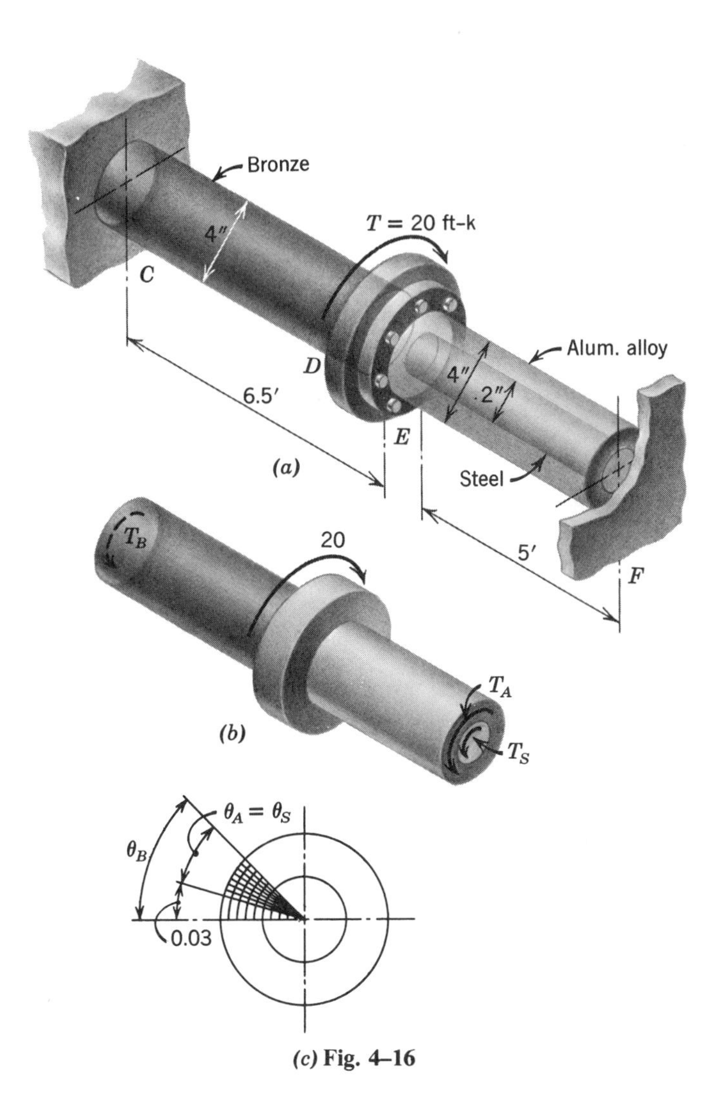

(c) **Fig. 4–16**

plane and the diameter remaining straight; the strains are still proportional to the distance from the shaft axis. However, the stresses are not proportional to the radii throughout the entire cross section because G is not single-valued.

From Fig. 4–16c, the distortion equations are

$$\theta_B = \theta_A + 0.03 \tag{m}$$

and

$$\theta_A = \theta_S \tag{n}$$

Eqs. (k), (m), and (n) can be written in terms of the same three unknowns (torque, angle, or stress) and solved simultaneously. Since the maximum stress is required, Eqs. (m) and (n) will be written in terms of the maximum stress in each material; thus,

$$\theta = \frac{\gamma L}{c} = \frac{\tau L}{Gc}$$

$$\frac{\tau_B(6.5)(12)}{6.5(10^3)(2)} = \frac{\tau_A(5)(12)}{4(10^3)(2)} + 0.03 \tag{m}$$

from which

$$\tau_B = (5/4)\tau_A + 5 \qquad (\text{therefore } \tau_B > \tau_A) \tag{o}$$

and

$$\frac{\tau_A(5)(12)}{4(10^3)(2)} = \frac{\tau_S(5)(12)}{11.6(10^3)(1)} \tag{n}$$

from which

$$\tau_S = 1.45\tau_A = 1.45(4\tau_B/5 - 4) \tag{p}$$

Writing Eq. (k) in terms of maximum stresses in each material and using the torsion formula gives

$$20(12) = \frac{\tau_B(\pi/2)(2^4)}{2} + \frac{\tau_A(\pi/2)(2^4 - 1^4)}{2} + \frac{\tau_S(\pi/2)(1^4)}{1}$$

or

$$960/\pi = 16\tau_B + 15\tau_A + 2\tau_S \tag{q}$$

Solving Eqs. (o), (p), and (q) simultaneously gives the results

$$\tau_B = 12.44 \text{ ksi}$$

$$\tau_A = 5.95 \text{ ksi}$$

and

$$\tau_S = 8.63 \text{ ksi}$$

Assuming for the steel and the bronze that the elastic strengths in shear are approximately six-tenths the elastic strengths in tension and, recognizing that the proportional limits may be somewhat lower than the elastic strengths (which may be yield strengths), the above stresses indicate that the elastic solution is valid, and the maximum shearing stress is at the surface of the bronze. Thus

$$\tau_{max} = \underline{12.44 \text{ ksi}} \qquad \text{Ans.}$$

PROBLEMS

Note. In the following problems, the action is known to be elastic unless (1) it is specifically stated to be inelastic, or (2) data are provided that will permit a judgement to be made on whether the action is elastic or plastic or both.

4-63* For the steel shaft of Fig. P4-63, determine the maximum shearing stress and the angle of rotation of the section where the given torque is applied.

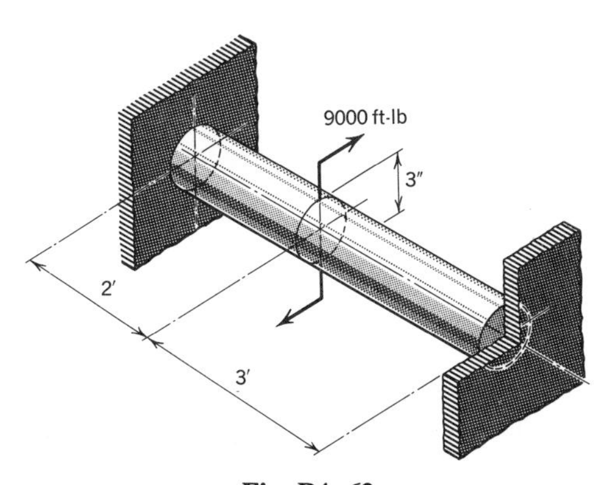

Fig. P4-63

4-64* The steel shaft of Fig. P4-64 is rigidly secured at the ends. The right 3 m of the shaft is hollow, having an inside diameter of 50 mm. Determine the maximum shearing stress in the shaft.

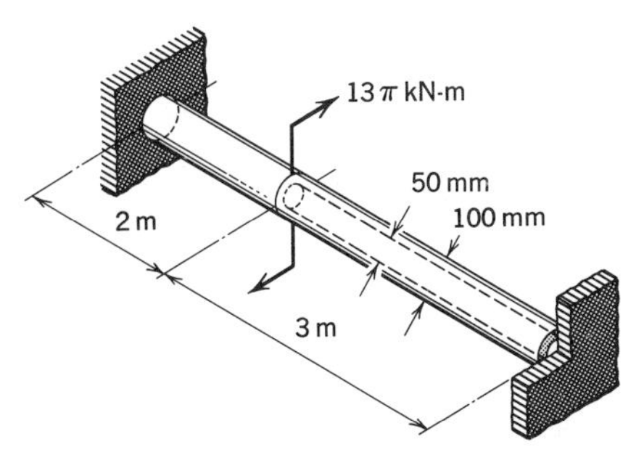

Fig. P4-64

4–65* Shown in Fig. P4–65 is a hollow cold-rolled bronze shaft ($G = 6000$ ksi) with a 6-in. outside diameter and a 4-in. inside diameter. Inside and concentric with this bronze shaft is a solid steel shaft with a 2-in. diameter. The two shafts are rigidly connected to a bar at the right end and to the wall at the left end. Determine the rotation of the bar AB due to the couple shown.

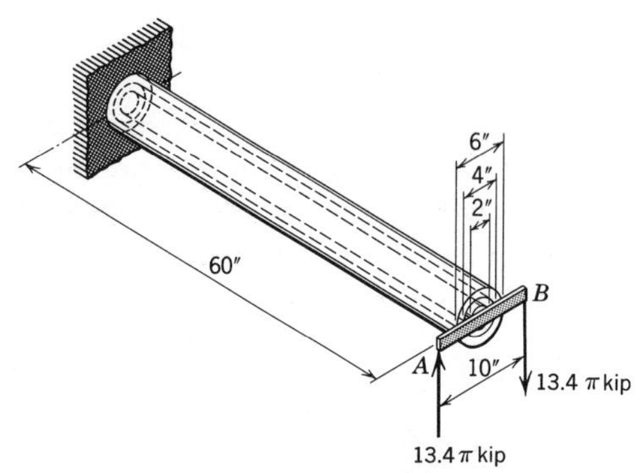

Fig. P4–65

4–66* A composite shaft is to be identical in construction and dimensions with that of Fig. P4–65, but the 6-in. hollow portion is to be of elastoplastic steel (see Fig. P4–66 for stress-strain diagram), and the 2-in. solid portion is to be of cold-rolled brass ($G = 5000$ ksi and proportional limit $= 30$ ksi). The torque applied to bar AB is unknown. Determine the torque necessary to rotate AB through an angle of 0.24 rad.

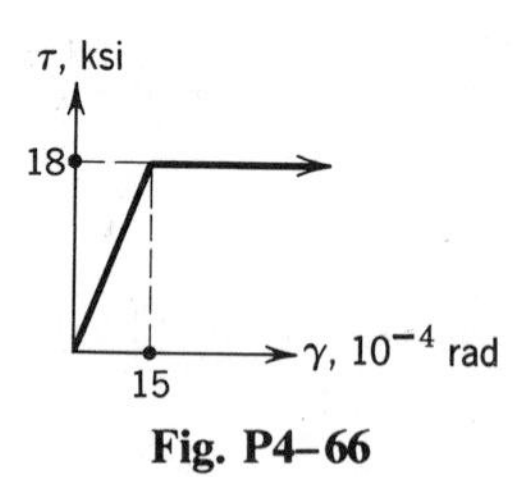

Fig. P4–66

4–67* Two 80-mm diameter solid circular shafts are rigidly connected and supported as shown in Fig. P4–67. An unknown torque T is applied at the junction of the two shafts as indicated. The allowable shearing stresses are 130 MN/m^2 for the steel and 40 MN/m^2 for the bronze. The modulus of rigidity for the bronze is 40 GN/m^2. Determine the maximum allowable magnitude for T.

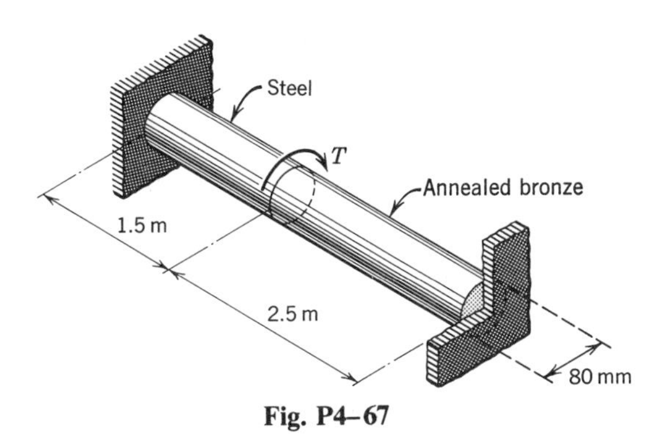

Fig. P4–67

4–68 The shaft of Fig. P4–68 is composed of an aluminum alloy 7075-T6 (G=4000 ksi and proportional limit in shear=33 ksi) segment AB securely connected to the steel (see Fig. P4–66 for stress-strain diagram) segment BC. Determine:

(a) The torque T that will induce a maximum shearing stress of 8 ksi in the aluminum alloy.

(b) The rotation of the section at B under the torque evaluated in part (a).

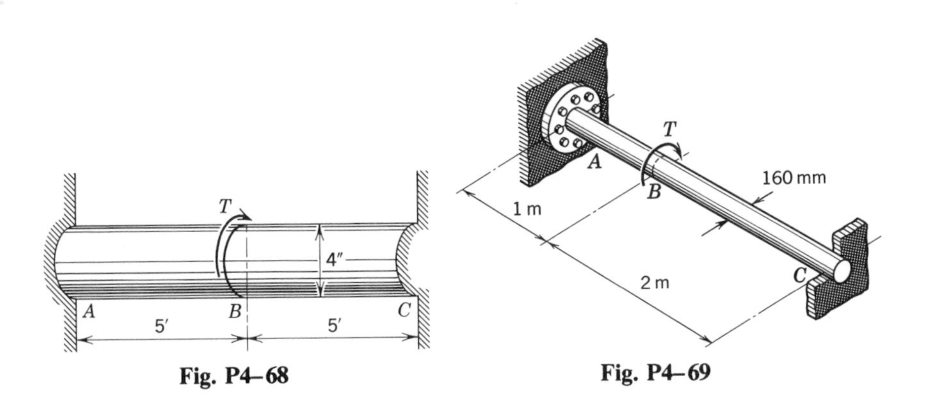

Fig. P4–68 **Fig. P4–69**

4–69* The aluminum alloy (G=28 GN/m^2) shaft of Fig. P4–69 is rigidly fixed to the wall at C, but the flange at A allows the left end of the shaft to rotate 0.010 rad before the bolts provide rigid support. Determine the maximum torque that can be applied at section B if the shearing stress is not to exceed 50 MN/m^2 in either section of the shaft.

4–70 A disk and two circular shafts are connected and supported between the two rigid walls of Fig. P4–70. What maximum permissible torque may be applied to the disk, which is securely attached to B but rotates through 0.005 rad before engaging A?

Shaft	Material	Diameter (in.)	Length (in.)	Modulus of Rigidity (ksi)	Allowable Shearing Stress (ksi)
A	Bronze	2	30	6,000	9
B	Steel	3	45	12,000	16

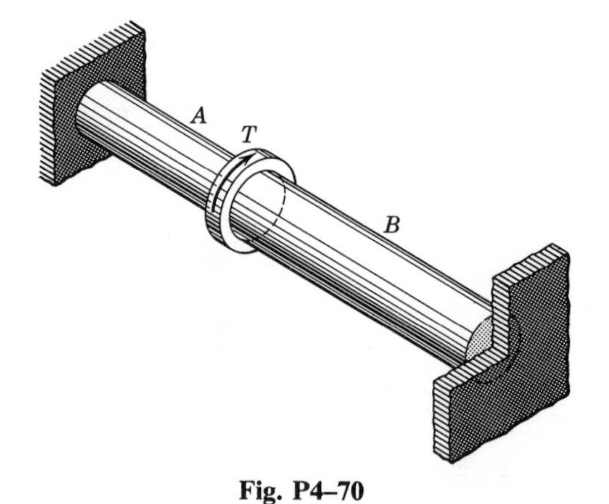

Fig. P4–70

4–71* A torque T of 8000 N-m is applied to the steel shaft of Fig. P4–71 without the brass shell. The brass shell is then slipped in place and attached to the steel, after which the torque T is released. Determine the torque in the brass shell after release of the original torque. The modulus of rigidity of the brass is 40 GN/m^2.

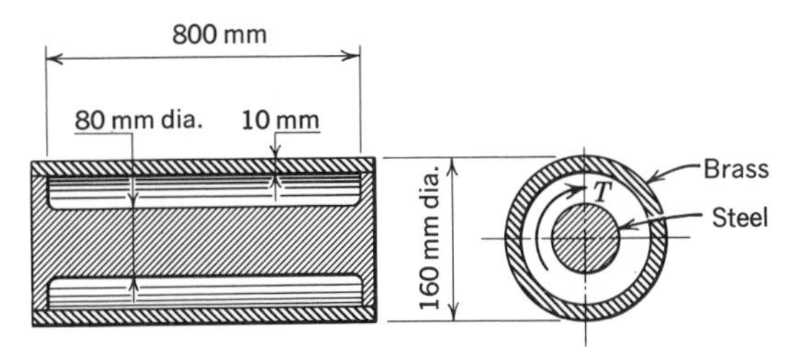

Fig. P4–71

4–72* In Fig. P4–72, the shaft segment AB is initially not connected to segment BC. The initial torque T_o is applied at D, and a secure connection between the two segments is then made at B, after which the torque T_o is removed. Determine the resulting maximum shearing stress in segment BC after torque T_o is removed. The moduli of rigidity are 6000 ksi for AB and 12,000 ksi for BC.

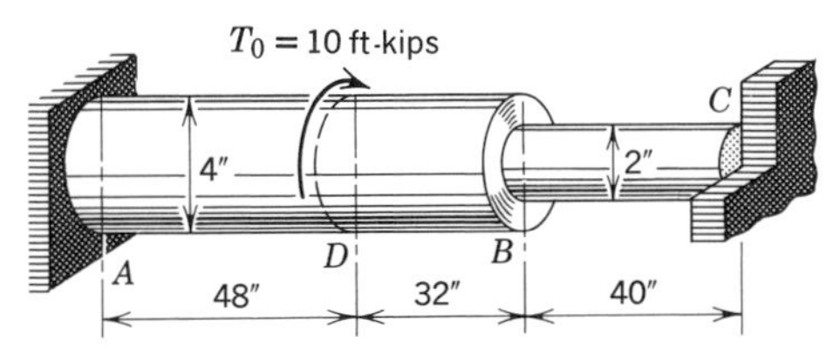

Fig. P4–72

4–73 The shaft of Fig. P4–73 is made up of a 2-m hollow steel section AB and a 1.5-m solid section of aluminum alloy CD ($G = 28\ \text{GN/m}^2$). The torque T_o of 14π kN-m is applied initially only to the steel section AB. Section CD is then connected and the torque T_o is released. When the torque is released, the connection slips 0.010 rad before the aluminum section CD takes any load. Determine the final maximum shearing stress in the aluminum alloy.

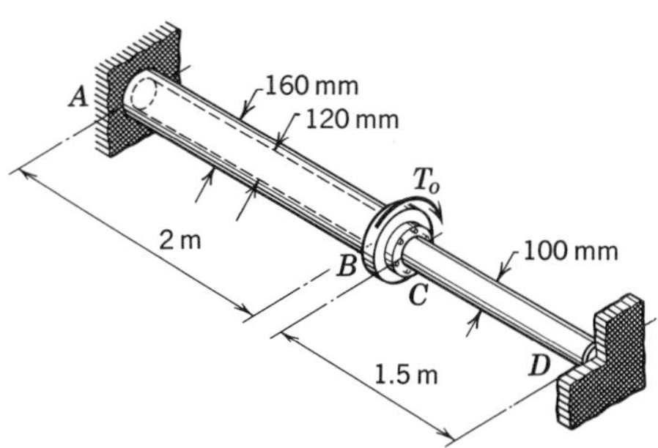

Fig. P4–73

4–74* In Fig. P4–64, change the length of the solid segment from 2 m to 15 ft and the length of the hollow segment from 3 m to 10 ft. Also change the diameters from 100 mm and 50 mm to 4 in. and 2 in. Let the torque be unknown. The steel has the shear stress-strain diagram of Fig. P4–66. Determine the minimum torque, applied as shown, that will produce completely plastic action in the hollow segment.

4–75 A steel shaft 7 ft long extends through, and is attached to, a hollow bronze shaft 4 ft long, as shown in Fig. P4–75. Both shafts are fixed at the wall. Use 6000 ksi for the modulus of rigidity of bronze. Determine:

(a) The maximum shearing stress in each material.
(b) The angle of twist of the right end.

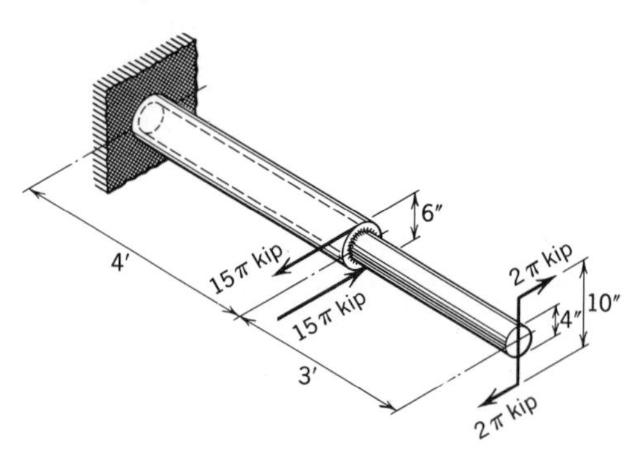

Fig. P4–75

4–76* The inner surface of the aluminum alloy ($G = 27\ \text{GN/m}^2$) sleeve A and the outer surface of the steel shaft B of Fig. P4–76 are smooth. Both the sleeve and the shaft are rigidly fixed to the wall at D. The 10-mm diameter pin C fills a hole drilled completely through a diameter of the sleeve and shaft. The shearing deformation in the pin and the bearing deformation between the pin and shaft can be neglected. Calculate the maximum torque T that can be applied to the steel shaft as shown without exceeding an average shearing stress of $20\ \text{MN/m}^2$ on the cross-sectional area of the pin at E, the interface between the shaft and sleeve.

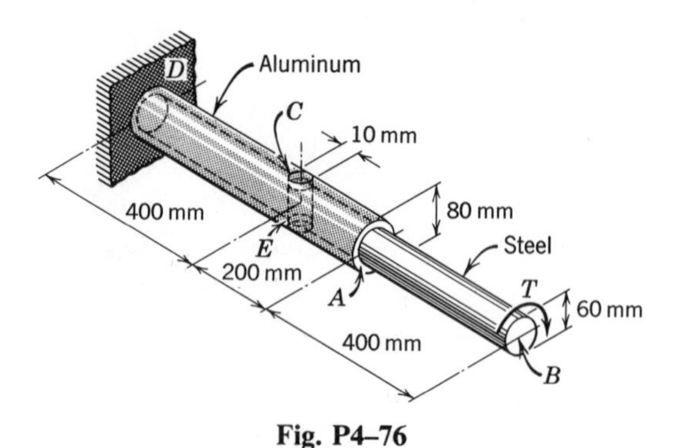

Fig. P4–76

4–77 In the assembly of Prob. 4–76, the hole in the sleeve is drilled oversize so that the pin moves 0.04 mm before it makes contact at the outside surface of the aluminum. Determine:

(a) The maximum torque T, as requested in Prob. 4–76.

(b) The angle of twist of the right end of shaft B relative to the wall.

4–78* The 6-in. diameter steel shaft of Fig. P4–78 has a 4-in. diameter bronze core ($G=6000$ ksi) inserted in 4 ft of the right end and securely bonded to the steel. When the torques of $13{,}000\pi$ and 6000π ft-lb are applied as shown, determine:

(a) The maximum shearing stress in the steel.

(b) The rotation of the free end of the shaft.

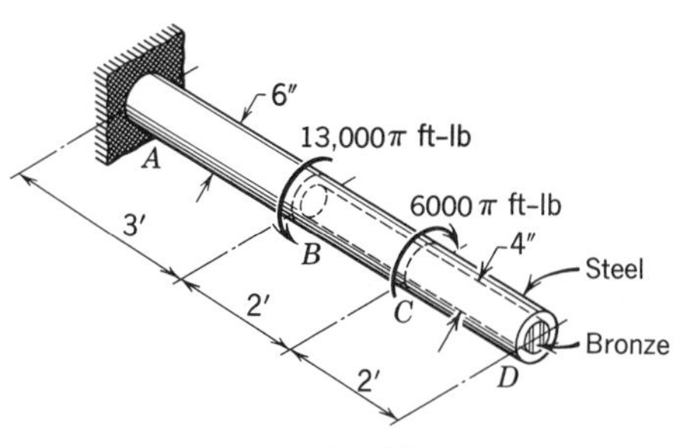

Fig. P4–78

4–79 Solve Prob. 4–78 with the following changes: change 6000π to $25{,}000\pi$ and $13{,}000\pi$ to $43{,}000\pi$, both torques applied as shown. The steel has the shear stress-strain diagram of Fig. P4–66, and the bronze is cold rolled with a proportional limit of 35 ksi.

4–80 The 100-mm diameter shaft of Fig. P4–80 is composed of a brass segment ($G=40$ GN/m^2) and a steel segment. Determine the maximum permissible magnitude for the torque T applied at C if the allowable shearing stresses are 35 MN/m^2 for the brass and 90 MN/m^2 for the steel.

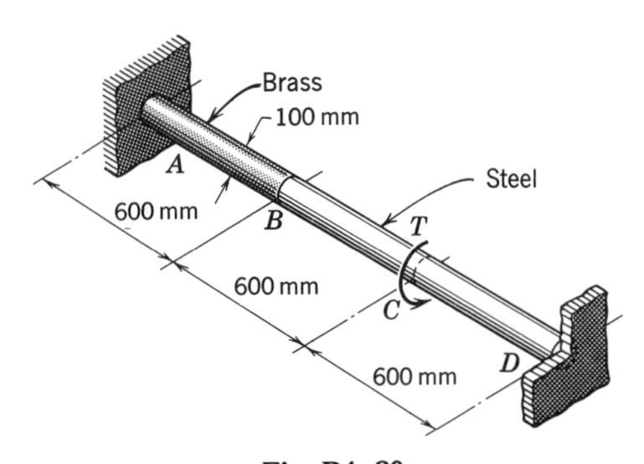

Fig. P4–80

4–81* Solve Prob. 4–80 with a second torque of magnitude $2T$ applied at B with the same sense as that of the torque shown.

4–82 The torsional assembly of Fig. P4–82 consists of an aluminum alloy ($G = 4000$ ksi) segment AB securely connected to a steel segment BCD by means of a flange coupling with four bolts. The diameters of both segments are 3 in., the cross-sectional area of each bolt is 0.2 sq in., and the bolts are located 2.5 in. from the center of the shaft. If the cross shearing stress in the bolts is limited to 8000 psi, determine:

(a) The maximum permissible magnitude for the torque T applied at C.
(b) The maximum shearing stress in the steel.

Neglect the deformation in the coupling.

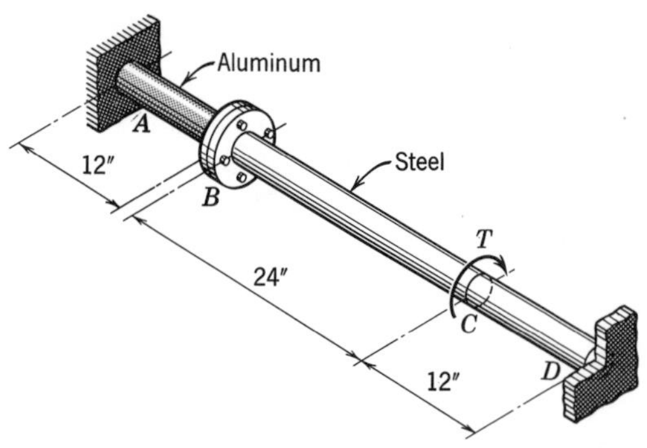

Fig. P4–82

4–83* The solid circular shafts A (aluminum alloy, $G = 28$ GN/m^2) and B and C (steel) are fixed at the ends and connected by the linkage at E (see Fig. P4–83). The

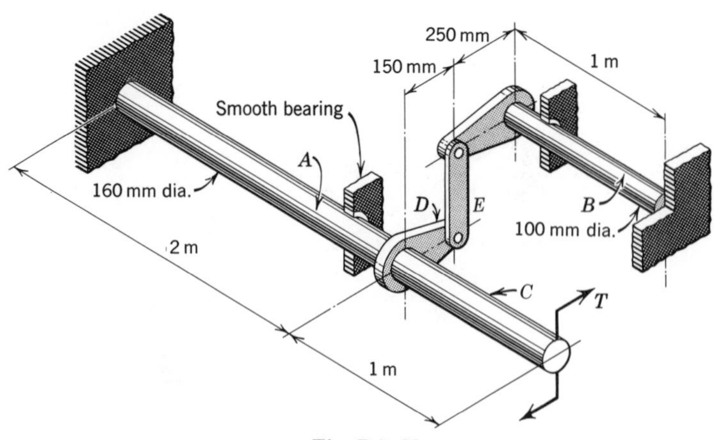

Fig. P4–83

play in the connection permits arm D to rotate through 0.005 rad before the slack is taken up. Neglect the elastic deformation of the linkage and compute the maximum permissible torque T if the torsional shearing stresses are not to exceed 60 MN/m^2 in the aluminum alloy and 90 MN/m^2 in the steel.

4–84 The circular shaft of Fig. P4–84 consists of a steel segment ABC securely connected to a bronze ($G = 6000$ ksi) segment CD, with ends A and D fixed. Determine the maximum shearing stresses in the steel and the bronze when T has a magnitude of 42π in-kips.

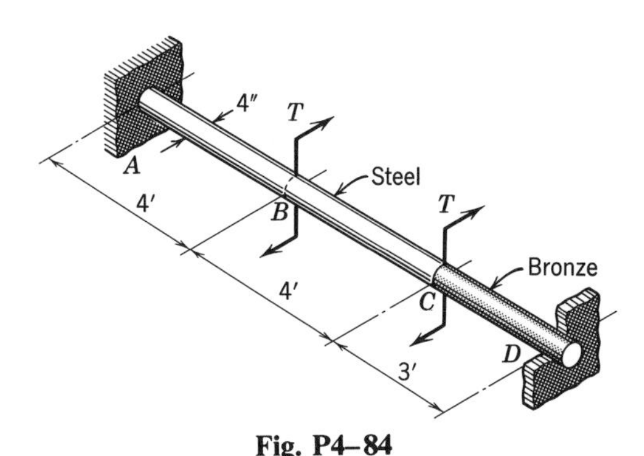

Fig. P4–84

4–85* For the shaft of Prob. 4–84, let T be unknown; let the steel have the shear stress-strain diagram of Fig. P4–66 and the bronze ($G = 6000$ ksi) be cold rolled with a proportional limit of 35 ksi. Under the applied torques, the rotation of the section at B is 0.072 rad in the direction of T. Determine the maximum shearing stresses in the steel and in the bronze.

4–86* The composite shaft of Fig. P4–86a is composed of a hollow steel shell and a solid core of cold-rolled bronze. The steel has the shear stress-strain diagram of Fig. P4–86b, and the bronze has a shearing proportional limit of 240 MN/m^2 and a modulus of rigidiy of 40 GN/m^2. The two components are fastened securely at the left end to a rigid support and at the right end to a rigid bar, to which the

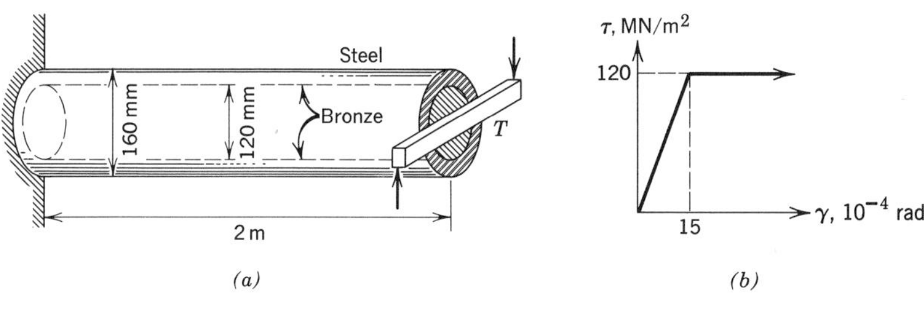

Fig. P4–86

torque T of $40{,}000\pi$ N-m is applied. Determine:

(a) The maximum shearing stresses in the steel and in the bronze.
(b) The angle of rotation of the right end.

4–9 NONCIRCULAR SECTIONS

Prior to 1820, when A. Duleau published experimental results to the contrary, it was thought that the shearing stresses in any torsionally loaded member were proportional to the distance from its axis. Duleau proved experimentally that this is not true for rectangular cross sections. An examination of Fig. 4–17 will verify Duleau's conclusion. If the stresses in the rectangular bar were proportional to the distance from its axis, the maximum stress would occur at the corners. However, if there were a stress of any magnitude at the corner, as indicated in Fig. 4–17*a*, it could be resolved into the components indicated in Fig. 4–17*b*; if these components existed, the two components shown dashed would also exist. These last components cannot exist, since the surfaces on which they are shown are free boundaries. Therefore, the shearing stresses at the corners of the rectangular bar must be zero.

The first correct analysis of the torsion of a prismatic bar of noncircular cross section was published by Saint Venant in 1855; however, the scope of

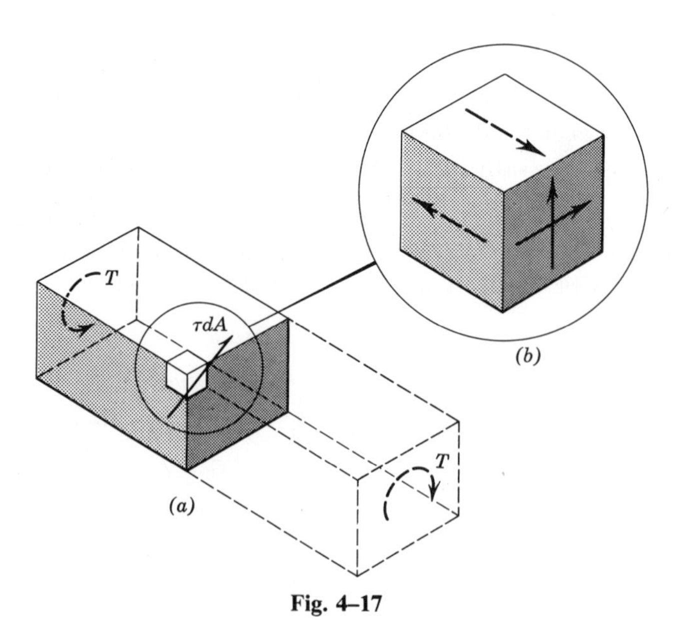

Fig. 4–17

this analysis is beyond the elementary discussions of this book.[3] The results of Saint Venant's analysis indicate that, in general, *except for members with circular cross sections*, every section will warp (not remain plane) when the bar is twisted.

For the case of the rectangular bar shown in Fig. 4–4*d*, the distortion of the small squares is maximum at the midpoint of a side of the cross section and disappears at the corners. Since this distortion is a measure of shearing strain, Hooke's law requires that the shearing stress be maximum at the midpoint of a side of the cross section and zero at the corners. Equations for the maximum shearing stress and angle of twist for a rectangular section obtained from Saint-Venant's theory are

$$\tau_{\max} = \frac{T}{\alpha a^2 b}$$

$$\theta = \frac{TL}{\beta a^3 bG}$$

where a and b are the lengths of the short and long sides of the rectangle, respectively. The numerical factors α and β can be obtained from Fig. 4–18.

Since the mathematical analysis of torsional stresses in noncircular solid sections is relatively difficult, other means such as the *membrane analogy*

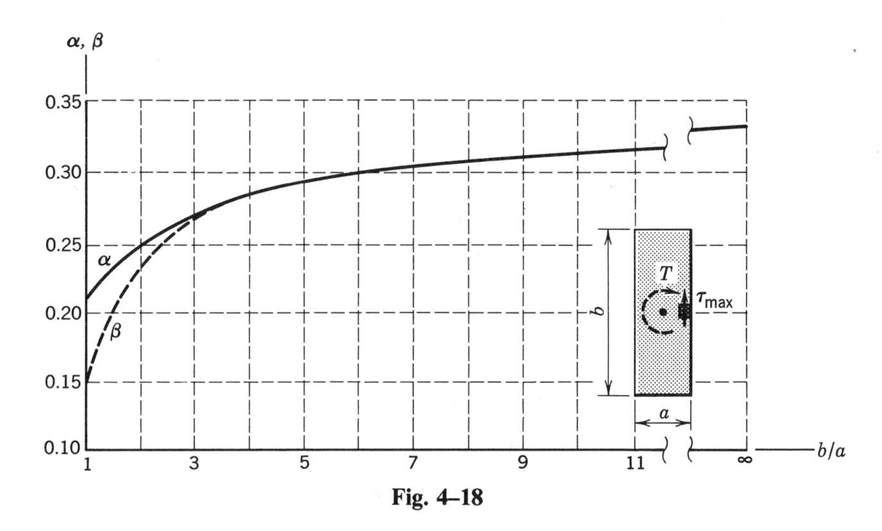

Fig. 4–18

[3]A complete discussion of this theory is presented in various books, such as *Mathematical Theory of Elasticity*, I. S. Sokolnikoff, Second Edition, McGraw-Hill, New York, 1956, pp. 109–134.

are frequently used to identify regions of large shearing stress. The membrane analogy is derived from the observation that the differential equation of the deflection of a membrane has the same form as the theoretical equation (from the theory of elasticity) for the shearing stresses in any uniform rod subjected to twisting moments.

When a thin membrane (such as a soap film) is stretched over a plane opening having the shape of the cross section of the bar, a slight pressure difference p applied to it will cause the membrane to distend. It can be shown that the volume between the membrane and the plane of the opening is proportional to the torque required to produce a given angle of twist per unit length of the shaft; and the slope of the membrane at any point is proportional to the shearing stress in the cross section at that point, the direction of the stress being normal to the direction in which the slope is measured. This last observation makes the method very useful in estimating stress distributions, since for many cross sections the shape of the distended membrane may be visualized; hence, points of maximum and minimum shearing stress may be located. For example, Fig. 4–19 indicates several sections through a membrane for a rectangular cross section, and these sections indicate that the shearing stress is zero (where the slope of the membrane must be zero) at the corners and at the center and is a maximum at points a and a. Thus, one observes that for a rectangular section, the maximum shearing stress is in the boundary material closest to the axis of the shaft rather than in the most distant material. In like manner, one can visualize the location of maximum torsional stresses in shaft keyways and how the addition of fillets in the reentrant corners of the keyway reduces the magnitude of the stress. The

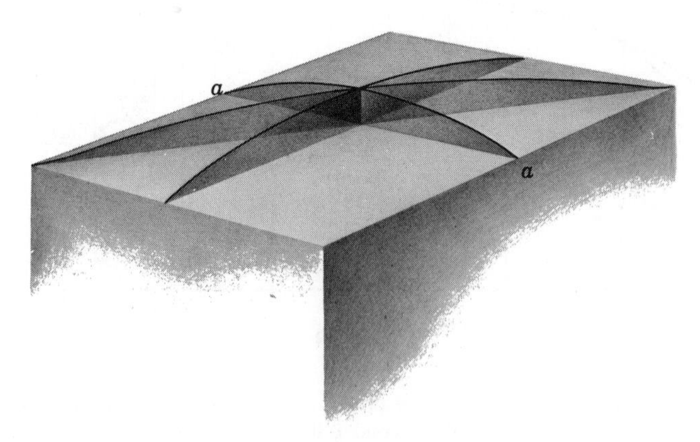

Fig. 4–19

discussion of the techniques involved in the evaluation of stresses by means of the membrane analogy will be left to others.[4]

4–10 SHEAR FLOW

Although the elementary torsion theory presented in Art. 4–1 is limited to circular sections, one class of noncircular sections that can be readily analyzed by elementary methods is the thin-walled section such as the one illustrated in Fig. 4–20*a* which represents a noncircular section with a wall of variable thickness (t varies).

A useful concept associated with the analysis of thin-walled sections is *shear flow* q, defined as internal shearing force per unit of length of the thin section. Typical units for q are pounds per inch. In terms of stress, q equals $\tau(1)(t)$, where τ is the average shearing stress across the thickness t. It will be demonstrated that the shear flow on a cross section is constant even though the thickness of the section wall varies. Figure 4–20*b* shows a block cut from the member of Fig. 4–20*a* between A and B, and as the member is subjected to pure torsion, the shear forces $V_1 \cdots V_4$ alone (no normal forces) are necessary for equilibrium. Summing forces in the x direction gives

$$V_1 = V_3$$

or

$$q_1\,dx = q_3\,dx$$

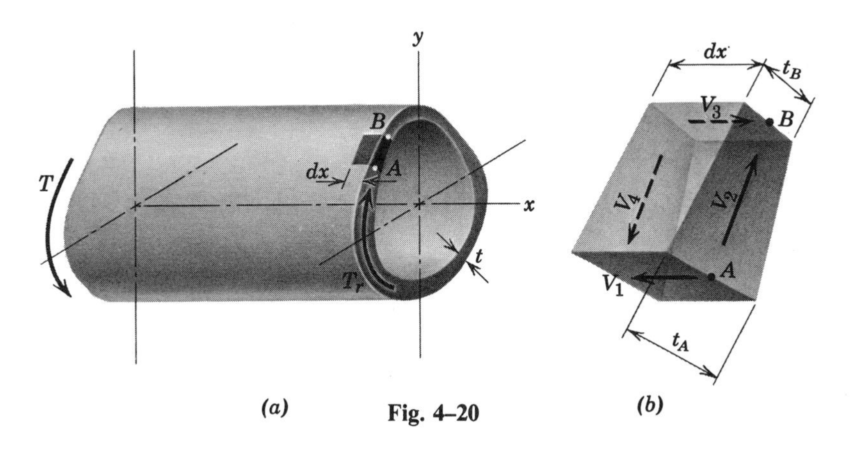

(*a*) Fig. 4–20 (*b*)

[4]*Handbook of Experimental Stress Analysis*, M. Hetenyi, Wiley, New York, 1950.

from which

$$q_1 = q_3$$

and, as $q = \tau(t)$,

$$\tau_1 t_A = \tau_3 t_B \tag{a}$$

The shearing stresses at point A on the longitudinal and transverse planes have the same magnitude; likewise, the shearing stresses at point B have the same magnitude on the two orthogonal planes; hence, Eq. (a) may be written

$$\tau_A t_A = \tau_B t_B$$

or

$$q_A = q_B$$

which was to be proved.

To develop the expression for the resulting torque on the section, consider the force dF acting through the center of a differential length of perimeter ds, as shown in Fig. 4–21. The resisting torque T_r is the resultant of the forces dF; that is,

$$T_r = \int (dF)\rho = \int (q\,ds)\rho = q\int \rho\,ds$$

This integral may be difficult to evaluate by formal calculus; however, the quantity $\rho\,ds$ is twice the area of the triangle shown shaded in Fig. 4–21, which makes the integral equal to twice the area A *enclosed by the median*

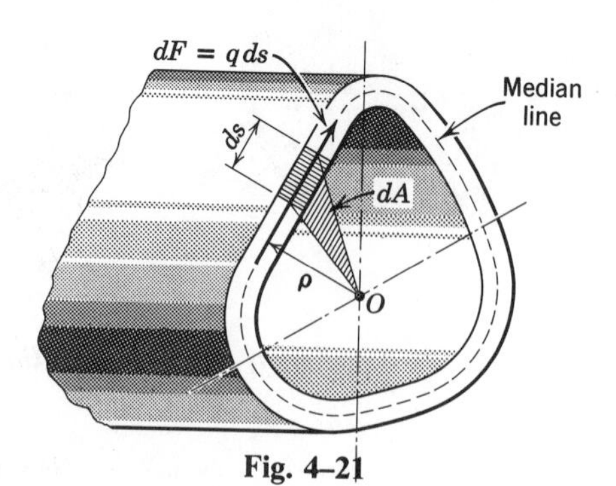

Fig. 4–21

line. The resulting expression is

$$T_r = q(2A)$$

or, in terms of stress,

$$\tau = T/(2At) \tag{4–4}$$

where τ is the *average* shearing stress across the thickness t (and tangent to the perimeter), and is reasonably accurate when t is relatively small. For example, in a round tube with a diameter to wall thickness ratio of 20, the stress as given by Eq. 4–4 is 5 percent less than that given by the torsion formula. It must be emphasized that Eq. 4–4 applies only to "closed" sections; that is, sections with a continuous periphery. If the member were slotted longitudinally (see, for example, Fig. 4–22), the resistance to torsion would be diminished considerably from that for the closed section.

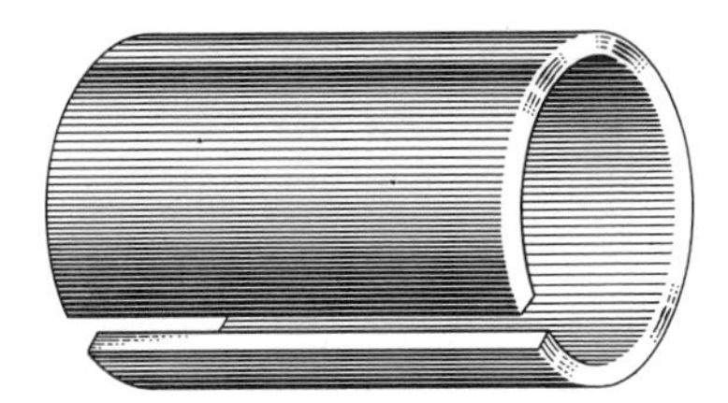

Fig. 4–22

It is sometimes necessary to evaluate the twist of a thin-walled member subjected to torsion. Since the ratio γ/ρ for a section such as that of Fig. 4–21 is not a constant as it is for a circular section, Eq. 4–2 ($\theta = L\gamma/\rho$) is not easily adapted to such a section. Instead, the principle of strain energy will be found most useful in the development of an expression for the angle of twist.

Assuming that energy is conserved, the work done by the torque in twisting a member through an angle θ is equal to the strain energy stored in the member, and if the torque-twist relation is linear (which limits stresses to values within the proportional limit), the expression becomes

$$W_k = U = \frac{T\theta}{2} = \int_{\text{Vol.}} u\,dv \tag{b}$$

where u is the strain energy per unit volume of the material in the member. All elements of the member are subjected to pure shear (see Fig. 4–20);

therefore, the expression for u is $u=\tau^2/(2G)$ (see Art. 2–9) and Eq. (b) becomes

$$U=\int_{\text{Vol.}}\frac{\tau^2}{2G}\,dv \tag{c}$$

When τ in Eq. (c) is replaced by Eq. 4–4 and dv by $t\,dx\,ds$ (see Figs. 4–20 and 4–21), the expression becomes

$$U=\int_{\text{Vol.}}\frac{T^2}{8A^2tG}\,dx\,ds \tag{d}$$

In general, t is a function of s, but for many practical problems, the integral may be replaced by a summation of n elements, each having a constant thickness t_i and a length L; thus,

$$U=\sum_{i=1}^{n}\frac{T^2L}{8A^2G}\left(\frac{\Delta s_i}{t_i}\right) \tag{e}$$

Substituting Eq. (e) into Eq. (b) yields

$$\theta=\frac{TL}{4A^2G}\sum_{i=1}^{n}\frac{\Delta s_i}{t_i} \tag{4–5a}$$

or, in terms of the shear flow q,

$$\theta=\frac{qL}{2AG}\sum_{i=1}^{n}\frac{\Delta s_i}{t_i} \tag{4–5b}$$

where $\Delta s_i/t_i$ is the ratio of plate width (measured along the periphery of the section) to thickness. Note that L and A have been treated as constants; hence, the shape and size of the cross section must be maintained constant (possibly by diaphragms, bulkheads, or cross bracing), and all the elements must have the same length. The following example will illustrate the use of Eqs. 4–4 and 4–5.

Example 4–10 A rectangular box section of aluminum alloy, 2024-T4, has outside dimensions 4×2 in. The plate thickness is 0.06 in. for the 2-in. sides and 0.10 in. for the 4-in. sides. Determine the maximum permissible torque to which the section may be subjected if a factor of safety of 2 with respect to failure by slip is specified. For this torque, determine the angle of twist in a 40-in. length of the member.

Solution The allowable stress is $28/2 = 14$ ksi, and this will occur in the thinnest plate; therefore, $q = 14(0.06) = 0.84$ kips/in. The torque is

$$T = 2qA = 2(0.84)(4-0.06)(2-0.10) = \underline{12.58 \text{ in-kips}} \qquad \text{Ans.}$$

The angle of twist is

$$\theta = \frac{LT}{4GA^2} \sum \frac{s_i}{t_i}$$

$$= \frac{40(12.58)}{4(4)(10^3)[3.94(1.90)]^2}\left(\frac{3.94(2)}{0.1} + \frac{1.90(2)}{0.06}\right)$$

$$\underline{= 0.0797 \text{ rad}} \qquad \text{Ans.}$$

PROBLEMS

Note. In all the following problems, assume that buckling of the shell is prevented and that the shape of the section is maintained by ribs, diaphragms, and stiffeners.

4–87* A 20-in. wide by 0.10-in. thick steel sheet is to be formed into a hollow section by bending through 360° and welding the long edges together (butt joint). The shape may be (1) circular, (2) square, or (3) a 6×4-in. rectangle. Assume a median length of 20 in. (no stretching of the sheet due to bending), and square corners for noncircular sections. The allowable shearing stress is 10 ksi. For each of the shapes listed, determine the magnitude of;

(a) The maximum permissible torque.
(b) The angle of twist in a 100-in. length.

4–88* A 600-mm wide by 3-mm thick steel sheet is to be formed into a hollow section by bending through 360° and welding the long edges together (butt joint). The shapes to be considered are (1) circular, (2) square, and (3) an equilateral triangle. Assume a median length of 600 mm (no stretching of the sheet due to bending) and sharp corners for noncircular sections. The allowable shearing stress is 85 MN/m^2. For each of the shapes listed, determine the magnitude of:

(a) The maximum permissible torque.
(b) The angle of twist in a 3-m length.

4–89 Figure P4–89 shows a cross section of the leading edge of an airplane wing. The length along the curve is 26 in. and the enclosed area is 82 sq in. Sheet thicknesses are shown on the diagram; the material is 2024-T4 aluminum alloy. For a torque of 100 in-kips, determine the magnitude of:

(a) The maximum shearing stress.
(b) The angle of twist per foot of length.

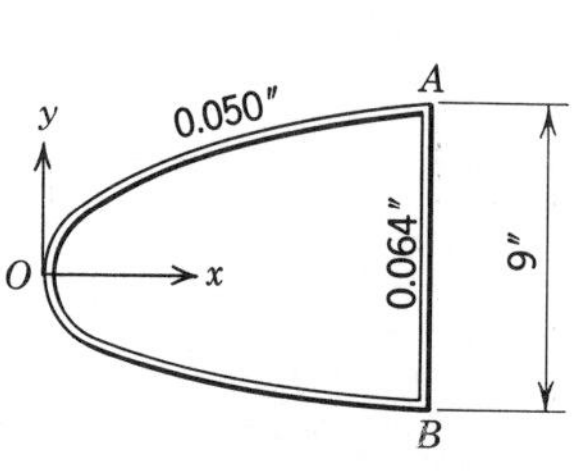

Fig. P4–89

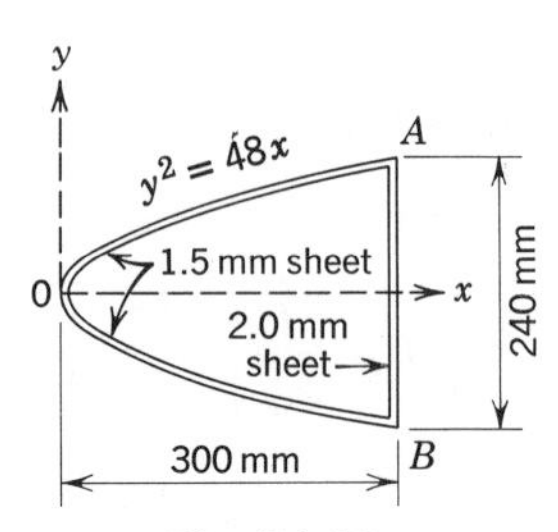

Fig. P4–90

4–90* A torque box of an aluminum alloy ($G = 27$ GN/m^2) has the parabolic cross section of Fig. P4–90. For an allowable shearing stress of 85 MN/m^2, determine the magnitude of:

(a) The maximum permissible torque.
(b) The angle of twist per metre of length.

4–91 Figure P4–91 represents the cross section of an airplane fuselage made of 2024-T4 aluminum alloy. The sheet thicknesses are 0.064 in. from A to B and C to D, 0.050 in. from B to C, and 0.040 in. from D to A. The fuselage is subjected to a torque of 1500 in-kips. Determine:

(a) The maximum shearing stress.
(b) The angle of twist per foot of length of fuselage.

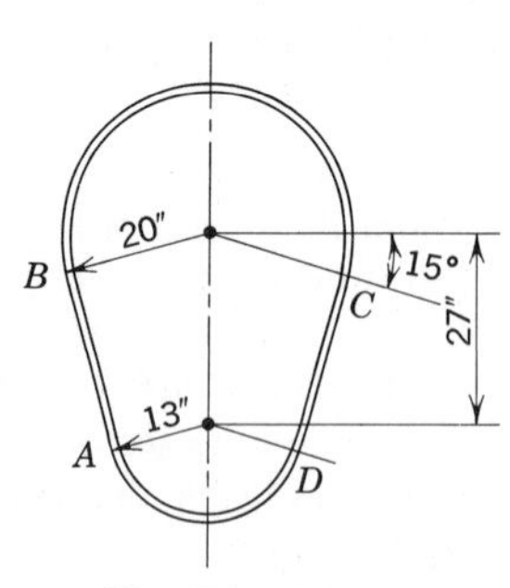

Fig. P4–91

4–92* Figure P4–92 represents the cross section of an airplane fuselage made of an aluminum alloy ($G = 27\ \mathrm{GN/m^2}$). The sheet thicknesses are 1.5 mm from A to B and C to D, 1.2 mm from B to C, and 1.0 mm from D to A. For an allowable shearing stress of 85 $\mathrm{MN/m^2}$, determine the magnitude of:

(a) The maximum permissible torque.
(b) The angle of twist per metre of length.

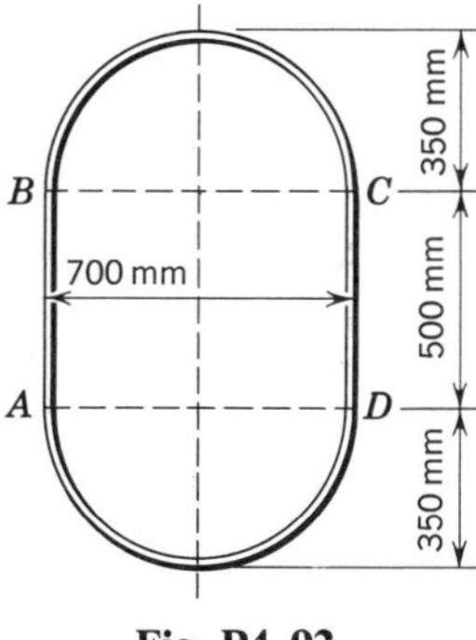

Fig. P4–92

4–93* Figure P4–93 shows the cross section of a torque box for an airplane wing. The top and bottom sheets are 40 in. long and the mean depth of the box is 11.2 in. Sheet thicknesses are shown on the figure; the material is 2024-T4 aluminum alloy. For a shear flow of 500 lb/ in., determine the magnitude of:

(a) The torque applied to the box.
(b) The maximum shearing stress in each sheet.
(c) The angle of twist per foot of wing.

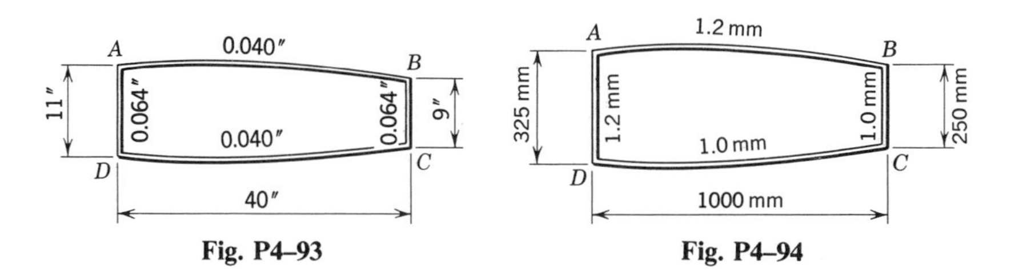

Fig. P4–93 **Fig. P4–94**

4–94 Figure P4–94 shows the cross section of an airplane wing torque box. The lengths of the top and bottom sheets are 1005 mm and the mean depth of the box is 350 mm. Sheet thicknesses are shown on the figure; the material is an aluminum alloy ($G = 27\ \mathrm{GN/m^2}$). For an applied torque of 50 kN-m, determine the magnitude of:

(a) The maximum shearing stress in each sheet.
(b) The angle of twist per metre of wing.

4–95 If an airplane fuselage with the cross section of Fig. P4–91 is subjected to a torque of 1000 in-kips, and the allowable shearing stress is 10 ksi, what is the minimum thickness of sheet (constant for the entire periphery) needed to resist the torque? For the given data, what is the magnitude of the angle of twist per foot of fuselage if the material is 2024-T4 aluminum alloy?

5 Flexural Loading: Stresses

5–1 INTRODUCTION

The beam, or flexural member, is frequently encountered in structures and machines, and its elementary stress analysis constitutes one of the more interesting facets of mechanics of materials. A beam is a member subjected to loads applied transverse to the long dimension, causing the member to bend. For example, Fig. 5–1 is a photograph of an aluminum alloy I-beam, AB, simply supported in a testing machine and loaded at the one-third points, and Fig. 5–2 depicts the shape (exaggerated) of the beam when loaded.

Before proceeding with a discussion of stress analysis for flexural members, it may be well to classify some of the various types of beams and loadings encountered in practice. Beams are frequently classified on the basis of the supports or reactions. A beam supported by pins, rollers, or smooth surfaces at the ends is called a *simple beam*. A simple support will develop a reaction normal to the beam but will not produce a couple. If either or both ends of a beam project beyond the supports, it is called a *simple beam with overhang*. A beam with more than two simple supports is a *continuous beam*. Figure 5–3*a*, *b*, and *c* shows, respectively, a simple

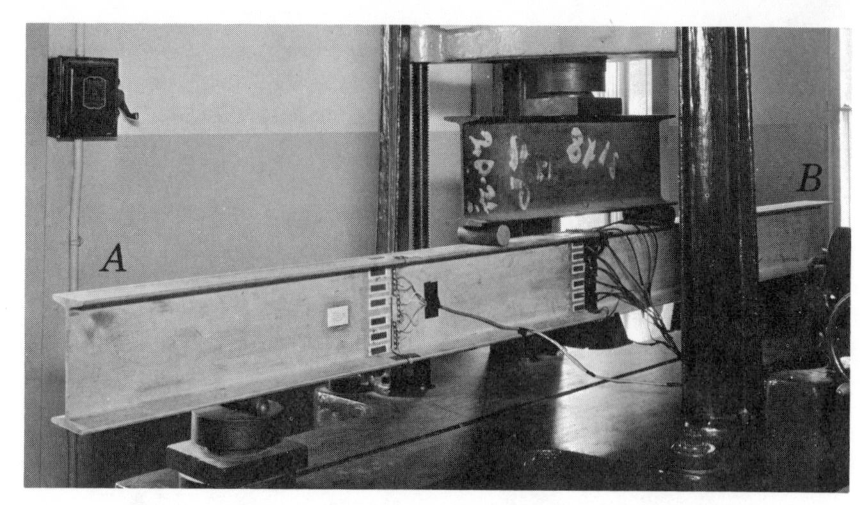

Fig. 5–1 Setup for measuring longitudinal strains in a beam.

beam, a beam with overhang, and a continuous beam. A *cantilever beam* is one in which one end is built into a wall or other support so that the built-in end cannot move transversely or rotate. The built-in end is said to be *fixed* if no rotation occurs and *restrained* if a limited amount of rotation occurs. The supports shown in Fig. 5–3d and e represent fixed ends unless otherwise stated. The beams in Fig. 5–3d, e, and f are, in order, a cantilever beam, a beam fixed (or restrained) at the left end and simply supported near the other end (which has an overhang), and a beam fixed (or restrained) at both ends.

Cantilever beams and simple beams have only two reactions (two forces or one force and a couple), and these reactions can be obtained from a free-body diagram of the beam by applying the equations of equilibrium. Such beams are said to be *statically determinate* since the reactions can be obtained from the equations of equilibrium. Continuous and other beams, with only transverse loads, with more than two reaction components are

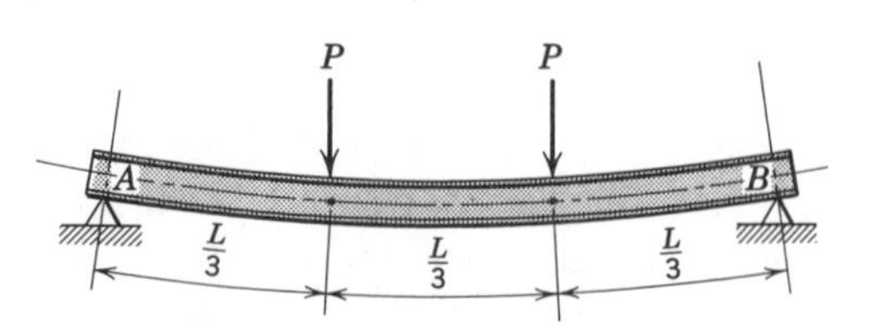

Fig. 5–2

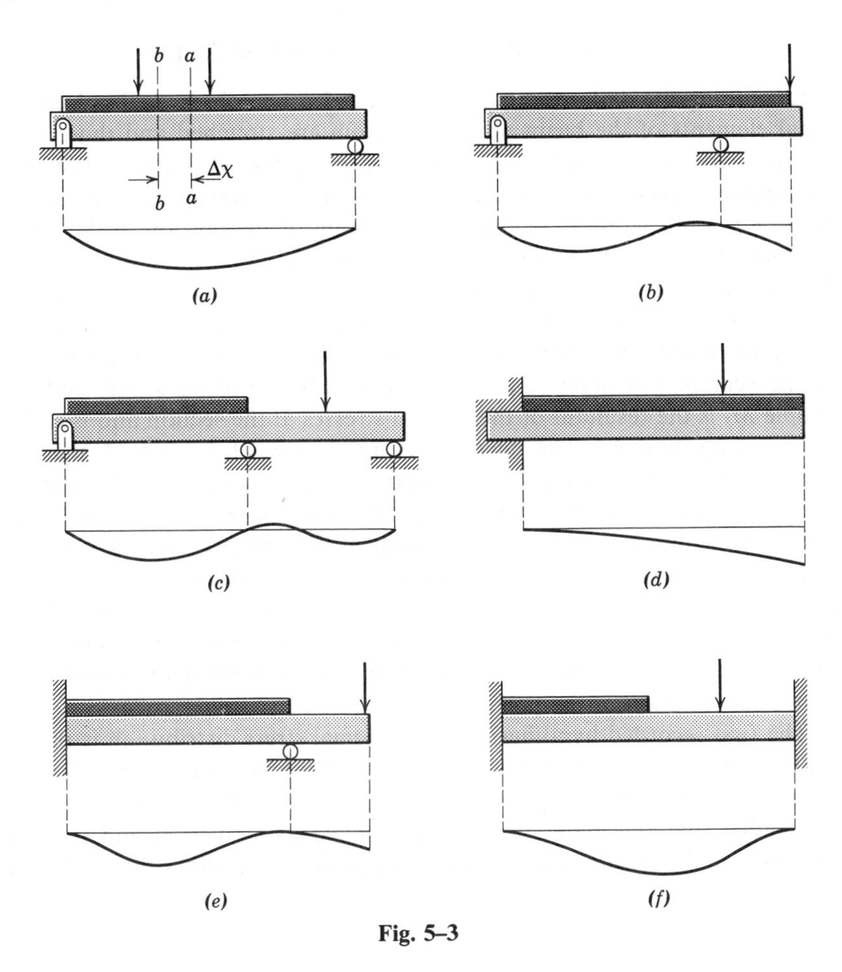

Fig. 5–3

called *statically indeterminate* since there are not enough equations of equilibrium to determine the reactions. Statically indeterminate beams are discussed in Chap. 7.

The beams shown in Fig. 5–3 are all subjected to uniformly distributed loads and to concentrated loads, and, although shown as horizontal, they may have any orientation. Distributed loads will be shown on the side of the beam on which they are acting; that is, if drawn on the bottom of the beam, the load is pushing upward, and if drawn on the right side of a vertical beam, the load is pushing to the left. The deflection curves shown beneath the beams are greatly exaggerated to assist in visualizing the shape of the loaded beam.

5–2 FLEXURAL STRESSES—CONCEPTS

Examining the deflection curve of Fig. 5–3a, one observes that longitudinal elements of the beam near the bottom are stretched and those near the top are compressed, indicating the existence of both tensile and compressive stresses on transverse planes. These stresses are designated *fiber* or *flexural* stresses. A free-body diagram of the portion of the beam between the left end and plane a–a is shown in Fig. 5–4. A study of this diagram reveals that a transverse force V_r and a couple M_r at the cut section and a force R (a reaction) at the left support are needed to maintain equilibrium. The force V_r is the resultant of the shearing stresses at the section (on plane a–a) and is called the resisting shear, and the couple M_r is the resultant of the normal stresses at the section and is called the resisting moment. The magnitudes and senses of V_r and M_r may be obtained from the equations of equilibrium $\Sigma F_y = 0$ and $\Sigma M_O = 0$ where O is any axis perpendicular to plane xy (the reaction R must be evaluated first from a free-body of the entire beam). For the present, the shearing stresses will be ignored while the normal stresses are studied. The magnitudes of the normal stresses can be computed if M_r is known and also if the law of variation of the normal stresses on the plane a–a is known. At this time, a brief résumé of the history of this problem is presented for a better appreciation of the problem.

The first recorded hypothesis on the normal stress distribution is that of Galileo Galilei (1564–1642) who, possibly observing that beams of stone when subjected to bending loads broke somewhat as indicated in Fig. 5–5, concluded that the material at point A acted as a fulcrum and that the resultant of the tensile stresses was at the center of the cross section (as for

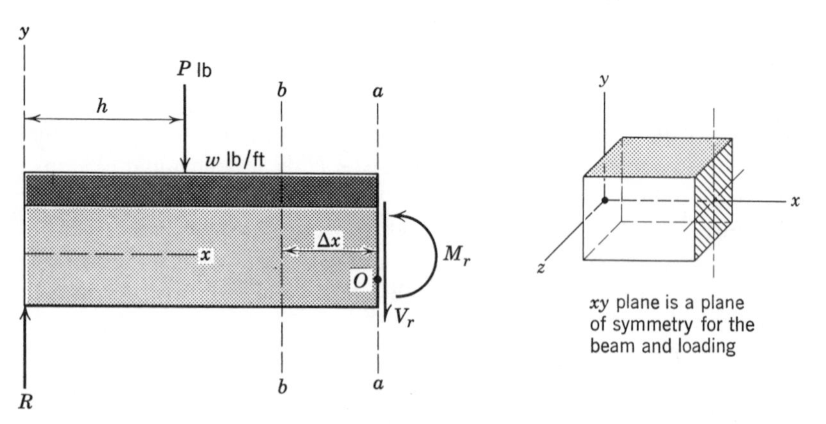

Fig. 5–4

axial loading).[1] This theory would lead to the stress distribution diagram of Fig. 5–6*a* in which F_T is the resultant of the tensile stresses. About fifty years after Galileo's observations, E. Mariotte (1620–1684), a French physicist, still retaining the concept of the fulcrum at the compression surface of the beam, but reasoning that the extensions of the longitudinal elements of the beam (fibers) would be proportional to the distance from the fulcrum, suggested that the tensile stress distribution was as indicated in Fig. 5–6*b*. Mariotte later rejected the concept of the fulcrum and observed that part of the beam on the compression side was subjected to compressive stress having a triangular distribution; however, his expression for the ultimate load was still based on his original concept and was accepted by famous scientists such as Jacob Bernoulli, Leonard Euler, and others until C. A. Coulomb (1736–1806), a French military engineer, in a paper published in 1773 discarded the fulcrum concept and proposed the

Fig. 5–5 Failure of plaster-of-paris beam subjected to flexure.

[1] *Two New Sciences*, Galileo Galilei, 1638. Actually, Galileo's published work contains no record of test results; therefore, the manner in which he arrived at this conclusion is mere speculation.

distribution of Fig. 5–6*c* in which both the tensile and compressive stresses have the same linear distribution.[2]

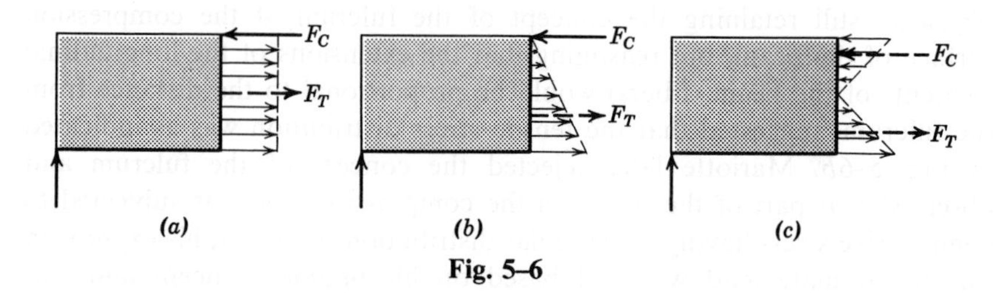

Fig. 5–6

The correctness of the Coulomb theory can be demonstrated as follows. A segment of the beam of Fig. 5–4 between the planes *a–a* and *b–b* is shown in Fig. 5–7 with the distortion greatly exaggerated. The assumption is made that *a plane section before bending remains plane after bending*. For this to be strictly true, it is necessary that the beam be bent only with couples (no shear on transverse planes); the beam must be so proportioned that it will not buckle and the loads applied so that no twisting occurs (this last limitation will be satisfied if the loads are applied in a plane of symmetry—a sufficient though not a necessary condition). One observes in Fig. 5–7 that at some distance c above the bottom of the beam, the longitudinal elements (sometimes referred to as fibers) undergo no change in length. The curved surface formed by these elements is referred to as the *neutral surface*, and the intersection of this surface with any cross section is called the *neutral axis* of the section. All elements (fibers) on one side of the neutral surface are compressed, and those on the opposite side are elongated. The assumption is made that *all longitudinal elements have the same length*. This assumption imposes the restriction that the beam be initially straight and of constant cross section; however, in practice, considerable deviation from these last restrictions is often tolerated. If Fig. 5–7 is referred to,

$$\frac{\delta_y}{y} = \frac{\delta_c}{c}$$

[2]Sixty years before Coulomb's paper, Parent (1666–1716), a French mathematician, proposed the correct triangular stress distribution. However, his work remained in obscurity, and the majority of engineers continued to use formulas based on Mariotte's theory. For a more complete discussion of the history of the flexure problem, the reader is referred to the *History of Strength of Materials*, S. P. Timoshenko, McGraw-Hill, New York, 1953, from which most of the foregoing discussion is taken.

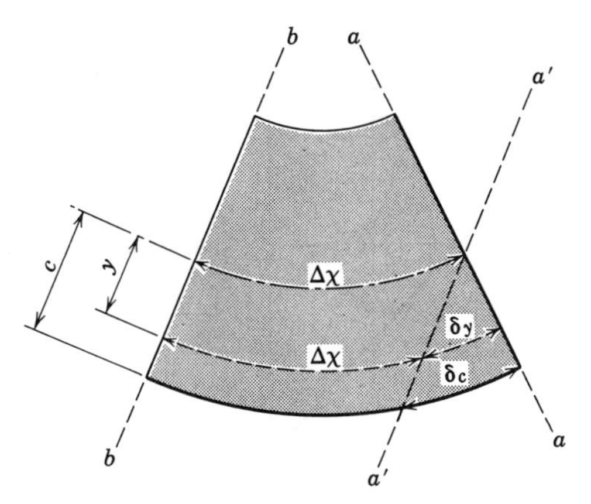

Fig. 5–7

from which

$$\delta_y = \frac{\delta_c}{c} y$$

Since all elements had the same initial length Δx, the strain of any element can be determined by dividing the deformation by the length of the element; the strain becomes

$$\epsilon = \frac{\epsilon_c}{c} y \tag{a}$$

where ϵ is the strain of a longitudinal element at a distance y from the neutral surface, and ϵ_c is the strain of a longitudinal element at the outer surface of the beam. On the basis of the general notation presented in Art. 1–5, both ϵ and ϵ_c should be designated ϵ_x. However, for simplicity, the subscript x is deleted on ϵ_x and σ_x throughout the chapter, except where the general notation is necessary. The preceding equation indicates that the strain of any fiber is directly proportional to the distance of the fiber from the neutral surface. This variation can be demonstrated experimentally by means of strain gages attached to a beam, as shown in Fig. 5–1. The strains, as measured by gages on two different sections, are plotted against the vertical position of the gages on the beam in Fig. 5–8. Curve 1 represents strains on a section at the center of the beam where pure bending occurs (no transverse shear), and curve 2 shows strains at a section near one end of the beam where both fiber stresses and transverse

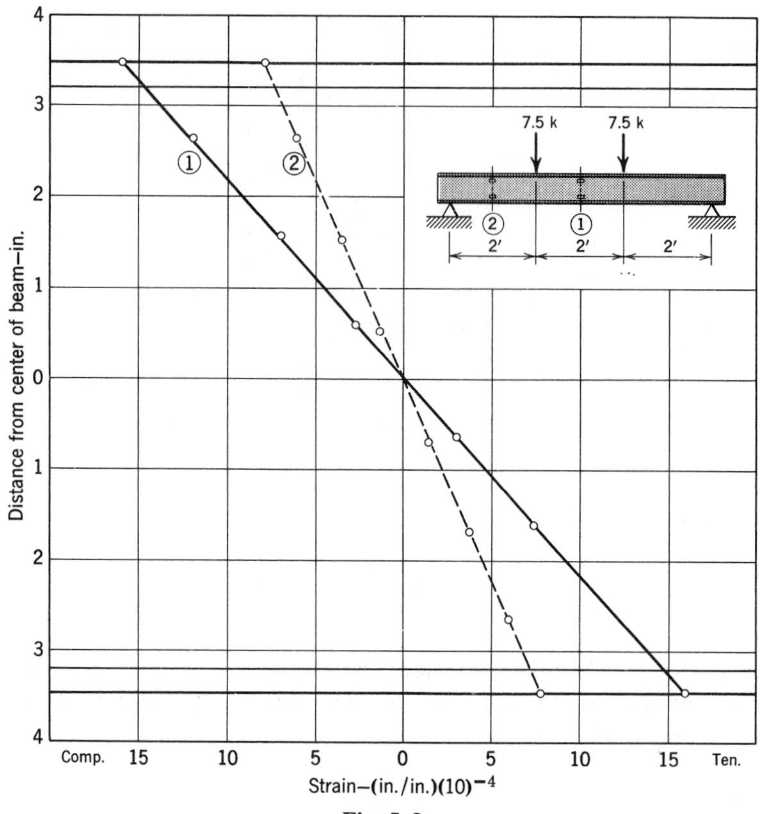

Fig. 5–8

shearing stresses exist. These curves are both straight lines within the limits of the accuracy of the measuring equipment.[3]

Note that Eq. (a) is valid for elastic or inelastic action so long as the beam does not twist or buckle and the transverse shearing stresses are relatively small. Problems in this book will be assumed to satisfy these restrictions.

With the acceptance of the premise that the longitudinal strains ϵ_x are proportional to the distance from the neutral surface, the law of variation

[3]A more exact analysis using principles developed in the theory of elasticity indicates that curve 2 should be curved slightly. *Note*: other experiments indicate that a plane section of an initially curved beam will also remain plane after bending and that the deformations will still be proportional to the distance of the fiber from the neutral surface. The strain, however, will not be proportional to this distance since each deformation must be divided by a different original length.

of the normal stress σ_x on the transverse plane can be determined by referring to a tensile-compressive stress-strain diagram and making the appropriate substitutions into Eq. (a) to obtain the stress distribution diagram σ vs y. For most real materials, the tension and compression stress-strain diagrams are identical in the elastic range; although the diagrams may differ somewhat in the inelastic range, the differences can be neglected for most real problems. For beam problems in this book, the compressive stress-strain diagram will be assumed to be identical to the tensile diagram. As an aid to the thought sequence, it is suggested that sketches be drawn of ϵ vs y, then σ vs ϵ, and finally σ vs y, as illustrated for some typical examples in Fig. 5–9. Note that the y axis is taken positive downward so that a combination of positive moment (defined in Art. 5–6) and positive y gives a positive, or tensile, stress. *The downward positive direction for* y *is used only for computation of stresses. In all other discussions and applications,* y *is positive upward.* Since no load is applied in the z direction, σ_z is zero throughout the beam. Because of loads applied in the y direction, σ_y is not zero throughout the beam; however, for the elementary solution considered here, σ_y is small enough to neglect (see Art. 5–12).

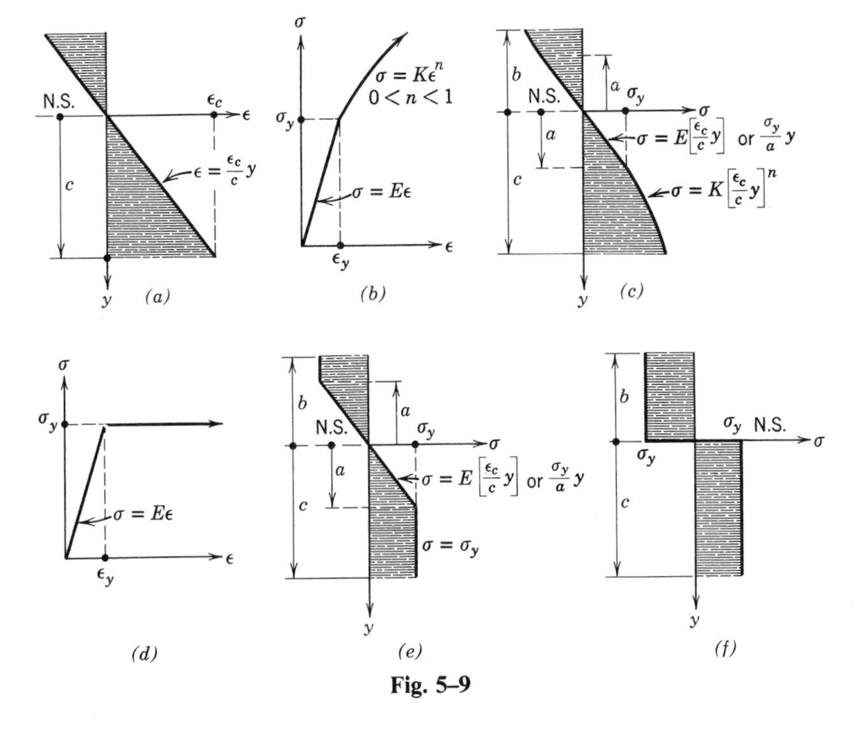

Fig. 5–9

Also, σ_x is uniformly distributed in the z direction because the plane cross section remains plane.

Since the stress-strain diagrams of Fig. 5–9 are for specific materials, the stress distribution diagrams are for the same materials throughout the beam. This imposes the limitation that *the beam be of a homogeneous material*. Figures 5–9*c* and *e* are self explanatory; however, Fig. 5–9*f* might require some explanation. This diagram indicates complete plastic action throughout the section for an ideal elastoplastic material such as that of Fig. 5–9*d*. Actually, there must be some elastic strain in the material near the neutral surface; however, the contribution of the stresses near the neutral surface to the resisting moment is so small that the assumption of plastic action as indicated in Fig. 5–9*f* is valid.

With the law of variation of flexural stress known, Fig. 5–4 can now be redrawn as in Fig. 5–10. The vectors F_C and F_T are the resultants of the compressive and tensile flexural stresses, and, because the sum of the forces in the x direction must be zero, F_C is equal to F_T; hence, they form a couple of magnitude M_r.

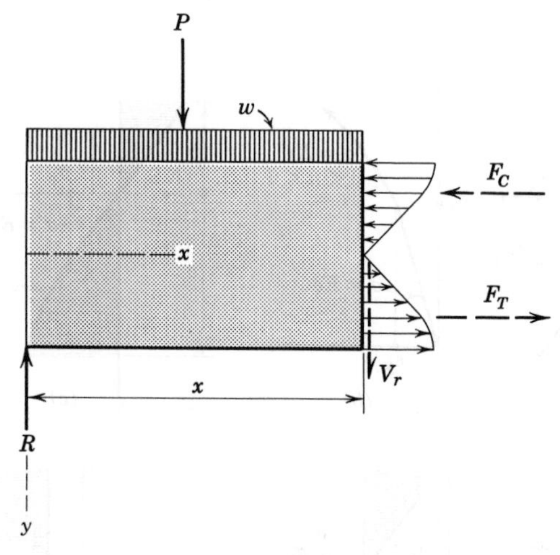

Fig. 5–10

5–3 FLEXURAL STRESSES—SYMMETRICAL BENDING

Referring to Fig. 5–11, the resisting moment M_r (which equals the applied moment M) is

$$M_r = \int_A dF y = \int_A \sigma \, dA \, y \tag{a}$$

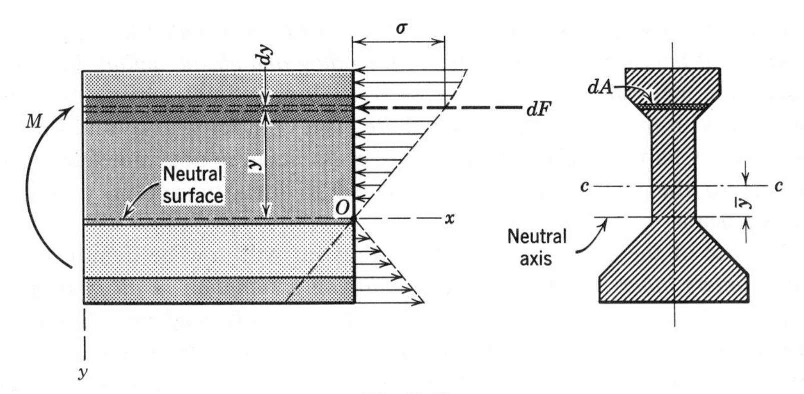

Fig. 5–11

and since y is measured from the neutral surface, it is first necessary to locate this surface by means of the equilibrium equation $\Sigma F_x = 0$, which gives

$$\int_A \sigma\, dA = 0 \tag{b}$$

where σ is some function of y, as indicated in Fig. 5–9c, e, or f.

For the special case of elastic action (which in this book always means *linearly* elastic action) over the entire cross section, the stress distribution diagram would be an extension of the elastic part of Fig. 5–9c or e to the top and bottom of the beam. The relation between σ and y can be obtained from generalized Hooke's law

$$\sigma_x = \frac{E}{(1+\nu)(1-2\nu)}[(1-\nu)\epsilon_x + \nu(\epsilon_y + \epsilon_z)] \tag{2–12}$$

which, since $\epsilon_y = \epsilon_z = -\nu\epsilon_x$, reduces to $\sigma_x = E\epsilon_x$: thus,

$$\sigma = E\left(\frac{\epsilon_c}{c}\right)y = \frac{\sigma_c}{c}y \tag{c}$$

When Eq. (c) is substituted into Eq. (b), the result is

$$\frac{\sigma_c}{c}\int_A y\, dA = \frac{\sigma_c}{c}A\bar{y} = 0$$

where $\bar{y}$ is the distance from the neutral axis to the centroidal axis (c–c) of the cross section. Since σ_c/c is not zero, $\bar{y}$ must be zero, indicating that the *neutral axis coincides with the centroidal axis (perpendicular to the plane of bending) of the cross section for flexural loading and elastic action.* Instances where the neutral axis (the line of zero longitudinal strain) does not coincide with the centroidal axis include (1) a combined axial and flexural load (see Chap. 8), (2) a beam with cross section unsymmetrical with respect to the neutral axis and subjected to inelastic action, and (3) a curved beam.

The use of the concepts established thus far will now be demonstrated. *Note: unless otherwise stated, as a simplifying assumption in all beam problems in this book, the weight of the beam itself is to be neglected.* Obviously this would not, as a rule, be done in professional practice. When recording normal stresses as answers to problems, it is necessary to indicate whether the stress is tension or compression.

Example 5–1 The beam of Fig. 5–3*a* has a rectangular cross section 4 in. wide by 12 in. deep and a span of 12 ft. It is subjected to a uniformly distributed load of 100 lb/ft and concentrated loads of 600 lb and 1200 lb at 2 and 7 ft, respectively, from the left support. Assuming elastic action, determine the maximum fiber stress at the center of the span by computing the magnitudes and positions of the vectors F_T and F_C in terms of $\sigma_{\max}$ and substituting into the moment equilibrium equation.

Solution The reactions are evaluated from a free-body diagram of the entire beam and moment equations with respect to each of the two reactions. By this method, the reactions are evaluated independently, and summation of forces may be used to check the results. The left reaction R_L is 1600 lb, and the right reaction R_R is 1400 lb, both upward. Figure 5–12*a* is a free-body diagram of the left half of the beam. The equation $\Sigma M_O = 0$ gives

$$\circlearrowleft + \qquad M_r + 600(4) + 100(6)(3) - 1600(6) = 0$$

from which $M_r = 5400$ ft-lb or 5400(12) in-lb as shown. The moment M_r is the resultant of the flexural stresses indicated in Fig. 5–12*b* and *c* in which F_C and F_T are the resultants of the compressive and tensile stresses, respectively. The resultants F_C and F_T (which act through the centroids of the wedge-shaped stress distribution diagram of Fig. 5–12*c*) are located through the centroids of the triangular stress distribution diagrams of Fig. 5–12*b*. The magnitude of the couple M_r equals $F_C(8)$ or $F_T(8)$, and $F_C = (\frac{1}{2})(\sigma_c)(6)(4) = 12(\sigma_c)$; therefore,

$$M_r = 8(12)\sigma_c = 5400(12) \text{ in-lb}$$

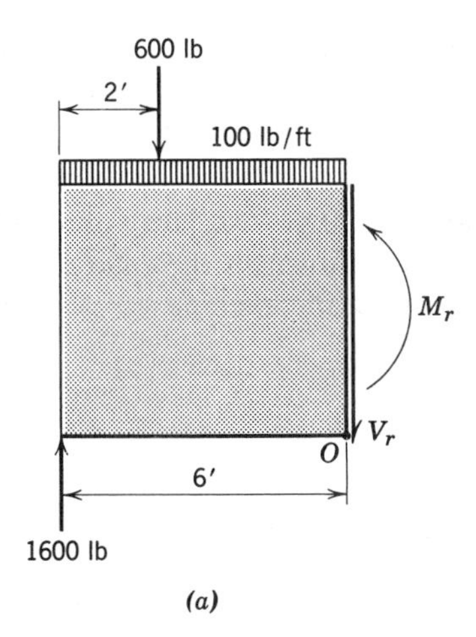

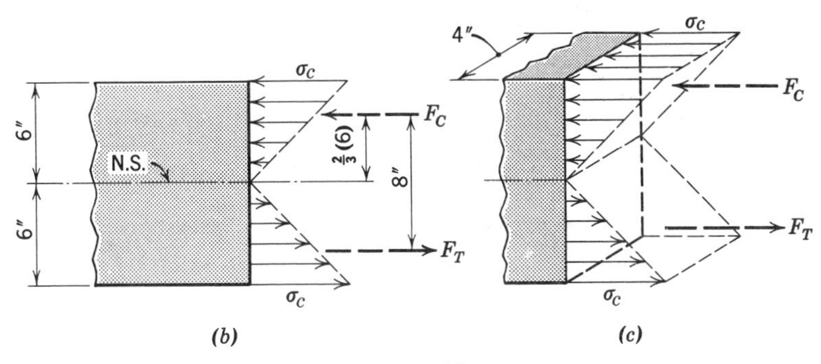

Fig. 5-12

from which

$$\sigma_{max} = \sigma_c = \underline{675 \text{ psi } C \text{ at top and } T \text{ at bottom}} \qquad \text{Ans.}$$

Example 5-2 Solve Exam. 5-1 if the beam is composed of two 2×10-in. timbers[4] securely fastened together to form a symmetrical T-section with the flange on top. Use the method of Exam. 5-1 and check the result by integration of Eq. (a).

Solution The horizontal centroidal axis of the cross section is first determined to be 4 in. down from the top surface (see any text on statics).

[4]Structural timbers specified as rough sawn will be approximately full size. Cross-sectional dimensions of dressed timbers will be from $\frac{3}{8}$ to $\frac{1}{2}$ in. smaller than nominal. For convenience, all timbers in this book will be considered as full size.

The fiber stress distribution diagrams are shown in Fig. 5–13. For convenience, the compression diagram is subdivided into three parts composed of two triangular distribution (wedge-shaped) diagrams and one uniform distribution diagram. This is done because the locations of the centroids of these diagrams are known. The values of the force components and their locations are shown in the figure. It will be noted that each component F_1, F_2, F_3, being the average stress multiplied by the area

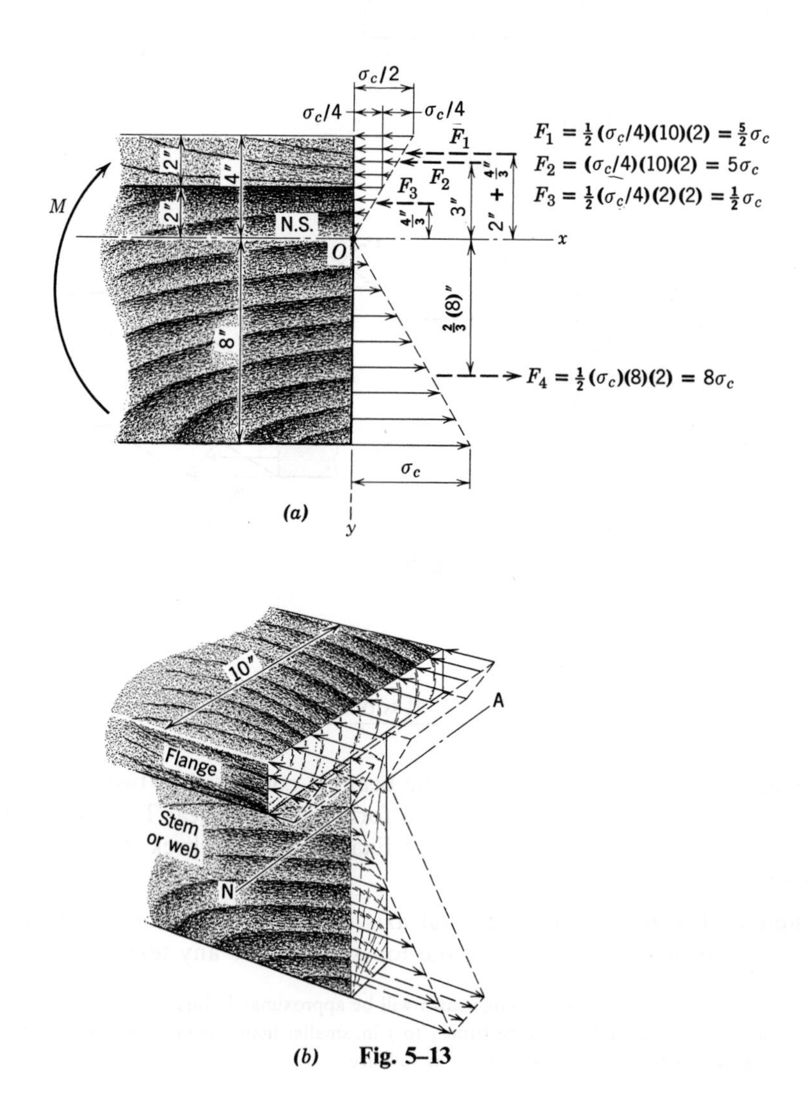

(b) **Fig. 5–13**

on which the stress acts is equal to the volume of the corresponding part of the three-dimensional diagram of Fig. 5–13*b*. The moment of the component forces with respect to O is

$$M_r = F_1\left(\frac{10}{3}\right) + F_2\left(\frac{9}{3}\right) + F_3\left(\frac{4}{3}\right) + F_4\left(\frac{16}{3}\right)$$

which from the preceding example is equal to 5400(12); therefore,

$$\left(\frac{5}{2}\right)\sigma_c\left(\frac{10}{3}\right) + 5\sigma_c\left(\frac{9}{3}\right) + \left(\frac{1}{2}\right)\sigma_c\left(\frac{4}{3}\right) + 8\sigma_c\left(\frac{16}{3}\right) = 5400(12)$$

from which

$$\sigma_{\text{max}} = \sigma_c = \underline{972 \text{ psi } T \text{ at bottom}} \qquad \text{Ans.}$$

A more direct solution can be obtained by integration of Eq. (a); thus,

$$M_r = \int_A \sigma\, dA\, y = \int_{-4}^{-2} \left(\frac{\sigma_c}{8}y\right)(10dy)(y) + \int_{-2}^{8} \left(\frac{\sigma_c}{8}y\right)(2dy)(y)$$

$$= \left[\frac{5\sigma_c y^3}{12}\right]_{-4}^{-2} + \left[\frac{\sigma_c y^3}{12}\right]_{-2}^{8} = 5400(12)$$

from which

$$\sigma_c = \sigma_{\text{max}} = \underline{972 \text{ psi } T \text{ at bottom}} \qquad \text{Ans.}$$

Example 5–3 A beam having the cross section of Fig. 5–14*a* is made of magnesium alloy having the approximate stress-strain diagram of Fig. 5–14*b*. Determine the magnitude of the bending moment applied in the vertical plane of symmetry necessary to produce a flexural stress of magnitude 14.4 ksi at point A, which is at the lower surface of the top flange.

Solution Because of symmetry, the neutral axis coincides with the centroidal z axis, and the strain and stress distribution diagrams are as shown in Fig. 5–14*c* and *d*. Since the specified stress is 14.4 ksi at $y = 2$ in., the corresponding strain from Fig. 5–14*b* is 3.5 (in/in)$(10)^{-3}$, and the strain at $y = 3$ in. is 5.25 which is less than 6; therefore, the second stress-strain function of Fig. 5–14*b* is valid for the entire flange. The first function is valid for the web. Substituting the ϵ vs y function from Fig.

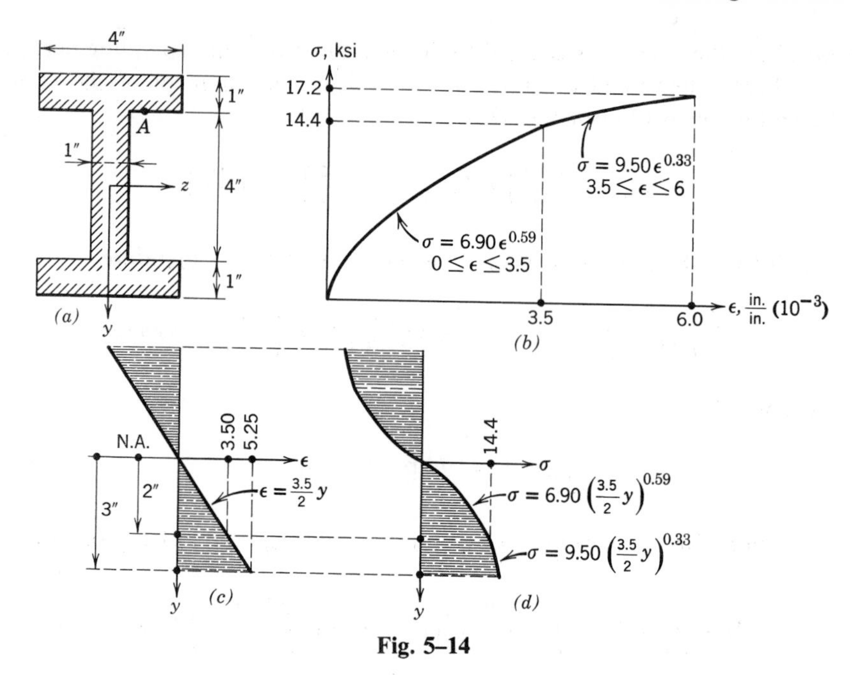

Fig. 5–14

5–14c into the σ vs ϵ functions gives the σ vs y functions of Fig. 5–14d. The solution of Eq. (a) follows.

$$M_r = \int_A \sigma\, dA\, y = 2\left\{ \int_0^2 \left[6.90(1.75)^{0.59} y^{0.59}\right][1\,dy]\,y + \int_2^3 \left[9.50(1.75)^{0.33} y^{0.33}\right][4\,dy]\,y \right\}$$

$$= 2\left[\frac{6.90(1.391)\,y^{2.59}}{2.59}\right]_0^2 + 2\left[\frac{38(1.203)\,y^{2.33}}{2.33}\right]_2^3 = 354.8$$

from which

$$M = M_r = \underline{355 \text{ in-kips}} \qquad \text{Ans.}$$

Example 5–4 A beam having the T cross section of Fig. 5–15 is made of elastoplastic steel having a proportional limit of 36 ksi. Determine the bending moment applied in the vertical plane of symmetry that will

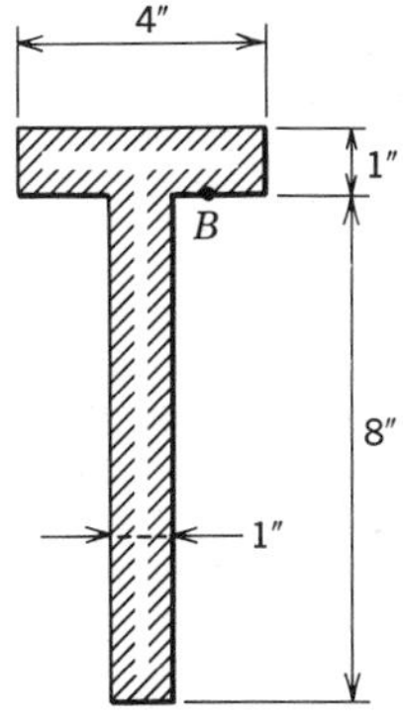

Fig. 5–15

produce a longitudinal strain of −0.0012 at point B at the lower face of the flange.

Solution The strain distribution, stress-strain, and stress distribution diagrams are shown in Fig. 5–16a, b, and c. Because of the inelastic action and unsymmetrical section, the location of the neutral axis must first be determined from Eq. (b). Thus,

$$\int_A \sigma\, dA = -36(1)(4) - \frac{36}{2}(a)(1) + \frac{36}{2}(a)(1) + 36(8-2a)(1) = 0$$

from which

$$a = 2 \text{ in.}$$

To find the moment, the method of Exam. 5–1 will be most convenient.

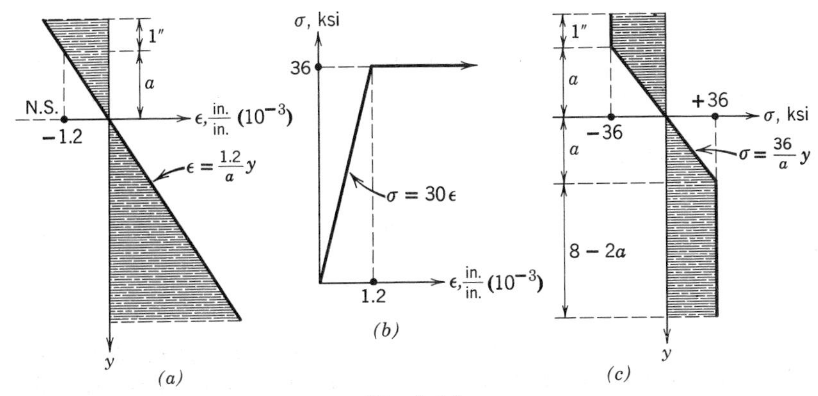

Fig. 5–16

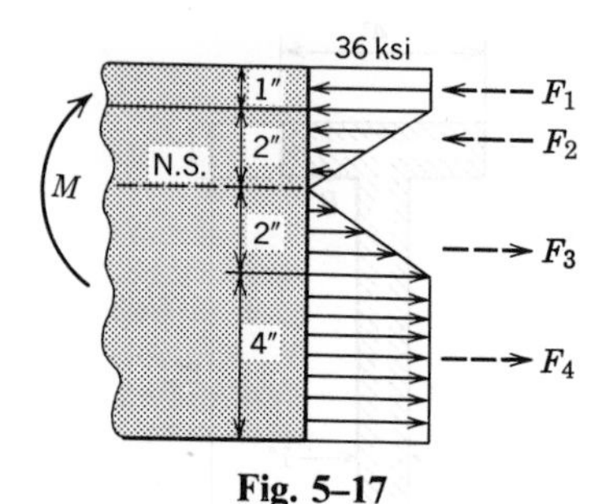

Fig. 5–17

Referring to Fig. 5–17,

$$M_r = 36(4)(1)\left(2+\frac{1}{2}\right)+\frac{36}{2}(2)(1)\left(\frac{2}{3}\right)(2)$$

$$+\frac{36}{2}(2)(1)\left(\frac{2}{3}\right)(2)+36(4)(2+2)$$

from which

$$M = M_r = \underline{1032 \text{ in-kips as shown}} \qquad \text{Ans.}$$

Example 5–5 Determine the minimum bending moment necessary to produce completely plastic action in the beam of Exam. 5–4. This moment is called the *plastic moment*.

Solution In this case, the stress distribution diagram would be like Fig. 5–9*f*. Letting the distance from the neutral surface to the bottom of the stem be designated a, Eq. (b) would yield

$$\int_A \sigma\, dA = \sigma(1)(a) - \sigma(1)(8-a) - \sigma(4)(1) = 0$$

from which $a = 6$ in.; therefore, the neutral surface is 3 in. below the top of the flange. Following the solution of Exam. 5–4, the moment becomes

$$M_r = 36(4)(1)\left(\frac{5}{2}\right) + 36(2)(1)(1) + 36(6)(1)(3)$$

$$M_p = M_r = \underline{1080 \text{ in-kips}} \qquad \text{Ans.}$$

PROBLEMS

Note. In the following problems, all beams are loaded in a plane of symmetry parallel to the largest cross-sectional dimension except as noted otherwise. Also, assume elastic action unless noted otherwise.

5–1* The maximum fiber stress at a certain section in a rectangular beam 4 in. wide by 8 in. deep is 1500 psi. Determine by the free-body diagram method of Exam. 5–1:

(a) The resisting moment in the beam at that section.
(b) The percentage decrease in this moment caused by removing the central (dotted portion) of the beam (see Fig. P5–1).

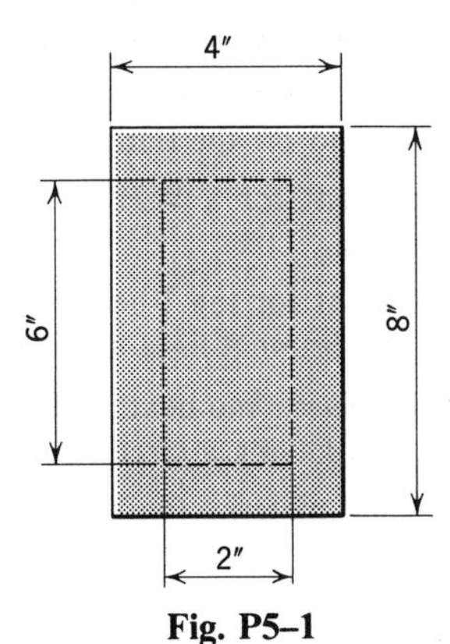

Fig. P5–1

5–2* For the beam of Fig. P5–2, loaded in the vertical plane of symmetry, determine:

(a) The resisting moment in terms of the maximum fiber stress.
(b) The magnitude of the resisting moment on this cross section if the fiber stress at point A is 6000 psi T.

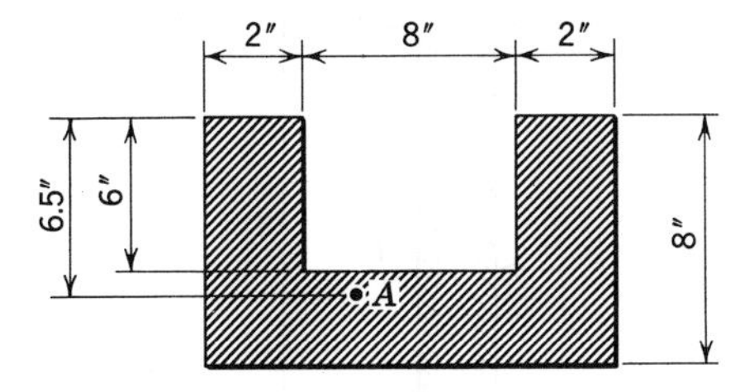

Fig. P5–2

5–3* A timber beam is made of three 50×150-mm planks fastened together to make an I-beam 150 mm wide by 250 mm deep. The allowable fiber stress is 10 MN/m^2. Determine the maximum allowable bending moment to which the beam may be subjected.

5–4 A timber beam is made of three 2×4-in. members fastened together to make an I-beam 4 in. wide by 8 in. deep. Determine the maximum fiber stress on a section where the bending moment is 6000 ft-lb.

5–5 A timber beam is fabricated of four 25×100-mm boards fastened together to form a box cross section with outside dimensions 100 mm wide by 150 mm deep. Determine the magnitude of the maximum permissible bending moment that may be applied in a plane parallel to the 150-mm side, when the fiber stress is limited to 9 $\mathrm{MN/m^2}$ T or C.

5–6 An I-beam is fabricated by welding two 400×50-mm flange plates to a 600×25-mm web plate. The beam is loaded in the plane of symmetry parallel to the web. Determine the percentage of the bending moment carried by the flanges.

5–7 Determine the percentage of the bending moment carried by the flanges of a W36×160 beam (see App. C). Round off the flange thickness to 1.00 in.

5–8* Determine the plastic moment for a W21×112 steel beam (see App. C) having a proportional limit of 36 ksi.

5–9* Determine the plastic moment for the I-beam of Prob. 5–6 if the beam is fabricated using steel having a proportional limit of 280 $\mathrm{MN/m^2}$.

5–10 Determine the ratio of the plastic moment to the maximum elastic moment for a beam of elastoplastic material having a circular cross section.

5–11* For a beam of elastoplastic material (with a yield point of 36 ksi) having the cross section of Fig. P5–11:

(a) Locate the neutral axis when the action is plastic over the entire section.
(b) Determine the plastic moment.
(c) Determine the ratio of the plastic to the maximum elastic moment.

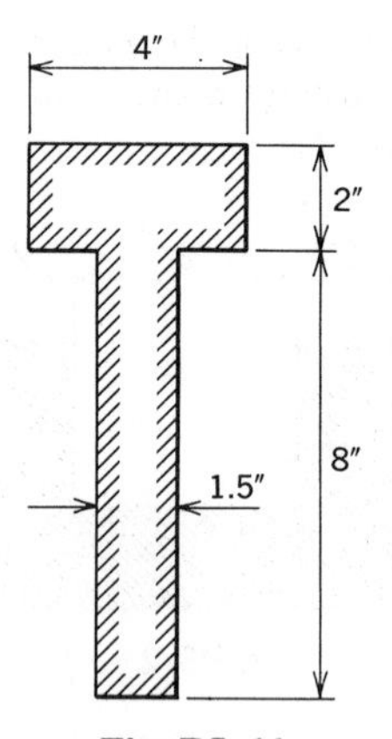

Fig. P5–11

5–12 For the beam of Prob. 5–11:

(a) Locate the neutral axis when the stress in the outer fibers of the flange reaches the yield point.
(b) Determine the resisting moment when the condition in part (a) is reached.

5–13* A beam is fabricated of two 150×25-mm and two 75×25-mm plates welded together to form a box section 75 mm wide by 200 mm deep. The material is aluminum alloy having the stress-strain diagram represented by the solid line of Fig. P5–13*a*. The dashed lines represent an approximation to the stress-strain curve. Using the approximate curve:

(a) Verify that the curve of Fig. P5–13*c* represents the aproximate stress distribution over the half cross section shown in Fig. P5–13*b* for a maximum longitudinal strain of 0.010.

(b) Determine the corresponding bending moment in the beam.

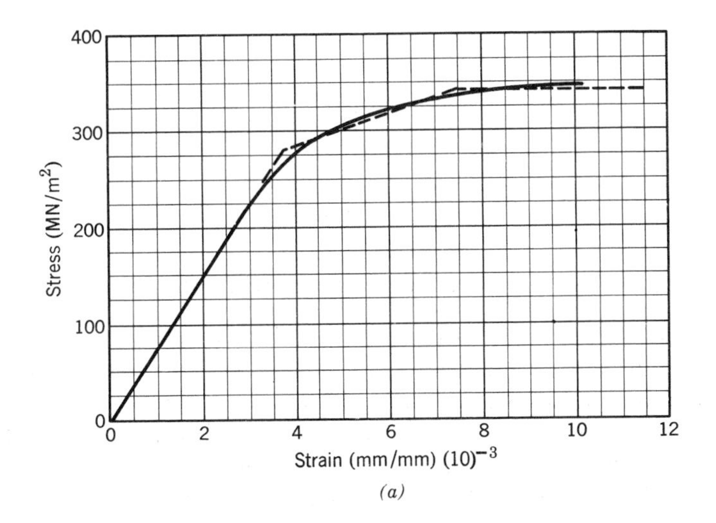

(*a*)

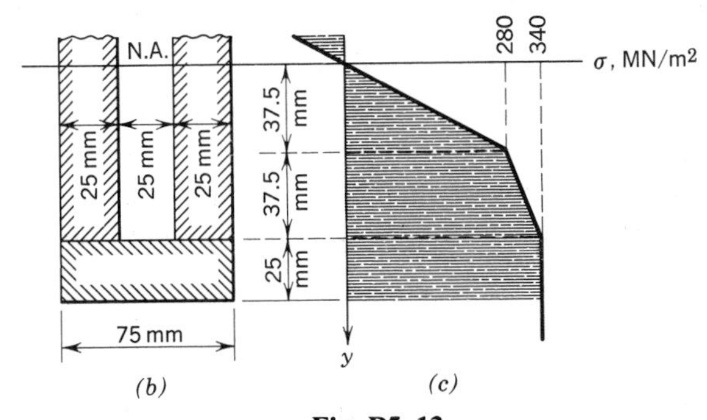

(*b*) (*c*)

Fig. P5–13

5–14 A beam having the T cross section of Fig. P5–14 is made of the magnesium alloy having the approximate stress-strain diagram of Fig. 5–14*b* (Exam. 5–3). The beam is subjected to a bending moment in the *xy* plane of symmetry (*x* is the longitudinal axis of the beam) of such magnitude that the maximum flexural stress is 14.4 ksi *T* and the neutral surface is located at the junction of the flange and stem. Determine:

(a) The dimension *c*.
(b) The bending moment in the beam.

Hint: Assume that *c* is greater than 2 in. and verify the assumption.

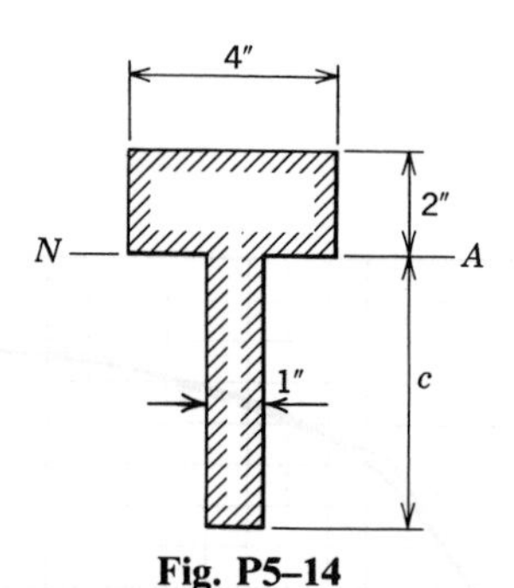

Fig. P5–14

5–4 THE ELASTIC FLEXURE FORMULA

As the sophistication of structural and machine design increases, problems involving inelastic action are receiving increased attention. However, the design of beams on the basis of linearly elastic action is still a very important area of engineering activity; for such design, the fundamental approach outlined in Art. 5–3 becomes too time-consuming. Therefore, a formula will be developed that has general applicability within the limitations noted in Arts. 5–2 and 5–3.

Substitution of Eq. (c) into Eq. (a) of Art. 5–3 yields the following:

$$M_r = \int_A \sigma \, dA \, y = \frac{\sigma_c}{c} \int_A y^2 \, dA = \frac{\sigma}{y} \int_A y^2 \, dA$$

The integral is the familiar second moment of the cross-sectional area, commonly called "moment of inertia," (see App. B), *with respect to the centroidal axis*. When this integral is replaced by the symbol I, the elastic flexure formula is obtained as

$$\sigma = \frac{My}{I} \tag{5–1}$$

where σ is the normal stress (σ_x in general notation) at a distance y from the neutral surface and on a transverse plane, and M is the resisting moment of the section.

At any section of the beam, the fiber stress will be maximum at the surface farthest from the neutral axis, and Eq. 5–1 becomes

$$\sigma_{max} = \frac{Mc}{I} = \frac{M}{Z}$$

where $Z = I/c$ is called the section modulus of the beam. Although the section modulus can be readily calculated for a given section, values of the modulus are often included in tables to simplify calculations. Observe that for a given area, Z becomes larger as the shape is altered to concentrate more of the area as far as possible from the neutral axis. Commercial rolled shapes such as I- and H-beams and the various built-up sections are intended to optimize the area-section modulus relation. Note that in the problems involving reference to rolled shapes in App. C, the beam (unless otherwise noted) is so oriented that the maximum section modulus applies.

Example 5–6 Solve Exam. 5–2 by means of the flexure formula.

Solution The second moment of the area can be obtained by integration or by composite areas, (see App. B) and is 1600/3 in.^{4} The maximum fiber stress is

$$\sigma_{max} = \frac{5400(12)(8)}{1600/3} = \underline{972 \text{ psi } T \text{ at bottom}} \qquad \text{Ans.}$$

PROBLEMS

Note. The action is linearly elastic in all of the following problems.

5–15 Solve Prob. 5–1 by means of the flexure formula.

5–16 Solve Prob. 5–2 by means of the flexure formula.

5–17 Solve Prob. 5–3 by means of the flexure formula.

5–18 Solve Prob. 5–6 by means of the flexure formula.

5–19* The beam loaded as shown in Fig. P5–19*a* has the cross section shown in Fig. P5–19*b*. At a point 9 ft from the left end and 4 in. above the neutral surface, determine the fiber stress.

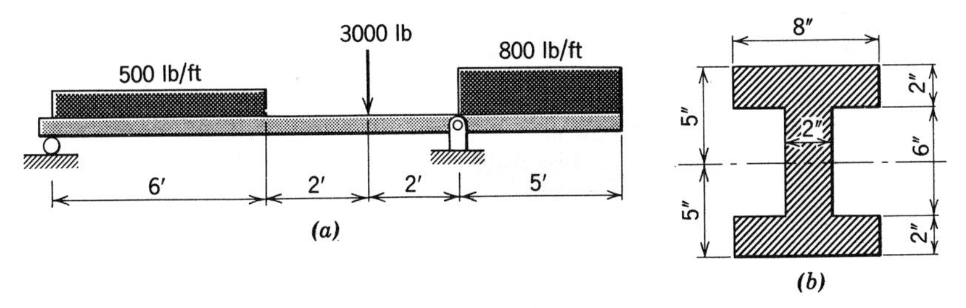

Fig. P5–19

5–20* For the beam of Prob. 5–19, determine the fiber stress at a point 3 in. below the neutral surface on a section 3 ft from the right end.

5–21* Figure P5–21*a* shows the cross section of a beam loaded and supported as shown in Fig. P5–21*b*. Determine the fiber stress at a point 100 mm below the neutral surface on a section 1.3 m from the right end.

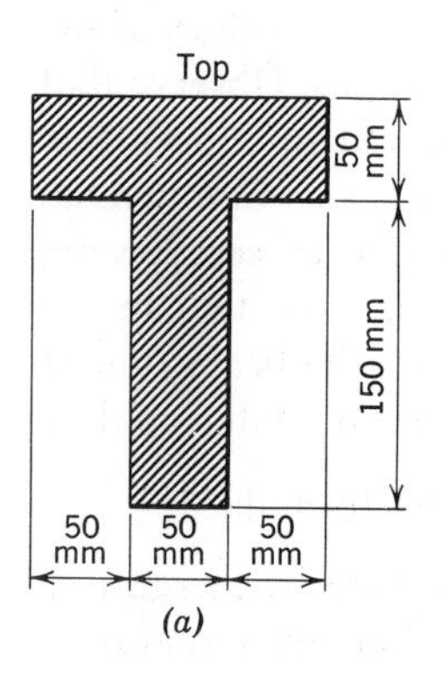

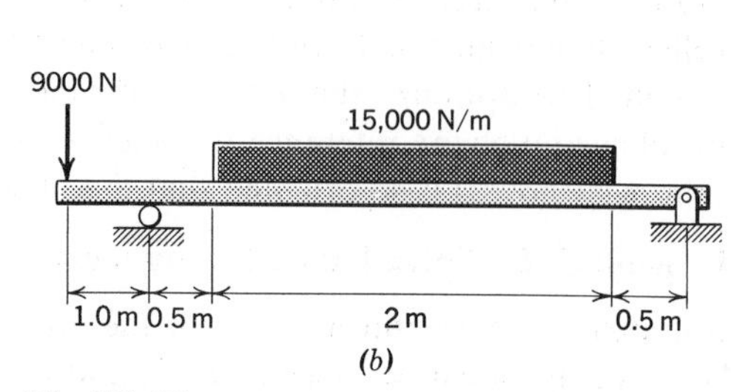

Fig. P5–21

5–22* For the beam of Prob. 5–21, determine the maximum fiber stress on a section 1 m from the left end.

5–23 An S7×15.3 steel beam (see App. C) is loaded and supported as shown in Fig. P5–23. Determine the maximum fiber stress on a section 3 ft from the left end.

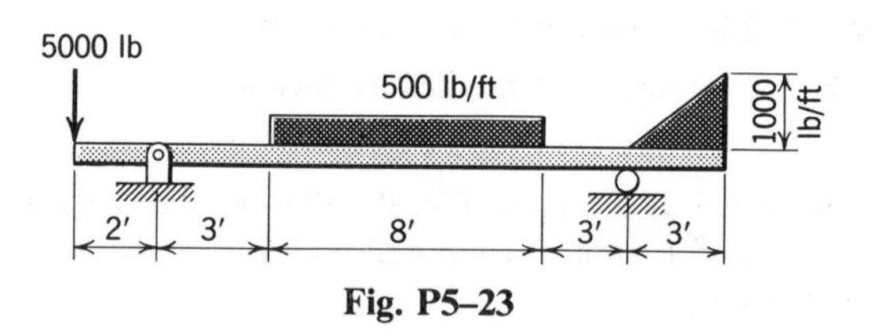

Fig. P5–23

5–24* A 200-mm diameter steel bar is loaded and supported as shown in Fig. P5–24. Determine the maximum fiber stress on a section 1.5 m from the wall.

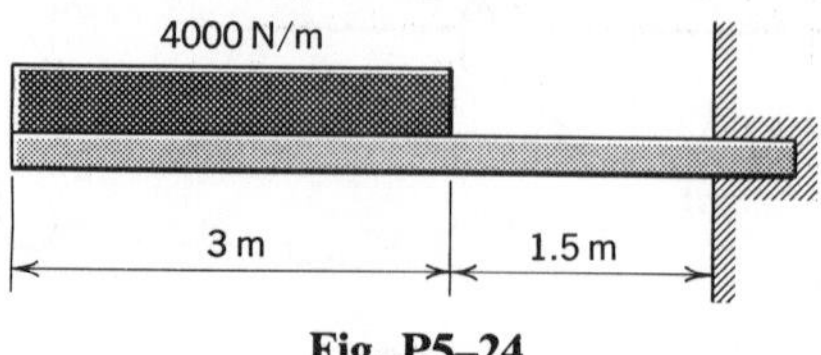

Fig. P5–24

5–25 An S6×12.5 steel beam (see App. C) is loaded and supported as shown in Fig. P5–25. Determine the maximum fiber stress on a section 7 ft to the right of B.

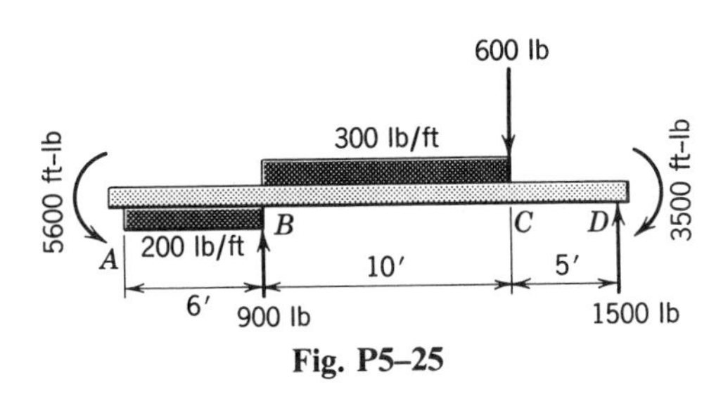

Fig. P5–25

5–26 For the beam shown in Fig. P5–26, the allowable fiber stresses on the section at the load are 42 MN/m^2 T and 70 MN/m^2 C. Determine the maximum allowable load P.

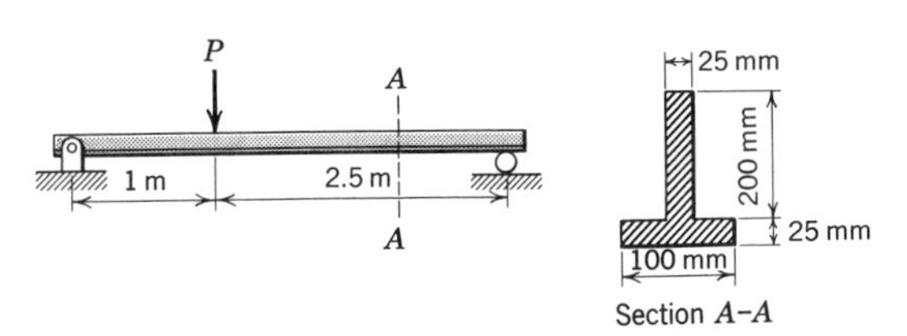

Fig. P5–26

5–27 Solve Prob. 5–26 with an additional load P placed 1 m to the left of the right support.

5–5 FLEXURAL STRESSES—UNSYMMETRICAL BENDING

The discussion of flexural stresses to this point has been limited to beams with at least one longitudinal plane of symmetry and with the load applied in the plane of symmetry. The work will now be extended to cover pure bending (bent with couples only; no transverse forces) of (1) beams with a plane of symmetry but with the load (couple) applied not in or parallel to the plane of symmetry, or (2) beams with no plane of symmetry. Additional discussion needed for the problem of unsymmetrical beams loaded with transverse forces will be found in Art. 5–13.

Figure 5–18 depicts a beam of unsymmetrical cross section loaded with a couple M in a plane making an angle α with the xy plane, where the origin of coordinates is at the centroid of the cross section. The neutral axis, which passes through the centroid for linearly elastic action (see Art. 5–3), makes an unknown angle β with the z axis. The beam is straight and of uniform cross section, and a plane cross section is assumed to remain plane after bending. Also the following development is restricted to linearly elastic action.

Since the orientation of the neutral axis is not known, the flexural stress distribution function cannot be expressed in terms of one variable as in Eq. (c) of Art. 5–3. However, since the plane section remains plane, the stress variation can be written thus

$$\sigma = k_1 y + k_2 z \tag{a}$$

Then the resisting moments with respect to the z and y axes can be written as follows:

$$M_{rz} = \int_A \sigma \, dA \, y = \int_A k_1 y^2 \, dA + \int_A k_2 zy \, dA = k_1 I_z + k_2 I_{yz}$$

$$M_{ry} = \int_A \sigma \, dA \, z = \int_A k_1 yz \, dA + \int_A k_2 z^2 \, dA = k_1 I_{yz} + k_2 I_y$$

where I_z and I_y are the moments of inertia of the cross-sectional area with respect to the z and y axes, respectively, and I_{yz} is the product of inertia

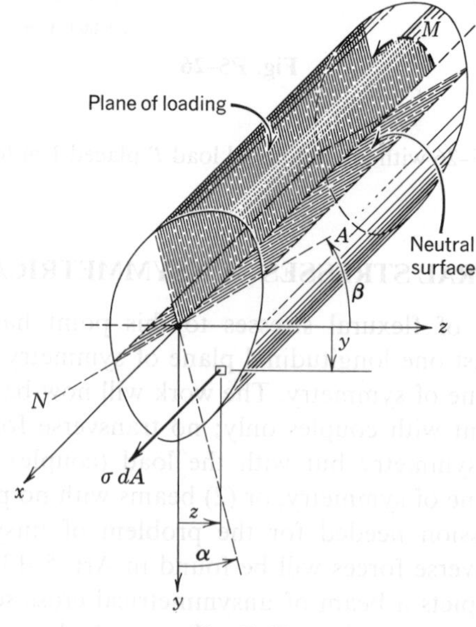

Fig. 5–18

(see App. B) with respect to these two axes. It will be convenient to let the y and z axes be *principal axes* Y and Z; then I_{YZ} is zero. Equating the applied moment to the resisting moment and solving for k_1 and k_2 gives

$$M_{rZ} = k_1 I_Z = M\cos\alpha, \qquad k_1 = \frac{M\cos\alpha}{I_Z}$$

and

$$M_{rY} = k_2 I_Y = M\sin\alpha, \qquad k_2 = \frac{M\sin\alpha}{I_Y}$$

Substituting these expressions for k into Eq. (a) gives the elastic flexure formula for unsymmetrical bending

$$\sigma = \frac{M\cos\alpha\, Y}{I_Z} + \frac{M\sin\alpha\, Z}{I_Y} \tag{5–2}$$

Since σ is zero at the neutral surface, the orientation of the neutral axis is found by setting Eq. 5–2 equal to zero, from which

$$\frac{\cos\alpha\, Y}{I_Z} = -\frac{\sin\alpha\, Z}{I_Y}$$

or

$$Y = -\tan\alpha \frac{I_Z}{I_Y} Z$$

which is the equation of the neutral axis in the YZ plane. The slope of the line is dY/dZ, and since $dY/dZ = \tan\beta$, the orientation of the neutral axis is given by the expression

$$\tan\beta = -\frac{I_Z}{I_Y}\tan\alpha \tag{5–3}$$

The negative sign indicates that the angles α and β are in adjacent quadrants.

It will be observed that the neutral axis is *not* perpendicular to the plane of loading *unless* (1) the angle α is zero, in which case the plane of loading is (or is parallel to) a principal plane, or (2) the two principal moments of inertia are equal; this reduces to the special kind of symmetry where all centroidal moments of inertia are equal (square, circle, etc). The following examples will illustrate the application of Eqs. 5–2 and 5–3. *A positive moment is defined as a moment for which the z component* ($M \cos \alpha$) will produce tension in the bottom of the beam.

Example 5–7 A beam having the T cross section of Fig. 5–19*a* is loaded with a couple of +13,600 in-lb applied in the plane *a–a* making an angle α with the yx plane (x is the longitudinal axis of the beam).

(a) Determine the maximum flexural stress by using Eq. 5–2.

(b) Determine, and show on a sketch, the orientation of the neutral axis.

(c) Check the stress obtained in part (a) by using Eq. 5–1 and the orientation of the neutral axis as determined in part (b).

Solution (a) Because of symmetry, the y axis is a principal axis (see App. B) of the cross section and the z axis is the other principal axis. The principal moments of inertia are $I_Z = I_z = 136$ in.^{4} and $I_Y = I_y = 40$ in.^{4}. The point of maximum flexural stress is not apparent by inspection, but the two likely points are A and B because for each of these points on the boundary of the section, the two components of stress are additive. Applying Eq. 5–2 yields

$$\sigma_A = \frac{0.8(13600)(-3)}{136} + \frac{0.6(13600)(-3)}{40} = -852$$

$$\sigma_B = \frac{0.8(13600)(5)}{136} + \frac{0.6(13600)(1)}{40} = +604$$

therefore, the maximum flexural stress is 852 psi C Ans.

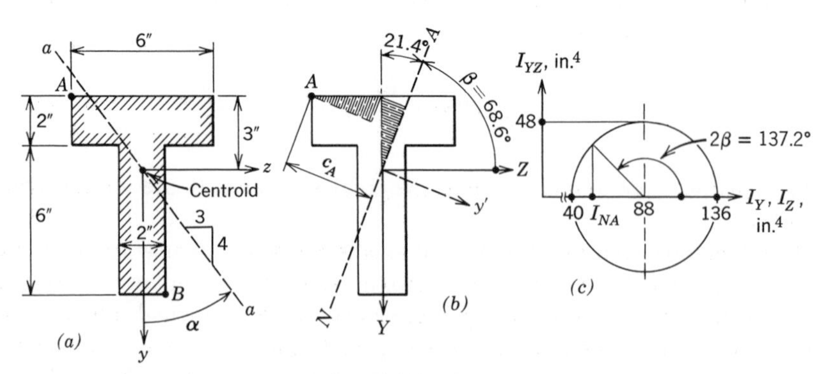

Fig. 5–19

(b) The angle β is obtained from Eq. 5–3; thus,

$$\tan\beta = -\frac{I_Z}{I_Y}\tan\alpha = -\frac{136}{40}\left(\frac{3}{4}\right) = -2.550$$

therefore,

$$\underline{\beta = 68.6^\circ \text{ as shown in Fig. 5–19}b} \qquad \text{Ans.}$$

(c) The flexure formula $\sigma = My/I$ applies when M is the moment in the plane perpendicular to the neutral axis, y is the perpendicular distance c_A from the neutral axis to point A, and I is I_{NA}. The distance c_A from geometry (see shaded triangles in Fig. 5–19b) is

$$c_A = -3\sin 21.4 - 3\cos 21.4 = -3.888 \text{ in.}$$

The moment of inertia is found from Mohr's circle (Fig. 5–19c); thus,

$$I_{NA} = 88 + 48\cos 137.2 = 88 - 35.2 = 52.8 \text{ in.}^4$$

Then the flexural stress at A is

$$\sigma_A = \frac{M\cos(\beta - \alpha)c_A}{I_{NA}} = \frac{13600\cos(68.6 - 36.87)(-3.888)}{52.8}$$

$$\sigma_A = -852 = \underline{852 \text{ psi } C} \qquad \text{Ans.}$$

Example 5–8 A beam having the cross section of Fig. 5–20a is loaded with a bending moment of $+300$ in-kips in the yx plane (where x is the

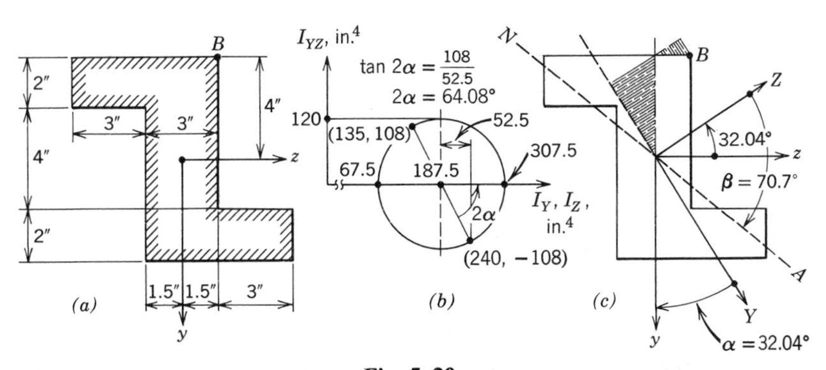

Fig. 5–20

longitudinal axis of the beam). The moments of inertia with respect to the z and y axes are: $I_z = 240$ in.^{4}, $I_y = 135$ in.^{4}, and $I_{yz} = +108$ in.^{4}. Determine:

(a) The flexural stress at point B.
(b) The orientation of the neutral axis; show result on a sketch.

Solution The application of Eqs. 5–2 and 5–3 requires principal moments of inertia; since the product of inertia I_{yz} is not zero, the z and y axes are not principal axes. The principal moments of inertia and the angle 2α are found from the Mohr's circle of Fig. 5–20*b* (Note that, for the axes shown, I_{yz} is always plotted with I_y). The principal axes are indicated in Fig. 5–20*c*.

(a) The coordinates Y_B and Z_B are found from geometry (see Fig. 5–20*c*); thus,

$$Y_B = -4\cos 32.04 + 1.5\sin 32.04 = -2.59 \text{ in.}$$

$$Z_B = +4\sin 32.04 + 1.5\cos 32.04 = +3.39 \text{ in.}$$

and the application of Eq. 5–2 yields

$$\sigma_B = \frac{M\cos\alpha\ Y_B}{I_Z} + \frac{M\sin\alpha\ Z_B}{I_Y}$$

$$= \frac{300(\cos 32.04)(-2.59)}{307.5} + \frac{300(-\sin 32.04)(3.39)}{67.5} = -10.12$$

$\sigma_B = \underline{10.12 \text{ ksi } C}$ Ans.

Note that in the development of Eq. 5–2, the angle α was positive counterclockwise from the Y axis; therefore, α for this problem is $-32.04°$, for which $\cos\alpha$ is positive and $\sin\alpha$ is negative.

(b) The angle β from Eq. 5–3 is

$$\tan\beta = -\frac{I_Z}{I_Y}\tan\alpha = -\frac{307.5}{67.5}\tan(-32.04) = +2.850$$

from which

$\beta = \underline{70.7° \text{ as shown in Fig. 5–20}c.}$ Ans.

PROBLEMS

Note. In the following problems, the *x* axis refers to the longitudinal centroidal axis of the beam.

5–28* A beam having the I cross section of Fig. P5–28 is subjected to a bending moment of 6160 in-lb applied in the *a–a* plane. Determine:

(a) The magnitude of the maximum flexural stress.
(b) The orientation of the neutral axis; show the result on a sketch.

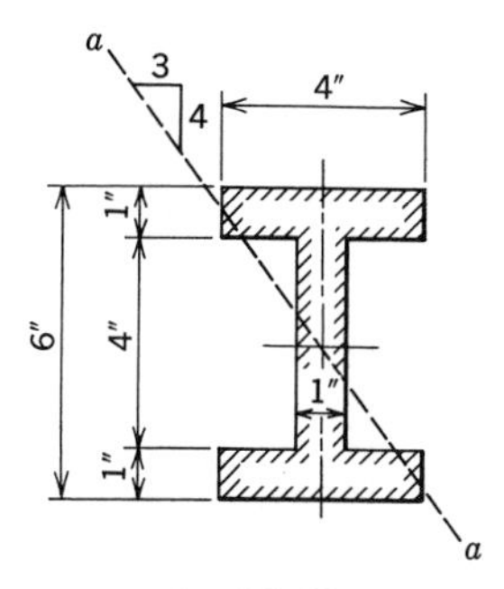

Fig. P5–28

5–29* A beam having the T cross section of Fig. P5–29 is subjected to a bending moment of 100 in-kips applied in the *a–a* plane. Determine:

(a) The magnitude of the maximum flexural stress.
(b) The orientation of the neutral axis; show the result on a sketch.

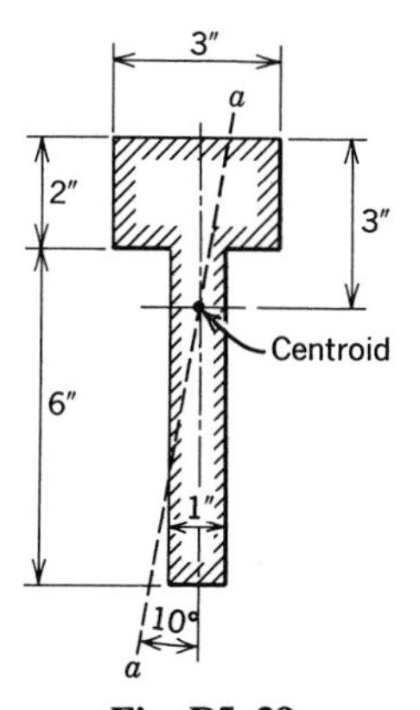

Fig. P5–29

5–30* A beam having a rectangular cross section 100 mm wide by 200 mm deep is subjected to a bending moment of +3000 N-m applied in a plane making an angle of 30° with the yx plane, as shown in Fig. P5–30. Determine:

(a) The maximum flexural stress.
(b) The orientation of the neutral axis; show the location on a sketch.

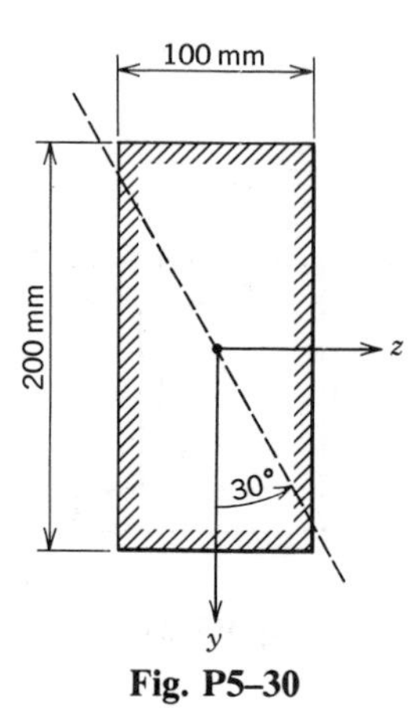

Fig. P5–30

5–31* A 200×200×24-mm angle is used for a beam that supports a bending moment of +10,000 N-m applied in the yx plane. The moments of inertia from a structural steel handbook are $I_z = I_y = 33.3(10^6)$ mm^4, and $I_{yz} = +19.5(10^6)$ mm^4 (see Fig. P5–31). Determine:

(a) The flexural stress at point A.
(b) The maximum flexural stress and its location on the cross section.
(c) The orientation of the neutral axis; show the location on a sketch.

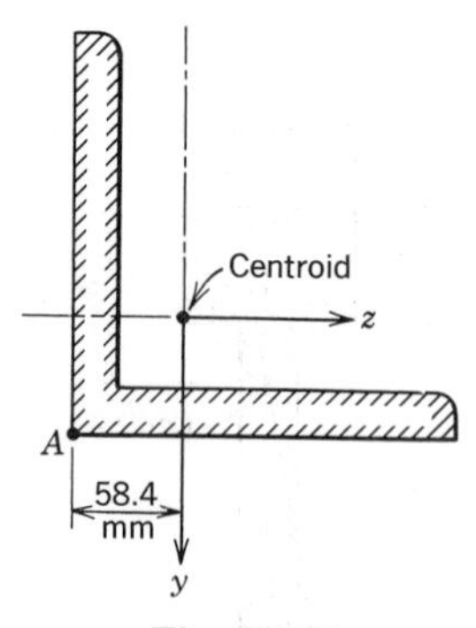

Fig. P5–31

5–32 A beam having the rectangular cross section of Fig. P5–32 is subjected to a bending moment M applied in a plane making an angle α with the yx plane.

(a) Develop an expression for the maximum flexural stress in terms of b, h, M, and α.
(b) Determine α in terms of b and h to produce the maximum flexural stress for a given moment M.
(c) Using the expression for α obtained in part (b), determine, and show on a sketch, the orientation of the neutral axis when $h=2b$.

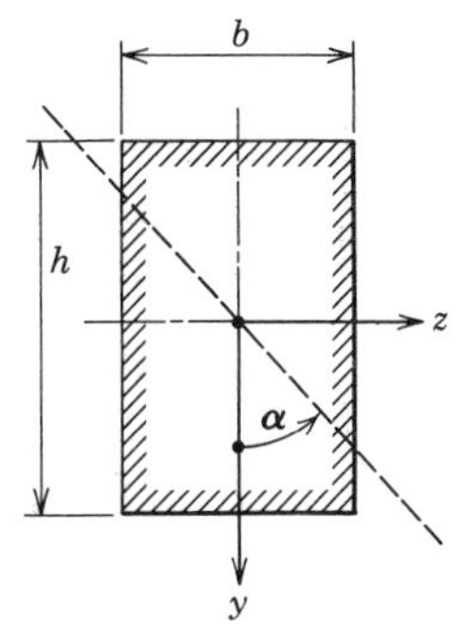

Fig. P5–32

5–33* A beam having the triangular cross section of Fig. P5–33 is subjected to a bending moment of $+10$ in-kips applied in the yx plane. Determine:
(a) The flexural stress at point A.
(b) The orientation of the neutral axis; show the location on a sketch.

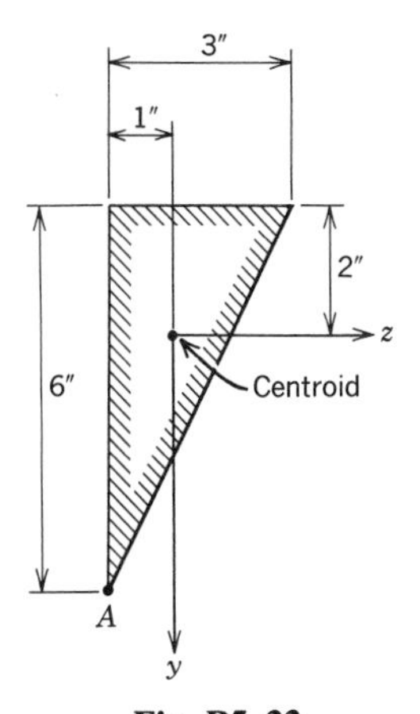

Fig. P5–33

5–34 A beam having the Z cross section of Fig. P5–34 is loaded with a bending moment of $+6000$ N-m applied in the yx plane. The moments of inertia from a structural steel handbook are $I_z = 16.64(10^6)$ mm^4, $I_y = 4.02(10^6)$ mm^4, and $I_{yz} = -6.05(10^6)$ mm^4. Determine:

(a) The flexural stresses at points A and B.

(b) The orientation of the neutral axis; show the location on a sketch.

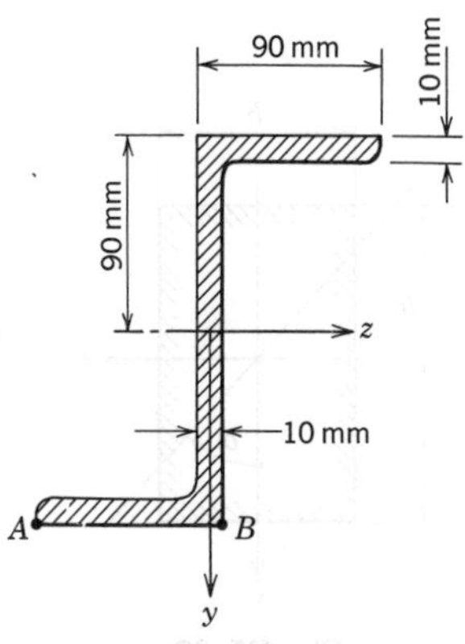

Fig. P5–34

5–35* A beam having the angle cross section of Fig. P5–35 is loaded with a bending moment of $+100$ in-kips applied in the yx plane. Determine:

(a) The flexural stress at point A.

(b) The orientation of the neutral axis; show the location on a sketch.

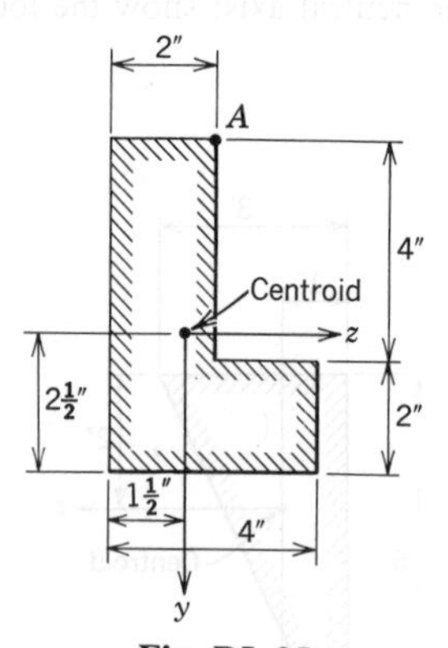

Fig. P5–35

5–36 A beam having the Z cross section of Fig. P5–36 is loaded with a bending moment of $+50$ in-kips applied in the yx plane. The moments of inertia from the *Alcoa Structural Handbook* are: $I_z = 25.40$ in^4, $I_y = 8.83$ in^4, and $I_Y = 3.08$ in^4. The value of I_Z may be obtained from the fact that the sum of the moments of inertia with respect to any pair of orthogonal axes intersecting at a common point is a constant. Determine:

(a) The flexural stress at point A.

(b) The orientation of the neutral axis; show the location on a sketch.

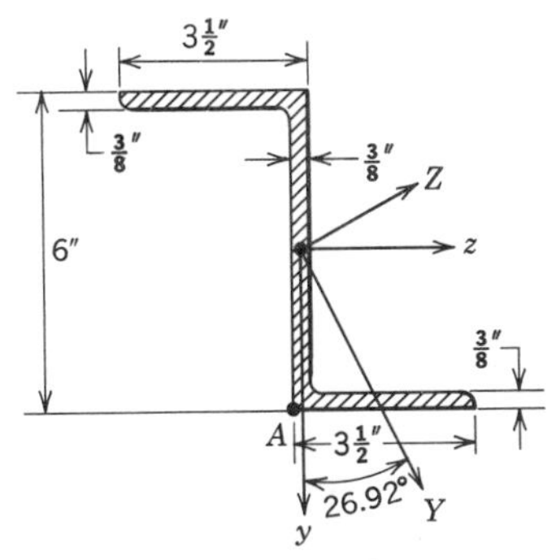

Fig. P5–36

5–6 SHEAR AND MOMENT IN BEAMS

The procedures for determining flexural stresses outlined in Arts. 5–3 and 5–4 are quite adequate if one wishes to compute the stresses at any specified section in a beam. However, if it is necessary to determine the maximum fiber (flexural) stress in a given beam subjected to loading producing a bending moment that varies with position along the beam, it is desirable to have a convenient method for determining the maximum moment.

In Art. 5–9 a relation will be developed between the resisting shear (V_r of Fig. 5–4) and the longitudinal and transverse shearing stresses at the section. Since the maximum transverse shearing stress will occur at the section at which V_r is maximum, a convenient method of determining such sections is likewise desirable.

When the equilibrium equation $\Sigma F_y = 0$ is applied to the free-body diagram of Fig. 5–4, the result can be written as

$$R - wx - P = V_r$$

or

$$V = V_r$$

where V *is defined as the resultant of the external transverse forces acting on the part of a beam to either side of a section and is called the transverse shear or just the shear at the section.*

As seen from the definitions of V and V_r, the shear is equal in magnitude and opposite in sense to V_r. Since these shear forces are always equal in magnitude, they are frequently treated as though they were identical. For simplicity the symbol V will be used henceforth to represent both the transverse shear and the resisting shear. The sign convention selected will have sufficient generality to apply to both quantities.

The resultant of the fiber stresses on any transverse section has been shown to be a couple (if only transverse loads are considered) and has been designated as M_r. When the equilibrium equation $\Sigma M_O = 0$ (where O is any axis parallel to the neutral axis of the section) is applied to the free-body diagram of Fig. 5–4, the result can be written as

$$Rx - wx^2/2 - P(x-h) = M_r$$

or

$$M = M_r$$

where M *is defined as the algebraic sum of the moments of the external forces, acting on the part of the beam to either side of the section, with respect to an axis in the section and is called the bending moment or just the moment at the section.*

As seen from the definitions of M and M_r, the bending moment is equal in magnitude and opposite in sense to M_r. Since these moments are always equal in magnitude, they are frequently treated as though they were identical. For simplicity the symbol M will be used henceforth to represent both the bending moment and the resisting moment. The sign convention selected will have sufficient generality to apply to both quantities.

The bending moment and the transverse shear are not normally shown on a free-body diagram. The usual procedure is to show each external force individually as indicated in Fig. 5–4. The variation of V and M along the beam can be shown conveniently by means of equations or shear and moment diagrams.

A sign convention is necessary for the correct interpretation of results obtained from equations or diagrams for shear and moment. The following convention will give consistent results regardless of whether one proceeds from left to right or from right to left. By definition, the shear at a section is positive when the portion of the beam to the left of the section (for a horizontal beam) tends to move upward with respect to the portion to the right of the section as shown in Fig. 5–21*a*. Also by definition, the bending moment in a horizontal beam is positive at sections for which the top of the beam is in compression and the bottom is in tension as shown in Fig. 5–21*b*. Observe that the signs of the terms in the preceding equations for V and M agree with this convention.

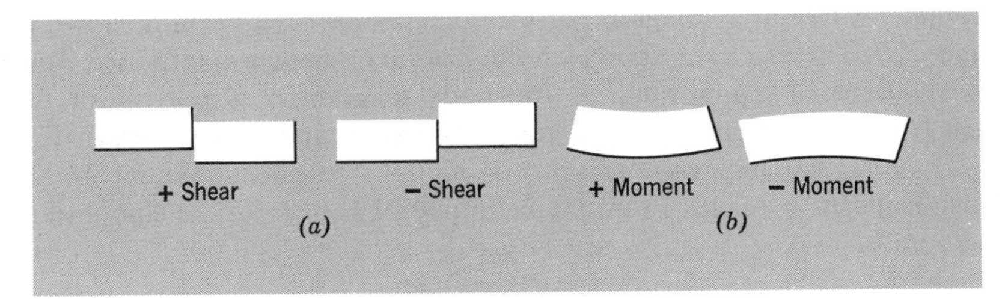

Fig. 5–21

Since M and V vary with x, they are functions of x, and equations for M and V can be obtained from free-body diagrams of portions of the beam. The procedure is illustrated in Exam. 5–9.

Example 5–9: A beam is loaded and supported as shown in Fig. 5–22*a*. Write equations for the shear and bending moment for any section of the beam in the interval BC.

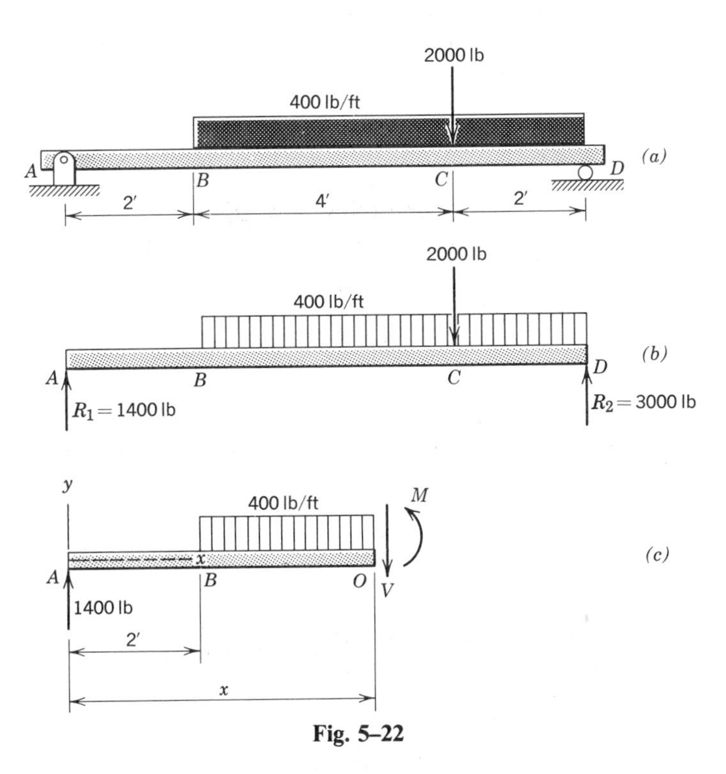

Fig. 5–22

Solution A free-body diagram, or load diagram, for the beam is shown in Fig. 5–22*b*. The reactions shown on the load diagram are determined from the equations of equilibrium. A free-body diagram of a portion of the beam from the left end to any section between B and C is shown in Fig. 5–22*c*. Note that the resisting shear V and the resisting moment M are shown as positive values. From the definition of V or from the equilibrium equation $\Sigma F_y = 0$,

$$V = 1400 - 400(x-2) = \underline{2200 - 400x} \qquad 2 \leqslant x \leqslant 6 \qquad \text{Ans.}$$

From the definition of M, or from the equilibrium equation $\Sigma M_O = 0$,

$$M = 1400x - 400(x-2)\left(\frac{x-2}{2}\right) = -\underline{200x^2 + 2200x - 800} \qquad 2 \leqslant x \leqslant 6 \qquad \text{Ans.}$$

The equations for V and M in the other intervals can be determined in a similar manner.

PROBLEMS

5–37* A beam is loaded and supported as shown in Fig. P5–37. Using the coordinate axes shown, write equations for the shear and bending moment for any section of the beam in the interval $2 \leqslant x \leqslant 8$.

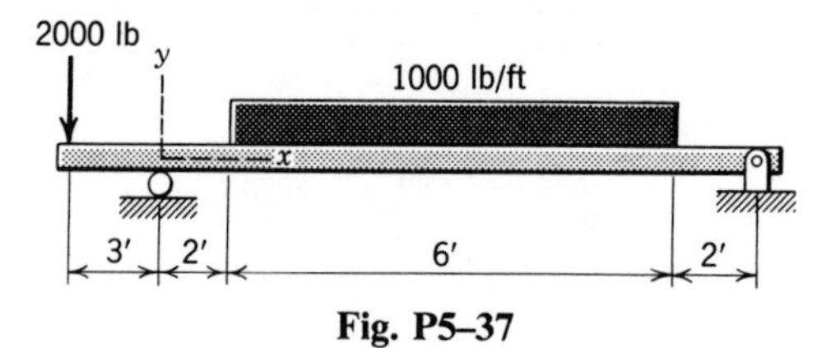

Fig. P5–37

5–38* A beam is loaded and supported as shown in Fig. P5–38. Using the coordinate axes shown, write equations for the shear and bending moment for any section of the beam in the interval $1 \leqslant x \leqslant 3$.

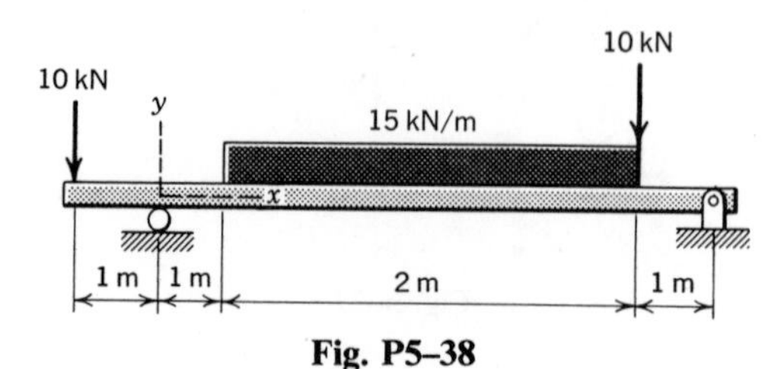

Fig. P5–38

5–39 A beam is loaded and supported as shown in Fig. P5–39. Using the coordinate axes shown, write equations for the shear and bending moment for any section of the beam in the interval $0 \leqslant x \leqslant 4$.

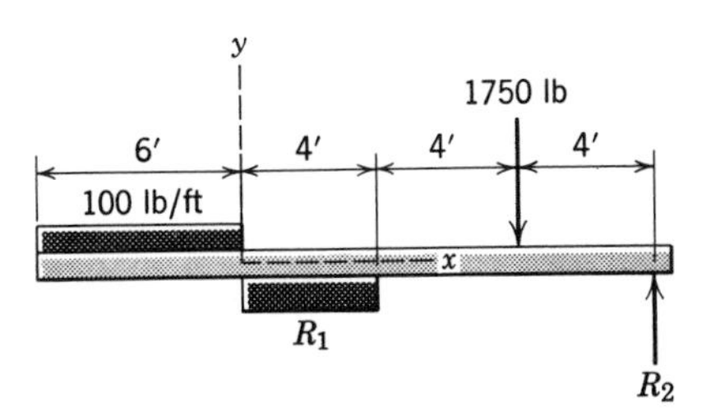

Fig. P5–39

5–40 A beam is loaded and supported as shown in Fig. P5–40. Using the coordinate axes shown, write equations for the shear and bending moment for any section of the beam in the interval $1 \leqslant x \leqslant 4$.

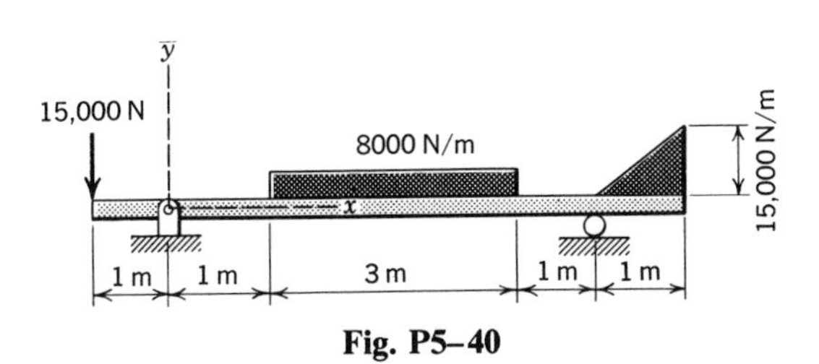

Fig. P5–40

5–41* For the beam in Fig. P5–41, and using the coordinate axes shown, write equations for the shear and bending moment for any section of the beam in the interval from B to C.

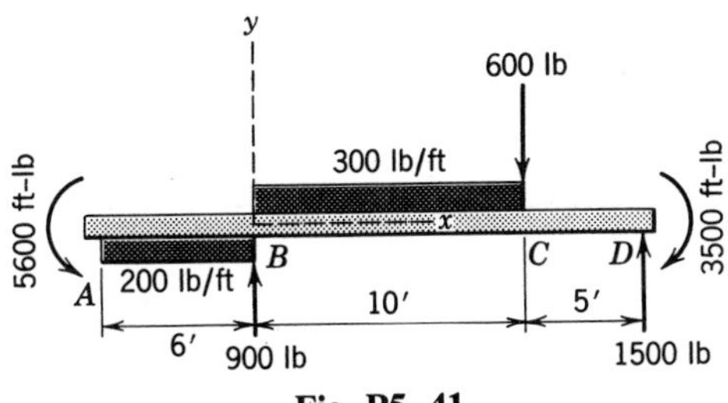

Fig. P5–41

5–42* For the beam in Fig. P5–42, and using the coordinate axes shown, write equations for the shear and bending moment for any section of the beam in the interval from A to B.

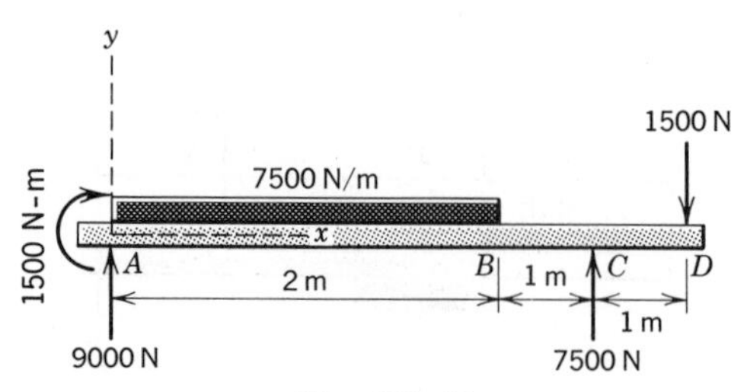

Fig. P5–42

5–43* A barge weighing 1000 lb/ft of length (not to be neglected) is loaded as shown in Fig. P5–43.

(a) Determine the length L if the barge is to float level (i.e., the water pressure must be uniform on the bottom of the barge).

(b) Write equations for the shear and bending moment for any section of the barge in the interval BC.

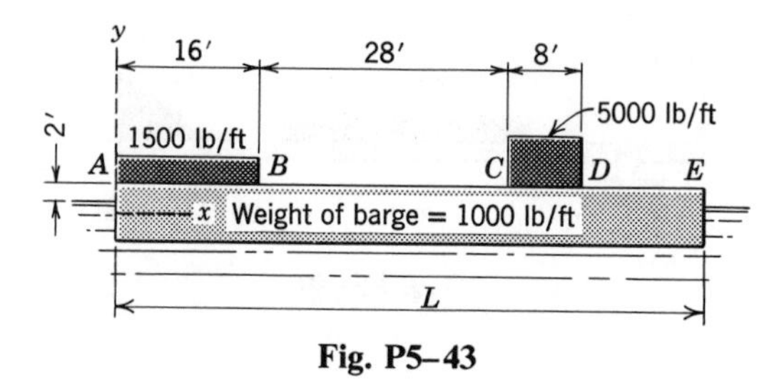

Fig. P5–43

5–44 A beam is loaded and supported as shown in Fig. P5–44. Using the coordinate axes shown, write equations for the shear and bending moment for any section of the beam in the interval $2 \leqslant x \leqslant 4.5$.

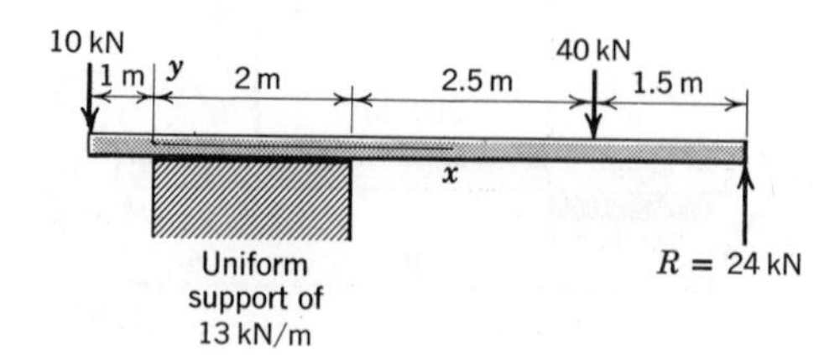

Fig. P5–44

5–45 For the beam in Fig. P5–45, and using the coordinate axes shown, write equations for the shear and bending moment for any section of the beam in the interval from C to D.

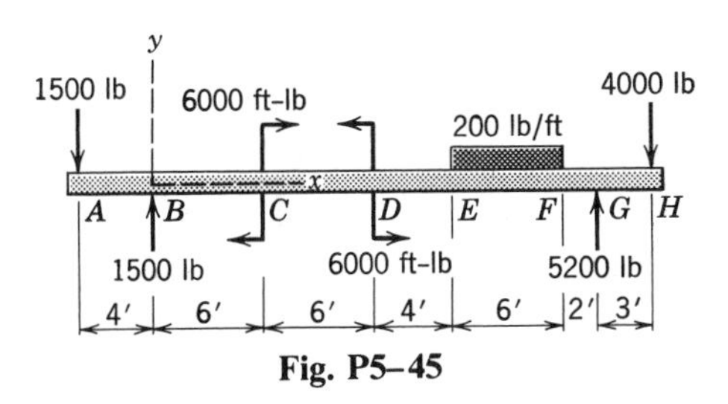

Fig. P5–45

5–46 For the beam in Fig. P5–46, and using the coordinate axes shown, write equations for the shear and bending moment for any section of the beam in the interval from C to D.

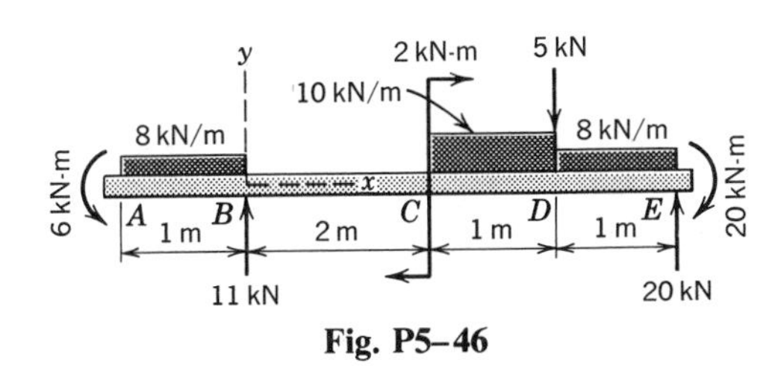

Fig. P5–46

5–7 LOAD, SHEAR, AND MOMENT RELATIONSHIPS

The mathematical relationships between loads and shears, and between shears and moments, can be used to facilitate construction of shear and moment diagrams. The relationships can be developed from a free-body diagram of an elemental length of a beam as shown in Fig. 5–23, in which the upward direction is considered positive for the applied load w and the shears and moments are shown as positive according to the sign convention established in Art. 5–6. The element must be in equilibrium, and the equation $\Sigma F_y = 0$ gives

$$(\uparrow +) \qquad \sum F_y = V + w\,dx - (V + dV) = 0$$

from which

$$dV = w\,dx \quad \text{or} \quad \frac{dV}{dx} = w$$

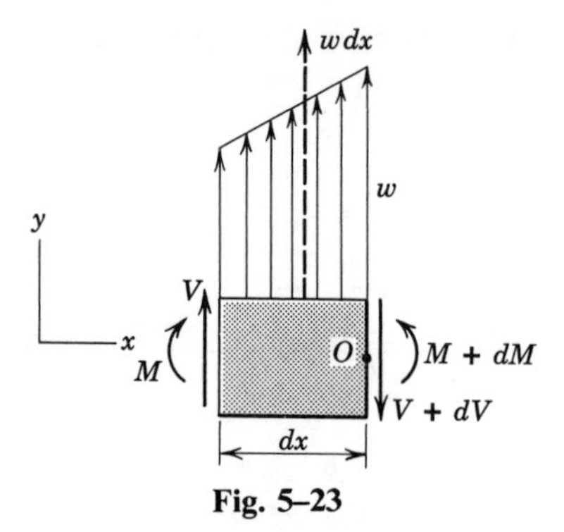

Fig. 5–23

The preceding equation indicates that at any section in the beam *the slope of the shear diagram is equal to the intensity of loading*. When w is known as a function of x, the equation can be integrated between definite limits as follows:

$$\int_{V_1}^{V_2} dV = \int_{x_1}^{x_2} w\,dx = V_2 - V_1$$

That is, *the change in shear between sections at x_1 and x_2 is equal to the area under the load diagram between the two sections.*

The equation $\Sigma M_O = 0$, applied to Fig. 5–23, gives

$$(\curvearrowleft +) \qquad \sum M_O = M + V\,dx + w\,dx(dx/2) - (M + dM) = 0$$

from which

$$dM = V\,dx + w(dx)^2/2$$

This expression reduces to

$$dM = V\,dx \quad \text{or} \quad V = \frac{dM}{dx}$$

when second-order differentials are considered as being negligible compared with first-order differentials. The preceding equation indicates that at any section in the beam *the slope of the moment diagram is equal to the shear*. The equation can be integrated between definite limits to give

$$\int_{M_1}^{M_2} dM = \int_{x_1}^{x_2} V\,dx = M_2 - M_1$$

Thus the change in moment between sections at x_1 and x_2 is equal to the area under the shear diagram between the two sections.

Note that the equations in this article were derived with the x axis positive to the right, the applied loads positive upward, and the shear and moment signs as indicated in Fig. 5–21. If one or more of these assumptions are changed, the algebraic signs in the equation may need to be altered.

The relations just developed will be useful in drawing shear and moment diagrams and in computing values of shear and moment at various sections along a beam, as illustrated in the next article.

5–8 SHEAR AND MOMENT DIAGRAMS

As stated in Art. 5–6, shear and moment diagrams provide a convenient method of obtaining maximum values of shear and moment and other information, to be discussed later. *A shear diagram is a graph in which abscissas represent distances along the beam and ordinates represent the transverse shear at the corresponding sections. A moment diagram is a graph in which abscissas represent distances along the beam and ordinates represent the bending moment at the corresponding sections.*

Shear and moment diagrams can be drawn by calculating values of shear and moment at various sections along the beam and plotting enough points to obtain a smooth curve. Such a procedure is rather time-consuming, and, although it may be desirable for graphic solutions of certain structural problems, other more rapid methods will be developed.

A convenient arrangement for constructing shear and moment diagrams is to draw a free-body diagram of the entire beam and construct shear and moment diagrams directly below. Two methods of procedure are presented in this article.

The first method consists of writing algebraic equations for the shear V and the moment M and constructing curves from the equations. This method has the disadvantage that unless the load is uniformly distributed or varies according to a known equation along the entire beam, no single elementary expression can be written for the shear V or the moment M, which applies to the entire length of the beam. Instead, it is necessary to divide the beam into intervals bounded by the abrupt changes in the loading. An origin should be selected (different origins may be used for different intervals); positive directions should be shown for the coordinate axes, and the limits of the abscissa (usually x) should be indicated for each interval.

Complete shear and moment diagrams should indicate values of shear and moment at each section where the load changes abruptly and at sections where they are maximum or minimum (negative maximum values). Sections where the shear and moment are zero should also be located.

The second method consists of drawing the shear diagram from the load diagram and the moment diagram from the shear diagram by means of the mathematical relationships developed in Art. 5–7. The latter method, though it may not produce a precise curve, is less time-consuming than the first and it does provide the information usually required.

When all loads and reactions are known, the shear and moment at the ends of the beam can be determined by inspection. Both shear and moment are zero at the free end of a beam unless a force or a couple or both is applied there; in this case, the shear is the same as the force and the moment the same as the couple. At a simply supported or pinned end, the shear must equal the end reaction and the moment must be zero. At a built-in or fixed-end beam, the reactions are the shear and moment values.

Once a starting point for the shear diagram is established, the diagram can be sketched by using the definition of shear and the fact that the slope of the shear diagram can be obtained from the load diagram. When positive directions are chosen as upward and to the right, a positive distributed load, one acting upward, will result in a positive slope on the shear diagram, and a negative load will give a negative slope. A concentrated force will produce an abrupt change in shear. The *change* in shear between any two sections is given by the area under the load diagram between the same two sections. The change of shear at a concentrated force is equal to the concentrated force.

The moment diagram is drawn from the shear diagram in the same manner. The slope at any point on the moment diagram is given by the shear at the corresponding point on the shear diagram, a positive shear representing a positive slope and a negative shear representing a negative slope, *when upward and to the right are positive*. The *change* in moment between any two sections is given by the area under the shear diagram between the corresponding sections. A couple applied to a beam will cause the moment to change abruptly by an amount equal to the moment of the couple. Examples 5–10 and 5–11 illustrate these two methods for construction of shear and moment diagrams.

Example 5–10 A beam is loaded and supported as shown in Fig. 5–24*a*.

(a) Write equations for the shear and bending moment for any section of the beam in the interval BC.

(b) Draw complete shear and moment diagrams for the beam.

Solution A free-body diagram, or load diagram, for the beam is shown in Fig. 5–24*b*. The reactions at C are computed from the equations of equilibrium. It is not necessary to compute the reactions on a cantilever

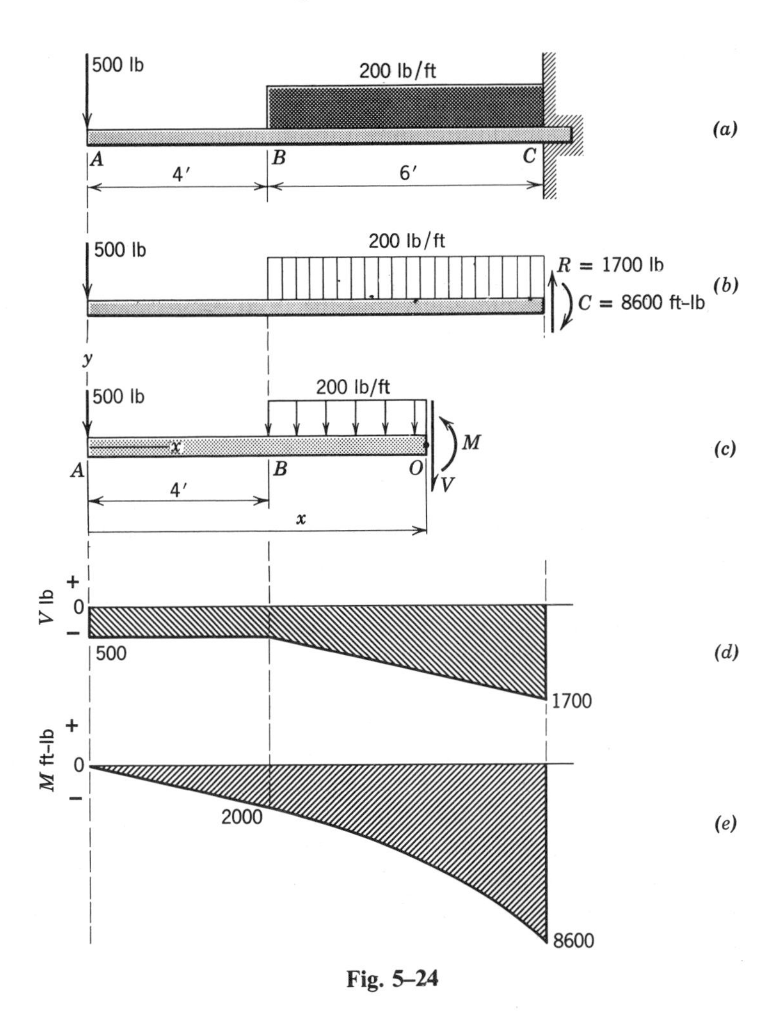

Fig. 5–24

beam in order to write equations for the shear and bending moment or to draw shear and moment diagrams from the load diagram, but the reactions provide a convenient check.

(a) A free-body diagram of a portion of the beam from the left end to any section between B and C is shown in Fig. 5–24c. Note that the resisting shear V and the resisting moment M are shown as positive values. From the definition of V or from the equilibrium equation $\Sigma F_y = 0$,

$$V = -500 - 200(x-4) = \underline{300 - 200x} \qquad 4 \leqslant x \leqslant 10 \qquad \text{Ans.}$$

From the definition of M, or from the equilibrium equation $\Sigma M_O = 0$,

$$M = -500x - 200(x-4)(x-4)/2$$

$$= \underline{-100x^2 + 300x - 1600} \qquad 4 \leqslant x \leqslant 10 \qquad \text{Ans.}$$

(b) The equations for V and M in the interval AB can be written in the same manner as in part (a) after which the two equations for V can be plotted in the appropriate intervals to give the shear diagram in Fig. 5–24*d*. Likewise, the two equations for M can be plotted in the appropriate intervals to give the moment diagram in Fig. 5–24*e*.

The shear and moment diagrams can also be drawn, working directly from the load diagram without writing the shear and moment equations.

The shear diagram is drawn below the load diagram (Fig. 5–24*d*). The shear just to the right of the 500-lb load is -500 lb. The slope of the shear diagram is equal to the load, and, since no load is applied between A and B, the slope of the diagram is zero. From B to C the load is uniform; therefore, the slope of the shear diagram is constant. The *change* of shear from B to C is equal to the applied load (1200 lb) from B to C, and the shear is -1700 lb at C. This shear is the same as the reaction at C, which provides a check.

The moment diagram is drawn below the shear diagram as Fig. 5–24*e*. The moment is zero at the free end A. From A to B the shear is constant (-500); therefore, the slope of the moment diagram is constant, and the moment diagram is a straight line. The shear increases (negatively) from -500 lb at B to -1700 lb at C; therefore, the slope of the moment diagram is negative from B to C and increases uniformly in magnitude. The change in moment from A to B is equal to the area under the shear diagram and is $500(4) = 2000$ ft-lb. The area under the shear diagram from B to C is $(\frac{1}{2})(500 + 1700)(6) = 6600$ ft-lb, which represents the change in moment from B to C. Thus, the moment at C is $-2000 + (-6600) = -8600$ ft-lb.

Example 5–11 A beam is loaded and supported as shown in Fig. 5–25*a*.

(a) Write equations for the shear and bending moment for any section of the beam in the interval CD.

(b) Draw complete shear and moment diagrams for the beam.

Solution The reactions shown on the load diagram (Fig. 5–25*b*) are determined from the equations of equilibrium.

(a) A free-body diagram of a portion of the beam from the left end to any section between C and D is shown in Fig. 5–25*c*. From the definition

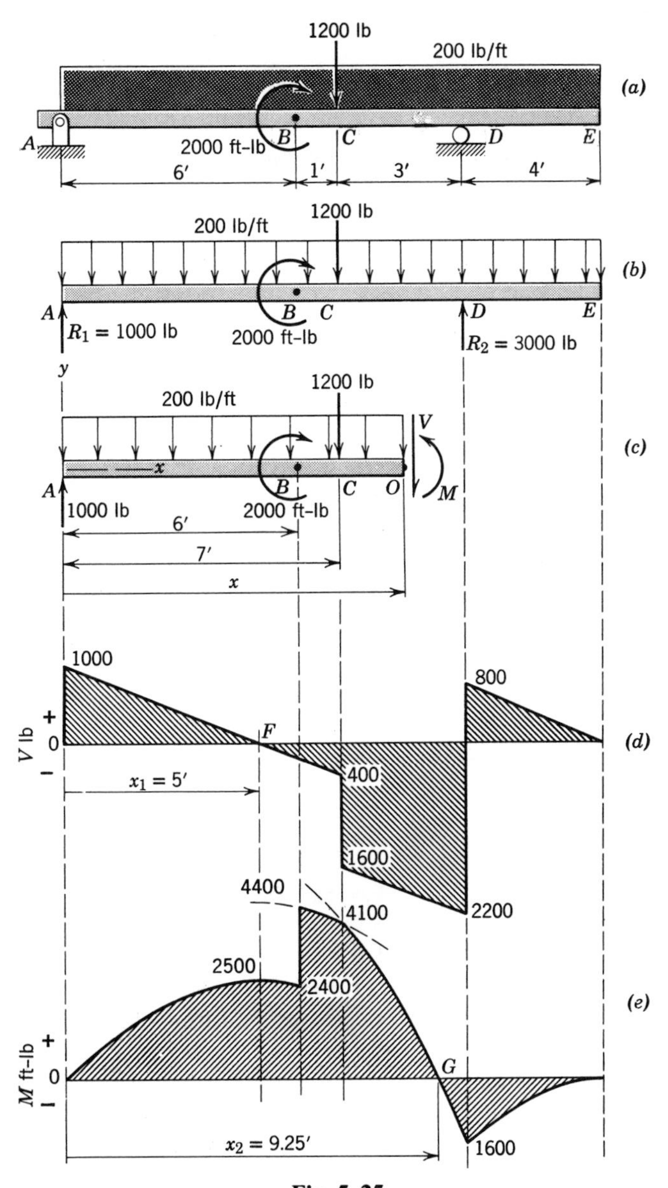
1200 lb
200 lb/ft
(a)
A
2000 ft-lb
B
C
D
E
6′
1′
3′
4′
1200 lb
200 lb/ft
(b)
A
R_1 = 1000 lb
B
C
2000 ft-lb
D
E
R_2 = 3000 lb
y
1200 lb
200 lb/ft
V
(c)
x
A
1000 lb
B
C
O
M
2000 ft-lb
6′
7′
x
1000
800
V lb
+
0
–
F
(d)
400
x_1 = 5′
1600
2200
4400
4100
2500
2400
(e)
M ft-lb
+
0
–
G
x_2 = 9.25′
1600

Fig. 5–25

of V, or from the equilibrium equation $\Sigma F_y = 0$,

$$V = 1000 - 200x - 1200 = -\underline{200x - 200} \qquad 7 \leqslant x \leqslant 10 \qquad \text{Ans.}$$

From the definition of M, or from the equilibrium equation $\Sigma M_O = 0$,

$$M = 1000x - 200x(x/2) + 2000 - 1200(x-7)$$

$$= \underline{10{,}400 - 200x - 100x^2} \qquad 7 \leqslant x \leqslant 10 \qquad \text{Ans.}$$

(b) The equations for V and M in the other intervals can be written and the shear and moment diagrams obtained by plotting these equations. In this example, the second method is used and the shear diagram is drawn directly from the load diagram.

The shear just to the right of A is 1000 lb and decreases at the rate of 200 lb each foot to -400 lb at C. The concentrated downward load at C causes the shear to change suddenly from -400 lb to -1600 lb. The shear changes uniformly to -2200 lb just to the left of D. The reaction at D changes the shear by 3000 lb (from -2200 lb to $+800$ lb). The shear decreases uniformly to zero at E. Note that the distributed load is uniform over the entire beam; consequently, the slope of the shear diagram is constant. Points of zero shear, such as F of Fig. 5–25d, are located from the geometry of the shear diagram. For example, the slope of the left portion of the shear diagram is 200 lb/ft. Therefore,

$$\text{slope} = 200 = 1000/x_1$$

and

$$x_1 = 5.00 \text{ ft}$$

The moment is zero at A, and the slope of the moment diagram (equal to the shear) is 1000 ft-lb per ft. From A to C the shear and, hence, the slope of the moment diagram, decreases uniformly to zero at F and to -400 at C. The abrupt change of shear at C indicates a sudden change of slope of the moment diagram; thus, the two parts of the moment diagram at C are not tangent. The slope of the moment diagram changes from -1600 at C to -2200 just to the left of D. From D to E the slope changes from $+800$ to zero.

The change of moment from A to F is equal to the area under the shear diagram from A to F and is

$$\Delta M = 1000(5)/2 = 2500 \text{ ft-lb}$$

which is the moment at F since the moment at A is zero. From F to B,

$\Delta M = -100$ ft-lb and $M_B = +2400$ ft-lb. The 2000-ft-lb couple (point moment) is applied at B. Since the moment just to the left of B is positive and the couple contributes an additional positive moment to sections to the right of B, the moment changes abruptly from +2400 ft-lb to +4400 ft-lb. Moments at C, D, and E can be determined from the shear diagram areas that give the changes in moment from section to section. If the moment at E is not equal to zero, it indicates that an error has occurred.

The point of inflection G, where the moment is zero, can be determined by setting the expression for the moment from part (a) equal to zero and solving for x. The result is

$$x_2 = -1 \pm \sqrt{105} = 9.247 = \underline{9.25 \text{ ft}} \qquad \text{Ans.}$$

Note in the example that maximum and minimum moments may occur at sections where the shear curve passes through zero. In general, the shear curve may pass through zero at a number of points along the beam, and each such crossing indicates a point of *possible* maximum moment (in engineering, the moment with the largest absolute value is the maximum moment). It should be emphasized that the shear curve does not indicate the presence of abrupt discontinuities in the moment curve; hence, the maximum moment may occur where a couple is applied to the beam, not where the shear passes through zero. All possibilities should be examined to determine the maximum moment.

Sections where the bending moment is zero, called points of inflection or contraflexure, can be located by equating the expression for M to zero. The fiber stress is zero at such sections, and if a beam must be spliced, the splice should be located at or near a point of inflection if there is one.

PROBLEMS

5–47 Draw complete shear and bending moment diagrams for the beam of Fig. P5–37.

5–48 Draw complete shear and bending moment diagrams for the beam of Fig. P5–38.

5–49* For the beam of Fig. P5–49:

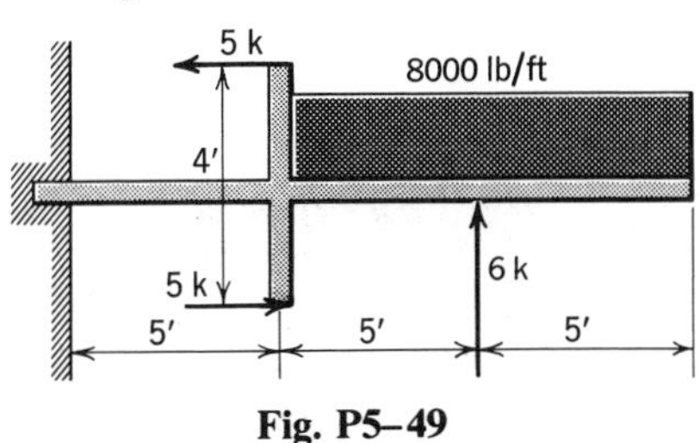

Fig. P5–49

(a) Write equations for the shear and bending moment for any section in the center 5-ft interval.

(b) Draw complete shear and bending moment diagrams for the beam.

5–50* For the beam of Fig. P5–50:

(a) Write the equations for the shear and bending moment at any section of the beam.

(b) Draw complete shear and bending moment diagrams for the beam.

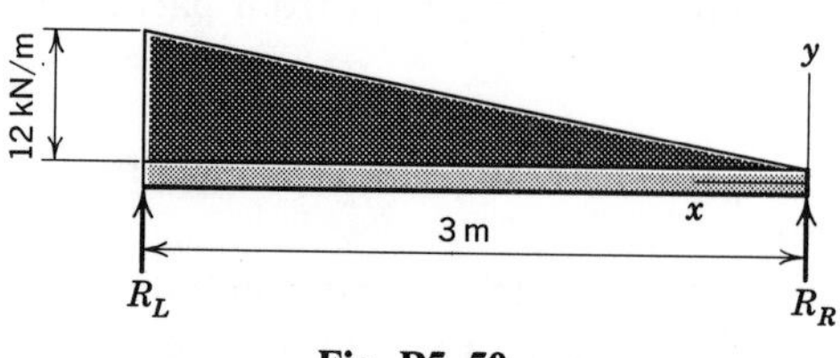

Fig. P5–50

5–51* For the beam of Fig. P5–51, and using the coordinate axes shown:

(a) Write the equations for the shear and moment at any section of the beam in the interval AB.

(b) Draw complete shear and moment diagrams for the beam.

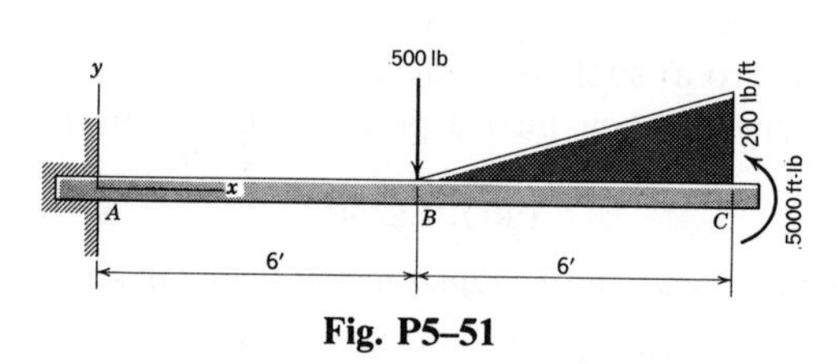

Fig. P5–51

5–52* For the beam of Fig. P5–52, and using the coordinate axes shown:

(a) Write equations for the shear and moment for any section between the supports.

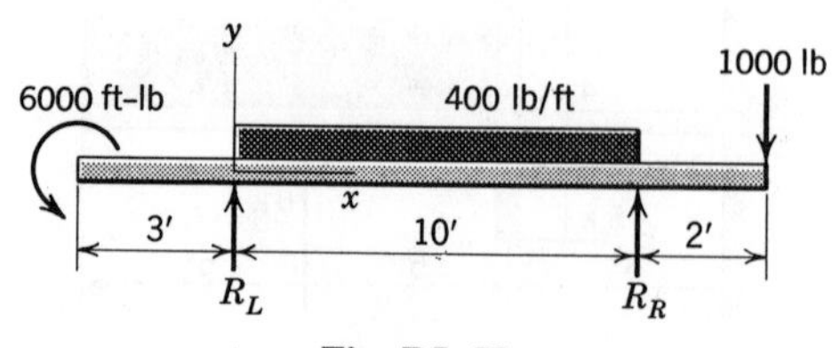

Fig. P5–52

(b) Draw the complete shear diagram and indicate sections of possible maximum moment.

(c) Without drawing a moment diagram, determine the maximum moment in the beam.

5–53* For the beam of Fig. P5–53, and using the coordinate axes shown:

(a) Write the equations for the shear and moment for any section of the beam in the interval from B to C.

(b) Draw the complete shear diagram and indicate sections of possible maximum moment.

(c) Without drawing a moment diagram, determine the maximum moment in the beam.

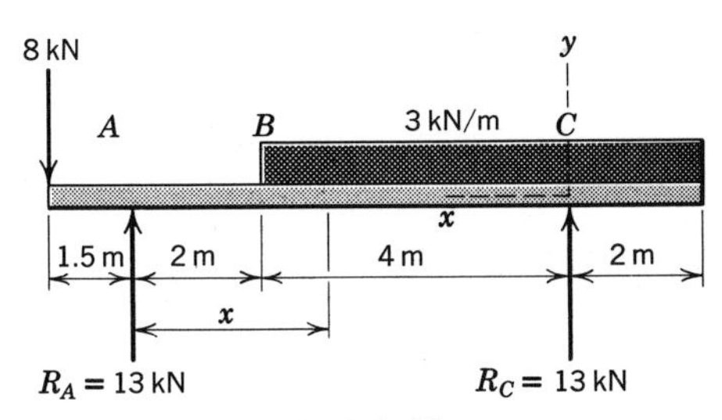

Fig. P5–53

5–54 Draw complete shear and moment diagrams for the beam of Fig. P5–39.

5–55 Draw complete shear and moment diagrams for the beam of Fig. P5–40.

5–56 Draw complete shear and moment diagrams for the beam of Fig. P5–41.

5–57 Draw complete shear and moment diagrams for the beam of Fig. P5–42.

5–58 Draw complete shear and moment diagrams for the beam of Fig. P5–58.

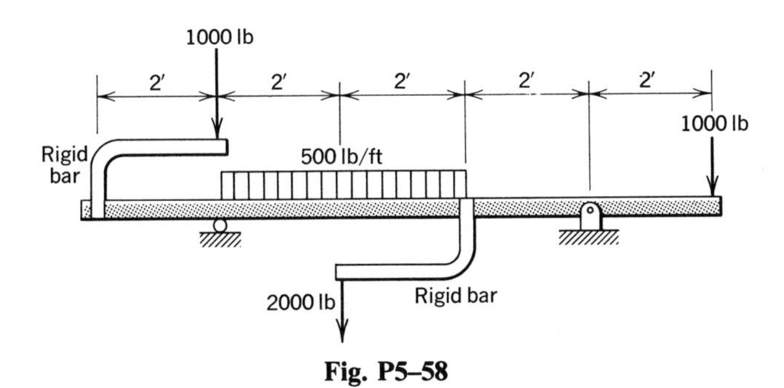

Fig. P5–58

5–59 Draw complete shear and moment diagrams for the beam of Fig. P5–59.

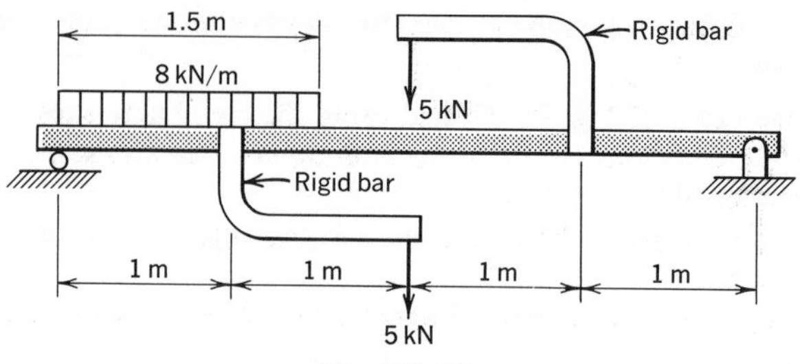

Fig. P5–59

5–60* Draw complete shear and moment diagrams for the beam of Fig. P5–60, and write the moment equation for the interval BC.

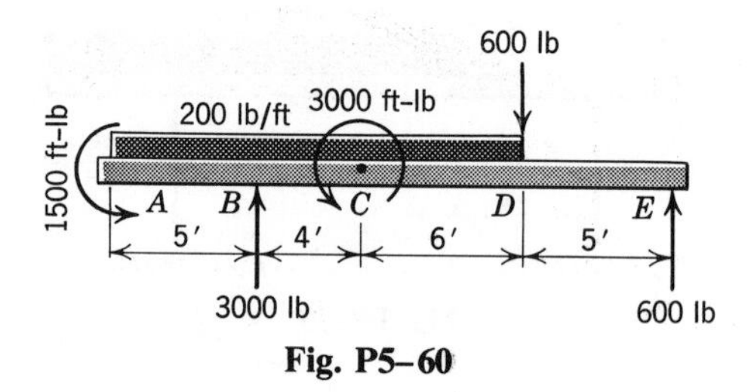

Fig. P5–60

5–61* Draw complete shear and moment diagrams for the beam of Fig. P5–61, and write the moment equation for the interval CD.

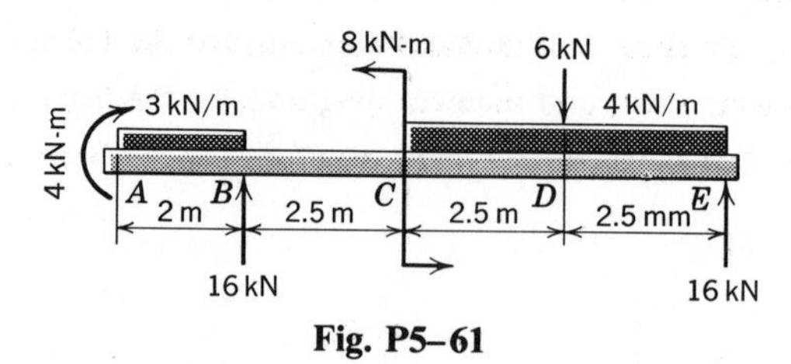

Fig. P5–61

5–62 Draw complete shear and moment diagrams for the beam of Fig. P5–62, and write moment equations for (1) the interval BC and (2) the interval EF.

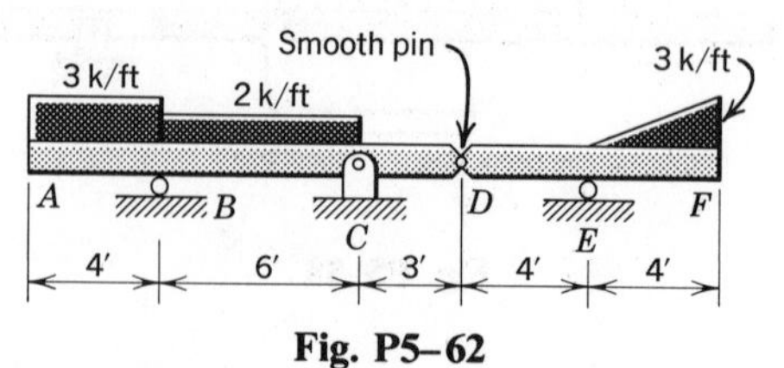

Fig. P5–62

5-63 The shear and moment at end A of the beam segment (in equilibrium) of Fig. P5-63 are $+3$ kN and -2 kN-m, respectively, and the shear and moment at end D are unknown. Draw complete shear and moment diagrams for the beam segment, and write the moment equation for the interval CD.

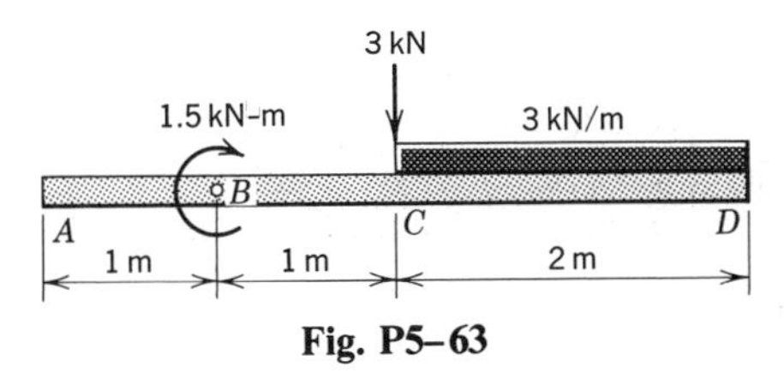

Fig. P5-63

5-64 In Fig. P5-64 one span of a continuous beam is shown. The moment at support A is $-wL^2/20$ ft-lb and the moment at support B is $-wL^2/10$ ft-lb. Draw complete shear and moment diagrams for the portion of the beam between supports A and B.

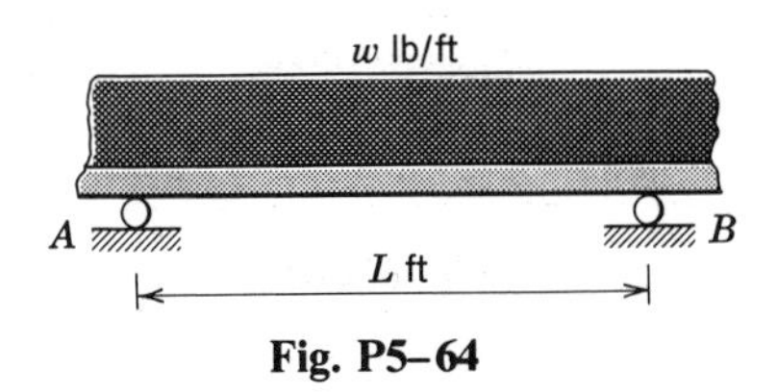

Fig. P5-64

5-65 The beam of Fig. P5-65 has a pin-connected joint at C. The moment and shear at A are $+12$ kN-m and $+6$ kN, respectively. Draw complete shear and moment diagrams for the length $ABCD$ of the beam, and write the moment equation for the interval BCD.

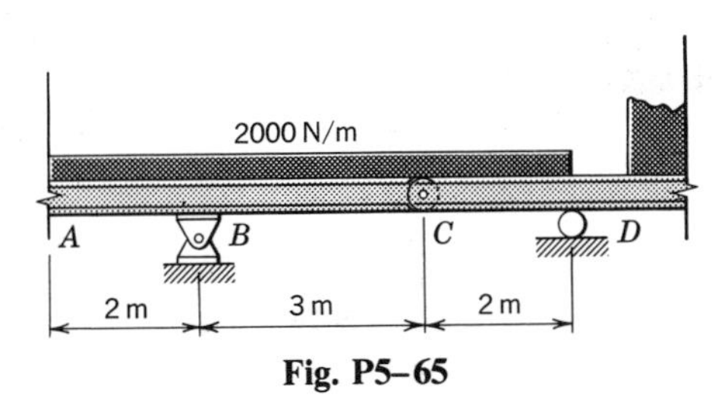

Fig. P5-65

5–66 Draw complete shear and moment diagrams for the beam of Fig. P5–45.

5–67 Draw complete shear and moment diagrams for the beam of Fig. P5–46.

5–68 Select the lightest steel wide-flange beam that may be used for the beam of Fig. P5–68. Use a working fiber stress of 18,000 psi.

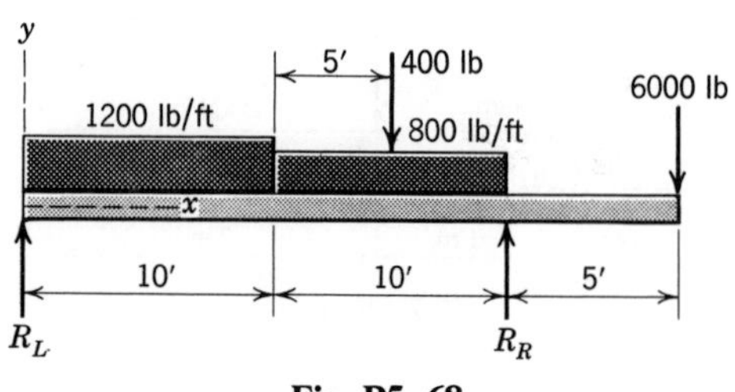

Fig. P5–68

5–69 Select the lightest steel wide-flange or American standard beam that may be used for the beam of Fig. P5–69. Use a working fiber stress of 18,000 psi.

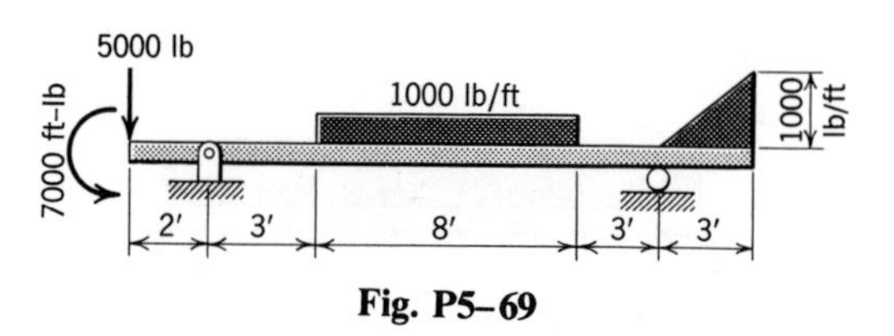

Fig. P5–69

5–70* The beam shown in Fig. P5–70*a* has the cross section shown in Fig. P5–70*b*. Determine the maximum fiber stress between sections *A* and *C*, and state where it occurs.

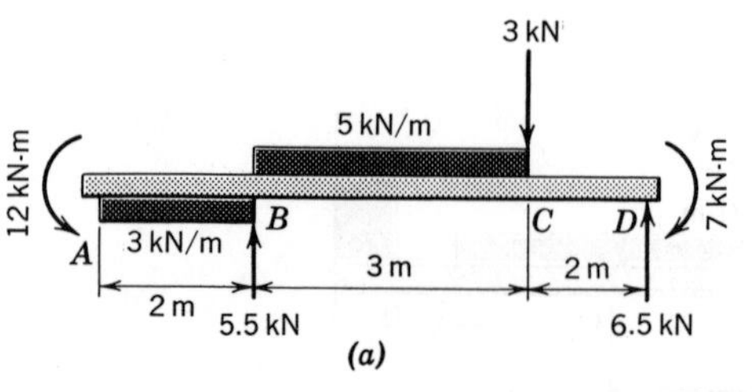

(*a*)

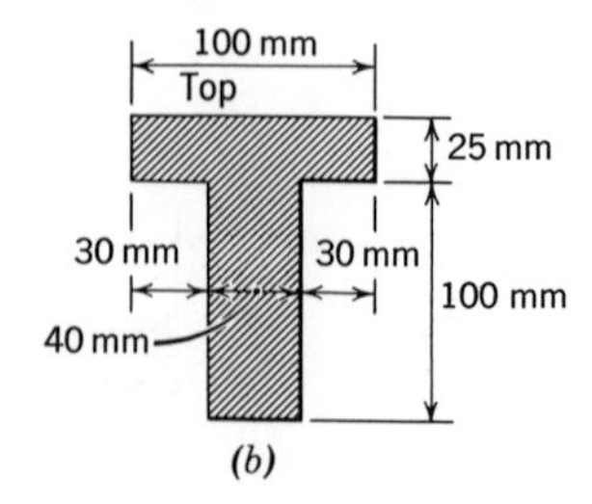

(*b*)

Fig. P5–70

5-71* The beam shown in Fig. P5-71*a* has the cross section shown in Fig. P5-71*b*. Determine:

(a) The maximum tensile fiber stress in the beam and state where it occurs.
(b) The maximum compressive fiber stress in the beam and state where it occurs.

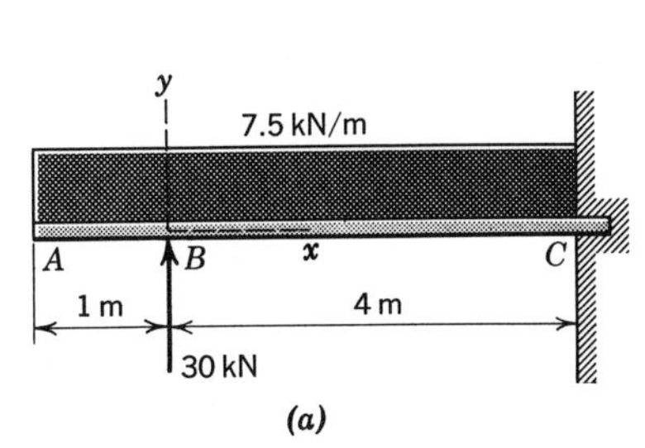

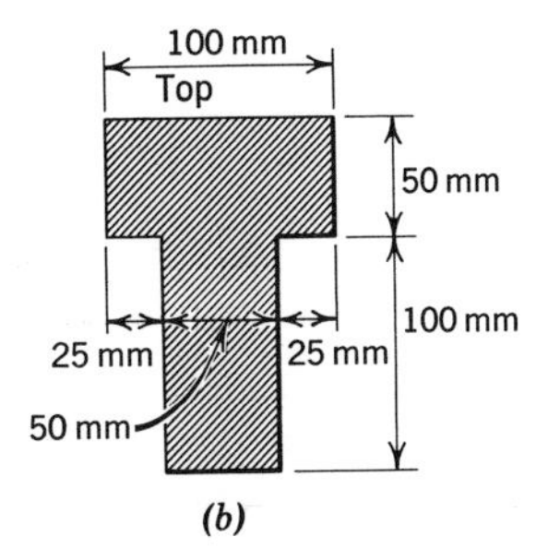

Fig. P5-71

5-72* The beam shown in Fig. P5-72*a* has the cross section shown in Fig. P5-72*b*. Determine:

(a) The maximum tensile fiber stress in the beam and state where it occurs.
(b) The maximum compressive fiber stress in the beam and state where it occurs.

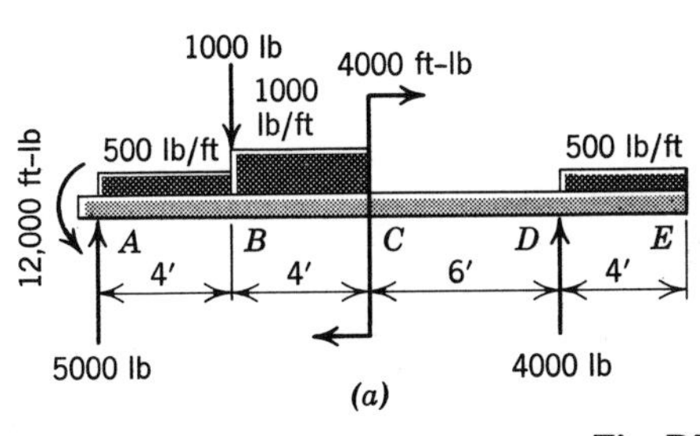

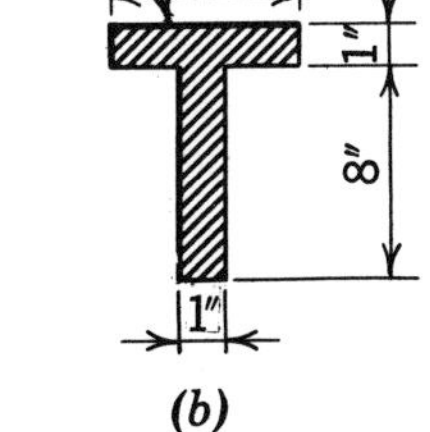

Fig. P5-72

5–73 The shear diagram for a beam, with zero moment on the left end and with no couples applied between the ends, is shown in Fig. P5–73. Draw the free-body diagram and the moment diagram for the beam.

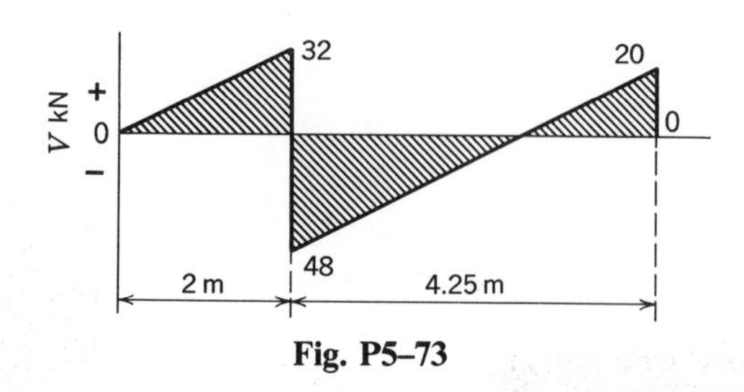

Fig. P5–73

5–74 The shear diagram for a beam fixed at the right end is shown in Fig. P5–74. There are no applied couples anywhere else along the beam. Draw the load and moment diagrams for the beam.

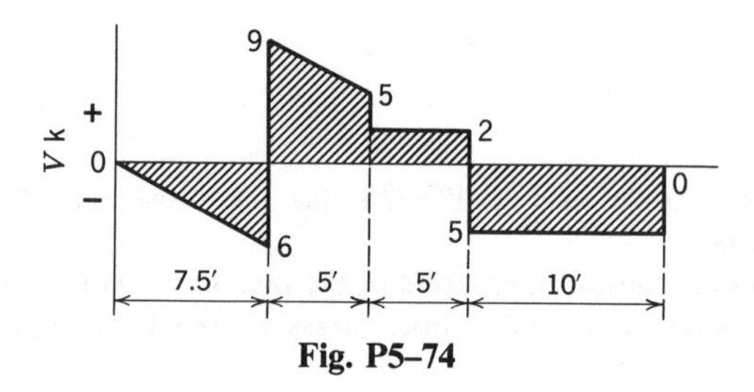

Fig. P5–74

5–75 The moment diagram for a beam is shown in Fig. P5–75. Construct the shear and load diagrams for the beam.

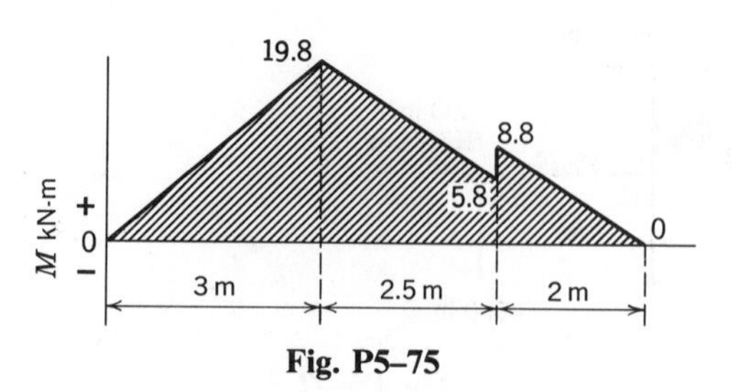

Fig. P5–75

5–76 Draw shear and load diagrams for the beam which has the moment diagram in Fig. P5–76.

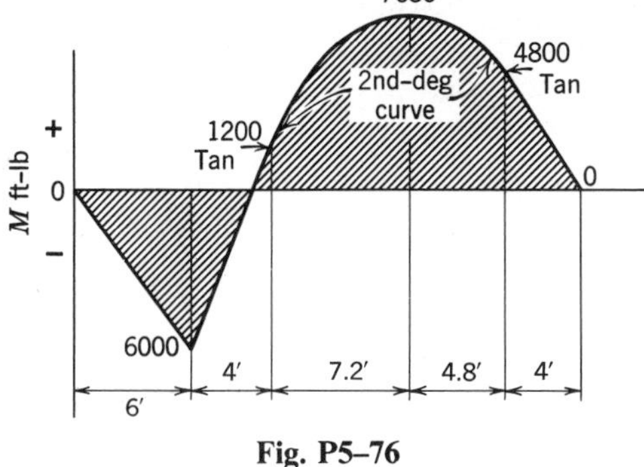

Fig. P5–76

5–9 SHEARING STRESSES BY EQUILIBRIUM

In Art. 5–2, the discussion of shearing stresses in beams was bypassed while flexural stresses were studied. This procedure seems to be in keeping with the historical record on the study of beam stresses. From the time of Coulomb's paper, which contained the correct theory of distribution of flexural stresses, approximately seventy years elapsed before the Russian engineer D. J. Jourawski (1821–1891), while designing timber railroad bridges in 1844–1850, developed the elementary shear stress theory used today. Saint-Venant in 1856 developed a rigorous solution for beam shearing stresses, but the elementary solution of Jourawski, being much easier to apply and quite adequate, is the one in general use today by engineers and architects and the one that will be presented here. Note that the method, because it involves the elastic flexure formula, is limited to elastic action. This limitation need be no cause for concern because, for practical engineering problems, the shearing stress evaluation discussed in this and following articles is important only in (1) the evaluation of principal elastic stresses at interior points in certain metallic beams (see Art. 5–11), and (2) the design of timber beams, because of the longitudinal plane of low shear resistance. The allowable stresses for timber are always in the elastic range. In beams of elastoplastic steel designed on the basis of plastic theory, methods for evaluation of shear stresses are available[5]; however, this area is beyond the scope of this book.

If one constructs a beam by stacking flat slabs one on top of another without fastening them together, and then loads this beam in a direction normal to the surface of the slabs, the resulting deformation will appear

[5]See, for example, *Commentary on Plastic Design in Steel*, A.S.C.E. Manual of Engineering Practice No. 41, 1961.

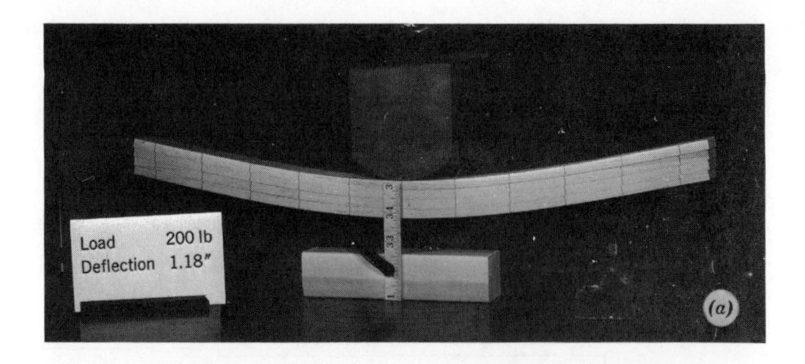

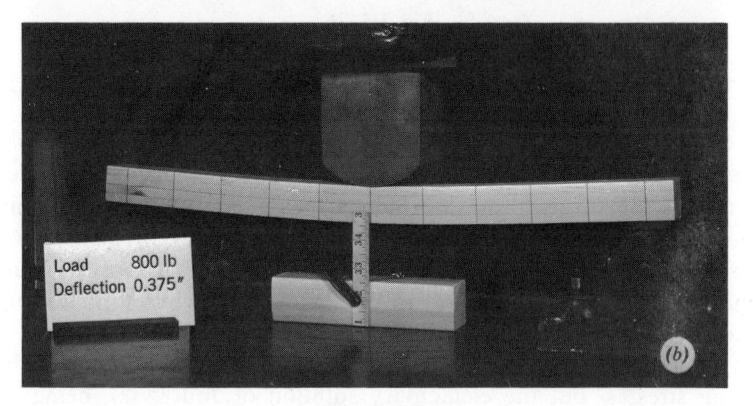

Fig. 5–26

somewhat like that in Fig. 5–26*a*. The same type of deformation can be observed by taking a pack of cards and bending them, noting the relative motion of the ends of the cards with respect to each other. The fact that a solid beam does not exhibit this relative movement of longitudinal elements (see Fig. 5–26*b*, in which the beam is identical to that of Fig. 5–26*a* except that the layers are glued together) indicates the presence of shearing stresses on longitudinal planes. The evaluation of these shearing stresses will now be studied by means of the free-body diagram and equilibrium approach, as outlined in the following example.

Example 5–12 A 12-in. long segment of a beam having the *T* cross section of Fig. 5–27*b* is subjected to a moment of +5400 ft-lb at the left end and a constant shear of +900 lb. Determine the average shearing stress on a horizontal plane 4 in. above the bottom of the beam. Assume elastic action.

Solution A free-body diagram of the segment is shown in Fig. 5–27*a*, where *AB* designates the plane on which the shearing stress is to be

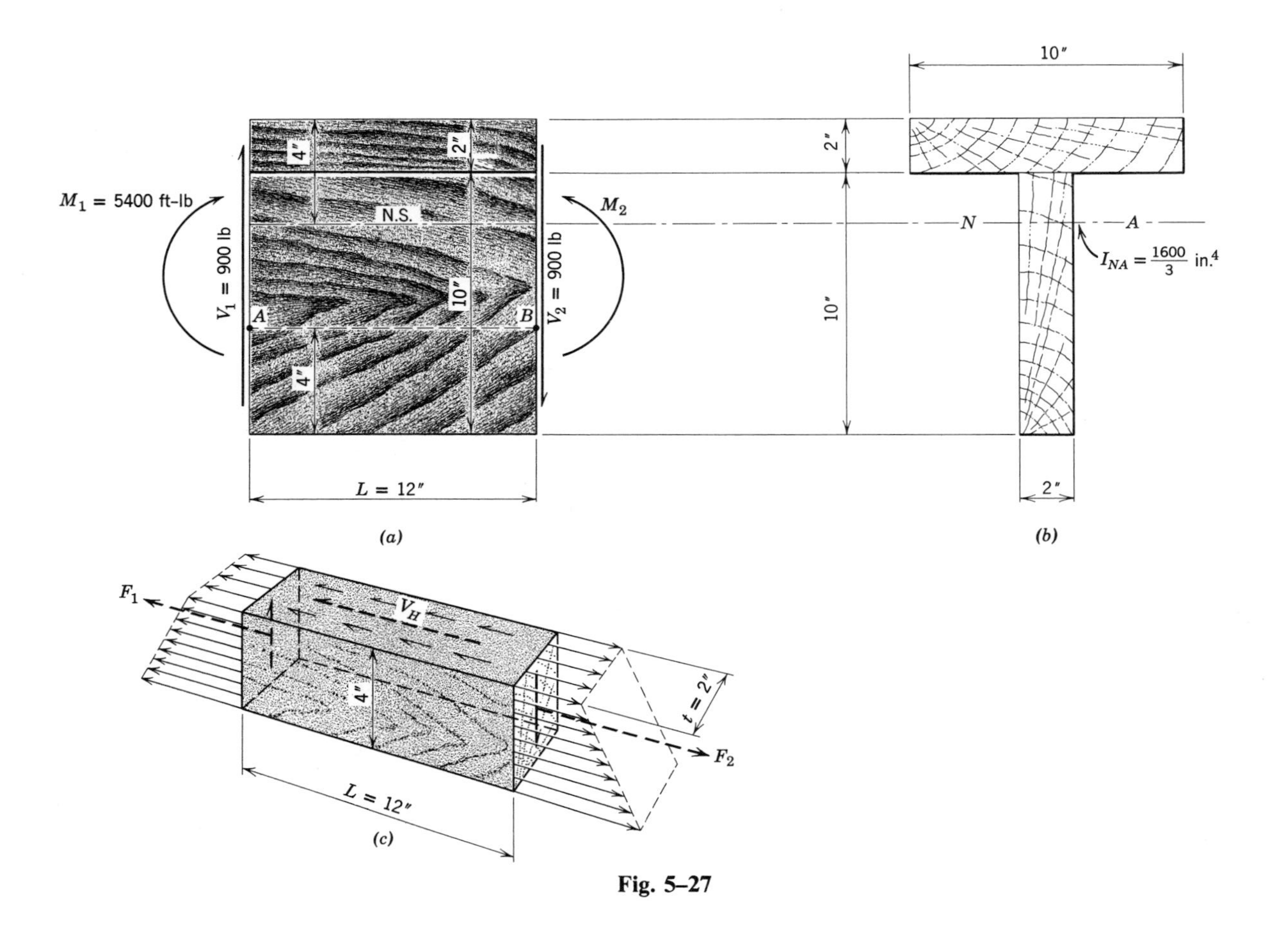

M_1 = 5400 ft-lb
V_1 = 900 lb
A
4"
2"
N.S.
10"
4"
B
V_2 = 900 lb
M_2
L = 12"
(a)
10"
2"
10"
N
A
$I_{NA} = \frac{1600}{3}$ in.^{4}
2"
(b)
F_1
V_H
4"
t = 2"
F_2
L = 12"
(c)

Fig. 5–27

evaluated. The moment M_2, obtained from a moment equilibrium equation, is +6300 ft-lb. A free-body diagram of the portion of the beam segment below plane AB is shown in Fig. 5–27c. The force V_H is the resultant of the shearing stresses on plane AB. The forces F_1 and F_2 are the resultants of the normal stresses on the ends of the segment and are equal to the average fiber stress (which occurs 6 in. below the neutral surface) on the end area multiplied by the area; that is,

$$F_1 = \frac{M_1 y}{I} A = \frac{5400(12)(6)}{1600/3}(2)(4) = 108(54) \text{ lb}$$

and

$$F_2 = \frac{M_2 y}{I} A = \left(\frac{M_2}{M_1}\right) F_1 = \frac{63}{54}(108)(54) = 108(63) \text{ lb}$$

Summing forces in the longitudinal direction in Fig. 5–27c gives

$$V_H = F_2 - F_1 = 108(63) - 108(54) = 972 \text{ lb}$$

The average shearing stress on plane AB is

$$\tau = \frac{V_H}{A_H} = \frac{972}{12(2)} = \underline{40.5 \text{ psi}} \qquad \text{Ans.}$$

In this example, the shearing stress on plane AB does not vary with the length; therefore, since the shearing stresses on orthogonal planes have the same magnitude, the shearing stresses on the transverse (vertical) planes at A and B are also 40.5 psi in the directions shown.

PROBLEMS

Note. Assume elastic action for the following problems.

5–77* A 300-mm section of a 200-mm deep by 80-mm wide timber beam is shown in Fig. P5–77a, and a free-body diagram of the bottom 60 mm of the beam is indicated in Fig. P5–77b.

(a) Determine values for M_2, F_1, and F_2.
(b) By applying an equilibrium equation to Fig. P5–77b, determine the resultant of the shearing stresses indicated on the area $ABCD$.
(c) Compute the average shearing stress on the area $ABCD$.

5–78* A timber T-beam is made up of two 2×6-in. planks glued together at $ABCD$, as shown in Fig. P5–78a. This glued joint must offer shearing resistance, as shown in Fig. P5–78b. The vertical shear at sections 1–1 and 2–2 is +1020 lb. The bending moment is +2040 ft-lb at section 1–1 and +3060 ft-lb at section 2–2.

(a) After computing values for F_1 and F_2, apply an equilibrium equation to Fig. P5–78b, and determine the total resistance offered by the glue.
(b) Determine the average shearing stress at the joint.

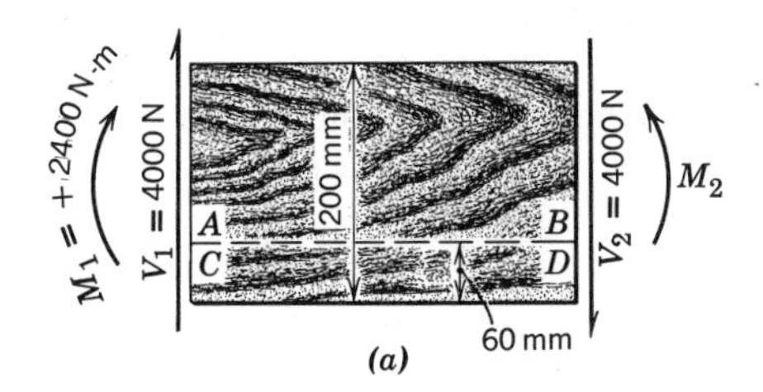

(a)

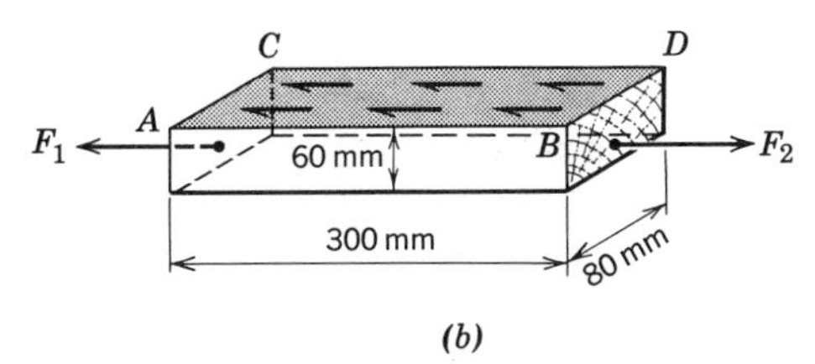

(b)

Fig. P5–77

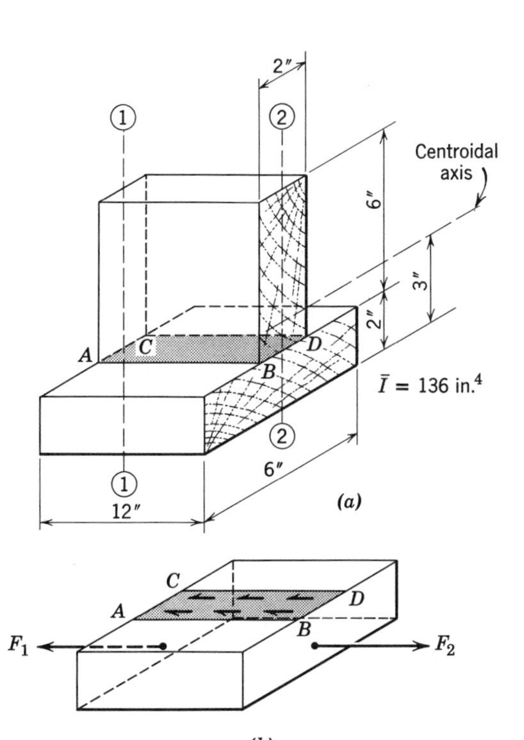

(a)

(b)

Fig. P5–78

5–79* If the beam of Prob. 5–78 is made up of one 2×8-in. plank and two 2×2-in. sticks as shown in Fig. P5–79, determine:

(a) The total resistance offered by the glue on surface $ABCD$.
(b) The average shearing stress at the joint.

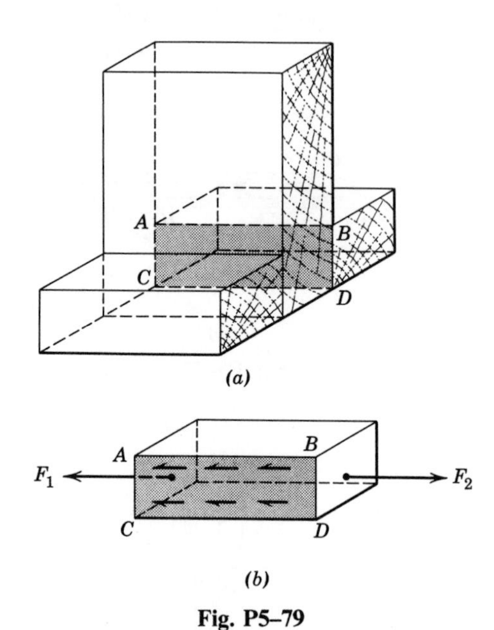

Fig. P5–79

5–80 A timber beam is to be fabricated of 50-mm planks to form the cross section of Fig. P5–80. A 400-mm length of the beam for which the shear is constant will be subjected to bending moments of +15 and +25 kN-m at the ends. Which method of fabrication, (*a*) or (*b*) of Fig. P5–80, will result in the least average shearing stress in the glued joint? Compute these stresses by the free-body diagram method.

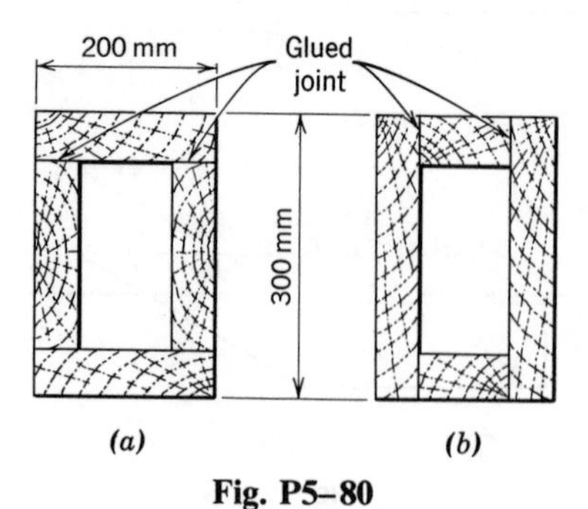

Fig. P5–80

5–81* A timber beam is fabricated of two 2×6-in. and one 2×4-in. pieces of lumber glued together to form the cross section of Fig. P5–81. In a region of constant shear, the two ends of a 12-in. segment of the beam are subjected to negative bending moments of 10 and 8 ft-kips, respectively. Determine, by the free-body diagram method, the average shearing stresses in each of the two glued joints between the flanges and the web.

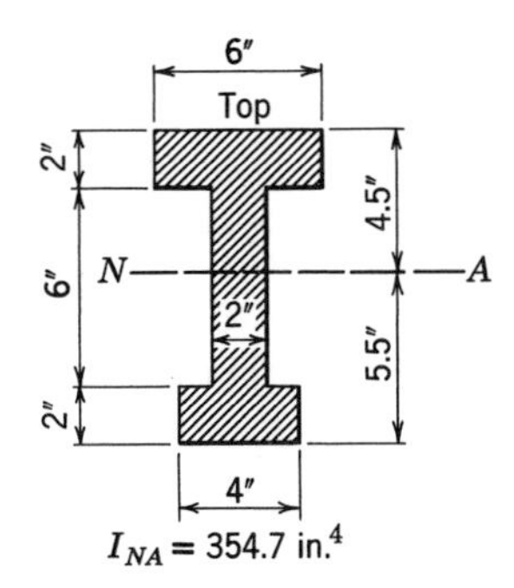

Fig. P5–81

5–10 THE SHEARING STRESS FORMULA

The shearing stress as determined in Exam. 5–12 is reasonably accurate only if the width of the section on plane AB is not large compared to the depth of the beam and the transverse shear in the beam is constant throughout the length of the segment AB. If the transverse shear varies, the length L must be reduced to a short length Δx. The method outlined in the example, though excellent for emphasizing the concept of horizontal shear, is too laborious for general use; therefore, a formula will be developed that will be generally applicable within limits imposed by the assumptions. For this development refer to Fig. 5–28, which is a free-body diagram similar to that of Fig. 5–27c. The force dF_1 is the normal force acting on a differential area dA and is equal to $\sigma\, dA$. The resultant of these differential forces is F_1 (not shown). Thus, $F_1 = \int \sigma\, dA$ integrated over the shaded area of the cross section, where σ is the fiber stress at a distance y from the neutral surface and is given by the expression $\sigma = My/I$. When the two expressions are combined, the force F_1 becomes

$$F_1 = \frac{M}{I}\int y\, dA = \frac{M}{I}\int_h^c ty\, dy$$

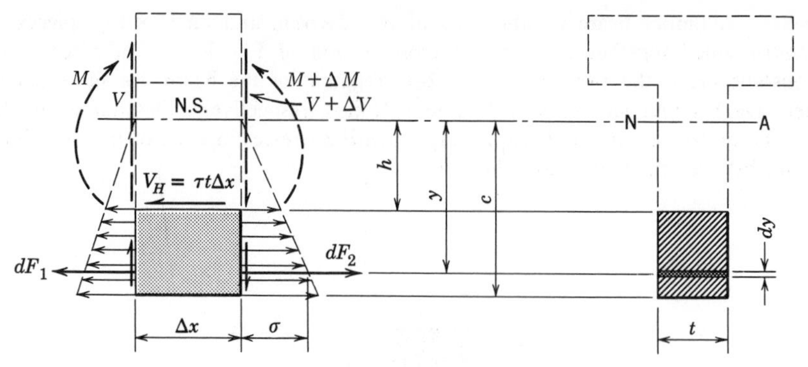

Fig. 5–28

Similarly, the resultant force on the right side of the element is

$$F_2 = \frac{(M+\Delta M)}{I} \int_h^c ty\,dy$$

The summation of forces in the horizontal direction on Fig. 5–28 yields

$$V_H = F_2 - F_1 = \frac{\Delta M}{I} \int_h^c ty\,dy$$

The average shearing stress is V_H divided by the area, from which

$$\tau = \lim_{\Delta x \to 0} \frac{\Delta M}{\Delta x}\left(\frac{1}{It}\right)\int_h^c ty\,dy = \frac{dM}{dx}\left(\frac{1}{It}\right)\int_h^c ty\,dy \qquad \text{(a)}$$

and from Art. 5–7, dM/dx equals V, the shear at the beam section where the stress is to be evaluated. Also, the integral is the first moment of that portion of the cross-sectional area between the transverse line where the stress is to be evaluated and the extreme fiber of the beam. The integral is designated Q, and when V and Q are substituted into Eq. (a), the formula for horizontal (or longitudinal) shearing stress becomes

$$\tau = \frac{VQ}{It} \qquad (5\text{–}4)$$

Because the flexure formula was used in the derivation, Eq. 5–4 is subject to the same assumptions and limitations. Although the stress given by Eq. 5–4 is associated with a particular point in a beam, it is averaged across the thickness t and hence is accurate only if t is not too great. For a rectangular section having a depth twice the width, the maximum stress as computed by Saint-Venant's more rigorous method is about 3 percent greater than that given by Eq. 5–4. If the beam is square, the error is about 12 percent. If the width is four times the depth, the error is almost 100 percent, from which one must conclude that, if Eq. 5–2 were applied to a point in the flange of the beam of Exam. 5–12, the result would be worthless. Furthermore, if Eq. 5–4 is applied to sections where the sides of the beam are not parallel, such as a triangular section, the average stress is subject to additional error because the transverse variation of stress is greater when the sides are not parallel.

As noted previously, at each point in a beam, the horizontal (longitudinal) and vertical (transverse) shearing stresses have the same magnitude; hence, Eq. 5–4 gives the vertical shearing stress at a point in a beam (averaged across the width). The variation of the transverse shearing stress in a beam will be demonstrated for the T-shaped beam of Exam. 5–12. Note that in Eq. 5–4 V and I are constant for any section, and only Q and t vary for different points in the section. The transverse shearing stress at any point in the stem of the T-section a distance y_O from the neutral axis is, from Fig. 5–29a and Eq. 5–4,

$$\tau = \frac{V}{It_1}\int_{y_O}^{8} yt_1\,dy = \frac{V}{2I}(8^2 - y_O{}^2) \qquad (-2 \leqslant y_O \leqslant 8)$$

Note that y_0 is shown as positive downward. An expression for the average

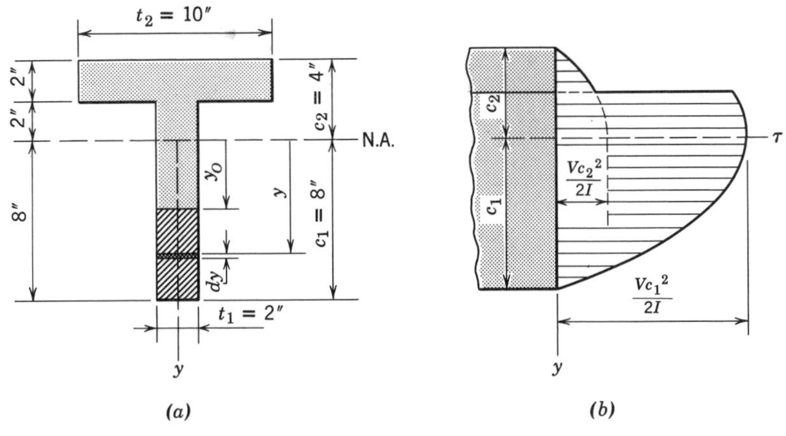

(*a*) (*b*)

Fig. 5–29

shearing stress in the flange can be written in a similar manner and is

$$\tau = \frac{V}{2I}(4^2 - y_O{}^2) \qquad (-4 \leqslant y_O \leqslant -2)$$

These are parabolic equations for the theoretical stress distribution, and the results are shown by Fig. 5–29*b*. The diagram has a discontinuity at the junction of the flange and stem because the thickness of the section changes abruptly. As pointed out previously, the distribution in the flange is fictitious, since the stress at the bottom of the flange must be zero (a free surface). Also, because of the abrupt change in section, a stress concentration will exist, and the maximum shearing stress at the junction of flange and stem will be more than the average stress (see Art. 1–10). From Fig. 5–29*b* and Eq. 5–4, one may conclude that in general the maximum[6] longitudinal and transverse shearing stress occurs at the neutral surface at a section where the transverse shear is maximum. There may be some exceptions such as a beam with a cross section in the form of a triangle or a Greek cross with Q/t at the neutral surface less than the value some distance from the neutral surface. The following example illustrates the use of the shearing stress formula.

Example 5–13 For the beam segment of Exam. 5–12, determine:

(a) The vertical shearing stress 4 in. from the bottom.

(b) The maximum vertical shearing stress.

Solution (a) The moment of inertia taken from Fig. 5–27*b* is 1600/3 in.^{4}. With reference to Fig. 5–30 the first moment Q is the shaded area multiplied by the distance from the centroid of the area to the neutral axis, which is 4(2)(6) in.^{3}

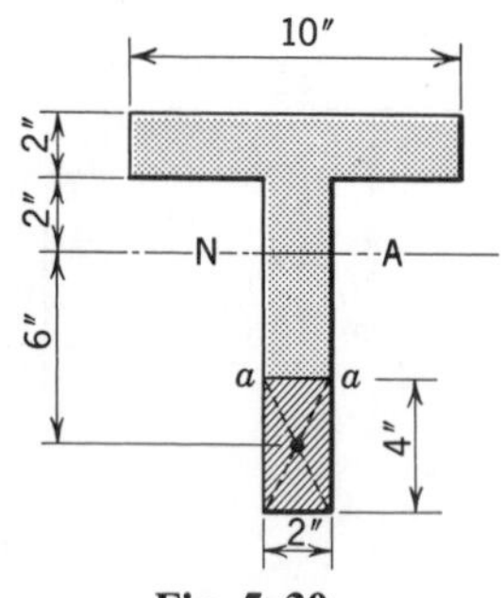

Fig. 5–30

[6]In this book the term *maximum*, as applied to longitudinal and transverse shearing stress, will mean the average stress across the thickness t at a point where such average has the maximum value.

Then

$$\tau = \frac{VQ}{It} = \frac{900(4)(2)(6)}{(1600/3)(2)} = \underline{40.5\text{ psi}} \qquad \text{Ans.}$$

(b) The maximum stress will occur at the neutral surface; it is

$$\tau_{\text{max}} = \frac{900(8)(2)(4)}{(1600/3)(2)} = \underline{54.0\text{ psi}} \qquad \text{Ans.}$$

Note that, in the computation of the value for Q, it is immaterial whether one takes the area above or below a transverse line such as a–a in Fig. 5–30. For example, in part b, Q for the area above the neutral axis is $Q_{\text{flange}} + Q_{\text{web}}$ equals $10(2)(2+1)+2(2)(1)$ equals $60+4$, which is the same as that used in part (b) above.

PROBLEMS

Note. Assume elastic action for the following problems.

5–82* The vertical shear at a certain section of a beam is 4000 lb. The beam has the cross section shown in Fig. P5–82. Determine:

(a) The maximum horizontal shearing stress, and indicate where it occurs within the cross section.

(b) The vertical shearing stress 3 in. below the top.

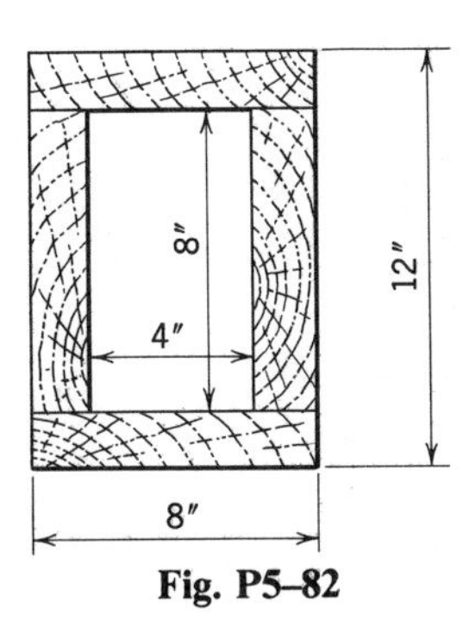

Fig. P5–82

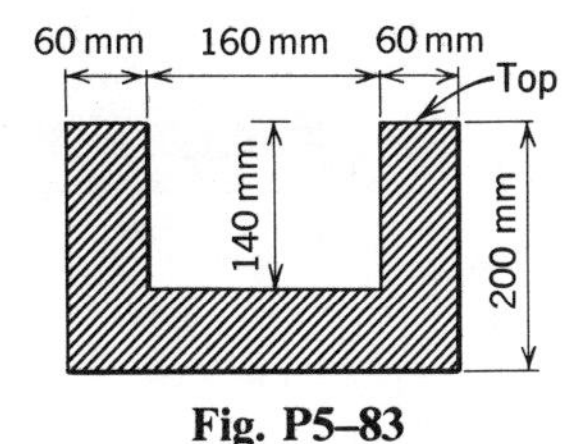

Fig. P5–83

5–83* A beam 6 m long is simply supported at its ends and has a cross section as shown in Fig. P5–83. The beam carries a uniformly distributed load of 5 kN/m over its entire length. Determine:

(a) The vertical shearing stress at a point 0.5 m from the right end and 100 mm below the top surface of the beam.

(b) The maximum horizontal and vertical shearing stresses, and state where each occurs.

5–84* A beam 16 ft long has the cross section shown in Fig. P5–84. It is simply supported at the ends and carries a uniformly distributed load of 300 lb/ft over its entire length. At a point 2 ft from the left end and 2 in. below the neutral surface, determine:

(a) The fiber stress.
(b) The horizontal shearing stress.
(c) The vertical shearing stress.

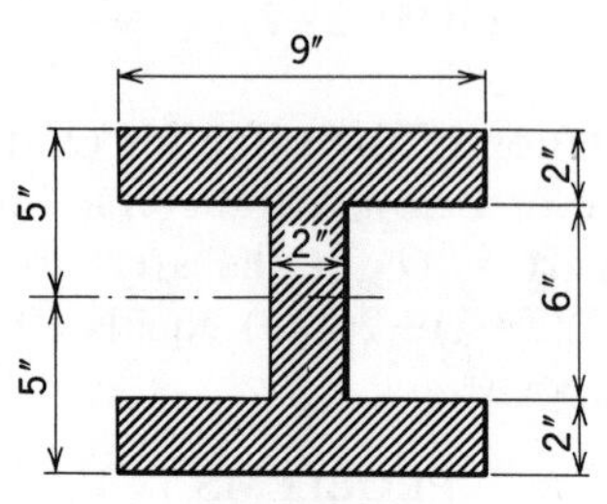

Fig. P5–84

5–85* The beam of Fig. P5–85*a* has the cross section shown in Fig. P5–85*b*. At section *a–a*, the shear is +5 kN, and the bending moment is −4 kN-m. Determine the fiber stress and the vertical shearing stress 40 mm below the neutral surface at section *b–b*.

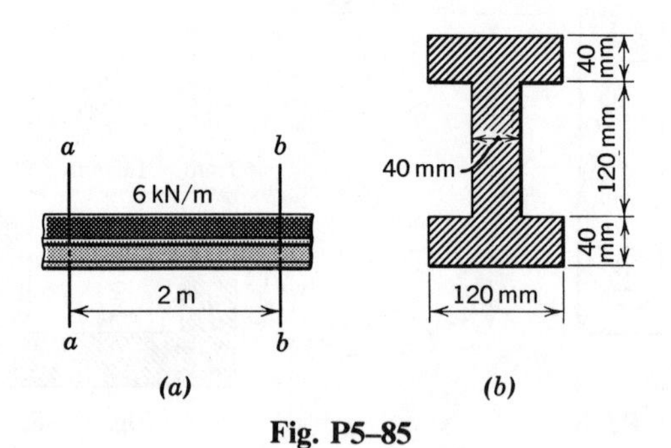

Fig. P5–85

5–86 Figure P5–86 shows the elevation and cross section of part of a beam. At section *b–b*, the vertical shear is −1600 lb, and the bending moment is +2000 ft-lb. Determine:

(a) The vertical shearing stress at point *A* in the section at *a–a*.
(b) The maximum compressive fiber stress at section *a–a*.

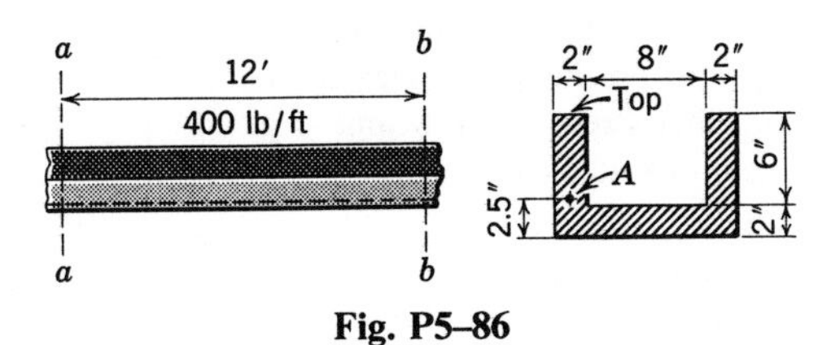

Fig. P5–86

5–87* A beam simply supported at its ends is 200 mm wide by 400 mm deep by 4 m long and carries a uniformly distributed load over its entire length. The allowable fiber stress in the beam is 12 MN/m^2 in tension or compression, and the allowable horizontal shearing stress is 0.8 MN/m^2. Determine the maximum allowable total load on the beam.

5–88 The cross section of the timber beam in Fig. P5–88 is a rectangle 6 in. wide by 12 in. deep. The allowable fiber stress is 1600 psi, and the allowable vertical shearing stress is 140 psi. Determine the maximum allowable value of w.

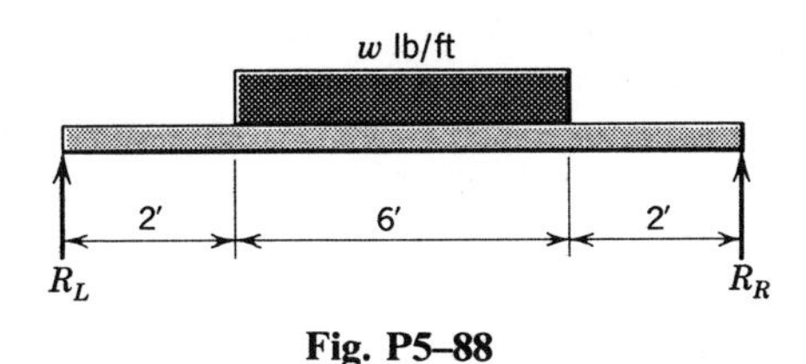

Fig. P5–88

5–89 The simple beam of Fig. P5–89a has the cross section shown in Fig. P5–89b. Determine the maximum permissible value for the load w (N/m) if the compressive flexural stress is not to exceed 9 MN/m^2 and the horizontal shearing stress is not to exceed 0.7 MN/m^2.

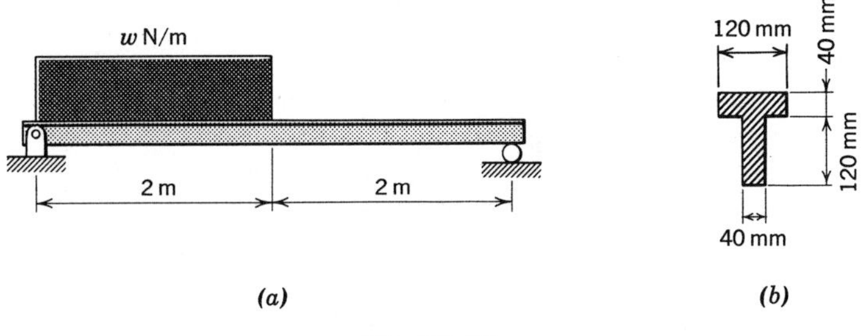

Fig. P5–89

5–90* The beam of Fig. P5–90*a* has the cross section of Fig. P5–90*b*. The left reaction is 1400 lb upward on the beam. Determine:

(a) The maximum fiber stress in the beam.
(b) The maximum horizontal shearing stress.

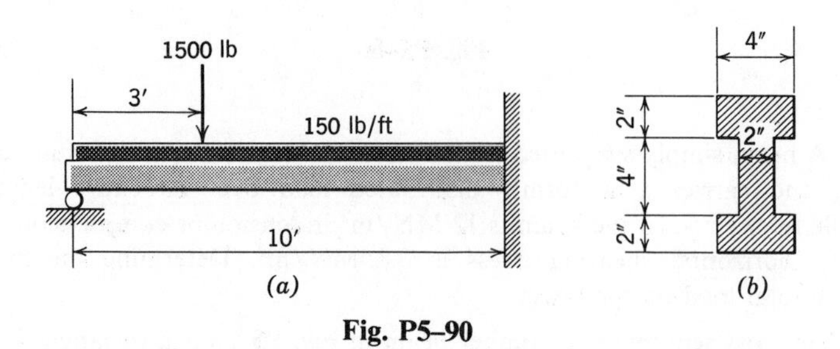

Fig. P5–90

5–91 Under the loading shown on the beam of Fig. P5–91, the left reaction is 300 lb upward on the beam. Determine for this loading:

(a) The maximum compressive fiber stress (neglect stress concentration and the thickness of the rigid loading yoke).
(b) The vertical shearing stress at a point 2 in. above the bottom of the beam in a section 5 ft from the right support.

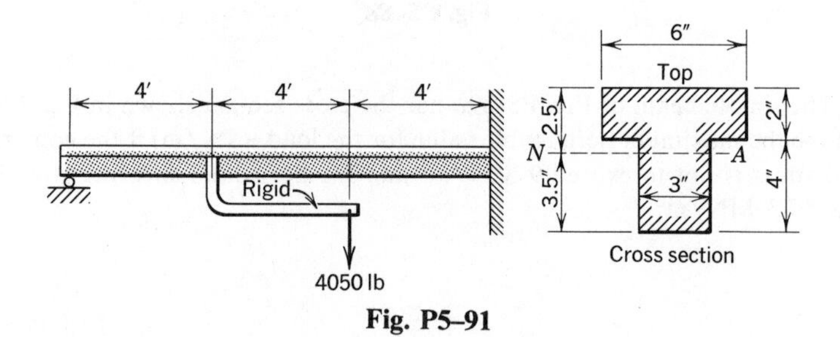

Fig. P5–91

5–92* The beam in Fig. P5–92 supports the indicated loads and has the cross section shown. Determine:

(a) The maximum compressive fiber stress.
(b) The maximum horizontal shearing stress.
(c) The vertical shearing stress 3.4 m from the left end at a point 60 mm above the bottom of the beam.
(d) The flexural stress at a point 1.5 m from the right end and 50 mm above the bottom of the beam.

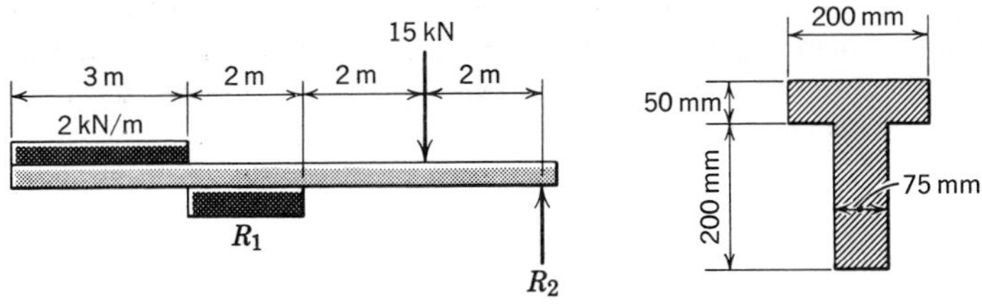

Fig. P5–92

5–93 A horizontal wooden beam with the cross section of Fig. P5–93 is subjected to maximum moments of +800 N-m and −1000 N-m, and a maximum shear of 2000 N—all in a vertical plane. Determine

(a) The maximum horizontal shearing stress.
(b) The maximum tensile fiber stress.

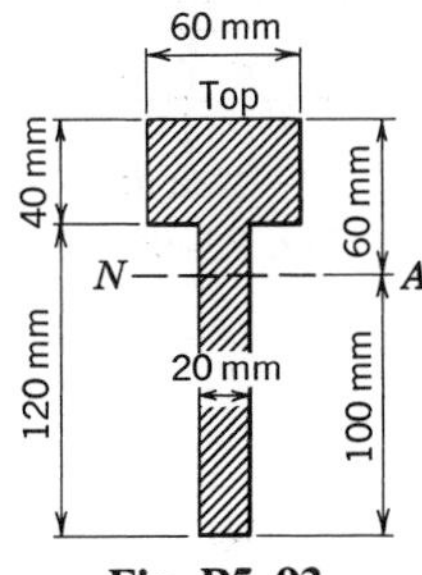

Fig. P5–93

5–94* A horizontal beam having the cross section of Fig. P5–94 is subjected to vertical loads producing the following maximum values: positive moment 10,000 ft-lb, negative moment 8000 ft-lb, and shear −1600 lb. Determine:

(a) The maximum compressive fiber stress.
(b) The maximum horizontal shearing stress.
(c) The fiber stress at a point 2 in. above the bottom of the beam at a section where the bending moment is +8000 ft-lb.

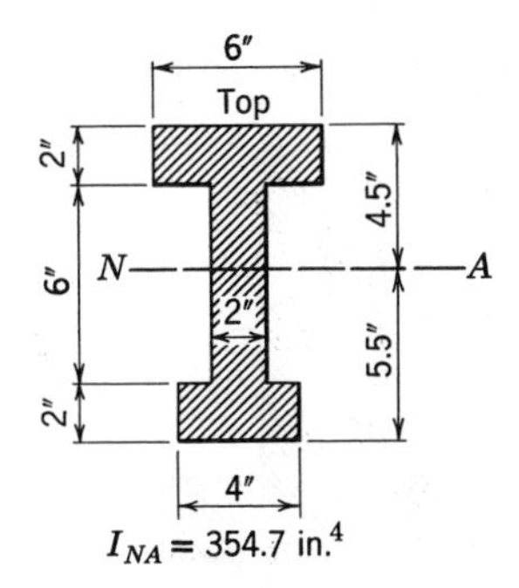

$I_{NA} = 354.7$ in.^{4}

Fig. P5–94

5–95 The beam of Fig. P5–95 is composed of two 1×6-in. and two 1×3-in. hard maple boards glued together to form the cross section shown. Determine the maximum fiber and the maximum horizontal shearing stresses for the loading shown, and state where these stresses occur.

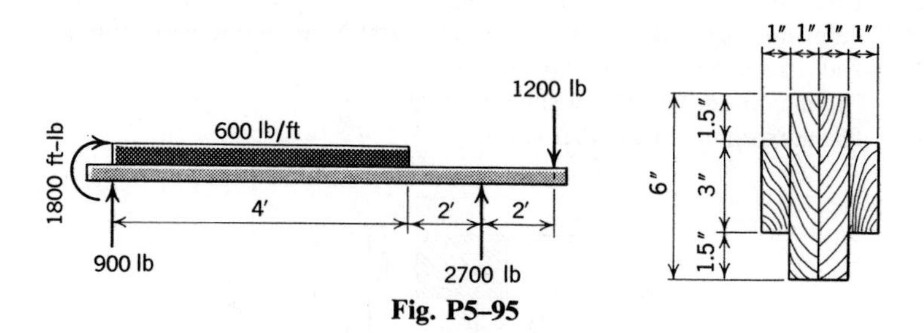

Fig. P5–95

5–96* A T-beam 5 m long is simply supported at the ends and has the cross section shown in Fig. P5–96. It is specified that the tensile fiber stress shall not exceed 12 MN/m^2 and that the horizontal shearing stress shall not exceed 0.7 MN/m^2. Determine the maximum concentrated downward load that may be placed 3 m from the right end.

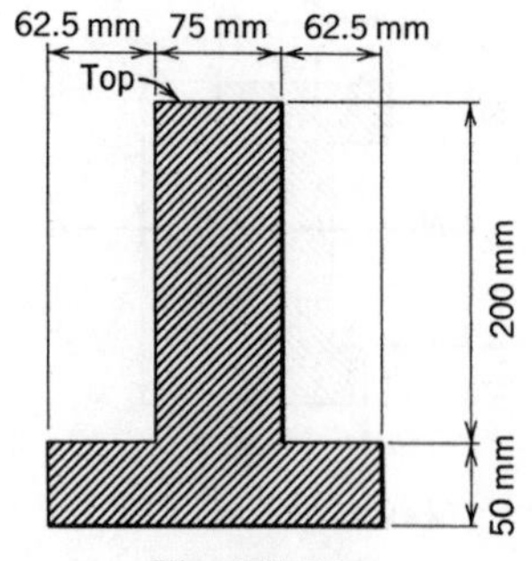

Fig. P5–96

5–97 The beam of Fig. P5–97 is composed of three pieces of timber glued together as shown. The maximum horizontal shearing stresses are not to exceed 100 and 120 psi in the glue joint and the wood, respectively. The tensile and compressive fiber stresses are not to exceed 1800 psi anywhere in the beam. What is the maximum permissible value of P?

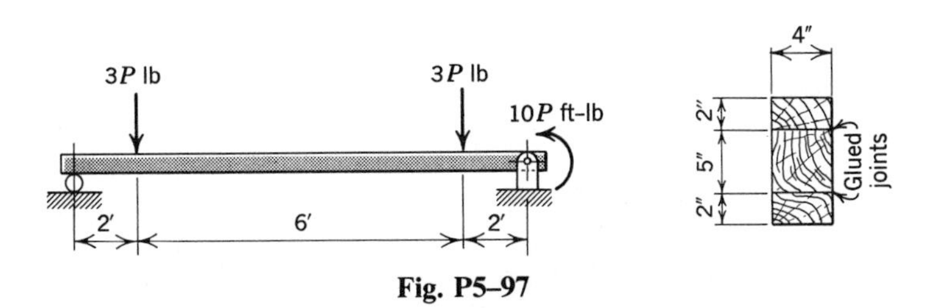

Fig. P5–97

5–98* A certain beam with restrained ends has a span L of 3 m and the cross section shown in Fig. P5–98. The beam is subjected to a loading that produces the following maximum values: positive moment $wL^2/18$, negative moment $wL^2/7$, and shear $5wL/2$, where L is in metres and w is in newtons per metre. The working stresses are: fiber 35 MN/m^2 T, 100 MN/m^2 C, and horizontal shear 15 MN/m^2. Determine the maximum allowable uniformly distributed load w for the beam.

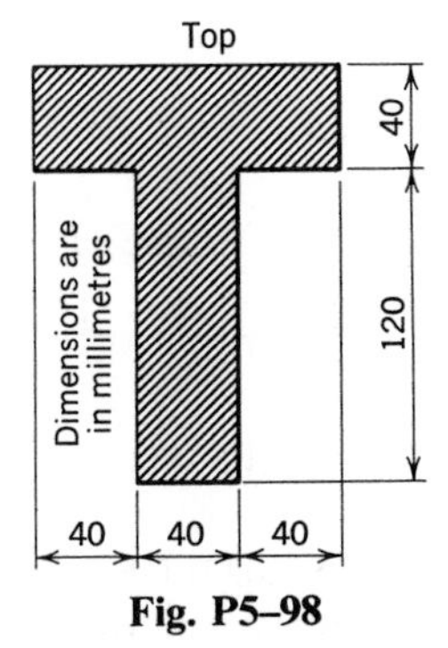

Fig. P5–98

5–99 In Fig. P5–99*a* are shown one span of a continuous beam and shear and moment diagrams for this span. The beam cross section is given in Fig. P5–99*b*. If the allowable fiber stress is 1800 psi *C* and the allowable horizontal shearing stress is 140 psi, determine the maximum permissible value for the uniformly distributed load *w*.

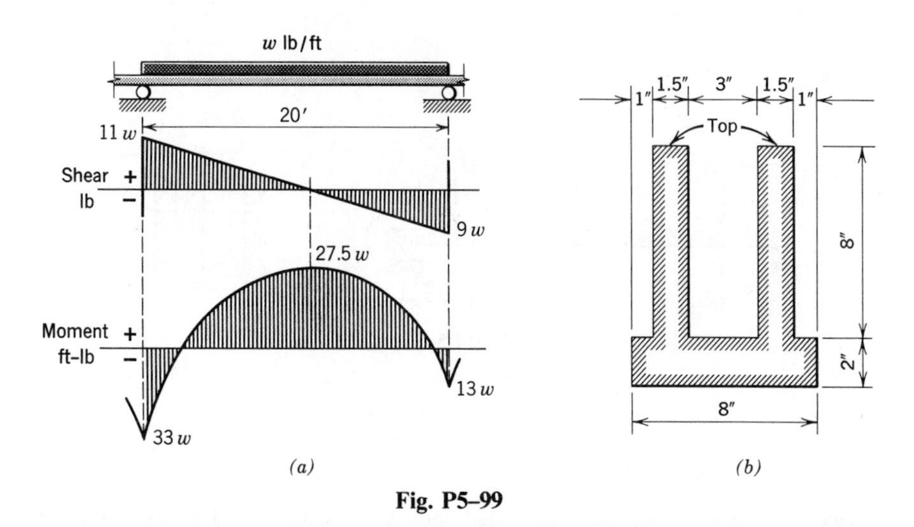

Fig. P5–99

5–100 The beam of Fig. P5–99*a* with a load *w* of 300 lb/ft is to be a rectangular timber stringer with the depth twice the width. The fiber and horizontal shearing stresses are not to exceed 1200 and 100 psi, respectively. Determine the minimum permissible dimensions of the cross section.

5–101* The wooden beam of Fig. P5–101*a* has the cross section of Fig. P5–101*b*. The allowable fiber and horizontal shearing stresses are 1200 and 120 psi, respectively. Determine the maximum permissible magnitude for *w*.

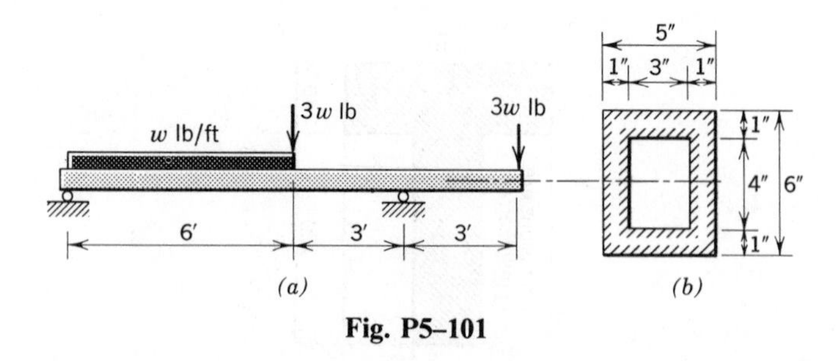

Fig. P5–101

5–102 The wooden beam of Fig. P5–102 is fabricated of four 25 × 150-mm boards glued together to form a box section 150 mm wide by 200 mm deep. The allowable fiber and horizontal shearing stresses in the wood are 10 and 0.8 MN/m^2, respectively. The allowable shearing stress in the glued joint is 0.6 MN/m^2. Determine the maximum permissible magnitude for the distributed load w.

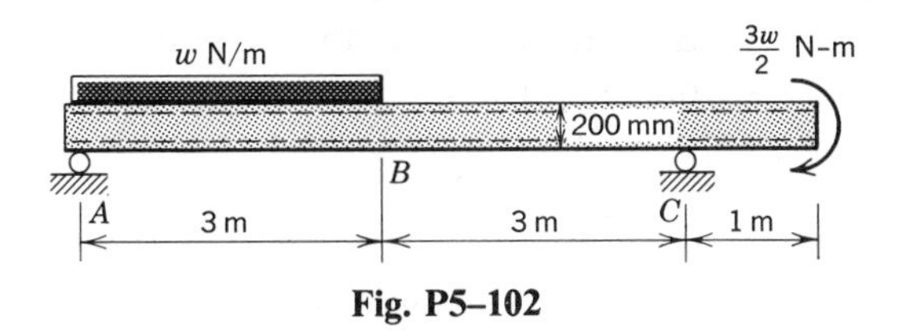

Fig. P5–102

5–103* The loaded beam of Fig. P5–103 is made of two 2×4-in. pieces of lumber glued together to form the T-section shown. Determine:

(a) The maximum permissible value for the load w for allowable stresses of 1600 psi compression and 120 psi horizontal shear.

(b) The maximum average shearing stress in the glued joint for the value of w determined in part (a).

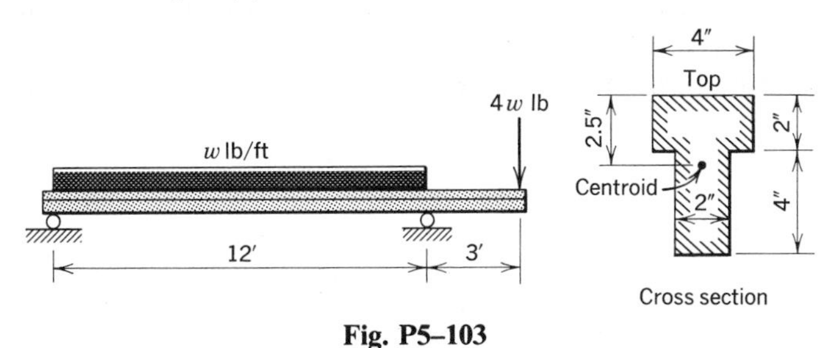

Fig. P5–103

5–11 PRINCIPAL STRESS FOR ELASTIC FLEXURE

Methods for finding the fiber stress at any point in a beam were presented in Arts. 5–3 and 5–4. Procedures for locating the critical sections of a beam (maximum M and V) were developed in Arts. 5–7 and 5–8. Methods for determining the transverse and longitudinal shearing stresses at any point in a beam were presented in Arts. 5–9 and 5–10. However, the discussion of stresses in beams is incomplete without careful consideration of the principal and maximum shearing stresses at carefully selected points on the sections of maximum shear and bending moment.

The fiber stress is maximum at the top or bottom edge of a section, and the transverse and longitudinal shearing stresses are zero at these points. Consequently, the maximum tensile and compressive fiber stresses on a given section are also the principal stresses at these points, and the corresponding maximum shearing stress is equal to one-half the fiber stress

$[\tau_{max} = (\sigma_p - 0)/2]$. The longitudinal and transverse shearing stresses are normally maximum for a given section at the neutral axis, where the fiber stress is zero. Thus, the evidence presented so far is limited to the extremes in a cross section. One might well wonder what the stress situation might be between these points. Unfortunately, the magnitude of the principal stresses throughout a cross section cannot be expressed for all sections as a simple function of position. However, in order to provide some insight into the nature of the problem, two typical sections will be discussed in the following paragraphs.

For a cantilever beam of rectangular cross section subjected to a concentrated load, the theoretical principal stress variation is indicated for two sections in Fig. 5–31. One observes from this figure that, at a distance from the load of one-fourth the depth of the beam, the maximum normal stress does not occur at the surface. However, at a distance of one-half the depth, the maximum fiber stress is the maximum normal stress. For either of these sections, Saint-Venant's principle (see Art. 1–10) indicates that the flexure (and transverse shearing stress) formula is inapplicable. Since such a small increase in bending moment is required to overcome the effect of the transverse shearing stress, the conclusion may be drawn that, for a

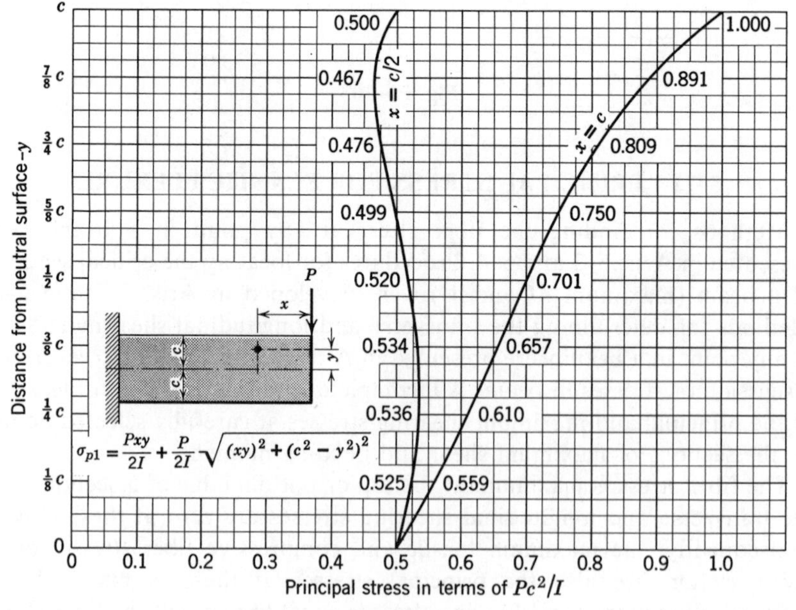

Fig. 5–31

rectangular cross section, in regions where the flexure formula applies, the maximum fiber stress is the maximum normal stress. Although for the rectangular cross section the maximum shearing stress will usually be one-half the maximum normal stress (at a surface of the beam), for materials having a longitudinal plane of weakness (for example, the usual timber beam), the longitudinal shearing stress may frequently be the significant stress; hence, the emphasis on this stress.

The other section to be discussed is the deep, wide flange section subjected to a combination of large shear and large bending moment. For this combination, the high fiber and transverse shearing stresses occurring simultaneously at the junction of flange and web sometimes yields a principal stress greater than the fiber stress at the surface of the flange (see Exam. 5–14). In general, at any point in a beam, a combination of large M, V, Q, and y and a small t should suggest a check on the principal stresses at such a point. Otherwise, the maximum flexural (fiber) stress will very likely be the maximum normal stress, and the maximum shearing stress will probably occur at the same point.

A knowledge of the directions of the principal stresses may aid in the prediction of the directions of cracks in a brittle material (concrete, for example) and thus may aid in the design of reinforcement to carry the tensile stresses. Curves drawn with their tangents at each point in the directions of the principal stresses are called *stress trajectories*. Since there are, in general, two nonzero principal stresses at each point (plane stress), there are two stress trajectories passing through each point. These curves will be perpendicular since the principal stresses are orthogonal; one set of curves will represent the maximum stresses, whereas the other represents the minimum stresses. The trajectories for a simply supported rectangular beam carrying a concentrated load at the midpoint are shown in Fig. 5–32,

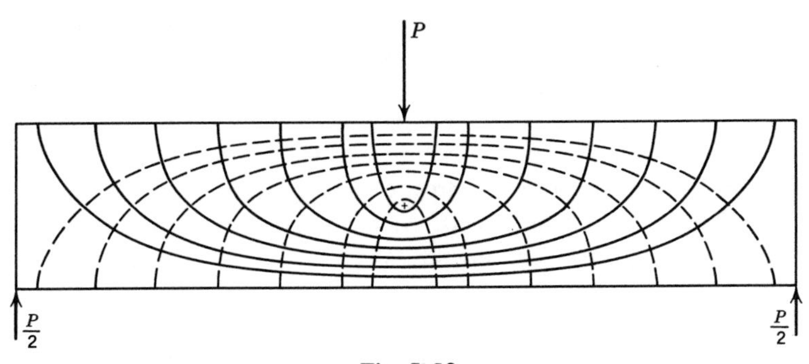

Fig. 5–32

with dashed lines representing the directions of the compressive stresses and solid lines showing tensile stress directions. In the vicinities of the load and reactions there are stress concentrations, and the trajectories become much more complicated. Figure 5–32 neglects all stress concentrations.

In order to determine the principal stresses and the maximum shearing stresses at a particular point in a beam, it is necessary to calculate the fiber stresses and transverse (or longitudinal) shearing stresses at the point. With the stresses on orthogonal planes through the point known, the methods of Art. 1–7 or 1–8 can be used to calculate the maximum stresses at the point. The following example illustrates the procedure and provides an example in which the principal stresses at some interior point are greater than the maximum fiber stresses.

Example 5–14 A W24×100 cantilever beam, supported at the left end, carries a uniformly distributed load of 11.60 kips per ft on a span of 8.00 ft. Determine the maximum normal and shearing stresses in the beam.

Solution For a cantilever beam with a uniformly distributed load, both the maximum bending moment and the maximum transverse shear occur on the section at the support. In this case they are

$$M = \frac{wL^2}{2} = \frac{(11.6)(8^2)(12)}{2} = 4454 \text{ in-kips}$$

and

$$V = wL = (11.6)(8) = 92.8 \text{ kips}$$

The upper half of the cross section of the beam at the wall is shown in Fig. 5–33*a*. The distribution of fiber stress for this half of the section is shown in Fig. 5–33*b*, and the distribution of the average transverse shearing stress is as shown in Fig. 5–33*c*. The vertical stresses σ_y due to the pressure of the load on the top of the beam are considered negligible.

Values will be calculated for three points—namely, at the neutral axis, in the web at the junction of the web and the top flange, and at the top surface. Both the fillets and the stress concentrations at the junction of the web and flange will be neglected. At the neutral axis, the fiber stress is zero, and the transverse shearing stress is

$$\tau = \frac{VQ}{It} = \frac{92.8[12(0.775)(11.61) + 11.225(0.468)(5.61)]}{2987(0.468)} = 9.12 \text{ ksi}$$

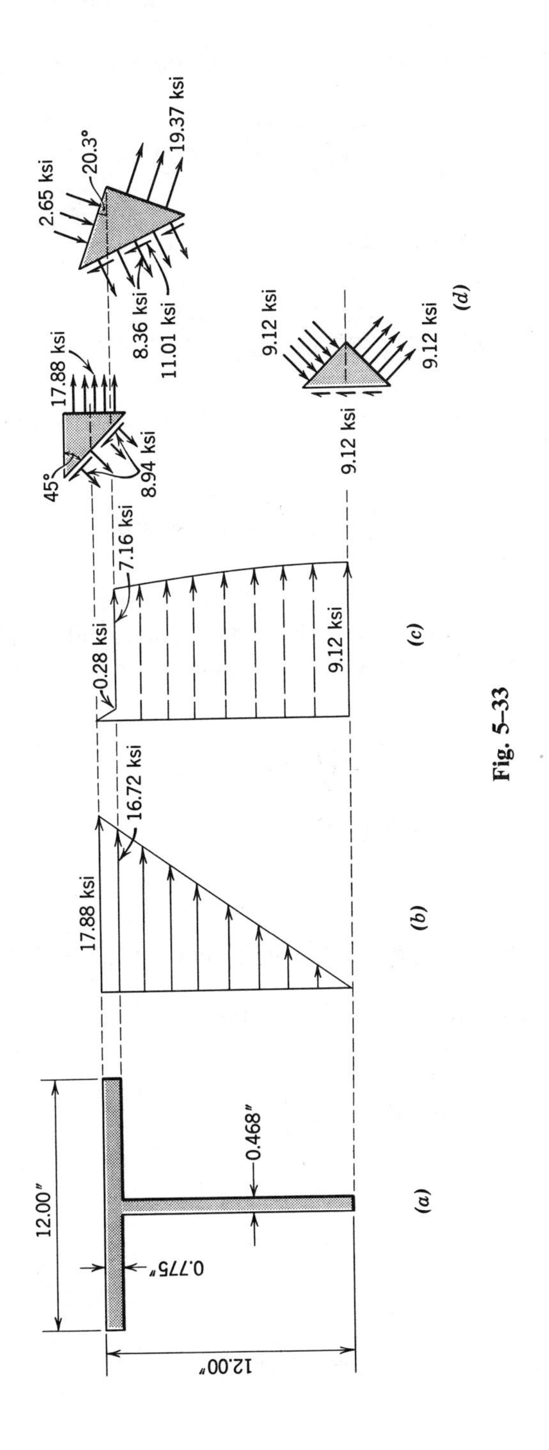
12.00″
0.775″
0.468″
12.00″
(a)
17.88 ksi
16.72 ksi
(b)
0.28 ksi
7.16 ksi
9.12 ksi
(c)
45°
8.94 ksi
17.88 ksi
9.12 ksi
9.12 ksi
9.12 ksi
2.65 ksi
20.3°
19.37 ksi
8.36 ksi
11.01 ksi
9.12 ksi
(d)

Fig. 5–33

In the web at the junction with the top flange, the fiber stress is

$$\sigma = \frac{My}{I} = \frac{4454(11.225)}{2987} = 16.72 \text{ ksi } T$$

and the transverse shearing stress is

$$\tau = \frac{VQ}{It} = \frac{92.8(12)(0.775)(11.61)}{2987(0.468)} = 7.16 \text{ ksi}$$

At the top surface, the transverse shearing stress is zero, and the fiber stress is

$$\sigma = \frac{My}{I} = \frac{M}{Z} = \frac{4454}{248.9} = 17.88 \text{ ksi } T$$

The principal stresses and maximum shearing stresses for each of the three selected points are shown in Fig. 5–33*d*. The calculations, from the equations of Art. 1–7, for the point at the junction of web and flange are

$$\sigma_p = \frac{16.72}{2} \pm \sqrt{\left(\frac{16.72}{2}\right)^2 + 7.16^2}$$
$$= 8.36 \pm 11.01$$

which gives

$$\sigma_{p1} = 19.37 \text{ ksi } T, \qquad \sigma_{p2} = 2.65 \text{ ksi } C, \quad \text{and} \quad \tau_{max} = 11.01 \text{ ksi}$$

The angle is not particularly important in this case, but it can be obtained from the equation

$$\tan 2\theta_p = -\frac{7.16}{8.36} = -0.857$$

The result is

$$2\theta_p = -40.6°$$

or

$$\theta_p = 20.3° \curvearrowright$$

The other stresses of Fig. 5–33*d* are obtained by inspection. It should be noted that the maximum tensile stress of 19.37 ksi is 8.32 percent above the maximum tensile fiber stress of 17.88 ksi and that the maximum shearing stress of 11.01 ksi is 20.7 percent above the maximum transverse shearing stress of 9.12 ksi.

PROBLEMS

Note. In the following problems neglect both the fillets and the stress concentrations at the junction of the web and flange. The action is elastic in all problems.

5–104* A 5-ft steel beam with the cross section shown in Fig. P5–104*a* is loaded and supported as shown in Fig. P5–104*b*. At point *A*, which is just to the right of the 10,000-lb load and just above the flange, determine, and show on a sketch, the principal stresses and the maximum shearing stress.

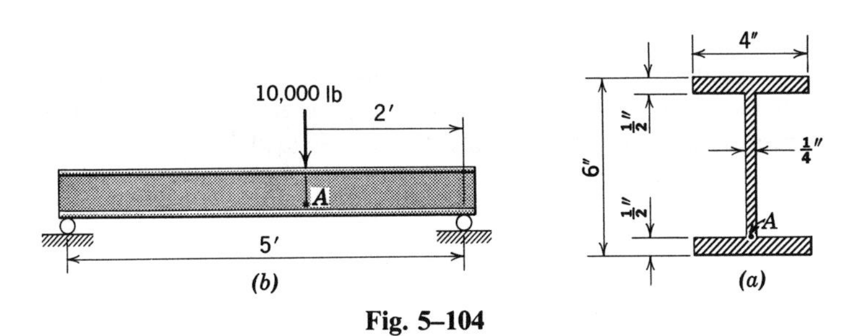

Fig. 5–104

5–105* A hollow rectangular timber beam with the cross section in Fig. P5–105*a* is loaded as shown in Fig. P5–105*b*. Determine, and show on a sketch, the principal stresses and the maximum shearing stress at point *A*.

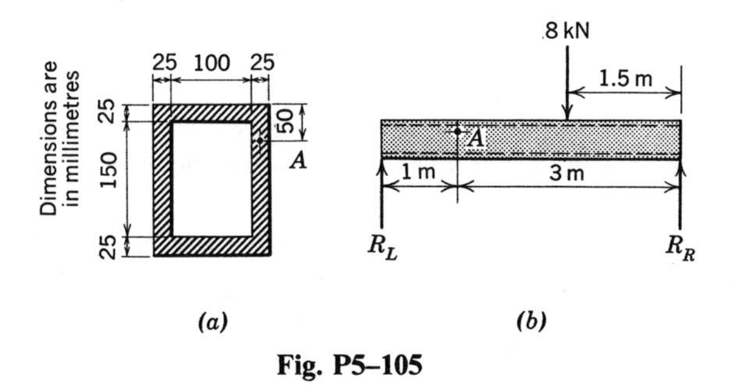

Fig. P5–105

5–106 The cantilever beam in Fig. P5–106a has the cross section shown in Fig. P5–106b. If the allowable stresses are 10,000 psi shear and 16,000 psi tension and compression at point A (just below the flange), determine the maximum allowable load P.

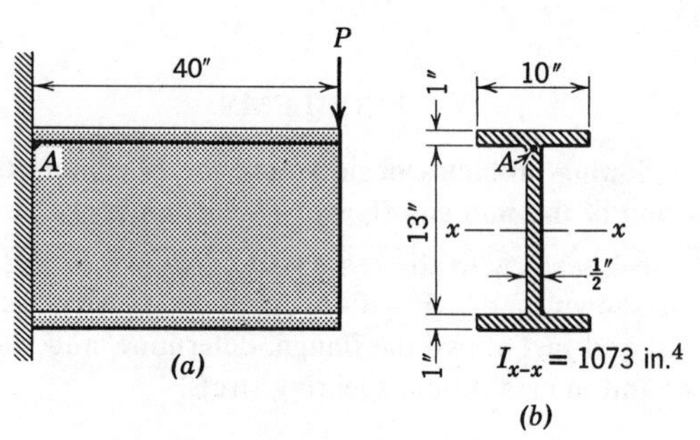

Fig. P5–106

5–107 The simple beam in Fig. P5–107a has the cross section shown in Fig. P5–107b. If the allowable stresses are 70 MN/m^2 shear and 110 MN/m^2 tension and compression, determine the maximum allowable load P.

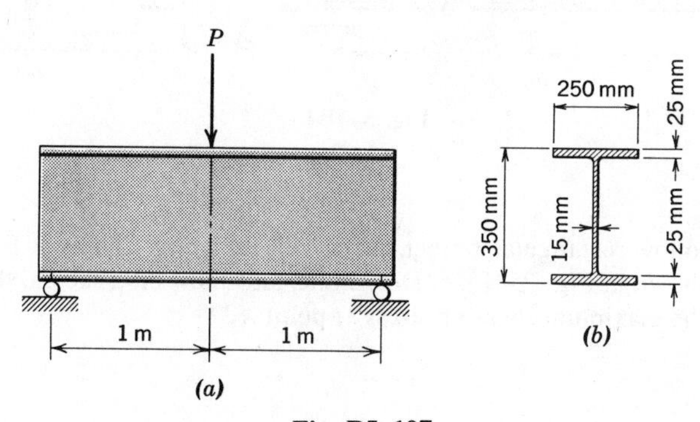

Fig. P5–107

5–108 A W24×100 cantilever beam with a span of 9 ft carries a uniformly distributed load of 15.0 kips/ft. Determine the maximum principal and shearing stresses in the beam.

5–109 A continuous beam having the cross section shown in Fig. P5–107b has an overhanging end 2 m long. The allowable normal and shearing stresses are 130 and 80 MN/m^2, respectively. Determine the maximum allowable uniformly distributed load per metre for the overhang.

5–110 The cantilever beam of Fig. P5–110 is made up of four 1×6-in. boards fastened together to form a box cross section 6 in wide by 8 in. deep. Determine, and show on a sketch the principal stresses and the maximum shearing stress at point A.

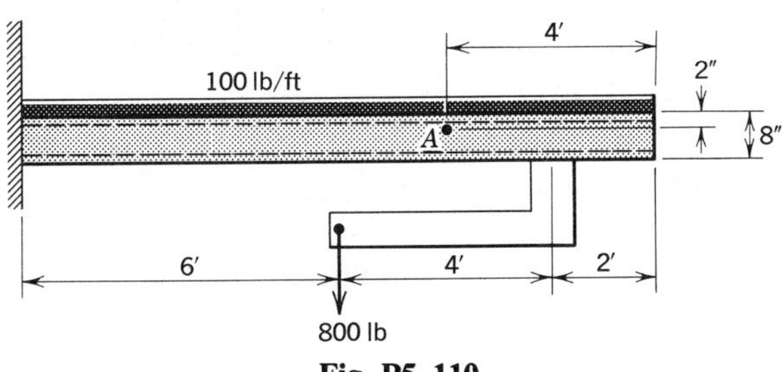

Fig. P5–110

5–12 NEGLECTED EFFECTS OF DISTRIBUTED LOADING

In Arts. 5–3 and 5–4 the flexure formula was developed by assuming that plane sections remain plane and that the beam is subjected to couples only. The formula for transverse shearing stresses was developed in Art. 5–10 for use with the flexure formula when transverse shearing forces are present. If shearing stresses exist, plane sections cannot remain plane, as Fig. 5–34 indicates, because the element b at the left section will undergo the maximum shearing strain, while elements a and c will experience no shearing strain. The result is a warped surface, as indicated. If the segment is in a region of constant shear, the warping of section *def* will match that of *abc*; consequently, the ends of any longitudinal element will be displaced the same amount, and the normal strains will be proportional to the distance from the neutral surface just as though the section had remained plane. In a region of varying shear (for example, the segment of a beam subjected to distributed loading), the warping of section *def* will not be the same as that of *abc*, and the normal strains will not be proportional to the distance from the neutral surface. For most engineering applications, the effects of cross-sectional warping can be neglected. Observe, for example, that curve 2 in Fig. 5–8, which represents measured longitudinal elastic strains in a region including flexural and transverse shearing stress, is apparently a straight line although the discussion above indicates it should not be. From the mathematical theory of elasticity, the error involved in using the elastic flexure formula to compute the maximum fiber stress in the cantilever beam of Fig. 5–35 is less than 0.2 percent for a span-depth ratio of 6.

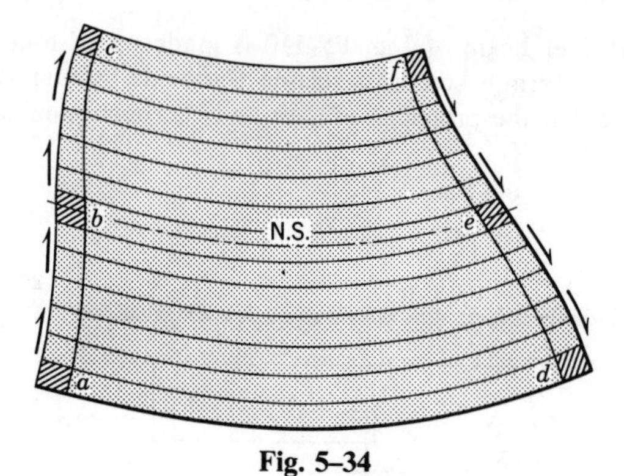

Fig. 5–34

Also neglected in previous articles are normal stresses in the transverse direction σ_y due to distributed loading. In order to complete the elementary picture of the state of stress in a beam, an expression for σ_y will now be developed for a rectangular cantilever beam of width b and depth $2c$ carrying a uniformly distributed load w over its entire length. The beam cross section and a free-body diagram of a portion of the free end of the beam are shown in Fig. 5–35a. Passing a horizontal plane through the beam segment results in the free-body diagram of Fig. 5–35b, from which the force equilibrium equation in the y direction gives the following:

$$\sigma_y bx + \Delta V + wx = 0 \tag{a}$$

where σ_y is uniformly distributed over the top of the beam (in both the x

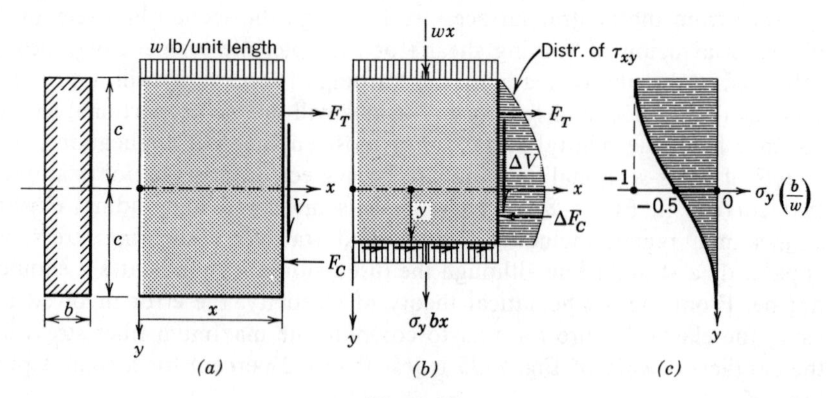

Fig. 5–35

and z directions). Also, the resultant of the shearing stresses is

$$\Delta V = \int_{-c}^{y} \tau_{xy} b\, dy$$

and from Art. 5–10,

$$\tau_{xy} = \frac{V}{Ib}\int_{y}^{c} yb\, dy = \frac{V}{2I}(c^2 - y^2)$$

therefore,

$$\Delta V = \frac{Vb}{2I}\int_{-c}^{y}(c^2 - y^2)dy = \frac{Vb}{2I}\left[c^2y - \frac{y^3}{3}\right]_{-c}^{y} = \frac{Vb}{6I}(2c^3 + 3c^2y - y^3)$$

From Fig. 5–35a, $V = -wx$, which when substituted into the expression for ΔV yields

$$\Delta V = -\frac{wxb}{6I}(2c^3 + 3c^2y - y^3) \tag{b}$$

Substituting (b) into (a), replacing I with $2bc^3/3$, and solving for σ_y gives

$$\sigma_y = \frac{w}{4b}\left(-\frac{y^3}{c^3} + \frac{3y}{c} + 2\right) - \frac{w}{b} \tag{5–5}$$

The graph of Eq. 5–5 is given in Fig. 5–35c, which shows that σ_y decreases rapidly from a maximum at the surface where the load is applied to zero at the opposite surface.

It will be of interest to compare the magnitudes of σ_x and σ_y at some point where both have maximum values. For the cantilever beam of Fig. 5–35, the maximum fiber stress σ_x at any section in the beam is at the top surface and is

$$\sigma_x = \frac{M(-c)}{I} = -\frac{3M}{2bc^2}$$

and since $M = -wx^2/2$,

$$\sigma_x = \frac{3wx^2}{4bc^2}$$

If x is written as Kc,

$$\sigma_x = \frac{3K^2}{4}\left(\frac{w}{b}\right) = -\frac{3K^2}{4}\sigma_y$$

If K is taken as 10, which would mean a span-depth ratio of 5 for a cantilever beam (since σ_x is maximum at the fixed end),

$$\sigma_{x\,\max} = -\frac{3(100)}{4}\sigma_y = -75\sigma_{y\,\max}$$

or

$$\sigma_{y\,\max} = 1.33 \text{ percent of } \sigma_{x\,\max}$$

The above discussion does not complete the picture. For a beam section composed of flanges connected by a thin web, the stress σ_y might have some effect on the maximum stresses. For example, refer to Exam. 5–14 in which the beam has a span depth ratio of 4. The stress σ_y in the web at the bottom of the flange is -2.03 ksi, and when this value is included in the computations for the maximum stresses, they would be changed from 19.37 ksi T and 11.01 ksi S to 19.15 ksi T and 11.80 ksi S, or -1.14 percent for $\sigma_{\max}$ and $+7.18$ percent for $\tau_{\max}$. A common specification for span-depth ratios (to put a practical limit on deflections) for simple beams and girders is 20 to 25. Such ratios would result in even smaller variations than those noted above; hence, the conclusion may be drawn that for real problems, neglecting σ_y is reasonable.

5–13 SHEARING STRESSES IN THIN-WALLED OPEN SECTIONS—SHEAR CENTER

One of the assumptions made in the development of the elastic flexure formula was that the loads were applied in a plane of symmetry. When this assumption is not satisfied, the beam will, in general, twist about a longitudinal axis. It is possible, however, to place the loads in such a plane that the beam will not twist. When any load is applied in such a plane, the line of action of the load will pass through the *shear center* (also known as the center of flexure or center of twist). The shear center is defined as the point in the plane of the cross section through which the resultant of the transverse shearing stresses due to flexure (no torsion) will pass for any orientation of transverse loads. For cross-sectional areas having two axes of symmetry, the shear center coincides with the centroid of the cross-sectional area. For those sections with one axis of symmetry, the shear center is always located on the axis of symmetry. If a beam having a cross section with one axis of symmetry is positioned such that the plane of symmetry is the neutral axis for flexural stresses, the plane of the loads must be perpendicular to the neutral axis but cannot pass through the centroid if bending is to occur without twisting. The channel is a common structural shape with one axis of symmetry. It is normally used with the axis of symmetry as the neutral axis since the section modulus is relatively large in this position.

Thin-walled open sections such as channels, angles, wide-flange sections, etc., develop significant shearing stresses during bending. Since the shearing stresses can be assumed to be uniform through the thickness, procedures based on longitudinal equilibrium similar to those of Art. 5–10 can be used to establish shearing stress distributions and to locate the shear centers for these sections.

Consider a cantilever beam of arbitrary cross section but constant thickness (see Fig. 5–36*a*) loaded in a plane parallel to one of its principal planes. The normal stress σ in any longitudinal fiber is then given by the elastic flexure formula as $\sigma = My/I$. The shearing stresses on section *a–a* can be determined by considering longitudinal equilibrium of the small block *B* shown in Fig. 5–36*a*. An enlarged free-body diagram of the block

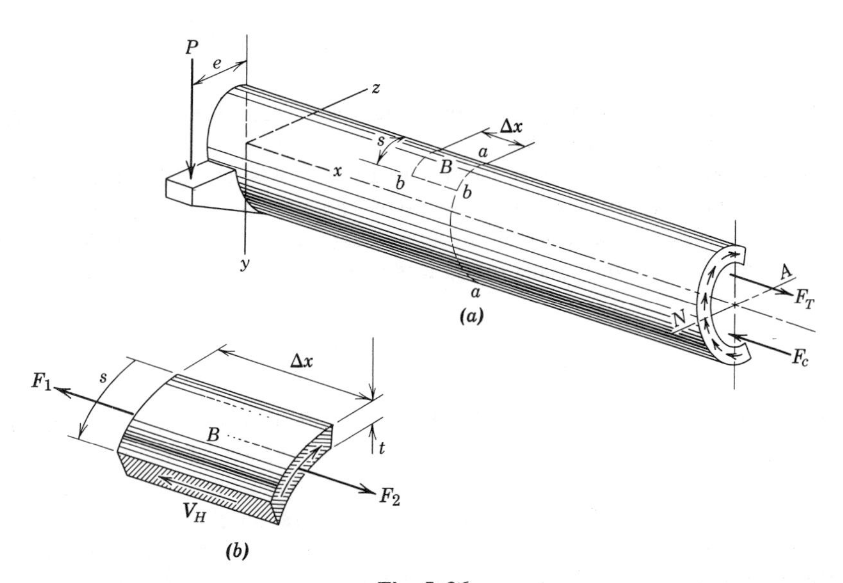

Fig. 5–36

is shown in Fig. 5–36*b*. Since the moments on sections at the ends of the block are M and $M+\Delta M$, the resultant forces F_1 and F_2 are

$$F_1=\int_{A_s}\sigma\,dA=\int_{A_s}\frac{My}{I}\,dA=\int_0^s\frac{My}{I}(t\,ds)$$

and

$$F_2=\int_{A_s}\sigma\,dA=\int_{A_s}\frac{(M+\Delta M)y}{I}\,dA=\int_0^s\frac{(M+\Delta M)y}{I}(t\,ds)$$

where A_s is the area of the cross section between the free edge of the block and the longitudinal plane b–b at a distance s from the free edge. A summation of forces in the longitudinal direction on the block of Fig. 5–36*b* yields

$$V_H=F_2-F_1=\int_{A_s}\frac{(\Delta M)y}{I}(dA)=\frac{\Delta M}{I}\int_0^s y(t\,ds)$$

The average shearing stress on the longitudinal plane b–b is V_H divided by the area. Thus

$$\tau=\lim_{\Delta x\to 0}\frac{\Delta M}{\Delta x}\left(\frac{1}{It}\right)\int_{A_s}y\,dA=\frac{dM}{dx}\left(\frac{1}{It}\right)\int_{A_s}y\,dA=\frac{dM}{dx}\left(\frac{1}{It}\right)\int_0^s y(t\,ds) \quad \text{(a)}$$

where dM/dx equals V, the shear at the beam section where the shearing stresses are to be evaluated.

Equation (a) is identical in form with Eq. (a) of Art. 5–10 and the terms have identical meanings. The integral is the first or static moment with respect to the neutral axis of the cross-sectional area to one side of the longitudinal plane b–b. This integral is usually referred to as Q_s. Thus

$$\tau=\frac{VQ_s}{It} \quad \text{(5–6)}$$

which is identical in form to Eq. 5–4. The shearing stresses are uniform through the thickness and act tangent to the surface of the beam. This same shearing stress acts on the transverse cross section at a distance s from the free edge of the section. The stresses on the transverse cross section "flow" in a continuous direction, as shown in Fig. 5–36*a*, and at the neutral axis they have the same sense as the shearing force V. The shearing force per unit length of cross section is frequently referred to as the shear flow q. Shear flow was previously discussed in Art. 4–10.

The procedure for locating a shear center will be illustrated by considering a channel section loaded as a cantilever beam (see Fig. 5–37). Since all fibers in a flange can be considered to be located at a distance $h/2$ from the neutral axis, the shearing stress distribution obtained from Eq. (a) is

$$\tau = \frac{V}{It}\int_0^s y(t\,ds) = \frac{V}{It}\int_0^s \left(\frac{h}{2}\right)(t\,ds) = \frac{Vhs}{2I}$$

This result shows that the horizontal shearing stress in the flange varies linearly from zero at the outer edge to $Vhb/(2I)$ at the web. The resultant shearing force on the top flange is

$$F_3 = (\tau_{\text{avg}})(\text{area}) = \left(\frac{Vhb}{4I}\right)(bt_1) = \frac{Vhb^2t_1}{4I}$$

as shown in Fig. 5–37. A similar analysis indicates that the force F_4 on the lower flange is equal in magnitude and opposite in direction to F_3. These two forces constitute a clockwise twisting moment that must be balanced by the resultant force in the web and the applied load if the beam is not to twist.

The resultant of the shearing forces in the web is V and must be equal to the applied load P for equilibrium. It is assumed that the vertical shearing force transmitted by the flanges is small enough to neglect (see Art. 5–10). The forces V and P will provide a counterclockwise couple to balance F_3 and F_4 if P is located a distance e from the center of the web as indicated in Fig. 5–37. When the moments of these two couples are equated, the result is

$$Ve = hF_3 = \frac{Vh^2b^2t_1}{4I}$$

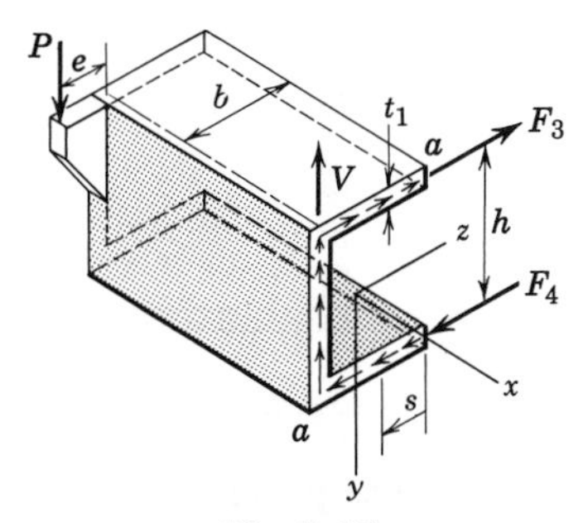

Fig. 5–37

from which

$$e = \frac{h^2 b^2 t_1}{4I}$$

If the load P of Fig. 5–37 were applied in the z direction, the forces corresponding to F_3 and F_4 (entirely different magnitudes) would be equal, and both would be directed to oppose the load P. There would also be two shearing forces in the web (parallel to the web) of equal magnitude and oppositely directed; hence, a moment equation would establish the fact that the shear center lies on the z axis of symmetry. If there is to be no twisting of the beam (buckling is assumed not to occur), any load must pass through the intersection of this axis of symmetry and the line of action of force P of Fig. 5–37. If the load is inclined, it may be resolved into components parallel to these axes, and the stresses may be determined using the principle of superposition.

PROBLEMS

5–111* A channel is 18 in. deep overall and has flanges 3.95 in. wide and 0.625 in. thick. The web is 0.450 in. thick. Locate the shear center of the section with respect to the center of the web. Assume that all of the section is effective in resisting fiber stresses and that only the web resists vertical shearing stresses.

5–112* An eccentric H-section is made by welding two 25×250-mm steel flanges to a 25×300-mm steel web, as indicated in Fig. P5–112. The section is used as a cantilever beam having a span of 2 m and carrying a concentrated load of 500 kN at the end, parallel to the web. Determine the magnitude of the resultant shearing force in the top flange, and locate the shear center of the section with respect to the center of the web. Assume that all of the section is effective in resisting fiber stresses and that only the web resists vertical shearing stresses.

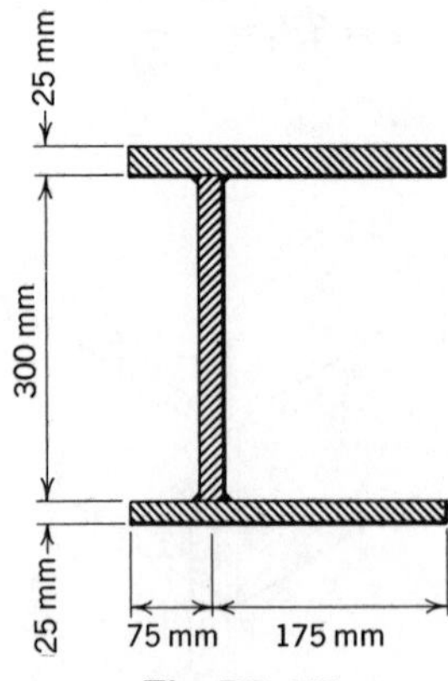

Fig. P5–112

5–113 A thin-walled cylindrical tube cut longitudinally to make a semicylinder is used as a cantilever beam. The load acts parallel to the cut section. Demonstrate that the shear center is $4R/\pi$ from the cut section. Use the element shown in Fig. P5–113.

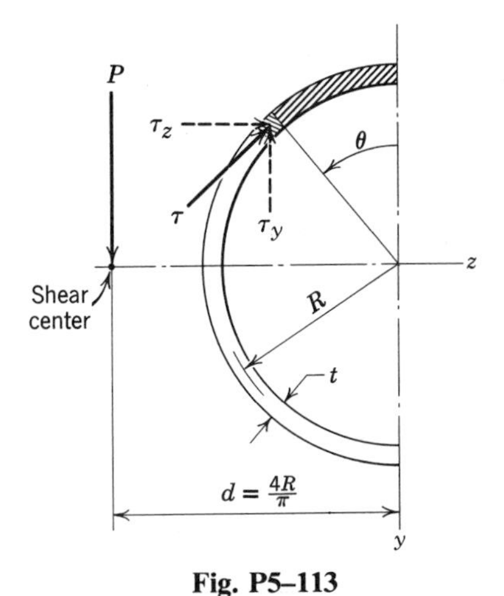

Fig. P5–113

5–114* Two $\frac{1}{2}$-in. thick plates are welded to the flanges of the channel in Prob. 5–111, as shown in Fig. P5–114. Determine the width of these plates that will cause the shear center to be located at the center of the web.

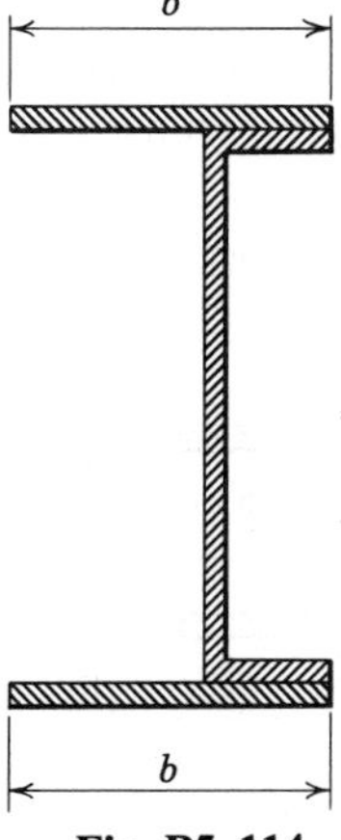

Fig. P5–114

5–115* A channel is to be fabricated with a depth of 150 mm, a web thickness of 5 mm, and a flange width of 90 mm. Structural considerations require the shear center of the channel to be located 40 mm from the center of the web. Determine the required thickness of the flanges.

5–116 A cantilever beam with the cross section shown in Fig. P5–116 is subjected to a vertical shear of 1000 lb. The section has a constant thickness of 0.200 in.

(a) Locate the shear center with respect to the center of the web.

(b) Make a sketch showing the distribution of shearing stress throughout the cross section when the 1000-lb load is applied through the shear center. Use the dimensions shown in Fig. P5–116 as centerline dimensions for the flanges, web, and extensions. Assume that all of the section is effective in resisting fiber stresses but that the vertical shearing force resisted by the flanges is negligible. The vertical shearing force in the extensions is not negligible.

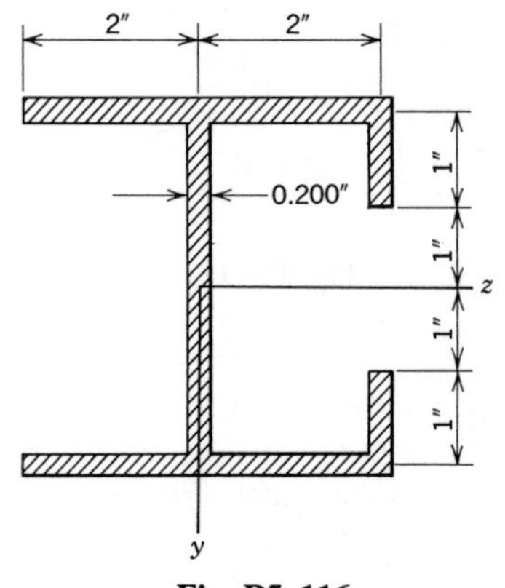

Fig. P5–116

5–117 A cantilever beam with the cross section shown in Fig. P5–117 is subjected to a vertical shear of 2.5 kN. The section has a constant thickness of 2.5 mm.

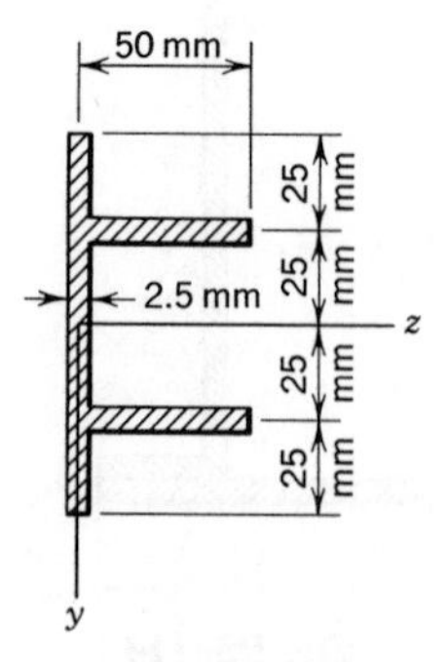

Fig. P5–117

(a) Locate the shear center with respect to the center of the web.
(b) Make a sketch showing the distribution of shearing stress throughout the cross section when the 2.5-kN load is applied through the shear center. Use the dimensions shown on Fig. P5–117 as centerline dimensions for the flanges and web. Assume that all of the section is effective in resisting fiber stresses and that the vertical shearing force resisted by the flanges is negligible.

5–118* A thin-walled slotted tube (see Fig. P5–118) is to be used as a beam loaded in a plane parallel to the *xy* plane. Assume the tube shape is maintained by suitable diaphragms. Locate the shear center (distance *d*).

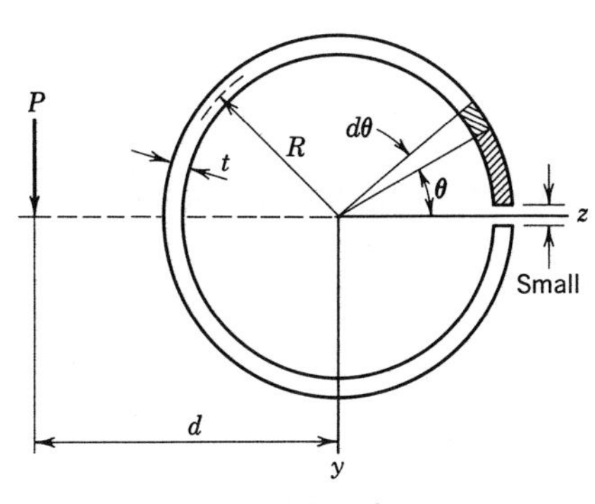

Fig. P5–118

5–119* A box beam with the rectangular cross section shown in Fig. P5–119 is loaded in a plane parallel to the *xy* plane. The shape is maintained by suitable stiffeners. Locate the shear center with respect to the center of the web.

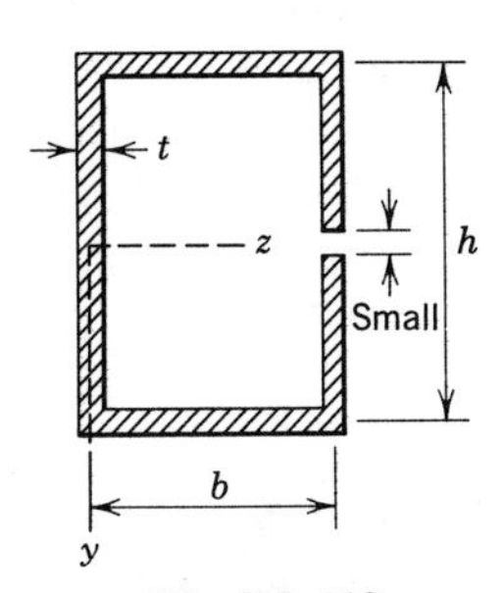

Fig. P5–119

5–120 All metal in the beam cross section of Fig. P5–120 is $\frac{1}{4}$ in. thick. Use the dimensions shown as centerline dimensions for the flanges and web. Locate the shear center with respect to the center of the web.

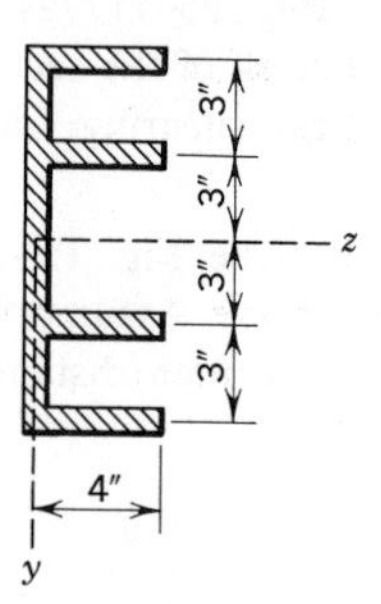

Fig. P5–120

5–14 BEAMS OF TWO MATERIALS BY TRANSFORMED SECTION

The method of fiber stress computation covered in Art. 5–3 is sufficiently general to cover symmetrical beams composed of longitudinal elements (layers) of different materials. However, for many real beams of two materials (often referred to as reinforced beams), a method can be developed to allow the use of the elastic flexure formula, thus reducing the computational labor involved. Of course, the method is applicable to elastic design only.

The assumption of the plane section remaining plane is still valid, provided the different materials are securely bonded together to provide the necessary resistance to longitudinal shearing stresses. Therefore, the usual linear transverse distribution of longitudinal strains is valid.

The beam of Fig. 5–38, composed of a central portion of material A and two outer layers of material B, will serve as a model for the development of the stress distribution. The section is assumed to be symmetrical with respect to the xy and xz planes, and the moment is applied in the xy plane. As long as neither material is subjected to stresses above the proportional limit, Hooke's law applies and the strain relation

$$\epsilon_b = \epsilon_a \frac{b}{a}$$

becomes

$$\frac{\sigma_b}{E_B} = \frac{\sigma_a}{E_A}\left(\frac{b}{a}\right)$$

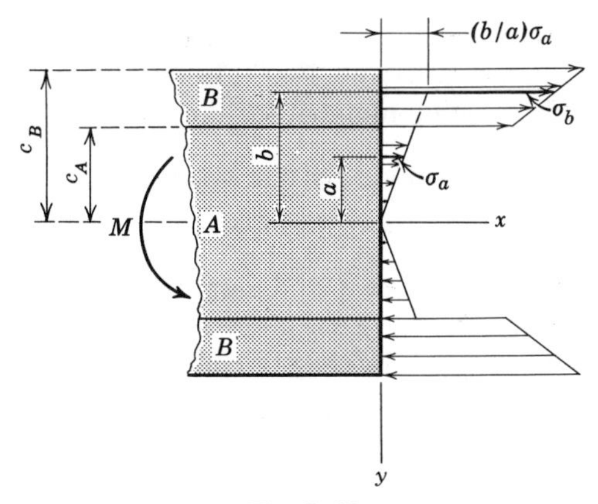

Fig. 5–38

or

$$\sigma_b = \sigma_a \left(\frac{E_B}{E_A} \right) \left(\frac{b}{a} \right) \tag{a}$$

From this relation it is evident that, at the junction between the two materials where distances a and b are equal, there is an abrupt change in stress determined by the ratio $n = E_B / E_A$ of the two moduli. Using Eq. (a), the normal force on a differential end area of element B is given by the expression

$$dF_B = \sigma_b \, dA = n \frac{b}{a} \sigma_a \, dA = \left(\frac{b}{a} \sigma_a \right) (nt) dy \qquad c_A \leqslant y \leqslant c_B$$

where t is the width of the beam at a distance b from the neutral surface. The first term in parentheses represents the linear stress distribution in a homogeneous material A. The second term in parentheses may be interpreted as the extended width of the beam from $y = c_A$ to $y = c_B$ if material B were replaced by material A, thus resulting in an equivalent or transformed *cross section* for a beam of homogeneous material. The transformed section is obtained by replacing either material by an equivalent amount of the other material as determined by the ratio n of their elastic moduli. The method is not limited to two materials; however, the use of more than two materials in one beam would be unusual. The method is illustrated by the following example.

Example 5–15 A timber beam 4 in. wide by 8 in. deep has an aluminum alloy plate with a net section $\frac{10}{3}$ in. wide by $\frac{1}{2}$ in. deep securely fastened to its bottom face. The working stresses in the timber and aluminum alloy are

1200 and 16,000 psi, respectively. The moduli of elasticity for the timber and aluminum alloy are 1250 and 10,000 ksi, respectively. Determine the maximum allowable resisting moment for the beam.

Solution The ratio of the moduli of the aluminum and timber is

$$n = 10{,}000/1250 = 8$$

The actual cross section (timber A and aluminum B) and the transformed timber cross section are shown in Fig. 5–39 a and b. The neutral axis of the transformed section is located by the principle of moments as

$$\bar{y} = \frac{\Sigma M}{\Sigma A} = \frac{(40/3)(1/4) + 32(9/2)}{(40/3) + 32} = 3.25 \text{ in.}$$

above the bottom of the aluminum plate. The moment of inertia of the cross-sectional area with respect to the neutral axis is

$$I_c = \frac{1}{12}\left(\frac{80}{3}\right)\left(\frac{1}{2}\right)^3 + \frac{40}{3}(3)^2 + \frac{1}{12}(4)(8)^3 + 32\left(\frac{5}{4}\right)^2 = 341 \text{ in.}^4$$

The stress relation

$$\sigma_b = \sigma_a\left(\frac{E_B}{E_A}\right)\left(\frac{b}{a}\right)$$

will be used to determine which working stress controls the capacity of the section. From Fig. 5–39 it is seen that the distance to the outermost timber fiber is 5.25 in. and that the distance to the outermost aluminum fiber is 3.25 in. With 1200 psi for σ_a the maximum stress in the aluminum is

$$\sigma_b = 1200(8)(3.25/5.25) = 5940 \text{ psi}$$

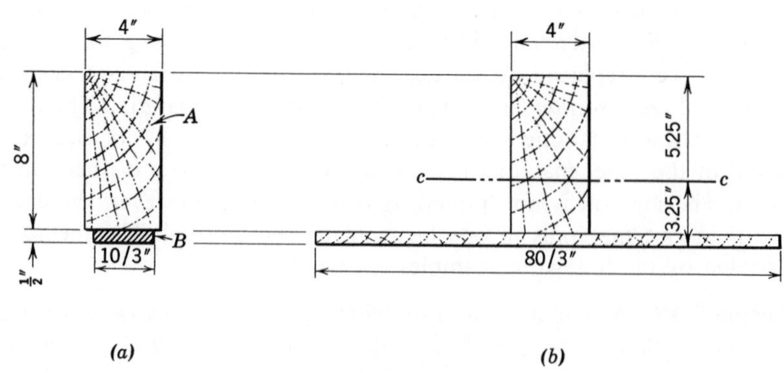

Fig. 5–39

which is well below the allowable stress. Therefore, the stress in the timber is the controlling stress. From the flexure formula

$$M = \frac{\sigma I}{c} = \frac{1200(341)}{5.25} = \underline{77{,}900 \text{ in-lb}} \qquad \text{Ans.}$$

PROBLEMS

5–121* A 4-in. wide by 6-in. deep timber cantilever beam 6 ft long is reinforced by two $\frac{1}{2}$-in. wide by 6-in. deep structural aluminum plates cemented securely to the sides of the timber beam. The flexural moduli of elasticity of this timber and aluminum are 1000 and 10,000 ksi, respectively. Determine the maximum tensile fiber stresses in each of the two materials when a static load of 1500 lb is placed on the free end of the beam. The 6-in. dimension is vertical.

5–122* A composite beam 225 mm wide by 300 mm deep by 5 m long is made by fastening two timber planks, 100 mm × 300 mm, to the sides of a steel plate, 25 mm wide by 300 mm deep. The flexural modulus of elasticity of the timber is 8 GN/m^2. The composite beam, simply supported at the ends, acts as a single unit in supporting a uniformly distributed load of 15 kN/m. Determine the maximum fiber stresses in the steel and wood.

5–123 A timber beam 4 in. wide by 10 in. deep has a $\frac{1}{2}$-in. deep by 4-in. wide steel plate cemented securely to its bottom face. The modulus of elasticity of this timber is 1500 ksi. Determine the maximum fiber stress in the timber when that in the steel is 10,000 psi T.

5–124 A 100-mm wide by 300-mm deep timber beam has a 100-mm wide by 15-mm deep steel plate cemented securely to its bottom face. The flexural modulus of elasticity of the timber is 10 GN/m^2. Determine the maximum fiber stress in the steel when the maximum fiber stress in the timber is 10 MN/m^2 C.

5–125* A cantilever beam 8 ft long carries a load of 3000 lb on the free end. The beam consists of a wooden section 4 in. wide by 8 in. deep to which are cemented 4-in. wide by $\frac{1}{2}$-in. deep steel plates top and bottom. The modulus of elasticity for this wood is 1800 ksi. Determine the maximum fiber stress in the wood and in the steel.

5–126* A 100-mm wide by 300-mm deep timber beam 5 m long is reinforced with 75-mm wide by 15-mm deep steel plates on the top and bottom faces. The beam is simply supported and carries a uniformly distributed load of 20 kN/m over its entire length. The modulus of elasticity of the timber is 10 GN/m^2. Determine the maximum fiber stress in the timber and in the steel.

5–127 The elastic design of reinforced concrete beams is based on the assumption that none of the concrete is effective in tension. This results in an effective cross section, as shown by the shaded areas of Fig. P5–127, where the rectangular area represents the concrete in compression with the usual linear stress distribution, and the lower circular areas represent the cross section (A_s) of the steel bars (assumed to be concentrated on a line).

(a) For allowable stresses of 1200 and 18,000 psi in the concrete and steel, respectively, determine the maximum permissible bending moment for a beam having the cross section shown. Use 3000 and 30,000 ksi for the moduli of elasticity of the concrete and steel, respectively.

(b) Based on a comparison of the concrete and steel stresses under the moment determined in part (a), would you consider this to be a good design?

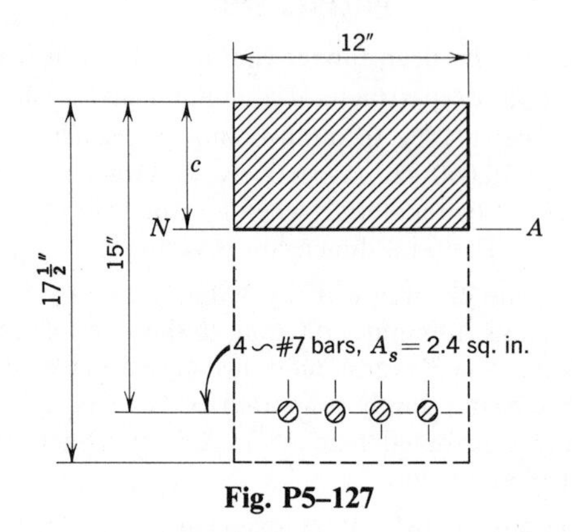

Fig. P5–127

5–128 Solve Prob. 5–127 for the beam section shown in Fig. P5–128.

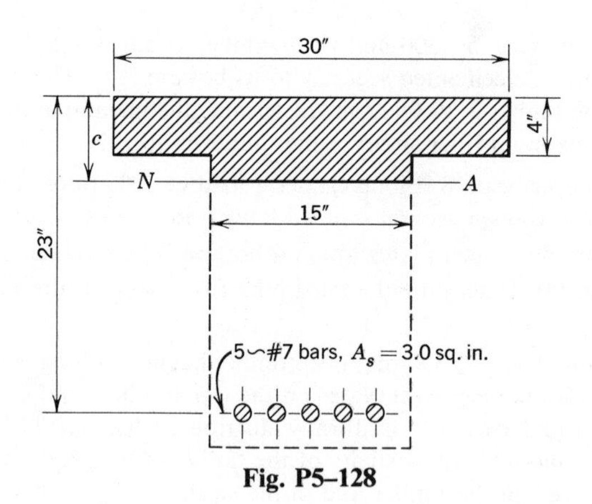

Fig. P5–128

5–129* A simple reinforced concrete beam carries a uniformly distributed load of 25 kN/m on a span of 5 m. The beam has a rectangular cross section 300 mm wide by 550 mm deep, and 1250 mm^2 of steel reinforcing rods are placed with their centers 70 mm from the bottom of the beam. The modulus of elasticity for the concrete is 20 GN/m^2. Determine the maximum fiber stress in the concrete and the average fiber stress in the steel. Assume that none of the concrete is effective in tension.

5–15 CURVED BEAMS

One of the assumptions made in the development of the flexural stress theory in Art. 5–2 was that all longitudinal elements of a beam have the same length, thus restricting the theory to initially straight beams of constant cross section. Although considerable deviation from this restriction can be tolerated in real problems, when the initial curvature of the beam becomes significant, the linear variation of strain over the cross section is no longer valid, even though the assumption of the *plane cross section remaining plane is valid.* A theory will now be developed for a beam, subjected to pure bending, having a constant cross section and a constant or slowly varying initial radius of curvature in the plane of bending. The development will be limited to linearly elastic action.

Figure 5–40*a* is the elevation of part of such a beam. The *xy* plane is the plane of bending and a plane of symmetry. Since a plane section before bending remains plane after bending, the longitudinal deformation of any element will be proportional to the distance of the element from the neutral surface, as indicated in Fig. 5–40*a*, from which

$$\delta = \frac{\delta_o}{b} y \tag{a}$$

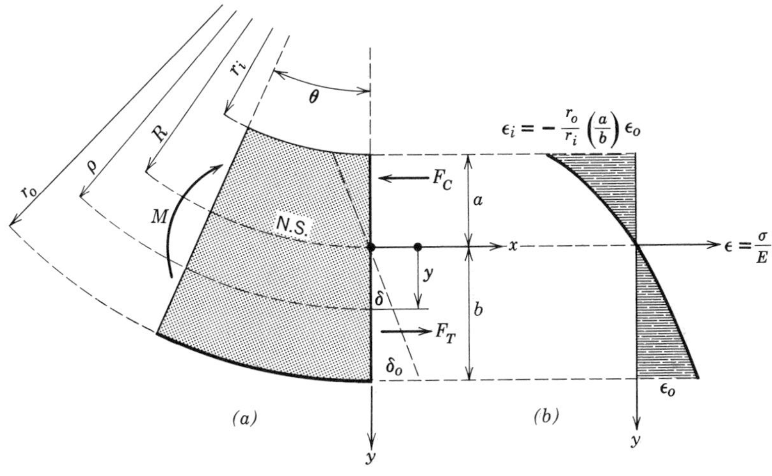

Fig. 5–40

The unstrained length of any longitudinal element is $\rho\theta$; therefore, Eq. (a) in terms of strain becomes

$$\rho\theta\epsilon = \frac{r_o\theta\epsilon_o}{b}y$$

from which the expression for the longitudinal strain distribution becomes

$$\epsilon = \frac{r_o\epsilon_o}{b}\frac{y}{\rho} \tag{b}$$

The plot of Eq. (b) is shown in Fig. 5–40*b*. Since the action is elastic, Hooke's law applies; therefore, $\epsilon = \sigma/E$ (for $\sigma_y = \sigma_z = 0$), which when substituted into Eq. (b) yields

$$\sigma = \frac{r_o\sigma_o}{b}\frac{y}{\rho} \tag{c}$$

Thus the curve in Fig. 5–40*b* indicates the flexural stress distribution as well as the strain distribution.

The location of the neutral surface is obtained from the equation $\Sigma F_x = 0$, thus

$$\int_A \sigma\, dA = \frac{r_o\sigma_o}{b}\int_A \frac{y}{\rho}\, dA = 0$$

Since $y = \rho - R$, where R is the radius of the neutral surface,

$$\frac{r_o\sigma_o}{b}\int_A \frac{\rho - R}{\rho}\, dA = 0 \tag{5–7}$$

In general, Eq. 5–7 should be solved for R for each specific problem; however, the general solution for a *rectangular cross section* of width t is easily obtained as follows:

$$\int_{r_i}^{r_o} \frac{\rho - R}{\rho} t\, d\rho = \int_{r_i}^{r_o} t\, d\rho - \int_{r_i}^{r_o} t\frac{R\, d\rho}{\rho} = 0$$

Since R is a constant,

$$r_o - r_i - R\ln(r_o/r_i) = 0$$

from which the radius of the neutral surface is

$$R = \frac{r_o - r_i}{\ln(r_o/r_i)} \tag{d}$$

The resisting moment in terms of flexural stress is obtained from the equilibrium equation $\Sigma M = 0$, thus

$$M_r = \int_A \sigma \, dA \, y = \frac{r_o \sigma_o}{b} \int_A \frac{y^2 dA}{\rho} = \frac{r_o \sigma_o}{b} \int_A \frac{(\rho - R)^2}{\rho} dA \tag{e}$$

The value of R for a given problem, obtained from Eq. 5–7, can be substituted in Eq. (e) and a solution for M_r thus obtained. However, it will, in general, be found more convenient to write Eq. (e) in the following form:

$$M_r = \frac{r_o \sigma_o}{b} \left[\int_A (\rho - R) dA - R \int_A \frac{\rho - R}{\rho} dA \right]$$

From Eq. 5–7, the second integral in the brackets is zero, and when $\rho - R$ is replaced by y in the first integral, the resisting moment is given by the expression

$$M_r = \frac{r_o \sigma_o}{b} \int_A y \, dA = \frac{r_o \sigma_o}{b} A \bar{y} \tag{f}$$

where $\bar{y}$ is the y coordinate of the centroid of the cross-sectional area A measured from N. A. Since r_o, b, σ_o, and A are all positive for a positive moment (defined as a moment producing tension at the boundary farthest from the center of curvature), $\bar{y}$ must always be positive, indicating that the neutral axis of the cross section is always displaced from the centroid toward the center of curvature. Replacing $(r_o \sigma_o)/b$ in Eq. (f) by $(\rho\sigma)/y$ from Eq. (c) and solving for σ gives

$$\sigma = \frac{My}{\rho A \bar{y}} = \frac{My}{(R + y) A \bar{y}} \tag{5–8}$$

which is the expression for the elastic flexural stress at any point in an initially curved beam.

The preceding development is for pure bending and neglects radial compressive stresses that occur within the material. These compressive stresses are usually very small. If the beam is loaded with forces (instead of couples), additional stresses will occur on the radial planes. Because the action is elastic, the principle of superposition applies, and the additional normal stresses can be added to the flexural stresses obtained from Eq. 5–8. Such problems are covered in Art. 8–2.

The application of Eqs. 5–7 and 5–8 is illustrated in the following examples.

Example 5–16 The segment of a curved beam (see Fig. 5–41) of high-strength steel (porportional limit = 95 ksi) having a trapezoidal cross section is subjected to a moment M of -100 in-kips.

(a) Determine the flexural stresses at the top and bottom surfaces.
(b) Sketch the flexural stress distribution in the beam.
(c) Determine the percentages of error if the flexure formula for a straight beam (Eq. 5–1) were used for part (a).

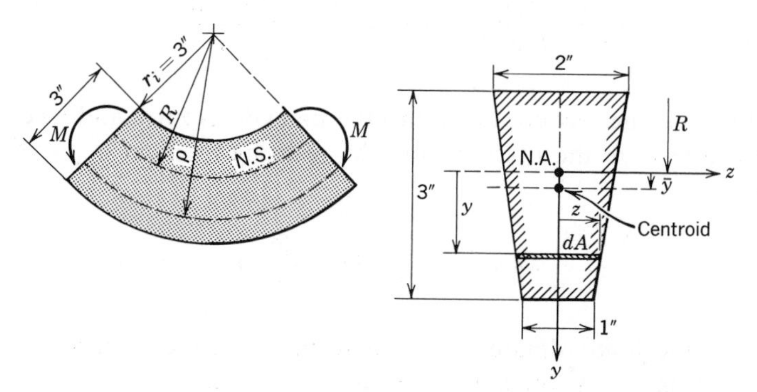

Fig. 5–41

Solution The first step will be to locate the neutral surface by setting the integral of Eq. 5–7 equal to zero, thus

$$\int_A \frac{\rho - R}{\rho}\, dA = \int_{r_i}^{r_o} \frac{\rho - R}{\rho}(2z)\, d\rho = 0$$

and from the cross section in Fig. 5–41,

$$2z = 2 - \frac{2-1}{3}(\rho - r_i) = 2 - \frac{1}{3}(\rho - 3) = 3 - \frac{\rho}{3}$$

which when substituted into the integral above yields

$$\int_3^6 \frac{\rho - R}{\rho}\left(3 - \frac{\rho}{3}\right) d\rho = 0$$

Integrating and substituting limits gives

$$3(6-3) - 3R\ln(6/3) - (36-9)/6 + (R/3)(6-3) = 0$$

from which

$$R = 4.1688 \text{ in.}$$

In order to apply Eq. 5–8, it is necessary to locate the centroid of the cross section, which is found to be 4/3 in. below the top edge. Therefore,

$$\bar{y} = 1.3333 - (4.1688 - 3) = 0.1645 \text{ in.}$$

(a) The required stresses from Eq. 5–8 are

$$\sigma_1 = \frac{-100(6-4.1688)}{6(4.5)(0.1645)} = \underline{41.2 \text{ ksi } C \text{ at bottom}}$$

$$\sigma_2 = \frac{-100(3-4.1688)}{3(4.5)(0.1645)} = \underline{52.6 \text{ ksi } T \text{ at top}} \qquad \text{Ans.}$$

These stresses are well below the proportional limit of the material.

(b) Plotting the two stresses from part (a) and zero stress at the neutral surface will indicate that the curve must be shaped as in Fig. 5–42.

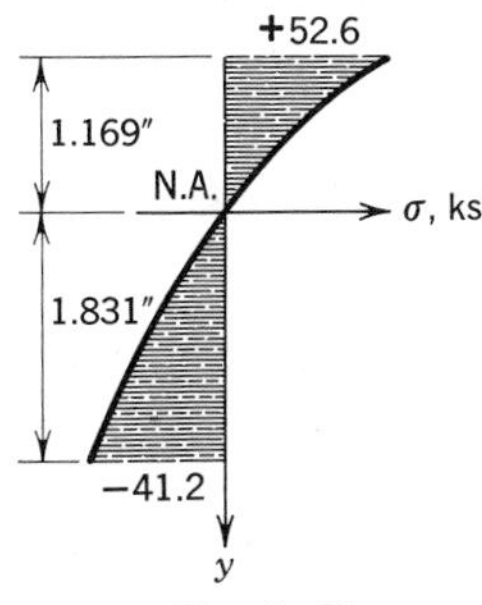

Fig. 5–42

(c) The cross-sectional moment of inertia with respect to the centroidal axis parallel to the neutral axis is found to be 13/4 in.^{4} and Eq. 5–1 yields

$$\sigma_1 = \frac{My}{I} = \frac{-100(5/3)}{(13/4)} = 51.3 \text{ ksi } C \text{ at bottom}$$

$$\sigma_2 = \frac{-100(-4/3)}{(13/4)} = 41.0 \text{ ksi } T \text{ at top}$$

Therefore the errors are

$$\frac{51.3-41.2}{41.2} = \underline{24.5\% \text{ high at bottom}}$$

$$\frac{52.6-41.0}{52.6} = \underline{22.0\% \text{ low at top}} \qquad \text{Ans.}$$

Example 5–17 The elevation and cross section of a segment of a gray cast-iron (proportional limit = 15 ksi T) punch press frame are shown in Fig. 5–43. Determine the maximum flexural tensile and compressive stresses due to a moment M of -500 in-kips.

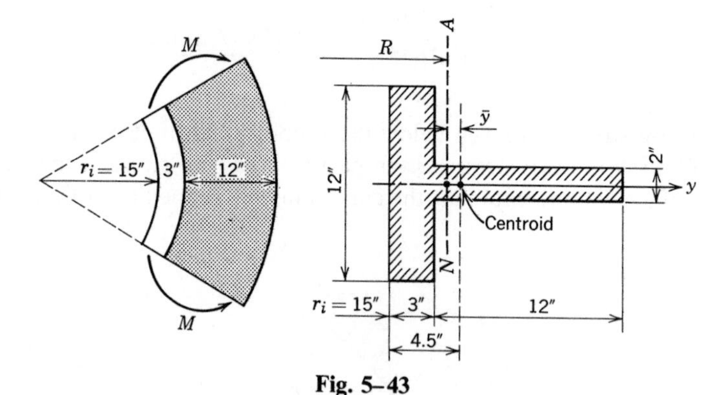

Fig. 5–43

Solution Assume that the neutral axis of the cross section is to the right of the flange and apply Eq. 5–7; thus,

$$\int_A \frac{\rho - R}{\rho} dA = \int_{15}^{18} \frac{\rho - R}{\rho}(12)d\rho + \int_{18}^{30} \frac{\rho - R}{\rho}(2)d\rho = 0$$

Integrating and substituting limits gives

$$12(18-15)-12R\ln(18/15)+2(30-18)-2R\ln(30/18)=0$$

from which

$$R=18.695 \text{ in.}$$

therefore, the assumption about the location of the neutral axis is correct. The centroid is 4.5 in. from the left edge of the section, and the distance from the neutral axis to the centroid is

$$\bar{y}=4.5-3.695=0.805 \text{ in.}$$

From Eq. 5–8 the stresses are

$$\sigma=\frac{My}{\rho A\bar{y}}=\frac{-500(11.305)}{30(60)(0.805)}=\underline{3.90 \text{ ksi } C \text{ at the outside of stem}} \qquad \text{Ans.}$$

and

$$\sigma=\frac{-500(-3.695)}{15(60)(0.805)}=\underline{2.55 \text{ ksi } T \text{ at inside of flange}} \qquad \text{Ans.}$$

The low stresses indicate elastic action.

PROBLEMS

Note. The action is elastic in the following problems, and moments are applied in a plane of symmetry.

5–130 A curved rectangular beam of width t and depth d (in the radial direction) and an inside radius r_i of $10d$ is subjected to a moment M in the plane of curvature. Demonstrate that the error in computing the maximum flexural stress by Eq. 5–1 instead of Eq. 5–8 is approximately 3.7 percent low.

5–131* A curved beam having an inside radius r_i of 20 in. has a rectangular cross section 3 in. wide by 6 in. deep (in the radial direction). The beam is subjected to a moment M of +300 in-kips in the plane of curvature. Determine the flexural stresses at the inside and outside curved surfaces of the beam.

5–132* A curved beam having an inside radius of 150 mm has the cross section shown in Fig. P5–132. The beam is subjected to a moment M in the plane of curvature. Determine the dimension b needed to make the flexural stress at the inside curved surface equal in magnitude to the flexural stress at the outside curved surface of the beam.

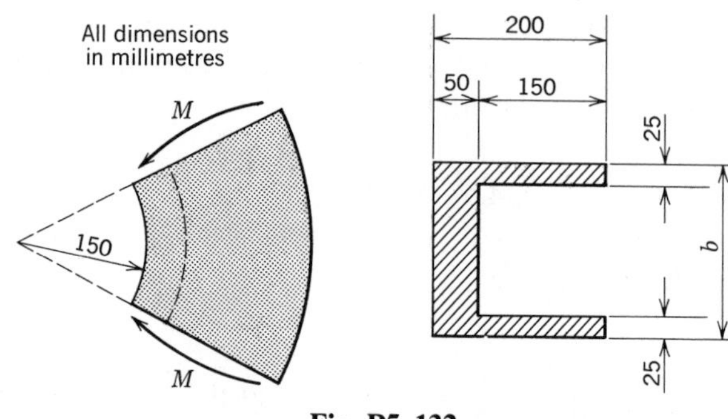

Fig. P5–132

5–133* The curved beam of Fig. P5–133 is subjected to the moment M of $+200$ in-kips. Determine the maximum tensile and compressive flexural stresses in the beam.

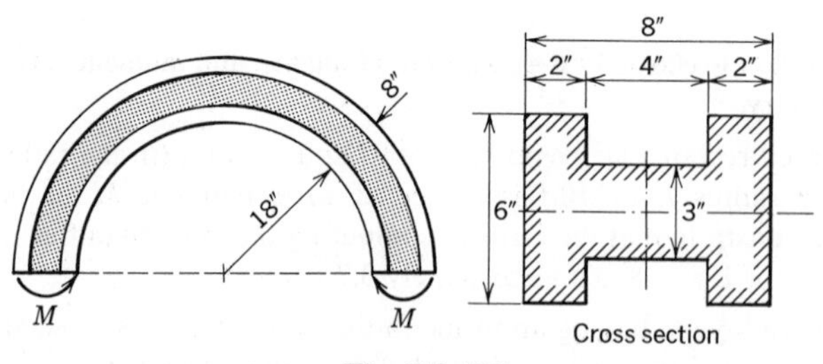

Fig. P5–133

5–134* The curved beam of Fig. P5–134 is subjected to the moment M of -50 kN-m. Determine the maximum tensile and compressive flexural stresses in the beam.

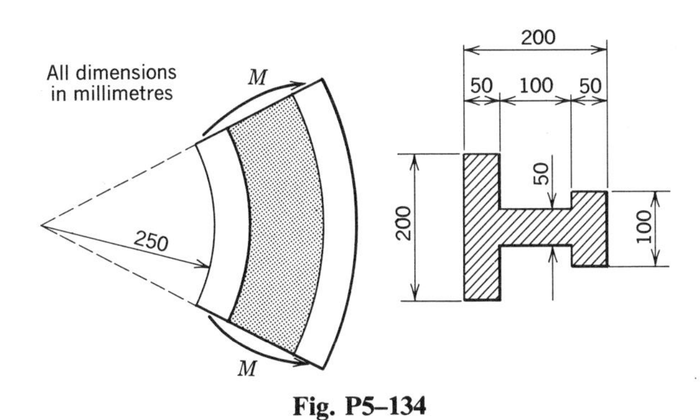

Fig. P5–134

5–135* Solve Exam. 5–17 with the depth of the stem of the T cross section changed from 12 to 9 in. and the moment changed from -500 to $+810$ in-kips. All other given data remain unchanged except the 4.5 in. to the centroidal axis, which now must be determined for the new section.

5–136 The curved beam of Fig. P5–136 is subjected to a positive moment M. Determine the maximum permissible magnitude for M if the flexural stress is not to exceed 35 MN/m^2 T or 140 MN/m^2 C.

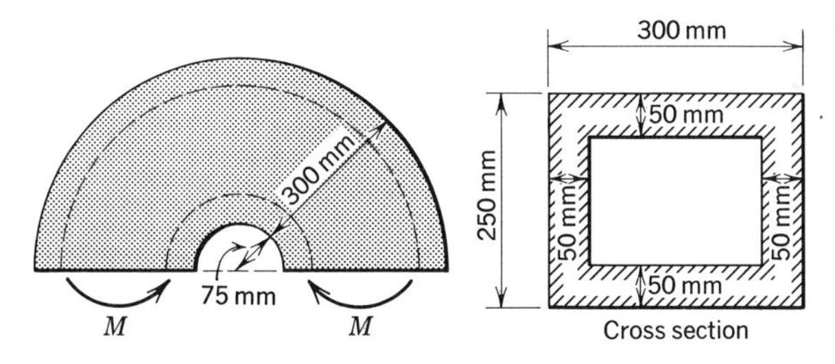

Fig. P5–136

5–137 The cross section of a segment of a crane hook is approximated by Fig. P5–137. The curvature and loading are in the *xy* plane. The inner radius r_i is 2.25 in. Determine the flexural stresses at the top and bottom surfaces due to a moment M of -80 in-kips.

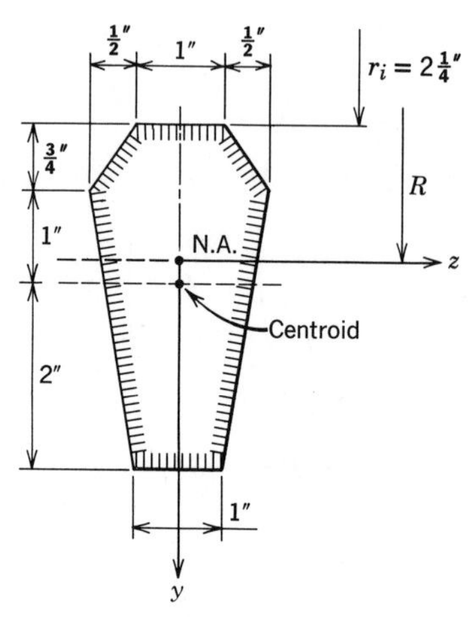

Fig. P5–137

6 Flexural Loading: Deflections

6–1 INTRODUCTION

The important relations between applied load and stress (fiber and shear) developed in a beam were presented in Chap. 5. However, a beam design is frequently not complete until the amount of deflection has been determined for the specified load. Failure to control beam deflections within proper limits in building construction is frequently reflected by the development of cracks in plastered walls and ceilings. Beams in many machines must deflect just the right amount for gears or other parts to make proper contact. In innumerable instances the requirements for a beam involve a given load-carrying capacity with a specified maximum deflection.

The deflection of a beam depends on the stiffness of the material and the dimensions of the beam as well as on the applied loads and supports. Four common methods for calculating beam deflections due to fiber stresses are presented here: (1) the double integration method, (2) the area-moment method, (3) the superposition method, and (4) an energy method. The material is presented so that methods 1 and 2 can be studied independently of each other. Deflections due to shearing stresses are also discussed.

6–2 THE RADIUS OF CURVATURE

When a straight beam is loaded and the action is elastic, *the longitudinal centroidal axis of the beam becomes a curve defined as the elastic curve*. In regions of constant bending moment, the elastic curve is an arc of a circle of radius ρ as indicated in Fig. 6–1, in which the portion AB of a beam is bent only with couples. Therefore, the plane sections A and B remain plane (see Art. 5–2), and the deformation (elongation and compression) of the fibers is proportional to the distance from the neutral surface, which is unchanged in length. From Fig. 6–1,

$$\theta = \frac{L}{\rho} = \frac{L+\delta}{\rho + c}$$

from which

$$\frac{c}{\rho} = \frac{\delta}{L} = \epsilon = \frac{\sigma}{E} = \frac{Mc}{EI}$$

Therefore

$$\frac{1}{\rho} = \frac{M}{EI} \tag{6–1}$$

which relates the radius of curvature of the neutral surface of the beam to the bending moment M, the stiffness of the material E, and the moment of inertia of the cross-sectional area I.

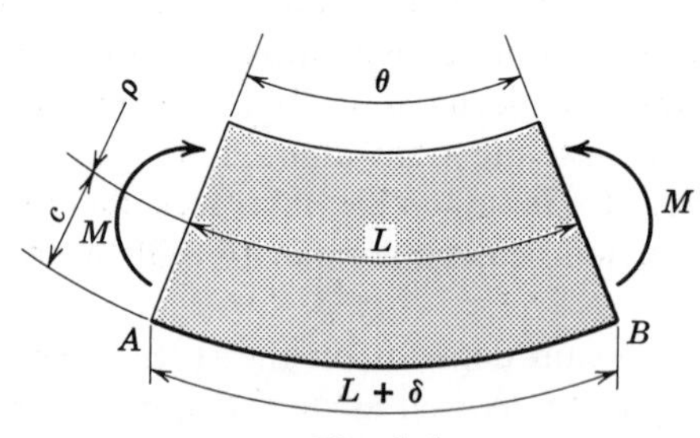

Fig. 6–1

PROBLEMS

6–1* The beam shown in Fig. P6–1*a* is made of two steel bars each 1×4 in. The bars are to be securely fastened together as shown in either Fig. P6–1*b* or *c*.

Determine:

(a) The radius of curvature and maximum fiber stress for configuration *b*.
(b) The radius of curvature and maximum fiber stress for configuration *c*.

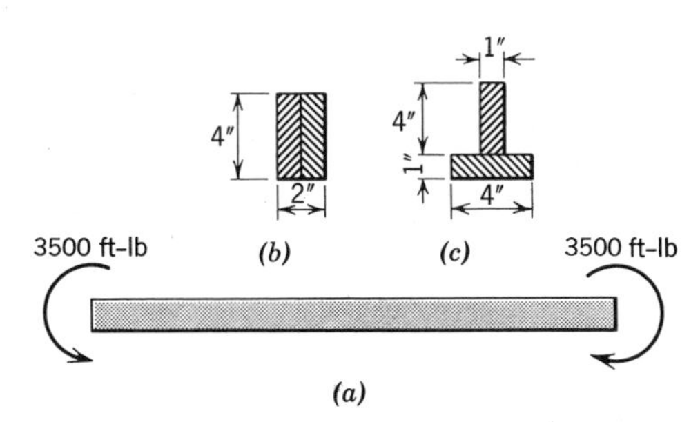

Fig. P6–1

6–2* The 4-m steel beam of Fig. P6–2*a* has the cross section shown in Fig. P6–2*b*.

(a) Determine the radius of curvature at any point between the supports.
(b) Determine the maximum fiber stress in the beam.
(c) What is the shape of the elastic curve?

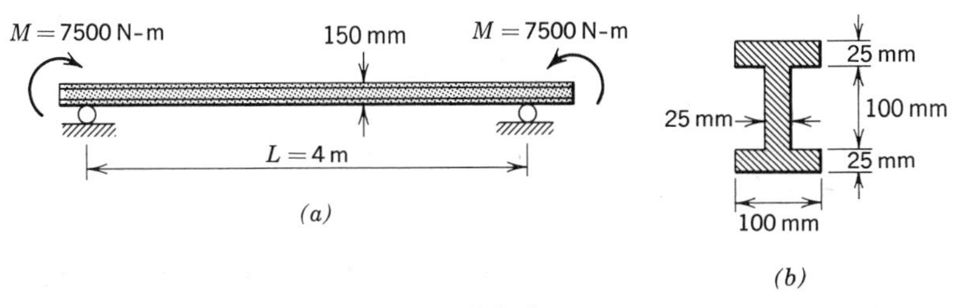

Fig. P6–2

6–3 A high-strength steel tube having an outside diameter of 3 in. and a wall thickness of 0.10 in. is bent into a circular curve of 75-ft radius. Determine the maximum fiber stress in the tube.

6–4 The boards for a concrete form are to be bent to a circular curve of 5-m radius. What maximum thickness can be used if the stress is not to exceed 15 MN/m^2? The timber is saturated, and Young's modulus may be taken as 6 GN/m^2.

6–5* Determine the radius of curvature of the aluminum cantilever beam of Fig. P6–5 and from it the deflection of the left end of the beam when loaded as shown. Use 10^7 psi for Young's modulus.

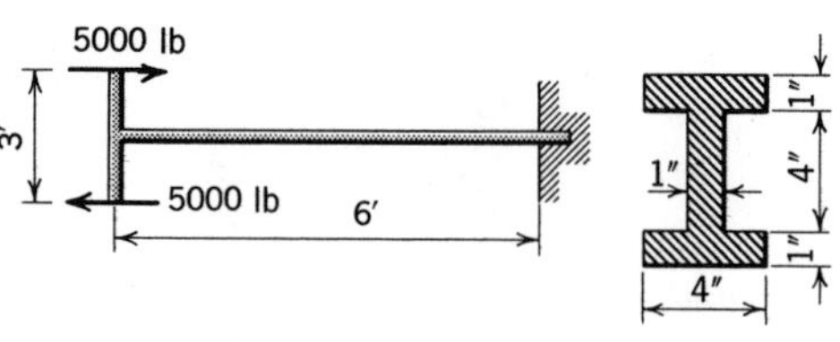

Fig. P6–5

6–6* For the steel beam loaded and supported as shown in Fig. P6–6, determine the maximum fiber stress and the deflection midway between the supports.

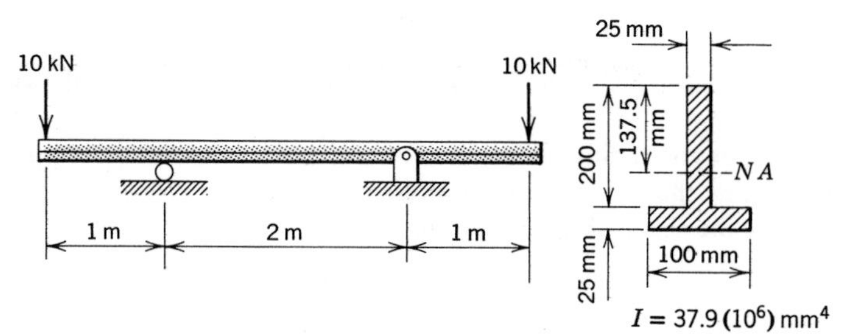

Fig. P6–6

6–7 Determine the minimum diameter required for a bandsaw pulley if the maximum fiber stress due to bending in a 0.025-in. thick steel bandsaw blade is not to exceed 30,000 psi.

6–8 A 25-mm thick board is simply supported at the ends of a span of 3 m. Loads are applied such that the elastic curve between supports is an arc of a circle and the deflection at the midpoint of the span is 100 mm. Use 6 GN/m^2 for Young's modulus. Determine the maximum fiber stress in the board.

6–3 THE DIFFERENTIAL EQUATION OF THE ELASTIC CURVE

The expression for the curvature of the elastic curve derived in Art. 6–2 is useful only when the bending moment is constant for the interval of the beam involved. For most beams the bending moment is a function of the position along the beam and a more general expression is required.

The curvature from the calculus (see any standard calculus textbook) is

$$\frac{1}{\rho}=\frac{d^2y/dx^2}{\left[1+(dy/dx)^2\right]^{3/2}}$$

For actual beams the slope dy/dx is very small, and its square can be neglected in comparison to unity. With this approximation,

$$\frac{1}{\rho}=\frac{d^2y}{dx^2}$$

and Eq. 6–1 becomes

$$EI\frac{d^2y}{dx^2}=M_x \tag{6–2}$$

which is the differential equation for the elastic curve of a beam where the subscript x on M is used as a reminder that M is a function of x.

The differential equation of the elastic curve can also be obtained from the geometry of the bent beam as shown in Fig. 6–2, where it is evident that $dy/dx=\tan\theta=\theta$ (approximately) for small angles and that $d^2y/dx^2 = d\theta/dx$. Again from Fig. 6–2

$$d\theta=\frac{dL}{\rho}=\frac{dx}{\rho}$$

(approximately) for small angles. Therefore

$$\frac{d^2y}{dx^2}=\frac{d\theta}{dx}=\frac{1}{\rho}=\frac{M_x}{EI}$$

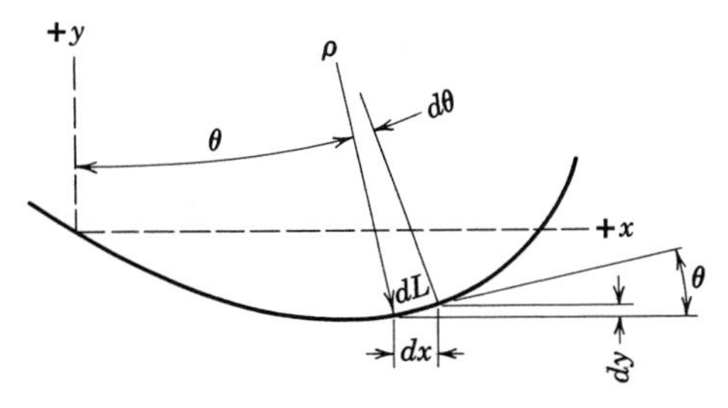

Fig. 6–2

or

$$EI\frac{d^2y}{dx^2} = M_x \tag{6–2}$$

The sign convention for bending moments established in Art. 5–6 will be used for Eq. 6–2. Both E and I are always positive; therefore, the signs of the bending moment and the second derivative must be consistent. With the coordinate axes selected as shown in Fig. 6–3, the slope changes from positive to negative in the interval from A to B; therefore, the second derivative is negative, which agrees with the sign convention of Art. 5–6. For the interval BC, both d^2y/dx^2 and M_x are seen to be positive.

Careful study of Fig. 6–3 reveals that the signs of the bending moment and the second derivative are also consistent when the origin is selected at the right with x positive to the left and y positive upward. Unfortunately, the signs are inconsistent when y is positive downward. *In this book, for horizontal beams,* y *will always be selected positive upward for beam deflection problems.*

Before proceeding with the solution of Eq. 6–2, it is desirable to correlate the successive derivatives of the deflection y of the elastic curve with the physical quantities that they represent in beam action. They are:

$$\text{deflection} = y$$

$$\text{slope} = \frac{dy}{dx}$$

$$\text{moment} = EI\frac{d^2y}{dx^2} \text{ (from Eq. 6–2)}$$

$$\text{shear} = \frac{dM}{dx} \text{ (from Art. 5–7)} = EI\frac{d^3y}{dx^3} \text{ (for } EI \text{ constant)}$$

$$\text{load} = \frac{dV}{dx} \text{ (from Art. 5–7)} = EI\frac{d^4y}{dx^4} \text{ (for } EI \text{ constant)}$$

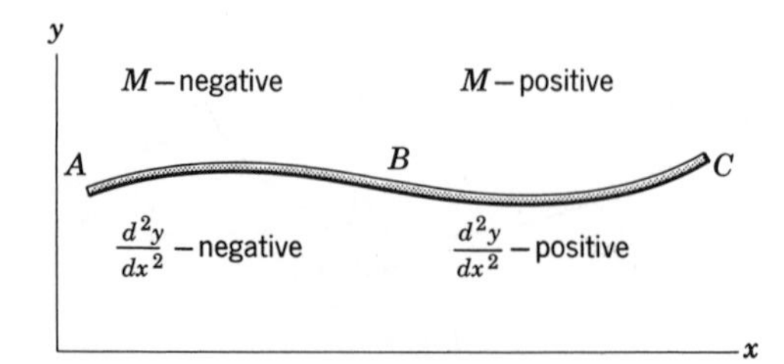

Fig. 6–3

where the signs are as given in Arts. 5–6 and 5–7.

In Art. 5–8 a method based on these differential relations was presented for starting from the load diagram and drawing first the shear diagram and then the moment diagram. This method can readily be extended to construction of the slope diagram from the relation

$$M = EI\frac{d\theta}{dx}$$

from which

$$\int_{\theta_A}^{\theta_B} d\theta = \int_{x_A}^{x_B} \frac{M}{EI}\,dx$$

This relation shows that except for a factor EI, the area under the moment diagram between any two points along the beam gives the change in slope between the same two points. Likewise, the area under the slope diagram between two points along a beam gives the change in deflection between these points. These relations have been used to construct the complete series of diagrams shown in Fig. 6–4 for a simply supported beam with a

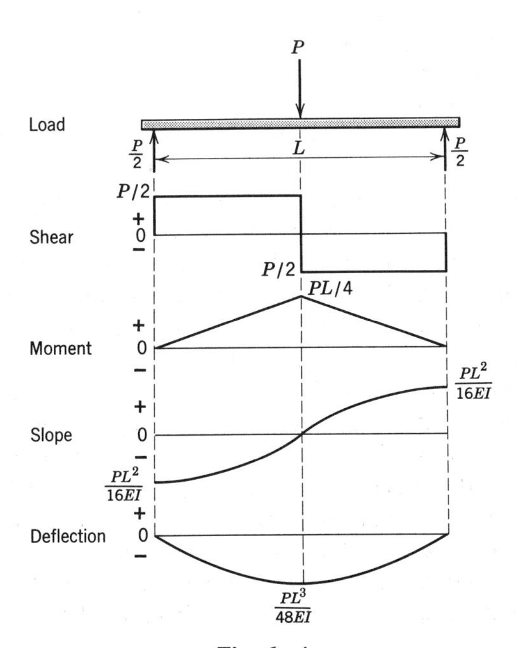

Fig. 6–4

concentrated load at the center of the span. The geometry of the beam was used to locate the points of zero slope and deflection, required as starting points for the construction. More commonly used methods for calculating beam deflections will be developed in succeeding articles.

Before proceeding with specific methods for calculating beam deflections, it is advisable to consider the assumptions used in the development of the basic relation Eq. 6–2. All of the limitations that apply to the flexure formula apply to the calculation of deflections because the flexure formula was used in the derivation of Eq. 6–2. It is further assumed that:

1. The square of the slope of the beam is negligible compared to unity.

2. The beam deflection due to shearing stresses is negligible (a plane section is assumed to remain plane). See Art. 6–12 for further information.

3. The values of E and I remain constant for any interval along the beam. In case either of them varies and can be expressed as a function of the distance x along the beam, a solution of Eq. 6–2 that takes this variation into account may be possible.

6–4 DEFLECTIONS BY INTEGRATION

Whenever the assumptions of the previous paragraph are essentially correct and the bending moment can be readily expressed as an integrable function of x, Eq. 6–2 can be solved for the deflection y of the elastic curve of a beam at any point x along the beam. The constants of integration can be evaluated from the applicable boundary or matching conditions.

A boundary condition is defined as a known set of values for x *and* y, *or* x *and* dy/dx, *at a specific location along the beam.* One boundary condition (one set of values) can be used to determine one and only one constant of integration.

Many beams are subjected to abrupt changes in loading along the beam, such as concentrated loads, reactions, or even distinct changes in the amount of a uniformly distributed load. Since the expressions for the bending moment on the left and right of any abrupt change in load are different functions of x, it is impossible to write a single equation for the bending moment in terms of ordinary algebraic functions that is valid for the entire length of the beam. This can be resolved by writing separate bending moment equations for each interval of the beam. Although the intervals are bounded by abrupt changes in load, the beam is continuous at such locations and, consequently, the slope and the deflection at the junction of adjacent intervals must match. *A matching condition is defined as the equality of slope or deflection, as determined at the junction of two intervals from the elastic curve equations for both intervals.* One matching condition (for example, at x equals $L/3$, y from the left equation equals y

from the right equation) can be used to determine one and only one constant of integration.

The procedure for obtaining beam deflections where matching conditions are required is lengthy and tedious. A method is presented in the next article in which *singularity functions* are used to write a single equation for the bending moment that is valid for the entire length of the beam; this eliminates the need for matching conditions and, accordingly, reduces the labor involved.

Calculating the deflection of a beam by the double integration method involves four definite steps, and the following sequence for these steps is strongly recommended.

---

1. Select the interval or intervals of the beam to be used; next, place a set of coordinate axes on the beam *with the origin at one end of an interval*, and then indicate the range of values of x in each interval. For example, two adjacent intervals might be

$$0 \leqslant x \leqslant L \quad \text{and} \quad L \leqslant x \leqslant 3L$$

2 List the available boundary and matching (where two or more adjacent intervals are used) conditions for each interval selected. Remember that two conditions are required to evaluate the two constants of integration for each interval used.

3. Express the bending moment as a function of x for each interval selected, and equate it to $EI\, d^2y/dx^2$.

4. Solve the differential equation or equations from item 3, and evaluate all constants of integration. Check the resulting equations for dimensional homogeneity. Calculate the deflection at specific points where required.

A roller or pin at any point in a beam (see Fig. 6–5*a* and *b*) represents a simple support at which the beam cannot deflect (unless otherwise stated in the problem) but can rotate. At a fixed end, as represented by Fig. 6–5*c* and *d*, the beam can neither deflect nor rotate unless otherwise stated.

The following examples illustrate the use of the double integration method for calculating beam deflections.

Example 6–1 For the beam loaded and supported as shown in Fig. 6–6*a* locate the point of maximum deflection between the supports.

Solution From a free-body diagram of the beam and the equations of equilibrium, the left reaction is found to be $7wL/12$ upward. As indicated in Fig. 6–6*b*, the origin of coordinates is selected at the left support, and the interval to be used is $0 \leqslant x \leqslant L$.

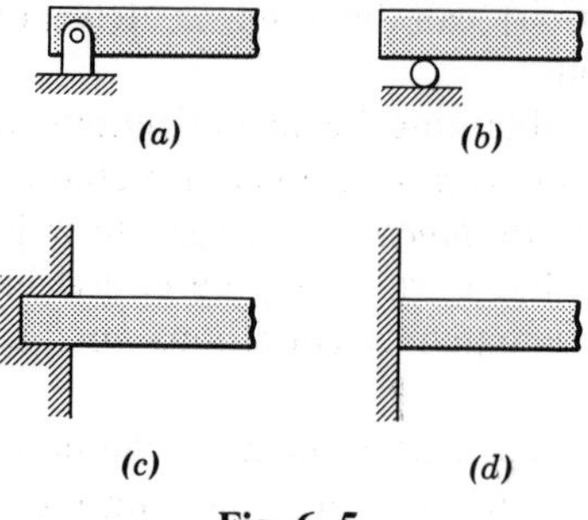

Fig. 6–5

The two required boundary conditions are $y=0$ when $x=0$ and $y=0$ when $x=L$. From the free-body diagram of a portion of the beam as shown in Fig. 6–6*b*, the expression for the bending moment at any point in the selected interval is seen to be

$$EI\frac{d^2y}{dx^2} = M_x = \frac{7wL}{12}x - \frac{wL^2}{12} - wx\left(\frac{x}{2}\right)$$

Successive integration gives

$$EI\frac{dy}{dx} = \frac{7wL}{24}x^2 - \frac{wL^2}{12}x - \frac{w}{6}x^3 + C_1$$

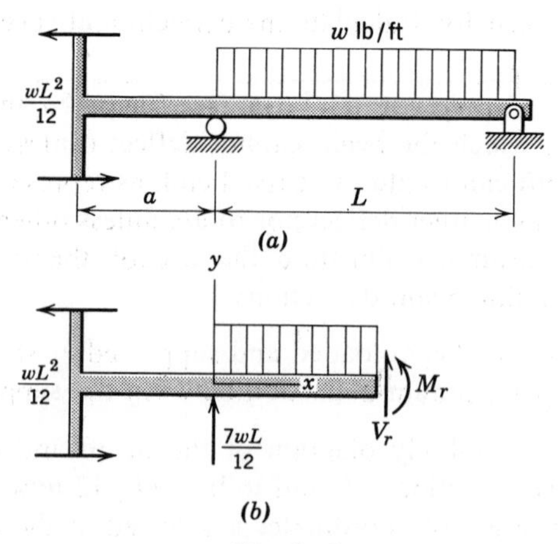

Fig. 6–6

and

$$EIy=\frac{7wL}{72}x^3-\frac{wL^2}{24}x^2-\frac{w}{24}x^4+C_1x+C_2$$

Substitution of the boundary condition $y=0$ when $x=0$ yields $C_2=0$. Substitution of the remaining boundary condition $y=0$ when $x=L$ gives $C_1=-wL^3/72$, and therefore the elastic curve equation is

$$EIy=\frac{7wL}{72}x^3-\frac{wL^2}{24}x^2-\frac{w}{24}x^4-\frac{wL^3}{72}x$$

The maximum deflection occurs where the slope dy/dx is zero or

$$0=\frac{7wL}{24}x^2-\frac{wL^2}{12}x-\frac{w}{6}x^3-\frac{wL^3}{72}$$

The solution of this cubic equation in x gives the point of maximum deflection at

$x=$ <u>$0.541L$ to the right of the left support</u>. Ans.

Although not requested here, the amount of the maximum deflection can readily be obtained by substituting $0.541L$ for x in the elastic curve equation. The result is

$$y=\frac{-7.88wL^4}{10^3EI}=\frac{7.88wL^4}{10^3EI}\text{ downward}$$

Example 6–2 For the beam loaded and supported as shown in Fig. 6–7*a* determine the deflection of the right end.

Solution From the free-body diagram of Fig. 6–7*b* the equations of equilibrium give the shear at the wall as $wL/3$ upward and the moment at the wall as $5wL^2/18$ counterclockwise. With the sign conventions established in Art. 5–6 these results are

$$V_W=+\frac{wL}{3}\quad\text{and}\quad M_W=-\frac{5wL^2}{18}$$

In Fig. 6–7*b* there are two intervals to be considered, namely, the loaded and the unloaded portions of the beam. The loaded portion of the beam must be used because it contains the point where the deflection is required.

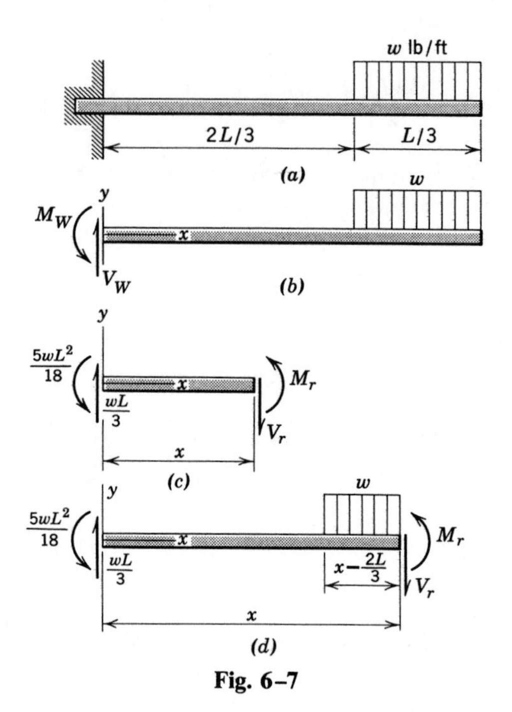

Fig. 6–7

However, a quick check reveals the absence of boundary conditions in this interval. It therefore becomes necessary to use both intervals as well as matching and boundary conditions; hence, the origin of coordinates is selected, as shown. For clarity, the two intervals are written side by side with available boundary conditions, matching conditions, and equations listed under the appropriate interval. The intervals are

$0 \leqslant x \leqslant 2L/3$	$2L/3 \leqslant x \leqslant L$

The available boundary conditions are

$$\frac{dy}{dx}=0 \quad \text{when } x=0$$

and

$$y=0 \quad \text{when } x=0$$

The available matching conditions are

$\frac{dy}{dx}$ from the left equation $= \frac{dy}{dx}$ from the right equation

when $x = 2L/3$

and

y from the left equation $= y$ from the right equation

when $x = 2L/3$

Four conditions (two boundary and two matching) are sufficient for the evaluation of the four constants of integration (two in each of the two elastic curve differential equations), therefore, the problem can be solved in this manner.

From the free-body diagram of an unloaded portion of the beam in Fig. 6–7*c*, the expression for the bending moment at any point in the left interval is seen to be

$$EI\frac{d^2y}{dx^2} = M_x = \frac{wL}{3}x - \frac{5wL^2}{18}$$

From the free-body diagram of Fig. 6–7*d* where the beam is cut in the loaded interval, the bending moment at any point in the right interval is found to be

$$EI\frac{d^2y}{dx^2} = M_x = \frac{wL}{3}x - \frac{5wL^2}{18} - w\left(x - \frac{2L}{3}\right)\left(\frac{x - 2L/3}{2}\right)$$

Integration of each differential equation gives

$$EI\frac{dy}{dx} = \frac{wL}{6}x^2 - \frac{5wL^2}{18}x + C_1$$

$$EI\frac{dy}{dx} = \frac{wL}{6}x^2 - \frac{5wL^2}{18}x - \frac{w}{6}\left(x - \frac{2L}{3}\right)^3 + C_3$$

Substitution of the boundary condition

$$\frac{dy}{dx}=0 \quad \text{when} \quad x=0$$

gives

$$C_1=0$$

Substitution of the matching condition

$$\frac{dy}{dx} \text{ from left Eq} = \frac{dy}{dx} \text{ from right Eq when } x=\frac{2L}{3}$$

gives

$$C_1=C_3=0$$

Integration of the resulting differential equations gives

$$EIy=\frac{wL}{18}x^3-\frac{5wL^2}{36}x^2+C_2 \qquad EIy=\frac{wL}{18}x^3-\frac{5wL^2}{36}x^2-\frac{w}{24}\left(x-\frac{2L}{3}\right)^4+C_4$$

Use of the remaining boundary condition yields $C_2=0$, and use of the final matching condition yields $C_2=C_4=0$.

The deflection of the right end of the beam can now be obtained from the elastic curve equation for the right interval by replacing x by its value L. The result is

$$EIy=\frac{wL}{18}L^3-\frac{5wL^2}{36}L^2-\frac{w}{24}\left(L-\frac{2L}{3}\right)^4$$

from which

$$y=-\frac{163}{1944}\frac{wL^4}{EI}=\underline{\frac{163}{1944}\frac{wL^4}{EI} \quad \text{downward}} \qquad \text{Ans.}$$

PROBLEMS

6–9* For the beam of Prob. 6–2:

(a) Derive the equation of the elastic curve in terms of E, I, L, and M by the double integration method.

(b) What is the shape of the elastic curve?

(c) Explain any discrepancy between the answers to Probs. 6–2*b* and 6–6*b*.

6–10* For the beam loaded and supported as shown in Fig. P6–10, derive the equation for the elastic curve in terms of *E*, *I*, *w*, and *L*. Use the designated axes.

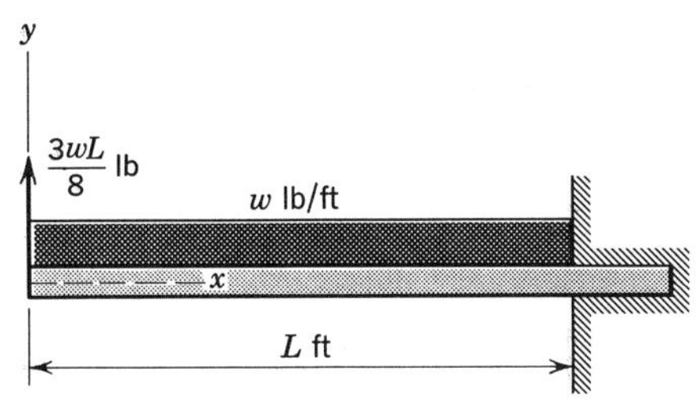

Fig. P6–10

6–11* A beam of length *L* with a uniformly distributed load of *w* lb per unit length over the entire span is simply supported at the left end and restrained at the right end. The ends remain on the same horizontal line, and a load cell shows that the left reaction is one-fourth of the total load. Determine the equation of the elastic curve of the beam in terms of *E*, *I*, *w*, and *L*.

6–12* For the beam in Fig. P6–12, determine the equation of the elastic curve for the portion of the beam between the supports in terms of *E*, *I*, *w*, and *L*.

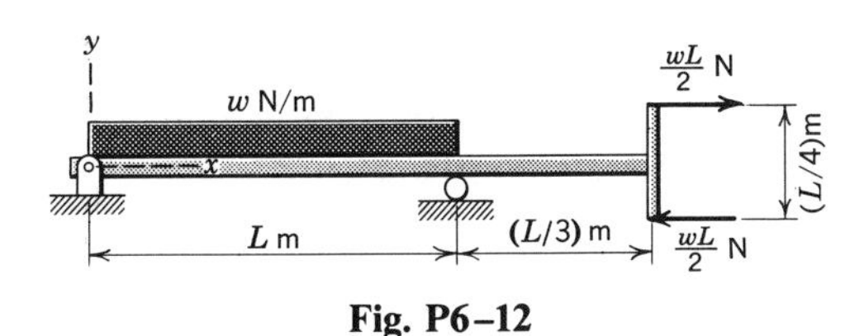

Fig. P6–12

6–13* For the beam loaded and supported as shown in Fig. P6–13, determine the equation of the elastic curve for the interval of the beam between the supports in terms of *E*, *I*, *w*, and *L*. Use the designated axes.

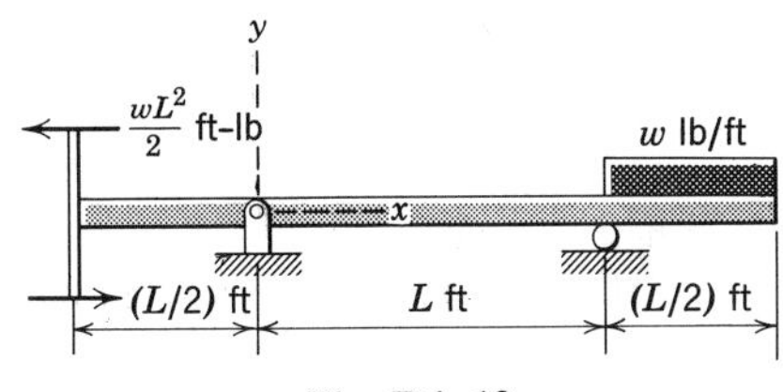

Fig. P6–13

6–14* Derive, in terms of w, L, E, I, x, and y, the equation of the elastic curve for span BC of the beam in Fig. P6–14. Use the indicated origin.

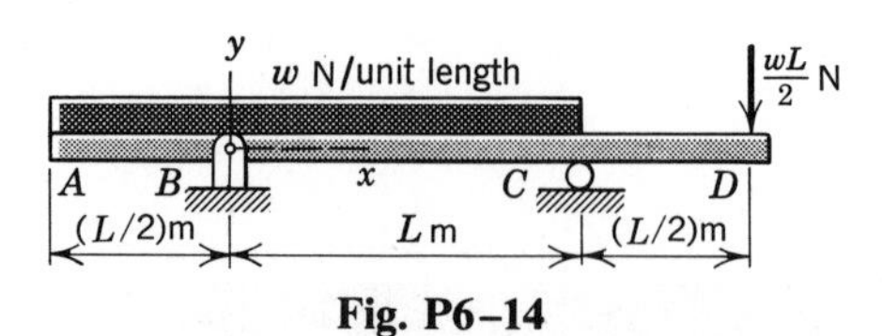

Fig. P6–14

6–15 Determine, in terms of w, L, E, and I, the deflection of the free end of the cantilever beam loaded as shown in Fig. P6–15.

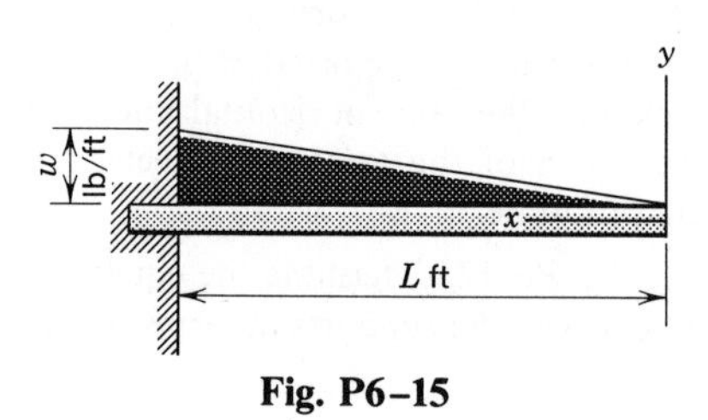

Fig. P6–15

6–16 Determine, in terms of w, L, E, and I, the deflection of the free end of the cantilever beam loaded as shown in Fig. P6–16.

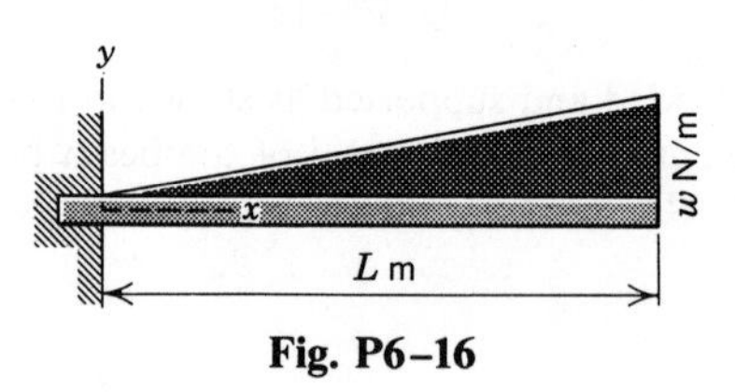

Fig. P6–16

6–17* A 4 × 10-in. timber having a modulus of elasticity of 1200 ksi is used as a post, as shown in Fig. P6–17. Compute the maximum horizontal movement of the top of the post in inches.

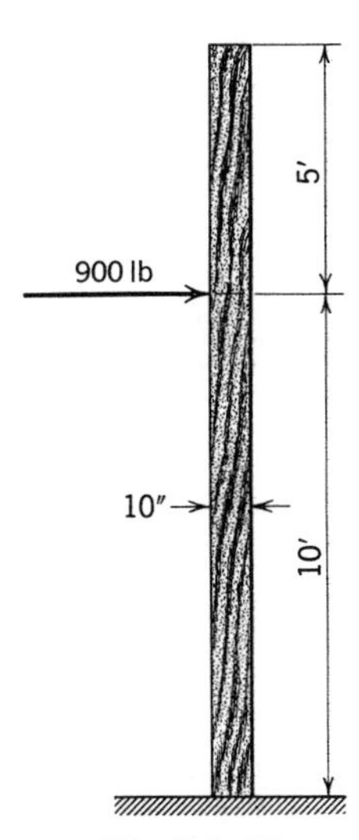

Fig. P6–17

6–18* A timber beam 150 mm wide by 300 mm deep is simply supported at the ends of a span of 6 m. It carries a concentrated load of 10 kN at the center of the span. The modulus of elasticity of the timber is 8 GN/m^2. A pointer 2 m long is attached to the right end, as shown in Fig. P6–18. Determine:

(a) The movement of the right end of the pointer in millimetres.
(b) The maximum deflection of the beam.

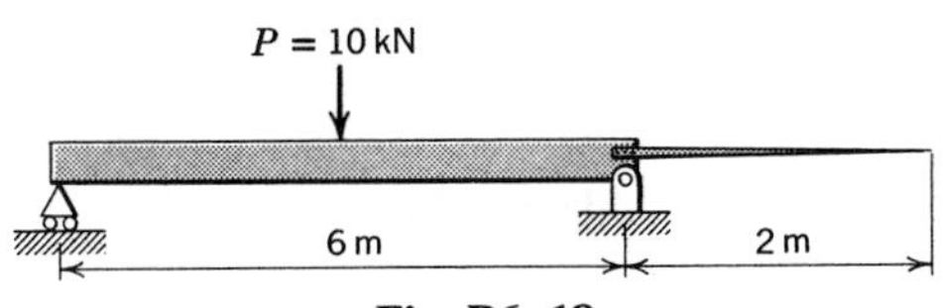

Fig. P6–18

6–19 An S12×50 steel beam 40 ft long is simply supported at its ends and carries a concentrated load P at midspan. Determine the value of P that will produce deflections of 0.9 in. at points 10 ft from the supports. The web of the cross section is vertical.

6–20 A beam of length L m is fixed at the left end and carries a uniformly distributed load of w N/m. The right end is subjected to a moment of $+3wL^2/8$. Determine:

(a) The equation of the elastic curve.
(b) The deflection of the right end when $I=2.5(10^6)$ mm^4, $E=200$ GN/m^2, $L=3$ m, and $w=1500$ N/m.

6–21* For the cantilever beam loaded as shown in Fig. P6–21, determine the deflection of the free end in terms of w, L, E, and I.

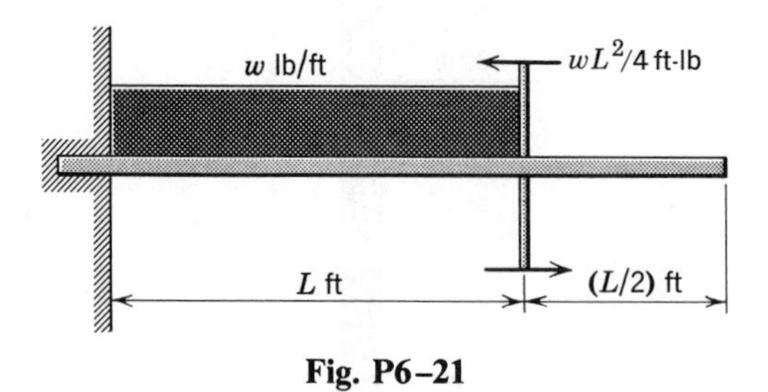

Fig. P6–21

6–22* For the continuous beam in Fig. P6-22, the bending moment at the second support is $-wL^2/12$, and the vertical shear just to the right of the second support is $+5wL/12$. Determine the equation of the elastic curve for the portion of the beam between the second and third supports.

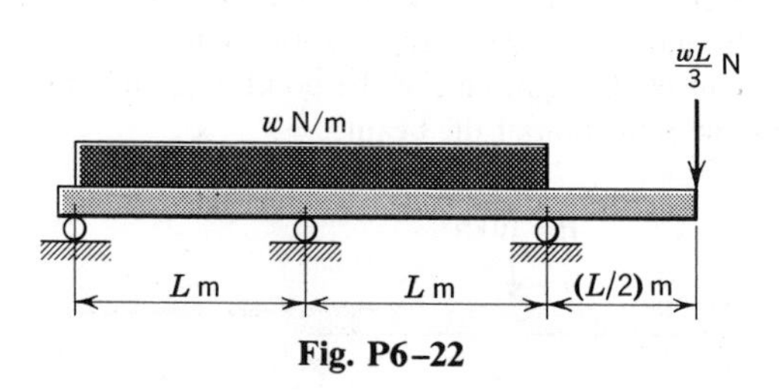

Fig. P6–22

6–23* Determine, for the beam of Fig. P6–23, the distance from the left reaction to the point of maximum deflection between the supports.

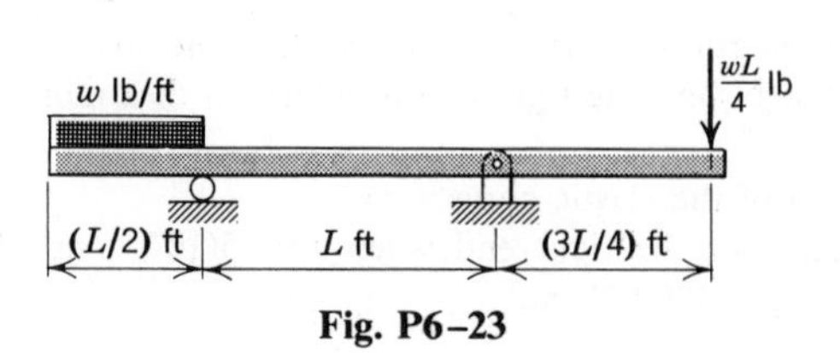

Fig. P6–23

6–24* For the beam loaded and supported as shown in Fig. P6–24, determine the maximum deflection between the supports in terms of P, L, E, and I.

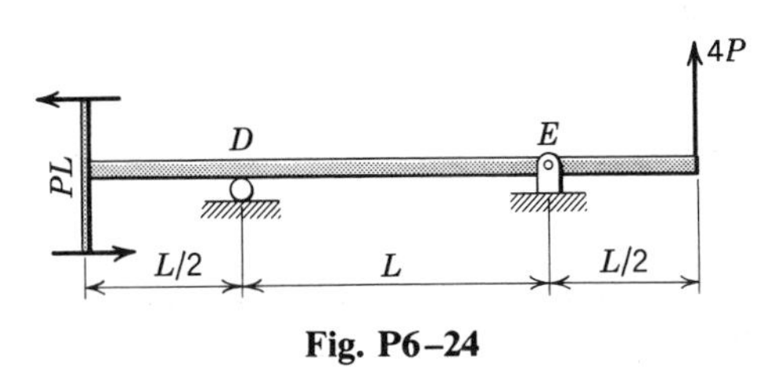

Fig. P6–24

6–25 Determine the maximum deflection in the left half of the beam loaded and supported as shown in Fig. P6–25 in terms of M, L, E, and I. The deflection at the center of the beam is zero due to symmetry.

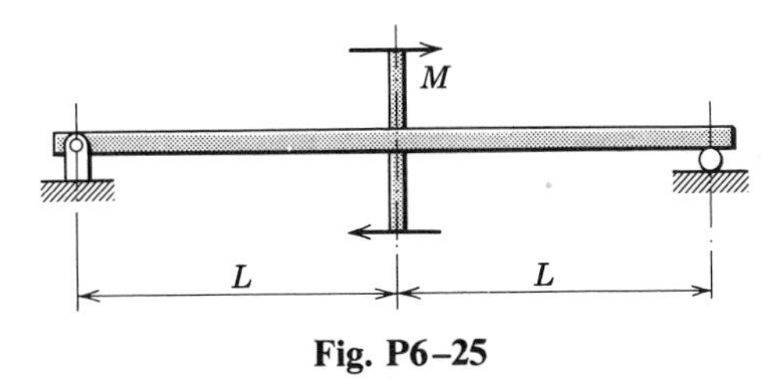

Fig. P6–25

6–26 Determine the maximum deflection in the interval AB of the beam loaded and supported as shown in Fig. P6–26 in terms of P, L, E, and I.

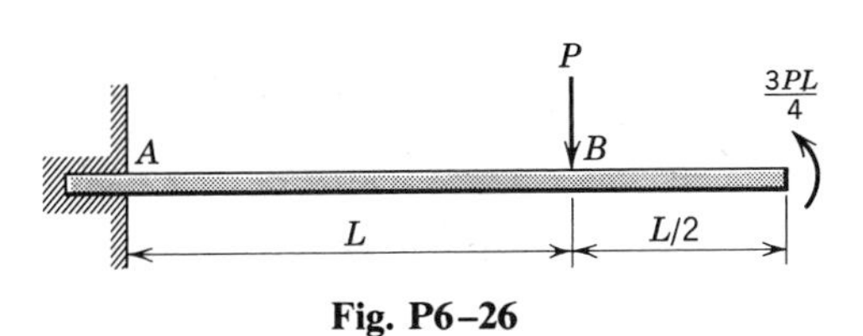

Fig. P6–26

6–27* For the beam loaded and supported as shown in Fig. P6–27, determine in terms of P, L, E, and I:

(a) The elastic curve equations.
(b) The location of the point of maximum deflection.
(c) The maximum deflection.

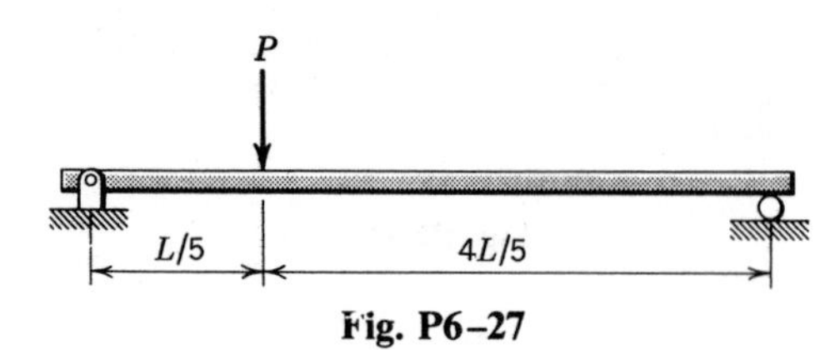

Fig. P6–27

6–28* A cantilever beam carries two concentrated loads as shown in Fig. P6–28. Determine, in terms of P, L, E, and I:

(a) The location of the point of maximum deflection.
(b) The deflection of the beam at the load P.

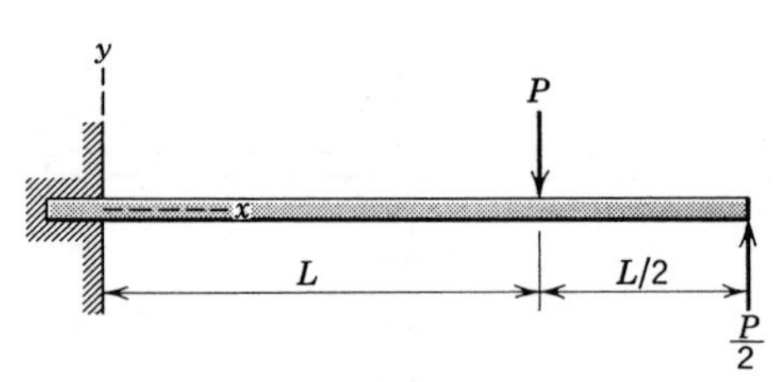

Fig. P6–28

6–29 Develop in terms of w, L, E, I, x, and y the elastic curve equation for segment BC of the beam in Fig. P6–29.

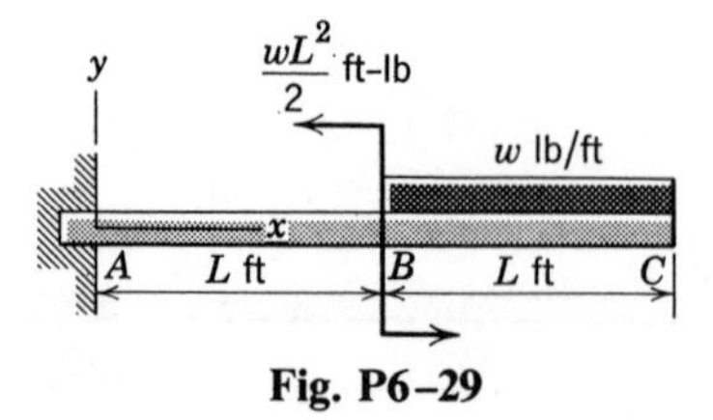

Fig. P6–29

6–30 For the beam in Fig. P6–30, determine the deflection of the beam at $x = L$ in terms of w, L, E, and I.

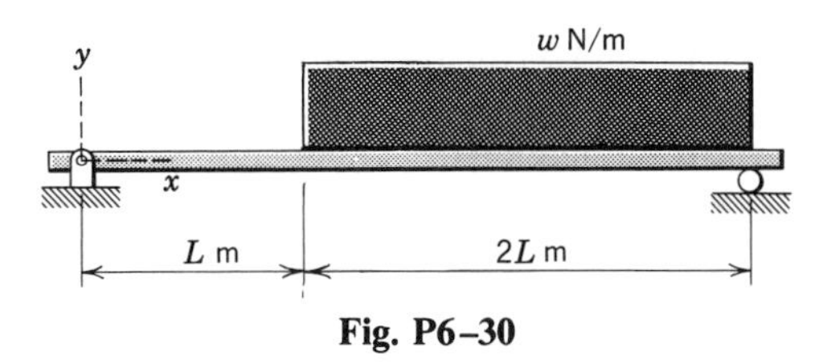

Fig. P6–30

6–31* For the beam in Fig. P6–31:

(a) Derive the elastic curve equation for the loaded portion of the beam in terms of E, I, w, and a. Use the designated axes.

(b) Determine the maximum deflection when $I =$ [illegible] in.^{4}, $E = 12(10^6)$ psi, $w =$ 1000 lb/ft, and $a = 3$ ft.

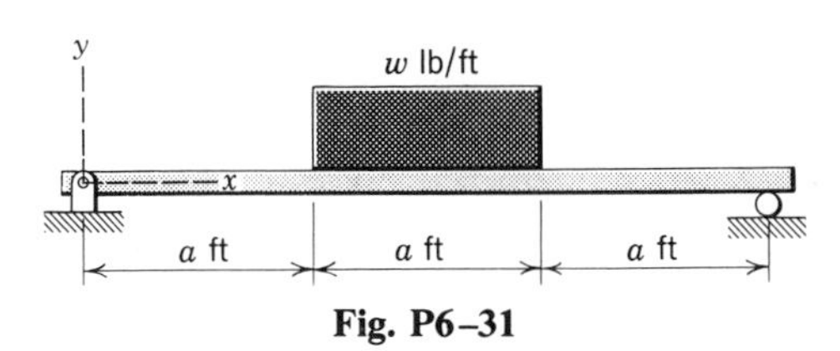

Fig. P6–31

6–32* For the beam in Fig. P6–32:

(a) Derive the elastic curve equation for the loaded portion in terms of E, I, w, and L. Use the designated axes.

(b) Determine the maximum deflection when $I = 36(10^6)$ mm^4, $E = 80$ GN/m^2, $w = 15$ kN/m, and $L = 1$ m.

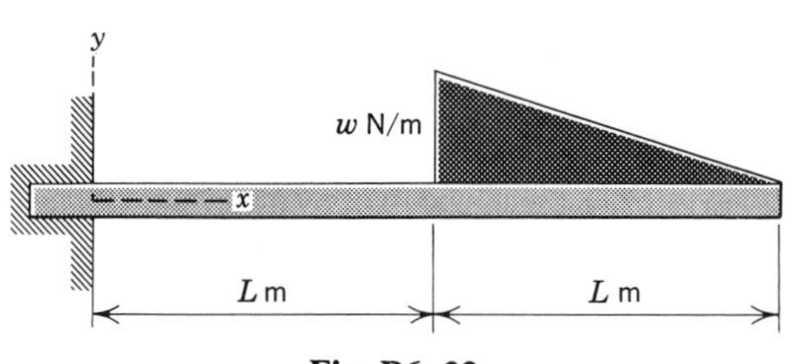

Fig. P6–32

6–5 SINGULARITY FUNCTIONS FOR BEAM DEFLECTIONS

The double integration method of Art. 6–4 becomes extremely tedious and time-consuming when several intervals and several sets of matching conditions are required. The labor involved in solving problems of this type can be diminished by making use of *singularity functions* following the method originally developed in 1862 by the German mathematician A. Clebsch (1833–1872).[1]

Singularity functions are closely related to the unit step function used by the British physicist O. Heaviside (1850–1925) to analyze the transient response of electrical circuits. Singularity functions will be used here for writing one bending moment equation that applies in all intervals along a beam, thus eliminating the need for matching conditions.

A singularity function of x is written as $\langle x-x_0\rangle^n$, where n is any *integer* (positive or negative) including zero, and x_0 is a constant equal to the value of x at the initial boundary of a specific interval along a beam. Selected properties of singularity functions required for beam-deflection problems are listed here for emphasis and ready reference.

$$\langle x-x_0\rangle^n = \begin{cases} (x-x_0)^n & \text{when } n>0 \text{ and } x \geqslant x_0 \\ 0 & \text{when } n>0 \text{ and } x<x_0 \end{cases}$$

$$\langle x-x_0\rangle^0 = \begin{cases} 1 & \text{when } x \geqslant x_0 \\ 0 & \text{when } x<x_0 \end{cases}$$

$$\int \langle x-x_0\rangle^n \, dx = \frac{1}{n+1}\langle x-x_0\rangle^{n+1} + C \quad \text{when } n \geqslant 0$$

$$\frac{d}{dx}\langle x-x_0\rangle^n = n\langle x-x_0\rangle^{n-1} \quad \text{when } n \geqslant 1$$

Several examples of singularity functions are shown in Fig. 6–8.

By making use of these properties of singularity functions, one is able to write a single equation for the bending moment for a beam and obtain the correct value of the moment in any interval along the beam. To amplify this statement consider the beam of Fig. 6–9. By the methods of Arts. 5–8

[1] For a rather complete history on the Clebsch method and the numerous extensions thereof, see "Clebsch's Method for Beam Deflections," Walter D. Pilkey, *Journal of Engineering Education*, Jan. 1964, p. 170.

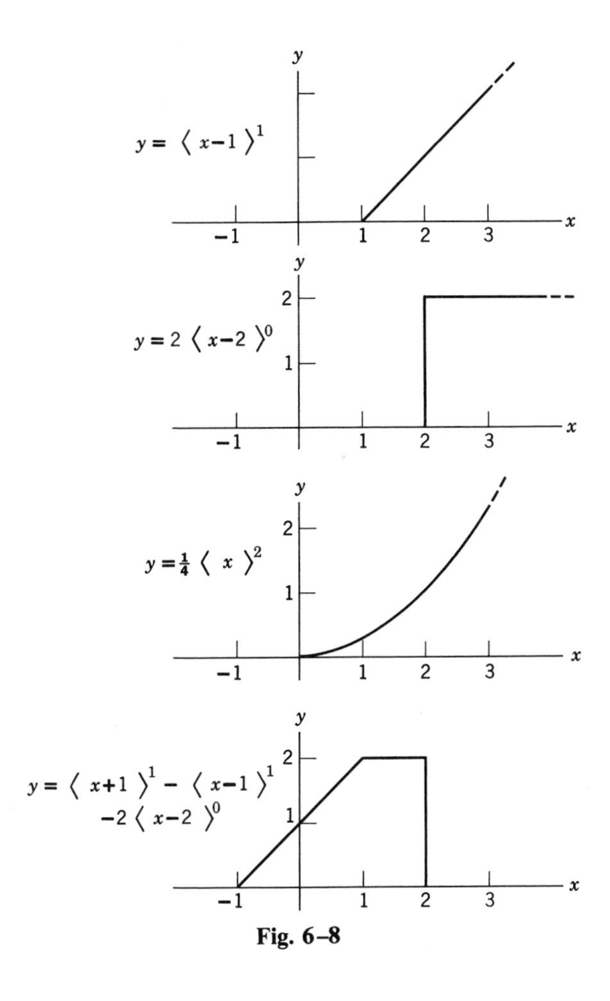
y
$y = \langle x-1 \rangle^1$
-1
1
2
3
x
$y = 2\langle x-2 \rangle^0$
$y = \frac{1}{4}\langle x \rangle^2$
$y = \langle x+1 \rangle^1 - \langle x-1 \rangle^1$
$-2\langle x-2 \rangle^0$

Fig. 6–8

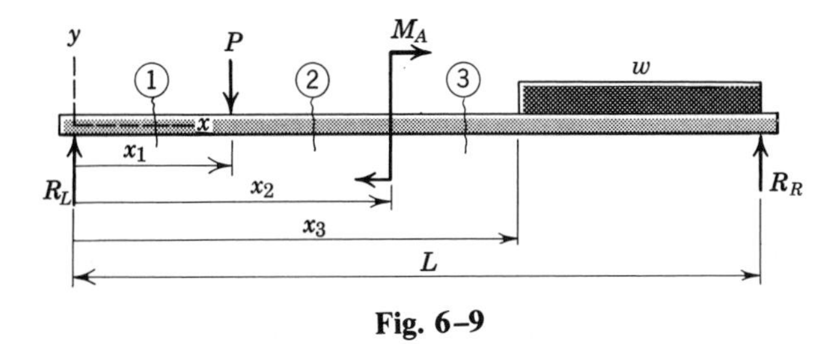
y
P
M_A
w
1
2
3
x
x_1
x_2
x_3
L
R_L
R_R

Fig. 6–9

and 6–4, the moment equations at the four designated sections are:

$$M_1 = R_L x \qquad 0 \leqslant x \leqslant x_1$$
$$M_2 = R_L x - P(x - x_1) \qquad x_1 \leqslant x \leqslant x_2$$
$$M_3 = R_L x - P(x - x_1) + M_A \qquad x_2 \leqslant x \leqslant x_3$$
$$M_4 = R_L x - P(x - x_1) + M_A - (w/2)(x - x_3)^2 \qquad x_3 \leqslant x \leqslant L$$

These four moment equations can be combined into a single equation by means of singularity functions to give

$$M_x = R_L x - P\langle x - x_1\rangle^1 + M_A\langle x - x_2\rangle^0 - \frac{w}{2}\langle x - x_3\rangle^2 \qquad 0 \leqslant x \leqslant L$$

Distributed beam loadings that are sectionally continuous (the distributed load cannot be represented by a single function of x for all values of x) require special consideration. Moment equations for such loads can be readily obtained by superposition, as illustrated in the following example. Further applications of singularity functions to the determination of beam deflections are given in Exams. 6–4 and 6–5.

Example 6–3 Use singularity functions to write a single equation for the bending moment at any section of the beam of Fig. 6–10*a*.

Solution The loading on the beam of Fig. 6–10*a* can be considered as a combination of the loadings shown in Fig. 6–10*b*, *c*, *d*, and *e*, where downward-acting loads are shown on top of the beam and upward-acting loads on the bottom. The moment of the linearly varying load can be obtained from the geometry of the load diagram. The moment equation is

$$M_x = R_L x - \frac{w_1}{6(x_2 - x_1)}\langle x - x_1\rangle^3 + \frac{w_1}{6(x_2 - x_1)}\langle x - x_2\rangle^3$$
$$+ \frac{w_1 - w_2}{2}\langle x - x_2\rangle^2 + \frac{w_2}{2}\langle x - x_3\rangle^2 \qquad \text{Ans.}$$

Example 6–4 Solve Exam. 6–2 (Art. 6–4) by use of singularity functions.

Solution The boundary conditions, in the interval $0 \leqslant x \leqslant L$, are $dy/dx = 0$ when $x = 0$ and $y = 0$ when $x = 0$ with the axes as shown in Fig. 6–7*b*. From the free-body diagram of Fig. 6–7*d*, the expression for the bending

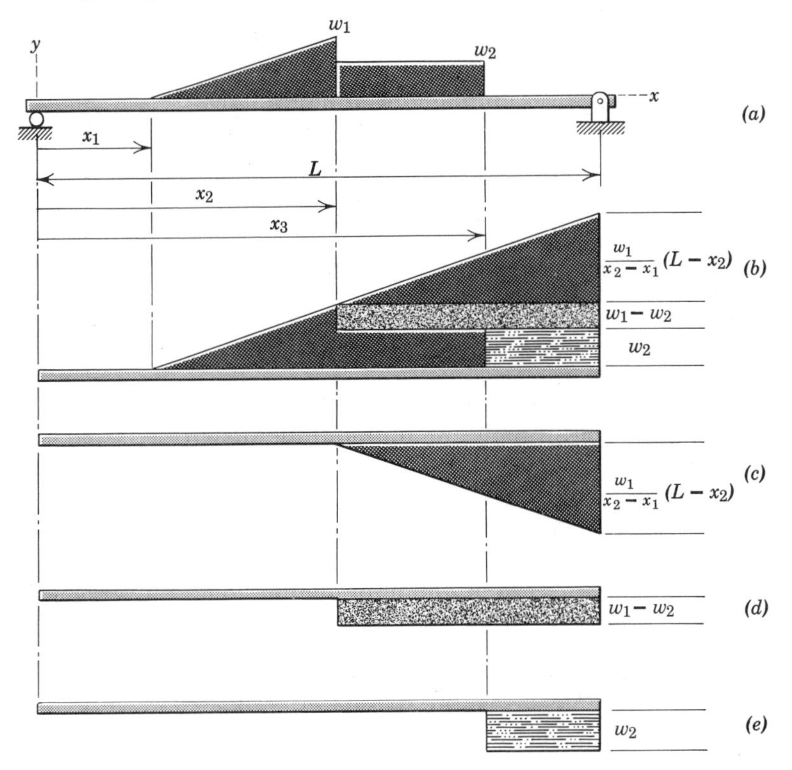

Fig. 6-10

moment is

$$EI\frac{d^2y}{dx^2} = -\frac{5wL^2}{18} + \frac{wLx}{3} - \frac{w}{2}\left\langle x - \frac{2L}{3}\right\rangle^2$$

The first integration gives

$$EI\frac{dy}{dx} = -\frac{5wL^2x}{18} + \frac{wLx^2}{6} - \frac{w}{6}\left\langle x - \frac{2L}{3}\right\rangle^3 + C_1$$

Substituting the boundary condition $dy/dx = 0$ when $x = 0$, and noting that for $x < 2L/3$, the term in the brackets is zero, gives $C_1 = 0$. Integrating again gives

$$EIy = -\frac{5wL^2}{36}x^2 + \frac{wL}{18}x^3 - \frac{w}{24}\left\langle x - \frac{2L}{3}\right\rangle^4 + C_2$$

The boundary condition $y=0$ when $x=0$ gives $C_2=0$. The deflection at the right end is obtained by substituting $x=L$ in the above elastic curve equation. The result is

$$y=-\frac{163}{1944}\frac{wL^4}{EI}=\underline{\frac{163}{1944}\frac{wL^4}{EI}\quad\text{downward}}\qquad\text{Ans.}$$

It will be observed that the result agrees with that of Exam. 6–2.

Example 6–5 Determine the deflection at the left end of the beam of Fig. 6–11*a*.

Solution The equations of equilibrium, applied to the free-body diagram of Fig. 6–11*b*, give the reaction R_L as $3wL/8$ upward. By use of singularity functions and the superposed loads of Fig. 6–11*c*, the bending-moment equation becomes

$$EI\frac{d^2y}{dx^2}=-\frac{w}{2}(x+L)^2+\frac{w}{2}\left\langle x+\frac{L}{2}\right\rangle^2+\frac{3wL}{8}\langle x\rangle^1+\frac{wL^2}{2}\left\langle x-\frac{L}{2}\right\rangle^0$$

The boundary conditions are $y=0$ when $x=0$, and $y=0$ when $x=L$. Two integrations of the moment equation give

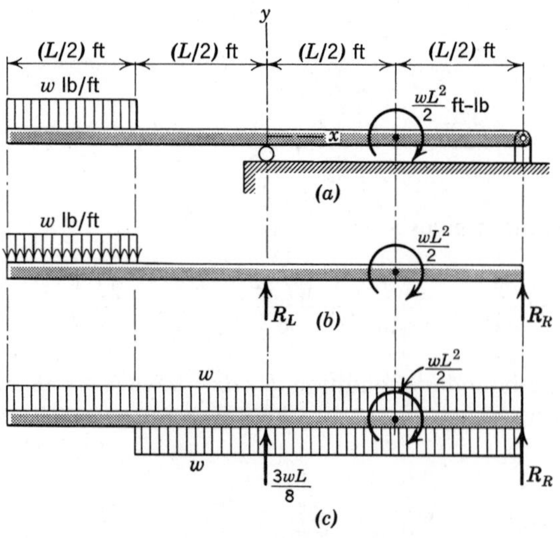

Fig. 6–11

$$EI\frac{dy}{dx} = -\frac{w}{6}(x+L)^3 + \frac{w}{6}\left\langle x+\frac{L}{2}\right\rangle^3 + \frac{3wL}{16}\langle x\rangle^2 + \frac{wL^2}{2}\left\langle x-\frac{L}{2}\right\rangle^1 + C_1$$

and

$$EIy = -\frac{w}{24}(x+L)^4 + \frac{w}{24}\left\langle x+\frac{L}{2}\right\rangle^4 + \frac{wL}{16}\langle x\rangle^3 + \frac{wL^2}{4}\left\langle x-\frac{L}{2}\right\rangle^2 + C_1x + C_2$$

The first boundary condition, $y=0$ when $x=0$, gives

$$0 = -\frac{wL^4}{24} + \frac{wL^4}{384} + 0 + 0 + 0 + C_2$$

from which

$$C_2 = +\frac{5}{128}wL^4$$

The second boundary condition, $y=0$ when $x=L$, gives

$$0 = -\frac{16wL^4}{24} + \frac{81wL^4}{384} + \frac{wL^4}{16} + \frac{wL^4}{16} + C_1L + \frac{5wL^4}{128}$$

from which

$$C_1 = +\frac{7}{24}wL^3$$

The deflection at $x=-L$ is found, by substitution, to be

$$y = -\frac{97}{384}\frac{wL^4}{EI} = \underline{\frac{97}{384}\frac{wL^4}{EI} \quad \text{downward}} \qquad \text{Ans.}$$

PROBLEMS

6–33* Use singularity functions to determine the deflection, in terms of P, L, E, and I, at the right end of the cantilever beam shown in Fig. P6–33.

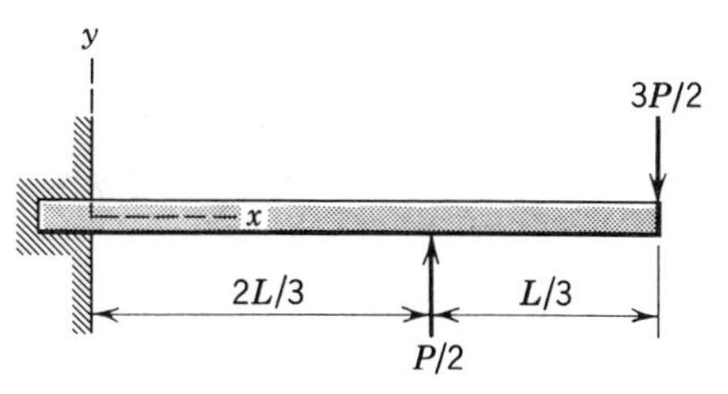

Fig. P6–33

6–34* Use singularity functions to determine the deflection, in terms of w, L, E, and I, at the left end of the cantilever beam shown in Fig. P6–34.

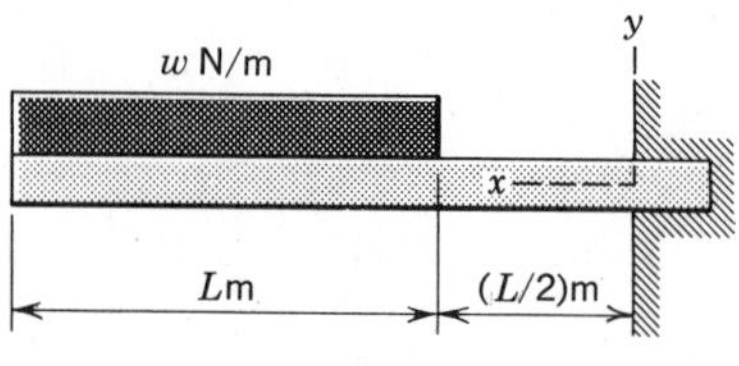

Fig. P6–34

6–35* Use singularity functions to solve Prob. 6–27.

6–36* Use singularity functions to solve Prob. 6–28.

6–37 Use singularity functions to solve Prob. 6–29.

6–38 Use singularity functions to solve Prob. 6–30.

6–39* Use singularity functions to solve Prob. 6–31.

6–40* Use singularity functions to solve Prob. 6–32.

6–41 Determine, in terms of w, E, I, and L, the deflection at the free end of the beam of Fig. P6–41.

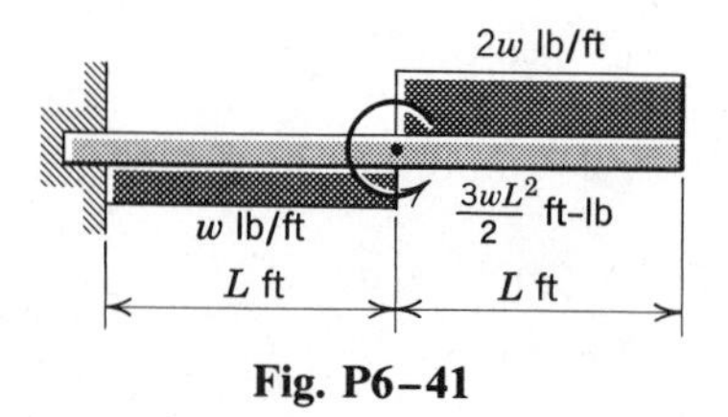

Fig. P6–41

6–42 For the beam loaded and supported as shown in Fig. P6–42, determine the deflection at the midpoint in terms of w, L, E, and I.

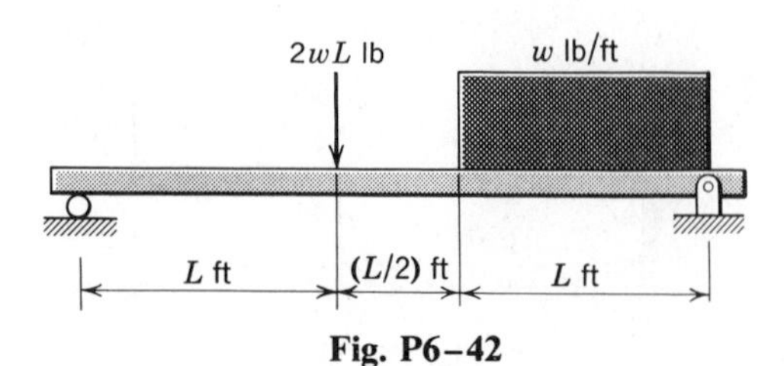

Fig. P6–42

6–43* A steel rod 25 mm in diameter and 500 mm long is used as a cantilever beam and is loaded as shown in Fig. P6–43. Determine the deflection of the free end of the beam.

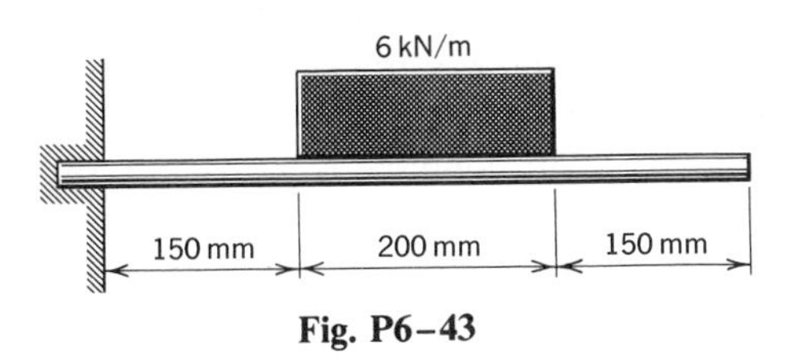

Fig. P6–43

6–44* For the beam in Fig. P6–44 determine the deflection of the left end with $E = 10(10^6)$ psi and $I = 100$ in.^{4}.

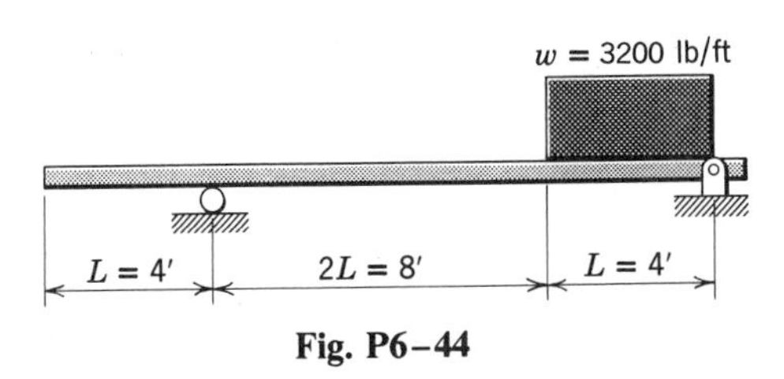

Fig. P6–44

6–45* For the beam loaded and supported as shown in Fig. P6–45, determine the deflection of the left end with $E = 70$ GN/m^2 and $I = 20(10^6)$ mm^4.

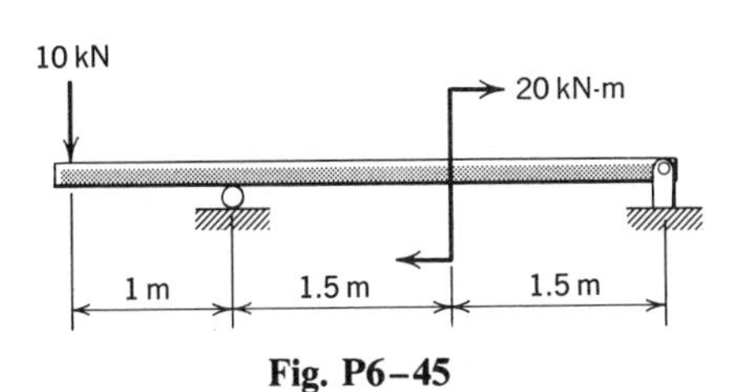

Fig. P6–45

6–6 THE DEVELOPMENT OF THE AREA-MOMENT METHOD

The area-moment method of determining the deflection at any specified point along a beam is a semigraphic method utilizing the relations between successive derivatives of the deflection y and the moment diagram. For problems involving several changes in loading, the area-moment method is usually much faster than the double-integration method; consequently, it is widely used in practice. The method is based on two propositions or theorems derived as follows.

From Fig. 6–2 it is seen that for any arc length dL along the elastic curve of a loaded beam

$$d\theta = dL/\rho$$

and from Eq. 6–1 this becomes

$$d\theta = \frac{M}{EI}\,dL$$

For actual beams the curvature is so small that dL can be replaced by the distance dx along the unloaded beam, and the relation becomes

$$d\theta = \frac{M}{EI}\,dx$$

Reference to Fig. 6–12 (which consists of a loaded beam, its moment diagram, and a sketch, greatly exaggerated, of the elastic curve of the deflected beam) shows that $d\theta$ is also the angle between tangents to the elastic curve at two points a distance dL apart. When any two points A and B along the elastic curve are selected as shown in Fig. 6–12c and the relation for $d\theta$ is integrated between these limits, the result is

$$\int_{\theta_A}^{\theta_B} d\theta = \theta_B - \theta_A = \int_{x_A}^{x_B} \frac{M}{EI}\,dx$$

The integral on the left, designated θ_{AB}, is the angle between tangents to the elastic curve at A and B. The theorem can be written as

$$\theta_{AB} = \int_{x_A}^{x_B} \frac{M\,dx}{EI} \qquad (6\text{–}3)$$

From Fig. 6–12b, the integral is clearly equal to $1/(EI)$ times the area under the moment diagram when E and I are constants (as they are in

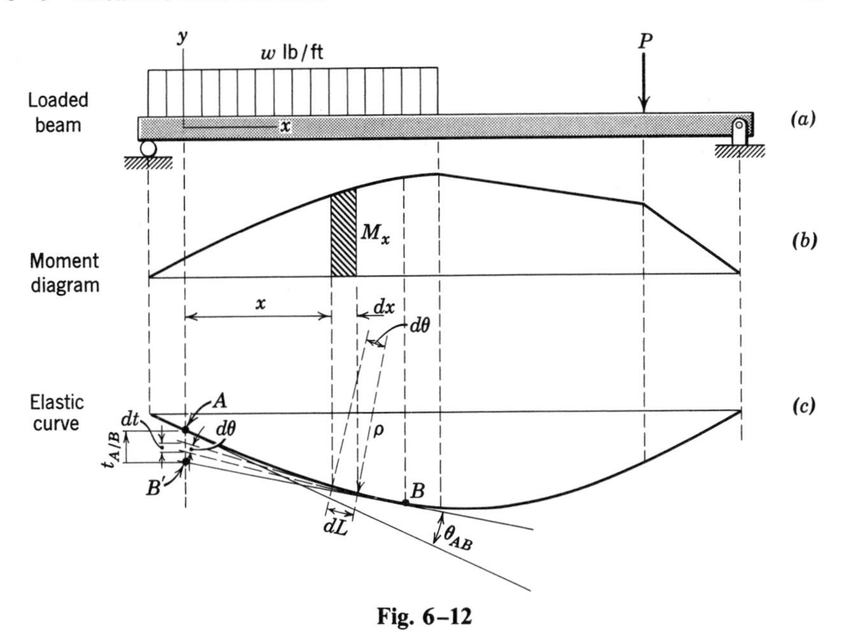

Fig. 6–12

many beams). In case either E or I vary along the beam, it is advisable to construct an $M/(EI)$ diagram instead of a moment diagram. The first area-moment theorem can be expressed in words as follows: *the angle between the tangents to the elastic curve of a loaded beam at any two points* A *and* B *along the elastic curve equals the area under the* M/(EI) *diagram of the beam between ordinates at* A *and* B.

The correct interpretation of numerical results frequently depends on a reliable sign convention, and the following convention is recommended. Areas under positive bending moments are considered positive. A positive area means that the angle is positive or counterclockwise when measured from the tangent to the point on the left to the other tangent, as shown in Fig. 6–13.

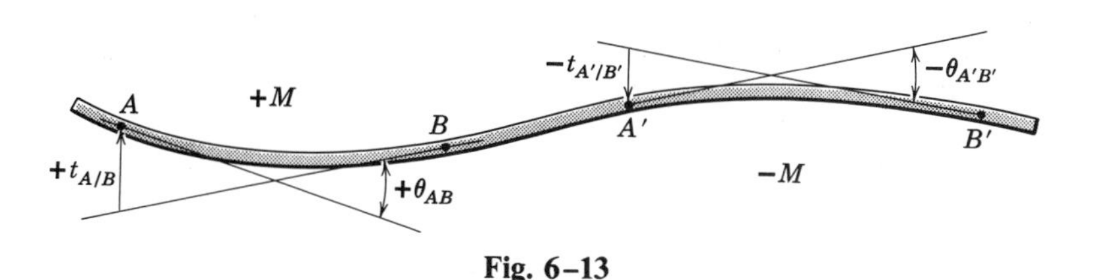

Fig. 6–13

In order to obtain the second area-moment theorem, let B' in Fig. 6–12*c* be the intersection of the tangent to the elastic curve at B with a vertical from point A on the elastic curve. Since the deflections of a beam are assumed to be small, it can be seen from Fig. 6–12*c* that

$$dt = x\,d\theta = \frac{xM}{EI}\,dx$$

Integration gives

$$\int_0^{t_{A/B}} dt = t_{A/B} = \int_{x_A}^{x_B} \frac{Mx}{EI}\,dx \qquad (6\text{–}4)$$

A study of Fig. 6–12*c* shows that the left integral is the vertical (actually perpendicular to the beam) distance $t_{A/B}$ of any point A on the elastic curve from a tangent drawn to any other point B on the elastic curve. This distance is frequently called the *tangential deviation* to distinguish it from the beam deflection y. The integral on the right, as seen from Fig. 6–12*b*, is the first moment, with respect to an axis through A parallel to the y axis, of the area under the $M/(EI)$ diagram between ordinates at A and B.

The second area-moment theorem (Eq. 6–4) can be stated in words as follows: *the vertical distance (tangential deviation) of any point* A *on the elastic curve of a beam from a tangent drawn at any other point* B *on the elastic curve equals the first moment, with respect to an axis at* A, *of the area under the* M/(EI) *diagram between ordinates at* A *and* B.

A sign convention is required for those instances where the shape of the elastic curve is not immediately evident from the given loads. The following sign convention is recommended. The first moment of the area under a positive $M/(EI)$ diagram is considered positive, and from the second area-moment theorem a positive first moment gives a positive tangential deviation. A positive tangential deviation means that point A on the beam (the moment center) is above the tangent drawn from the other point B. Examples of the sign conventions for both the first and second area-moment theorems are shown in Fig. 6–13.

The effectiveness of the area-moment method for determining beam deflections depends, to a considerable extent, on the ease of calculating areas and first moments of areas under the $M/(EI)$ diagrams. Naturally, these values can be obtained by integration of the general expressions from Eqs. 6–3 and 6–4. In general, however, this procedure offers little or no advantage over the double integration method of Arts. 6–3 and 6–4. The device of drawing a separate moment diagram for each load and reaction on a beam, as explained in the next article, greatly simplifies the finding of areas and moments of areas because in this way nearly all areas ıre triangular, rectangular, or parabolic.

6–7 MOMENT DIAGRAMS BY PARTS

In Chap. 5 a composite or complete moment diagram was drawn that included the contribution or effect of all loads and reactions acting on a beam. In the application of the area-moment method for obtaining deflections, it is sometimes easier and more practical to draw a series of diagrams showing the moment due to each load or reaction on a separate sketch. For this purpose, each moment diagram is drawn from the load to an arbitrarily selected section called the reference section. It is preferable to select the reference section at one end of the beam or at one end of a uniformly distributed load. With a reference section at one end, start with the load or reaction at the opposite end of the beam, and draw the bending moment diagram immediately below the beam as if no other loads or reactions were acting on the beam except at the reference section. Proceed toward the right (or left), drawing the moment diagram for each load and reaction below the previous one. The composite moment diagram can be obtained, where necessary, by superposition (addition) of the component diagrams. The actual or total moment on the beam at any section is the algebraic sum of the moments from the separate diagrams, except at an interior reference section where the sum of the moments on either side of the reference section gives the total moment. The following example illustrates the procedure for drawing moment diagrams by parts.

Example 6–6 For the beam in Fig. 6–14*a* construct the moment diagram by parts.

Solution From a free-body diagram of the beam, the equations of equilibrium give the left reaction as 2500 lb upward and the right reaction as 3500 lb upward. The reference section can be selected anywhere along the beam, and each different reference section yields a different series of component moment diagrams. Three different solutions of the example are given to emphasize the importance of a careful selection of the reference section.

(a) The reference section is selected at the right end of the beam. The bending moment due to the 2500-lb left reaction is positive and equal to 2500 times the distance from the left end, as shown in Fig. 6–14*b*. The bending moment due to the 2400-lb concentrated load is negative and equal to 2400 times the distance from the load, as shown in Fig. 6–14*c*. The bending moment due to the distributed load is negative and equal to 300 times the square of the distance from C (the left edge of the uniform

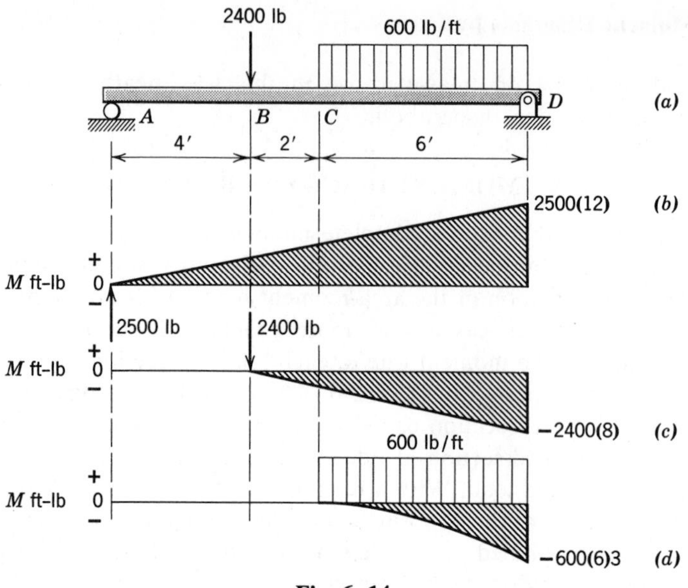

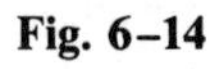
Fig. 6–14

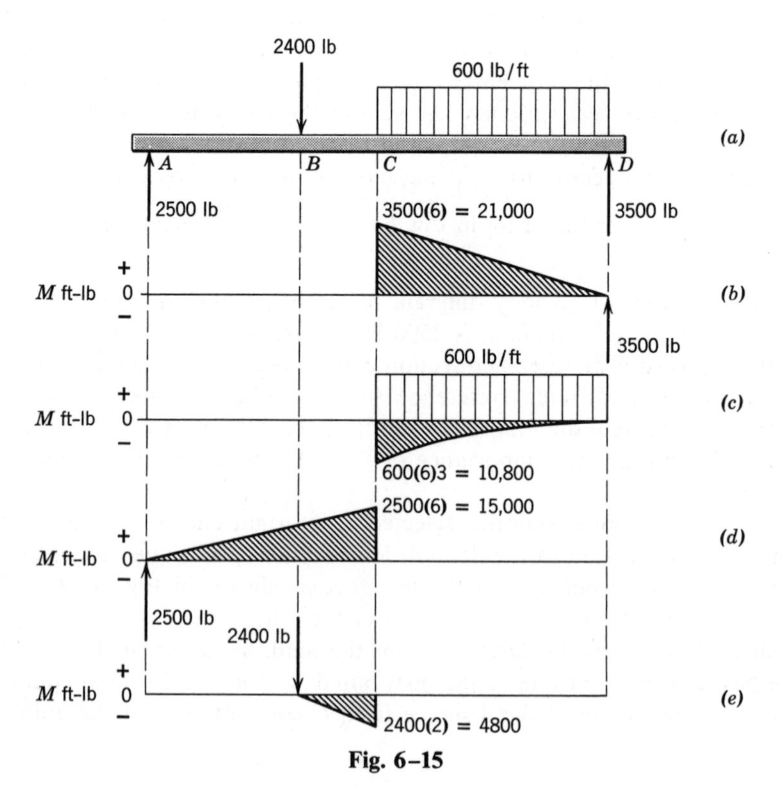

Fig. 6–15

load), as shown in Fig. 6–14*d*. Since the diagrams were drawn from left to right, the right reaction makes no contribution to the moment. It is in the reference section.

(b) The reference section is selected at *C*, the left edge of the distributed load, The bending moment due to the 3500-lb right reaction is positive, varies directly as the distance from the right end, and is shown in Fig. 6–15*b*. The bending moment due to the distributed load is negative, equal to 300 times the square of the distance from the right end, as shown in Fig. 6–15*c*. Likewise, the moments due to the left reaction and the concentrated load are shown in Fig. 6–15*d* and *e*. The total bending moment at the reference section *C* is the sum of the individual moments just to the left or right of the section, namely, +10,200 ft-lb.

(c) The reference section is selected at the left end. The bending moment due to the 3500-lb right reaction is shown in Fig. 6–16*b*. The bending moment due to the distributed load is shown in Fig. 6–16*c*, and the area is subdivided into a triangle, a rectangle, and a parabolic area. This shows the disadvantage of selecting the reference section away from

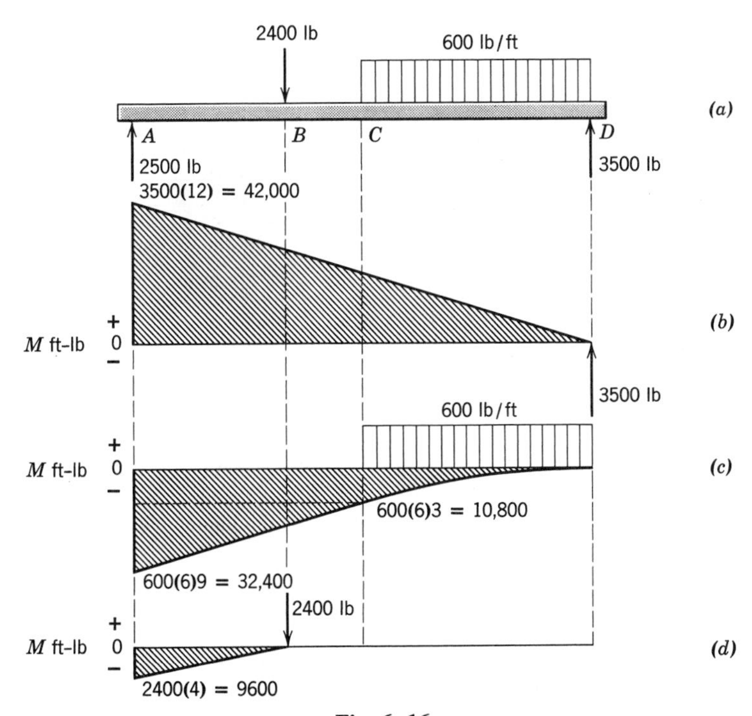

Fig. 6–16

one end of a distributed load. The bending moment due to the concentrated load is shown in Fig. 6–16*d*.

Study of the three solutions reveals that the amount of numerical work involved in finding areas and first moments is greatly altered by the choice of reference section to be used. In solution (a) there are three component areas, whereas solution (b) involves four component areas, and solution (c) is the least desirable with five component areas, three of which are due to the distributed load.

PROBLEMS

6–46 For the beam of Fig. P6–46, construct the moment diagram by parts. Use a reference section at the right support.

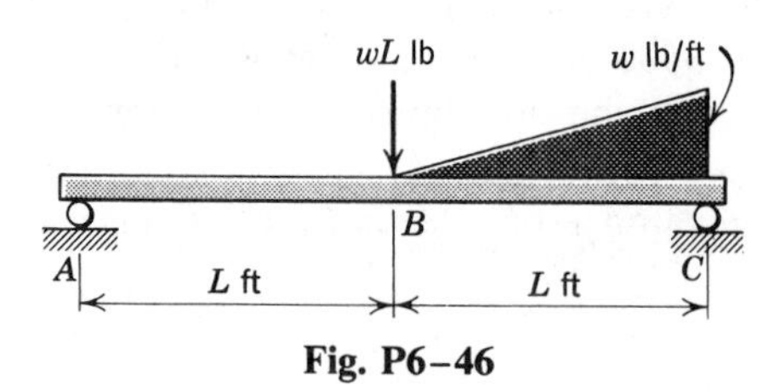

Fig. P6–46

6–47 For the beam of Fig. P6–47, construct the moment diagram by parts. Use a reference section at the right support.

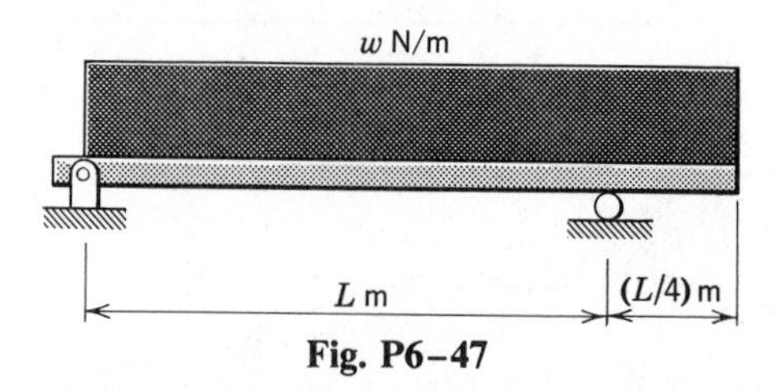

Fig. P6–47

6–48 Construct the moment diagram by parts for the beam of Fig. P6–31. Use a reference section at the right end of the distributed load.

6–49 Construct the moment diagram by parts for the beam of Fig. P6–43. Use a reference section at the right end of the distributed load.

6–50 Construct the moment diagram by parts for the beam of Fig. P6–12. Use a reference section at the right support.

6–51 Construct the moment diagram by parts for the beam of Fig. P6–13. Use a reference section at the right support.

6–52 Construct the moment diagram by parts for the beam of Fig. P6–14. Use a reference section at the right support.

6–53 Construct the moment diagram by parts for the beam of Fig. P6–21. Use a reference section at the left end of the beam.

6–54 Construct the moment diagram by parts for the beam of Fig. P6–29. Use a reference section at the left end of the distributed load.

6–55 Construct the moment diagram by parts for the beam of Fig. P6–32. Use a reference section at the left end of the distributed load.

6–8 DEFLECTIONS BY AREA MOMENTS

The deflection of a specific point along the elastic curve of a beam can normally be determined by one or more applications of the second area-moment theorem. The amount of time and effort required for a given problem depends to a considerable extent on the judgment used in selecting the moment center for the second theorem and in drawing the moment diagram by parts. The ability to construct the correct moment diagram and sketch the elastic curve is essential to the application of the method.

A few general suggestions are presented here to facilitate the application of the method.

---

1. It is frequently easier to use a composite moment diagram instead of a moment diagram by parts for those cases where the beam carries only one load.

2. It will be found that selection of the reference section of a cantilever beam at the fixed end for construction of the moment diagram by parts usually gives the minimum number of areas.

3. For overhanging beams it is advisable to avoid taking a tangent to the elastic curve at the overhanging end.

4. It is sometimes advantageous to solve the problem with symbols and substitute numerical values in the last step.

The list of steps given below for the solution of deflection problems by the area-moment method may prove helpful. There is no point in memorizing these steps as they are not a substitute for an understanding of the principles involved. They are:

---

1. Sketch the loaded beam, the moment [or $M/(EI)$] diagrams (either by parts or composite diagram, depending on the complexity of the problem), and the elastic curve (if the shape is not known, it may be assumed).

2. Visualize which tangent lines may be most helpful, and draw such lines on the elastic curve. When a point of zero slope is known, either from symmetry or from supports, a tangent drawn at this point will frequently be useful.

3. By application of the second area-moment theorem, determine the tangential deviation at the point where the beam deflection is desired and at any other points required.

4. From geometry, determine the perpendicular distance from the unloaded beam to the tangent line at the point where the beam deflection is desired, and, using the results of step 3, solve for the required deflection.

For essential data on properties of curves see App. B. Examples 6–7 and 6–8 illustrate the method.

Example 6–7 Determine the deflection of the free end of the cantilever beam of Fig. 6–17*a* in terms of w, L, E, and I.

Solution Since E and I are constant, a bending moment diagram is used instead of an $M/(EI)$ diagram. The moment diagram is shown in Fig. 6–17*b*, and the area under it has been divided into rectangular, triangular, and parabolic parts for ease in calculating areas and moments. The elastic curve is shown in Fig. 6–17*c* with the deflections greatly exaggerated. Points A and B are selected at the ends of the beam because the beam has a horizontal tangent at B and the deflection at A is required. The vertical distance to A from the tangent at B ($t_{A/B}$ in Fig. 6–17*c*) equals the deflection of the free end of the beam y_A. For this reason, the second area-moment theorem can be used directly to obtain the required deflection. The area of each of the three portions of the area under the moment

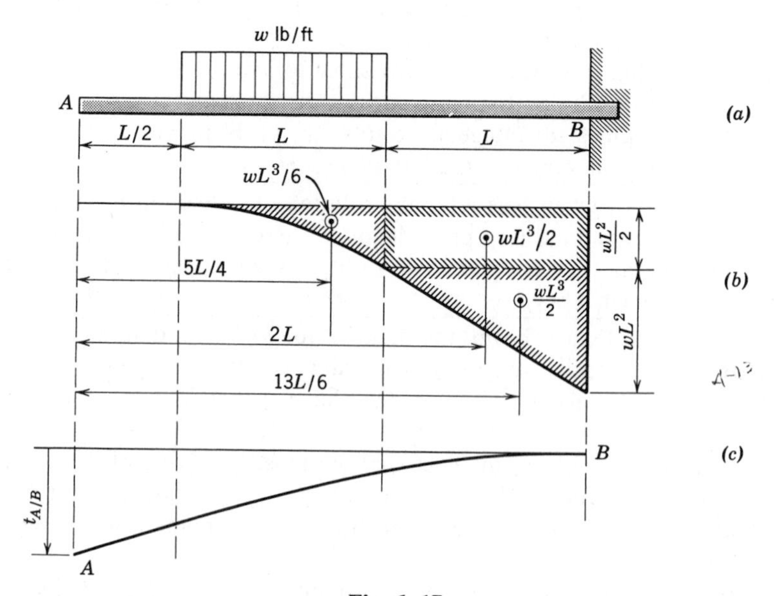

Fig. 6–17

diagram is shown in Fig. 6–17b along with the distance from the centroid of each part to the moment axis at the free end A. The second area-moment theorem gives

$$EIt_{A/B} = -\frac{wL^3}{6}\left(\frac{5L}{4}\right) - \frac{wL^3}{2}(2L) - \frac{wL^3}{2}\left(\frac{13L}{6}\right)$$

from which

$$y_A = t_{A/B} = -\frac{55wL^4}{24EI} = \underline{\frac{55wL^4}{24EI}\text{ downward}} \qquad \text{Ans.}$$

Example 6–8 Determine the deflection of the left end of the beam loaded and supported as shown in Fig. 6–18a, in terms of P, L, E, and I.

Solution The equations of equilibrium yield the left reaction equal to $4P/3$ upward and the right reaction equal to $P/3$ downward. Since E and I are constant, a moment diagram (instead of an $M/(EI)$ diagram) is drawn, and, with only one applied load, a composite moment diagram as shown in Fig. 6–18b is used.

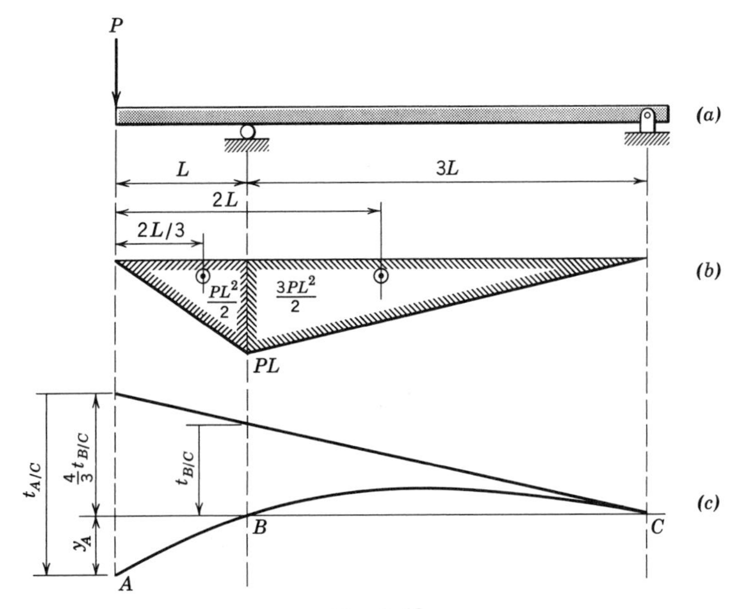

Fig. 6–18

The elastic curve with a tangent drawn at C is shown in Fig. 6–18c. The tangent line could have been drawn at B for an equally satisfactory solution. The second area-moment theorem gives the tangential deviation at B as

$$EIt_{B/C} = -\frac{3PL^2}{2}(L) = -\frac{3PL^3}{2}$$

where the negative sign indicates that the moment center B is below the tangent drawn from C. Another application of the second area-moment theorem gives the tangential deviation at A as

$$EIt_{A/C} = -\frac{3PL^2}{2}(2L) - \frac{PL^2}{2}\left(\frac{2L}{3}\right) = -\frac{10PL^3}{3}$$

where the sign indicates A is below the tangent from C. Finally,

$$EIy_A = EI\left(t_{A/C} - \frac{4}{3}t_{B/C}\right) = -\frac{10PL^3}{3} - \frac{4}{3}\left(-\frac{3PL^3}{2}\right) = -\frac{4}{3}PL^3$$

and

$$y_A = \frac{4}{3}\frac{PL^3}{EI} \text{ downward} \qquad \text{Ans.}$$

PROBLEMS

6–56* Use area moments to determine the deflection at the right end of the beam of Fig. P6–56 in terms of P, L, E, and I.

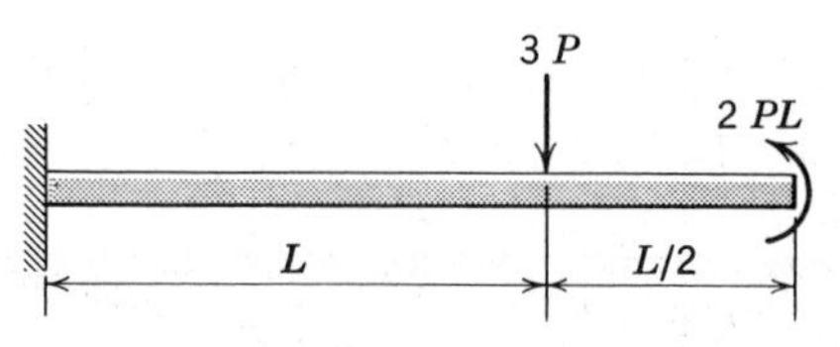

Fig. P6–56

6–57* Use area moments to determine the deflection at the point of application of the load $P/2$ for the beam of Fig. P6–57 in terms of P, L, E, and I.

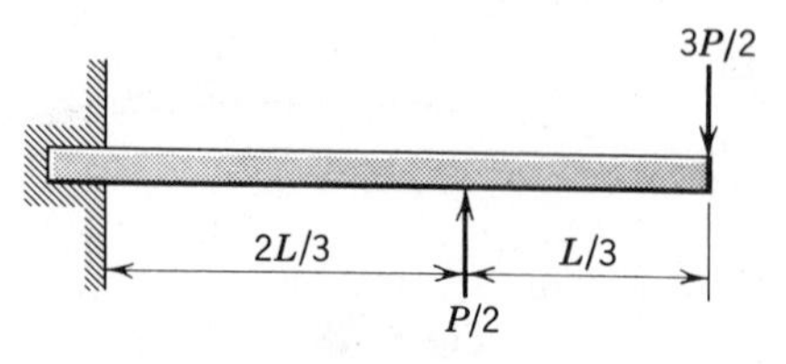

Fig. P6–57

6–58* For the beam loaded and supported as shown in Fig. P6–58, use area moments to determine the deflection at the right end in terms of P, L, E, and I.

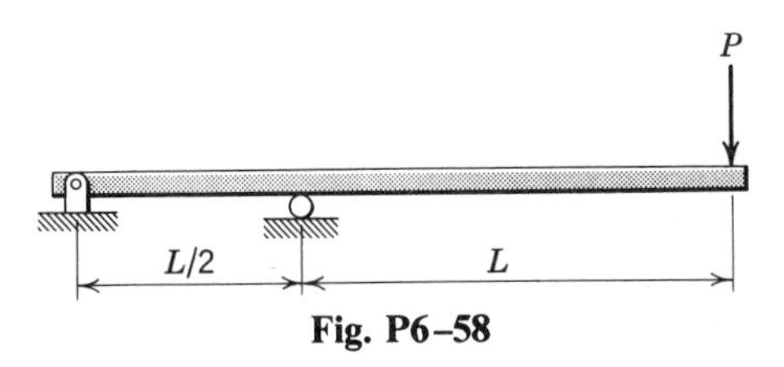

Fig. P6–58

6–59* For the beam loaded and supported as shown in Fig. P6–59, use area moments to determine in terms of P, L, E, and I:

(a) The deflection at the left load.
(b) The deflection at the center of the span.

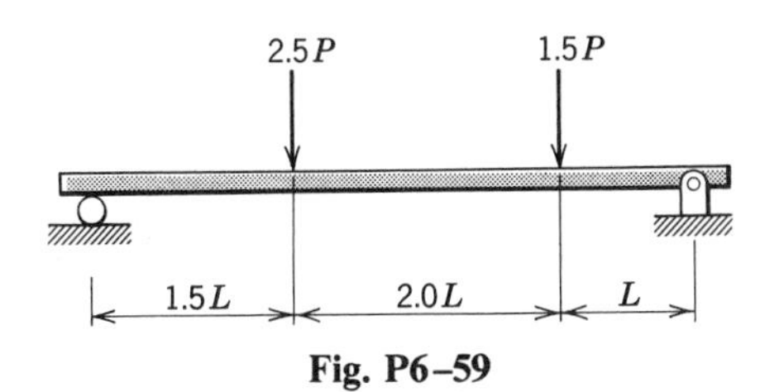

Fig. P6–59

6–60 Solve Prob. 6–21 by area moments.

6–61* Solve Prob. 6–16 by area moments.

6–62 For the beam of Fig. P6–29, use area moments to determine the deflection at the right end in terms of w, L, E, and I.

6–63 For the beam of Fig. P6–32, use area moments to determine the deflection at the right end in terms of w, L, E, and I.

6–64* For the beam loaded and supported as shown in Fig. P6–64, use area moments to determine the deflection at the right end in terms of w, L, E, and I.

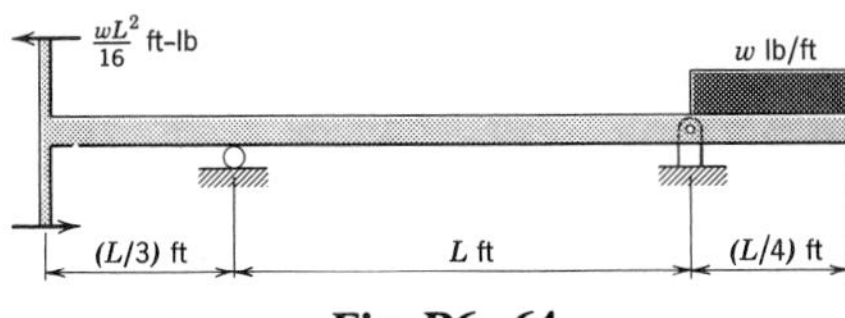

Fig. P6–64

6–65* A UB254×28 steel beam is shown in Fig. P6–65. When the deflection at the free end is 10 mm downward, determine:

(a) The load P.
(b) The maximum fiber stress in the beam.

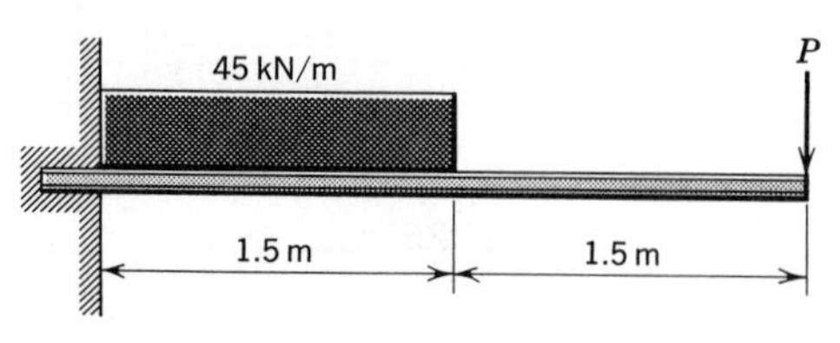

Fig. P6–65

6–66* Select the lightest allowable steel wide-flange section for the beam of Fig. P6–66 with $L = 6$ ft, $w = 1000$ lb/ft, an allowable fiber stress of 18,000 psi, and an allowable deflection at the free end of 0.90 in.

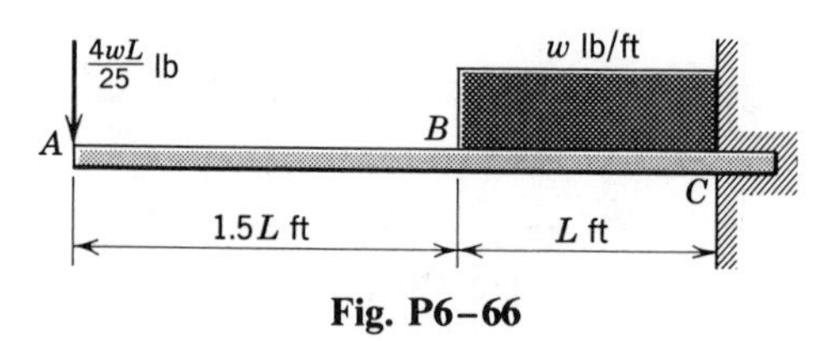

Fig. P6–66

6–67* A UB305×48 steel beam is loaded and supported as shown in Fig. P6–67. Determine the maximum allowable value of P for an allowable fiber stress of 150 MN/m^2 and a maximum allowable deflection at midspan of 15 mm.

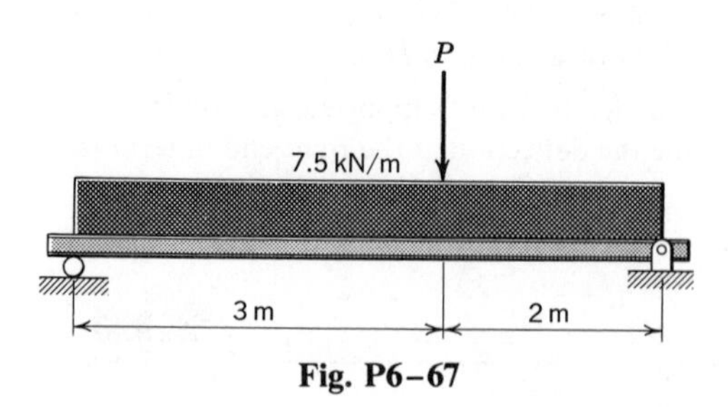

Fig. P6–67

6–68 For the beam loaded and supported as shown in Fig. P6–68, use area moments to determine the deflection at the concentrated load in terms of w, L, E, and I.

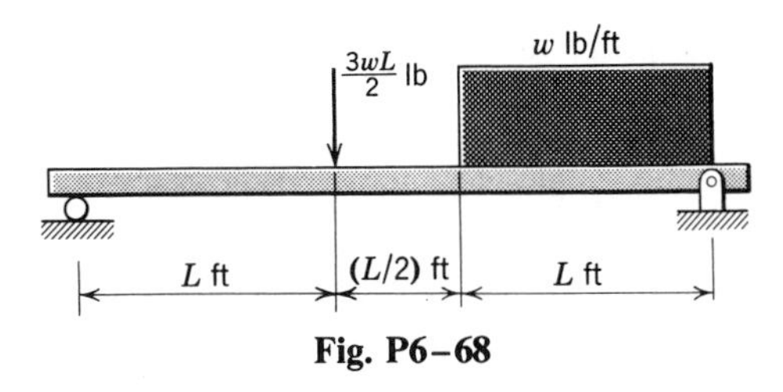

Fig. P6–68

6–69* For the beam loaded and supported as shown in Fig. P6–69, use area moments to determine the deflection at the left end in terms of w, L, E, and I.

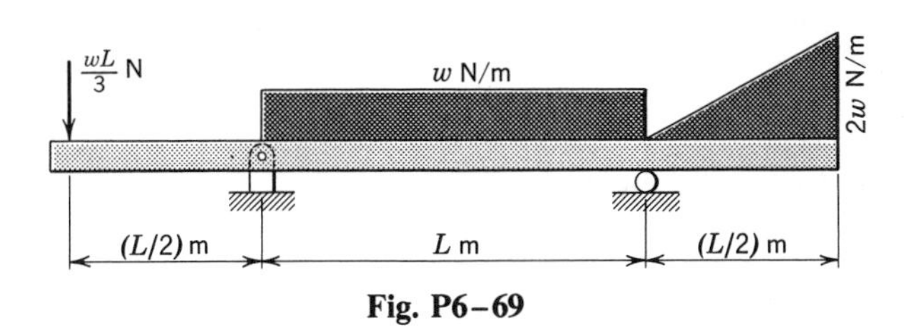

Fig. P6–69

6–70 For the beam of Fig. P6–70, use area moments to determine the deflection at the left end in terms of w, L, E, and I.

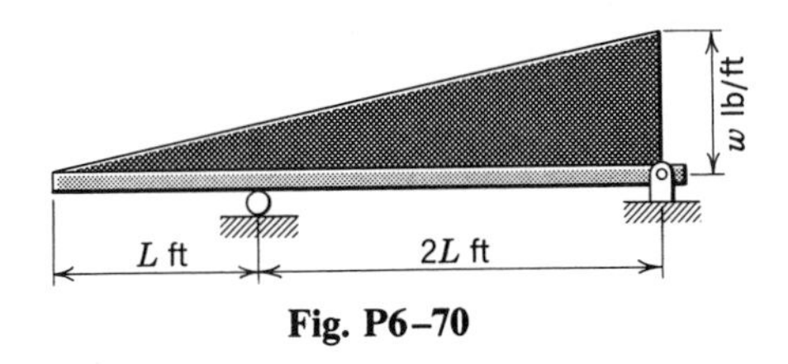

Fig. P6–70

6–71* Solve Prob. 6–42 by area moments.

6–72* Solve Prob. 6–30 by area moments.

6–73 Solve Prob. 6–41 by area moments.

6–74 Solve Prob. 6–20*b* by area moments.

6–75* Solve Prob. 6–31*b* by area moments.

6–9 MAXIMUM DEFLECTIONS BY AREA MOMENTS

In most cantilever beams and in symmetrically loaded simply supported beams, the point of maximum deflection can be determined by inspection. In such instances the procedure outlined in Art. 6–8 is usually adequate for finding the maximum deflection. When the location of the point of maximum deflection is unknown, it is necessary to use both the first and second area-moment theorems for determining the maximum deflection.

Figure 6–19 shows a typical situation where the maximum deflection, at point B on the elastic curve, is required. The deflection y_B, or the tangential deviation $t_{A/B}$, is wanted, but the distance x along the beam is unknown and must be determined before the deflection can be obtained. Once $t_{C/A}$ is determined from the second area-moment theorem, $t_{C/A}/L$ gives $\tan\theta_A$ which is approximately equal to θ_A. The angle θ_A can also be determined, in terms of x, from the first area-moment theorem because the tangent to the elastic curve at B is horizontal and $\theta_A = \theta_{AB}$. When these two expressions for θ_A are equated, the distance x is found. Finally, $t_{A/B}$ is determined from the second area-moment theorem. The following example illustrates the procedure for a specific case.

Example 6–9 For the beam in Fig. 6–20a, determine the maximum deflection in terms of w, L, E, and I.

Solution The equations of equilibrium give the left reaction as $wL/2$ upward and the right reaction as wL upward. The moment diagram by parts is shown in Fig. 6–20b, c, and d. The elastic curve, very much exaggerated, of the loaded beam is shown in Fig. 6–20e with tangents drawn at A and B. The second area-moment theorem, with the moment axis at C, gives the vertical distance to C from the tangent at A as

$$EIt_{C/A} = L\left(\frac{9wL^3}{4}\right) - \frac{2L}{3}(wL^3) - \frac{L}{4}\left(\frac{wL^3}{6}\right) = \frac{37}{24}wL^4$$

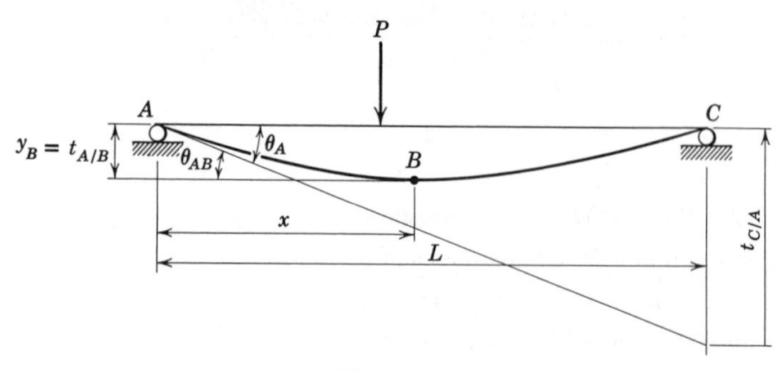

Fig. 6–19

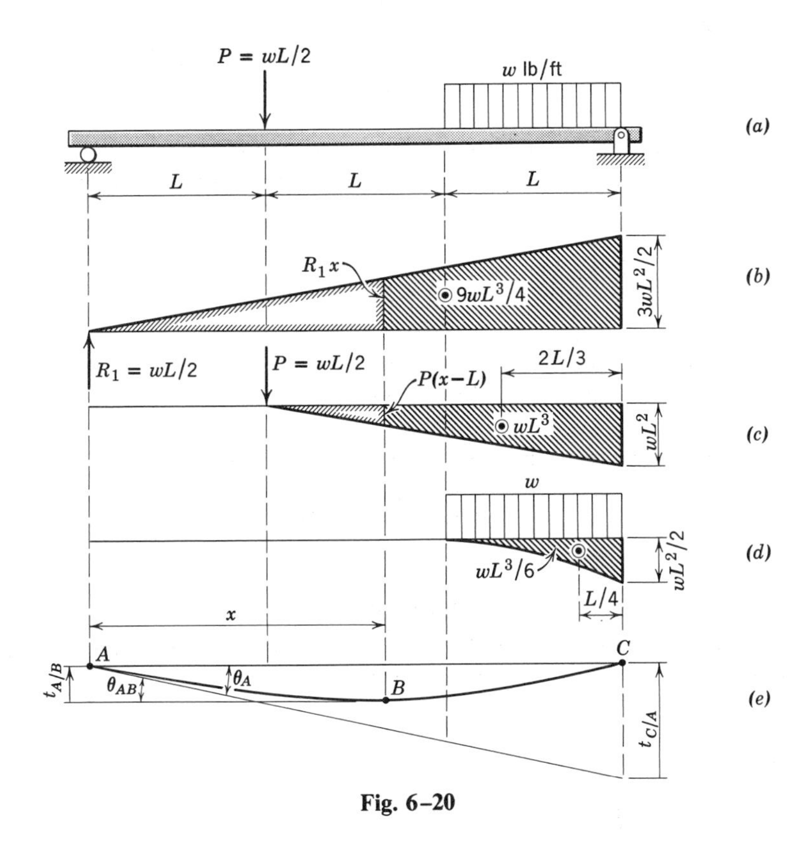

Fig. 6–20

Since the deflections are small,

$$\theta_A = \tan\theta_A = \frac{t_{C/A}}{3L} = \frac{37wL^3}{72EI}$$

Because the tangent at B is horizontal, the first area-moment theorem applied to the segment AB gives

$$\theta_A = \theta_{AB} = \frac{1}{EI}\left[\frac{wL}{2}x\left(\frac{x}{2}\right) - \frac{wL}{2}(x-L)\frac{x-L}{2}\right] = \frac{wL}{4EI}(2Lx - L^2)$$

Equating the two expressions for θ_A gives

$$\frac{37wL^3}{72EI} = \frac{wL}{4EI}(2Lx - L^2)$$

from which

$$x = 55L/36$$

Application of the second theorem to the segment AB, with the moment axis at A, gives the vertical distance to A from the horizontal tangent at B as

$$|y_B| = t_{A/B} = \frac{1}{EI}\left\{\left(\frac{2x}{3}\right)\left(\frac{wLx}{2}\right)\left(\frac{x}{2}\right) - \left[L + \frac{2}{3}(x-L)\right]\frac{wL}{2}\frac{(x-L)^2}{2}\right\}$$

$$= \frac{wL^4}{2EI}\left[\frac{1}{3}\left(\frac{55}{36}\right)^3 - \frac{1}{2}\left(\frac{19}{36}\right)^2 - \frac{1}{3}\left(\frac{19}{36}\right)^3\right]$$

where the positive result means that A is above the tangent line. Therefore

$$y_B = 0.500\frac{wL^4}{EI} \text{ downward} \qquad \text{Ans.}$$

PROBLEMS

6–76* For the beam of Fig. P6–76, use area moments to determine the maximum deflection between the supports in terms of P, L, E, and I.

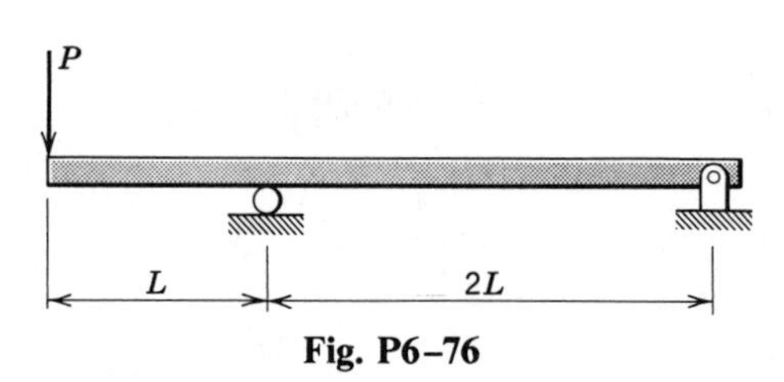

Fig. P6–76

6–77* For the beam of Fig. P6–77, use area moments to determine the maximum deflection between the supports in terms of P, L, E, and I.

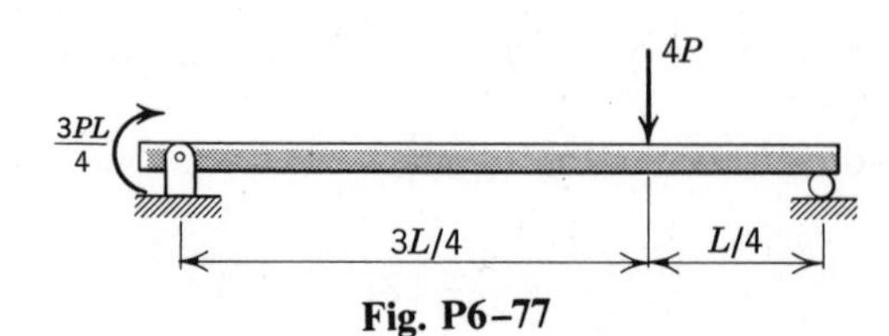

Fig. P6–77

6–78* An S12×50 steel beam is loaded and supported as shown in Fig. P6–78. Use area moments to determine:

(a) The deflection at the right end.

(b) The maximum deflection between the supports.

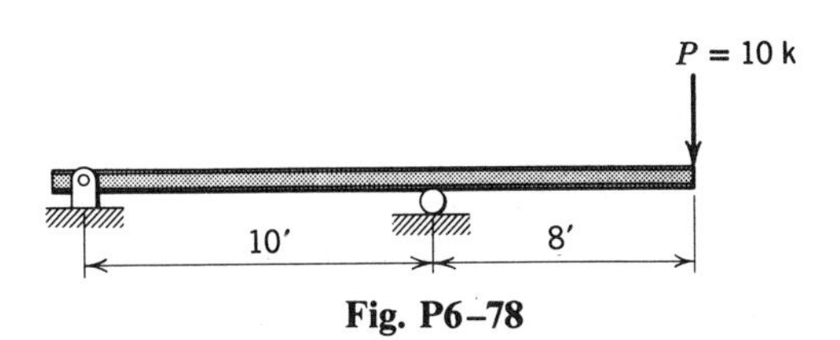

Fig. P6–78

6–79* A 200-mm wide by 300-mm deep timber beam is simply supported on a span of 7 m. A 15 kN load is applied 2 m from the right support. Use area moments to determine:

(a) The maximum deflection for a modulus of elasticity of 7 GN/m^2.

(b) The deflection of the beam at the 15-kN load.

6–80* Solve Prob. 6–28*a* by area moments.

6–81* Solve Prob. 6–27*b* by area moments.

6–82 For the beam of Fig. P6–26, use area moments to determine:

(a) The deflection of the beam at the load P.

(b) The location of the point of maximum deflection.

(c) The maximum deflection.

6–83* For the beam of Fig. P6–47, use area moments to determine the maximum deflection between the supports in terms of w, L, E, and I.

6–84* For the beam of Fig. P6–13, use area moments to determine the maximum deflection between the supports in terms of w, L, E, and I.

6–10 DEFLECTIONS BY SUPERPOSITION

As explained in Art. 3–3, the concept or method of superposition consists of finding the resultant effect of several loads acting on a member simultaneously as the sum of the contributions from each of the loads applied individually. The results for the separate loads are frequently available from previous work or readily determined by a previous method. In such instances the superposition method becomes a powerful concept or tool for finding stresses, deflections, etc. The method is applicable in all cases where a linear relation exists between the stresses or deflections and the applied loads.

To show that beam deflections can be accurately determined by superposition, consider the cantilever beam of Fig. 6–21 with loads Q, w, and

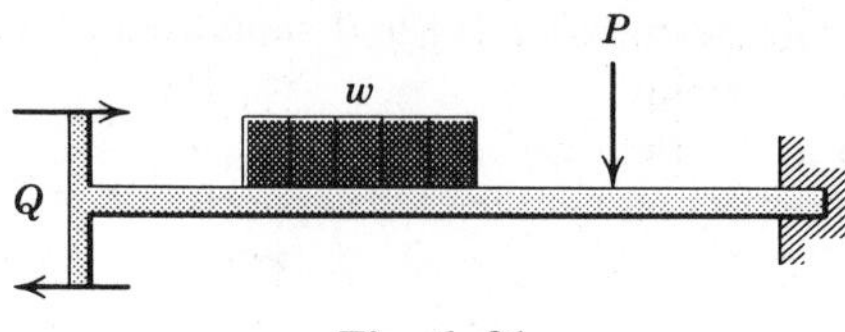

Fig. 6–21

P. In order to determine the deflection at any point of this beam by the double integration method, it is necessary to express the bending moment in terms of the applied loads. For each interval along the beam, the value of M is the algebraic sum of the moments due to the separate loads. After two successive integrations, the solution for the deflection at any point will still be the algebraic sum of the contributions from each applied load. Furthermore, for any given value of x, the relation between applied load and resulting deflection will be linear. It is evident, therefore, that the deflection of a beam is the sum of the deflections produced by the individual loads. Once the deflections produced by a few typical individual loads have been determined by one of the methods already presented, the superposition method provides a means of rapidly solving a wide variety of more complicated problems by various combinations of known results. As more data become available, a wider range of problems can be solved by superposition.

The data in App. D are provided for use in mastering the superposition method. No attempt is made to give a large number of results because such data are readily available in various handbooks. The data given and the illustrative examples are for the purpose of making the concept and methods clear.

Example 6–10 For the beam in Fig. 6–22a, determine the maximum deflection when E is $12(10^6)$ psi and I is 81 in.^{4}.

Solution From symmetry and $\Sigma F_y = 0$, the reactions are each 4500 lb upward. Because of the symmetrical loading, the slope of the beam is zero at the center of the span, and the right (or left) half of the beam can be considered as a cantilever beam with two loads. As shown in Fig. 6–22b, the cantilever with two loads can be replaced by two beams (designated 1 and 2), each carrying one of the two loads.

The elastic curve (exaggerated) for part 1, as shown in Fig. 6–22c, gives the deflection at the right end as $y_{4.5} + y_3$, where $y_{4.5}$ is the deflection at the end of the uniformly distributed load and y_3 is the additional deflection of the unloaded 3 ft. From case 2 of App. D

$$y_{4.5} = -\frac{1000(4.5^4)(12^3)}{8(12)(10^6)(81)} = -\frac{9^3}{8(10^3)} = -0.0911 \text{ in.}$$

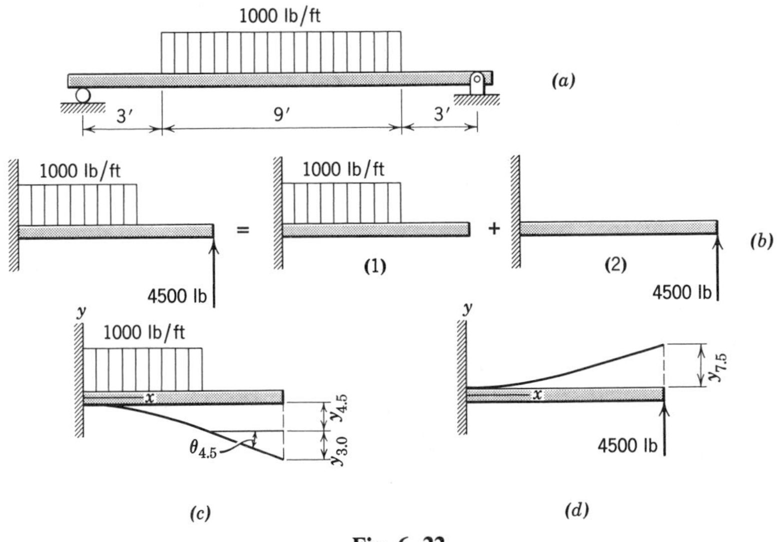

Fig. 6-22

and

$$\theta_{4.5} = -\frac{1000(4.5^3)(12^2)}{6(12)(10^6)(81)} = -\frac{9}{4(10^3)} = -0.00225 \text{ rad}$$

from which

$$y_{3.0} = 3(12)(0.00225) = -0.0810 \text{ in.}$$

consequently, the total deflection of the right end is 0.1721 in. downward.

The elastic curve (exaggerated) for part 2 is shown in Fig. 6-22*d*. From case 1 of App. D

$$y_{7.5} = +\frac{4500(7.5^3)(12^3)}{3(12)(10^6)(81)} = 1.125 \text{ in. upward}$$

The algebraic sum of the deflections for parts 1 and 2 is +0.953 in., which means that the right end of the beam is 0.953 in. above the center. Obviously the right end does not move, and the maximum deflection is at the center and is

$$y_{max} = \underline{0.953 \text{ in. downward}} \qquad \text{Ans.}$$

Example 6–11 Determine the deflection at mid-span for the steel beam ($I = 180$ in.^{4}) of Fig. 6–23a.

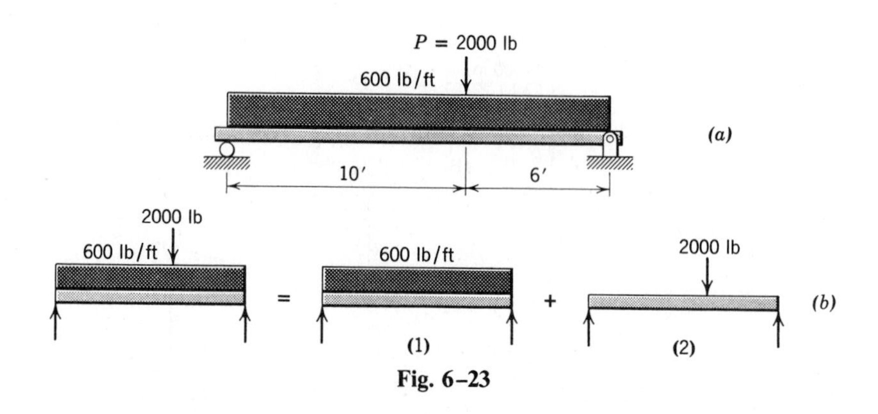

Fig. 6–23

Solution The given loading is equivalent to the two loads, parts 1 and 2, shown in Fig. 6–23b. For part 1, the deflection at the center of the span is given as case 7 of App. D. It is

$$(y_C)_1 = -\frac{5(600)(16^4)(12^3)}{384(30)(10^6)(180)} = -0.1638 \text{ in.}$$

For part 2, the deflection at the center is given as a part of case 5 of App. D. It is

$$(y_C)_2 = -\frac{2000(6)\left[3(16^2) - 4(6^2)\right]12^3}{48(30)(10^6)(180)} = -0.0499 \text{ in.}$$

The algebraic sum of the deflections for midspan is

$$y_C = -0.1638 + (-0.0499) = -0.2137 \text{ in.} = \underline{0.214 \text{ in. downward}} \qquad \text{Ans.}$$

PROBLEMS

Note. Use the method of superposition to determine deflections in the following problems.

6–85* Determine the deflection of the free end of the cantilever beam in Fig. P6–85 when w is 800 lb/ft, L is 10 ft, E is $12(10^6)$ psi, and I is 20 in.^{4}.

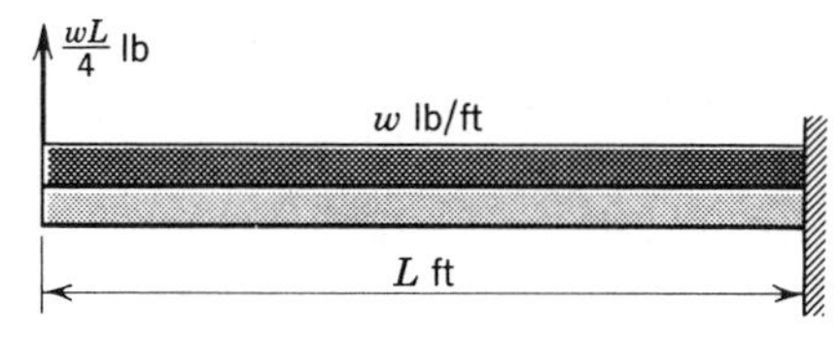

Fig. P6–85

6–86* A 7-m simply supported UB254×31 steel beam carries a uniformly distributed load of 6 kN/m over the entire span and a concentrated load of 8 kN at midspan. Determine the maximum deflection of the beam.

6–87* A 10×10-in. timber beam 14 ft long is simply supported at its ends. The beam carries a uniformly distributed load of 1000 lb/ft over its entire length and a concentrated load of 2000 lb at the center. The modulus of elasticity of this timber is 1000 ksi. Determine the maximum deflection of the beam.

6–88 A timber cantilever beam 100 mm wide by 150 mm deep is 4 m long. If an upward pull of 3 kN is applied at the free end, determine the uniformly distributed load W, equal to wL, required to bring the free end to a point 10 mm above its unloaded position. Young's modulus is equal to 10 GN/m^2.

6–89* For the beam in Fig. P6–89, determine the maximum deflection in the right half in terms of P, L, E, and I.

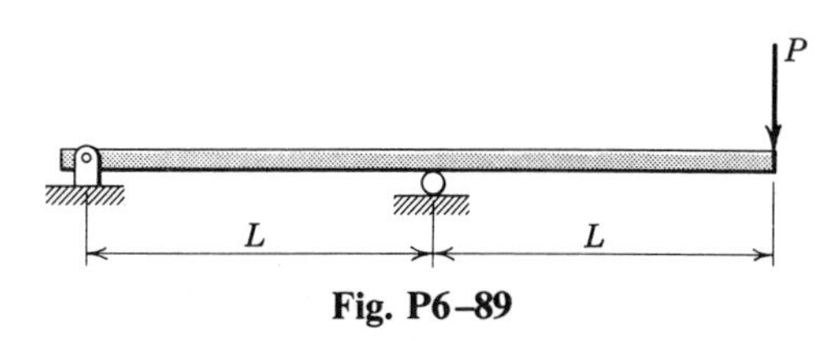

Fig. P6–89

6–90 A 100-mm wide by 300-mm deep timber cantilever beam supports a concentrated load of 6 kN at the free end and a uniformly distributed load of w N/m over the entire 3-m span. For a modulus of elasticity of 8 GN/m^2, an allowable fiber stress of 13 MN/m^2, an allowable horizontal shearing stress of 0.7 MN/m^2, and a maximum allowable deflection of 35 mm, determine the maximum permissible value of w.

6–91* Determine the deflection of the free end of the beam in Fig. P6–91 in terms of w, L, E, and I.

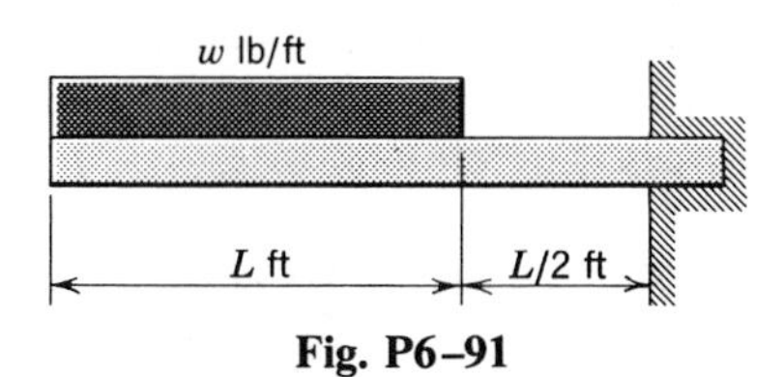

Fig. P6–91

6–92 Solve Prob. 6–91 with an added couple of wL^2 counterclockwise applied at the right end of the distributed load.

6–93 For the beam of Fig. P6–93, determine the midspan deflection if w is 90 lb/ft, L is 12 ft, E is $1.44(10^6)$ psi, and I is 120 in.^{4}.

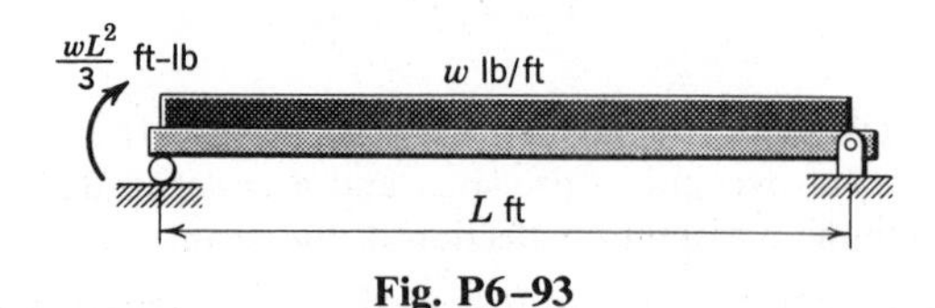

Fig. P6–93

6–94* Determine the deflection at the right end of the beam in Fig. P6–13 in terms of w, L, E, and I.

6–95* Determine the deflection at midspan for the beam of Fig. P6–13 in terms of w, L, E, and I.

6–96* Determine the deflection at the right end of the beam in Fig. P6–14 in terms of w, L, E, and I.

6–97* Determine the deflection at midspan for the beam of Fig. P6–14 in terms of w, L, E, and I.

6–98 Determine the deflection at the right end of the beam in Fig. P6–29 in terms of w, L, E, and I.

6–99 Determine the deflection at the right end of the beam in Fig. P6–41 in terms of w, L, E, and I.

6–11 DEFLECTIONS BY ENERGY METHODS—CASTIGLIANO'S THEOREM

Strain energy techniques are frequently used to analyze the deflections of beams and structures. Of the many available methods, the application of Castigliano's theorem, to be developed here, is one of the most widely used. It was presented in 1873 by the Italian engineer Alberto Castigliano (1847–1884). Although the theorem will be derived by considering the strain energy stored in beams, it is applicable to any structure for which the force-deformation relations are linear.

If the beam shown in Fig. 6–24 is slowly and simultaneously loaded by the two forces P_1 and P_2 with resulting deflections y_1 and y_2, the strain energy U of the beam is, by Clapeyron's theorem (see Art. 2–9), equal to the work done by the forces. Therefore,

$$U = \frac{1}{2}P_1y_1 + \frac{1}{2}P_2y_2 \tag{a}$$

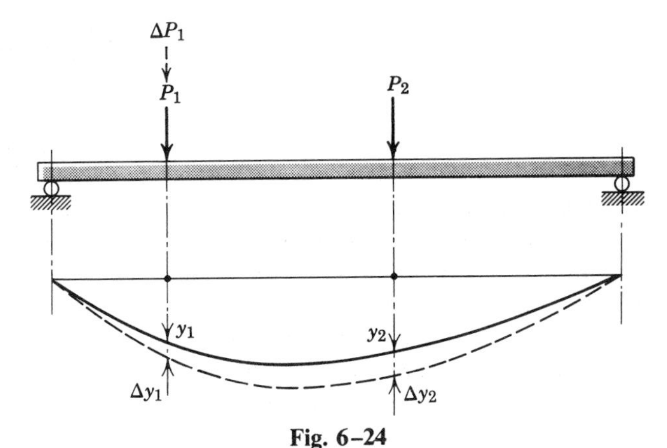

Fig. 6-24

Let the force P_1 be increased by a small amount ΔP_1, and Δy_1 and Δy_2 be the changes in deflection due to this incremental load. Since the forces P_1 and P_2 are already present, the increase in the strain energy is

$$\Delta U = \frac{1}{2}\Delta P_1 \Delta y_1 + P_1 \Delta y_1 + P_2 \Delta y_2 \tag{b}$$

If the order of loading were reversed so that the incremental force, ΔP_1, were applied first, followed by P_1 and P_2, the resulting strain energy would be

$$U + \Delta U = \frac{1}{2}\Delta P_1 \Delta y_1 + \Delta P_1 y_1 + \frac{1}{2}P_1 y_1 + \frac{1}{2}P_2 y_2 \tag{c}$$

The resulting strain energy must be independent of the order of loading; hence, by combining Eqs. (a), (b), and (c)

$$\Delta P_1 y_1 = P_1 \Delta y_1 + P_2 \Delta y_2 \tag{d}$$

Equations (b) and (d) can be combined to give

$$\frac{\Delta U}{\Delta P_1} = y_1 + \frac{1}{2}\Delta y_1$$

or upon taking the limit as ΔP_1 approaches zero[2]

$$\frac{\partial U}{\partial P_1} = y_1 \tag{e}$$

[2]The partial derivative is used because the strain energy is a function of both P_1 and P_2.

For the general case in which there are many loads involved, Eq. (e) is written as

$$\frac{\partial U}{\partial P_i} = y_i \tag{6–5a}$$

The following is a statement of Castigliano's theorem. *If the strain energy of a linearly elastic structure is expressed in terms of the system of external loads, the partial derivative of the strain energy with respect to a concentrated external load is the deflection of the structure at the point of application and in the direction of that load.* By a similar development, Castigliano's theorem can also be shown to be valid for applied moments and the resulting rotations (or changes in slope) of the structure. Thus

$$\frac{\partial U}{\partial M_i} = \theta_i \tag{6–5b}$$

If the deflection is required either at a point where there is no unique point load or in a direction not aligned with the applied load, a dummy load is introduced at the desired point acting in the proper direction. The deflection is obtained by first differentiating the strain energy with respect to the dummy load and then taking the limit as the dummy load approaches zero. Also, for the application of Eq. 6–5*b*, either a unique point moment or a dummy moment must be applied at point i. The moment will be in the direction of rotation at the point. Note that if the loading consists of a number of point loads, all expressed in terms of a single parameter (e.g., P, $2P$, $3P$, wL, $2wL$), and if the deflection is wanted at one of the applied loads, one must either write the moment equation with this load as a separate identifiable term or add a dummy load at the point, so that the partial derivative can be taken with respect to this load only.

It was shown in Art. 2–9 that the strain energy per unit volume for uniaxial stress is $\sigma^2/(2E)$; hence, the total strain energy under uniaxial stress is

$$U = \int_{\text{vol.}} \frac{\sigma^2}{2E} dV$$

For a beam of constant (or slowly varying) cross section subjected to pure bending, the principal stresses are parallel to the axis of the beam; therefore, using Eq. 5–1,

$$U = \frac{1}{2E} \int_{\text{vol.}} \left(\frac{My}{I} \right)^2 dV \tag{f}$$

By writing dV as $dA\,dx$ (where x is measured along the axis of the beam), Eq. (f) becomes

$$U = \frac{1}{2E}\int_0^L \left(\frac{M^2}{I^2}\int_{\text{area}} y^2\,dA\right)dx = \frac{1}{2E}\int_0^L \frac{M^2}{I}\,dx \qquad (6\text{–}6)$$

Equation 6–6 was developed for a beam loaded in pure bending. However, most real beams will be subjected to transverse loads that induce shearing stresses and, in the case of distributed loading, transverse normal stresses. In Art. 5–12, it was shown that the transverse normal stress is small enough to neglect, and in the following Art. 6–12, it is shown that, except for short, deep beams, the deflection due to shearing stresses is also small enough to neglect. Hence, Eq. 6–6, which neglects strain energy due to these stresses, is applicable to the usual real beams. In applying Eq. 6–5 to Eq. 6–6, it is usually much simpler to apply Leibnitz's rule[3] to differentiate under the integral sign so that

$$y_i = \frac{\partial U}{\partial P_i} = \frac{1}{E}\int_0^L \frac{M}{I}\frac{\partial M}{\partial P_i}\,dx \qquad (6\text{–}7)$$

Example 6–12 Determine the deflection at the free end of a cantilever beam of constant cross section and length L, loaded with a force P at the free end and a distributed load that varies linearly from zero at the free end to w at the support.

Solution From the free-body diagram of Fig. 6–25, the moment equation is

$$M = -Px - \frac{wx^3}{6L}$$

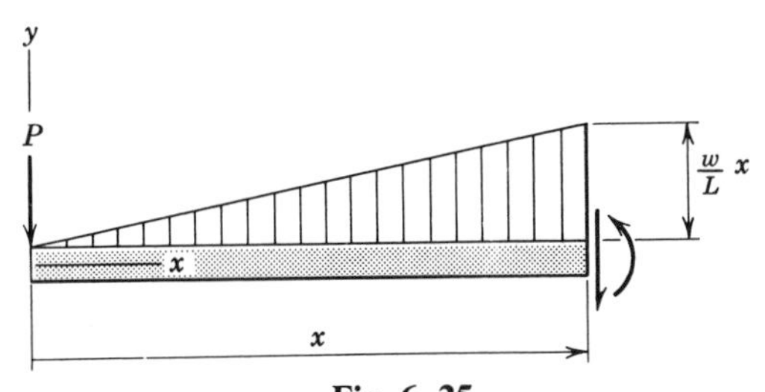

Fig. 6–25

[3]*Advanced Calculus*, W. Kaplan, Addison-Wesley Publishing Company, Reading, Mass., 1953, p. 219.

The partial derivative of M with respect to P is $-x$, and Eq. 6–7 becomes

$$EIy = \int_0^L \left(Px^2 + \frac{wx^4}{6L}\right)dx = +\frac{PL^3}{3} + \frac{wL^4}{30}$$

The positive signs indicate deflection in the direction of the force P; hence,

$$\underline{y = \frac{PL^3}{3EI} + \frac{wL^4}{30EI} \text{ downward}} \qquad \text{Ans.}$$

The result may be verified by reference to App. D.

Example 6–13 Determine the deflection at the center of a simply supported beam of constant cross section and span L carrying a uniformly distributed load w lb/ft over its entire length.

Solution In the free-body diagram of Fig. 6–26, the dashed force P represents a fictitious load applied at the location and in the direction of the desired deflection. The moment equation is

$$M = M_w + M_P = \frac{wLx}{2} - \frac{wx^2}{2} + \frac{Px}{2} - P\left\langle x - \frac{L}{2}\right\rangle^1$$

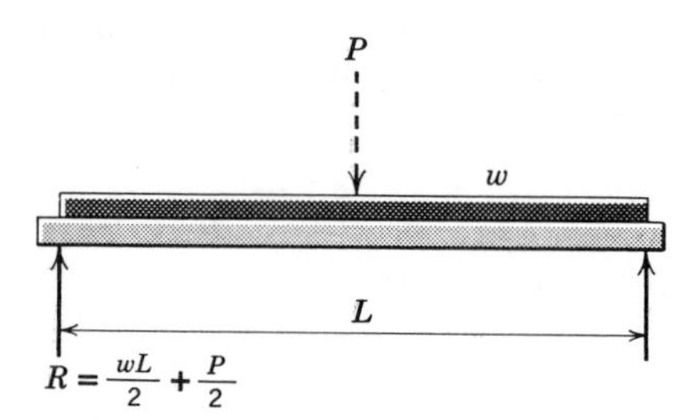

Fig. 6–26

where the quantity $\langle x - L/2\rangle^1$ is zero for all $x \leqslant L/2$ (see Art. 6–5). The partial derivative of M with respect to P is

$$\frac{\partial M}{\partial P} = \frac{x}{2} - \left\langle x - \frac{L}{2}\right\rangle^1$$

The dummy force P is equated to zero and the deflection is given by

$$EIy = \int_0^L \left(\frac{wLx}{2} - \frac{wx^2}{2}\right)\left(\frac{x}{2} - \left\langle x - \frac{L}{2}\right\rangle^1\right)dx$$

which, for ease of integration, can be written as

$$EIy = \frac{w}{4}\int_0^L (Lx^2 - x^3)dx + \frac{w}{4}\int_{L/2}^L (L^2x - 3Lx^2 + 2x^3)dx = \frac{5wL^4}{384EI}$$

Since the deflection is positive in the direction of the force, the deflection is downward and

$$y = \frac{5wL^4}{384EI} \text{ downward} \qquad \text{Ans.}$$

PROBLEMS

Note. Solve the following problems by means of Castigliano's theorem.

6-100* For the beam of Fig. P6-100, determine the deflection at P in terms of P, L, E, and I.

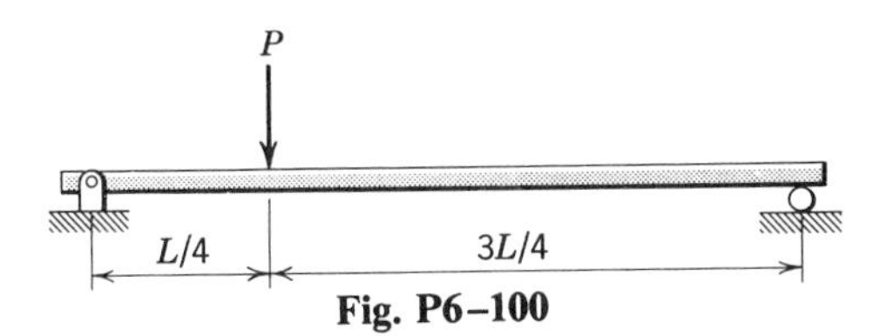

Fig. P6-100

6-101* For the beam of Fig. P6-101, determine the deflection at the concentrated load in terms of P, L, E, and I.

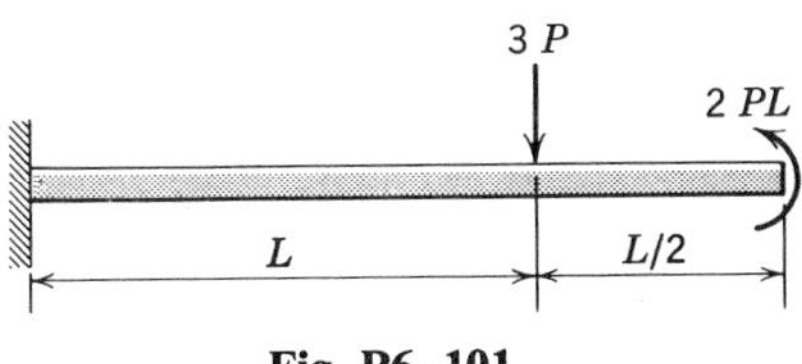

Fig. P6-101

6-102 For the beam of Fig. P6-102, determine the deflection at the right end in terms of w, L, E, and I.

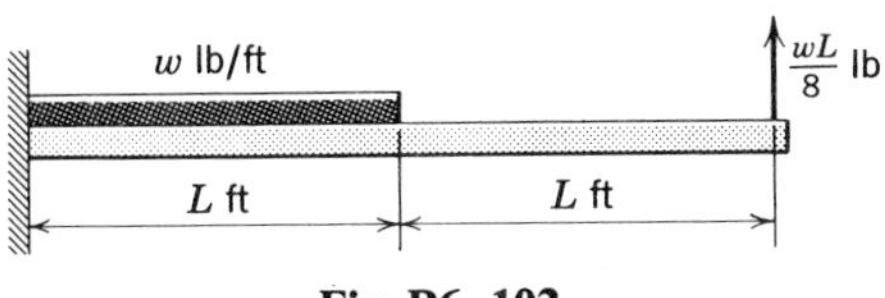

Fig. P6-102

6–103 For the beam of Fig. P6–103, determine the deflection at P in terms of P, L, E, and I.

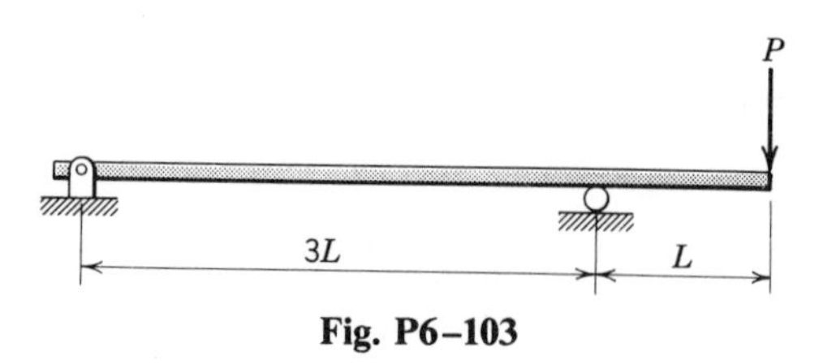

Fig. P6–103

6–104* For the beam of Fig. P6–104, determine the slope at the section where the couple is applied in terms of M, L, E, and I.

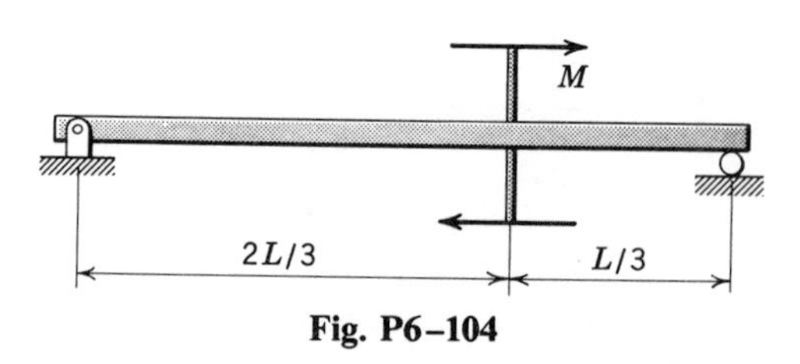

Fig. P6–104

6–105* A cantilever beam is loaded as shown in Fig. P6–105. The moment Q is equal to $3PL/4$. Determine the slope at the left end of the beam in terms of P, L, E, and I.

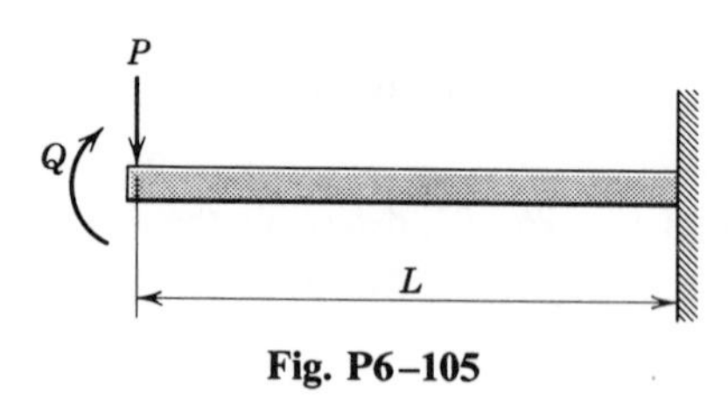

Fig. P6–105

6–106 An S15×42.9 steel beam is loaded as shown in Fig. P6–106. When the deflection at the free end is 0.50 in. downward, determine the force P.

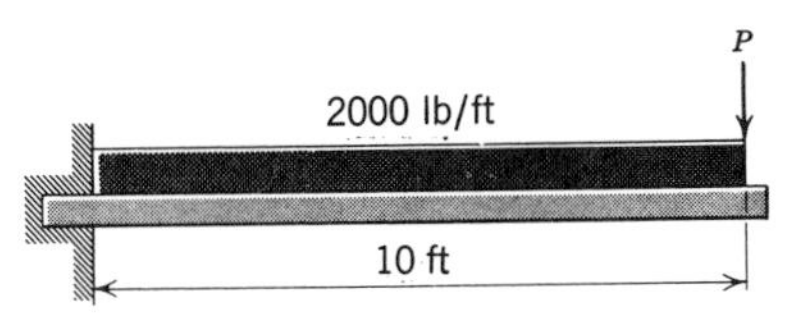

Fig. P6–106

6–107* For the beam of Fig. P6–93, determine the midspan deflection if w is 800 N/m, L is 4 m, E is 10 GN/m^2, and I is 30(10^6) mm^4.

6–108 Prob. 6–44.

6–109 Prob. 6–45.

6–110 For the beam of Fig. P6–110, determine the deflection at B in terms of w, L, E, and I.

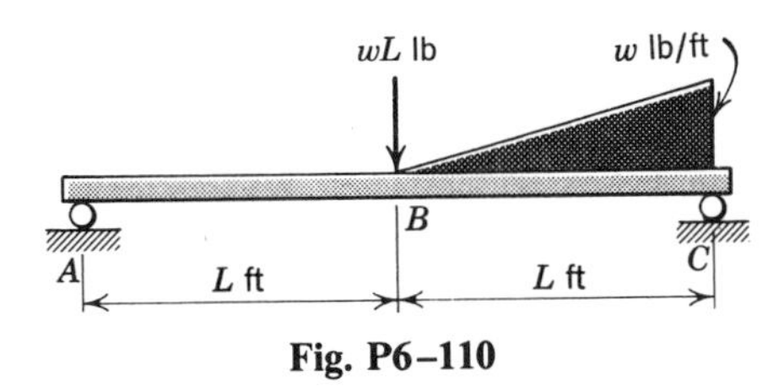

Fig. P6–110

6–111* For the curved bar of Fig. P6–111, determine the vertical component of the deflection at B due to flexure. Assume that the radius of curvature is sufficient for the elastic flexure formula to apply.

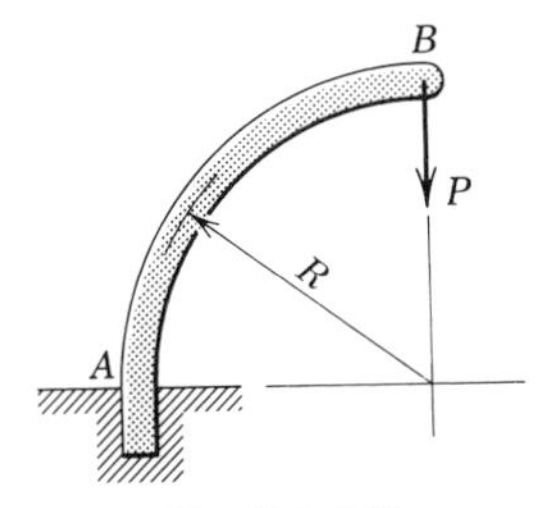

Fig. P6–111

6–112* For the curved bar of Fig. P6–112, determine the vertical component of the deflection at B due to flexure. Assume that the radius of curvature is sufficient for the elastic flexure formula to apply.

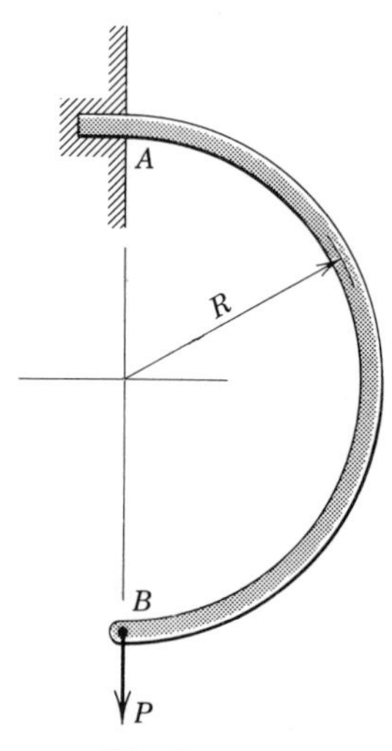

Fig. P6–112

6–113* For the curved bar of Fig. P6–112, determine the horizontal component of the deflection at B due to flexure. Assume that the radius of curvature is sufficient for the elastic flexure formula to apply.

6–114* For the curved bar of Fig. P6–112, determine the slope at B.

6–12 DEFLECTIONS DUE TO SHEARING STRESS

As mentioned in Art. 6–3, the beam deflections calculated so far neglect the deflection produced by the shearing stresses in the beam. For short, heavily loaded beams this deflection can be significant, and an approximate method for evaluating such deflections will now be developed. The deflection of the neutral surface dy due to shearing stresses in the interval dx along the beam of Fig. 6–27 is

$$dy = \gamma\, dx = \frac{\tau}{G}\, dx = \frac{VQ}{GIt}\, dx$$

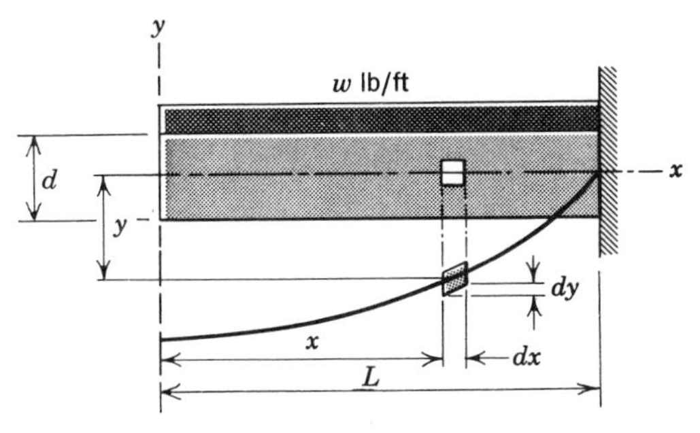

Fig. 6–27

from which, since the shear in Fig. 6–27 is negative,

$$\frac{GIt}{Q}\frac{dy}{dx} = -V \tag{6–8}$$

Since the vertical shearing stress varies from top to bottom of a beam, the deflection due to shear is not uniform. This nonuniform deflection due to shear is reflected in the slight warping of the sections of a beam (see Art. 5–12). Equation 6–8 gives values too high because the maximum shearing stress (at the neutral surface) is used and also because the rotation of the differential shear element (see Art. 5–12) is ignored.

In order to obtain an idea of the relative amount of beam deflection due to shearing stress, consider a rectangular cross section for the beam of Fig. 6–27, and use the maximum stress for which

$$\frac{Q}{It} = \frac{3}{2A}$$

The expression for dy becomes

$$dy = \frac{3wx\,dx}{2AG}$$

where V is replaced by its value $-wx$. Integration along the entire beam gives the change in y due to shear as

$$\Delta y = \frac{3w}{2AG}\int_0^L x\,dx = \frac{3wL^2}{4AG}$$

which equals the deflection at the left end. For this same beam the magnitude of the deflection at the left end due to flexural stresses is

$$\frac{wL^4}{8EI} = \frac{3wL^4}{2EAd^2}$$

and the magnitude of the total deflection at the left end becomes

$$y = \frac{3wL^4}{2EAd^2} + \frac{3wL^2}{4AG} = \frac{3wL^2}{4AG}\left(\frac{2L^2G}{d^2E} + 1\right) \tag{a}$$

For a material such as steel the ratio G/E is approximately 0.4, and, with the ratio L/d selected as 6 (a very short, deep beam), the deflection due to flexural stress is 28.8 times the deflection due to shear. On the other hand, for a W36×160 steel beam and the same span-depth ratio, the flexural stress deflection is 4.85 times the approximate shear deflection. For a working fiber stress of 18 ksi, the two deflections are 0.389 and 0.080 in., respectively. Equation (a) indicates that the ratio of the two deflections increases as the square of L/d, which means that the deflection due to shear is of importance only in the case of very short, deep beams.

6–13 CLOSURE

Four methods for determining the deflection of a beam were presented in this chapter. The integration method consists of solving the elastic curve differential equation and evaluating the constants of integration from the boundary and matching conditions. Singularity functions are quite helpful if more than one interval is involved. The area-moment method consists of applying one or both of the area-moment theorems to the $M/(EI)$ diagram and using the geometry of the beam to obtain the required deflections. Note that the integration method can be adapted to the area-moment method by replacing the indefinite integral with a definite integral, whereby, upon integration, one obtains the change in slope and the change in tangential deviation (measured from the tangent drawn to the curve at the origin) between the limits selected. The superposition method consists of subdividing the applied loads and supports into an equivalent set of less involved components, for which solutions are readily available, and obtaining the required deflections as the algebraic sum of the deflections for the component parts. An energy method—Castigliano's theorem—consists of writing an expression for the energy U stored in the beam due to all applied loads and then evaluating the partial derivative of U with respect to a real or dummy load P applied at the point where the deflection is to be obtained.

The selection of the best method for a given problem requires judgment and experience.

7 Statically Indeterminate Beams

7–1 INTRODUCTION

As explained in Art. 5–1, a beam, subjected only to transverse loads, with more than two reaction components is statically indeterminate because the equations of equilibrium are not sufficient to determine all the reactions. In such cases the geometry of the deformation of the loaded beam is utilized to obtain the additional relations needed for evaluation of the reactions (or other unknown forces). For problems involving elastic action, each additional constraint on a beam provides additional information concerning slopes or deflections. Such information, when utilized with appropriate slope or deflection equations, yields expressions that supplement the independent equations of equilibrium.

Any of the methods for determining beam deflections outlined in Chap. 6 can be utilized to provide the additional equations needed to solve for the unknown reactions or loads in a statically indeterminate beam provided the action is linearly elastic. In this book no attempt is made at extensive coverage of beams subjected to inelastic action. However, one type of analysis, because of its extensive use in modern design of structures, is presented in Art. 7–6. Once the reactions and loads are known, the methods of Chaps. 5 and 6 will give the required stresses and deflections.

7–2 THE INTEGRATION METHOD

For statically determinate beams, known slopes and deflections were used to obtain boundary and matching conditions, from which the constants of integration in the elastic curve equation could be evaluated. For statically indeterminate beams, the procedure is identical. However, the moment equations will contain reactions or loads that cannot be evaluated from the available equations of equilibrium, and one additional boundary condition is needed for the evaluation of each such unknown. For example, if a beam is subjected to a force system for which there are two independent equilibrium equations and if there are four unknown reactions or loads on the beam, two boundary or matching conditions are needed in addition to those necessary for the determination of the constants of integration. These extra boundary conditions, when substituted in the appropriate elastic curve equations (slope or deflection), will yield the necessary additional equations. The following example illustrates the method.

Example 7–1 For the beam loaded and supported as shown in Fig. 7–1a, determine the reactions in terms of w and L.

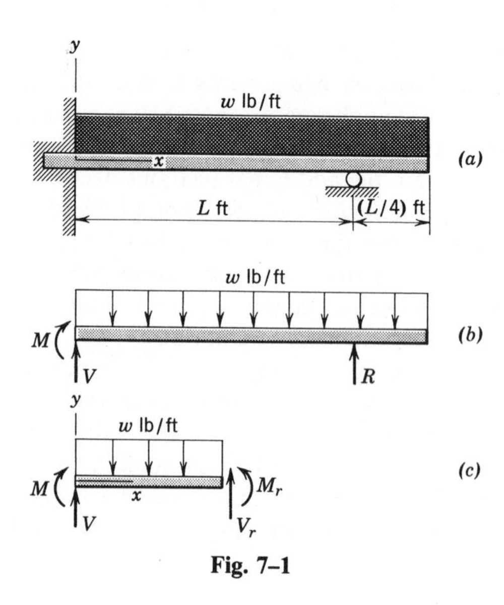

Fig. 7–1

Solution From the free-body diagram of Fig. 7–1*b* it is seen that there are three unknown reaction components (M, V, and R) and that only two independent equations of equilibrium are available. The additional unknown requires the use of the elastic curve equation, for which one extra boundary condition is required in addition to the two required for the constants of integration. Because three boundary conditions are available in the interval between the supports, only one elastic curve equation need be written. The origin of coordinates is arbitrarily placed at the wall, and, for the interval $0 \leqslant x \leqslant L$, the boundary conditions are: when $x=0$, $dy/dx=0$; when $x=0$, $y=0$; and when $x=L$, $y=0$. From Fig. 7–1*c* the bending moment equation is

$$EI\frac{d^2y}{dx^2} = M_x = Vx + M - \frac{wx^2}{2}$$

Integration gives

$$EI\frac{dy}{dx} = \frac{Vx^2}{2} + Mx - \frac{wx^3}{6} + C_1$$

The first boundary condition gives $C_1 = 0$. A second integration yields

$$EIy = \frac{Vx^3}{6} + \frac{Mx^2}{2} - \frac{wx^4}{24} + C_2$$

The second boundary condition gives $C_2 = 0$, and the last boundary condition gives

$$0 = \frac{VL^3}{6} + \frac{ML^2}{2} - \frac{wL^4}{24}$$

which reduces to

$$4VL + 12M = wL^2$$

The equation of equilibrium $\Sigma M_R = 0$ for the free-body diagram of Fig. 7–1*b* yields

$$VL + M = w\left(\frac{5L}{4}\right)\left(\frac{5L}{8} - \frac{L}{4}\right) = \frac{15wL^2}{32}$$

The simultaneous solution of these equations gives

$$M = -\frac{7wL^2}{64} = \underline{\frac{7wL^2}{64} \text{ ft-lb counterclockwise}} \qquad \text{Ans.}$$

and

$$V = \frac{37wL}{64} \text{ lb upward} \qquad \text{Ans.}$$

Finally, the equation $\Sigma F_y = 0$ for Fig. 7–1*b* gives

$$R + V = 5wL/4$$

from which

$$R = \frac{43wL}{64} \text{ lb as shown} \qquad \text{Ans.}$$

An alternate solution would be to place the origin of coordinates at the right support and write the moment equation for the interval $0 \leqslant x \leqslant L$. This equation would involve only one unknown, the reaction R; upon integration and evaluation of the constants, the third boundary condition would yield the value of R directly. The two independent equilibrium equations could then be used to evaluate M and V.

PROBLEMS

7–1* The beam in Fig. P7–1 has a couple Q, applied at the left end, of such magnitude that the slope at the left end is zero. Determine:

(a) The magnitude of the couple Q in terms of w and L.

(b) The deflection at the left end in terms of w, L, E, and I.

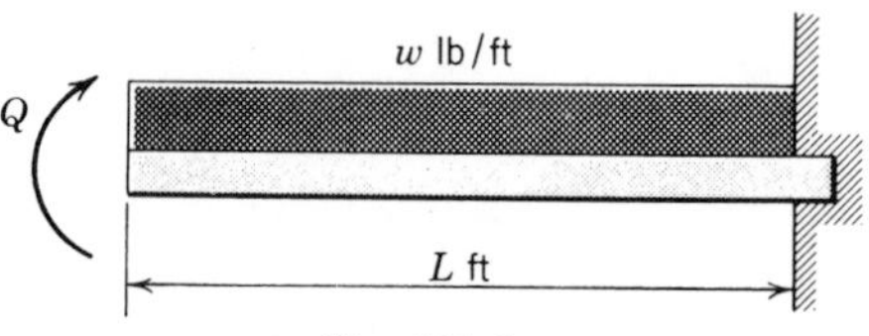

Fig. P7–1

7–2* The deflection of the left end of the cantilever beam in Fig. P7–2 is $wL^4/(24EI)$ upward. Determine Q in terms of w and L.

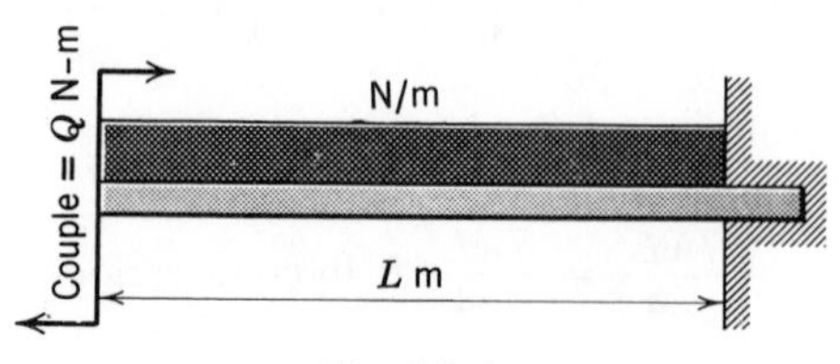

Fig. P7–2

7–3* A cantilever beam is loaded as shown in Fig. P7–2. Determine:

(a) The value of Q in terms of w and L such that the slope of the elastic curve will be zero at the center of the span.

(b) The deflection at the left end, in terms of w, L, E, and I when Q has the value determined in part (a).

7–4 For the beam in Fig. P7–4, determine:

(a) The magnitude of the force P, in terms of w and L, that will make the slope zero at point A.

(b) The deflection at point A in terms of w, L, E, and I when P has the value obtained in part (a).

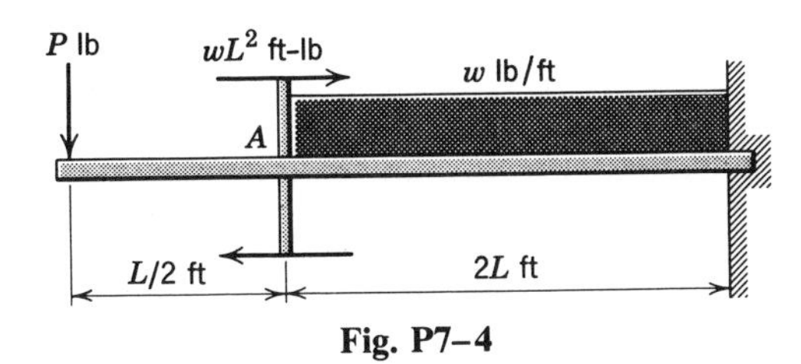

Fig. P7–4

7–5* A beam is supported and loaded as shown in Fig. P7–5.

(a) Determine the value of Q, in terms of w and L, that would make the slope zero over the left support.

(b) Locate the section of maximum deflection between the supports.

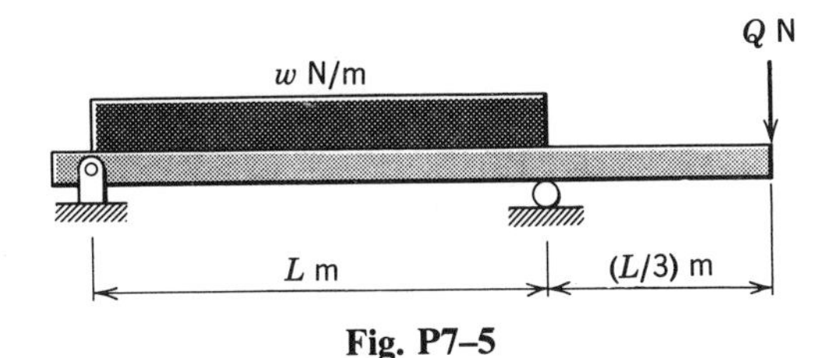

Fig. P7–5

7–6* For the beam in Fig. P7–6:

(a) Determine the load P that will make the slope at the left support be zero.

(b) Locate the section of maximum deflection between the supports.

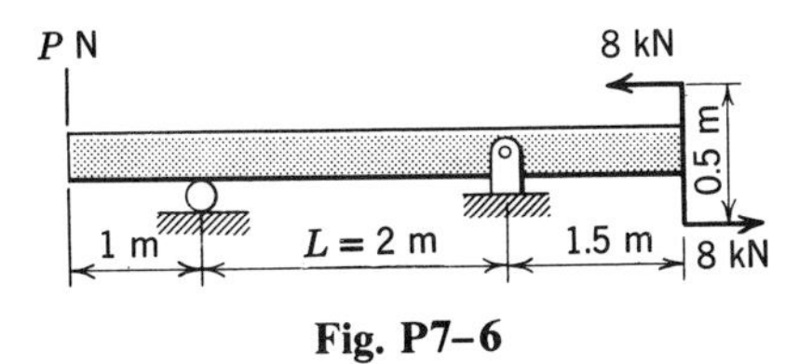

Fig. P7–6

7–7* For the beam loaded and supported as shown in Fig. P7–7, determine the left reaction.

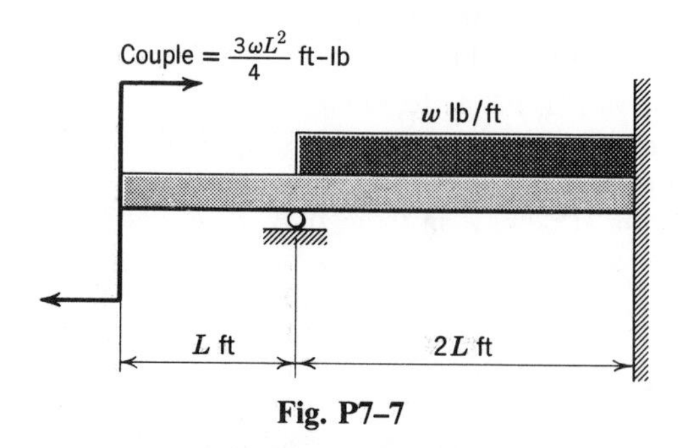

Fig. P7–7

7–8* Determine, in terms of w and L, the components of the reaction (shear and moment) at A on the beam ABC of Fig. P7–8.

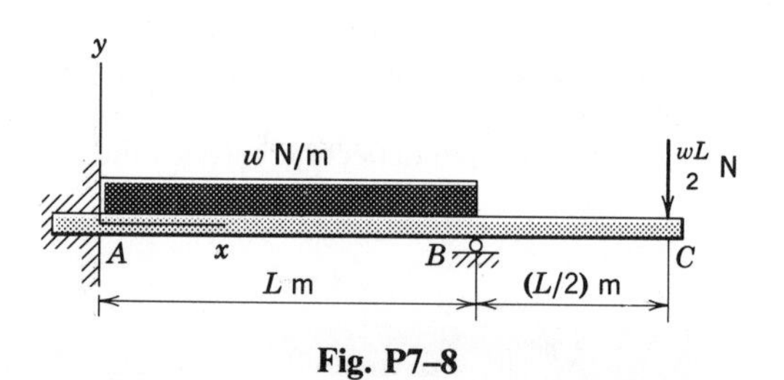

Fig. P7–8

7–9 The beam in Fig. P7–9 is partially fixed at the right end. The slope at this point is $wL^3/(48EI)$ upward to the right. Determine in terms of L and w:

(a) The left end reaction.
(b) The maximum bending moment in the beam.

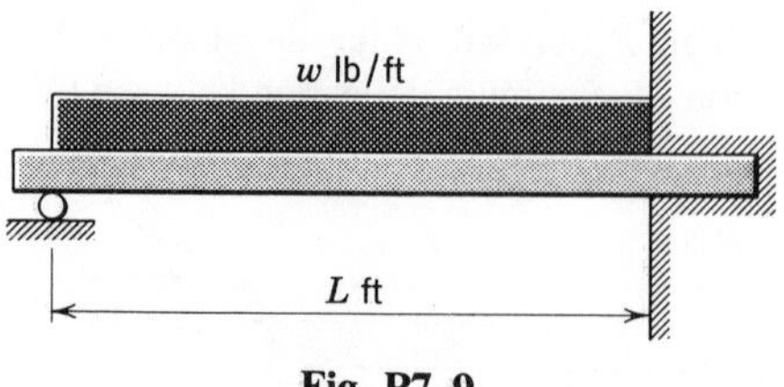

Fig. P7–9

7–10 For the beam shown in Fig. P7–10, determine in terms of M, L, E, and I:

(a) The reaction at the right support.

(b) The location of the section of maximum deflection between the supports.

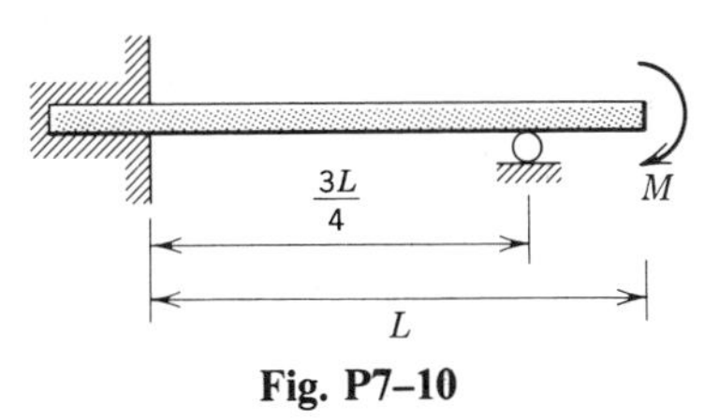

Fig. P7–10

7–11* Determine the right reaction in terms of w and L for the uniformly loaded beam shown in Fig. P7-11.

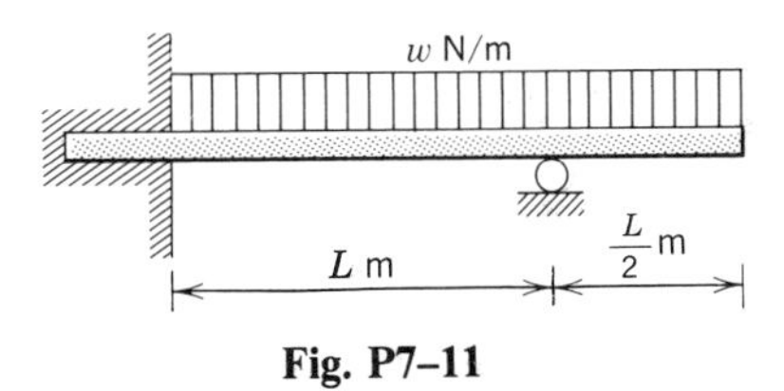

Fig. P7–11

7–12 Determine the maximum deflection between the supports for the beam in Fig. P7–12. Express the result in terms of P, L, E, and I.

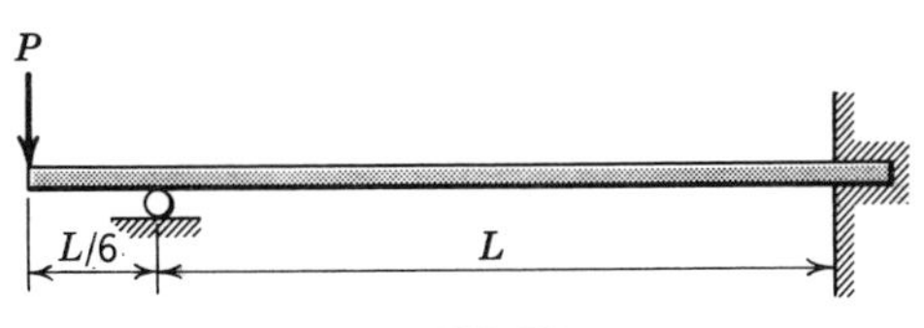

Fig. P7–12

7–13* For the beam in Fig. P7–13, determine the moment M_O in terms of w and L that will cause the slope of the deflected beam over the support at A to be $wL^3/(60EI)$ upward to the left.

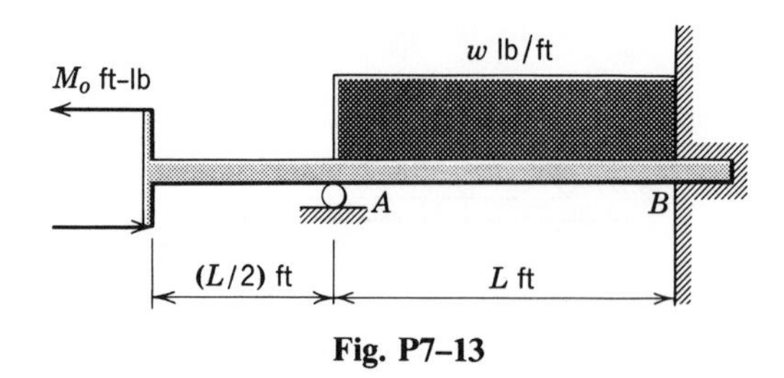

Fig. P7–13

7–14* In Fig. P7–14, the slope of the beam over the support B is $wL^3/(24EI)$ downward to the right. Determine all of the reactions in terms of w and L.

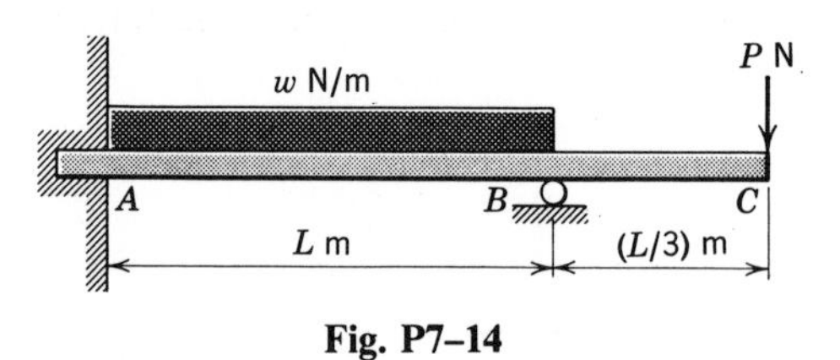

Fig. P7–14

7–15 Determine, in terms of w and L, the center reaction on the continuous beam of Fig. P7–15.

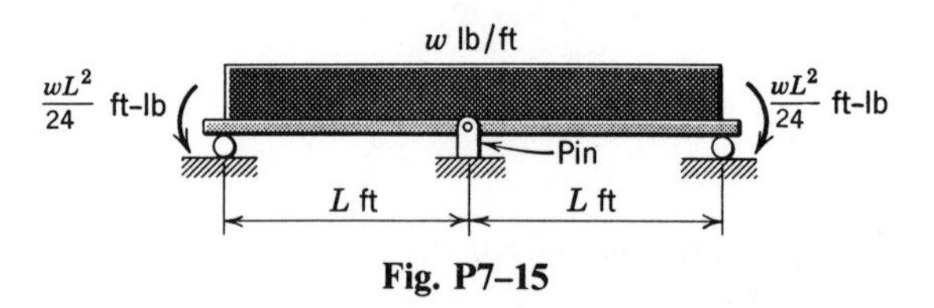

Fig. P7–15

7–16* Determine, in terms of w and L, the left reaction on the beam in Fig. P7–16.

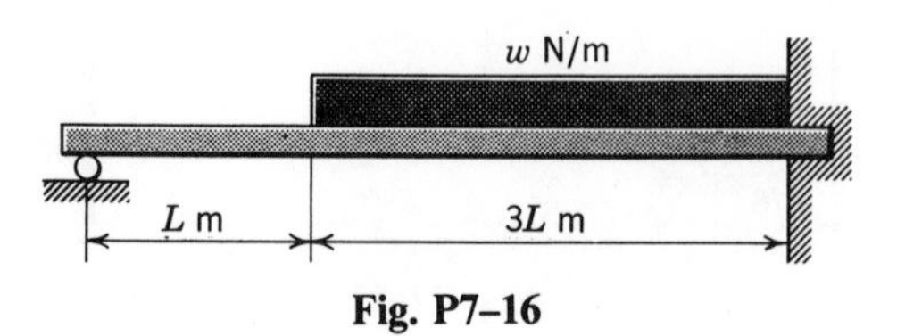

Fig. P7–16

7–17 Determine, in terms of P and L, the reaction at B on the beam of Fig. P7–17.

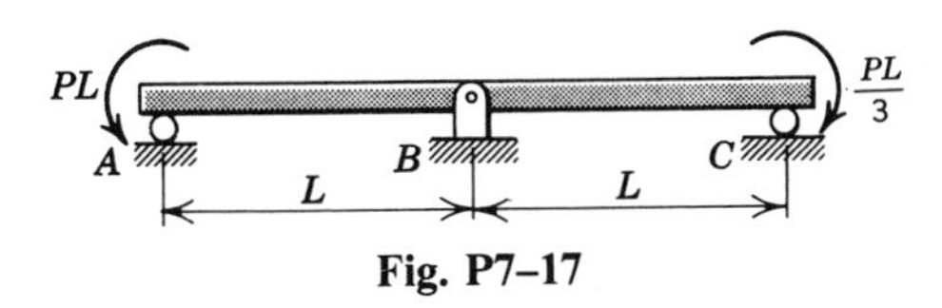

Fig. P7–17

7–18* Determine P, in terms of w and L, such that the slope of the elastic curve will be zero at B in the beam of Fig. P7–18.

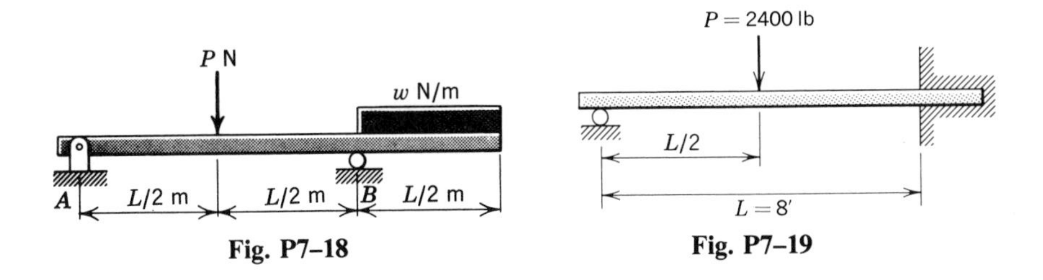

Fig. P7–18 **Fig. P7–19**

7–19 The uniform beam shown in Fig. P7–19 supports a load P of 2400 lb. Determine the reaction at the left end and draw complete shear and bending moment diagrams.

7–20* A beam of span L m is fixed at the left support and restrained at the right support. Under a uniformly distributed load of w N/m, the right support settles an amount equal to $wL^4/(60EI)$ and rotates until the slope of the elastic curve is $wL^3/(30EI)$ upward to the right. Determine the reaction components at the left support in terms of w and L.

7–21 Under the loading shown in Fig. P7–21, the right support rotates counterclockwise until the slope of the elastic curve at the support is $wL^3/(30EI)$. Both supports remain on the same level. Determine the shear and the moment at the left support in terms of w and L.

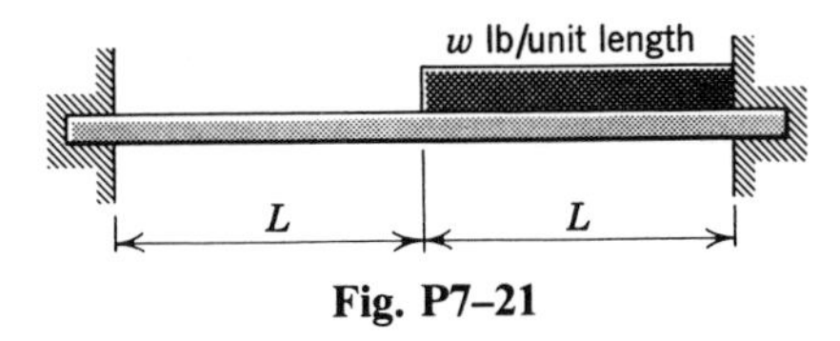

Fig. P7–21

7–22* The beam of Fig. P7–22 is simply supported at the left end and is framed into a column at the right end. The beam carries a distributed load varying linearly from zero to w N/m at the column. Under the load the ends of the beam remain at the same level, but the right end rotates (due to loading the adjacent span) clockwise until the slope of the elastic curve is $wL^3/(12EI)$ downward to the right. Determine the left reaction on the beam in terms of w and L.

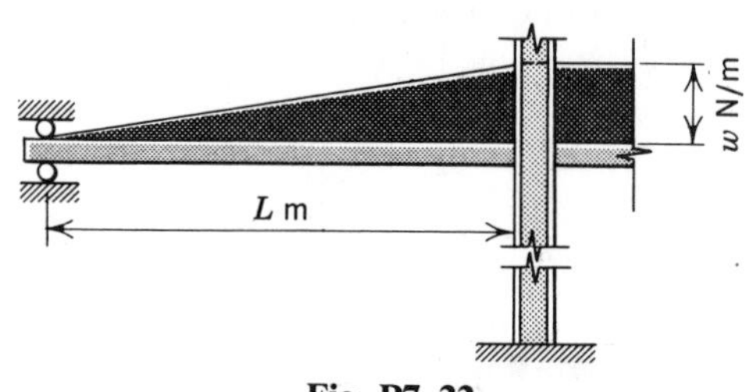

Fig. P7–22

7–3 THE AREA-MOMENT METHOD

For statically determinate beams, points with known slopes or deflections were used as tangent points and moment centers for the application of the area-moment theorems. For statically indeterminate beams, each additional constraint provides information regarding a slope or deflection, which makes it possible to write an area-moment equation to supplement the equilibrium equations. The following example illustrates the application of the concepts.

Example 7–2 For the beam of Fig. 7–2*a* determine the reactions in terms of w and L.

Solution There are four unknown reactions (V_A, M_A, V_B, and M_B) on the free-body diagram of Fig. 7–2*b*, and there are only two independent equations of equilibrium. Therefore, two area-moment relations are necessary to supplement the equations of equilibrium.

Inspection of the elastic curve (very much exaggerated) of Fig. 7–2*c* reveals that the angle between tangents drawn at A and B is zero. Also, the distance from a tangent drawn at A to point B is zero.

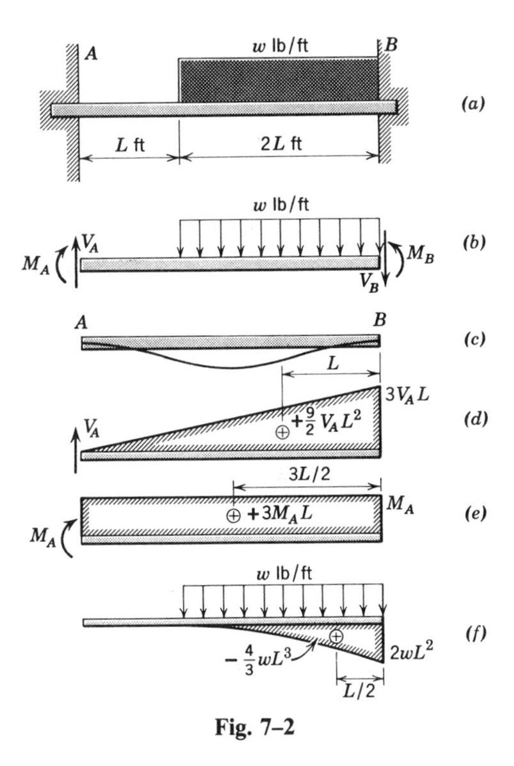

Fig. 7–2

These geometric statements can be expressed mathematically by the area-moment theorems. Thus, since the angle between the tangents at A and B is zero, the area under the moment diagram between A and B must be zero. Also, since the distance from the tangent drawn at A to point B is zero, the moment of the area under the $M/(EI)$ diagram between A and B, with respect to an axis at B, is zero. (In this particular problem the moment axis could also have been selected at A since the deflection of A from a tangent drawn at B is also zero.) These two area-moment relations give

$$EI\theta_{AB} = 0 = \frac{9}{2} V_A L^2 + 3M_A L - \frac{4}{3} wL^3$$

from which

$$27V_A L + 18M_A = 8wL^2$$

and

$$EIt_{B/A} = 0 = \frac{9}{2} V_A L^2(L) + 3M_A L\left(\frac{3L}{2}\right) - \frac{4wL^3}{3}\left(\frac{L}{2}\right)$$

from which

$$27V_AL + 27M_A = 4wL^2$$

The simultaneous solution of these equations yields

$$M_A = -\frac{4wL^2}{9} = \underline{\frac{4wL^2}{9}\text{ ft-lb counterclockwise}} \qquad \text{Ans.}$$

and

$$V_A = \underline{\frac{16wL}{27}\text{ lb upward}} \qquad \text{Ans.}$$

With two reactions known, the equations of equilibrium give the other two as

$$M_B = -\frac{2wL^2}{3} = \underline{\frac{2wL^2}{3}\text{ ft-lb clockwise}} \qquad \text{Ans.}$$

and

$$V_B = -\frac{38wL}{27} = \underline{\frac{38wL}{27}\text{ lb upward}} \qquad \text{Ans.}$$

PROBLEMS

7–23* A cantilever beam of length L ft carries a uniformly distributed load of w lb/ft over the entire span. Determine the moment in terms of w and L that must be applied at the free end in order to make the deflection zero at that end.

7–24* Given a beam loaded and supported as shown in Fig. P7–24, determine the force Q in terms of w and L that will cause the beam to be horizontal at the left support.

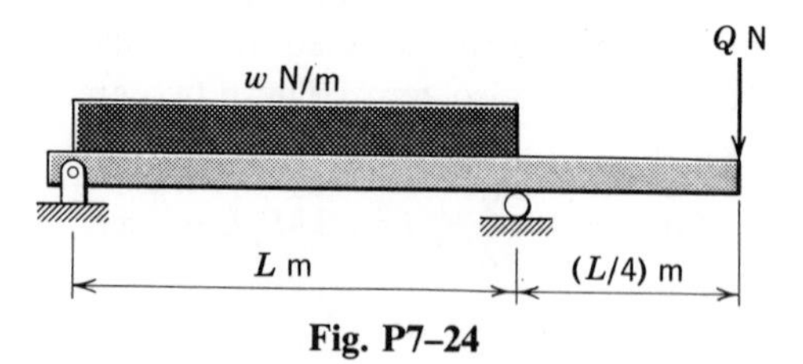

Fig. P7–24

7–25* The beam AB of Fig. P7–25 is fixed at the end B. When the couple is applied, the support A settles an amount equal to $PL^3/(24EI)$. Determine the reaction at A in terms of P.

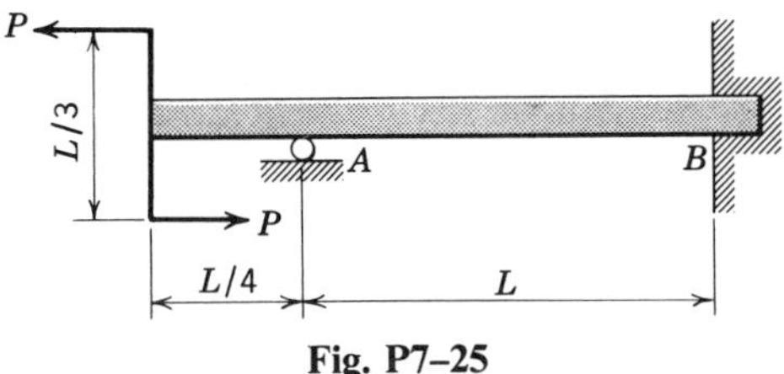

Fig. P7–25

7–26 Solve Prob. 7–16 by the area-moment method.

7–27 Solve Prob. 7–15 by the area-moment method.

7–28* Solve Prob. 7–17 by the area-moment method.

7–29 A beam is loaded and supported as shown in Fig. P7–29. Determine:

(a) The value of M, in terms of P and L, for which the slope of the elastic curve will be $PL^2/(12EI)$ upward to the right at the left support.

(b) The location, in terms of L, of the point of maximum deflection between the supports for the value of M determined from part (a).

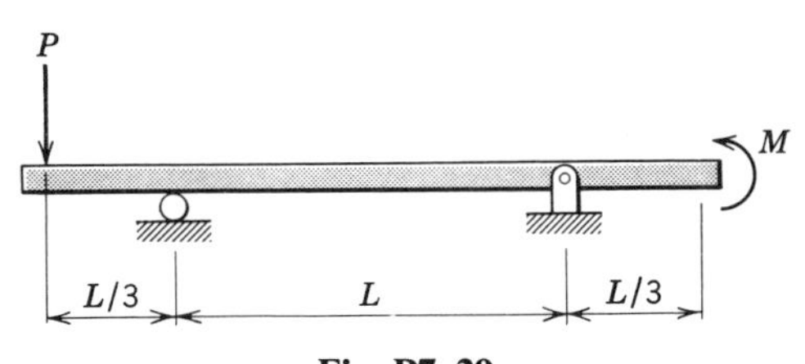

Fig. P7–29

7–30* For the beam in Fig. P7–30, determine:

(a) The force P that will give a zero slope of the elastic curve at the right support.

(b) The location of the point of maximum deflection between the supports for the value of P as determined in part (a).

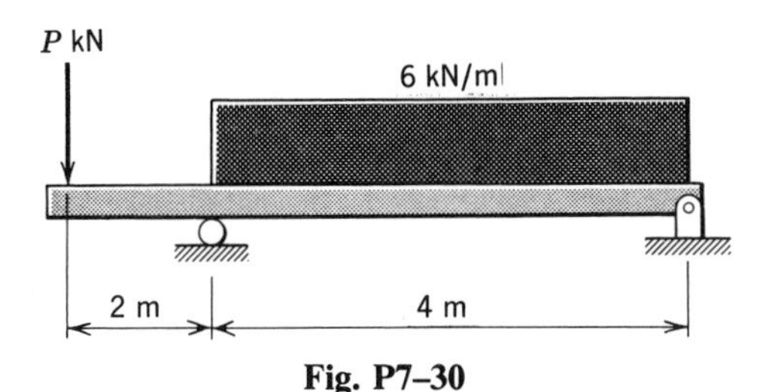

Fig. P7–30

7–31 For the beam in Fig. P7–31, determine:
(a) The left reaction.
(b) The deflection at B. Use 30,000 ksi and 800 in.^{4} for E and I.

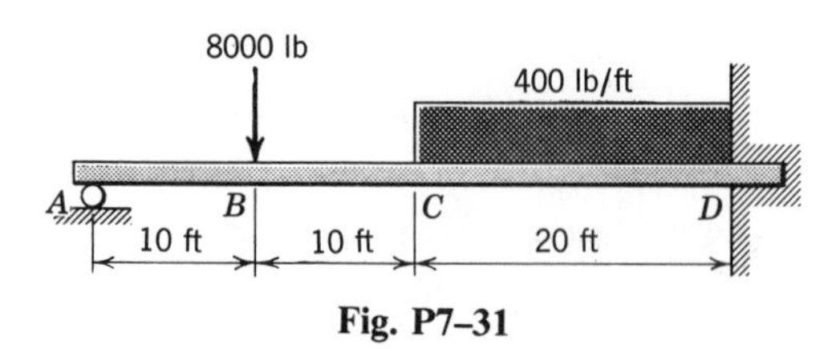

Fig. P7–31

7–32* Determine the reactions on the beam loaded and supported as shown in Fig. P7–32.

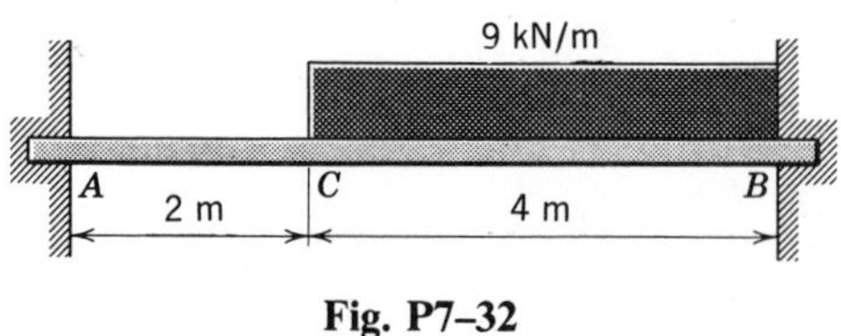

Fig. P7–32

7–33* For the uniform beam shown in Fig. P7–33, determine:
(a) The reaction at the right end in terms of M and L.
(b) The location of the point of maximum deflection.
(c) The value of the maximum deflection in terms of M, L, E, and I.

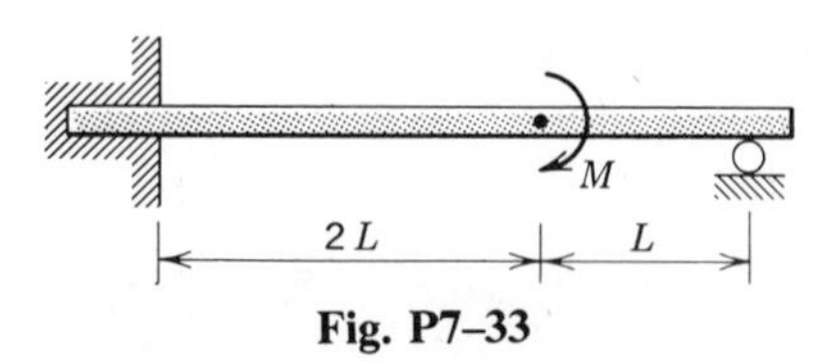

Fig. P7–33

7–34 For the uniform beam loaded as shown in Fig. P7–34, determine:
(a) The reactions at A, B, and C.
(b) The deflection at the center of span AB in terms of P, L, E, and I.

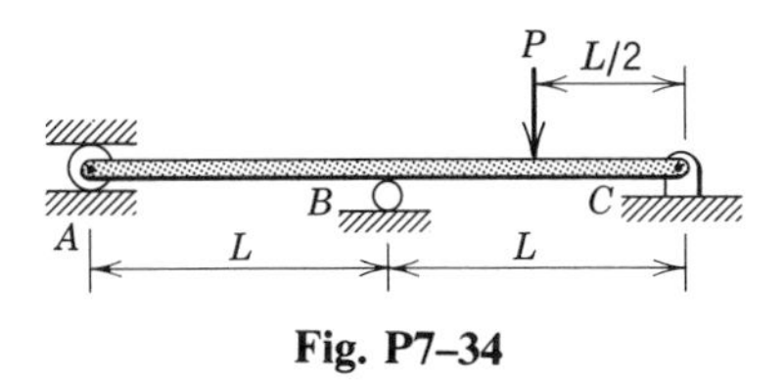

Fig. P7–34

7–35* For the uniform beam loaded and supported as shown in Fig. P7–35, determine the reactions at A, B, and C.

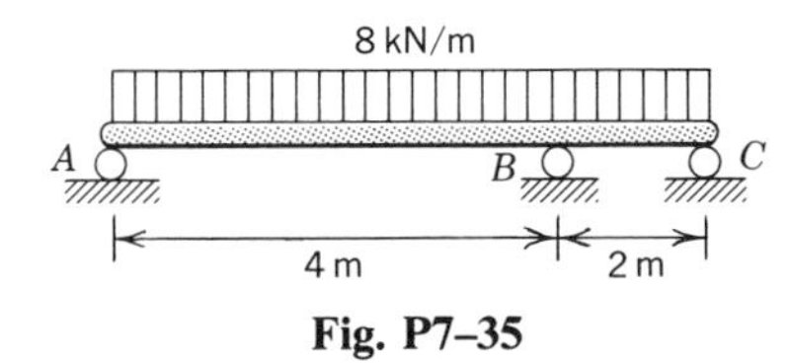

Fig. P7–35

7–36 When unloaded, the 40-ft timber beam shown in Fig. P7–36 has a 0.40-in. gap between the center support and the bottom of the beam. The cross section of the beam is 6 in. wide by 12. in. deep and E for the timber is 2000 ksi. Determine the center reaction when the beam is supporting a uniformly distributed load of 300 lb/ft.

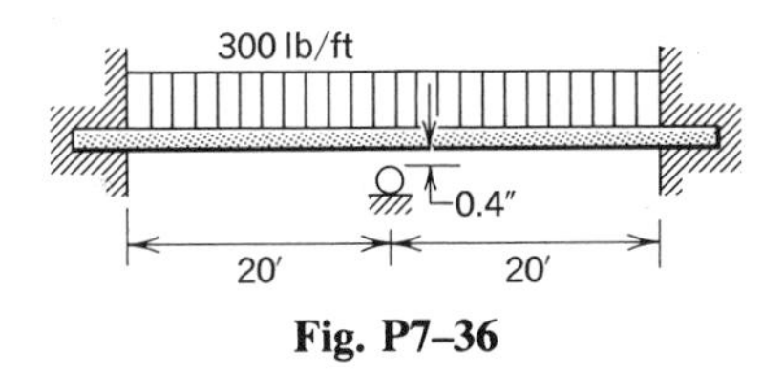

Fig. P7–36

7–4 THE SUPERPOSITION METHOD

The concept (discussed in Art. 6–10) that a slope or deflection due to several loads is the algebraic sum of the slopes or deflections due to each of the loads acting individually is frequently used to provide the deformation equations needed to supplement the equilibrium equations in the solution of statically indeterminate beam problems. The general method of attack is as outlined in Art. 3–3. To provide the necessary deformation equations, selected restraints are removed and replaced by unknown loads (forces and couples); the deformation diagrams corresponding to individual loads (both known and unknown) are sketched, and the component deflections or slopes are summed to produce the known configuration. The following examples illustrate the use of superposition for this purpose.

Example 7–3 A steel beam 20 ft long is simply supported at the ends and at the midpoint. Under a load of 400 lb/ft uniformly distributed over the entire beam, the center support is 0.12 in. above the end supports. Determine the reactions. The moment of inertia of the cross section with respect to the neutral axis is 100 in.^{4}.

Solution With three unknown reactions and only two equations of equilibrium available, the beam is statically indeterminate. Replace the center support with an unknown applied load. The resulting simply supported beam is equivalent to two beams with individual loads as shown in Fig. 7–3. The resulting deflection at the midpoint of the beam is the upward deflection y_R due to R_C, minus the downward deflection y_w due to the uniform load; that is,

$$y = y_R - y_w = 0.12 \text{ in.} \tag{a}$$

The deflection y_w can be obtained from App. D, case 7, and is

$$y_w = \frac{5wL^4}{384EI} = \frac{5[400(20)][20(12)]^3}{384(30)(10^6)(100)} = 0.48 \text{ in. downward}$$

The deflection y_R can be obtained in terms of R_C from App. D, case 6, and is

$$y_R = \frac{R_C L^3}{48EI} = \frac{R_C[20(12)]^3}{48(30)(10^6)(100)} = \frac{96R_C}{10^6} \text{ in. upward}$$

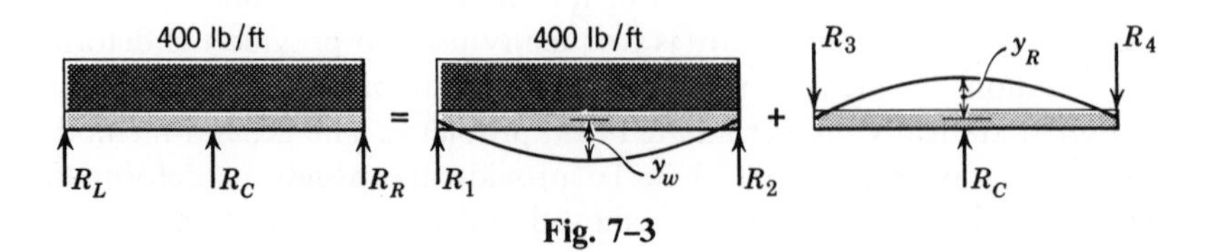

Fig. 7–3

When these values are substituted in Eq. (a), the result is

$$\frac{96R_C}{10^6} - 0.48 = 0.12$$

from which

$$R_C = \underline{6250 \text{ lb upward}} \qquad \text{Ans.}$$

The equilibrium equation $\Sigma F_y = 0$ and symmetry give

$$R_L = R_R = \underline{875 \text{ lb upward}} \qquad \text{Ans.}$$

Note that in determining y_w, w lb/ft times L ft gives the applied load in pounds. The remaining L^3 must be expressed in inches if E and I are in inches and the result is to be in inches. The arithmetic will frequently be simplified if the expressions for the deflections are substituted in the deflection equation in symbol form. In this example Eq. (a) becomes

$$\frac{R_C L^3}{48EI} - \frac{5wL^4}{384EI} = 0.12$$

which reduces (when multiplied by $48EI/L^3$) to

$$R_C - \frac{5wL}{8} = \frac{0.12(48EI)}{L^3}$$

or

$$R_C = \frac{5(400)(20)}{8} + \frac{0.12(48)(30)(10^6)(100)}{[20(12)]^3} = 6250 \text{ lb upward}$$

Example 7–4 Determine all reactions for the beam in Fig. 7–4*a* in terms of P, L, and a.

Solution There are four unknown reactions (a shear and moment at each end), and only two equations of equilibrium are available; therefore, the beam is statically indeterminate, and two deformation equations are necessary. Replace the constraint at the right end with an unknown force and couple. The resulting cantilever beam is equivalent to three beams with

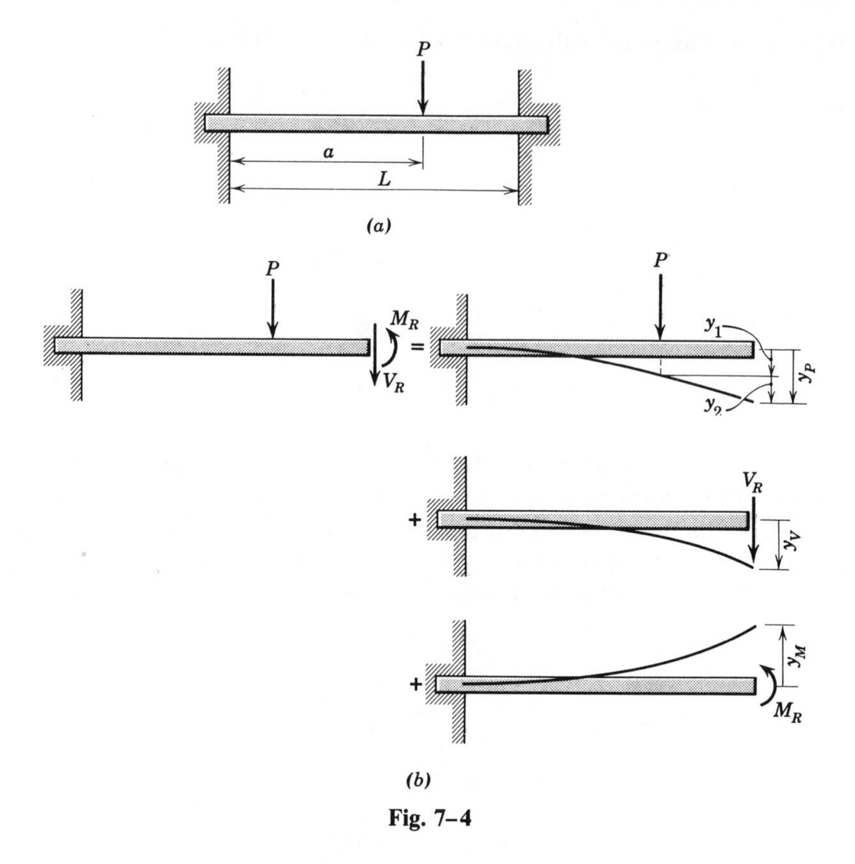

Fig. 7–4

individual loads, as shown in Fig. 7–4*b*. Note that the unknown shear and moment at the right end are both shown as positive values so that the algebraic sign of the result will be correct. From the geometry of the constrained beam, the resultant slope and the resultant deflection at the right end are both zero. The slope and deflection at the end of each of the three replacement beams can be obtained from the expressions in App. D. Thus the first beam with load P (see case 1, App. D) has a constant slope from P to the end of the beam, which is

$$\theta_P = -\frac{Pa^2}{2EI}$$

The deflection y_P at the end is made up of two parts: y_1 for a beam of length a, and y_2 the added deflection of the tangent segment (straight line) from P to the end of the beam. This deflection is

$$y_P = -y_1 - y_2 = -\frac{Pa^3}{3EI} - (L-a)\theta_P$$

$$= -\frac{Pa^3}{3EI} - (L-a)\frac{Pa^2}{2EI} = \frac{Pa^3}{6EI} - \frac{Pa^2L}{2EI}$$

The slope and deflection at the end of the beam due to the shear V_R are (also from case 1, App. D)

$$\theta_V = -\frac{V_R L^2}{2EI} \quad \text{and} \quad y_V = -\frac{V_R L^3}{3EI}$$

Finally, the slope and deflection at the right end of the beam due to M_R are (see case 4, App. D)

$$\theta_M = \frac{M_R L}{EI} \quad \text{and} \quad y_M = \frac{M_R L^2}{2EI}$$

Since the resultant slope is zero,

$$\theta_P + \theta_V + \theta_M = -\frac{Pa^2}{2EI} - \frac{V_R L^2}{2EI} + \frac{M_R L}{EI} = 0$$

Similarly

$$y_P + y_V + y_M = \frac{Pa^3}{6EI} - \frac{Pa^2L}{2EI} - \frac{V_R L^3}{3EI} + \frac{M_R L^2}{2EI} = 0$$

The simultaneous solution of these two equations yields

$$M_R = \frac{-Pa^2(L-a)}{L^2} \quad \text{and} \quad V_R = \frac{-Pa^2(3L-2a)}{L^3} \qquad \text{Ans.}$$

When the equations of equilibrium are applied to a free-body diagram of the beam, the shear and moment at the left end are found to be

$$M_L = \frac{-Pa(L-a)^2}{L^2} \quad \text{and} \quad V_L = \frac{+P(L^3 - 3a^2L + 2a^3)}{L^3} \qquad \text{Ans.}$$

PROBLEMS

7–37* When unloaded, the clearance between the center support and the bottom of the timber beam shown in Fig. P7–37 is 1.44 in. Determine the center reaction when the beam supports a uniformly distributed load of 800 lb/ft. The value of E is 1600 ksi.

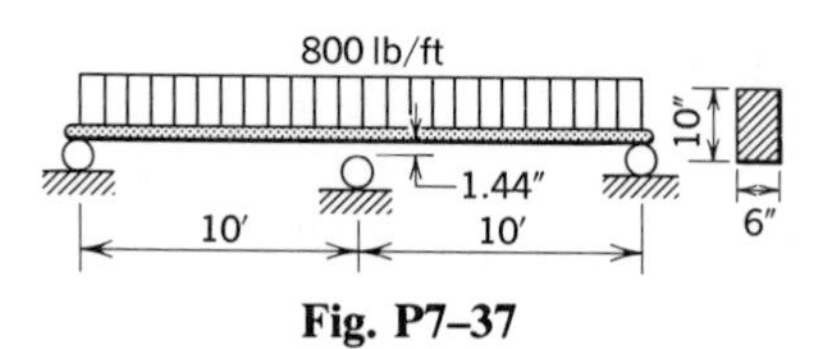

Fig. P7–37

7–38* A timber beam 100 mm wide by 200 mm deep is 4 m long. It is fixed at the left end and simply supported at the right end. It is subjected to a uniformly distributed load over the entire span. If the maximum allowable fiber stress is 12 MN/m^2 and the left support settles without rotation an amount equal to $wL^4/(120EI)$ where w is the load in newtons per metre, determine the maximum permissible total load. The flexural modulus of elasticity for this timber is 8 GN/m^2.

7–39 The aluminum alloy beam of Fig. P7–39 is fixed at the left end, and, when unloaded, there is a 0.60-in. clearance between the right end and the support. The moment of inertia of the beam cross section is 15 in.^{4} and E may be taken as 10^7 psi. Determine the maximum positive moment in the beam when the load of 300 lb/ft is applied.

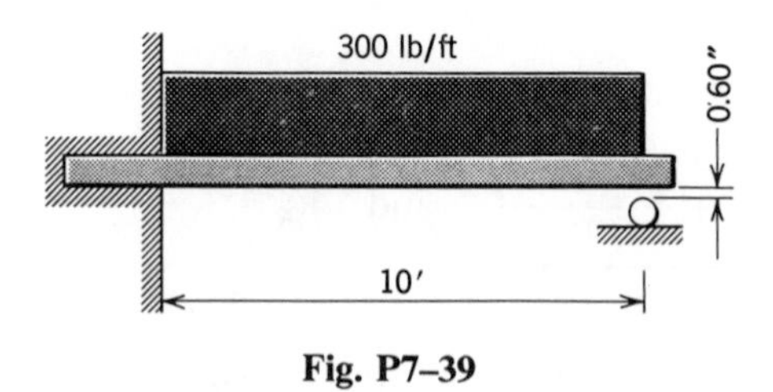

Fig. P7–39

7–40 With the beam loaded and supported as shown in Fig. P7–40, the slope at the wall was found to be $wL^3/(48EI)$ upward to the right, and the ends were both on the same horizontal line. Determine the reactions in terms of w and L.

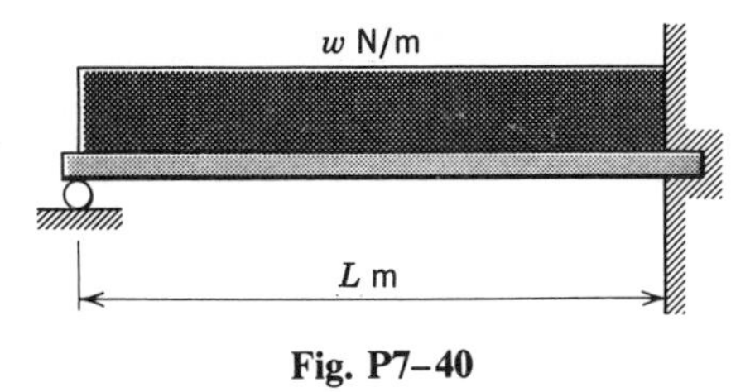

Fig. P7–40

7–41* Determine, in terms of w and L, the reaction at B on the beam of Fig. P7–41.

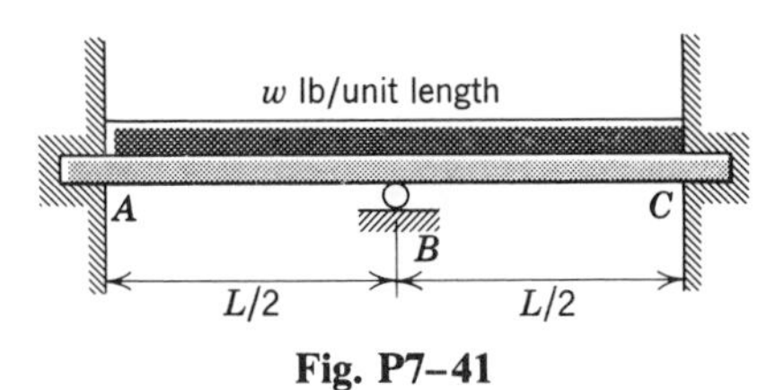

Fig. P7–41

7–42* The uniformly loaded timber beam of Fig. P7–42 is simply supported at the ends and has a high-strength steel vertical rod (area = 100 mm^2) support at midspan as shown. Consider the supports at A, D, and C to be rigid. Construct the shear diagram for the loaded beam. Use 10 GN/m^2 for the modulus of elasticity of the timber. The tension in rod BD is zero before the load is applied to the beam.

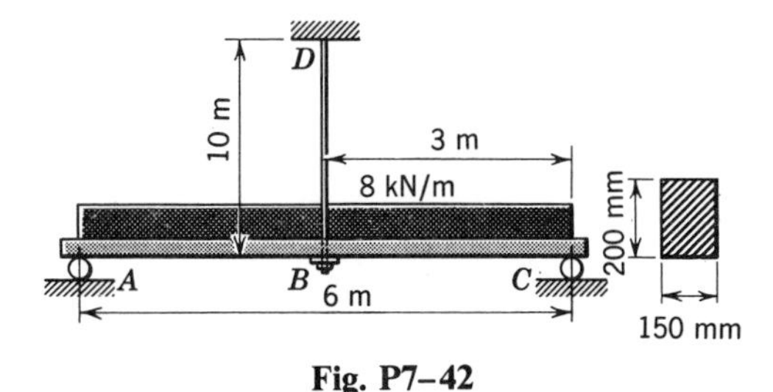

Fig. P7–42

7–43 Figure P7–43 shows a W8×40 steel beam supported at the ends and at the center. Consider the end supports and the support at D (but not the post) to be rigid. The post BD is a 4×6-in. timber (E = 1200 ksi), which is braced to prevent buckling. Determine the reaction at B due to the load of 600 lb/ft.

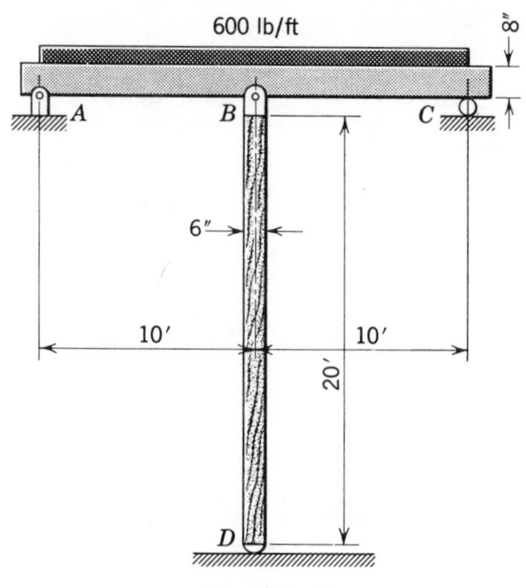

Fig. P7–43

7–44* Solve Prob. 7–43 if the post *BD* is replaced by two coil springs in series at *B* (one spring on top of the other and a rigid support under the springs). The spring moduli are 8000 and 12,000 lb/in. (i.e., one spring will deflect 1 in. under a load of 8000 lb and the other will deflect 1 in. under a load of 12,000 lb).

7–45 The 90-mm wide by 200-mm deep wooden beam of Fig. P7–45 is fixed at the left end and supported by an aluminum alloy tie rod having a cross-sectional area of 100 mm^2 at the right end. Use 8 and 72 GN/m^2, respectively, for the moduli of elasticity of the wood and the aluminum, and determine the tension in the tie rod if it is unstressed before the load is applied.

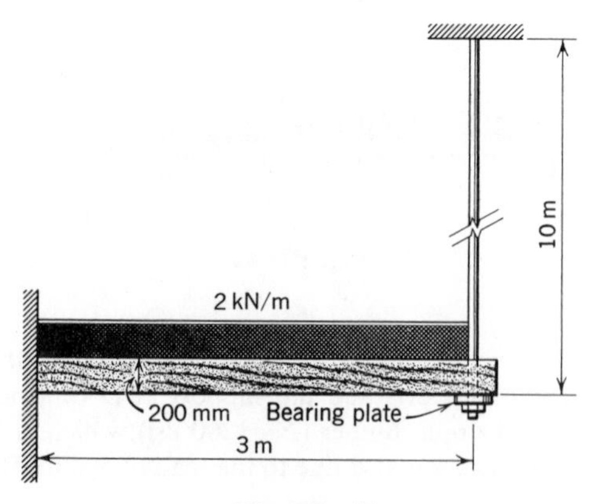

Fig. P7–45

7–46 The tie rod of Prob. 7–45 is replaced by a coil spring placed on a rigid support beneath the right end of the beam with the top of the spring 5 mm below the bottom of the beam. The spring modulus is $16(10^4)$ N/m (i.e., the spring deflects 1 mm under a load of 160 N). Determine the reaction of the spring on the beam.

7–47* The system shown in Fig. P7–47 is composed of two S3×7.5 steel beams. In the unloaded condition, the beams are parallel and the clearance between them at the center is 0.200 in. When the top beam is subjected to a uniform load of 300 lb/ft, determine the reaction at B on the upper beam and the center deflection of the upper beam.

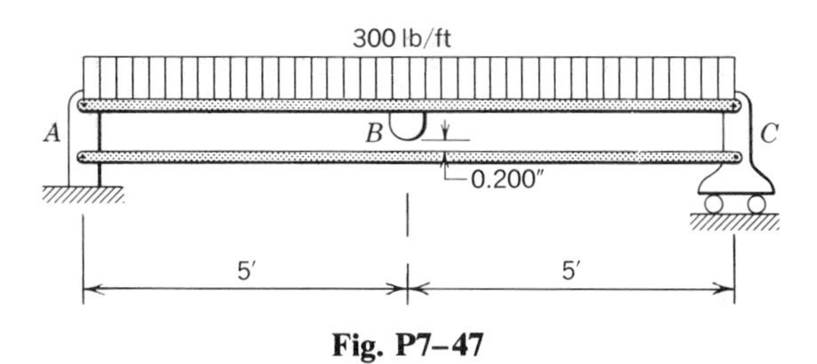

Fig. P7–47

7–48* In Fig. P7–48 beam A is of brass and beam B is of steel. The moment of inertia of B is twice that of A, and the modulus of elasticity of the steel is twice that of the brass. The reaction at C is zero before the load w is applied. Determine:

(a) The reaction at C on beam B in terms of w and L.

(b) The deflection of beam A at C when $L=4$ m, I for beam $A=32(10^6)$ mm^4, and $w=18$ kN/m.

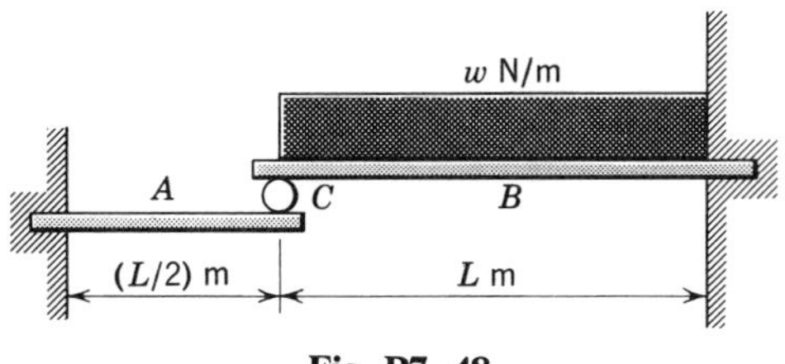

Fig. P7–48

7–49* In Fig. P7–49, the aluminum alloy tie rod passes through a hole in the aluminum alloy cantilever beam and through the coil spring positioned on the end of the tie rod. Before loading, there is 0.10-in. clearance between the beam and the top of the spring. The cross-sectional area of the tie rod is 0.15 sq in., the moment of inertia of the beam cross section is 100 in.^{4}, and the spring modulus is 6000 lb/in. Determine the maximum permissible magnitude for w applied with the moment M of 80,000 in-lb as shown for a limiting axial stress of 10 ksi in the tie rod.

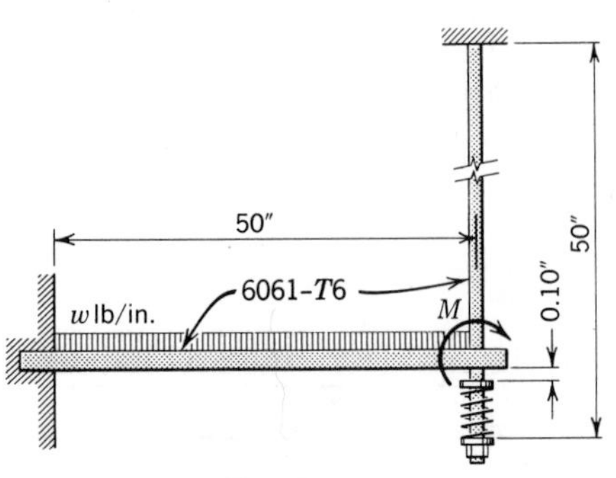

Fig. P7–49

7–50 The steel beam of Fig. P7–50 is fixed at the left end and, when loaded, is supported by the helical spring B at the right end. Determine the reaction of the spring B on the beam for the following numerical data: $I = 20(10^6)$ mm^4, $L = 5$ m, $w = 3$ kN/m, $y_o = 5$ mm (when the beam is unloaded), spring modulus $= 1200$ kN/m.

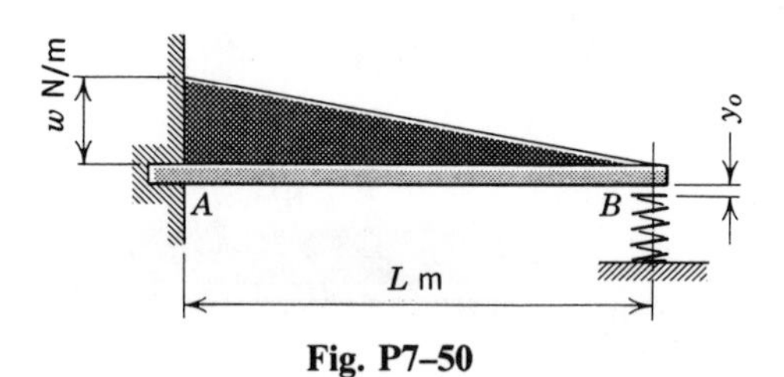

Fig. P7–50

7–51* The steel beam AB of Fig. P7–51 is fixed at the ends A and B and supported at the center by the pin-connected wooden struts CD and CE. The cross-sectional area of each strut is 20 sq. in. and the modulus of elasticity of the wood is 1000 ksi. Determine the force in each strut due to the distributed load of 35 lb/in. The constants to be used in the expression for maximum deflection of a beam with fixed ends are 1/384 for a uniformly distributed load and 1/192 for a concentrated load at the center.

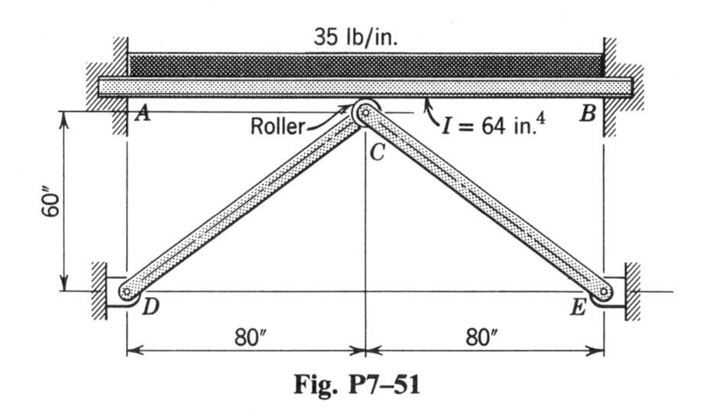

Fig. P7–51

7–52 Determine the reactions at A, B, and C for the uniform beam shown in Fig. P7–52.

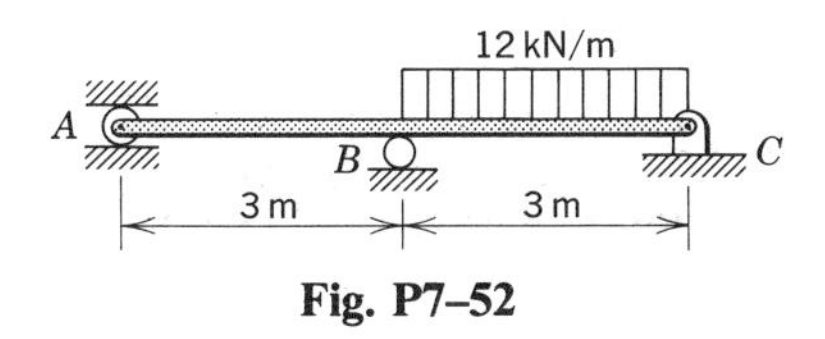

Fig. P7–52

7–53* In the unloaded condition, the S4×9.5 cantilever beam AB in Fig. P7–53

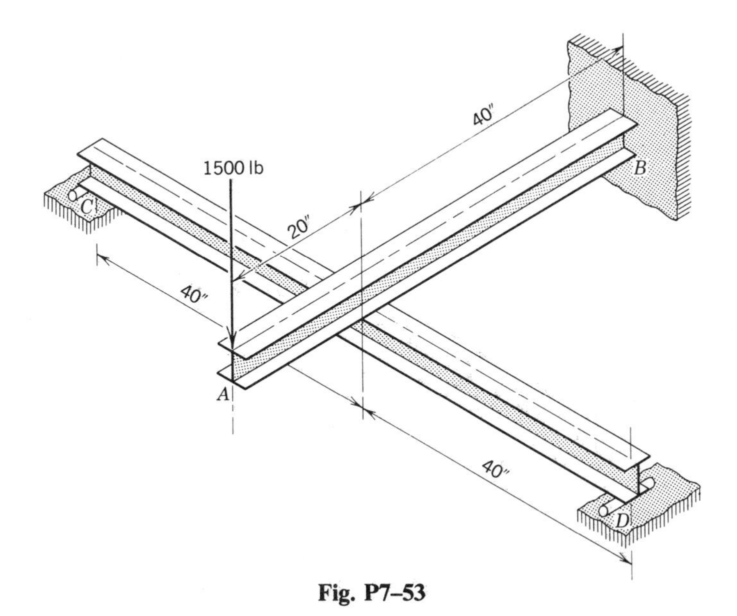

Fig. P7–53

touches but exerts no force on the S5×14.75 simply supported beam *CD*. Both beams are of steel. Determine the deflection of beam *AB* at the point of application of the 1500-lb force.

7–5 ENERGY METHOD—CASTIGLIANO'S THEOREM

Castigliano's theorem (see Art. 6–11) is an effective supplement to the equations of equilibrium in the solution of statically indeterminate structures. If the deflection at some point in a structure is known (a boundary condition), Eq. 6–5*a* may be applied at this point and set equal to the known deflection. The resulting equation will contain one or more unknown forces. If the equation contains more than one unknown, additional equations must be written, such as equilibrium equations or the application of Eq. 6–5*b*. However, there are certain restrictions necessary to ensure that the equations are independent; also, some remarks regarding procedure may be in order. The following outline may be found helpful:

---

1. For any system of loads and reactions, as many independent energy equations (Eq.6–5*a* or 6–5*b*) can be written as there are redundant unknowns—defined as unknowns not necessary to maintain equilibrium of the structure. This restriction is equivalent to the restriction of Art. 7–2 requiring one boundary condition for each extra unknown.

2. If only one independent energy equation can be written, and the moment equation contains two unknowns (forces or couples, including dummy loads), one of the unknowns must be expressed in terms of the other by means of an equilibrium equation. This case is illustrated in the alternate solution to Exam. 7–5 which follows. If two independent energy equations can be written, these can contain two unknowns which can be obtained from a solution of the two energy equations, as illustrated by Exam. 7–7.

3. If the loading changes along the beam, the moment equation can be written for the entire span using the singularity notation of Art. 6–5 and, after multiplying by $\partial M/\partial P_i$ or $\partial M/\partial M_i$ (also using singularity notation), integration can be performed over the entire span. However, this procedure requires integration by parts of such expressions as $x\langle x-L\rangle^1 dx$. Sometimes it may be more convenient to write an ordinary algebraic moment equation for each interval, multiply by $\partial M/\partial P_i$ or $\partial M/\partial M_i$ (different for each interval), and integrate over each interval; then add the results of all the integrations.

Example 7–5 Solve Exam. 7–1 of Art. 7–2 by the application of Castigliano's theorem. The beam is loaded and supported as shown in Fig. 7–5*a*.

Solution The free-body diagram of Fig. 7–5*b* indicates that the problem is statically indeterminate, since there are three unknown reaction components and only two independent equations of equilibrium available.

If the portion of the beam to the right of the reaction R were removed and replaced by an equivalent shearing force and bending moment at the transverse section above R, neither R nor the elastic curve in the interval $0 \leqslant x \leqslant L$ would be changed. Hence, it is necessary to deal with only the strain energy of the beam in the interval $0 \leqslant x \leqslant L$. With the coordinate system placed as shown, the resulting moment equation obtained from the free-body diagram in Fig. 7–5c is

$$M = Rx - \frac{w}{2}\left(x + \frac{L}{4}\right)^2$$

$$= Rx - \frac{wx^2}{2} - \frac{wLx}{4} - \frac{wL^2}{32}$$

From Eq. 6–5a, the deflection at the right support is given by

$$y_O = \frac{\partial U}{\partial R} = \frac{1}{EI}\int_0^L M\frac{\partial M}{\partial R}\,dx$$

Since the partial derivative of M with respect to R is x, the expression for the deflection is

$$y_O = \frac{1}{EI}\int_0^L \left(Rx^2 - \frac{wx^3}{2} - \frac{wLx^2}{4} - \frac{wL^2x}{32}\right)dx$$

which upon integration and substitution of limits becomes

$$y_O = \frac{1}{EI}\left(\frac{RL^3}{3} - \frac{wL^4}{8} - \frac{wL^4}{12} - \frac{wL^4}{64}\right)$$

Since the support is unyielding ($y = 0$ when $x = 0$), the above expression can be equated to zero and solved for R, yielding

$$\underline{R = \frac{43wL}{64} \text{ lb upward}} \qquad \text{Ans.}$$

The other reaction components (M_A and V_A) are obtained from the equilibrium equations $\Sigma M_A = 0$ and $\Sigma F_y = 0$, and are

$$\underline{M_A = \frac{7wL^2}{64} \text{ ft-lb counterclockwise}} \qquad \text{Ans.}$$

and

$$V_A = \frac{37wL}{64} \text{ lb upward} \qquad \text{Ans.}$$

Alternate solution: This solution will make use of the free-body diagram of Fig. 7–5*d*, from which the moment equation is

$$M = M_A + V_A x - \frac{wx^2}{2}$$

Since there are two unknowns in the moment equation and only one independent energy equation can be written, the equilibrium equation, $\Sigma M_B = 0$ will be used to obtain a relation between the two unknowns. Thus,

$$M_A + V_A L - \frac{wL^2}{2} + \frac{wL^2}{32} = 0$$

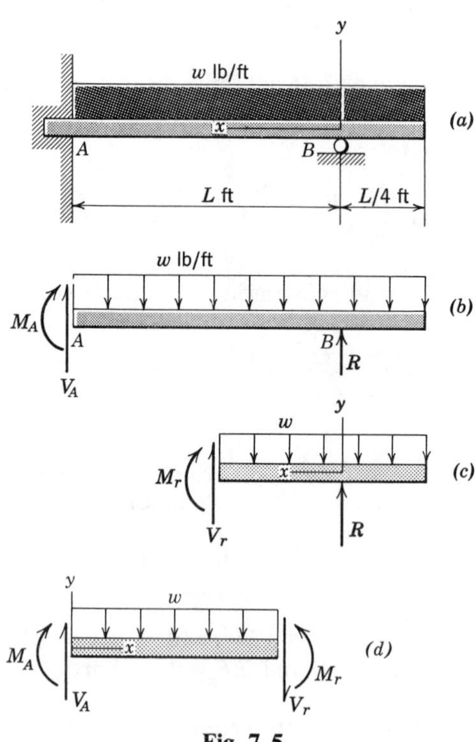

Fig. 7–5

from which

$$M_A = \frac{15wL^2}{32} - V_A L \quad \text{or} \quad V_A = \frac{15wL}{32} - \frac{M_A}{L}$$

Eliminating M_A from the moment equation gives

$$M = V_A(x-L) + \frac{15wL^2}{32} - \frac{wx^2}{2}$$

and

$$\frac{\partial M}{\partial V_A} = x - L$$

From Eq. 6–5a, the deflection at the left support is given by

$$y_A = \frac{\partial U}{\partial V_A} = \frac{1}{EI}\int_0^L M \frac{\partial M}{\partial V_A} dx$$

$$= \int_0^L \left[V_A(x-L)^2 - \frac{wx^2}{2}(x-L) + \frac{15wL^2}{32}(x-L) \right] dx = 0$$

from which

$$\underline{V_A = \frac{37wL}{64} \text{ lb upward}} \qquad \text{Ans.}$$

The other reaction components (M_A and R) can be obtained from equilibrium equations.

Instead of retaining V_A in the moment equation, M_A could have been retained, in which case the moment equation would be

$$M = M_A\left(1 - \frac{x}{L}\right) + \frac{15wLx}{32} - \frac{wx^2}{2}$$

and

$$\frac{\partial M}{\partial M_A} = 1 - \frac{x}{L}$$

From Eq. 6–5b, the rotation at the left support is given by

$$\theta_A = \frac{\partial U}{\partial M_A} = \frac{1}{EI}\int_0^L M\frac{\partial M}{\partial M_A}\,dx$$

$$= \int_0^L \left[M_A\left(1-\frac{x}{L}\right)^2 + \frac{15wLx}{32}\left(1-\frac{x}{L}\right) - \frac{wx^2}{2}\left(1-\frac{x}{L}\right)\right]dx = 0$$

from which

$$M_A = -\frac{7wL^2}{64}\ \text{ft-lb} \qquad \text{Ans.}$$

Again, the two remaining reactions can be obtained from equilibrium equations. The complete set of answers is given in the first solution.

Example 7–6 Beam BC of Fig. 7–6a is a rolled steel W8×15 beam having a cross-sectional moment of inertia of 48 in.^{4}. The beam is fixed at B and supported at C by an aluminum alloy tie rod having a cross-sectional area of 0.15 sq in. Determine the force in the tie rod due to the distributed load of 800 lb/ft. Use 30,000 and 10,000 ksi for the moduli of elasticity of the steel and aluminum, respectively.

Solution The free-body diagram of the beam and tie rod is shown in Fig. 7–6b with the origin of coordinates at the top of the tie rod where the boundary condition is $x=0$, $y=0$. The total strain energy in the system is the sum of the energies in the tie rod and the beam; that is,

$$U = \frac{\sigma^2}{2E_a}Al + \int_0^L \frac{M^2}{2E_sI}\,dx = \frac{P^2l}{2AE_a} + \int_0^L \frac{M^2}{2E_sI}\,dx$$

The deflection at D is

$$y_O = \frac{\partial U}{\partial P} = \frac{Pl}{AE_a} + \frac{1}{E_sI}\int_0^L M\frac{\partial M}{\partial P}\,dx = 0$$

The moment in the beam is

$$M = Px - \frac{wx^2}{2}$$

and

$$\frac{\partial M}{\partial P} = x$$

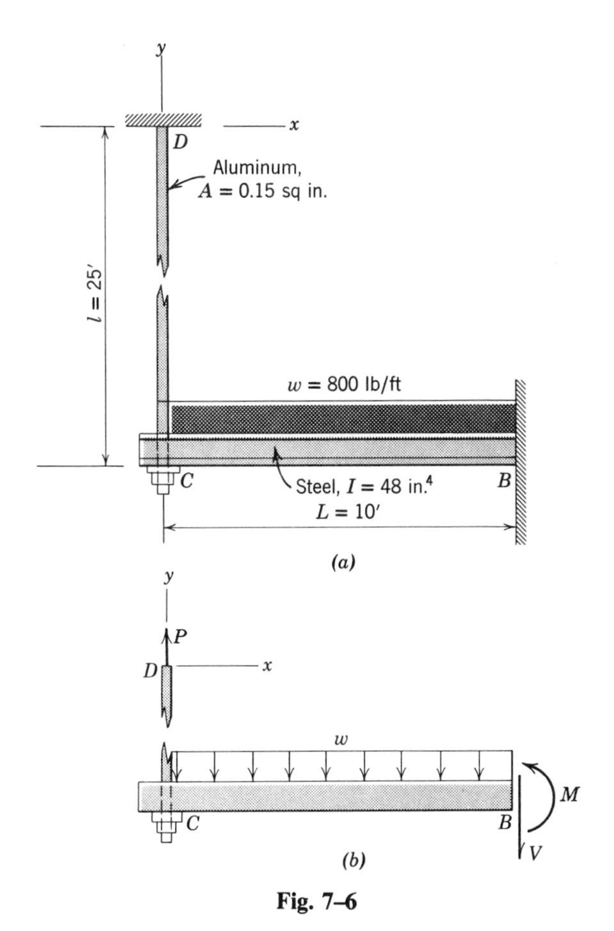

Fig. 7–6

Therefore,

$$y_O = \frac{Pl}{AE_a} + \frac{1}{E_s I}\int_0^L \left(Px^2 - \frac{wx^3}{2}\right)dx = \frac{Pl}{AE_a} + \frac{PL^3}{3E_s I} - \frac{wL^4}{8E_s I}$$

Since the structure is fixed at D, y_O is zero and substitution of the given data yields

$$\frac{P(25)12}{15(10^{-2})(10^7)} + \frac{P(10^3)(12^3)}{3(3)(10^7)48} = \frac{800(10)(10^3)(12^3)}{8(3)(10^7)48}$$

from which

$$P = \underline{2000 \text{ lb } T} \qquad \text{Ans.}$$

Example 7–7 Determine the reactions at the left end of the beam of Fig. 7–7*a* in terms of P, L, and a.

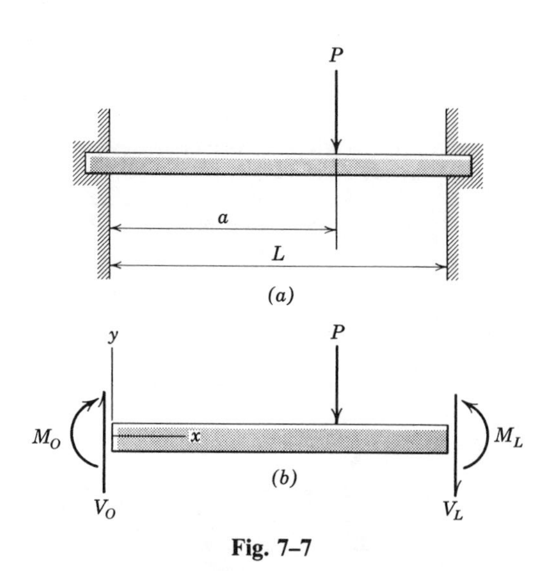

Fig. 7–7

Solution The free-body diagram of Fig. 7–7*b* indicates four unknown reaction components, and since there are only two independent equations of equilibrium, two supplementary equations are necessary. These are obtained by applying Eq. 6–5*a* and 6–5*b* and solving the resulting simultaneous equations for M_O and V_O. With the origin at the left support, the boundary conditions are $y=0$ and $\theta=0$ at $x=0$. Since the moment equation is

$$M = M_O + V_O x - P\langle x-a\rangle^1$$

the deflection at the left end is given by Eq. 6–5*a* as

$$y_O = \frac{\partial U}{\partial V_O} = \frac{1}{EI}\int_0^L M\frac{\partial M}{\partial V_O}\,dx$$

and

$$EIy_O = \left[\frac{M_O x^2}{2} + \frac{V_O x^3}{3} - P\left(\frac{x}{2}\langle x-a\rangle^2 - \frac{1}{6}\langle x-a\rangle^3\right) \right]_0^L$$

where the last term was integrated by parts. Substitution of the limits gives

$$EIy_O = \frac{M_O L^2}{2} + \frac{V_O L^3}{3} - \frac{PL}{2}(L-a)^2 + \frac{P}{6}(L-a)^3$$

The application of the boundary condition $y=0$ at $x=0$ gives

$$\frac{M_O L^2}{2} + \frac{V_O L^3}{3} - \frac{PL^3}{3} + \frac{PL^2 a}{2} - \frac{Pa^3}{6} = 0 \qquad \text{(a)}$$

When Eq. 6–5a is applied, the rotation at the left end is given by the expression

$$\theta_O = \frac{\partial U}{\partial M_O} = \frac{1}{EI}\int_0^L M \frac{\partial M}{\partial M_O}\,dx$$

and

$$EI\theta_O = \int_0^L (M_O + V_O x - P\langle x-a\rangle^1)\,dx$$

$$= \left[M_O x + \frac{V_O x^2}{2} - \frac{P}{2}\langle x-a\rangle^2 \right]_0^L$$

which reduces to

$$M_O L + \frac{V_O L^2}{2} - \frac{PL^2}{2} + PaL - \frac{Pa^2}{2} = 0 \qquad \text{(b)}$$

Equations (a) and (b) are solved to obtain the following results:

$$\underline{V_O = P - \frac{3Pa^2}{L^2} + \frac{2Pa^3}{L^3}} \qquad \text{Ans.}$$

and

$$\underline{M_O = -Pa + \frac{2Pa^2}{L} - \frac{Pa^3}{L^2}} \qquad \text{Ans.}$$

PROBLEMS

Note. The following problems are to be solved using Castigliano's theroem.

7–54* Solve Prob. 7–1.

7–55* Solve Prob. 7–2.

7–56* Solve Prob. 7–7.

7–57* Solve Prob. 7–5*a*.

7–58* Solve Prob. 7–13.

7–59* Solve Prob. 7–14.

7–60* Solve Prob. 7–8.

7–61* Solve Prob. 7–22.

7–62 Solve Prob. 7–45.

7–63 Solve Prob. 7–50.

7–64* Solve Prob. 7–16.

7–65* Solve Prob. 7–33*a*.

7–66 Solve Prob. 7–31*a*.

7–67* Solve Prob. 7–37.

7–68* Solve Prob. 7–44.

7–69* The curved beam of Fig. P7–69 has a constant radius and constant E and I. Assume that the ratio ρ/c is large enough for the elastic flexure formula to apply and neglect the axial stress. Determine the reaction at B on the beam in terms of P.

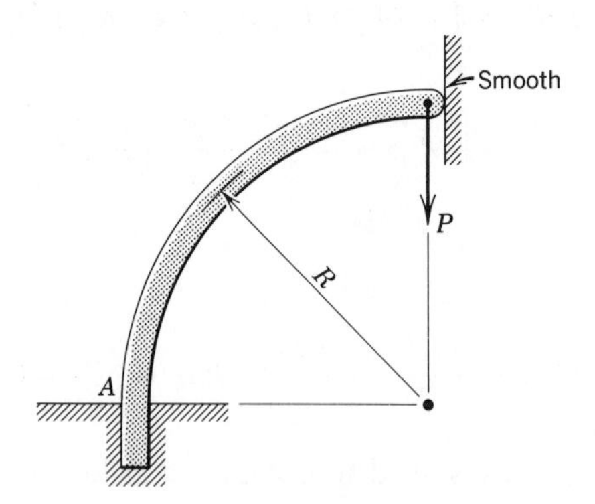

Fig. P7–69

7–70* The frame ABC of Fig. P7–70 is a continuous member of constant E and I. Assume the corner at B remains a right angle and neglect the energy due to the axial stress in the leg BC. Determine:

(a) The reaction at A on member AB in terms of w and L.

(b) The deflection of end A in terms of w, L, E, and I.

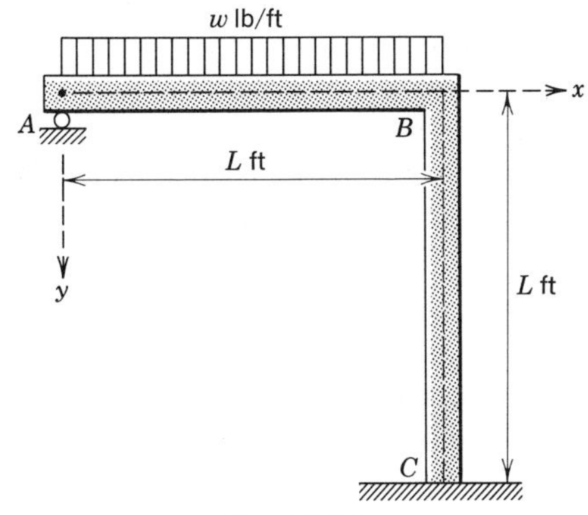

Fig. P7-70

7-71* A frame and loading are identical to those of Prob. 7-70 except that the end A is pinned as in Fig. P7-71. Neglecting the energy due to the axial stresses in AB and BC, determine the horizontal and vertical components of the pin reaction at A on member AB.

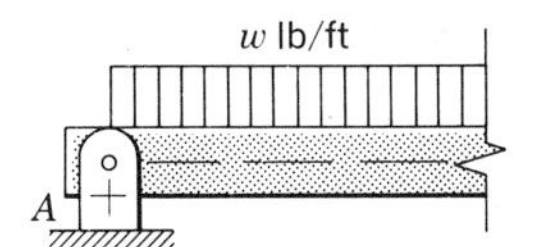

Fig. P7-71

7-6 PLASTIC ANALYSIS

Because of the increasing emphasis on structural analysis based on plastic action, a brief introduction to plastic analyses of statically indeterminate beams is presented below. The analysis will be limited to ideal elastoplastic materials as defined in Art. 2-7. For practical purposes, ordinary structural steel will be satisfactory, and for this material the following assumptions are reasonable and may be verified (within practical limits) by experiment:

1. As the load on the beam increases, the fiber stress will increase proportionately until the yield point for the material is attained at the section of maximum moment, after which yielding at this section will progress over the entire section (see Art. 5-2 and Fig. 5-9f). The resisting moment will increase with the load until the stress in each fiber has reached the yield point, after which the moment will remain constant. In other words, a hinge exerting a constant moment forms at the section producing a discontinuity in the slope of the elastic curve. The moment at the hinge is termed the *plastic moment* M_p.

2. As the loading is increased, a second hinge will form; and additional hinges will form until the beam is reduced to what has been termed a *mechanism*; that is, a system of members, connected at their ends by pins (each pin exerting a constant moment), which will continue to deflect with no additional load.

The load producing a mechanism is considered the *ultimate load*, although strain hardening of the material might necessitate the application of a larger load to produce complete collapse. The hinge, which actually extends over a finite length of beam, is assumed to form at a discrete point in the beam.

One procedure for analyzing statically indeterminate beams consists of arbitrarily assigning the plastic moment value to the redundant moments; this procedure makes the beam statically determinate. Additional hinges are introduced at points of high moment until a mechanism develops. Finally, a check of the moments should show that nowhere does the moment exceed the magnitude of the plastic moment M_p. This is not the only method used in plastic analysis,[1] but for the simpler cases it is probably the easiest to apply. The method is illustrated by the following example.

Example 7–8 The continuous beam of Fig. 7–8*a* is a steel W36×160 section. Consider the steel to be elastoplastic with a yield point of 36 ksi, and determine the ultimate load w (which for span BC includes the weight of the beam) for the beam.

Solution The free-body diagram of Fig. 7–8*b* indicates three unknown reactions, and for this force system there are only two independent equations of equilibrium; hence, there is one redundant quantity. If one moment, shear, or reaction is determined, the beam becomes statically determinate. However, the ultimate load is not attained until a mechanism forms; therefore, it becomes necessary to assume more than one possible hinge location. To assist in estimating the location of possible hinges, the shape of the moment diagram by parts and the composite moment diagram are shown in Fig. 7–8*c*, where the solid line indicates the composite diagram and the dashed lines represent the diagrams for two simple beams with a moment at B. This diagram indicates that hinges may possibly form at B, D, and E. Possible deformations are indicated (greatly exaggerated) in Fig. 7–8*d*, the top curve indicating the shape of the elastic curve when the action is entirely elastic. As the magnitude of the loading is

[1]For additional material see *Plastic Design of Steel Frames*, L. S. Beedle, Wiley, New York, 1958.

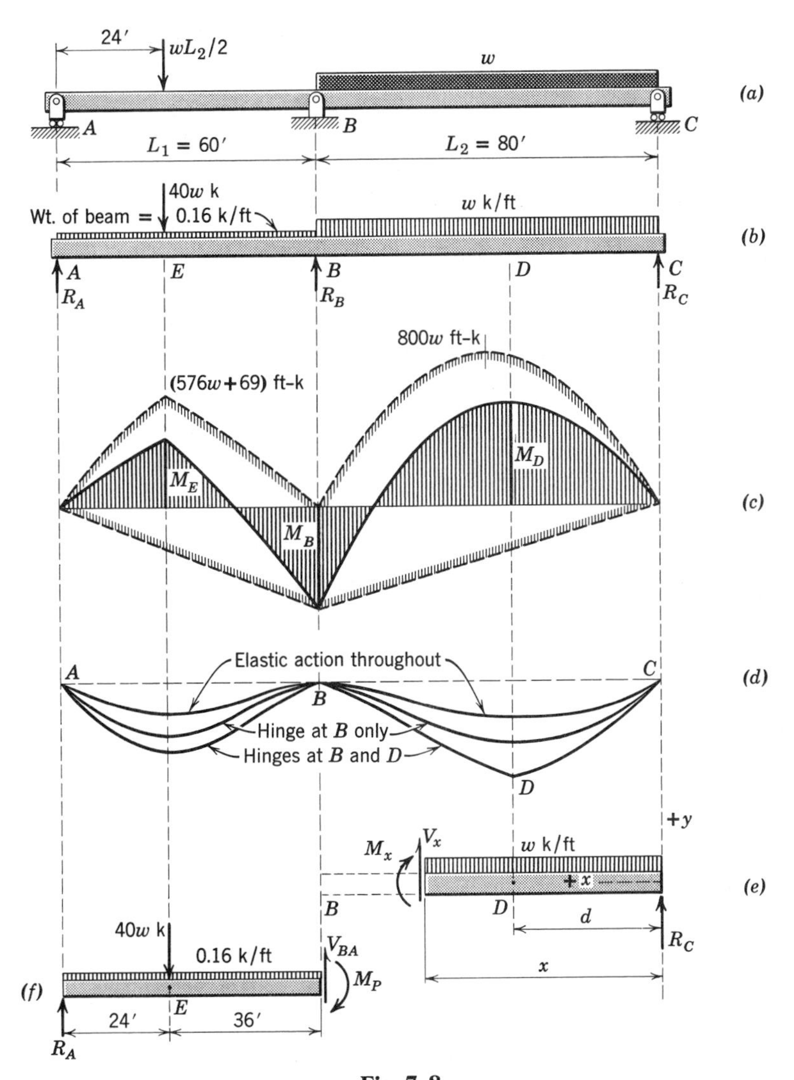
24′
wL2/2
w
(a)
A
B
C
L1 = 60′
L2 = 80′
40w k
Wt. of beam = 0.16 k/ft
w k/ft
(b)
A
E
B
D
C
RA
RB
RC
800w ft-k
(576w+69) ft-k
ME
MD
MB
(c)
Elastic action throughout
A
C
B
Hinge at B only
Hinges at B and D
D
(d)
+y
Vx
Mx
w k/ft
+x
B
D
d
RC
x
(e)
40w k
0.16 k/ft
VBA
MP
(f)
E
24′
36′
RA

Fig. 7–8

increased beyond that necessary to produce the yield point stress in the outer fibers, a hinge will form at the section where the flexural stress is the highest. The first hinge is assumed to form at B, and under this condition the beam would assume the shape indicated by the middle curve of Fig. 7–8d. Since a mechanism has not yet formed, the loading is increased until a second hinge forms. This hinge is assumed to form at D, and from the lower curve of Fig. 7–8d it is apparent that a mechanism has developed since, with hinges at B and D and with end C supported on rollers, deflection can continue in the span BC without an increase in the load. The second hinge may have formed at E instead of at D; hence, the moment at E must be evaluated as a check on the hinge assumptions.

The magnitude of the load necessary to produce hinges at B and D will first be determined, after which the moment at E will be checked. It should be noted that when the plastic moment M_p is used in either an equilibrium or a moment equation, it must have the correct sense or sign. With reference to the free-body diagram of Fig. 7–8e, which represents the right portion of span BC, the equation for the moment in span BC is

$$M_x = R_C x - wx^2/2 \tag{a}$$

When x is 80 ft,

$$M_B = -M_p = R_C(80) - w(80)^2/2$$

from which

$$R_C = 40w - M_p/80 \tag{b}$$

The negative sign on M_B is obtained from Fig. 7–8c. The section of maximum positive moment in span BC is at D where the shear is zero and is located from the shear equation

$$V_x = -R_C + wx$$

When x is equal to d;

$$V_x = 0 = -R_C + wd$$

from which

$$d = R_C/w \tag{c}$$

From Eq. (a) the moment at D (the plastic moment) is

$$M_D = +M_p = R_C d - wd^2/2$$

Since R_C is equal to wd from Eq. (c), the plastic moment becomes

$$M_p = wd^2 - \frac{wd^2}{2} = \frac{w}{2}\left(40 - \frac{M_p}{80w}\right)^2 \qquad \text{(d)}$$

where the plus sign on M_D is obtained from Fig. 7–8c and R_C is obtained from Eq. (b). Solving Eq. (d) for w gives

$$w = \frac{5.8284M_p}{3200} \quad \text{or} \quad \frac{0.1716M_p}{3200}$$

When these expressions are substituted into Eq. (b), only the larger value yields a positive result for d; therefore, this expression will give the correct magnitude for w unless the bending moment at E exceeds M_p. The plastic moment for the W36×160 section is

$$M_p = \Big[\overbrace{\sigma_{yp}(12)(1.02)(18-0.51)}^{\text{one flange}} + \overbrace{\sigma_{yp}(0.653)(18-1.02)^2(1/2)}^{\text{one-half web}}\Big](2) = 616\sigma_{yp}$$

$$= 616(36)/12 = 1848 \text{ ft-kips}$$

Substituting this value for M_p in the expression for w yields

$$w = 5.828(616)(3)/3200 = 3.366 \text{ kips/ft}$$

To check the bending moment at E, the reaction at A will first be obtained by summing moments with respect to B on the free-body diagram of Fig. 7–8f. The equation

$$60R_A - 40w(36) - 0.16(60)(30) + M_p = 0$$

gives

$$R_A = 24(3.366) + 4.80 - 616(3)/60 = 54.8 \text{ kips}$$

The bending moment at E is

$$M_E = 54.8(24) - 0.16(24)(12) = 1269 \text{ ft-kips}$$

which is less than 1848; therefore, a hinge does not form at E, the moment in the beam is nowhere greater than M_p, and the ultimate load is

$$w = \underline{3.37 \text{ kips/ft}} \qquad \text{Ans.}$$

Some economy in the design could be effected by reducing the size of the beam in the left span to such a section that a hinge might form at E; that is, a beam section could be selected for which the plastic moment is at least 1269 ft-kips. If a table of plastic section moduli were available,[2] the section could be determined from the expression $M_p = Z_p \sigma_{yp}$, where Z_p is the plastic section modulus.[3] If such a table is not available, the following method may be used. It has been shown[4] that the ratio Z_p/Z (where Z is the elastic section modulus) for wide flange and American standard sections ranges from 1.10 to 1.23, with 1.14 as the average for all wide flange sections and 1.18 as the average for American standard sections. Using the average value of 1.14 gives

$$Z = \frac{Z_p}{1.14} = \frac{M_E}{1.14\sigma_{yp}} = \frac{1269(12)}{1.14(36)} = 371 \text{ in.}^3$$

From App. C, a W33×130 with a section modulus of 404.8 appears to be a likely choice, and it will be checked as follows. When the cross-sectional dimensions given in App. C are used, the plastic moment is

$$\Big[\overbrace{11.51(0.855)(16.12)}^{\text{one flange}} + \overbrace{15.69(0.580)(15.69/2)}^{\text{one-half web}}\Big](2)(36/12) = 1379 \text{ ft-kips}$$

which is more than adequate. Therefore, one could splice the W33×130 to the W36×160 at some section between B and E. Theoretically, the splice could be made where the bending moment in span AB has a value of -1379 ft-kips (which is approximately 5.3 ft left of B). Practically, to

[2]See AISC *Manual of Steel Construction* and BCSA-CONSTRADO *Structural Steelwork Handbook* for standard metric sections.

[3]The plastic modulus is the arithmetic sum of the absolute values of the first moments, with respect to the neutral axis, of the cross-sectional areas on each side of the neutral axis.

[4]See Fig. 7.1 of *Plastic Design of Steel Frames*, L. S. Beedle, Wiley, New York, 1958.

avoid splicing at a section of large moment, the joint might be located some distance to the left, the distance depending on the judgement of the designer or a pertinent specification. The resulting beam will develop hinges at B and D and will just begin to yield at E under a load of 3.37 kips/ft.

The maximum value for w, based on elastic analysis and with the assumption that the I_{AB} for the full length of 60 ft is $(7/10)I_{BC}$, is found to be 2.76 kips/ft; thus, the plastic analysis for this problem gives a loading approximately 22 percent higher than that obtained from elastic analysis. Part of this increase is due to the plastic moment being greater (about 14 percent) than the elastic moment, and part is due to the increase in load necessary to cause formation of the second plastic hinge. These loads will be considered failure loads that, when divided by an appropriate factor of safety,[5] will yield working loads.

The discussion presented in this article is very limited in scope and hence does no more than present some of the fundamental concepts of plastic design involving an ideal elastoplastic material. A complete analysis would involve the rotation capacity of the beam sections used, to ascertain if the first hinge to form will sustain sufficient plastic deformation to permit the formation of successive hinges. Furthermore, shearing stresses or local buckling may be quite significant for some designs. These and other design problems are beyond the scope of this book.

PROBLEMS

Note. The following problems are to be solved by plastic analysis, and for simplification the weight of the beam is to be neglected. All beams are of structural steel having a yield point and proportional limit of 36 ksi or 250 MN/m². For wide flange or universal beam design, use $Z_p = 1.14\,Z$.

7–72* A two-span continuous beam with simple supports is subjected to a uniformly distributed load of 1.5 kips/ft on both spans. The span lengths are 30 and 40 ft, respectively, and the beam is to have a constant section throughout its length. Determine the lightest wide flange beam required if a factor of safety of 2 is specified.

7–73* A UB610×113 ($Z_p = 3283$ cm³) is used as a two-span, simply supported continuous beam with spans of 8 and 14 m, respectively. The beam carries uniformly distributed loads of $2w$ kN/m on the 8-m span and w kN/m on the 14-m span. Determine the ultimate load w for the beam.

[5]Here, the factor of safety is defined as the ratio of the ultimate (failure) load to the computed (or specified) applied load. This definition, which differs from that of Art. 3–2, will be used in all plastic analysis problems in this book.

7–74 The beam of Prob. 7–73 will develop a plastic hinge at the intermediate support when the value of w is 48.8 kN/m. For this loading, determine the lightest universal beam required for the 8-m span, and locate the theoretical splice point.

7–75 The beam of Fig. P7–75 has a span L of 20 ft; it is to carry a load of 2 kips/ft with a factor of safety of 2. Determine the lightest wide flange section required.

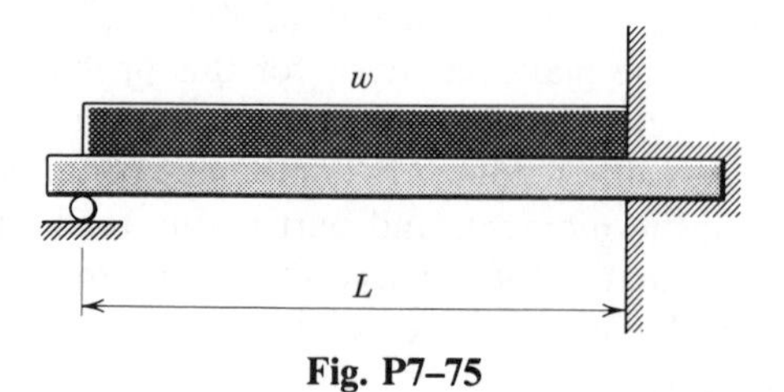

Fig. P7–75

7–76* Determine the magnitude of the ultimate load w, in N/m, on the beam of Fig. P7–75 when the section is a UB 305×48 ($Z_p = 704.9$ cm^3) and the span L is 8 m.

7–77* The beam of Fig. P7–77 is a UB 610×113 ($Z_p = 3283$ cm^3). Determine the ultimate load w.

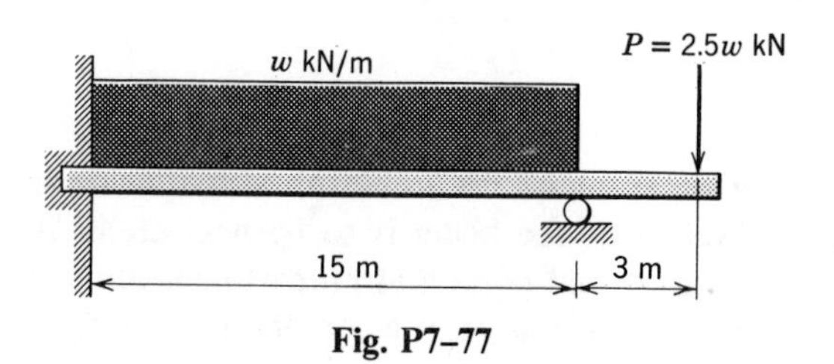

Fig. P7–77

7–78* The continuous beam of Fig. P7–78 is a W12×27 ($Z_p = 38$ in.^{3}). Determine the ultimate load w. Hint: assume three plastic hinges in span BC.

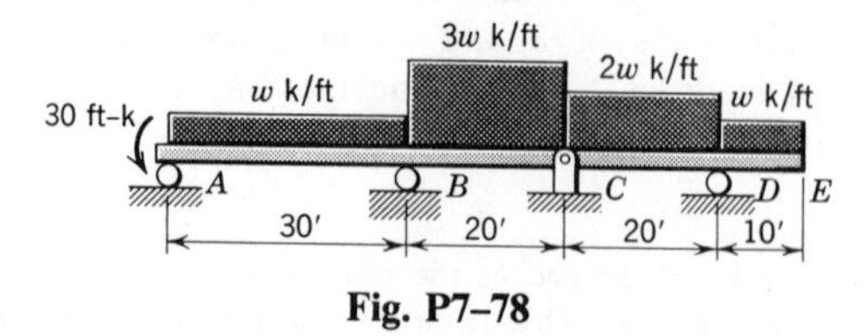

Fig. P7–78

7–79 Under a load w of 1.52 kips/ft applied as shown in Fig. P7–78, a plastic hinge forms at C. Determine the lightest wide flange section required for the portion of the beam from near C to E. Locate the theoretical splice point.

7–7 CLOSURE

In this chapter were presented four methods of elastic analysis and one method of plastic analysis of statically indeterminate beams. When the analysis is to be based on elastic action, the choice of method will be governed by the particular problem and the analyst's familiarity with certain of the four methods presented. As a possible aid to selection of the method, the following outline is submitted:

1. The integration method is the fundamental method and may be used to solve any problem although its application may sometimes be laborious. This method requires that, in addition to the boundary conditions necessary for the determination of constants of integration, there be additional boundary (or matching) conditions equal in number to the difference between the number of unknowns and the number of independent equations of equilibrium.

2. The area-moment method provides a neat and somewhat less laborious technique to supplement or replace the integration approach. The method requires the application of an area-moment theorem for each unknown in excess of the number of independent equations of equilibrium. The area-moment equations to be written will be dictated by the geometry of the deflected beam. If the expressions for the areas and centroids of the moment diagrams are not known, the areas and their first moments may be obtained by integration. The method is applicable to any problem and is readily applied to a beam with variable moment of inertia by drawing an M/I diagram and subdividing it into finite areas.

3. For many problems the superposition method provides the most rapid solution. The application of this method depends on the availability of a summary (or a memorization) of slope and deflection expressions for various spans and loadings and, except for the simpler cases, depends on the ingenuity of the analyst.

4. The energy method is preferred by many and is frequently much shorter than other methods. The method presented in this book requires one or more applications of Castigliano's theorem (Eq. 6–5*a* or 6–5*b*) at points where the deflections or slopes are known. The use of singularity functions to provide a single moment equation for the entire span is frequently advantageous.

8 Combined Static Loading

8–1 INTRODUCTION

The stresses and strains produced by three fundamental types of loads (centric, torsional, and flexural) have been analyzed in the preceding chapters. Combinations of centric and torsional loads were also presented in Art. 4–5. Many machine and structural elements are subjected to a combination of any two or all three of these types of loads, and a procedure for calculating the stresses on a given plane at a point is required. One method is to replace the given force system by an equivalent system of forces and couples, each of which is selected in such a manner that the stresses produced by it can be calculated by methods already developed. The composite effect can be obtained by the principle of superposition, as explained in Art. 3–3, provided that the combined stresses do not exceed the proportional limit. Various combinations of loads that can be analyzed in this manner are discussed in the following articles.

8–2 CENTRIC AND FLEXURAL LOADS COMBINED

A common example of a centric load combined with a bending load is a structural or machine part with a force acting parallel to but not along the axis of the member; for example, the vertical portion AB of the loading hook of Fig. 8–1. Such an eccentric load can be replaced by a centric load (a force acting along the centroidal axis of the part AB) and one or more couples. The stresses resulting from the axial load were discussed in Chap. 3. Procedures developed in Chap. 5 can be applied to evaluate the stresses produced by the couples.

The normal stress at any point on a transverse plane through AB is the algebraic sum of the normal stresses due to the centric load and that due to the bending load (the couples). Since the transverse shear is zero, this stress

Fig. 8–1 Five-ton Littell Coil Loading Hook. Courtesy: Steel Founders Society of America.

will be a principal stress. The procedure is illustrated in the following example.

Example 8–1 The cast-iron frame of a small press is shaped as shown in Fig. 8–2*a*. Figure 8–2*b* represents the cross section *a–a*, and the moment of inertia of the area is 1000/3 in.^{4} with respect to the centroidal axis *c–c*. For a load P of 16 kips, and assuming elastic action, determine:

(a) The normal stress distribution on section *a–a*.
(b) The principal stresses for the critical points on section *a–a*.

Solution (a) When the free-body diagram of the top portion of the frame was drawn, two equal opposite collinear forces P through the centroid of the section *a–a* were added as shown in Fig. 8–2*c*. The force P_1 is a centric load equal to the resisting force T on section *a–a*. The other two forces P and P_2 constitute an $18P$-in-kip couple equal to the resisting moment M on section *a–a*.

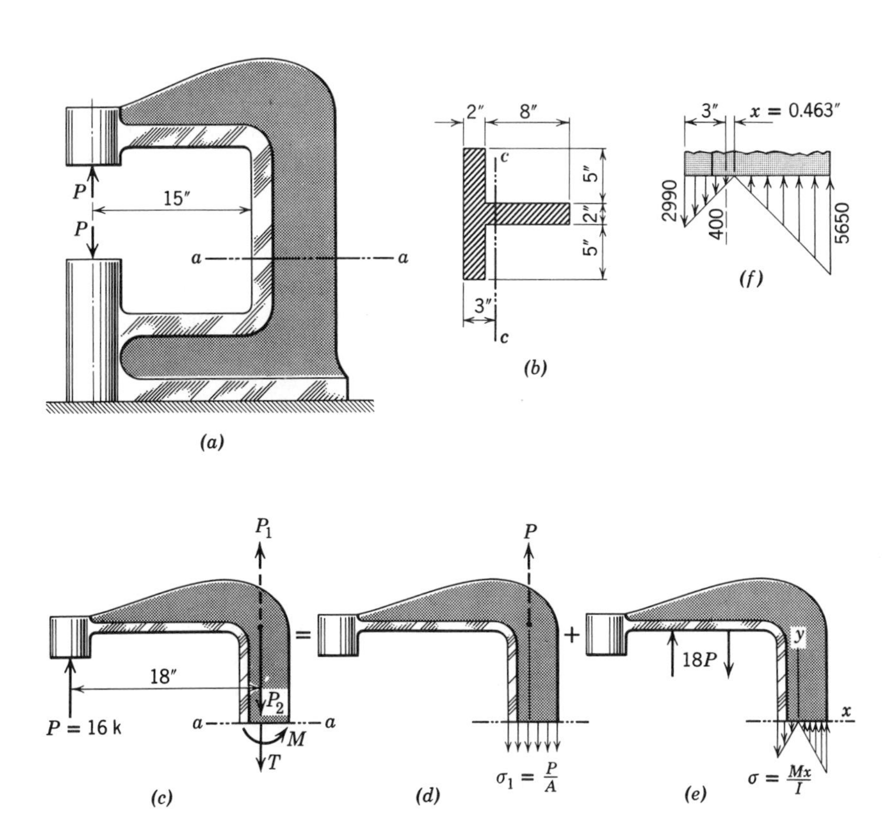

Fig. 8–2

The axial stress uniformly distributed over section *a–a*, as shown in Fig. 8–2*d*, is

$$\sigma_1 = \frac{P}{A} = \frac{16{,}000}{40} = 400\,\text{psi}\,T$$

The maximum tensile fiber stress occurs at the left edge of the section *a–a* and is

$$\sigma_2 = \frac{Mc}{I} = \frac{16{,}000(18)(3)}{1000/3} = 2592\,\text{psi}\,T$$

The maximum compressive fiber stress occurs at the right edge of the section *a–a* and is

$$\sigma_3 = \frac{Mc}{I} = \frac{16{,}000(18)(7)}{1000/3} = 6048\,\text{psi}\,C$$

The distribution of fiber stresses is shown in Fig. 8–2*e*.

When the principle of superposition is applied, the total normal stress at any point on section *a–a* is the algebraic sum of the component stresses at the point. When the stress distributions of Fig. 8–2*d* and *e* are added, the resultant stress distribution for section *a–a* is as shown in Fig. 8–2*f*, which is the answer for part (a).

(b) With no shearing stresses on section *a–a*, the normal stresses are principal stresses, and the critical points, as observed from the stress distribution of Fig. 8–2*f*, are the left and right edges of the section. The principal stresses for the right edge are

$$\sigma_p = \underline{5650\,\text{psi}\,C} \quad \text{and} \quad \underline{0} \qquad \text{Ans.}$$

The principal stresses for the left edge are

$$\sigma_p = \underline{0} \quad \text{and} \quad \underline{2990\,\text{psi}\,T} \qquad \text{Ans.}$$

Although not requested in this example, the location of the neutral axis (line of zero stress) can be determined as the place where the compressive fiber stress is 400 psi because this stress will just balance the axial tensile stress of 400 psi. The distance on Fig. 8–2*f* can be obtained by writing

$$400 = \frac{16{,}000(18)x}{1000/3}$$

from which

$$x = 0.463 \text{ in.}$$

to the right of the centroidal axis of the cross section.

Example 8–2 A gray cast-iron compression member is subjected to a vertical load P, as shown in Fig. 8–3a. The allowable normal stresses on plane $ABCD$ are 20,000 psi C and 5000 psi T. Determine the maximum allowable load P.

Solution The specified stresses indicate elastic action. When two vertical forces P (labeled P_1 and P_2 for clarity) of opposite sense are introduced at the center of the top section, $EFGH$, the eccentric force P is replaced by a centric force P_2 and a couple PP_1, as shown in Fig. 8–3b.

The stresses produced by the centric load are easily calculated; however, the couple does not act in a plane of symmetry of the member, and the flexural stresses produced are not readily computed for this orientation. This difficulty can be eliminated by replacing the resultant couple PP_1 by its components in or parallel to planes of symmetry of the compression member. When another pair of vertical forces equal to P (labeled P_3 and P_4 for clarity) of opposite sense are introduced at the center of the edge EH, the couple PP_1 is replaced by two component couples PP_3 and P_4P_1, as shown in Fig. 8–3c. The component couple P_4P_1 acts in a plane of symmetry, and the flexural stresses produced are given by the flexure formula. Since no transverse shearing stresses occur to cause twisting (see Art. 5–13), the flexural stress due to couple PP_3 can also be determined by means of the flexure formula, even though PP_3 is not in a plane of symmetry.

The centric load P_2 gives a uniformly distributed compressive stress on plane $ABCD$, and the value is

$$\sigma_1 = \frac{P}{A} = \frac{P}{28.8} \text{ psi}$$

The couple P_4P_1 gives a compressive stress along the edge CD of plane $ABCD$ and an equal tensile stress along the edge AB. These stresses are

$$\sigma_2 = \frac{Mc}{I} = \frac{2.4P(2.4)}{(6)(4.8)^3/12} = \frac{P}{9.6} \text{ psi}$$

The couple PP_3 gives a compressive stress along edge BC of plane

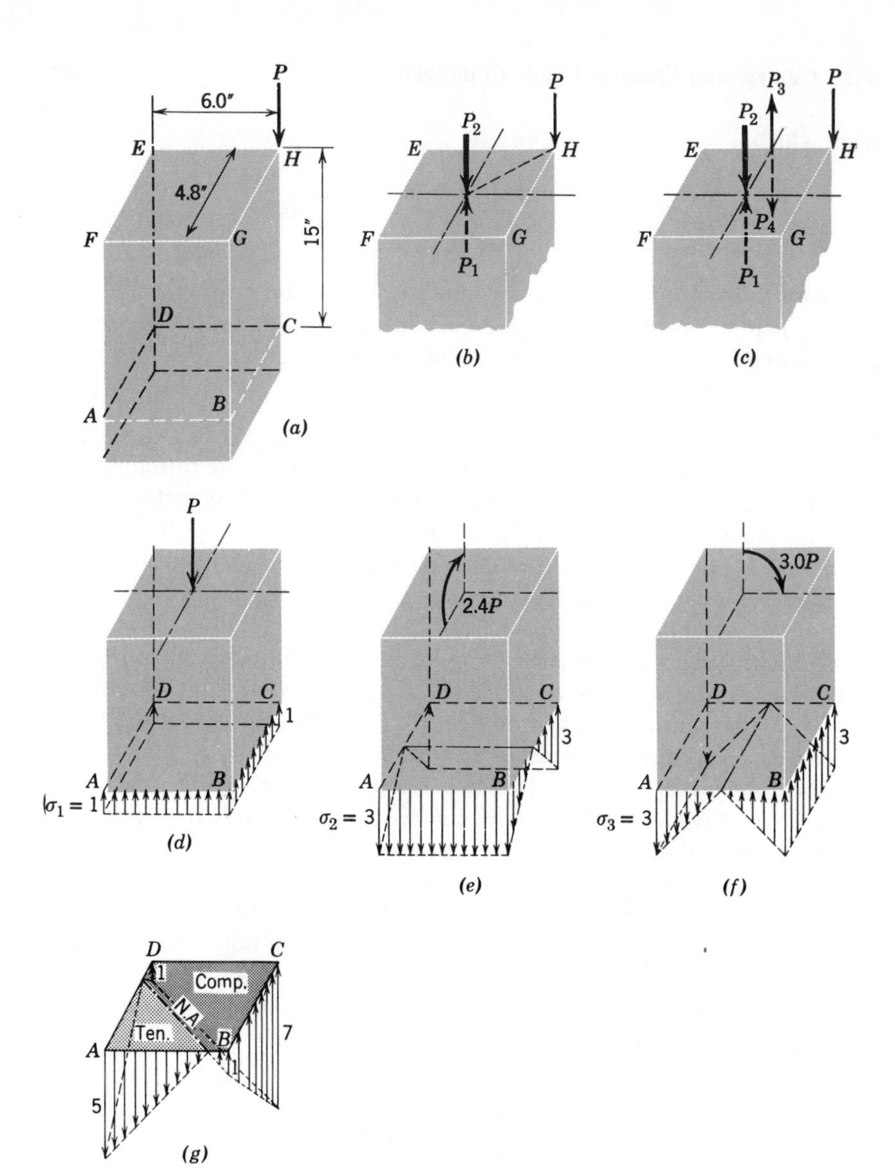

P
6.0″
E
H
4.8″
15″
F
G
D
C
A
B
(a)
P
P2
E
H
F
G
P1
(b)
P3
P
P2
E
H
P4
F
G
P1
(c)
P
D
C
1
A
B
σ1 = 1
(d)
2.4P
D
C
3
A
B
σ2 = 3
(e)
3.0P
D
C
3
A
B
σ3 = 3
(f)
D
C
1
Comp.
N.A.
Ten.
7
A
B
1
5
(g)

Fig. 8–3

$ABCD$ and an equal tensile stress along the edge AD equal to

$$\sigma_3 = \frac{Mc}{I} = \frac{3P(3)}{(4.8)6^3/12} = \frac{P}{9.6} \text{ psi}$$

The component stresses are shown in the three stress distribution diagrams of Fig. 8–3*d*, *e*, *f*, and from the diagrams one observes that the maximum compressive stress occurs at C and is

$$\sigma_C = \frac{P}{28.8} + \frac{P}{9.6} + \frac{P}{9.6} = \frac{7P}{28.8} \text{ psi}$$

and that the maximum tensile stress occurs at A and is

$$\sigma_T = \frac{P}{9.6} + \frac{P}{9.6} - \frac{P}{28.8} = \frac{5P}{28.8} \text{ psi}$$

When the allowable stresses are substituted in the above expressions, the values of P become

$$P_C = \frac{28.8}{7}(20{,}000) = 82{,}300 \text{ lb}$$

and

$$P_T = \frac{28.8}{5}(5000) = 28{,}800 \text{ lb}$$

Since P_T is less than P_C, the maximum allowable load is

$$P = \underline{28{,}800 \text{ lb}} \qquad \text{Ans.}$$

The distribution of the combined stresses is shown in Fig. 8–3*g* as an aid to the visualization of the stress pattern due to the given loading.

Another example involving combined centric and flexural loads is a beam with a force or a component of a load acting parallel to the axis of the beam. In this case the beam shearing stress is not zero everywhere and may be an important factor in some instances.

This type of combined load is complicated by having the deflection of the beam and the amount of bending moment dependent on each other. As a beam deflects, the centroid of a specific cross section of the beam moves relative to the line of action of any longitudinal force. This movement alters the moment or couple applied to the beam due to the eccentricity of the longitudinal force, as further explained in Exam. 8–3.

For this reason a complete solution suggests the use of successive approximations.[1] Fortunately, the change in stress due to the deflection of the beam can normally be neglected as shown by the following example.

Example 8–3 A 5-ft long timber cantilever beam with the cross section of Fig. 8–4*a* is loaded elastically, as shown in Fig. 8–4*b*. The modulus of elasticity of the timber is 1200 ksi. Determine:

(a) The normal stress distribution on a vertical plane through the beam at the wall.

(b) The principal stresses and maximum shearing stress at a point 2.50 in. above the centroidal axis on a cross section at the wall.

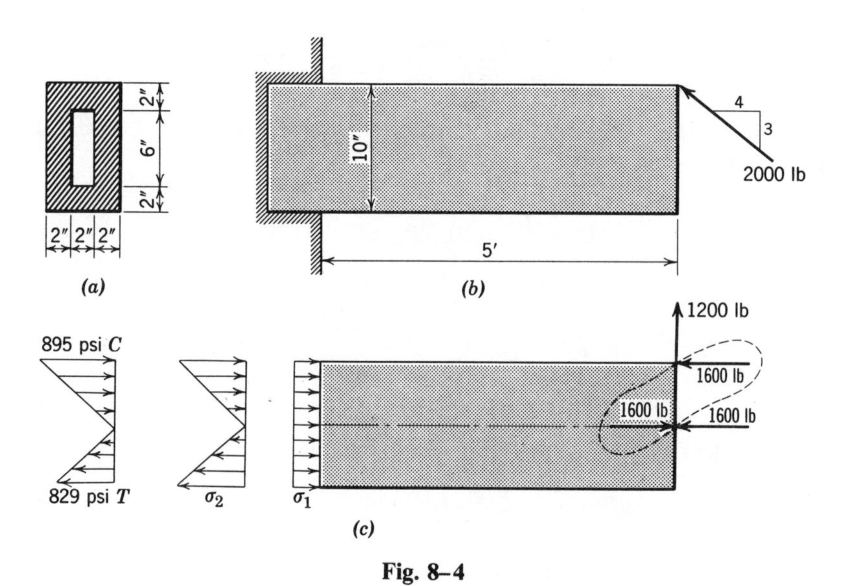

Fig. 8–4

Solution The 2000-lb load can be replaced by a 1200-lb vertical force, a 1600-lb horizontal force through the centroid of the cross section of the beam at the wall, and a couple with 1600-lb horizontal forces, as shown in Fig. 8–4*c*.

(a) The axial compressive stress produced by the 1600-lb horizontal

[1]It is recognized that the differential equation expressing the moment as a function of both x and y has a solution (see Art. 9–8). However, for many practical problems, the method of successive approximations is more efficient.

force on the section at the wall is

$$\sigma_1 = \frac{P}{A} = \frac{1600}{60-12} = 33.3\,\text{psi}\,C$$

as shown in Fig. 8–4*c*.

If the deflection at the right end of the beam is neglected, the bending moment at the support is $+1200(60)+1600(5)$, which is equal to $+80{,}000$ in-lb. If the deflection of the beam is appreciable, the couple at the free end will be increased since, for the load to be centric, the 1600-lb force must pass through the centroid of the resisting section (which is at the support for this problem). The deflection of the right end due to the 1200-lb load as given by case 1 of App. D is

$$y_1 = \frac{PL^3}{3EI} = \frac{1200(60^3)}{3(12)(10^5)\left[6(10^3)-2(6^3)\right]/12} = \frac{9}{58}\text{ in. upward}$$

The deflection at the right end due to the 8000-in-lb couple as given by case 4 of App. D is

$$y_2 = \frac{ML^2}{2EI} = \frac{1600(5)(60^2)}{2(12)(10^5)(464)} = \frac{1.5}{58}\text{ in. upward}$$

The sum of these deflections is 0.1810 in. upward; with this deflection included, the moment of the couple becomes 1600(5.18) or 8288, thus making the total moment 80,290 in-lb or 0.36 percent higher, which is so slight an increase that it will be neglected. Hence, the flexural stresses at the wall are

$$\sigma_2 = \frac{Mc}{I} = \frac{80{,}000(5)}{464} = 862\,\text{psi}\,T\text{ and }C$$

The distributions of the component stresses are shown in Fig. 8–4*c*. Also shown is the distribution of the combined stresses, which is the answer to part (a).

(b) The point specified is halfway from the centroidal axis to the outermost fiber; therefore, the normal stress on the transverse plane is

$$\sigma = \frac{862}{2} + 33 = 464\,\text{psi}\,C$$

The horizontal and vertical shearing stresses at the specified point are

$$\tau = \frac{VQ}{It} = \frac{1200[12(4)+2(2.75)]}{464(4)} = 34.6\,\text{psi}$$

From Eqs. 1–7 and 1–9, the principal stresses and maximum shearing stress at the specified point are

$$\begin{aligned}\sigma_p &= -232 \mp \sqrt{232^2 + 34.6^2} \\ &= -232 \mp 235 \\ &= \underline{467\,\text{psi}\,C} \quad \text{and} \quad \underline{3\,\text{psi}\,T}\end{aligned} \qquad \text{Ans.}$$

and

$$\tau_{\max} = \underline{235\,\text{psi}} \qquad \text{Ans.}$$

PROBLEMS

Note. In the following problems, the effect of deflection on the eccentricity is to be neglected and elastic action is to be assumed, unless otherwise noted.

8–1* For the member in Fig. P8–1, determine the vertical normal stresses at points *A*, *B*, *C*, and *D*.

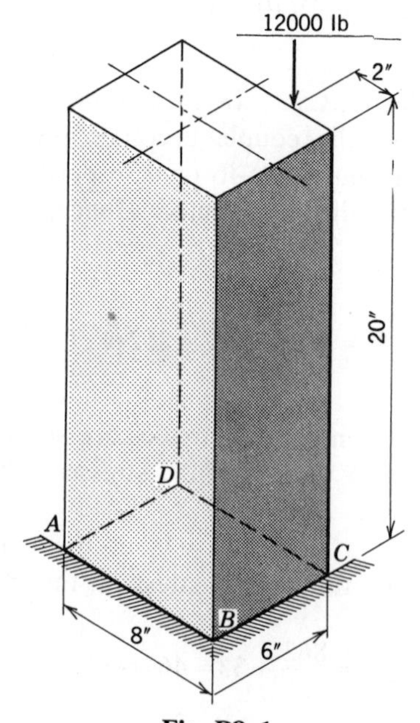

Fig. P8–1

8-2* For the member in Fig. P8-2, determine the distribution of normal stresses on section *AB* and show it on a sketch.

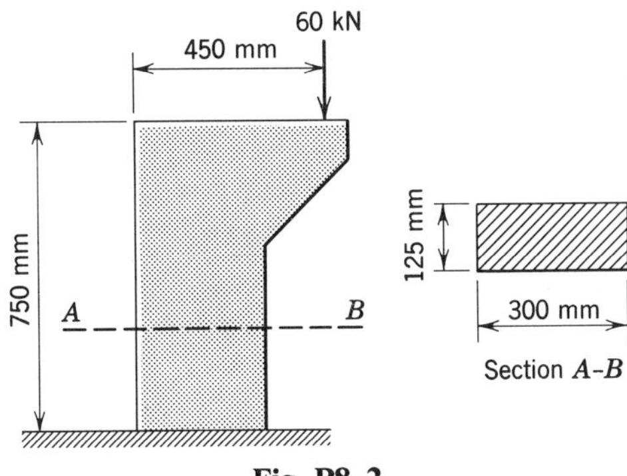

Fig. P8-2

8-3 Determine, and show on a sketch, the principal and maximum shearing stresses at point *A* of the cantilever beam in Fig. P8-3.

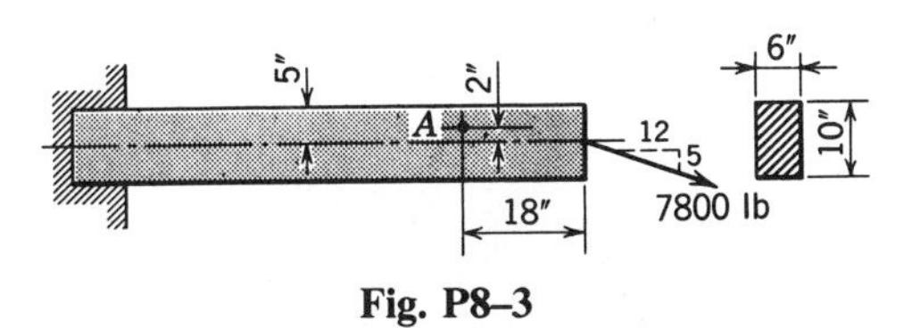

Fig. P8-3

8-4 A 75-mm wide by 150-mm deep cantilever beam is loaded as shown in Fig. P8-4. Determine, and show on a sketch, the principal and maximum shearing stresses at point *A*.

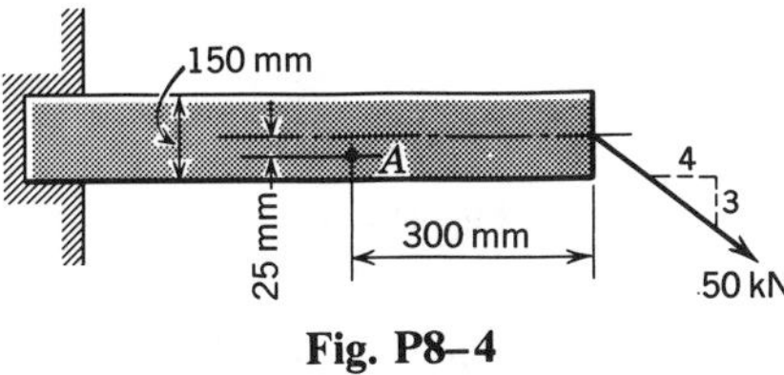

Fig. P8-4

8–5* Determine the maximum normal stress on a transverse plane in the straight portion of the structure in Fig. P8–5, and indicate clearly where it occurs. The member is braced in the plane perpendicular to the plane of symmetry.

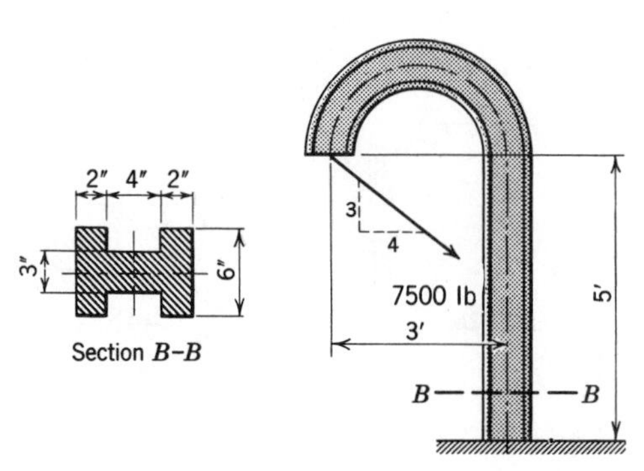

Fig. P8–5

8–6* The straight portion AB of the cast-steel machine part of Fig. P8–6 has a hollow rectangular cross section with outside dimensions 300×250 mm and walls 50 mm thick. For the indicated loading, determine the maximum normal stress on a transverse plane within the portion AB.

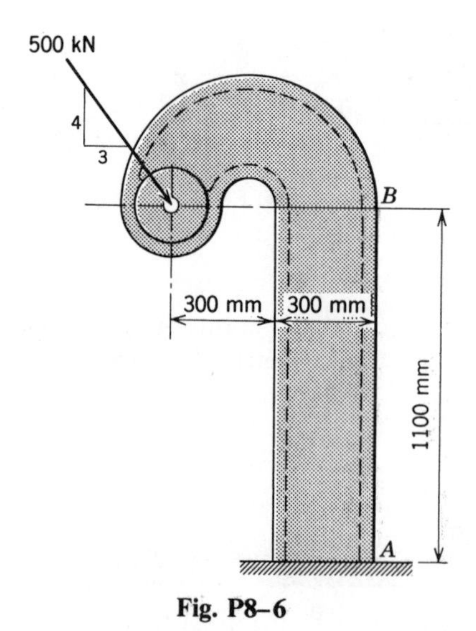

Fig. P8–6

8–7 The cast-iron frame of a small press is shaped as shown in Fig. P8–7. The allowable stresses on plane A–A are 3500 psi T and 15,000 psi C. Determine the capacity P of the press.

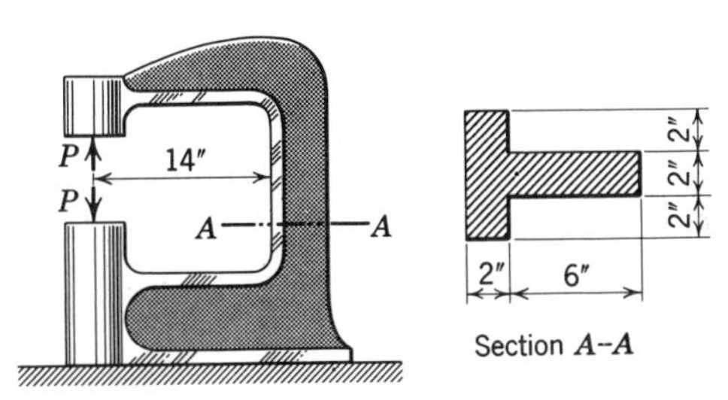

Fig. P8–7

8–8 The estimated maximum total force to be exerted on the C-clamp in Fig. P8–8 is 2500 N. If the normal stress on section A–A is not to exceed 140 MN/m^2, determine the minimum dimension h of the cross section indicated.

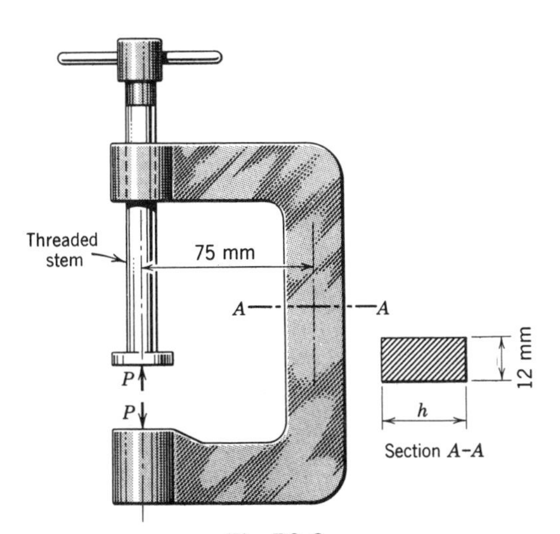

Fig. P8–8

8–9* The cross section of the aluminum beam ($E = 10^7$ psi) of Fig. P8–9 is a rectangle 4 in. wide by 6 in. deep. Determine the maximum normal stress on a vertical section at the wall:

(a) Neglecting the deflection of the member.

(b) Including the approximate increase in stress due to the deflection of the 50-in. horizontal portion of the member.

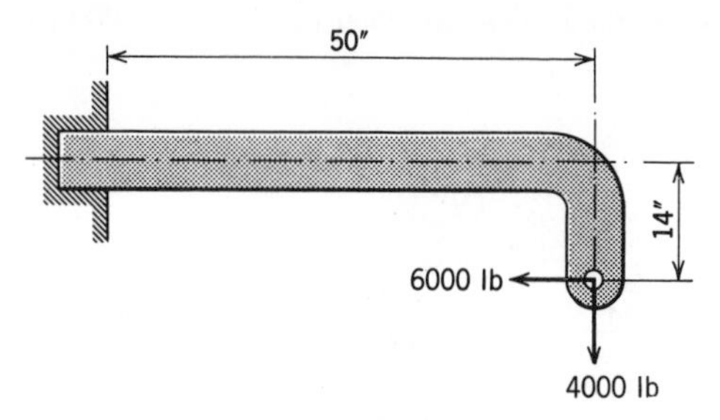

Fig. P8–9

8–10 For the member in Fig. P8–10, determine the maximum allowable value of P for a limiting tensile stress of 35 MN/m^2 and a limiting compressive stress of 80 MN/m^2 on plane AB.

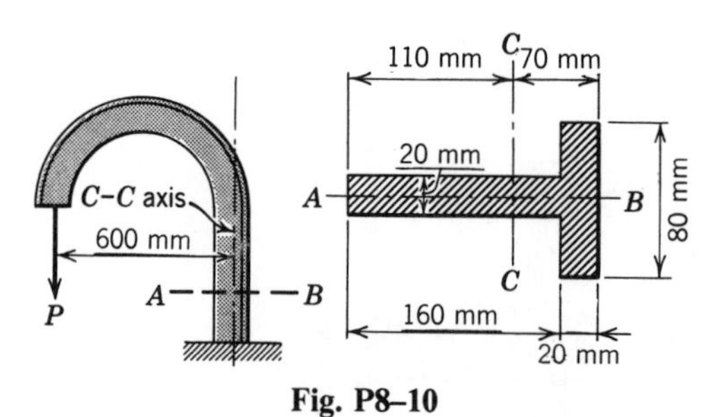

Fig. P8–10

8–11 The member in Fig. P8–11 has a solid circular cross section of radius r. Determine the minimum permissible radius of the member if the maximum tensile stress on section a–a is not to exceed 16,000 psi.

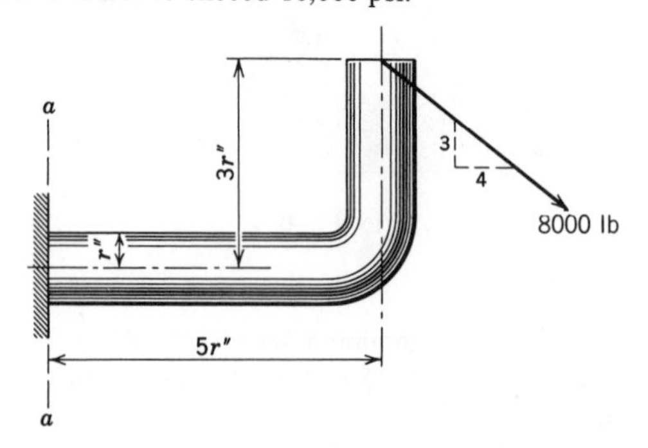

Fig. P8–11

8–12* A short post is loaded by a vertical force P of 25 kN and a horizontal force H of 5 kN, as shown in Fig. P8–12. Determine the vertical normal stresses at the corners A, B, C, and D and locate the line of zero normal stress in plane $ABCD$.

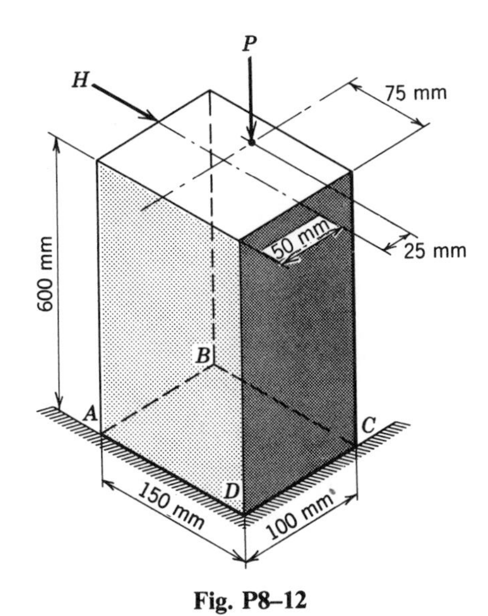

Fig. P8–12

8–13* For the beam of Fig. P8–13, determine the maximum normal stress on plane a–a.

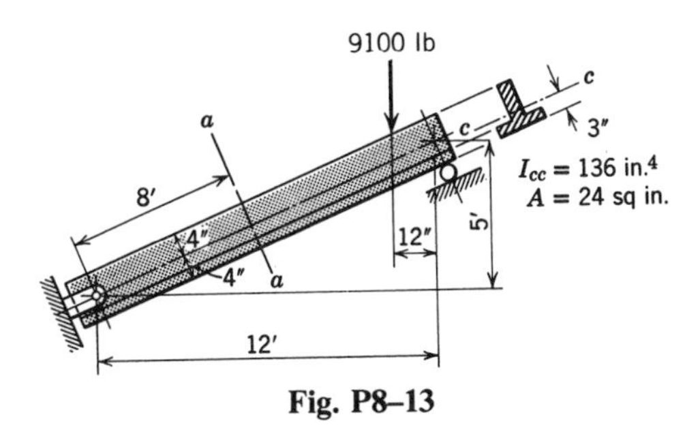

Fig. P8–13

8–14* The beam in Fig. P8–14 has an 80×200-mm rectangular cross section and is loaded in the plane of symmetry. Determine, and show on a sketch, the principal and maximum shearing stresses at point A.

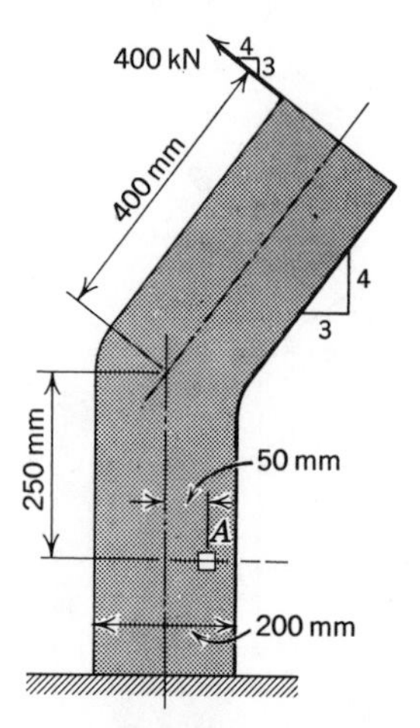

Fig. P8–14

8–15* The force P in Fig. P8–15 results in a maximum compressive stress on plane A–A of 20 ksi. Determine the corresponding average longitudinal shearing stress at the centroidal axis for this load.

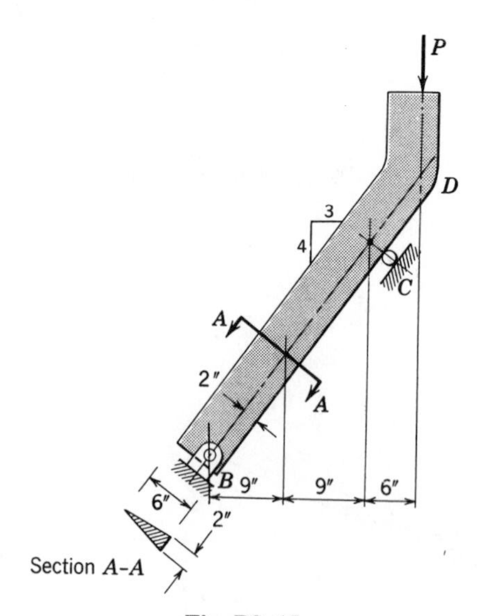

Fig. P8–15

8–16* The following data apply to the cantilever beam of Fig. P8–16: $P=25$ kN, $T=10$ kN, $e=150$ mm, and $a=600$ mm. Determine, and show on a sketch, the principal and maximum shearing stresses at point A (in the web).

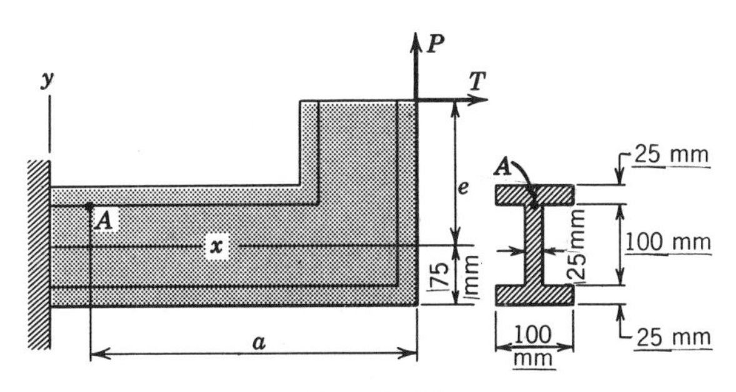

Fig. P8–16

8–17 In Fig. P8–16, $e=150$ mm, $a=500$ mm, and the stresses at point A (in the web) are $\sigma_x=35\ \text{MN/m}^2\ C$ and $\tau_{xy}=7\ \text{MN/m}^2$.

(a) Determine the magnitudes of P and T.

(b) Determine, and show on a sketch, the principal and maximum shearing stresses at point A.

8–18* The machine element of Fig. P8–18 is loaded in a plane of symmetry. Determine, and show on a sketch, the principal and maximum shearing stresses at point A, which is in the web at the junction of flange and web. Neglect stress concentrations.

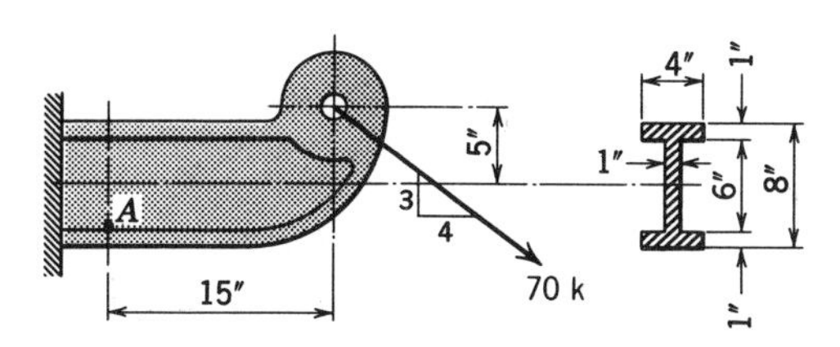

Fig. P8–18

8–19* Determine, and show on a sketch, the principal stresses and the maximum shearing stress at point A in the roof beam in Fig. P8-19.

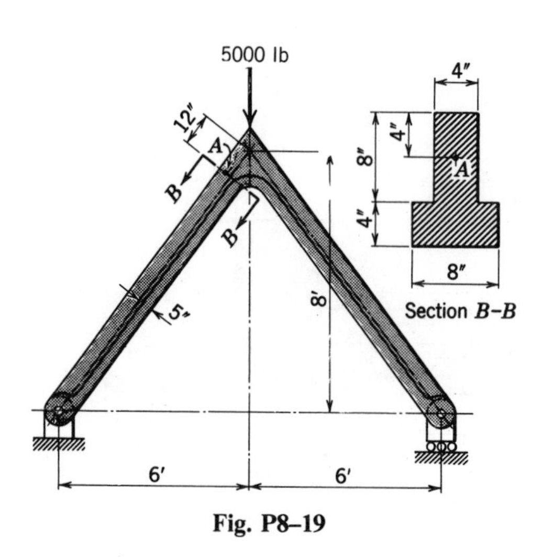

Fig. P8–19

8–20* The frame in Fig. P8–20 is constructed of 100×100-mm timbers. Determine, and show on sketches, the principal and maximum shearing stresses at points G and H.

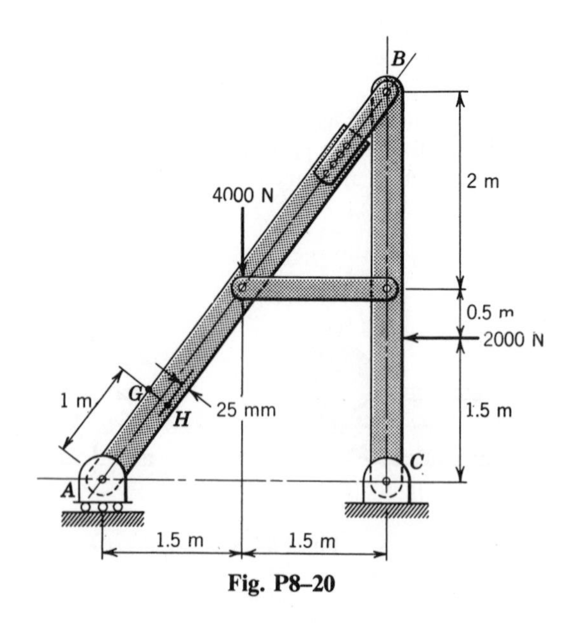

Fig. P8–20

8–21* A crane hook has the dimensions and cross section shown in Fig. P8–21. The allowable stresses on plane A–A are 15,000 psi T and 20,000 psi C. Determine the capacity P of the hook.

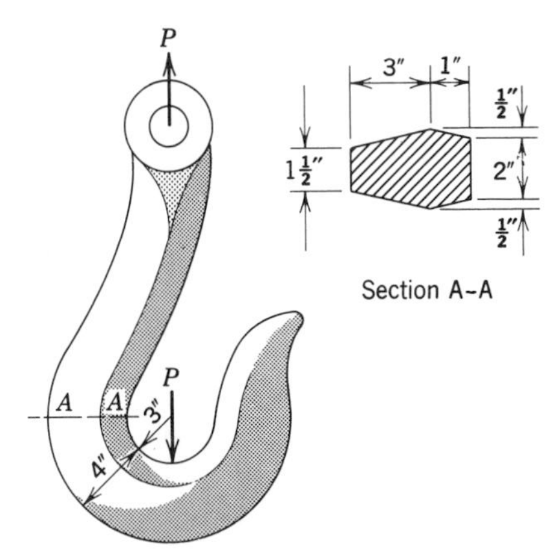

Fig. P8–21

8–22* A split ring has the dimensions and cross section shown in Fig. P8–22. Determine the maximum load P that can be applied to the ring if the normal stresses on plane A–A are not to exceed 100 MN/m² T or 150 MN/m² C.

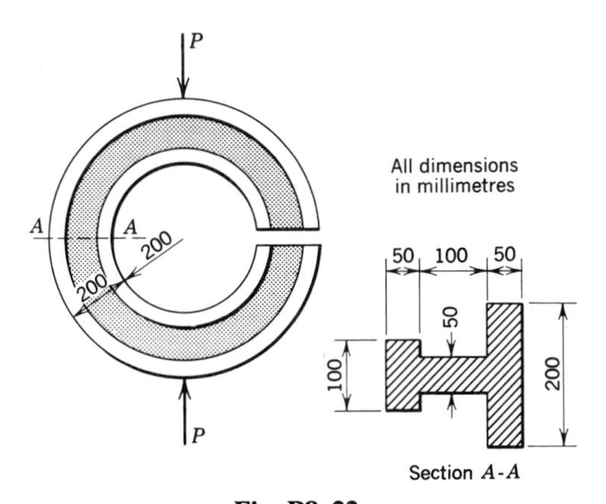

Fig. P8–22

8–23* The short rectangular post in Fig. P8–23 is made of an elastoplastic material with a yield point of σ_y.

(a) Locate the line of zero stress in the base of the post when the load P causes the maximum normal stress on the base to equal the yield point.

(b) Determine the maximum tensile stress on the base in terms of σ_y for the load in part (a).

(c) If the load P is increased beyond the value in part (a), a portion of the base, indicated by the distance c, will be plastically deformed. If the distance c is given by kd (where $0 \leqslant k < 1$), determine the value of k when the maximum tensile stress on the base is $2\sigma_y/3$.

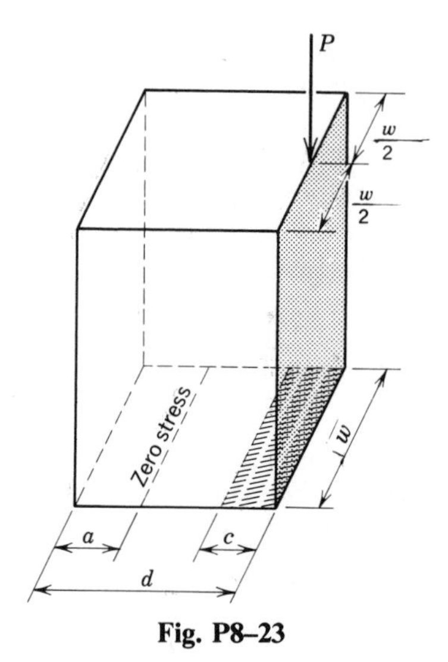

Fig. P8–23

8–24 Derive an expression for the distance a in Fig. P8–23 (the distance to the line of zero stress) in terms of the width d and the ratio $k = c/d$. Assume that the material is elastoplastic and that the load is sufficient to produce plastic deformation in compression but not large enough to make the maximum tensile stress exceed σ_y.

8–25 Determine the value of k in Prob. 8–24 at which the maximum tensile stress in the base will equal the yield point.

8–3 CENTRIC, TORSIONAL, AND FLEXURAL LOADS COMBINED

In numerous industrial situations, machine members are subjected to all three general types of loads studied so far. When a flexural load is combined with torsional and centric loads, it is frequently difficult to locate the point or points where the most severe stresses occur. The following review or summary statements may be helpful in this regard.

The longitudinal and transverse shearing stresses are maximum at the centroidal axis of the section where V is maximum. The flexural stress is maximum at the greatest distance from the centroidal axis on the section where M is maximum. The torsional shearing stress is maximum at the surface of a shaft at the section where T is maximum. With these facts in mind, it is normally possible to locate one or more possible points of high stress. Even so, it may be necessary to calculate the various stresses at more than one point of a member before locating the most severely stressed point. For stresses below the proportional limit of the material, the superposition method can be used to combine stresses on any given plane at any specific point of a loaded member.

After the stresses on a pair of mutually perpendicular planes at a specific point are determined, the methods of Arts. 1–7 and 1–8 can be used to find the principal stresses and the maximum shearing stress at the point.

The following example illustrates the procedure for the solution of elastic combined stress problems.

Example 8–4 The 4-in. diameter shaft of Fig. 8–5*a* is subjected to the loads shown. Determine the maximum permissible value of P for allowable stresses of 16,000 psi T and 9000 psi shear.

Solution When two equal opposite forces P are introduced along line EF as in Fig. 8–5*b*, it is seen that the shaft is subjected to a couple of $20P$ in-lb clockwise looking toward the wall and a concentrated load P acting at E. In other words, the shaft acts as a cantilever beam with a concentrated load P at the right end, a torsion member subjected to a $20P$-in-lb torque, and an axially loaded member with a load of $48P$ lb.

The couple produces a torsional shearing stress at all surface points of the shaft of

$$\tau_{xt} = \frac{Tc}{J} = \frac{20P(2)}{\pi(2^4)/2} = \frac{5P}{\pi}$$

as shown in Fig. 8–5*c*.

The axial tensile stress at all points of the shaft is

$$\sigma_{x_1} = \frac{P}{A} = \frac{48P}{\pi(2^2)} = \frac{12P}{\pi}$$

as shown in Fig. 8–5*d*.

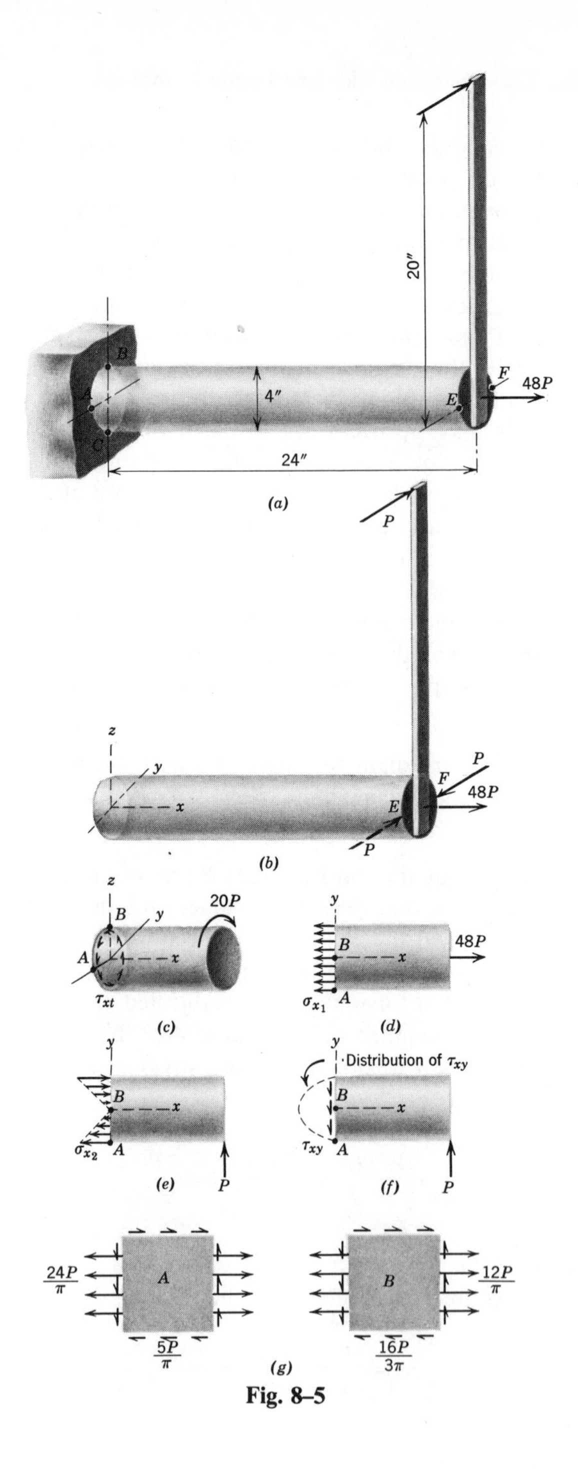
20″
B
A
C
4″
E
F
48P
24″
(a)
P
z
y
x
P
F
48P
E
P
(b)
z
B
y
20P
A
x
τ_{xt}
(c)
y
B
48P
x
σ_{x_1}
A
(d)
y
B
x
σ_{x_2}
A
(e)
P
y
Distribution of τ_{xy}
B
x
τ_{xy}
A
(f)
P
$\frac{24P}{\pi}$
A
$\frac{5P}{\pi}$
B
$\frac{12P}{\pi}$
$\frac{16P}{3\pi}$
(g)

Fig. 8–5

The cantilever beam load P produces a maximum tensile fiber stress at point A and an equal compressive stress on the opposite side of the shaft. The axial stress is tensile, and therefore the compressive flexural stress is not critical. The tensile fiber stress at A is (see Fig. 8–5e)

$$\sigma_{x_2} = \frac{Mc}{I} = \frac{24P(2)}{\pi(2^4)/4} = \frac{12P}{\pi}$$

The transverse shearing stress is equal to the longitudinal shearing stress, which is maximum on the xz plane including such surface points as B and C where the torsional shearing stress is also maximum. The maximum shearing stress is (see Fig. 8–5f)

$$\tau_{xy} = \frac{VQ}{It} = \frac{P\left[\frac{\pi(2^2)}{2}\right]\left[\frac{4(2)}{3\pi}\right]}{\frac{\pi(2^4)}{4}(4)} = \frac{P}{3\pi}$$

The torsional and transverse shearing stresses are in the same direction along the top of the shaft and in opposite directions along the bottom of the shaft. Therefore, any critical situation due to these two stresses will develop along the top of the shaft at some point such as B.

All four stresses are maximum at some point or points in the section of the shaft at the wall since the fiber stress is not maximum at any other section of the shaft. Therefore, a complete analysis for the member can be made on the section at the wall. A complete analysis including interior points of this section is beyond the scope of this book. However, a fairly good solution can be obtained by investigating the combined stresses at two surface points A and B. At A the fiber stress, axial stress, and torsional shearing stress are all maximum, whereas the transverse shearing stress is zero. At B the axial stress, torsional shearing stress, and transverse shearing stress are all maximum, whereas the fiber stress is zero. The fiber stress and transverse shearing stress are never maximum at the same point. Furthermore, the torsional shearing stress is never maximum at an interior point. Therefore, the chance of finding a more severe stress at an interior point is unlikely.

The stresses on orthogonal planes through A and B are as shown in Fig. 8–5g. When these values are substituted in Eq. 1–7 of Art. 1–7 and the resulting expressions compared, it is evident that the combination of stresses at A is more severe than that at B. The maximum stresses at A are

$$\sigma_A = \frac{12P}{\pi} + \sqrt{\left(\frac{12P}{\pi}\right)^2 + \left(\frac{5P}{\pi}\right)^2} = \frac{25P}{\pi}$$

and

$$\tau_A = \sqrt{\left(\frac{12P}{\pi}\right)^2 + \left(\frac{5P}{\pi}\right)^2} = \frac{13P}{\pi}$$

The allowable values of P from the allowable stresses and these relations are

$$16{,}000 = 25P_\sigma/\pi$$

or

$$P_\sigma = 640\pi = 2010 \text{ lb}$$

and

$$9000 = 13P_\tau/\pi$$

or

$$P_\tau = 9000\pi/13 = 2170 \text{ lb}$$

The smaller value controls and the maximum permissible value is

$$P = \underline{2010 \text{ lb}} \qquad \text{Ans.}$$

PROBLEMS

Note. In the following problems, stress concentrations and the effect of beam deflection on the eccentricity are to be neglected unless otherwise noted. Assume elastic action.

8–26* A solid shaft 4 in. in diameter is acted on by forces P and Q, as shown in Fig. P8–26. Determine, and show on a sketch, the principal stresses and the maximum shearing stress at point A on the surface of the shaft.

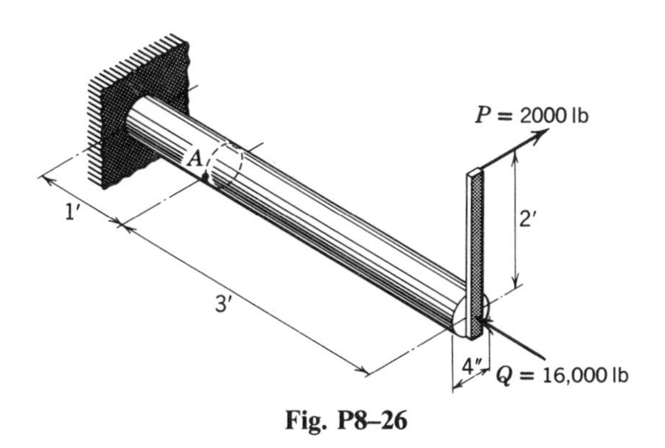

Fig. P8–26

8–27* A solid shaft 100 mm in diameter is acted on by forces P and Q, as shown in Fig. P8–27. Determine, and show on a sketch, the principal stresses and the maximum shearing stress at point A on the surface of the shaft.

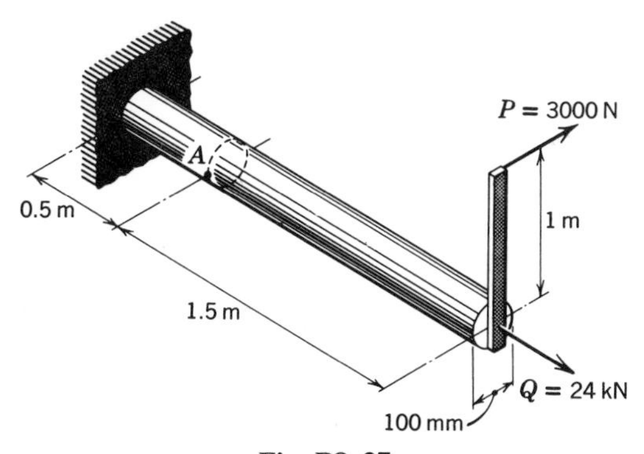

Fig. P8–27

8–28* A 2-in. diameter steel rod is loaded as shown in Fig. P8–28. Determine, and show on a sketch, the principal stresses and the maximum shearing stress at the top surface adjacent to the support.

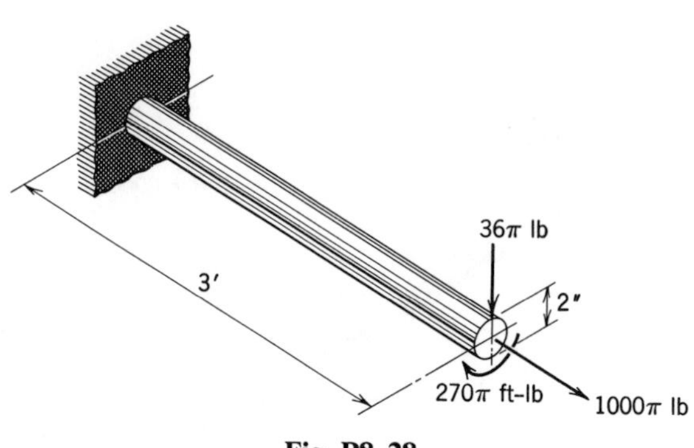

Fig. P8–28

8–29* Determine, and show on a sketch, the principal stresses and the maximum shearing stress at point P on the solid shaft in Fig. P8–29.

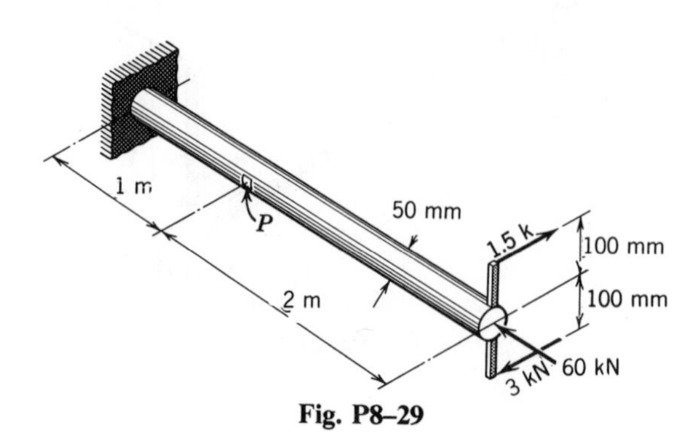

Fig. P8–29

8–30 The shaft in Fig. P8–30 is subjected to the loads shown. If the tensile and shearing stresses at point A (on the surface of the shaft) may not exceed 20,000 and 12,000 psi, respectively, determine the value of R.

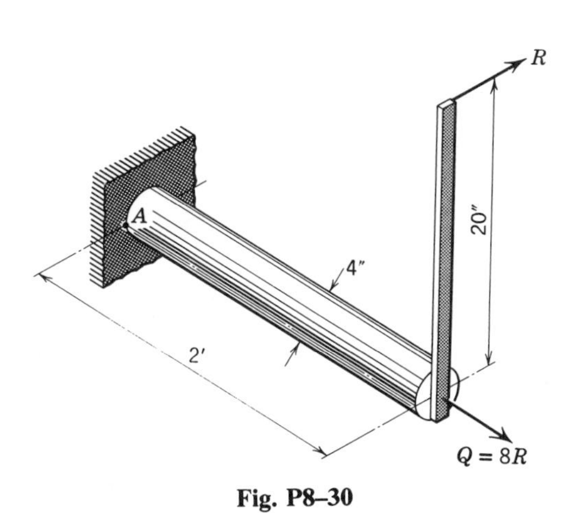

Fig. P8–30

8–31 Determine the minimum permissible diameter for the shaft of Fig. P8–31 if the shearing and tensile stresses at point A are not to exceed 75 and 130 MN/m^2, respectively.

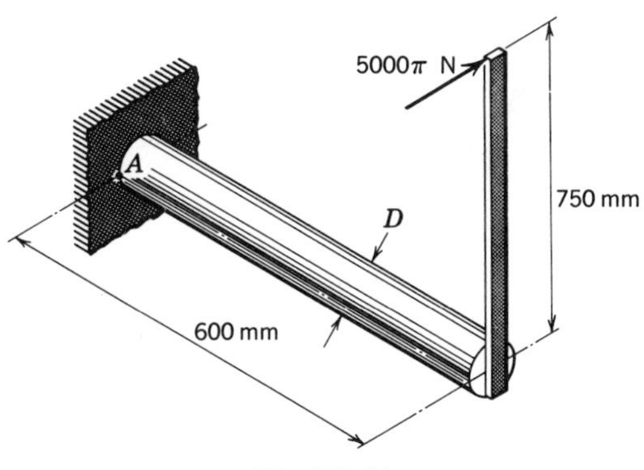

Fig. P8–31

8–32* A steel shaft 4 in. in diameter is supported in flexible bearings at its ends. Two pulleys each 2 ft in diameter are keyed to the shaft. The pulleys carry belts as shown in Fig. P8-32. Determine, and show on a sketch, the principal stresses and the maximum shearing stress at point *A*.

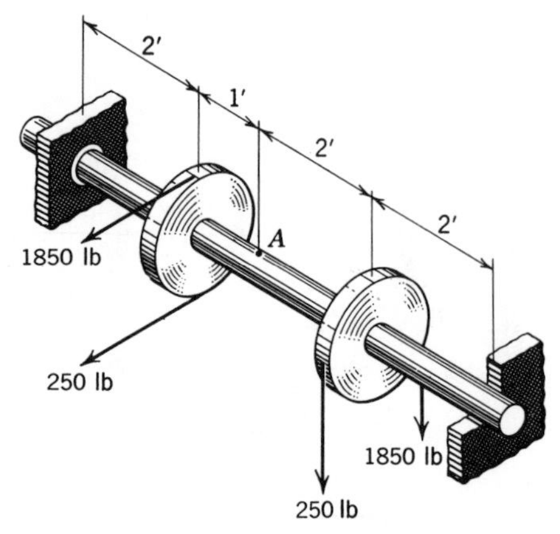

Fig. P8–32

8–33* A steel shaft 100 mm in diameter is supported in flexible bearings at its ends. Two pulleys each 600 mm in diameter are keyed to the shaft. The pulleys carry belts as shown in Fig. P8–33.

(a) At what point along the top line of the shaft is the compressive stress highest?

(b) Determine, and show on a sketch, the principal stresses and the maximum shearing stress at point *A*.

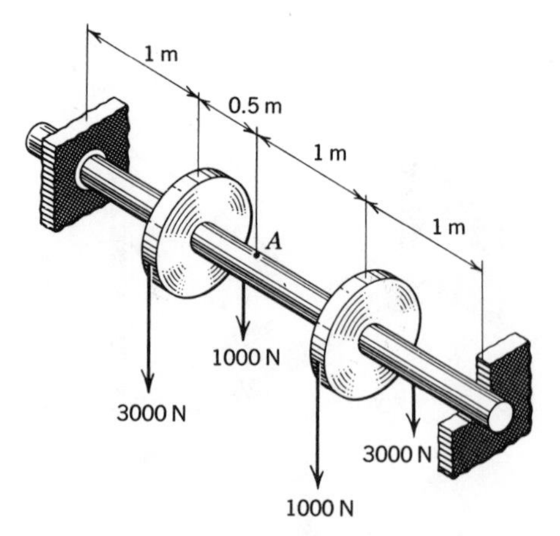

Fig. P8–33

8–34 A 2-in. diameter shaft is subjected to an axial tensile force of $8P$ kips, a bending moment of $22P$ in-kips, and a torque of $18P$ in-kips. The maximum normal stress in the shaft is 20 ksi T. Determine the maximum shearing stress and the magnitude of P.

8–35* A solid shaft 100 mm in diameter is subjected to a torque of 4000π N-m and a tensile force of 100π kN applied parallel to the axis of the shaft but with an eccentricity of 10 mm. Determine, and show on a sketch, the maximum normal stress and the maximum shearing stress in the shaft.

8–36* A short hollow shaft, having an outside diameter of 4 in. and an inside diameter of 2 in., is subjected to a torque of 2000π ft-lb and a compressive force of $12{,}000\pi$ lb applied parallel to the axis of the shaft but with an eccentricity of 0.50 in. Determine, and show on a sketch, the maximum normal stress and the maximum shearing stress in the shaft.

8–37* Figure P8–37 is a free-body diagram of a closed cylindrical pressure vessel with an inside diameter of 60 in. and a wall thickness of 0.25 in. The tank is subjected to an internal pressure of 125 psi, and the external loads shown are applied to rigid plates attached to the ends of the cylinder. Determine, and show on a sketch, the principal and maximum shearing stresses acting on an element at the inside of the tank at A. Neglect the weight of the tank and assume that the walls do not buckle.

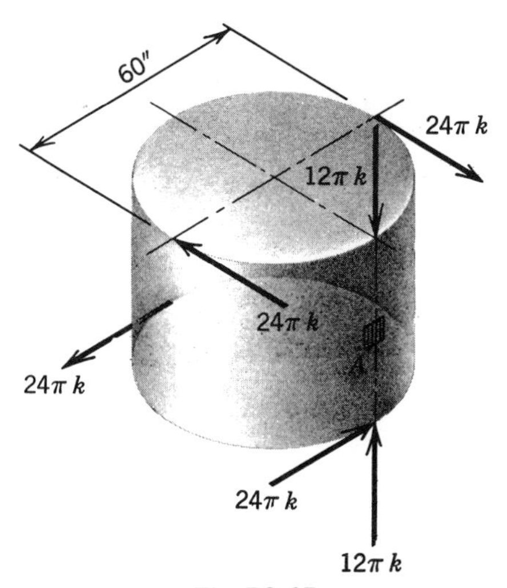

Fig. P8–37

8–38 Determine, and show on a sketch, the principal stresses and the maximum shearing stress at point A of the hollow shaft of Fig. P8–38. The shaft is in equilibrium.

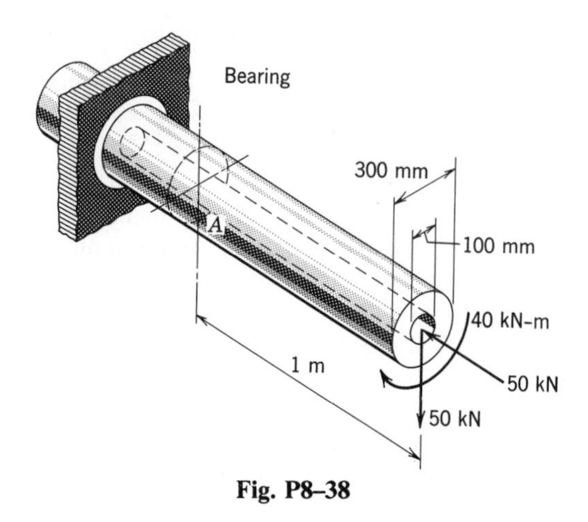

Fig. P8–38

8–39* The solid circular steel shaft of Fig. P8–39 is subjected to the torques and loads indicated. Determine, and show on a sketch, the principal stresses and the maximum shearing stress at point A.

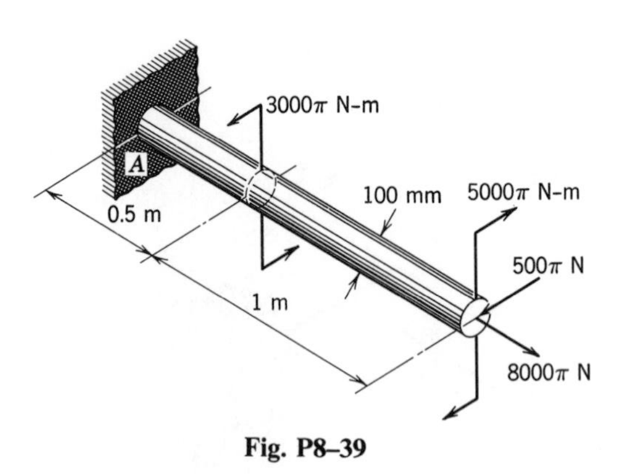

Fig. P8–39

8–40 Because of discontinuities at points A and B on the shaft of Fig. P8–40, the normal and shearing stresses at these points are limited to 15 ksi T and 10 ksi, respectively. Determine the maximum permissible value of P.

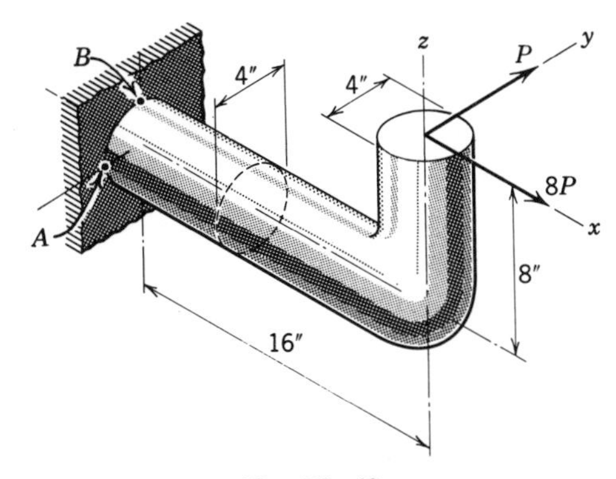

Fig. P8–40

8–4 THEORIES OF FAILURE

A tension test of an axially loaded member is easy to conduct, and the results, for many types of materials, are well known. When such a member fails, the failure occurs at a specific principal (axial) stress, a definite axial strain, a maximum shearing stress of one-half the axial stress, and a specific amount of strain energy per unit volume of stressed material. Since all of these limits are reached simultaneously for an axial load, it makes no difference which criterion (stress, strain, or energy) is used for predicting failure in another axially loaded member of the same material.

For an element subjected to biaxial or triaxial loading, however, the situation is more complicated because the limits of normal stress, normal strain, shearing stress, and strain energy existing at failure for an axial load are not all reached simultaneously. In other words, the cause of failure, in general, is unknown. In such cases it becomes important to determine the best criterion for predicting failure, because test results are difficult to obtain and the combinations of loads are endless. Several theories have been proposed for predicting failure of various types of material subjected to many combinations of loads. Unfortunately, none of the theories agree with test data for all types of materials and combinations of loads. Several of the more common theories of failure are presented and briefly explained in the following paragraphs.

MAXIMUM-NORMAL-STRESS THEORY[2] The maximum-normal-stress theory predicts failure of a specimen subjected to any combination

[2]Often called Rankine's theory after W. J. M. Rankine (1820–1872), an eminent engineering educator in England.

of loads when the maximum normal stress at any point reaches the axial failure stress as determined by an axial tensile or compressive test of the same material.

The maximum-normal-stress theory is presented graphically in Fig. 8–6*b* for an element subjected to biaxial principal stresses in the x and y directions, as shown in Fig. 8–6*a*. The limiting stress σ_f is the failure stress for this material when loaded axially. Any combination of biaxial principal stresses σ_x and σ_y represented by a point inside the square of Fig. 8–6*b* is safe according to this theory, whereas any combination of stresses represented by a point outside of the square will cause failure of the element on the basis of this theory.

MAXIMUM-SHEARING-STRESS THEORY[3] The maximum-shearing-stress theory predicts failure of a specimen subjected to any combination of loads when the maximum shearing stress at any point reaches the failure stress τ_f equal to $\sigma_f/2$, as determined by an axial tensile or compressive test of the same material. For ductile materials the shearing elastic limit, as determined from a torsion test (pure shear), is greater than one-half the tensile elastic limit (with an average value of τ_f about $0.57\sigma_f$). This means that the maximum-shearing-stress theory errs on the conservative side by being based on the limit obtained from an axial test.

The maximum-shearing-stress theory is presented graphically in Fig. 8–6*c* for an element subjected to biaxial (σ_z is equal to zero) principal stresses, as shown in Fig. 8–6*a*. In the first and third quadrants, σ_x and σ_y have the same sign, and the maximum shearing stress is half of the numerically larger principal stress σ_x or σ_y, as explained in Art. 1–7. In the second and fourth quadrants, where σ_x and σ_y are of opposite sign, the maximum shearing stress is half of the arithmetical sum of the two principal stresses. In the fourth quadrant the equation of the boundary, or limit stress, line is

$$\sigma_x - \sigma_y = \sigma_f$$

and in the second quadrant the relation is

$$\sigma_x - \sigma_y = -\sigma_f$$

MAXIMUM-NORMAL-STRAIN THEORY[4] The maximum-normal-strain theory predicts failure of a specimen subjected to any combination

[3]Sometimes called Coulomb's theory because it was originally stated by him in 1773. More frequently called Guest's theory or law because of the work of J. J. Guest in England in 1900.

[4]Often called Saint Venant's theory because of the work of Barre de Saint Venant (1797–1886), a great French mathematician and elastician.

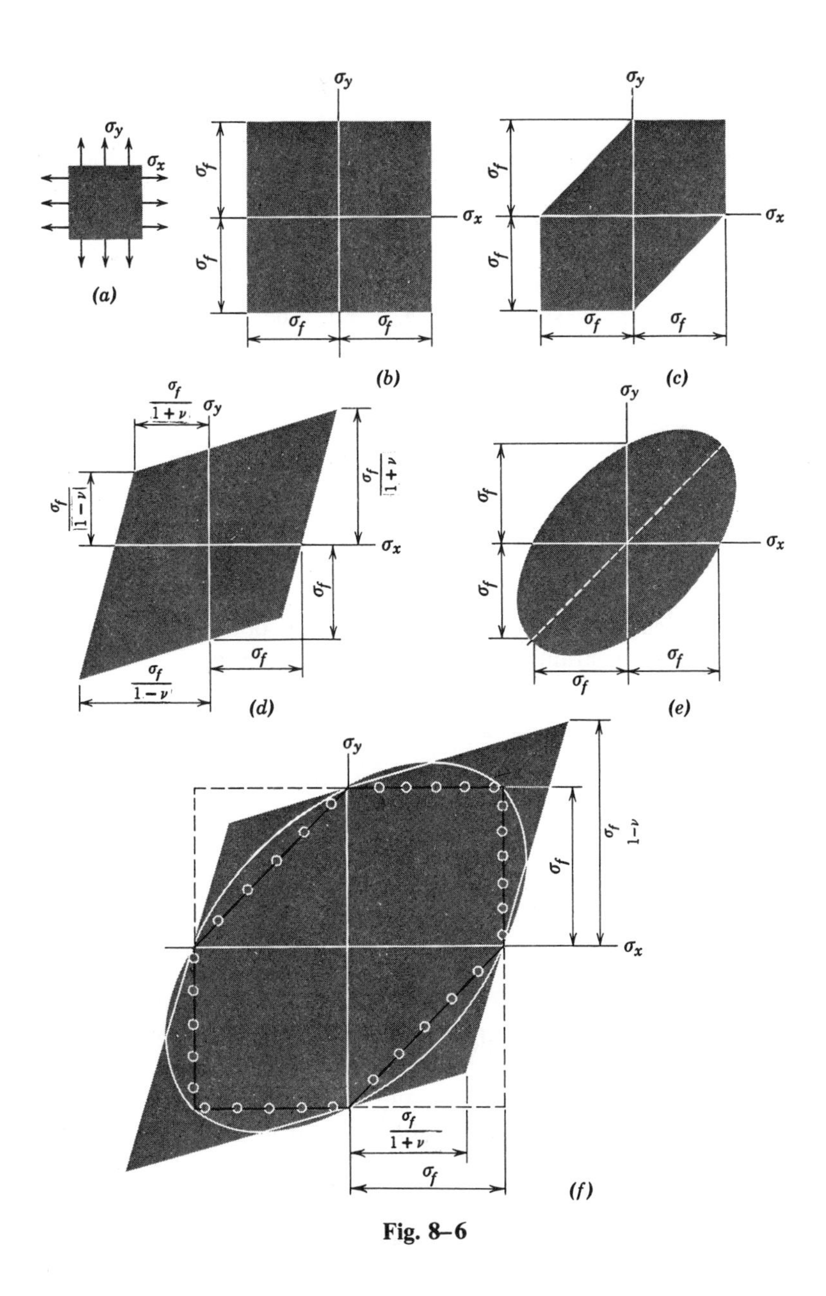

σ_y
σ_x
(a)
σ_f
(b)
(c)
$\frac{\sigma_f}{1+\nu}$
$\frac{\sigma_f}{1-\nu}$
(d)
(e)
(f)

Fig. 8–6

of loads when the maximum normal strain at any point reaches the failure strain σ_f/E at the proportional limit, as determined by an axial tensile or compressive test of the same material.

The maximum-normal-strain theory is presented graphically in Fig. 8–6*d* for an element subjected to biaxial principal stresses (not greater than the proportional limit), as shown in Fig. 8–6*a*. The limiting strain in the positive x direction is

$$\epsilon_x = \frac{\sigma_f}{E} = \frac{\sigma_x - \nu\sigma_y}{E}$$

from which

$$\sigma_y = \frac{1}{\nu}(\sigma_x - \sigma_f)$$

which is observed to be a straight line through the point $(\sigma_f, 0)$ with a slope of $1/\nu$. The limiting strain in the positive y direction is

$$\epsilon_y = \frac{\sigma_f}{E} = \frac{\sigma_y - \nu\sigma_x}{E}$$

from which

$$\sigma_y = \nu\sigma_x + \sigma_f$$

which is observed to be a straight line through the point $(0, \sigma_f)$ with a slope equal to ν.

MAXIMUM-STRAIN-ENERGY THEORY The maximum-strain-energy theory predicts failure of a specimen subjected to any combination of loads when the strain energy per unit volume (see Chap. 2 for the derivation of the relation between σ_f and the elastic strain energy per unit volume) of any portion of the stressed member reaches the failure value of the strain energy per unit volume as determined by an axial tensile or compressive test of the same material. The maximum-strain-energy theory has been largely replaced by the maximum-distortion-energy theory.

MAXIMUM-DISTORTION-ENERGY THEORY[5] This theory differs from the maximum-strain-energy theory in that the portion of the strain energy producing volume change is considered ineffective in causing failure by yielding. Supporting evidence comes from experiments showing

[5]Frequently called the Huber-Hencky-von Mises theory, because it was proposed by M. T. Huber of Poland in 1904 and independently by R. von Mises of Germany in 1913. The theory was further developed by H. Hencky and von Mises in Germany and the United States.

that homogeneous materials can withstand very high hydrostatic stresses without yielding. The portion of the strain energy producing change of shape of the element is assumed to be completely responsible for the failure of the material by inelastic action.

The strain energy of distortion is most readily computed by determining the total strain energy of the stressed material and subtracting the strain energy corresponding to the volume change. Article 2–9 defined the quantity $\sigma^2/(2E)$ as the strain energy per unit volume for a member subjected to a slowly applied axial load. This expression can also be written as

$$u=\frac{\sigma^2}{2E}=\frac{\sigma\epsilon}{2}$$

where u is the strain energy per unit volume and σ and ϵ are the slowly applied axial stress and strain. This equation assumes that the stress does not exceed the proportional limit.

When an elastic element is subjected to triaxial loading, the stresses can be resolved into three principal stresses such as σ_x, σ_y, and σ_z, where x, y, and z are the principal axes. These stresses will be accompanied by three principal strains related to the stresses by Eq. 2–10 (Art. 2–8). If it is assumed that the loads are applied simultaneously and gradually, the stresses and strains will increase in the same manner. The total strain energy per unit volume is the sum of the energies produced by each of the stresses (energy is a scalar quantity and can be added algebraically regardless of the directions of the individual stresses); thus,

$$u=\left(\frac{1}{2}\right)(\sigma_x\epsilon_x+\sigma_y\epsilon_y+\sigma_z\epsilon_z)$$

When the strains are expressed in terms of the stresses this equation becomes

$$u=[1/(2E)]\left[(\sigma_x^2+\sigma_y^2+\sigma_z^2)-2\nu(\sigma_x\sigma_y+\sigma_y\sigma_z+\sigma_z\sigma_x)\right]$$

The strain energy can be resolved into two components u_v and u_d, resulting from a volume change and a distortion, respectively, by considering the principal stresses to be made up of two sets of stresses as indicated in Fig. 8–7*a*, *b*, and *c*. The state of stress in Fig. 8–7*c* will result in distortion only (no volume change) if the sum of the three normal strains is

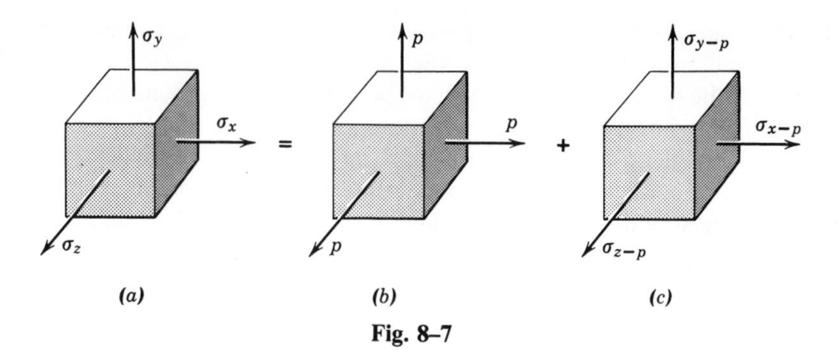

Fig. 8–7

zero. That is,

$$\begin{aligned} E(\epsilon_x+\epsilon_y+\epsilon_z)_d = &[(\sigma_x - p) - \nu(\sigma_y+\sigma_z-2p)] \\ &+[(\sigma_y - p) - \nu(\sigma_z+\sigma_x-2p)] \\ &+[(\sigma_z - p) - \nu(\sigma_x+\sigma_y-2p)] = 0 \end{aligned}$$

which reduces to

$$(1-2\nu)(\sigma_x+\sigma_y+\sigma_z-3p)=0$$

Therefore,

$$p=(\sigma_x+\sigma_y+\sigma_z)/3$$

The three normal strains due to p are, from Eq. 2–10,

$$\epsilon_v=(1-2\nu)p/E$$

and the energy resulting from the hydrostatic stress (the volume change) is

$$\begin{aligned} u_v &= 3\left(\frac{p\epsilon_v}{2}\right)=\frac{3}{2}\,\frac{1-2\nu}{E}\,p^2 \\ &= \frac{1-2\nu}{6E}(\sigma_x+\sigma_y+\sigma_z)^2 \end{aligned}$$

The energy resulting from the distortion (change of shape) is

$$u_d = u - u_v$$
$$= \frac{1}{6E}\Big[3(\sigma_x^2 + \sigma_y^2 + \sigma_z^2) - 6\nu(\sigma_x\sigma_y + \sigma_y\sigma_z + \sigma_z\sigma_x)$$
$$- (1-2\nu)(\sigma_x + \sigma_y + \sigma_z)^2\Big]$$

When the third term in the brackets is expanded, the expression can be rearranged to give

$$u_d = \frac{1+\nu}{6E}\Big[(\sigma_x^2 - 2\sigma_x\sigma_y + \sigma_y^2) + (\sigma_y^2 - 2\sigma_y\sigma_z + \sigma_z^2)$$
$$+ (\sigma_z^2 - 2\sigma_z\sigma_x + \sigma_x^2)\Big]$$
$$= \frac{1+\nu}{6E}\Big[(\sigma_x - \sigma_y)^2 + (\sigma_y - \sigma_z)^2 + (\sigma_z - \sigma_x)^2\Big] \qquad \text{(a)}$$

The maximum-distortion-energy theory of failure assumes that inelastic action will occur whenever the energy given by Eq. (a) exceeds the limiting value obtained from a tensile test. For this test, only one of the principal stresses will be nonzero. If this stress is called σ_f, the value of u_d becomes

$$(u_d)_f = \frac{1+\nu}{3E}\sigma_f^2$$

and when this value is substituted in Eq. (a) it becomes

$$2\sigma_f^2 = (\sigma_x - \sigma_y)^2 + (\sigma_y - \sigma_z)^2 + (\sigma_z - \sigma_x)^2 \qquad \text{(b)}$$

for failure by slip.

When a state of plane stress exists, assume σ_z equals zero, Eq. (b) becomes

$$\sigma_x^2 - \sigma_x\sigma_y + \sigma_y^2 = \sigma_f^2$$

This last expression is the equation of an ellipse with its major axis along the line σ_x equals σ_y, as shown in Fig. 8–6*e*.

An alternate approach to the distortion-energy theory of failure consists of determining the shearing stress on the octahedral planes, defined as any plane whose normal makes equal angles with the three principal axes.

Figure 8–8 is a free-body diagram of an element bounded by the three principal planes and the octahedral plane. Since

$$\theta_x = \theta_y = \theta_z$$

$$\cos^2\theta_x + \cos^2\theta_y + \cos^2\theta_z = 3\cos^2\theta_x = 1$$

or

$$\cos\theta_x = \cos\theta_y = \cos\theta_z = 1/\sqrt{3}$$

If the area of the octahedral plane is ΔA, the other three areas are $\Delta A/\sqrt{3}$. The equilibrium equation along the n axis gives

$$\Sigma F_n = \sigma_{\text{oct}}\Delta A - \frac{1}{\sqrt{3}}\left(\sigma_x \frac{\Delta A}{\sqrt{3}}\right) - \frac{1}{\sqrt{3}}\left(\sigma_y \frac{\Delta A}{\sqrt{3}}\right) - \frac{1}{\sqrt{3}}\left(\sigma_z \frac{\Delta A}{\sqrt{3}}\right) = 0$$

or

$$\sigma_{\text{oct}} = (\sigma_x + \sigma_y + \sigma_z)/3$$

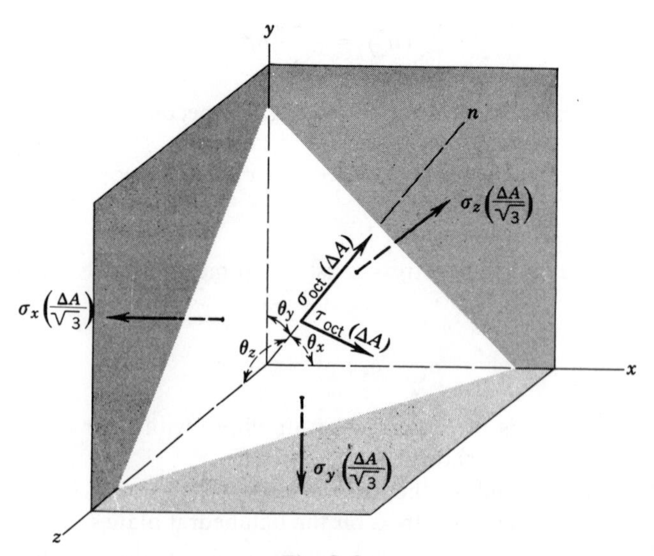

Fig. 8–8

Note that σ_{oct} is the same as p (the hydrostatic stress) in Fig. 8–7b. The resultant of the normal and shearing forces on the octahedral plane must equal the resultant of the forces on the other three planes; thus,

$$(\sigma_{oct}\Delta A)^2+(\tau_{oct}\Delta A)^2=\left(\sigma_x\frac{\Delta A}{\sqrt{3}}\right)^2+\left(\sigma_y\frac{\Delta A}{\sqrt{3}}\right)^2+\left(\sigma_z\frac{\Delta A}{\sqrt{3}}\right)^2$$

from which

$$\begin{aligned}(\tau_{oct})^2&=\frac{\sigma_x^2+\sigma_y^2+\sigma_z^2}{3}-\left(\frac{\sigma_x+\sigma_y+\sigma_z}{3}\right)^2\\&=\frac{1}{9}\left[3(\sigma_x^2+\sigma_y^2+\sigma_z^2)-(\sigma_x^2+\sigma_y^2+\sigma_z^2)\right.\\&\qquad\left.-2(\sigma_x\sigma_y+\sigma_y\sigma_z+\sigma_z\sigma_x)\right]\\&=\frac{1}{9}\left[(\sigma_x-\sigma_y)^2+(\sigma_y-\sigma_z)^2+(\sigma_z-\sigma_x)^2\right]\end{aligned} \tag{c}$$

When Eqs. (b) and (c) are compared, it is seen that

$$2\sigma_f^2=(3\tau_{oct})^2$$

or

$$\tau_{oct}=\frac{\sqrt{2}}{3}\sigma_f=0.471\sigma_f$$

The octahedral shearing stress is thus related directly to the tensile failure stress and can be used as a criterion of the distortion energy at any point.

The graphic representations of Fig. 8–6b, c, d, and e are superimposed in Fig. 8–6f for convenient comparison of the different theories. The maximum-distortion-energy theory has been found to agree best with available test data for ductile materials, and its use in design practice is increasing. The maximum-normal-strain theory is unsafe for ductile materials. The maximum-normal-stress theory is also unsafe for ductile materials with principal stresses of opposite sign. The maximum-shearing-stress theory is conservative for ductile materials and is used in some design codes. The maximum-normal-stress theory agrees with test data for brittle materials and is generally accepted in design practice for such materials.

The following example illustrates the application of the theories of failure in predicting the load-carrying capacity of a member.

Example 8–5 The solid circular shaft of Fig. 8–9 has a proportional limit of 64 ksi and a Poisson's ratio of 0.30. Determine the value of the load R for failure by slip as predicted by each of the four given theories of failure. Assume that point A is the most severely stressed point.

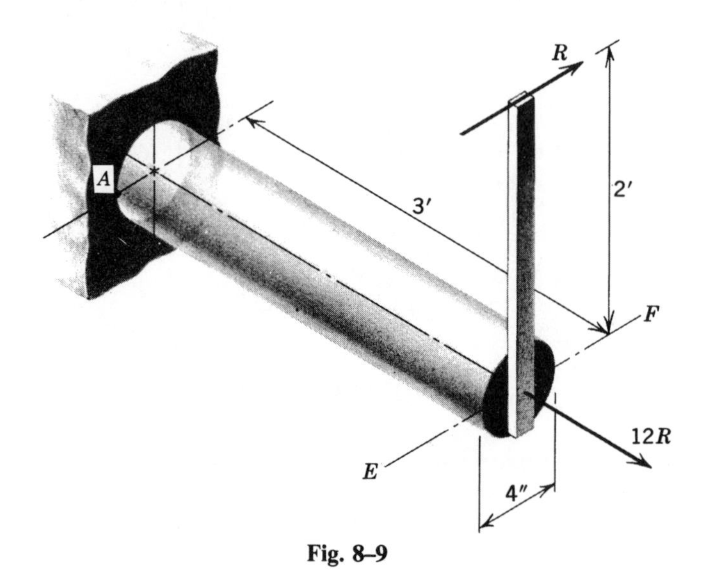

Fig. 8–9

Solution When two equal opposite forces R are introduced along line EF of Fig. 8–9, it is evident that the bending moment at point A is $36R$ in-kips, the torsional moment is $24R$ in-kips, and the axial load is $12R$ kips. The flexural stress at A is

$$\sigma_1 = \frac{Mc}{I} = \frac{36R\,(2)}{\pi(2^4)/4} = \frac{18R}{\pi}\,T$$

The direct stress at A is

$$\sigma_2 = \frac{P}{A} = \frac{12R}{\pi(2^2)} = \frac{3R}{\pi}\,T$$

and the sum of the two tensile stresses at A becomes

$$\sigma = \sigma_1 + \sigma_2 = \frac{21R}{\pi}$$

The torsional shearing stress at A is

$$\tau = \frac{Tc}{J} = \frac{24R(2)}{\pi(2^4)/2} = \frac{6R}{\pi}$$

The maximum stresses at A are

$$\tau_{max} = \sqrt{\left(\frac{10.5R}{\pi}\right)^2 + \left(\frac{6R}{\pi}\right)^2} = \frac{12.1R}{\pi}$$

$$\sigma_{p1} = \frac{10.5R}{\pi} + \frac{12.1R}{\pi} = \frac{22.6R}{\pi}T$$

and

$$\sigma_{p2} = \frac{10.5R}{\pi} - \frac{12.1R}{\pi} = -\frac{1.6R}{\pi} = \frac{1.6R}{\pi}C$$

According to the maximum-normal-stress theory

$$64 = \sigma_f = \sigma_{p1} = 22.6R/\pi$$

from which

$$R = 64\pi/22.6 = \underline{8.90 \text{ kips}}$$ Ans.

According to the maximum-shearing-stress theory

$$64/2 = \tau_f = \tau_{max} = 12.1R/\pi$$

from which

$$R = 32\pi/12.1 = \underline{8.31 \text{ kips}}$$ Ans.

The maximum-normal-strain theory gives

$$\frac{\sigma_f}{E} = \frac{\sigma_{p1} - \nu\sigma_{p2}}{E}$$

or

$$64 = \frac{22.6R}{\pi} - 0.3\left(-\frac{1.6R}{\pi}\right)$$

which gives

$$R = 64\pi/23.1 = \underline{8.71 \text{ kips}} \qquad \text{Ans.}$$

According to the maximum-distortion-energy theory

$$\sigma_f^{\,2} = \sigma_{p1}^{\,2} + \sigma_{p2}^{\,2} - \sigma_{p1}\sigma_{p2}$$

or

$$64^2 = \left(\frac{22.6R}{\pi}\right)^2 + \left(-\frac{1.6R}{\pi}\right)^2 - \frac{22.6R}{\pi}\left(-\frac{1.6R}{\pi}\right)$$

which gives

$$R = 64\pi/23.4 = \underline{8.59 \text{ kips}} \qquad \text{Ans.}$$

In this example, the maximum-shearing-stress theory is seen to be the most conservative, and the maximum-normal-stress theory gives the least conservative result.

PROBLEMS

8–41* A material with a proportional limit in tension and compression of 60 ksi and a Poisson's ratio of 0.30 is subjected to biaxial stress. The principal stresses are 35 ksi *T* and 40 ksi *C*. Determine which, if any, of the theories of failure will predict failure by slip in this instance.

8–42* At a critical point on the free surface of a structural steel member, the principal stresses are 160 MN/m^2 *T* and 125 MN/m^2 *C*. The material has a proportional limit in tension and compression of 250 MN/m^2 and a Poisson's ratio of 0.30. Determine which, if any, of the theories of failure will predict failure by slip at the critical point.

8–43* A material with a proportional limit of 40 ksi in tension and compression and a Poisson's ratio of 1/3 is subjected to principal stresses: $\sigma_x = 25$ ksi *T*, $\sigma_y = 15$ ksi *T* or *C*, and $\sigma_z = 0$. Determine the factor of safety with respect to slip failure by each of the four theories of failure if (1) the stress σ_y is tension and (2) σ_y is compression.

8–44* At a point on the free surface of a steel machine component (Poisson's ratio equals 0.30) the principal stresses are 350 MN/m^2 *T* and 200 MN/m^2 *C*. What minimum proportional limit is required according to each of the four theories if failure by slip is to be avoided?

8–45 Determine the minimum permissible diameter for the shaft of Fig. P8–45 if it is to be made of steel with the following properties: proportional limit 36 ksi in tension and 20 ksi in shear; ultimate tensile strength 60 ksi; and Poisson's ratio

0.30. Use the maximum-normal-stress, maximum-shearing-stress, and maximum-normal-strain theories of failure and a factor of safety of 2.5 with respect to failure by slip. Assume point A is critical.

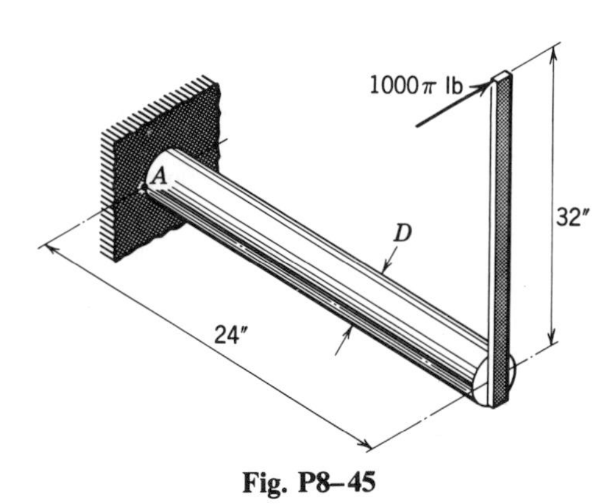

Fig. P8–45

8–46 Solve Prob. 8–45 using the maximum-distortion-energy theory of failure.

8–47 Determine the maximum permissible value for the axial load P of Fig. P8–47 if the member is made of steel with the following properties: proportional limit 250 MN/m^2 in tension or compression and 140 MN/m^2 in shear; ultimate tensile strength 420 MN/m^2; and Poisson's ratio 0.30. Use the maximum-normal-stress and maximum-shearing-stress theories of failure and a factor of safety of 2 with respect to failure by slip.

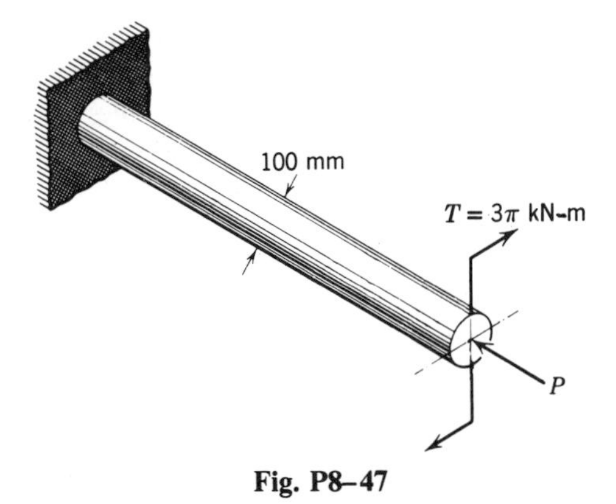

Fig. P8–47

8–48 Solve Prob. 8–47 using the maximum-normal-strain theory of failure.

8–49 Determine the maximum allowable value of R of Prob. 8–30 for a factor of safety of 2.5 with respect to failure by slip if the material has a proportional limit of 64 ksi. Use the maximum-distortion-energy theory of failure.

8–50 Solve Prob. 8–49 using the maximum-shearing-stress and the maximum-normal-strain theories. Poisson's ratio is 0.30 for this material.

8–51* A horizontal steel (yield point 420 MN/m^2) shaft of 75-mm diameter has a 1200-mm diameter pulley attached to the end of a 400-mm overhang. The shaft is not free to turn. A cable is attached to and wrapped around the pulley and supports a load of 10 kN. Determine:

(a) The maximum fiber and torsional shearing stresses on a transverse section of the shaft 400 mm from the pulley.
(b) The three principal stresses and the maximum shearing stress at the top of the section of part (a).
(c) For the stresses of part (b), the factor of safety, with respect to failure by slip, according to the maximum-shearing-stress theory.

8–52* A 3-ft cantilever beam with a solid circular cross section is subjected at the free end to a downward transverse load of 500π lb and a torque in a transverse plane of 1125π ft-lb. The critical point is at the top surface at the support. The material is Ni-Vee bronze with a proportional limit of 15 ksi in tension and compression. Determine the minimum permissible diameter for the beam. A factor of safety of 2.5 with respect to failure by slip is specified and the design must satisfy the maximum-normal-stress, the maximum-shearing-stress, and the maximum-distortion-energy theories of failure.

8–53* The bar of Fig. P8–40 is made of 8630 steel alloy (see stress strain diagram in App. A) and the loading is to be $8P$ in the x direction as shown, and $3P$ in the y direction (replacing the force P). The critical point is at the top of a vertical diameter at the wall. Using the maximum-normal-stress, the maximum-shearing-stress, and the maximum-distortion-energy theories of failure, determine the maximum permissible magnitude for P to give a minimum factor of safety of 2 with respect to failure by slip.

8–54 Determine the minimum permissible diameter for the shaft of Fig. P8–31 if it is to be made of steel with the following properties: proportional limit 250 MN/m^2 in tension or compression and 140 MN/m^2 in shear; ultimate tensile strength 420 MN/m^2; and Poisson's ratio 0.30. Use the maximum-normal-stress and the maximum-normal-strain theories of failure and a factor of safety of 2 with respect to failure by slip. Assume point A is critical.

8–55 Solve Prob. 8–54 using the maximum-shearing-stress and the maximum-distortion-energy theories of failure.

8–56* Determine the maximum allowable internal pressure to which a closed-end, thick-walled cylinder with an inside diameter of 6 in. and an outside diameter of 10 in. may be subjected if the cylinder is made of steel with a proportional limit of 80 ksi and a Poisson's ratio of 0.30. A factor of safety of 2.5 with respect to failure by slip according to the maximum-distortion-energy theory of failure is specified.

8–57* A thick-walled hydraulic cylinder with an inside diameter of 150 mm is required to operate under a maximum internal pressure of 50 MN/m^2. The cylinder is to be made of steel with a proportional limit of 280 MN/m^2 and a Poisson's ratio of 0.30. Determine the minimum outside diameter required if a factor of safety of 2 with respect to failure by slip according to the maximum-shearing-stress theory of failure is specified. The wall of the cylinder is not required to carry axial load.

8–58* A laminated thick-walled hydraulic cylinder consists of a sleeve with an inside diameter of 6 in. and an outside diameter of 8 in. shrunk over a tube with an inside diameter of 4 in. The maximum shrinkage stress developed in the sleeve is 6 ksi T. Determine the maximum permissible internal pressure to which the laminated cylinder may be subjected if a factor of safety of 3 with respect to failure by slip according to the maximum-distortion-energy theory of failure is specified. The cylinder is made of steel with a proportional limit of 40 ksi and a Poisson's ratio of 0.30. The wall of the cylinder is not required to carry axial load.

9 Columns

9–1 INTRODUCTION

In their simplest form, columns are long, straight, prismatic bars subjected to compressive, axial loads. As long as a column remains straight, it can be analyzed by the methods of Chap. 1; however, if a column begins to deform laterally, the deflection may become large and lead to catastrophic failure. This situation, called *buckling*, can be defined as the sudden large deformation of a structure due to a slight increase of an existing load under which the structure had exhibited little, if any, deformation before the load was increased. A yardstick illustrates this phenomenon; it will support a compressive load of several pounds without discernible lateral deformation, but once the load becomes large enough to cause the yardstick to "bow out" a slight amount, any further increase of load produces large lateral deflections.

Buckling of such a column is caused not by failure of the material of which the column is composed but by deterioration of what had been a stable equilibrium to an unstable one. The three states of equilibrium can be illustrated with a ball at rest on a surface, as shown in Figure 9–1. The ball in Fig. 9–1a is in a *stable equilibrium* position at the bottom of the pit

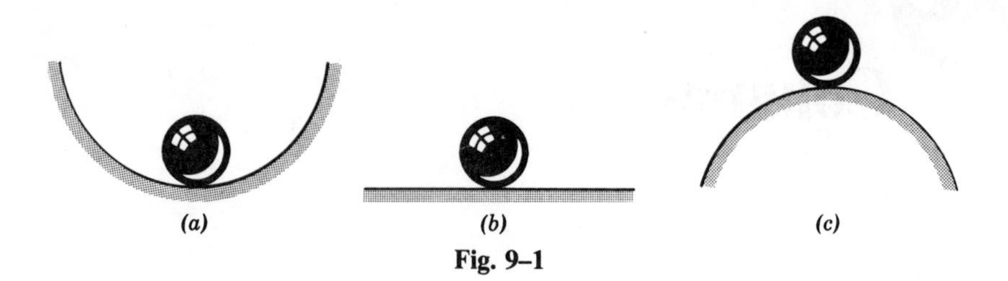

(a) (b) (c)

Fig. 9–1

because gravity will cause it to return to its equilibrium position if perturbed. The ball in Fig. 9–1*b* is in a *neutral equilibrium* position on the horizontal plane because it will remain at any new position to which it is displaced, tending neither to return to nor move farther from its original position. The ball in Fig. 9–*c*, however, is in an *unstable equilibrium* position at the top of a hill because, if it is perturbed, gravity will cause it to move even farther from its original location until it eventually finds a stable equilibrium position at the bottom of another pit.

As the compressive load on a column is gradually increased from zero, the column is at first in a state of stable equilibrium. During this state, if the column is perturbed by inducing small lateral deflections, it will return to its straight configuration when the loads are removed. As the load is increased further, a critical value is reached at which the column is on the verge of experiencing a lateral deflection so, if it is perturbed, it will not return to its straight configuration. The load cannot be increased beyond this value unless the column is restrained laterally; should the lateral restraints be removed, the slightest perturbation will trigger large lateral deflections. For long, slender columns, the *critical buckling load* (the maximum load for which the column is in stable equilibrium) occurs at stress levels much less than the proportional limit for the material. This indicates that this type of buckling is an elastic phenomenon.

9–2 STABILITY OF ELASTIC MECHANICAL SYSTEMS

Some insight into the problem of stability of a column can be gained by considering the stability of a simple elastic mechanical system such as the one shown in Fig. 9–2. The rigid prismatic bar is supported by a smooth pin at the bottom and is loaded by an axial force at the top. A torsional spring at the bottom offers a resisting couple $M_r = K\theta$ to rotation about the pin. The spring is adjusted initially so that the bar will stand in the vertical position. If the bar is displaced from the vertical position (by a momentary lateral force) to the position shown in the free-body diagram (Fig. 9–2*b*), the torsion spring at the bottom will exert a restoring couple that will

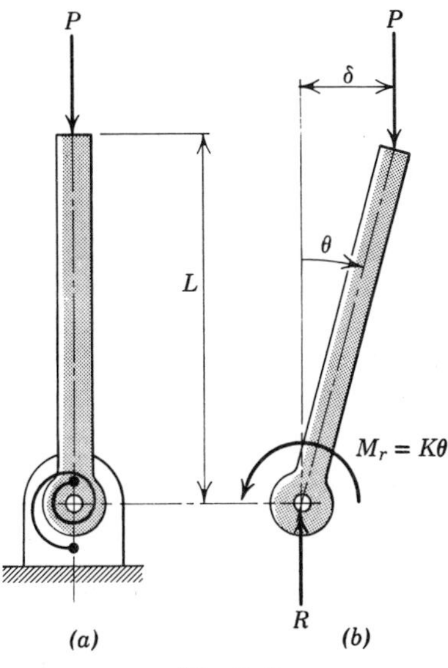

Fig. 9–2

return the bar to the original (vertical) position provided the restoring couple exceeds the moment of the force P with respect to the pin. Since the moment of P about the pin is $PL\sin\theta$, this condition for stable equilibrium can be expressed by the inequality

$$K\theta > PL\sin\theta$$

or

$$P < \frac{K}{L}\frac{\theta}{\sin\theta} \tag{a}$$

By applying L'Hospital's rule to Eq. (a) as $\theta \to 0$, it is clear that the vertical position ($\theta = 0$) is stable as long as $P < K/L$.

If the load P is increased beyond the value K/L, the vertical position becomes unstable and the bar, if perturbed, will rotate through an angle θ until it finds a new equilibrium position for which the restoring moment M_r balances the moment of the force P about the pin. In this new position

$$K\theta = PL\sin\theta \tag{b}$$

or

$$PL/K = \theta/\sin\theta \tag{c}$$

Figure 9–3 shows a plot of the quantity PL/K as a function of θ. For $P<K/L$ (or $PL/K<1$), the only equilibrium position is $\theta=0$. For $P>K/L$ (or $PL/K>1$), there are three possible equilibrium positions. Equation (b) can be satisfied by equating θ to zero; however, this is already known to be unstable. Thus, the other branches, which are graphs of the solution to Eq. (c), are the stable positions for $P>K/L$. There are two branches to the curve because the bar could rotate either right or left (the spring is assumed to be equally effective in either direction). Observe that an increase of only 11 percent in the magnitude of P above K/L will result in a rotation of 45° ($\theta=\pi/4$) before stable equilibrium is established. Displacements of this magnitude are typical of those encountered during elastic buckling. Thus, there is a *critical load* $P=K/L$ where the nature of the equilibrium of the elastic mechanical system changes markedly. The point ($\theta=0, P=K/L$) where the curve branches is called a bifurcation point.

If the bar and torsion spring system of Fig. 9–2 is replaced by an elastic bar fixed at the bottom as in Fig. 9–4, the analogy between the action of a column and the mechanical system is apparent. The entire column is now the spring, which resists lateral deflection. The lateral deflection can be initiated by accidental eccentricities due to (1) deviations from absolute straightness of the column, (2) nonuniformity of stress distribution within the member due to nonhomogeneity of the material or residual stresses, or (3) error in alignment when the load is applied. Only a small increase in load above the critical load will significantly increase the initial deflection —possibly enough to induce inelastic action in the material on the concave side of the deflected column, with the further possibility that such inelastic

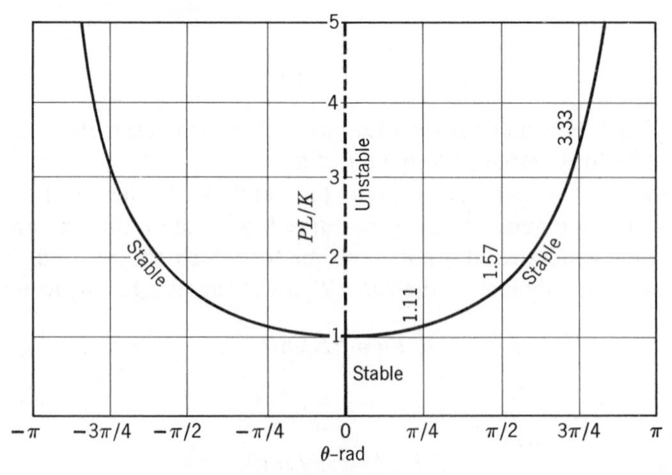

Fig. 9–3

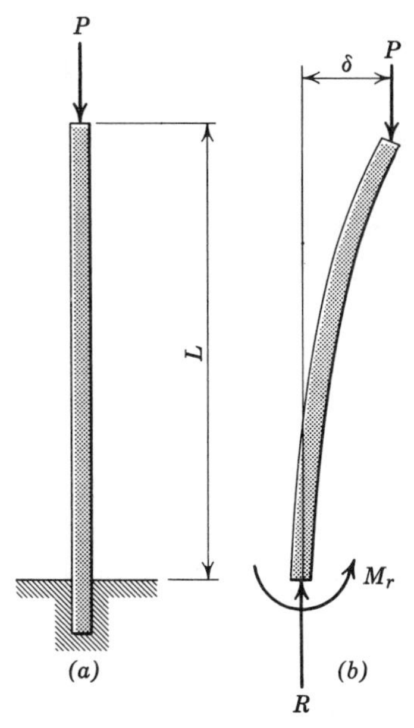

Fig. 9–4

action could be sufficient to cause collapse (stability failure).

In the following article an expression comparable to Eq. (a) is developed for the limiting compressive load that may be applied to a long, slender, elastic, prismatic bar (an axially loaded perfect column).

PROBLEMS

Note. All springs in the following problems can act in either tension or compression.

9–1* The 20-in. rigid bar shown in Fig. P9–1 is supported at the top by two

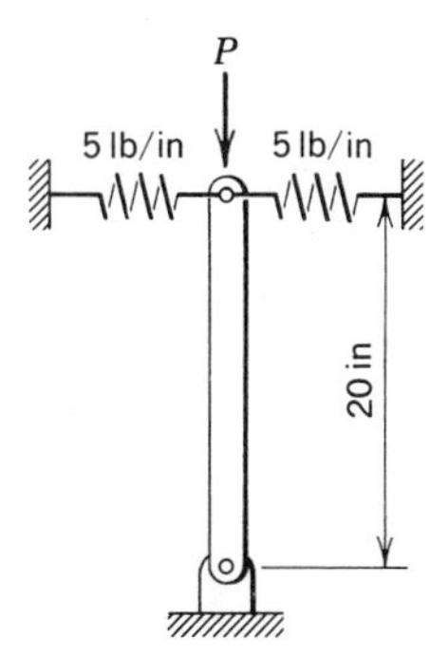

Fig. P9–1

springs adjusted to hold the bar in a vertical position. Each spring has a modulus of 5 lb/in. Determine the maximum load P for which the bar is in stable equilibrium in the vertical position by considering the behavior of the system under small angular rotations from the vertical.

9–2* The rigid T-shaped structure shown in Fig. P9–2 rests on two springs spaced 150 mm apart. Each spring has a modulus of 5 kN/m. Determine the maximum load P for which the structure is in stable equilibrium in the vertical position by considering the behavior of the system for small angular deviations from the vertical.

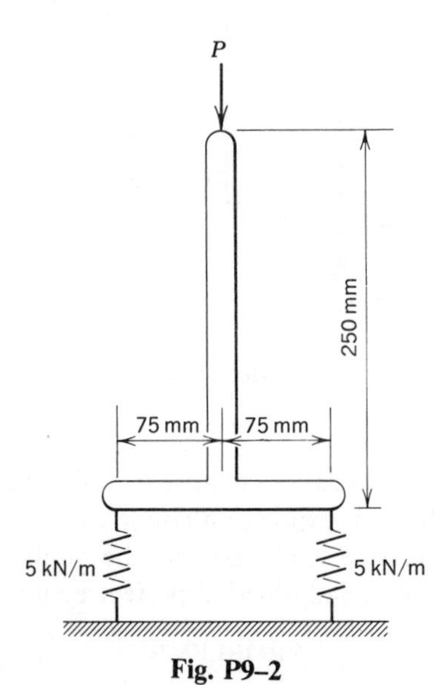

Fig. P9–2

9–3* The system of two rigid bars shown in Fig. P9–3 is restrained at B by two springs adjusted so that the system is in equilibrium when bars AB and BC are aligned. Each of the springs has a modulus of 20 lb/in. Determine the maximum load P that can be applied at C and have the system remain in stable equilibrium with the bars aligned. Perform the analysis by considering small angular rotations of bar AB.

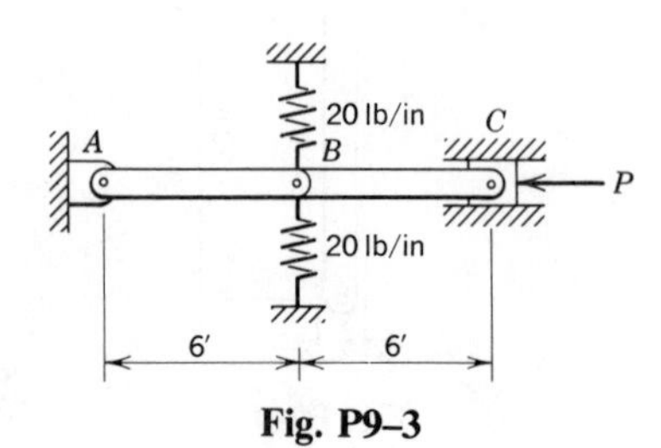

Fig. P9–3

9–4* The system of two rigid bars shown in Fig. P9–4 is restrained at B by two springs adjusted so that the system is in equilibrium when bars AB and BC are aligned. Each of the springs has a modulus of 20 kN/m. Determine the maximum load P that can be applied at C and have the system remain in stable equilibrium with the bars aligned. Perform the analysis by considering small angular rotations of bar AB.

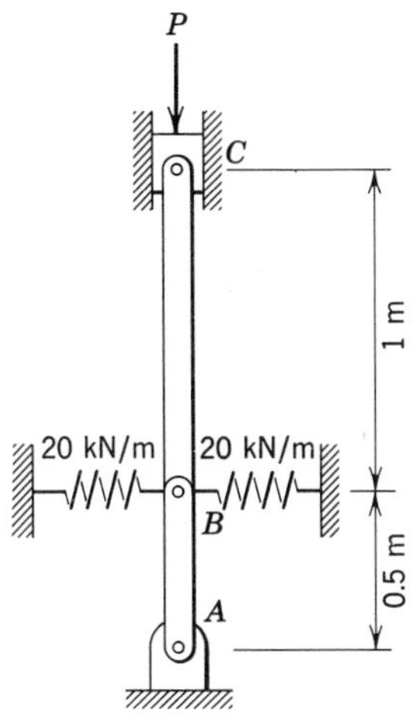

Fig. P9–4

9–3 BUCKLING OF LONG STRAIGHT COLUMNS

The first solution for the buckling of long slender columns was published in 1757 by the Swiss mathematician Leonhard Euler (1707–1783). Although the results of this article can be used only for long slender columns, the analysis, similar to that used by Euler, is mathematically revealing and helps explain the behavior of columns.

The purpose of this analysis is to determine the minimum axial compressive load for which a column will experience lateral deflections. A straight, slender, pivot-ended column centrically loaded by axial compressive forces P at each end is shown in Fig. 9–5a. A pivot-ended column is supported such that the bending moment and lateral movement are zero at the ends. In Fig. 9–5b, the load P has been increased sufficiently to cause a lateral deflection δ at the midpoint of the span. Axes are selected with the origin at the center of the span for convenience. If the column is sectioned at an arbitrary position x, a free-body diagram of the portion to the right of the section will appear as shown in Fig. 9–5c. The two forces constitute a couple of magnitude $P(\delta - y)$ that must equal the resisting moment M_r; thus, $M_r = P(\delta - y)$. The differential equation for the elastic curve, as given

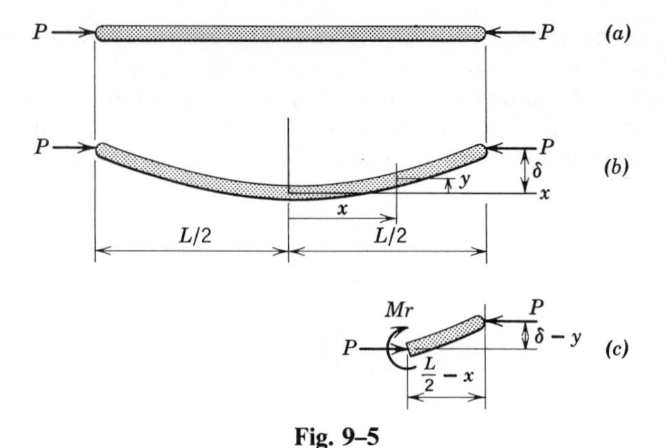

Fig. 9–5

by Eq. 6–2, becomes

$$EI\frac{d^2y}{dx^2} = M_r = P(\delta - y)$$

or

$$\frac{d^2y}{dx^2} + \frac{P}{EI}y = \frac{P\delta}{EI} \tag{a}$$

Equation (a) is an ordinary second-order linear differential equation with constant coefficients and a constant right-hand side. Established methods for the solution of such equations show it to be of the form

$$y = A\sin px + B\cos px + C \tag{b}$$

where A, B, C, and p are constants. By differentiating Eq. (b), substituting the results into Eq. (a), and collecting like terms, the following expression is obtained for evaluating p and C:

$$(-p^2 + P/EI)(A\sin px + B\cos px) + (P/EI)C = (P/EI)\delta$$

from which it follows that

$$p^2 = P/EI \quad \text{and} \quad C = \delta$$

The constants A and B can be obtained from the two boundary conditions that the slope and deflection of the elastic curve are zero at the origin; that is, at $x=0$, $y=0$ and $dy/dx=0$. Since the purpose of this analysis is to determine the minimum load P for which lateral deflections occur, it is necessary to have a third boundary condition; namely, at $x=L/2$, $y=\delta > 0$. The midspan deflection δ is required to be nonzero; otherwise, $\delta=0$, $y\equiv 0$ could be a solution of Eq. (a). When the boundary conditions are substituted in Eq. (b) and its first derivative, they give

$$0=B+C$$

$$0=Ap$$

from which

$$A=0 \quad \text{and} \quad B=-C=-\delta$$

The solution to Eq. (a) thus becomes

$$y=\delta\left(1-\cos\sqrt{\frac{P}{EI}}\,x\right) \tag{9–1}$$

In order to satisfy the condition that at $x=L/2$, $y=\delta$, the cosine term in Eq. 9–1 must vanish. Thus

$$\cos\sqrt{\frac{P}{EI}}\,\frac{L}{2}=0$$

This equation is satisfied when the argument of the cosine is an odd multiple of $\pi/2$ or

$$\sqrt{\frac{P}{EI}}\,\frac{L}{2}=\frac{\pi}{2},\frac{3\pi}{2},\frac{5\pi}{2},\ldots$$

Only the first value has physical significance since it determines the minimum value of P for a nontrivial solution. This value for P is called the critical buckling load, is designated by P_{cr}, and has the magnitude

$$P_{cr}=\pi^2 EI/L^2 \tag{9–2}$$

The term P_{cr}[1] is usually called the *Euler buckling load* in honor of Leonhard Euler.

[1]While the analysis predicts the buckling load, it does not determine the corresponding lateral deflection δ. This deflection can assume any nonzero value small enough that the nonlinear factor $[1+(dy/dx)^2]^{3/2}$ in the curvature expression is approximately unity.

The moment of inertia I in Eq. 9–2 refers to the axis about which bending occurs. When I is replaced by Ar^2, where r is the radius of gyration about the axis of bending, Eq. 9–2 becomes

$$P_{cr}/A = \pi^2 E/(L/r)^2 \tag{9–3}$$

The quantity L/r is the *slenderness ratio* and is determined for the axis about which bending tends to occur. For a pivot-ended, centrically loaded column with no intermediate bracing to restrain lateral motion, bending occurs about the axis of minimum moment of inertia (minimum radius of gyration).

The Euler buckling load as given by Eq. 9–2 or 9–3 agrees well with experiment only if the slenderness ratio is large, typically in excess of 140 for steel columns. Whereas short compression members can be treated as explained in Chap. 1, many columns lie between these extremes where neither solution is applicable. These intermediate length columns are analyzed by empirical formulas described in later sections.

Example 9–1 An 8-ft, pivot-ended, air-dried douglas fir column has a 2×4-in. rectangular cross section. Determine:

(a) The slenderness ratio.
(b) The Euler buckling load.
(c) The ratio of the axial stress under the action of the buckling load to the elastic strength of the material.
(d) The maximum load the column can support with a factor of safety of 2.5.

Solution (a) The length of the column is 8 ft or 96 in. To determine the slenderness ratio the minimum radius of gyration must be calculated. The moment of inertia of a rectangular cross section is $bh^3/12$ and the area is bh; therefore, the radius of gyration is $\sqrt{I/A}$ or $h/2\sqrt{3}$. The minimum radius of gyration is found by using the centroidal axis parallel to the longer side of the rectangle. Thus, $b=4$ in. and $h=2$ in., so that

$$r = h/2\sqrt{3} = 2/2\sqrt{3} = 0.5773 \text{ in.}$$

The slenderness ratio is then found to be

$$L/r = 96/0.5773 = \underline{166.3} \qquad \text{Ans.}$$

(b) From App. A, the modulus of elasticity for air-dried douglas fir is found to be 1.9×10^6 psi. Thus, by using Eq. 9–3, the Euler buckling load is found to be

$$P_{cr} = \pi^2 EA/(L/r)^2 = \pi^2(1.9)(10^6)(4)(2)/(166.3)^2 = \underline{5424\text{ lb}} \qquad \text{Ans.}$$

(c) The axial stress under the action of the buckling load is

$$\sigma_{cr} = P_{cr}/A = 5424/(4)(2) = 678 \text{ psi}$$

From App. A, the elastic strength of air-dried douglas fir is found to be 6400 psi. Thus, the stress ratio is

$$\sigma_{cr}/\sigma_y = 678/6400 = 0.1059 = \underline{10.59 \text{ percent}} \qquad \text{Ans.}$$

(This demonstrates that buckling can occur at stresses well below the elastic limit of a material for sufficiently slender columns.)

(d) The maximum load that the column can support with a factor of safety of 2.5 must be based on the Euler buckling load. Therefore,

$$P_{max} = P_{cr}/2.5 = 5424/2.5 = \underline{2170 \text{ lb}} \qquad \text{Ans.}$$

Example 9–2 Two 2×2×1/8-in. steel angles 9 ft long are to serve as a pivot-ended column. Determine the slenderness ratio and the Euler buckling load if:

(a) The two angles are not connected and each acts as an independent member.

(b) The two angles are fastened together as shown in Fig. 9–6 to act as a unit.

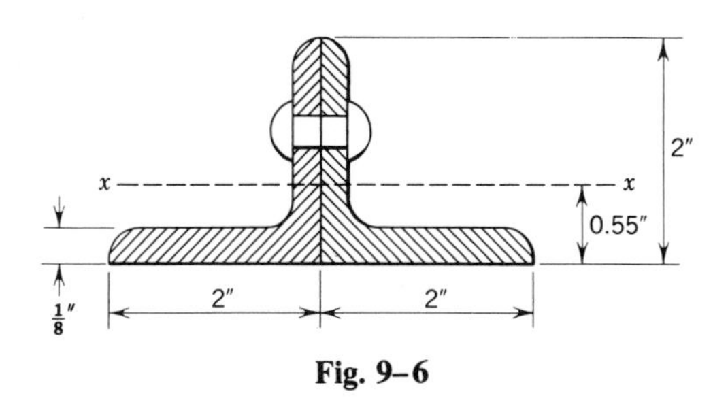

Fig. 9–6

Solution (a) If the angles are not connected and each acts independently, the slenderness ratio is determined by the least radius of gyration of the individual cross sections. From App. C, r for the Z–Z axis for this angle is 0.40 in. Thus, the slenderness ratio is

$$L/r = (9)(12)/0.40 = \underline{270} \qquad \text{Ans.}$$

The area for each angle is 0.48 sq in., making the total area 0.96 sq in. Using $30(10^6)$ psi for the modulus of elasticity for steel, the buckling load is found to be

$$P_{cr} = \pi^2 EA/(L/r)^2 = \pi^2(30)(10^6)(0.96)/(270)^2 = \underline{3900 \text{ lb}} \qquad \text{Ans.}$$

(b) With the two angles connected as shown in Fig. 9–6, buckling will occur about the X–X axis and the radius of gyration (see App. C) is 0.63 in. Thus, the slenderness ratio is

$$L/r = (9)(12)/0.63 = \underline{171.4} \qquad \text{Ans.}$$

The corresponding Euler buckling load is then

$$P_{cr} = \pi^2 EA/(L/r)^2 = \pi^2(30)(10^6)(0.96)/(171.4)^2 = \underline{9680 \text{ lb}} \qquad \text{Ans.}$$

PROBLEMS

9–5* A hollow circular aluminum column with pivot ends is 8 ft long and has inner and outer diameters of 2 and 3 in., respectively. Determine the minimum radius of gyration of the cross section and the slenderness ratio.

9–6* A hollow circular brass tube 1.5 m long is used as a pivot-ended column. The inner and outer diameters of the tube are 24 and 30 mm, respectively. Determine the minimum radius of gyration of the cross section and the slenderness ratio.

9–7* A 12-ft pivot-ended column is constructed of rough-cut (i.e. full dimension) 1×4-in. timbers, as shown in Fig. P9–7. The timbers are nailed together so that the four of them act as a unit. Determine the minimum radius of gyration of the cross section and the slenderness ratio.

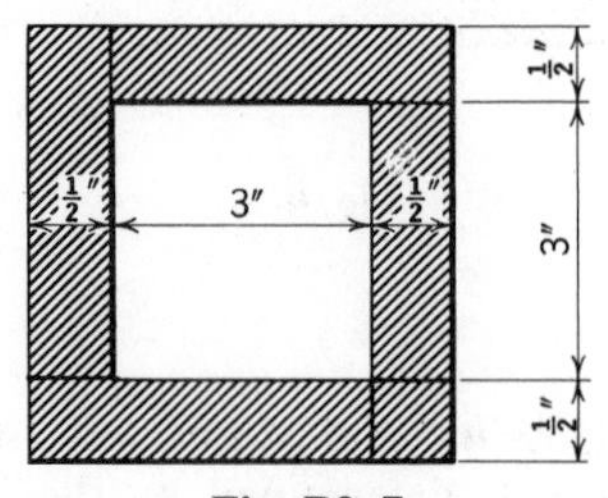

Fig. P9–7

9–8* A hollow circular aluminum tube 6 m long is used as a pivot-ended column. The slenderness ratio of the column is 120. The outside diameter of the tube is 150 mm. Determine the inside diameter of the tube.

9–9 The cross section of an 18-ft pivot-ended timber column is shown in Fig. P9–9. Determine the minimum radius of gyration of the cross section and the slenderness ratio.

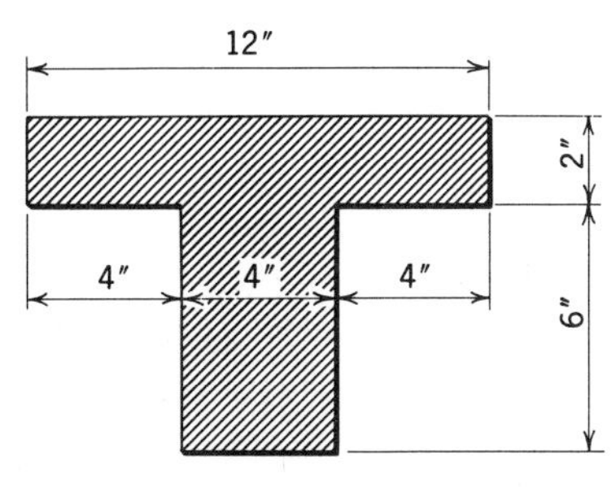

Fig. P9–9

9–10 A 4-m pivot-ended timber column is made of two 50×150-mm timber sections nailed together so that the two of them act as a unit with the cross section shown in Fig. P9–10. Determine the minimum radius of gyration of the cross section and the slenderness ratio.

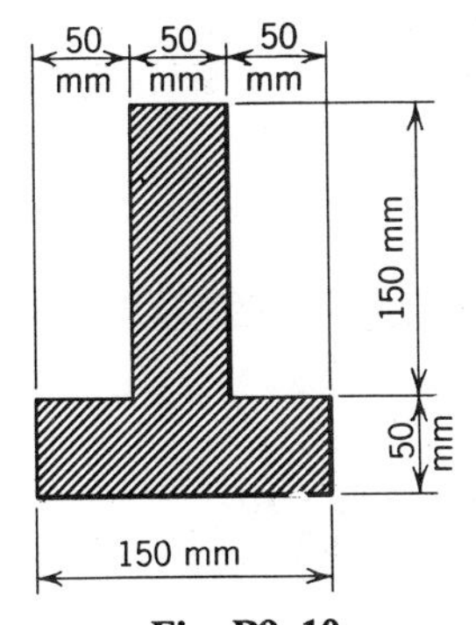

Fig. P9–10

9–11* A yardstick has a rectangular cross section $1\frac{1}{16} \times \frac{5}{32}$ in. Assuming Young's modulus for the wood to be 1.9 (10^6) psi, determine the slenderness ratio and the Euler buckling load if the yardstick is loaded as a pivot-ended column.

9–12* A 1/2-in. diameter steel rod 4 ft long is used as a pivot-ended column. Using 30 (10^6) psi for Young's modulus, determine:

(a) The slenderness ratio.
(b) The Euler buckling load.
(c) The axial stress in the rod under the action of the Euler buckling load.

9–13* A hollow circular aluminum tube 2 m long is used as a pivot-ended column. The inside and outside diameters of the tube are 30 and 40 mm, respectively, and Young's modulus for the aluminum is 70 GN/m^2. Determine:

(a) The slenderness ratio.
(b) The Euler buckling load.
(c) The mass the column could support with a factor of safety of 2.5.

9–14* A W8×28 steel member acts as a 25-ft pivot-ended column. Determine:

(a) The slenderness ratio.
(b) The Euler buckling load.
(c) The maximum load the column could support with a factor of safety of 1.8.

9–15 An 8-m pivot-ended steel column is composed of a single UB356×67 member. Determine:

(a) The slenderness ratio.
(b) The Euler buckling load.
(c) The mass the column could support with a factor of safety of 2.0.

9–16 A 4×4×3/8-in. aluminum angle is used as a 10-ft pivot-ended column. If Young's modulus for the aluminum is 10.6 (10^6) psi, determine:

(a) The slenderness ratio.
(b) The Euler buckling load.
(c) The maximum load the column can carry with a factor of safety of 2.5.

9–17* Two 75×50×8-mm steel angles serve as a pivot-ended column 4 m long. Determine the total load required to buckle the two members if:

(a) They act independently of each other.
(b) They are riveted together as a unit, as shown in Fig. P9–17.

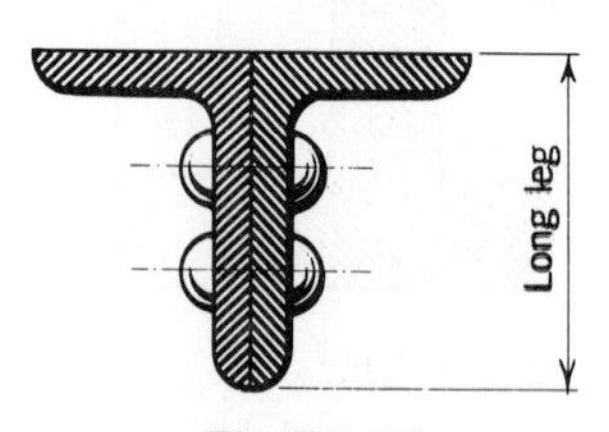

Fig. P9–17

9–18* An extruded aluminum pivot-ended column 4 m long has a solid cross section as shown in Fig. P9–18. Young's modulus for the aluminum is 70 GN/m^2.

Determine:
(a) The slenderness ratio.
(b) The Euler buckling load.
(c) The mass the column could support with a factor of safety of 2.25.

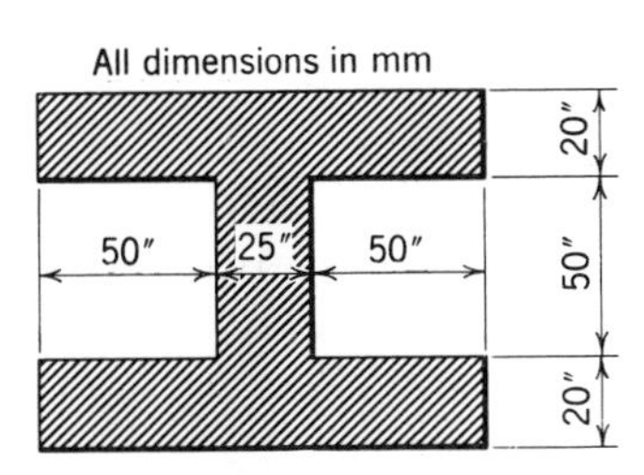

Fig. P9–18

9–19 A 600-kg mass is to be supported by a solid circular aluminum pivot-ended column 1.5 m long. If Young's modulus for the aluminum is 70 GN/m^2 and a factor of safety of 2 is specified, determine:
(a) The minimum diameter of the column.
(b) The slenderness ratio.

9–20 A 15-ft pivot-ended timber column with a solid square cross section must support a 40,000-lb load. If the modulus of elasticity of the timber is 1.9 (10^6) psi and a factor of safety of 2.5 is specified, determine:
(a) The size of the cross section.
(b) The slenderness ratio.

9–4 EFFECT OF END CONDITIONS

The Euler buckling formula, as expressed by either Eq. 9–2 or Eq. 9–3, was derived for a column with pivoted ends. The Euler equation changes for columns with different end conditions such as the four common ones shown in Fig. 9–7.

While it is possible to set up the differential equation with the appropriate boundary conditions to determine the Euler equation for each new case, a more common approach makes use of the concept of an effective length. The pivot-ended column, by definition, has zero bending moments at each end. The length L in the Euler equation, therefore, is the distance between successive points of zero bending moment. All that is needed to modify the Euler column formula for use with other end conditions is to replace L by L', where L' is defined as the *effective length* of the column (the distance between two successive inflection points or points of zero moment).

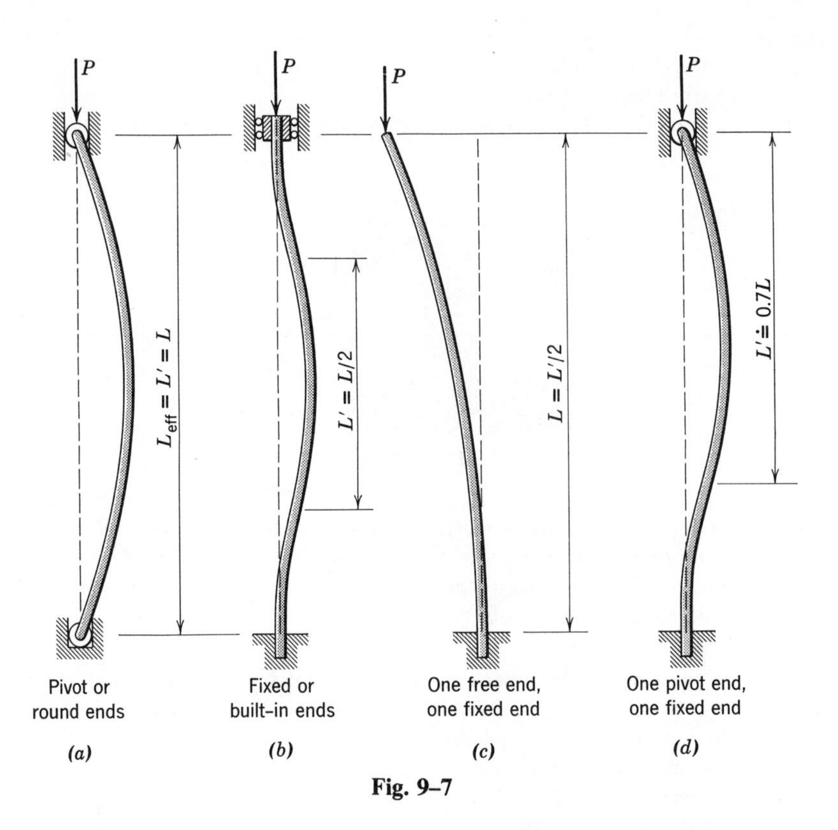

Fig. 9–7

The ends of the column in Fig. 9–7*b* are built in or fixed; since the deflection curve is symmetrical, the distance between successive points of zero moment (inflection points) is half the length of the column. Thus, the effective length L' of a fixed-end column for use in the Euler column formula is half the true length ($L' = 0.5L$). The column in Fig. 9–7*c*, being fixed at one end and free at the other end, has zero moment only at the free end. If a mirror image of this column is visualized below the fixed end, however, the effective length between points of zero moment is seen to be twice the actual length of the column ($L' = 2L$). The column in Fig. 9–7*d* is fixed at one end and pinned at the other end. The effective length of this column cannot be determined by inspection, as could be done in the previous two cases; therefore, it is necessary to solve the differential equation to determine the effective length. This procedure yields $L' = 0.7L$.

A pin-ended column is usually loaded through a pin which, as a result of friction, is not completely free to rotate; therefore, there will always be an indeterminate moment at the ends of a pin-connected column that will reduce the distance between the inflection points to a value less than L.

Also, it is impossible to support a column so that all rotation is eliminated, and so the effective length of the column in Fig. 9–7*b* will be somewhat greater than $L/2$. As a result, it is usually necessary to modify the effective column lengths indicated by the ideal end conditions. The amount of the corrections will depend on the individual application. In summary, the term L/r in all column formulas in this book is interpreted to mean the effective slenderness ratio L'/r. In the problems in this book, the length given for a member is assumed to be the effective length unless otherwise noted.

PROBLEMS

9–21 By solving the differential equation of the elastic curve for a column with an applied axial load and equal moments applied at each end, verify the effective length given for the column shown in Fig. 9–7*b*.

9–22 Figure P9–22 shows a free-body diagram of a column of the type shown in Fig. 9–7*d*. Derive an equation for the bending moment as a function of x and apply the boundary conditions to show that the critical buckling load is given by the solution to the transcendental equation $\tan pL = pL$, where $p^2 = P/EI$. Verify the effective length given for the column shown in Fig. 9–7*d*.

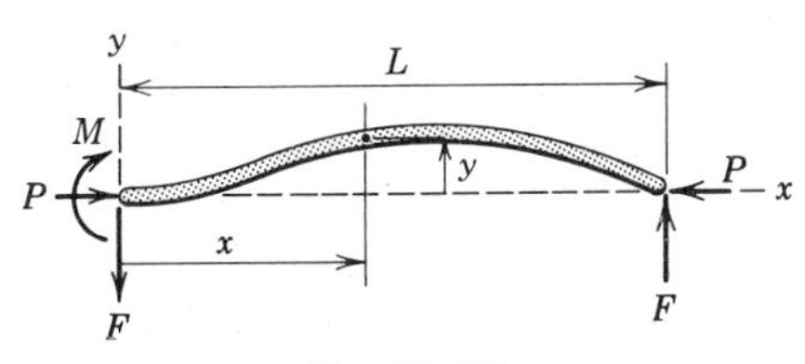

Fig. P9–22

9–23* Consider the rigid beam supported by two columns built in at both ends, as shown in Fig. P9–23. Assuming that buckling would occur as shown, determine the

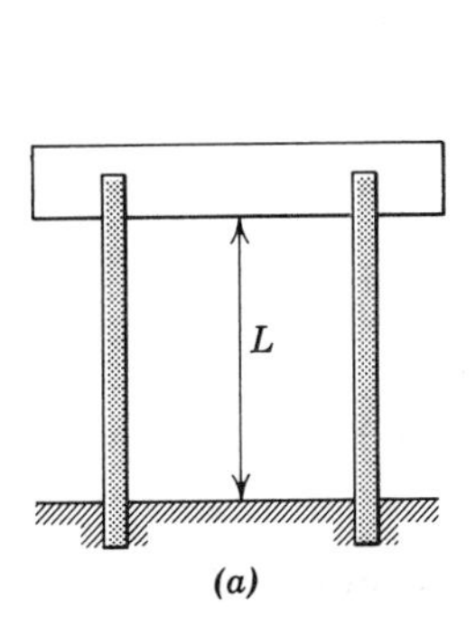

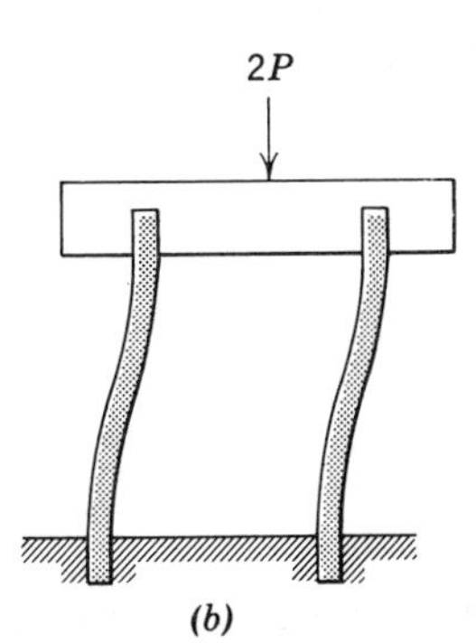

Fig. P9–23

effective lengths of the columns by solving the differential equation of the elastic curve.

9–24* A finished 2×4 (actual size $1\frac{5}{8}$ in.×$3\frac{1}{2}$ in.×8 ft long) is used as a fixed-end, free-end column. If Young's modulus for the timber is 1.9 (10^6) psi and a factor of safety of 2 is specified, determine:

(a) The slenderness ratio.
(b) The maximum safe load.

9–25* A 4×3×1/4-in. aluminum angle [$E=10.6$ (10^6) psi] is used as a fixed-ended column having an actual length of 12 ft. If a factor of safety of 1.5 is specified, determine:

(a) The slenderness ratio.
(b) The maximum safe load.

9–26* A UB203×30 steel section is used for a fixed-end, pinned-end column having an actual length of 6.5 m. Determine:

(a) The slenderness ratio.
(b) The maximum mass the column could support if a factor of safety of 2 is specified.

9–5 THE SECANT FORMULA

Many real columns do not behave as predicted by the Euler formula because of imperfections in the alignment of the loading. In this section, the effect of imperfect alignment is examined by considering an eccentric loading. The pivot-ended column shown in Fig. 9–8 is subjected to compressive forces acting at a distance e from the centerline of the undeformed column. As the loading is increased, the column deflects laterally, as shown in Fig. 9–8*b*. Proceeding as in Art. 9–3, axes are chosen with the origin at the center of the span. The bending moment at any section is

$$M = P(e+\delta-y)$$

and, if the stress does not exceed the proportional limit and deflections are small, the differential equation for the elastic curve becomes

$$EI\frac{d^2y}{dx^2} = P(e+\delta-y)$$

or

$$\frac{d^2y}{dx^2} + \frac{P}{EI}y = \frac{P}{EI}(e+\delta) \qquad \text{(a)}$$

This is the same differential equation as Eq. (a) of Art. 9–3 except that the

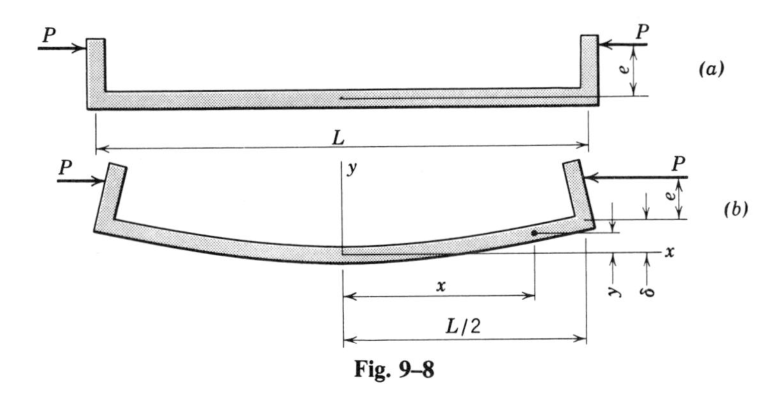

Fig. 9–8

constant on the right-hand side is $P(e+\delta)/EI$ rather than $P\delta/EI$. The solution, therefore, is of the form

$$y = A\sin px + B\cos px + C \tag{b}$$

where $p^2 = P/EI$, $C = (e+\delta)$, and A and B are constants to be determined from the boundary conditions, as was the case in Art. 9–3. Since the slope and deflection of the elastic curve are zero at the origin (that is, when $x=0$, $y=0$ and $dy/dx=0$), the constants A and B become

$$A = 0 \quad \text{and} \quad B = -(e+\delta)$$

The solution to Eq. (a) thus becomes

$$y = (e+\delta)\left[1 - \cos\sqrt{P/(EI)}\; x\right] \tag{9–4}$$

In this case a relationship can be found between the maximum deflection δ, which is the value of y at $x = L/2$, and the load P. Thus,

$$\delta = e + \delta - e\cos\sqrt{P/(EI)}\;(L/2) - \delta\cos\sqrt{P/(EI)}\;(L/2)$$

or

$$\delta = e\left[\frac{1-\cos\sqrt{P/(EI)}\;(L/2)}{\cos\sqrt{P/(EI)}\;(L/2)}\right] = e\left[\sec\sqrt{P/(EI)}\;(L/2) - 1\right] \tag{c}$$

Equation (c) indicates that, for a given column in which E, I, and L are fixed and $e > 0$, the column exhibits lateral deflection for even small values of the load P. For any value of e, the quantity

$$\sec\sqrt{P/(EI)}\,(L/2)$$

approaches positive or negative infinity as the argument $\sqrt{P/(EI)}\,(L/2)$ approaches $\pi/2$, $3\pi/2$, $5\pi/2, \ldots$, and the deflection δ increases without bound, indicating that the critical load corresponds to one of these angles. If $\pi/2$ is chosen, since this value yields the smallest load, then

$$\sqrt{P/(EI)}\,(L/2) = \pi/2$$

or

$$\sqrt{P/(EI)} = \pi/L$$

from which

$$P_{cr} = \frac{\pi^2 EI}{L^2} \tag{9–2}$$

which is the Euler formula discussed in Art. 9–3.

Since it has been assumed that the stresses do not exceed the proportional limit (when the elastic curve equation was written), the maximum compressive stress can be obtained by superposition of the axial stress and the maximum bending stress. The maximum bending stress occurs on a section at the midspan of the column where the bending moment assumes its largest value, $M_{max} = P(e+\delta)$. Thus, the maximum stress is

$$\sigma_{max} = \frac{P}{A} + \frac{M_{max}c}{I} = \frac{P}{A} + \frac{P(e+\delta)c}{Ar^2} \tag{d}$$

in which r is the radius of gyration of the column cross section. From Eq. (c)

$$e + \delta = e\sec\sqrt{\frac{P}{EI}}\,\frac{L}{2}$$

therefore,

$$\sigma_{max} = \frac{P}{A}\left(1 + \frac{ec}{r^2}\sec\sqrt{\frac{P}{EI}}\,\frac{L}{2}\right)$$

or

$$\frac{P}{A}=\frac{\sigma_{\max}}{1+\dfrac{ec}{r^2}\sec\left(\dfrac{L}{2r}\sqrt{\dfrac{P}{EA}}\right)} \tag{9–5}$$

Equation 9–5 is known as the secant formula and relates the average unit load P/A to the dimensions of the column, the properties of the column material, and the eccentricity e. The term L/r is the same slenderness ratio found in the Euler buckling formula, Eq. 9–3. For columns with different end conditions, the effective slenderness ratio L'/r should be used, as discussed in Art. 9–4. The quantity ec/r^2 is called the *eccentricity ratio* and is seen to depend on the eccentricity of the load and the dimensions of the column. If the column were loaded axially, e would presumably be zero, and the maximum stress would be equal to P/A. It is virtually impossible, however, to eliminate all eccentricity that might result from various factors such as the initial crookedness of the column, minute flaws in the material, and a lack of uniformity of the cross section as well as accidental eccentricity of the load. A committee of the American Society of Civil Engineers[2] made an extensive study of the results of a great many column tests and came to the conclusion that a value of 0.25 for ec/r^2 would give results with the secant formula that would be in good agreement with experimental tests on axially loaded columns of structural steel in ordinary structural sizes. A more complete discussion of eccentrically loaded columns is included in Art. 9–7. When Eq. 9–5 is used, $\sigma_{\max}$ is the proportional limit, although the yield point of structural steel may be used for convenience, and P is the load at which inelastic action would start to occur. This load should be divided by a factor of safety for design purposes. Note that P/A occurs on both sides of the equation, and thus it is not directly proportional to σ. Consequently, the load resulting from reducing σ to some proportion of the proportional limit will not be reduced by the same amount. In order to obtain a given factor of safety, the factor must be applied to the load each time it occurs in the formula and not to the value of σ; that is, in Eq. 9–5 kP_w should be substituted for P, where P_w is the working load (maximum safe load) and k is the factor of safety.

In order to make effective use of Eq. 9–5, curves showing P/A and L/r can be drawn for various values of ec/r^2 for any given material. Figure 9–9 is such a set of curves for a material with $\sigma=40{,}000$ psi and $E=30(10^6)$ psi (or $\sigma=280$ MN/m^2 and $E=200$ GN/m^2). For these material properties, graphic solutions can be obtained from Fig. 9–9. Digital computers

[2]Final Report of Special Committee on Steel Column Research, *Trans. Am. Soc. of Civil Engrs.*, Vol. 98, 1933.

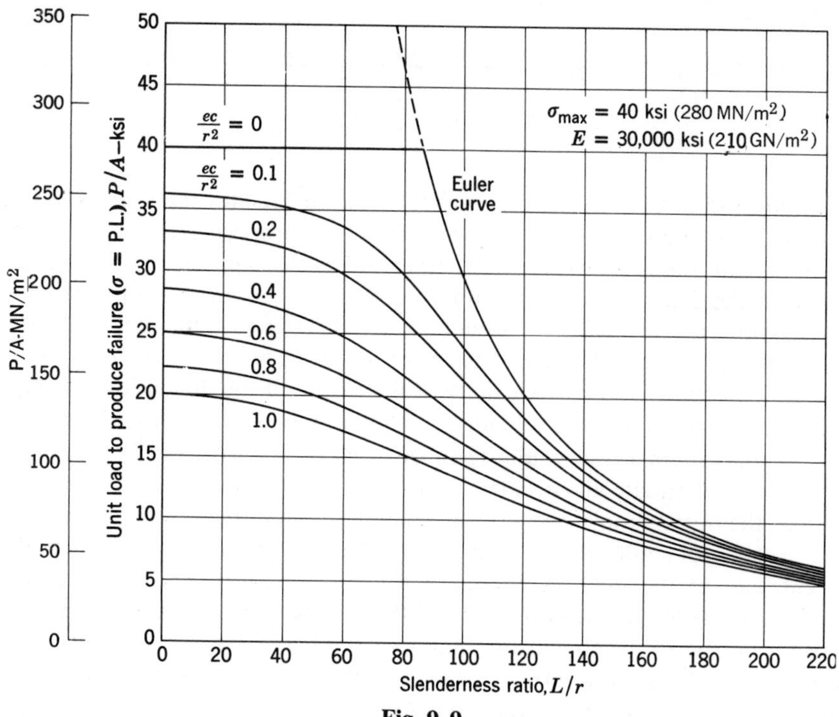

Fig. 9–9

can also be programmed to solve the secant formula directly using iterative techniques.[3]

The outer envelope of Fig. 9–9, consisting of the horizontal line $\sigma = 40$ ksi and the Euler curve, corresponds to zero eccentricity. The Euler curve is truncated at 40 ksi since this is the maximum allowable stress for the material. With Young's modulus equal to 30,000 ksi, the truncation occurs when $L/r = 86$. From the data presented in Fig. 9–9, it is seen that eccentricity of loading plays a significant role in reducing the working load (maximum safe load) in the short and intermediate column ranges (slenderness ratios less than 150 for the steel in Fig. 9–9). For large slenderness ratios, the curves for the various eccentricity ratios tend to merge with the Euler curve. Consequently, the Euler formula can be used to analyze columns with large slenderness ratios. For a given problem, *the slenderness ratio must be computed* to determine whether or not the Euler equation is valid.

[3]For a program for the secant formula, see *Numerical Methods in Fortran IV Case Studies*, W. S. Dorn and D. D. McCracken, Wiley, New York, 1972, pp. 43–49.

The quantities E, I, and L can be eliminated from Eq. (c) by using Eq. 9–2, in which case the maximum deflection becomes

$$\delta = e\left[\sec(\pi/2)\sqrt{P/P_{cr}} - 1\right] \qquad \text{(e)}$$

From this equation it would appear that the deflection δ becomes infinite as P approaches P_{cr}; however, under these conditions, the slope of the deflected shape of the column is no longer sufficiently small to be neglected in the curvature expression. As a result, accurate deflections can only be obtained by using the nonlinear form of the differential equation of the elastic curve. Graphs of Eq. (e) for various values of e are shown in Fig. 9–10. From these curves it can be seen that the deflection is extremely small as e approaches zero until the load is near P_{cr}, at which time the

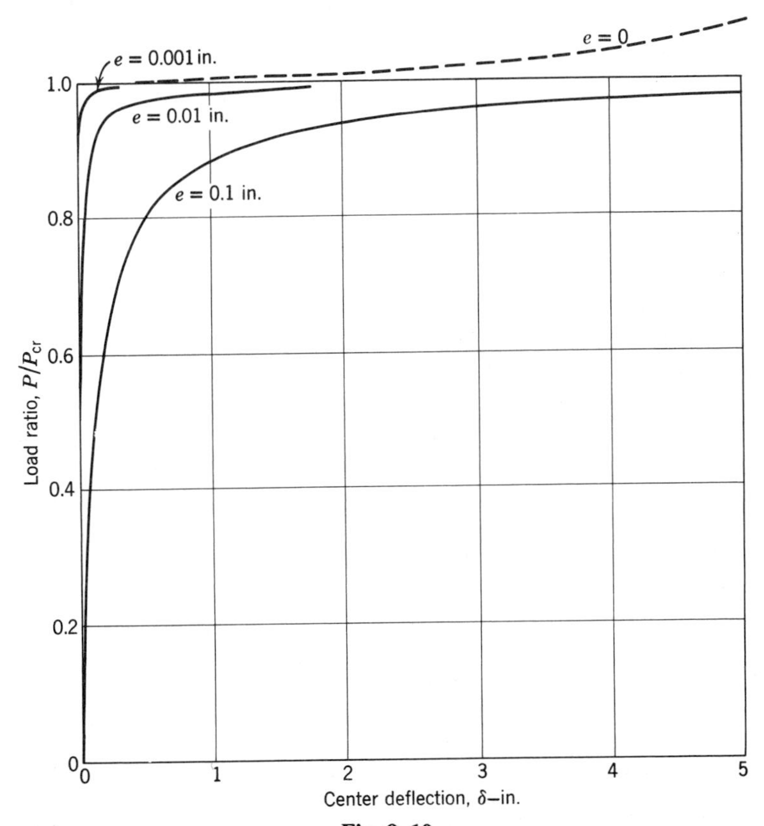

Fig. 9–10

deflection increases very rapidly. In Fig. 9–10 the curve shown in dashed lines for $e=0$ is a plot of the exact solution.[4] This limiting case represents an axially loaded column, and the remaining curves illustrate how the deflection under a supposedly axial load can be initiated by a small accidental eccentricity. The curve for $e=0$ indicates that a load only slightly greater than P_{cr} would probably cause the column to collapse because the proportional limit of the material would have been exceeded.[5] Note the similarity between this curve and that of Fig. 9–3.

PROBLEMS

Note. In the following problems, use the effective slenderness ratio for the particular end conditions and assume that Euler's equation is applicable for steel with L/r greater than 140 and for aluminum or timber with L/r greater than 80.

9–27* Use Fig. 9–9 to estimate the maximum safe compressive load for a pivot-ended steel strut having a 1×2-in. cross section and a 30-in. length. Use a factor of safety of 2.5 and assume ec/r^2 to be 0.1. What is the corresponding value of the eccentricity?

9–28* Use Fig. 9–9 to estimate the maximum safe compressive load for a fixed-ended solid steel bar having a 50-mm diameter and a 2-m length. The load acts 3 mm from the axis of the bar. A factor of safety of 2.5 is specified.

9–29* The American Society of Civil Engineers recommends using a value of 0.25 for the eccentricity ratio for structural steel columns. What eccentricity does this imply for:

(a) A solid circular bar of diameter d?
(b) A thin-walled tube of thickness t and diameter d?
(c) A W8$\times$17 wide flange section?

9–30 Compute the value of L/r corresponding to $P/A=20$ ksi and $ec/r^2=0.20$ for the data used to plot Fig. 9–9. Check this point to see that it is in reasonable agreement with Fig. 9–9.

9–31 A UB203$\times$25 steel universal beam 1.25 m long is used as a free-end, fixed-end column. Use Fig. 9–9 to estimate the maximum safe load this column could support if $ec/r^2=0.2$ and a factor of safety of 2 is specified.

9–32* A W10$\times$21 steel section 12 ft long was used as a pivot-ended column. The column failed under a load of 105 kips. Use Fig. 9–9 to estimate the probable eccentricity of the load.

9–33* An S12$\times$31.8 steel section 15 ft long is used as a fixed-ended, pivot-ended column. If the column carries a load of 50 kips with an assumed eccentricity ratio of 0.30, determine the factor of safety with the aid of Fig. 9–9.

9–34 A W14$\times$78 steel section 20 ft long is used for a fixed-ended column. Use

[4]See *Theory of Elastic Stability*, S. Timoshenko and J. Gere, Second Edition McGraw-Hill, New York, 1961, pp. 77–82.

[5]The curve is valid for any deflection only until the proportional limit of the material is exceeded, at which time the curve drops.

the secant formula to determine the allowable load P if it is assumed that the eccentricity ratio is 0.25, the modulus of elasticity is 30,000 ksi, the proportional limit is 36,000 psi, and the factor of safety is 2.

9–35 A UB356×67 steel section 2 m long is used for a fixed-ended, pivot-ended column. Use the secant formula to determine the allowable load P if it is assumed that the eccentricity ratio is 0.25, the modulus of elasticity is 200 GN/m^2, the proportional limit is 250 MN/m^2, and the factor of safety is 2.5.

9–6 EMPIRICAL COLUMN FORMULAS–CENTRIC LOADING

Referring to Fig. 9–9, one observes that the envelope of the family of curves is the horizontal line $P/A = 40$ ksi up to L/r of 86 and the plot of the Euler formula (Eq. 9–3) for L/r greater than 86. This curve represents the failure unit loads for *ideal* columns of any length under centric loading. However, for real columns (which will always have built-in eccentricities), the true curve will be an S-shaped curve similar to those shown in Fig. 9–9. Numerous column experiments indicate that the Euler formula (Eq. 9–2 or 9–3) is reliable for designing axially loaded columns, provided the slenderness ratio is within the range where the eccentricity has relatively little effect. This range is called the *slender* range, which in Fig. 9–9 would begin in the region of L/r equal to 120–140 (the region of transition depends on the material). At the low end of the slenderness scale, the column would behave essentially as a compression block, and the failure unit load would be the compressive strength of the material (for metals, usually the elastic strength σ_y). The extent of this range, the *compression block* range, is a matter of judgment, or is dictated by specifications. In Fig. 9–9, the small eccentricity curves seem to indicate that the range 0 to 40 could be considered the compression block range. The range between the compression block and the slender ranges is known as the *intermediate* range, and the only equation of a rational nature that applies to real columns in this range is the secant formula Eq. 9–5. Since the application of the secant formula to centric loading requires an estimate of the accidental eccentricity ratio, the equation acquires an empirical nature. Furthermore, the application is so involved that simpler empirical formulas have been developed that give results in reasonable agreement with experimental results within the intermediate range. Although a great many equations have been proposed, most of them come under one of three major classifications. These three general formulas are the *Gordon-Rankine*, the *straight-line*, and the *parabolic* formulas, described in the following paragraphs.

The *Gordon-Rankine* formula, one of the first empirical formulas, was developed by Gordon and Rankine, English engineers, during the middle of the nineteenth century. The formula is usually written as

$$\frac{P}{A} = \frac{\sigma_0}{1 + C_1(L/r)^2} \tag{a}$$

in which σ_0 and C_1 are experimentally determined constants. The equation plots as a curve similar to the secant curve.

The *straight-line* formula, proposed by T. H. Johnson about 1890, is written in general form as

$$\frac{P}{A} = \sigma_0 - C_2 \frac{L}{r} \tag{b}$$

where, again, σ_0 and C_2 are experimentally determined material-dependent constants.

At about the same time that T. H. Johnson proposed the straight-line formula, Professor J. B. Johnson proposed a *parabolic* formula of the general form

$$\frac{P}{A} = \sigma_0 - C_3 \left(\frac{L}{r}\right)^2 \tag{c}$$

where σ_0 and C_3 are the experimentally determined constants. This latter formula with the proper choice of constants can be made to agree quite closely in the intermediate range with the secant formula with certain prescribed numerical constants. See, for example, the close correlation between curves (2) and (7) in Fig. 9–11.

For design purposes, the entire range of unit loads for a given material is covered by an appropriate set of specifications known as a *column code*. The code will specify the empirical formula for the intermediate range and the limits of the intermediate range; it may also specify the Euler formula in the slender range. If the code is written for *safe* (or *allowable*) loads, the factor of safety k will be specified for the Euler formula and will either be specified or be included in the constants for the empirical formula. A few representative codes are listed in Table 9–1, in which the Euler formula is specified for the slender range, except for code 5. Some codes, such as 4 and 5, are for ultimate (buckling) loads. When these codes are used to determine safe loads, the values obtained from the formulas must be divided by an appropriate factor of safety. The data in Table 9–1 are taken or adapted from handbooks published by the American Institute of Steel Construction (A.I.S.C.), the American Association of State Highway Officials (A.A.S.H.O.), the Armed Forces Supply Support Center (Mil-

Hdbk-5), the Aluminum Company of America (Alcoa Handbook), the Aluminum Association (Aluminum Construction Manual), and the Forest Products Laboratory (U. S. D. A. Handbook #72). A graphic comparison of selected codes is provided in Fig. 9–11.

In order to provide a smooth transition from the intermediate to the slender range, the constants may be adjusted so that the intermediate curve becomes tangent to the Euler curve, thus defining the upper limit of the intermediate range. For safe loads, the factor of safety to be included in the Euler formula can be adjusted to make this curve tangent to any of the formulas for safe loads in the intermediate range. This is done by equating the two expressions for P/A and equating the two expressions for the first derivatives of P/A with respect to L/r. The resulting two equations can be solved for the factor of safety to be used in the Euler formula and the

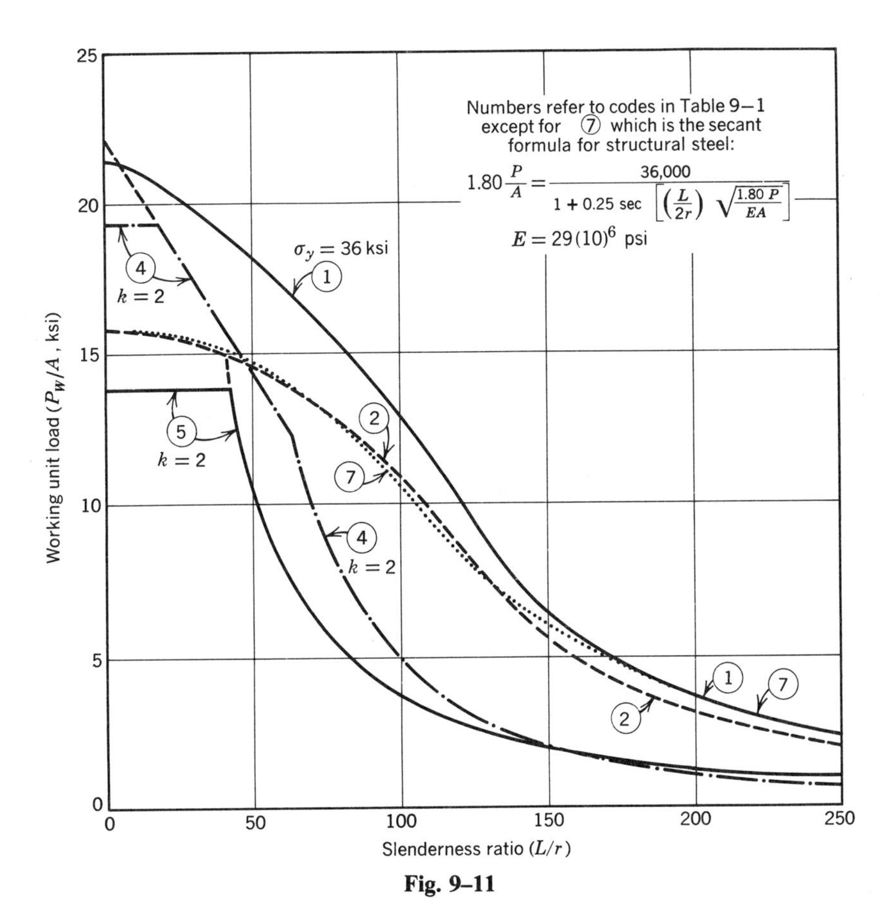

Fig. 9–11

TABLE 9–1 Some Representative Column Codes

Code No.	Source	Material and Modulus of Elasticity (ksi)	Intermediate Range Formulas and Limitations Note: $\frac{L}{r}$ refers to the effective ratio $\frac{L'}{r}$	Slender Range Factor of Safety
1	A.I.S.C. Specification 1970	Steel with Yield Point of σ_y 29(10^3)	$\frac{P_w}{A}=\frac{1}{k}\left[\sigma_y-\frac{\sigma_y}{2C_c^2}\left(\frac{L}{r}\right)^2\right]$ $C_c^2=\frac{2\pi^2 E}{\sigma_y}$ $k=\frac{5}{3}+\frac{3(L/r)}{8C_c}-\frac{(L/r)^3}{8C_c^3}$ $0\leqslant\frac{L}{r}\leqslant C_c$	23/12
2	Adapted from A.A.S.H.O. Specification 1973	A36 Structural Steel 29(10^3)	$\frac{P_w}{A}=16{,}000-\frac{1}{2}\left(\frac{L}{r}\right)^2$, psi[a] k included $0\leqslant\frac{L}{r}\leqslant 130$	2.24[b]
3	Aluminum Construction Manual	Aluminum Alloy 2014-T6 10.6(10^3)	$P_w/A=31{,}330-238(L/r)$, psi k included $11.3\leqslant(L/r)\leqslant 55$	1.95

[a]The formula as given in the A.A.S.H.O. specification is written in terms of the actual column length; changing the constant C_3 to 1/2 adjusts the formula for use with the effective length.

[b]Not in the A.A.S.H.O. specification, which specifies the secant formula (curve 7) for the slender range.

TABLE 9–1 (Continued)

Code No.	Source	Material and Modulus of Elasticity (ksi)	Intermediate Range Formulas and Limitations Note: $\frac{L}{r}$ refers to the effective ratio $\frac{L'}{r}$	Slender Range Factor of Safety
4	Adapted from Alcoa Structural Handbook 1960	Aluminum Alloy 2024-T4 $10.6(10^3)$	$\frac{P_u}{A} = 44{,}800 - 313\left(\frac{L}{r}\right)$, psi Ultimate load; use k as specified $18.5 \leqslant \frac{L}{r} \leqslant 64$	As specified
5	Adapted from Mil-Hdbk-5 1966	Magnesium Alloy ZK60A-T5 $6.5(10^3)$	$\frac{P_u}{A} = \frac{76(10^5)}{(L/r)^{1.5}}$, psi Ultimate load; use k as specified $43 < \frac{L}{r}$	No Euler Equation
6	Adapted from U.S.D.A. Wood Handbook #72 1955	Douglas Fir (coast type) $1.6(10^3)$	For square or round solid cross sections $\frac{P_w}{A} = 1450 - 1.656(10^{-5})\left(\frac{L}{r}\right)^4$, psi k included $38 \leqslant \frac{L}{r} \leqslant 73.5$ Use 1450 for $\frac{L}{r} < 38$	3

L/r at the transition from intermediate to slender range. This technique will be demonstrated later. The practice of making the intermediate and Euler curves tangent is not universal; many column codes plot as curves similar to that for curve ④ of Fig. 9–11.

The formulas in Table 9–1 are representative of a large number of column equations that have been incorporated in various design codes. The slenderness ratio in this table is always the effective slenderness ratio L'/r. If no end conditions are specified in the problems presented later, the stated length is the effective length. Note that the use of high-strength materials will increase the allowable (P/A) value for short columns but will have little, if any, effect on the strength of long columns, since the critical load (the Euler load) depends on Young's modulus, not on the elastic strength of the material. Note also that the use of fixed or restrained ends, which has the effect of reducing the length of the column, materially increases the strength of slender columns but has much less influence on short compression members.

Instead of working with sundry codes for particular materials, it has been suggested that, if only one or two general types of curves would be satisfactory for all materials, the equations of the curves could be presented in dimensionless form to apply to columns made of any material. The usual technique is to divide all values of P/A by a stress parameter σ_0, which is usually (but not always) taken as the elastic strength σ_y of a material, and all values of L/r by a slenderness ratio parameter, usually the slenderness ratio, which when applied to the Euler equation will give unity for the ratio $(P/A)/\sigma_0$ (a one-to-one ratio). The resulting curve is illustrated by Fig. 9–12 where, for convenience, the parameters are labeled R and S. The general equation for the Euler part of the curve is

$$R=\frac{1}{S^2} \qquad (S \geqslant S_t) \tag{9–6}$$

and the general equation for the parabola, which is tangent to the Euler curve, is

$$R=1-C_3S^2 \tag{d}$$

In order to find the values of C_3 and S_t, the technique previously described will be employed as follows. Equating the two expressions above yields

$$1-C_3S_t^2=1/S_t^2$$

and equating the first derivatives of these expressions yields

$$-2C_3S_t=-2/S_t^3$$

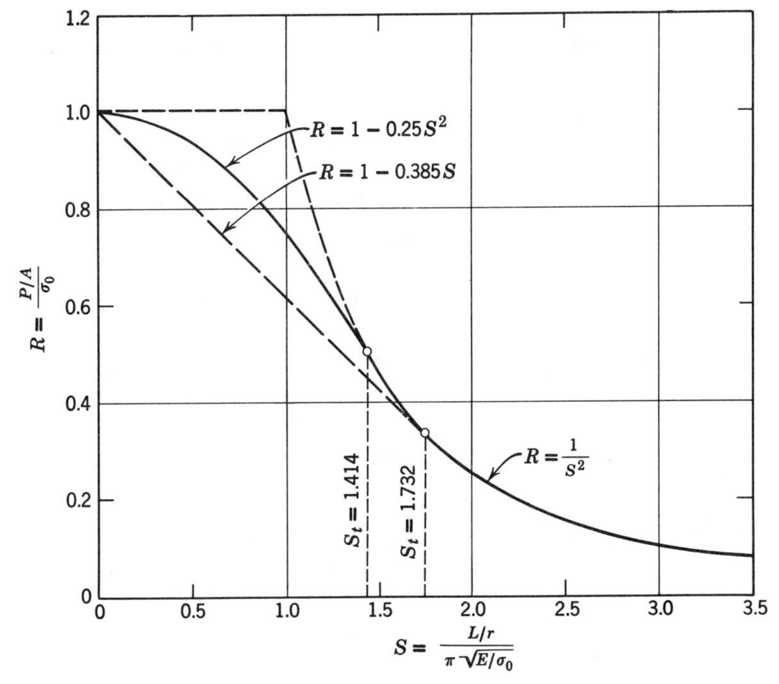

Fig. 9–12

The solution of these two equations for C_3 and S_t yields $C_3 = 0.250$ and $S_t = \sqrt{2} = 1.414$. Hence, the equation of the parabola is

$$R = 1 - 0.250S^2 \qquad (0 \leqslant S \leqslant 1.414) \tag{9–7}$$

In similar fashion, using the general straight-line equation,

$$R = 1 - C_2 S \tag{e}$$

and proceeding as before, the constants are evaluated as, $C_2 = 0.385$ and $S_t = \sqrt{3} = 1.732$. Hence, the equation of the straight line tangent to the Euler curve is

$$R = 1 - 0.385S \qquad (0 \leqslant S \leqslant 1.732) \tag{9–8}$$

To recapitulate, the nondimensional curves presented here apply to any material provided that the behavior of the column can be reliably predicted by one of the two curves illustrated. Note that for a straight-line formula [Eq. (b)] the value of σ_0 is often an experimentally determined stress larger than the compressive strength σ_y. In such cases, the straight-line curve in Fig. 9–12 will have a horizontal segment $R_0 = \sigma_y/\sigma_0$ at the low L/r end, so that P/A will never exceed σ_y (see, for example, curve ④ in Fig. 9–11). Note also that if the best straight line for a given material is not tangent to the Euler curve, Eq. 9–8 may not be reliable.

All of the discussion so far has been concerned with *primary* instability, where the column deflects as a whole into a smooth curve. No discussion of compression loading is complete without reference to *local instability* in which the member fails locally by crippling of thin sections. Thin open sections such as angles, channels, and H-sections are particularly sensitive to crippling failure. The design of such members to avoid crippling failure is usually governed by specifications controlling the width-thickness ratios of outstanding flanges. Closed section members of thin material (thin-walled tubes, for example) must also be examined for crippling failure when the members are short (see Mil-Hdbk-5).

Example 9–3 Two steel C10×25 channels are latticed 5 in. back to back, as shown in Fig. 9–13*a* and *b*, to form a column. Determine the maximum allowable axial load for effective lengths of 25 ft and 40 ft using:

(a) Code 1 for structural steel (see App. A).

(b) Equations 9–6 and 9–7 for structural steel. Use the same factors of safety as in part (a).

Solution (a) Both I_x and I_y (see Fig. 9–13) or r_x and r_y must be known in order to determine the minimum radius of gyration. Properties of the channel section are given in App. C. The value of I_y for two channels is obtained by using the parallel axis theorem; thus,

$$I_y = 2I_c + 2A\,d^2 = 2Ar_c^{\,2} + 2A\,d^2$$

and

$$r_y = \sqrt{I_y/(2A)} = \sqrt{\frac{2A\,(r_c^{\,2} + d^2)}{2A}} = \sqrt{r_c^{\,2} + d^2}$$

The above expression indicates that the radius of gyration for the two channels is the same as that for one channel, a fact obtained directly from the definition of radius of gyration. Then

$$r_y = \sqrt{(0.68)^2 + (2.50 + 0.62)^2} = 3.19 \text{ in.}$$

which is less than the tabular value of 3.52 in. for r_x. This means that the

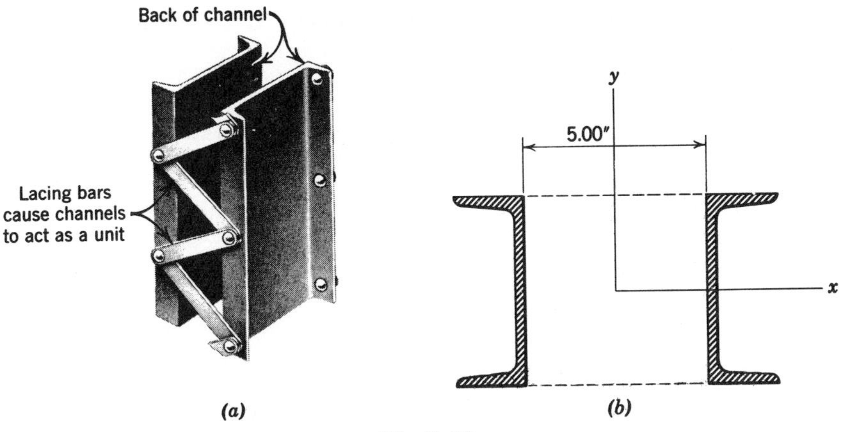

Fig. 9–13

column tends to buckle with respect to the y axis, and the slenderness ratios are

$$\frac{L}{r}=\frac{12(25)}{3.19}=94.0 \quad \text{and} \quad \frac{12(40)}{3.19}=150.5$$

From code 1,

$$C_c^2=2\pi^2(29)(10^3)/36=15{,}900 \qquad C_c=126$$

The slenderness ratios above indicate that the 25-ft column is in the intermediate range; hence,

$$k\frac{P}{A}=36-\frac{36(94^2)}{2(15{,}900)}=26.0$$

The factor of safety is given by

$$k=\frac{5}{3}+\frac{3(94/126)}{8}-\frac{(94/126)^3}{8}=1.89$$

Hence, the safe load for the 25-ft column is

$$P=\frac{26.0(2)(7.33)}{1.89}=\underline{202\text{ kips}} \qquad \text{Ans.}$$

The 40-ft column having a slenderness ratio of 150.5 is in the slender range; hence,

$$P=\frac{\pi^2 EA}{k(L/r)^2}=\frac{\pi^2(29)(10^3)(2)(7.33)}{(23/12)(150.5^2)}=\underline{96.7 \text{ kips}} \qquad \text{Ans.}$$

(b) Reference to Fig. 9–12 shows that the parameter $\pi\sqrt{E/\sigma_0}$ is evaluated as

$$\pi\sqrt{E/\sigma_y}=\pi\sqrt{29(10^3)/36}=89.16$$

For the 25-ft column, $S=94/89=1.055$ which is less than 1.414; therefore, the parabolic formula applies and

$$R=1-0.25(1.055^2)=0.722=(P/A)/\sigma_y$$

from which

$$P=0.722(36)(2)(7.33)/1.89=\underline{202 \text{ kips}} \qquad \text{Ans.}$$

For the 40-ft column, $S=150.5/89=1.69$ which is greater than 1.414; hence, the Euler formula applies and

$$R=\frac{1}{S^2}=\frac{1}{1.69^2}=0.350=\frac{P/A}{\sigma_y}$$

from which

$$P=0.350(36)(2)(7.33)/1.92=\underline{96.3 \text{ kips}} \qquad \text{Ans.}$$

Example 9–4 Determine the dimensions necessary for a 20-in. rectangular strut to carry an axial load of 1500 lb with a factor of safety of 2. The material is aluminum alloy 2024-T4, and the width of the strut is to be twice the thickness.

(a) Use code 4.
(b) Use Eqs. 9–6 and 9–8.

Solution (a) The code is represented by two different equations that depend on the value of L/r, which in turn will depend on the equation used. Thus, it will be necessary to assume that one of the equations applies and use it to obtain the dimensions of the column, after which the value of L/r must be calculated and used to check the validity of the equation

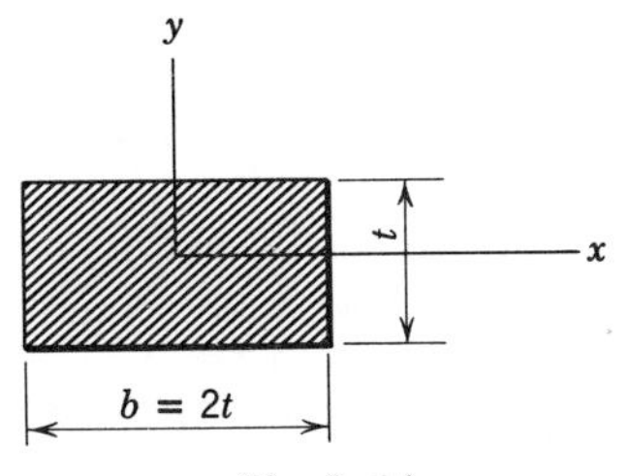

Fig. 9–14

used. Assume L/r is less than 64, in which case the straight-line equation is valid. The cross section is shown in Fig. 9–14, and the least moment of inertia I_x is equal to $bt^3/12$, the area A is equal to bt or $2t^2$ and the least radius of gyration is

$$r=\left(\frac{bt^3/12}{bt}\right)^{1/2}=\frac{t}{\sqrt{12}}$$

The slenderness ratio is

$$\frac{L}{r}=\frac{20\sqrt{12}}{t}$$

and, when this value and the expression for the area are substituted in the straight-line formula of code 4, it becomes

$$k\left(\frac{P}{A}\right)=\frac{2(1500)}{2t^2}=44{,}800-313\left(\frac{20\sqrt{12}}{t}\right)$$

from which

$$t=0.546 \text{ in.}$$

The value of L/r for this thickness is

$$\frac{L}{r}=\frac{20\sqrt{12}}{0.546}=127$$

which is greater than 64 and indicates that the straight-line formula is not valid. The problem must be solved again using the Euler equation. Thus

$$\frac{P}{A}=\frac{1500}{2t^2}=\frac{\pi^2E}{2(L/r)^2}=\frac{9.87(10.6)(10^6)}{2\left[(20)\sqrt{12}\,/t\right]^2}$$

from which

$$t^4 = 0.06882 \quad \text{and} \quad t = 0.512 \text{ in.}$$

The value of L/r is 135.3 for this thickness which confirms the use of the Euler formula. The dimensions of the cross section are

$$t = \underline{0.512 \text{ in.}} \quad \text{and} \quad b = \underline{1.024 \text{ in.}} \qquad \text{Ans.}$$

(b) Assume S is less than 1.732, in which case the straight-line formula applies. The parameter $\pi\sqrt{E/\sigma_0}$ becomes

$$\pi\sqrt{10.6(10^3)/44.8} = 48.3$$

consequently,

$$S = \frac{L/r}{\pi\sqrt{E/\sigma_0}} = \frac{20\sqrt{12}}{48.3t} = \frac{1.433}{t}$$

When this value is substituted into Eq. 9–8, R is found to be

$$R = 1 - 0.385\left(\frac{1.433}{t}\right) = \frac{k(P/A)}{\sigma_0} = \frac{2(1500)}{2t^2(44{,}800)} = \frac{0.0335}{t^2}$$

from which

$$t = 0.607 \text{ in.}$$

Upon substituting this value for t in the equation for S,

$$S = \frac{1.433}{0.607} = 2.36$$

which is greater than the 1.732 assumed, and therefore the slender-range formula must apply. Equation 9–6 yields

$$R = \frac{k(P/A)}{\sigma_0} = \frac{0.0335}{t^2} = \frac{1}{S^2} = \left(\frac{t}{1.433}\right)^2$$

from which

$$t^4 = 0.0688 \quad \text{and} \quad t = 0.512 \text{ in.}$$

Therefore, the dimensions of the cross sections are

$$t = \underline{0.512 \text{ in.}} \quad \text{and} \quad b = \underline{1.024 \text{ in.}} \qquad \text{Ans.}$$

Note that using Eq. 9–8 resulted in a larger value for t (0.607) than that obtained from code 4 (0.546). The reason for this is that the straight line in Fig. 9–12 is tangent to the Euler curve at (1.732, 0.333), whereas the straight line of code 4, if reduced to nondimensional form, would intersect the Euler curve at (1.333, 0.544). Thus, for any value of S less than 1.732 in Fig. 9–12, the value of R, hence P/A, would be more conservative than that obtained from code 4.

When a rolled section is to be selected to support a specified load, it is usually necessary to make several trial solutions since there is no direct relationship between areas and radii of gyration for different structural shapes. The best section is usually the section with the least area (lightest weight) that will support the load. A minimum area can be obtained by assuming $L/r = 0$. The load-carrying capacity of various sections with areas larger than this minimum can then be calculated, using the proper column formula, to determine the lightest one that will carry the specified load.

Example 9–5 Select the lightest wide flange steel section listed in App. C to support a centric load of 150 kips as a 15-ft column. Use code 2.

Solution If L/r were small, P/A would be 16,000 psi, and the area would be

$$A = 150{,}000/16{,}000 = 9.38 \text{ sq in.}$$

A column should be selected from App. C with an area greater than 9.38 sq in. for the first trial. In this case try a W8×40 section for which A is 11.76 sq in. and $r_{\min}$ is 2.04 in. The value of L/r for this column is

$$\frac{L}{r} = \frac{180}{2.04} = 88.2$$

and the load it can support is

$$P = A\left[16{,}000 - \frac{(L/r)^2}{2}\right] = 11.76\left[16{,}000 - \frac{88.2^2}{2}\right] = 142{,}400 \text{ lb}$$

This load is less than the design load; therefore, a column with either a larger area, a larger radius of gyration, or both must be investigated. As a second trial value use a W10×45 section for which A is 13.24 sq in. and r

is 2.00 in. For this section L/r is equal to 90, and the load is

$$P = 13.24(16{,}000 - 90^2/2) = 158{,}200 \text{ lb}$$

Since the 45-lb column is stronger than necessary and the 40-lb column is not strong enough, any other sections investigated should weigh between 40 and 45 lb/ft. The only wide flange section in App. C that might satisfy the requirements is a W14×43 section for which A is 12.65 sq in. and r is 1.89 in. The slenderness ratio and allowable load for this column are

$$(L/r) = 95.2 \quad \text{and} \quad P = 145{,}100 \text{ lb}$$

which is less than the design load.
Thus, the lightest column is a $W10\times45$ section. Ans.

The above is not necessarily the best procedure. Different designers have different approaches to the trial-and-error procedure, and for certain problems one approach may be better than another. The important point to make here is that the problem of design involving rolled shapes (other than simple geometric shapes) is, in general, solved by trial and error.

PROBLEMS

Note. Refer to App. A for necessary properties. When applying Eqs. 9–7 and 9–8, use the elastic strength for σ_0.

9–36* Use code 1 and structural steel for this problem.

(a) Determine the maximum allowable total load that three rectangular steel bars 1×4 in. by 8 ft long will carry as columns, each acting independently of the other two.

(b) Determine the maximum allowable load that these three bars can carry when they are welded together to form an H-column.

9–37* Four 3×3×1/4-in. aluminum alloy 2024-T4 angles 6.5 ft long are to serve as a column. Using code 4 with a factor of safety of 2, determine the maximum

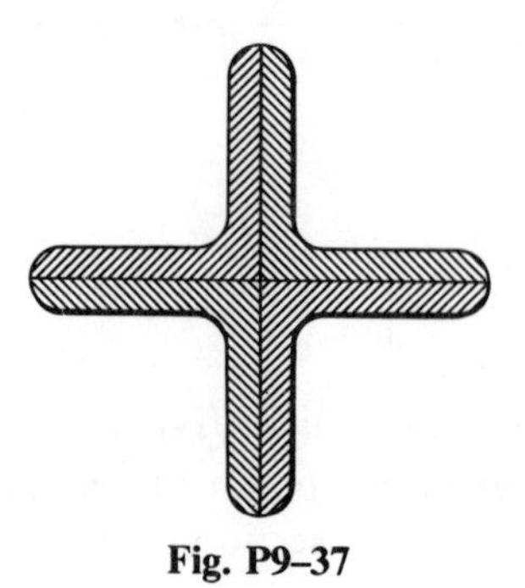

Fig. P9–37

allowable load:

(a) When the angles are fastened together as in Fig. P9–37 to act as a unit.

(b) When the angles are not connected and each acts as an independent axially loaded member.

9–38* A 7.5-m column is composed of two structural steel C254×35.74 channels. Using code 2 ($\sigma_0 = 110$ MN/m^2 and $C_3 = 0.00345$ MN/m^2), determine the maximum permissible load:

(a) If the channels are latticed 150 mm back to back as in Fig. 9–13 *a*.

(b) If the channels are not connected and each acts as an independent axially loaded member.

9–39* Two 125×75×12-mm steel angles 4 m long are to serve as a column. Using code 2 ($\sigma_0 = 110$ MN/m^2 and $C_3 = 0.00345$ MN/m^2), determine the maximum allowable load:

(a) When the angles are fastened together to form a T-section with the 125-mm legs connected.

(b) When the angles are not connected and each acts as an independent axially loaded member.

9–40 A connecting rod made of SAE 4340 heat-treated steel has the cross section of Fig. P9–40. The pins at the ends of the rod are parallel to the x axis and are 5 ft apart. Assume that the pins offer no restraint to bending about the x axis but provide complete fixity for bending about the y axis. Using code 1, determine the maximum allowable axial compression load for this rod.

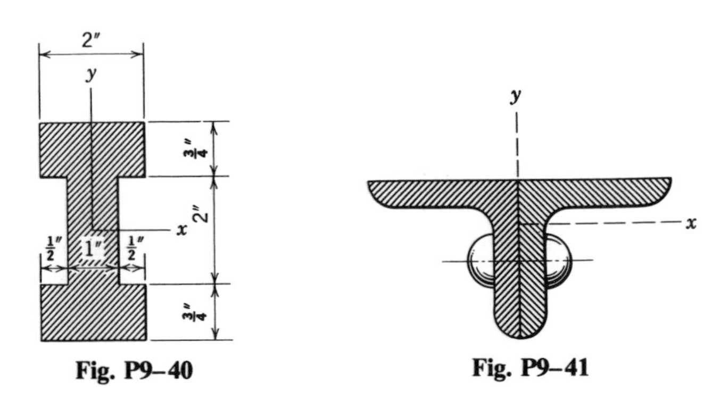

Fig. P9–40 **Fig. P9–41**

9–41* A column of aluminum alloy 2024-T4 is composed of two 5×5×3/4-in. angles riveted together as shown in Fig. P9–41. The length between end connections is 10 ft, and the end connections are such that there is no restraint to bending about the y axis; but restraint to bending about the x axis reduces the effective length to $10/\sqrt{2}$ ft. Using code 4 and a factor of safety of 2, determine the maximum permissible load.

9–42 Figure P9–42 shows a connecting rod made of SAE 4340 heat-treated steel ($\sigma_y = 1000$ MN/m^2 and $E = 200$ GN/m^2) that is to carry an axial compression load. The pins at the ends offer practically no restraint to bending about the axis of the pin, but the restraint about the perpendicular axis reduces the effective length to $L/\sqrt{3}$. Using code 1, determine the maximum allowable load.

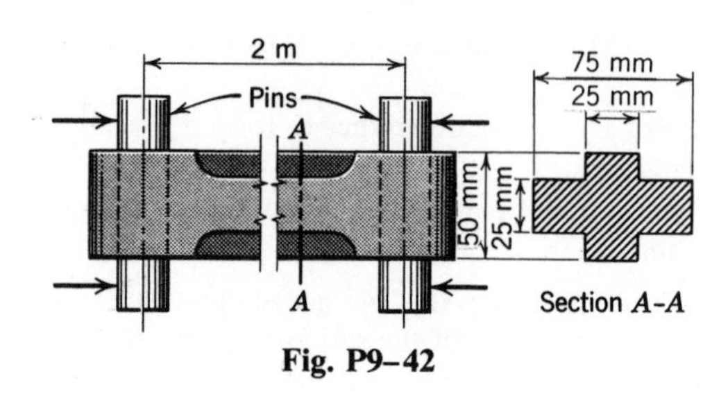

Fig. P9–42

9–43* Using code 1 for low-alloy steel ($\sigma_y = 50$ ksi) and a factor of safety of 23/12 (both ranges), determine the maximum allowable load that a W10×60 member can carry as a 30-ft column:

(a) If it is unbraced throughout its length.
(b) If it is laterally braced 20 ft from one end in the direction parallel to the x axis; that is, the effective length is 20 ft for bending about the y axis.

9–44 The compression member AB of the truss in Fig. P9–44 is a W10×45 steel section with the x–x axis lying in the plane of the truss. The member is continuous from A to B. Consider all connections to be the equivalent of pivot ends and use code 2.

(a) Determine the maximum permissible load in AB.
(b) Determine the maximum permissible load if the bracing member CD is removed.

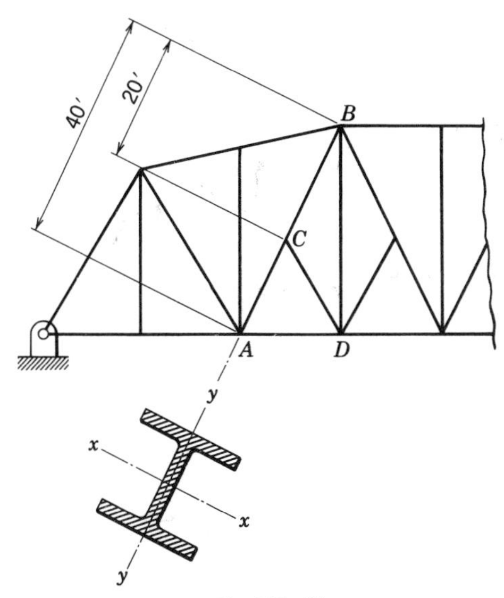

Fig. P9–44

9–45* Solve Prob. 9–44 assuming that the joints at A and B provide enough restraint that the effective length A to B is $40/\sqrt{1.5}$. Assume the connection at C offers no restraint to bending.

9–46* A 12-m column is composed of two C305×41.69 structural steel channels latticed together with the flanges turned out as in Fig. 9–13a. The column is fixed at the base, and the top is pinned to a roller that slides in a vertical slot.

(a) What should be the spacing of the channels (back to back) in order to provide a column with equal resistance to bending about either principal axis?

(b) Determine the maximum allowable load according to code 1 if $\sigma_y = 250$ MN/m^2 and $E = 200$ GN/m^2. Use a factor of safety of 23/12 for both ranges.

9–47 Using code 1, select a structural steel UB section to serve as a column 4 m long carrying an axial load of 1000 kN with a factor of safety of 23/12. For structural steel $\sigma_y = 250$ MN/m^2 and $E = 200$ GN/m^2.

9–48 Using code 2 ($\sigma_0 = 110$ MN/m^2 and $C_3 = 0.00345$ MN/m^2), select two steel channels to form a column 7.5 m long to carry an axial load of 525 kN. The channels are to be spaced and latticed so that the resistance to bending with respect to either axis of symmetry is the same.

9–49 A sand bin weighing 150 tons is to be supported by four square timber columns 12 ft long. What size dressed douglas fir (dry) timbers should be bought to comply with code 6? Assume that the load is to be equally divided among the columns and that the column load is axial. Dressed timber posts are 1/2 in. smaller than nominal.

9–50 A 5×5×3/4-in. aluminum alloy 2014-T6 angle is used as a fixed-ended column. Using code 3, determine the maximum allowable length of this column that will support a load of 120 kips.

9–51 A magnesium alloy (ZK60A-T5) strut 60 in. long is to transmit an axial compression load of 25 kips. Restraint at the ends of the strut make the effective length one-third greater than that for completely fixed ends. The strut is to be rectangular in cross section, one dimension being fixed at 2.50 in. Using code 5 and a factor of safety of 2, determine the other dimension.

9–52* Design a square column 60 in. long to carry a load of 5000 lb with a factor of safety of 3. The material is annealed 18–8 stainless steel. Use Eqs. 9–6 and 9–7.

9–53 Design a hollow circular column 1.5 m long to carry a load of 70 kN with a factor of safety of 2. The material is aluminum alloy 6061-T6 ($\sigma_y = 275$ MN/m^2 and $E = 70$ GN/m^2) and the ratio of inside to outside diameters is to be 0.8. Use Eqs. 9–6 and 9–8.

9–54 A cast-steel column ($\sigma_y = 250$ MN/m^2 and $E = 200$ GN/m^2) has the cross section of Fig. P9–54. The column is 3 m long and is fixed at the base and free at the top. Using Eqs. 9–6 and 9–7 and a factor of safety of 3, determine the maximum allowable load.

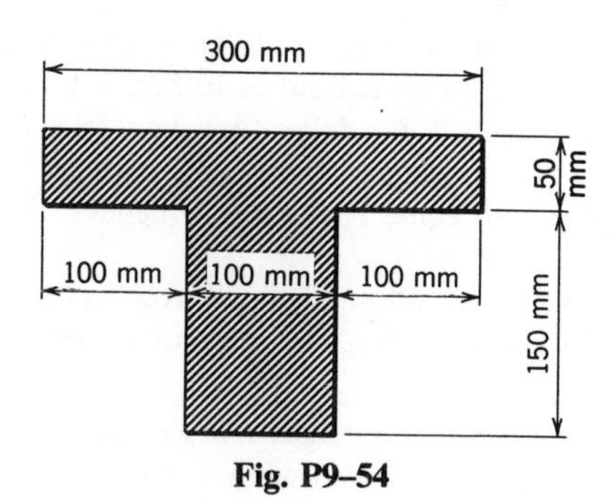

Fig. P9–54

9–55* Two $6 \times 3\frac{1}{2} \times \frac{1}{2}$-in. angles of aluminum alloy 2014-T4 are welded together, as shown in Fig. P9–55, to form a 15-ft pin-ended column. The pins provide no restraint to bending about the x axis but do reduce the effective length to 10 ft for bending about the y axis. The sectional properties of one angle are given in the figure where point C is the centroid of one angle. Using Eqs. 9–6 and 9–8, determine the maximum permissible axial load for a factor of safety of 2.

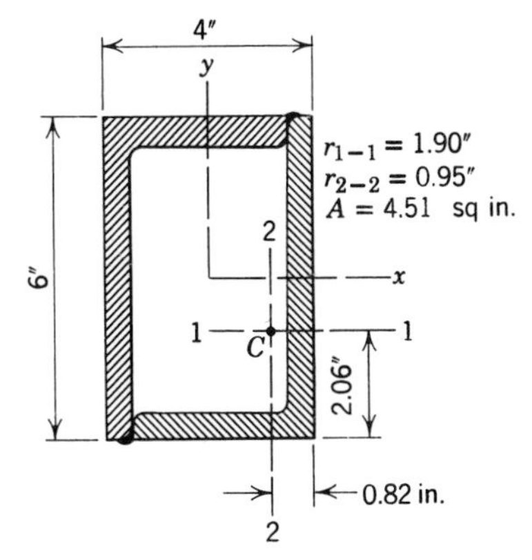

Fig. P9–55

9–56* Four 6061-T6 aluminum alloy C152×6.17 channels are fabricated as in Fig. P9–56 to form a pin-ended column 5 m long. The pins offer no restraint to bending about the y axis, but for bending about the x axis the pins provide restraint sufficient to reduce the effective length to 3 m. The cross-sectional properties of one channel are the same as those for a C152×17.88 steel channel in App. C. Using Eqs. 9–6 and 9–8, determine the maximum permissible axial load for a factor of safety of 2.5 if $\sigma_y = 275\ \text{MN/m}^2$ and $E = 70\ \text{GN/m}^2$.

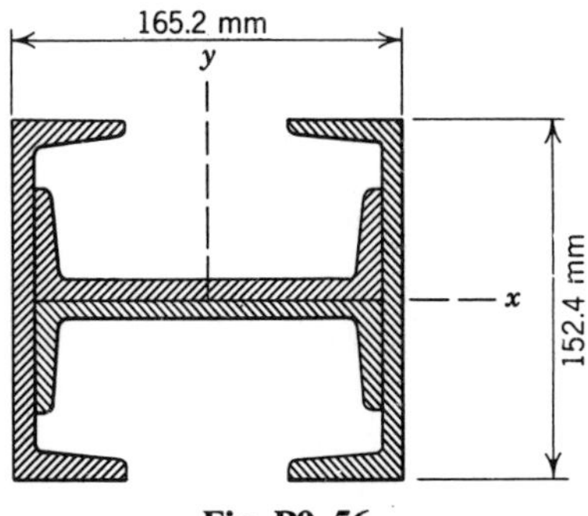

Fig. P9–56

9–57* A 5-ft strut having the cross section of Fig. P9–57 is made of T-1 steel having an elastic strength of 90 ksi. Assume the strut to be fixed at both ends. Using Eqs. 9–6 and 9–7, determine the maximum allowable axial compressive load that may be applied with a factor of safety of 2.

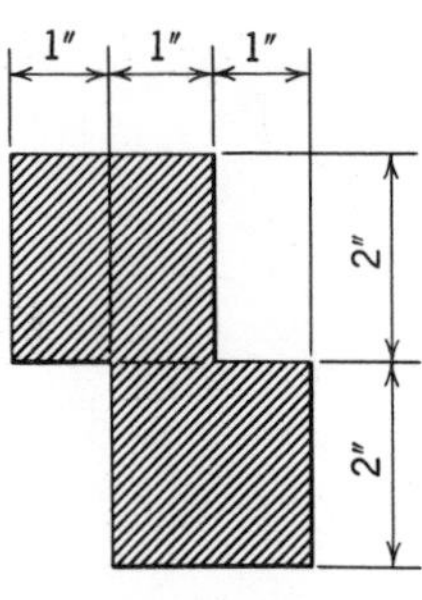

Fig. P9–57

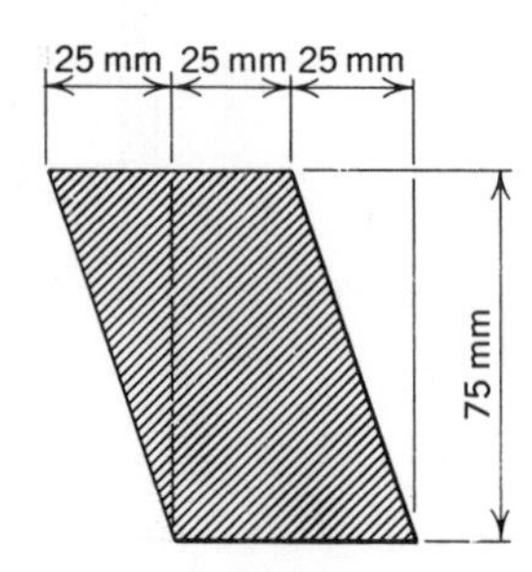

Fig. P9–58

9–58* A 2-m strut having the section of Fig. P9–58 is made of aluminum alloy 6061-T6 ($\sigma_y = 275$ MN/m^2 and $E = 70$ GN/m^2). Assume the strut to have fixed ends. Using Eqs. 9–6 and 9–8 and a factor of safety of 2, determine the maximum permissible axial compressive load.

9–7 ECCENTRIC LOADS

Although a given column will support the maximum load when the load is applied centrically, it is sometimes necessary to apply an eccentric load to a column. For example, a beam supporting a floor load in a building may in turn be supported by an angle riveted or welded to the side of a column, as shown in Fig. 9–15. Frequently, the floorbeam is framed into the column with a stiff connection, and the column is then subject to a bending moment due to the continuity of floorbeam and column. The eccentricity of the load, or the bending moment, will increase the stress in the column or reduce its load-carrying capacity. Three methods will be presented for computing the allowable load on a column subjected to an eccentric load (or an axial load combined with a bending moment).

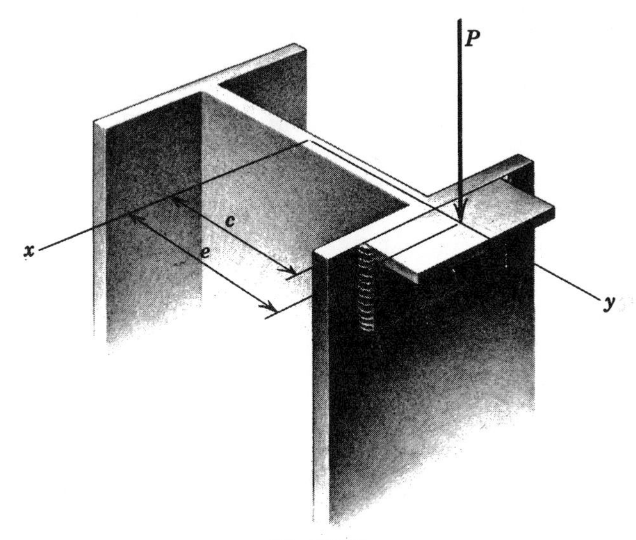

Fig. 9–15

THE SECANT FORMULA The secant formula (Eq. 9–5) was derived on the assumption that the applied load had an initial eccentricity e; this value can be substituted in the equation to determine the failure load (the load to cause incipient inelastic action). As mentioned previously, there is usually a small amount of unavoidable eccentricity which must be approximated when using this formula for centric loads. This unavoidable eccentricity should be added to the intentional eccentricity, particularly if the intentional eccentricity is small. The use of the secant formula is quite time-consuming, unless a series of curves, such as those in Fig. 9–9, is available.[6] The secant formula is specified by the American Railway Engineering Association and the American Association of State Highway Officials.

MODIFIED COLUMN FORMULAS The second solution is based on the arbitrary specification that the sum of the direct and bending stresses ($P/A + Mc/I$) shall not exceed the average unit load prescribed by the appropriate column formula. For example, when this approach is used with the parabolic formula, the equation becomes

$$\frac{P}{A} + \frac{Mc}{I} = \sigma_0 - C_3\left(\frac{L}{r}\right)^2 \tag{a}$$

[6]See, for example, App. C of the A.A.S.H.O. Standard Specifications for Highway Bridges.

In effect, this merely replaces the low L/r end of the secant curve with a parabola (see Fig. 9–11), and if the constants in the parabola are properly chosen, the agreement between the two curves is quite good. Since the parabola will deviate from the secant curve very rapidly beyond the point of reverse curvature, Eq. (a) is limited to the intermediate range. This leaves only the secant formula valid for columns in the slender range with eccentric loading. Although it is possible to use other types of intermediate formulas for the right side of Eq. (a), the use of the parabola only is recommended here, since it is an attempt to approximate a formula known to be rational.

The application of the secant formula or Eq. (a) to an eccentrically loaded column such as that of Fig. 9–15 must be accompanied by an investigation of the stability of the column as an axially loaded member. Referring to Fig. 9–15, when the load is applied in the plane containing the axis about which the moment of inertia is least (the y axis of Fig. 9–15), the column might fail by buckling about either axis, and both possibilities should be checked.

---

1. It can fail by buckling about the axis of least radius of gyration (the y axis in Fig. 9–15) because L/r is largest about this axis. In this case, all formulas developed for centric loading apply[7] and the radius of gyration is r_y.

2. The column can fail by buckling about the axis of maximum radius of gyration (the x axis) due to the combined effect of the bending moment and the direct load. In this case, the values required are r_x, I_x, and the distance c, as shown in Fig. 9–15. Note that c is the maximum distance to the compression side of the centroidal axis. The smaller of the two loads is, of course, the limiting load.

THE INTERACTION FORMULA One of the modern expressions for treating combined loads is known as the interaction formula, of which several types are in use. The analysis for compression members subjected to bending and direct stress may be derived as follows. It is assumed that the stress in the column can be written as

$$\frac{P}{A} + \frac{Mc}{I} = \sigma_w$$

[7]The load is assumed to be axial; that is, the eccentricity in the y direction does not decrease resistance to buckling about the y axis of Fig. 9–15. For columns with I-sections in the slender range, it can be shown that such an assumption may be unwarranted. See "Lateral Buckling of I-Section Columns with Eccentric End Loads in Plane of the Web," B. Johnston, *Journal of Applied Mechanics*, December 1941, p. A-176.

and, when this expression is divided by σ_w, it becomes

$$\frac{P/A}{\sigma_w}+\frac{Mc/I}{\sigma_w}=1 \tag{b}$$

When considering eccentrically loaded columns, the value of σ_w will, in general, be different for the two terms. In the first term P/A represents an axial unit load on a column; therefore, the value of σ_w should be the average working unit load on an axially loaded column as obtained by an empirical formula such as those presented in the preceding article or the Euler formula. In the second term, Mc/I represents the bending stress induced in the member as a result of the eccentricity of the load or an applied bending moment; therefore, the corresponding value of σ_w should be the allowable flexural stress. Since the two values of σ_w are different, one recommended form of Eq. (b) is the following interaction formula[8]

$$\frac{P/A}{\sigma_a}+\frac{Mc/I}{\sigma_b}=1 \tag{c}$$

in which P/A = the average unit load on the eccentrically loaded column, σ_a = the *allowable* average unit load for an axially loaded column (note that the greatest value of L/r should be used to calculate σ_a), Mc/I = the bending stress in the column, and σ_b = the *allowable* bending stress[9]

When Eq. (c) is used to select the most economic section, it will usually not be possible to obtain a section that will satisfy the equation exactly. Any section that makes the sum of the terms on the left side of the equation less than unity is considered to be safe, and the *safe* section giving the largest sum (less than unity) is the most efficient section.

Example 9–6 A 20-ft W18×96 steel column supports an eccentric load applied 5 in. from the center of the section measured along the web. Use code 1 with $\sigma_y = 42$ ksi, $C_c^2 = 13{,}630$, and $k = 1.90$ and 1.76 for maximum and minimum L/r, respectively. Determine the maximum safe load:

(a) According to the modified column formula.

(b) According to the interaction formula with σ_b equal to 28 ksi.

Solution (a) The following properties are obtained from App. C: $A =$

[8]The 1970 AISC specifications permit this formula if $(P/A)/\sigma_a \leqslant 0.15$; otherwise, the second term of the equation is amplified to provide for secondary moments due to lateral displacement of any cross section.

[9]The 1970 AISC specifications gives $\sigma_b = 0.66\sigma_y$ with a reduced value for compression flanges in some cases. See *The Steel Construction Manual.*

28.22 sq in., $r_{min}=2.71$ in., $r_{max}=7.70$ in., and $c=18.16/2=9.08$ in. Assume first that the column tends to buckle about the axis parallel to the web. Note that there is no eccentricity about this axis. The value of L/r is

$$L/r=20(12)/2.71=88.6$$

which is less than C_c of 116.8. The average unit load is

$$P/A=\frac{42}{1.90}\left[1-\frac{(88.6)^2}{2(13630)}\right]=15.74 \text{ ksi}$$

For bending about the axis perpendicular to the web, the radius of gyration is 7.70 in. (L/r will be much less than 88.6), and the modified column formula becomes

$$\frac{P}{A}+\frac{Mc}{I}=\frac{\sigma_y}{k}\left[1-\frac{(L/r)^2}{2C_c^{\,2}}\right]$$

The bending moment is Pe, and, when numerical values are substituted in the equation, it becomes

$$\frac{P}{A}+\frac{P(5)9.08}{A(7.70)^2}=\frac{42}{1.76}\left[1-\frac{(240/7.7)^2}{2(13630)}\right]$$

from which

$$P/A=13.03 \text{ ksi}$$

Since this last unit load is less than that for buckling about the web, the maximum allowable load is

$$P=13.03(28.22)=\underline{368 \text{ kips}} \qquad \text{Ans.}$$

(b) The interaction formula for eccentrically loaded columns is

$$\frac{P/A}{\sigma_a}+\frac{Mc/I}{\sigma_b}=\frac{P/A}{\sigma_a}+\frac{Pec/(Ar^2)}{\sigma_b}=1$$

The value of σ_a was found to be 15.74 ksi in part (a), and, when numerical data are substituted in the expression, it becomes

$$\frac{P}{A}\left[\frac{1}{15.74}+\frac{5.00(9.08)/(7.70)^2}{28}\right]=1$$

which gives

$$P/A = 11.00 \text{ ksi}$$

The allowable load using this specification is

$$P = 11(28.22) = \underline{310 \text{ kips}} \qquad \text{Ans.}$$

The two results are about 18 percent apart in this case.

PROBLEMS

Note. Use the table of codes of Art. 9–6 modified for eccentricity unless otherwise specified. Refer to App. C for properties. For σ_0 in Eq. 9–8, use the elastic strength unless specified otherwise.

9–59* A hollow square steel member 3 in. in outside dimension and with 1/2-in. thickness of walls functions as a column 7 ft long. Determine the maximum allowable load the column will carry when the load is applied with a known eccentricity of 1/2 in. along an axis of symmetry parallel to a pair of sides. Use Fig. 9–9 and a factor of safety of 2. Assume an accidental eccentricity ratio of 0.11.

9–60* A 60-mm diameter steel strut is subjected to an eccentric compression load P located as shown in Fig. P9–60. The effective length for bending about the x axis is 1500 mm but for bending about the y axis, the end conditions reduce the effective length to 900 mm. Determine the load P that may be applied with a factor of safety of 1.5. Use Fig. 9–9 and assume an accidental eccentricity ratio of 0.10.

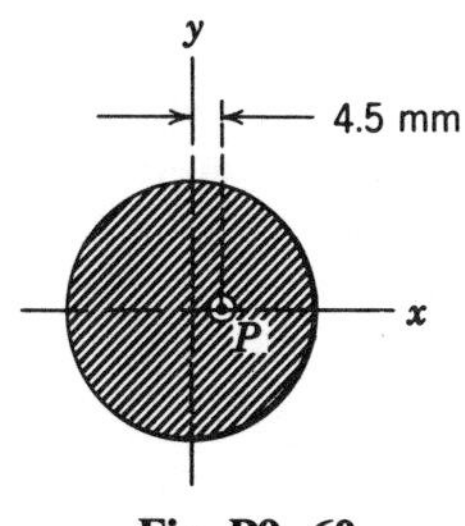

Fig. P9–60

9–61 A hollow square steel member with outside and inside dimensions of 4 in. and 3 in. (walls are 1/2 in. thick) functions as a column 13 ft long. The load is applied with a known eccentricity of 0.59 in. along one of the diagonals of the square, and the end conditions are such that for bending about the other diagonal the effective length is reduced to 8 ft. Determine the load that may be applied with a factor of safety of 2. Use Fig. 9–9 and assume an accidental eccentricity ratio of 0.20.

9–62* A UB 203×30 structural steel section acts as a 5-m column. Determine the maximum allowable load it can carry if the resultant load acts along the web 100 mm from the column axis. Use code 2 with $E=200$ GN/m^2, $\sigma_0=110$ MN/m^2, and $C_3=0.00345$ MN/m^2.

9–63* Two C9×15 structural steel sections 30 ft long are latticed 3 in. back to back as shown in Fig. P9–63 to make a column. Determine the maximum working load the column may carry if the load acts at point A. Use code 1, with $k=1.92$ for both ranges.

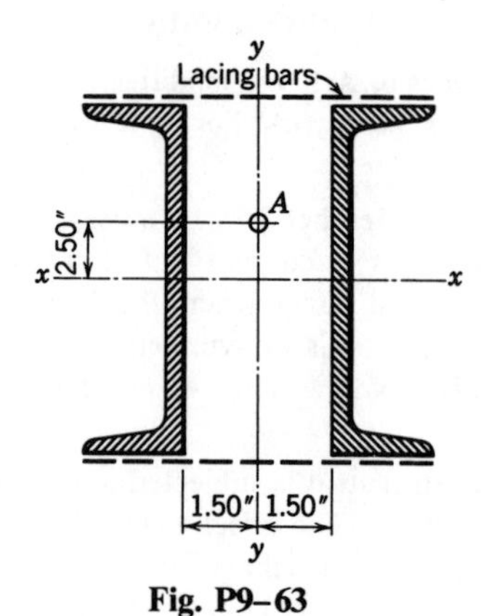

Fig. P9–63

9–64 A structural steel T-section having the properties given in Fig. P9–64 is to be used as a compression member with an effective length of 7 m. Using code 1 with $E=200$ GN/m^2, $\sigma_y=250$ MN/m^2, and $k=1.92$ for both ranges, determine:

(a) The maximum permissible axial load.

(b) The maximum permissible bending moment, applied to the ends of the member in the yz plane (the z axis is the axis of the member), that may be superimposed on the load of part (a) if the moment induces compression in the flange.

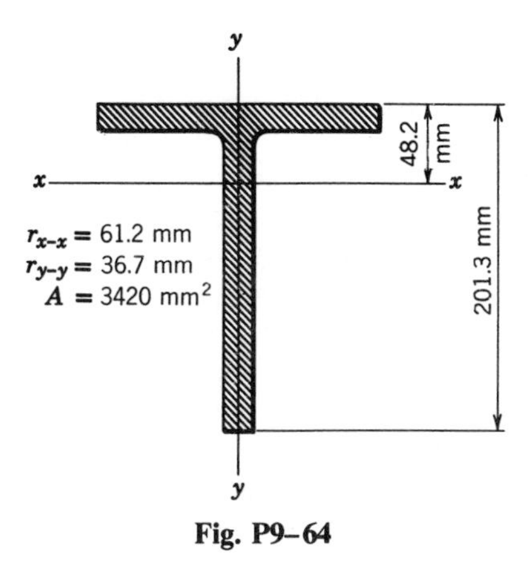

Fig. P9–64

9–65* A 2014-T6 aluminum alloy compression member with an effective length of 5 ft has the T cross section of Fig. P9-65. Using code 3, determine:

(a) The maximum permissible axial load.

(b) The maximum permissible bending moment in the *yz* plane that can be superimposed as shown on the axial load of part (a).

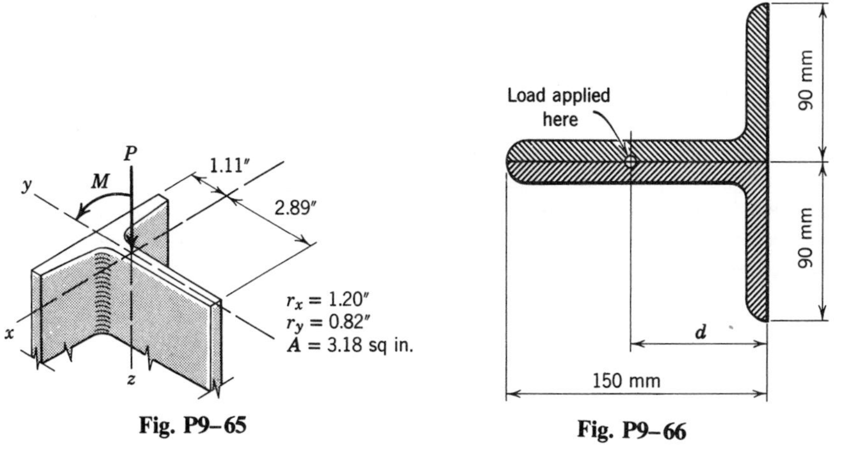

Fig. P9–65 **Fig. P9–66**

9–66 Two L150×90×10-mm structural steel angles are welded together, as shown in Fig. P9-66, to act as a column with an effective length of 5 m. Using code 2 with $E=200\ \text{GN/m}^2$, $\sigma_0=110\ \text{MN/m}^2$, and $C_3=0.00345\ \text{MN/m}^2$, determine:

(a) The maximum safe axial load.

(b) The maximum and minimum values for the distance *d* when the load of part (a) is applied as shown.

9–67* A W16×64 structural steel section is to function as a 30-ft column fixed at the base. The column is to be considered free at the top for bending with respect to the x axis (see App. C) and pivot-ended at the top for bending with respect to the y axis. The load is applied on the web 3 in. from the center of the cross section. Using code 2, determine the maximum permissible value for the eccentric load.

9–68* A UB203×25 low-alloy steel compression member has an effective length of 4 m and is subjected to an axial load P and a bending moment of 16 kN-m with respect to the x axis of the cross section (see App. C). Determine the maximum safe load P. Use code 1 with $E=200\ \text{GN/m}^2$ and $\sigma_y=290\ \text{MN/m}^2$ and the interaction formula with an allowable flexural stress of $0.66\sigma_y$.

9–69 Two C8×11.5 structural steel channels are latticed 4 in. back-to-back with the flanges turned out to form a column with a length of 32 ft. The load is to be located on the axis of symmetry parallel to the backs of the channels and with an eccentricity of 2 in. Using code 2 and the interaction formula for eccentric loads with an allowable flexural stress of 20 ksi, determine the maximum allowable load.

9–70* A UB406×74 steel column has an effective length of 10 m for bending about the x axis (see App. C). End restraint reduces the effective length for bending about the y axis to 6 m. The column carries a load of 350 kN located on the y axis at a distance e from the center of the section. Determine the maximum permissible value of e. Use code 1 with $E=200\ \text{GN/m}^2$ and $\sigma_y=250\ \text{MN/m}^2$ and the interaction formula for eccentric loads with an allowable flexural stress of $0.66\sigma_y$.

9–71 Determine the maximum load F that can be applied with a factor of safety of 5 to the gray cast-iron frame of Fig. P9–71. Use Eqs. 9–6 and 9–7 for the compressive strength, and since gray cast iron is a brittle material, use the ultimate compressive strength for σ_0. Note that the flange is in tension and the load may be limited by the tensile strength of the iron.

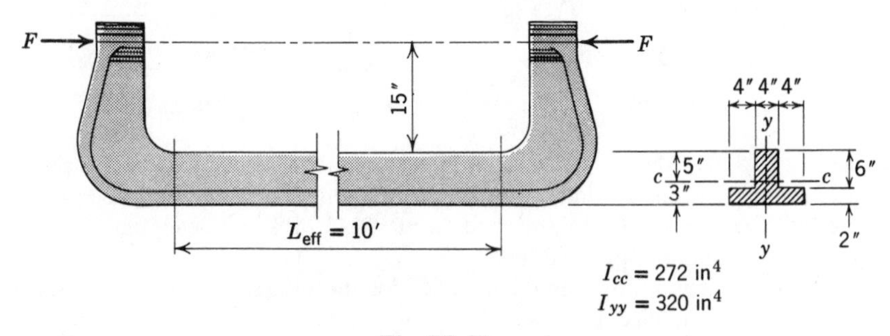

Fig. P9–71

9–72* The compression member of Fig. P9–72 has an effective length of 3 m and is subjected to a 7-kN-m bending moment as shown. The material is aluminum alloy 6061-T6 ($E = 70\ \text{GN/m}^2$ and $\sigma_y = 275\ \text{MN/m}^2$). Determine the maximum safe load P that may be applied (in addition to the bending moment). Use Eqs. 9–6 and 9–8 with a factor of safety of 2.

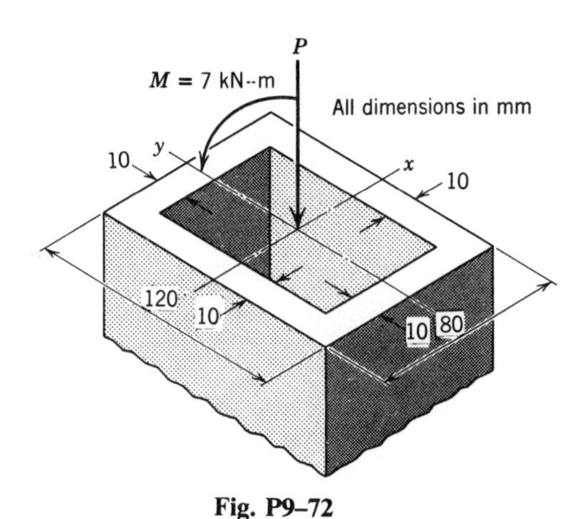

Fig. P9–72

9–73 Determine the maximum allowable load for a 3-m dry douglas fir column with the cross section and loading shown in Fig. P9–73. Use Eqs. 9–6 and 9–7 with $E = 13\ \text{GN/m}^2$, $\sigma_y = 45\ \text{MN/m}^2$, and a factor of safety of 3.

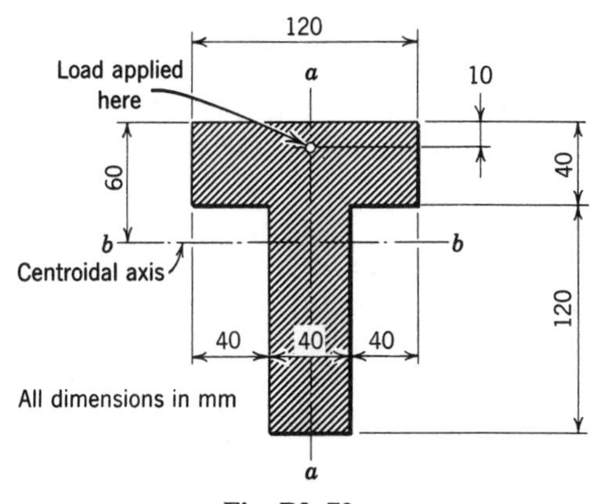

Fig. P9–73

9–74 A 3-ft strut with a rectangular cross section is to be subjected to an eccentric load P, as shown in Fig. P9–74. The material is SAE 4340 heat-treated steel. Determine the maximum permissible value of P and the value of the dimension b if the strut is to have equal resistance to buckling about either principal axis. Use Eqs. 9–6 and 9–7 and a factor of safety of 2.

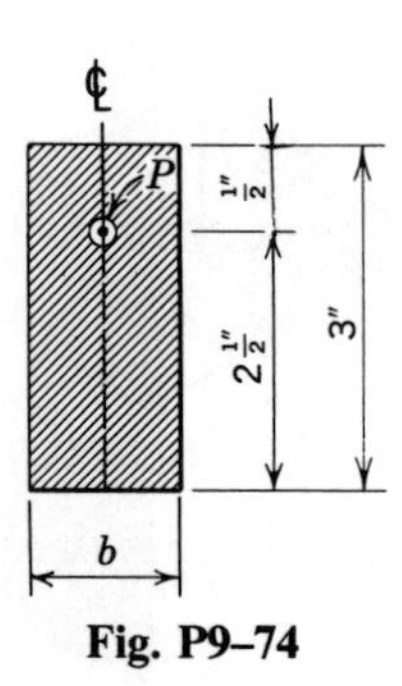

Fig. P9–74

9–8 SPECIAL COLUMN PROBLEMS

Although the secant formula and its various empirical approximations are adequate for treating a wide variety of engineering problems involving columns, there are certain types of problems involving column action for which the secant formula does not apply. One example is the class of problems known as beam colums—members subjected to lateral loading in addition to compressive axial loading. Unless the problem has been treated previously,[10] it is necessary to revert to the differential equation of the elastic curve to find a solution for the particular problem. It may be difficult to develop an explicit solution for P/A as a function of L/r (as was done in Arts. 9–3 and 9–5); in fact, such a solution may be unnecessary. Much of engineering design is done by iteration; hence, if an expression for the maximum deflection in a beam column can be found, the maximum moment can be determined and the maximum normal stress is $P/A + Mc/I$, where M is a function of the compressive as well as the

[10]There is considerable literature available on this subject. See, for example, Chap. 1, *Theory of Elastic Stability*, S. P. Timoshenko and J. M. Gere, Second Edition, McGraw-Hill, New York, 1961.

transverse loads. With an assumed trial section for the member, the maximum stress can be computed to check the adequacy of the section. The following example will illustrate one method of solving the differential equation.

Example 9–7 Develop the expression for the maximum deflection of the beam column of Fig. 9–16*a*. The column ends are free to rotate, but are restrained against lateral displacement.

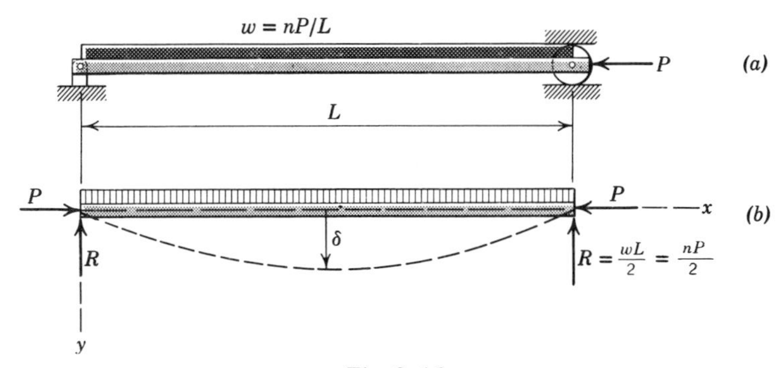

Fig. 9–16

Solution The free-body diagram is shown in Fig. 9–16*b*, with the y axis chosen positive downward to facilitate the solution. For this choice of axes, the curvature indicated is negative, and the differential equation is

$$EI\frac{d^2y}{dx^2} = -Py - \frac{nP}{2}x + \frac{nP}{2L}x^2$$

The boundary conditions are: $y=0$ at $x=0$ and, from symmetry, $dy/dx=0$ at $x=L/2$. By rearranging terms and letting $P/(EI)=k^2$, the equation becomes

$$\frac{d^2y}{dx^2} + k^2y = \frac{nk^2}{2L}x^2 - \frac{nk^2}{2}x \tag{a}$$

The complementary solution for this equation is

$$y_c = A\sin kx + B\cos kx$$

and the particular solution is of the form

$$y_p = Cx^2 + Dx + E$$

When this last equation is substituted into Eq. (a) and coefficients of like powers of x are matched, the constants are found to be

$$C = \frac{n}{2L} \qquad D = -\frac{n}{2} \qquad \text{and} \qquad E = -\frac{n}{Lk^2}$$

Combining the solutions gives

$$y = y_c + y_p = A\sin kx + B\cos kx + \frac{nx^2}{2L} - \frac{nx}{2} - \frac{n}{Lk^2} \tag{b}$$

Substituting the boundary conditions gives

$$y|_{x=0} = B - \frac{n}{Lk^2} = 0$$

$$\left.\frac{dy}{dx}\right|_{x=L/2} = Ak\cos\frac{kL}{2} - Bk\sin\frac{kL}{2} = 0$$

from which

$$B = \frac{n}{Lk^2} \qquad A = \frac{n}{Lk^2}\tan\frac{kL}{2}$$

The maximum deflection δ is found by substituting these constants into Eq. (b) and assigning x the value $L/2$; thus,

$$\delta = \frac{n}{Lk^2}\left(\sec\frac{kL}{2} - 1\right) - \frac{nL}{8}$$

or in terms of P and the distributed load w

$$\underline{\delta = \frac{wEI}{P^2}\left(\sec\frac{L}{2}\sqrt{\frac{P}{EI}} - 1 - \frac{PL^2}{8EI}\right)} \qquad \text{Ans.}$$

PROBLEMS

9–75* Develop the expression for the maximum deflection in the beam column of Fig. P9–75 in terms of P, E, I, L, and n.

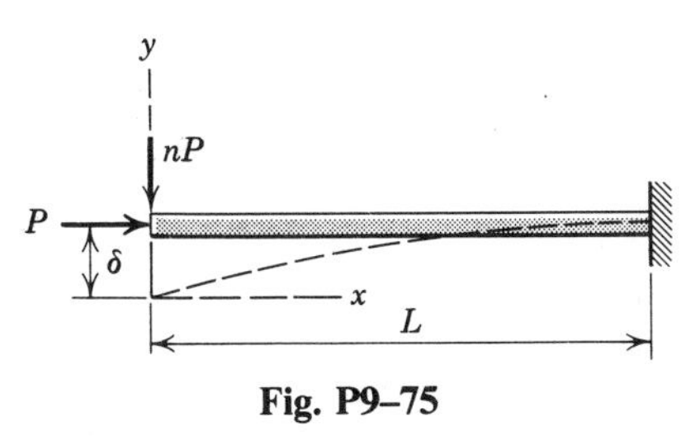

Fig. P9–75

9–76* Develop the expression for the maximum deflection in the beam column of Fig. P9–76 in terms of P, E, I, L, and n.

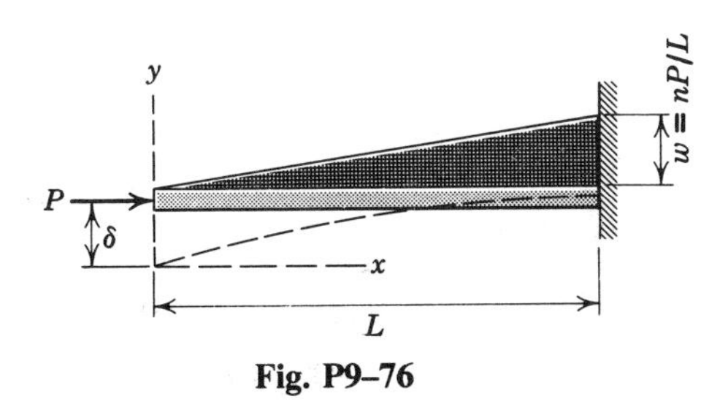

Fig. P9–76

9–77 Develop the equation of the elastic curve for the beam column of Fig. P9–77 in terms of n, P, L, E, I, x, and y.

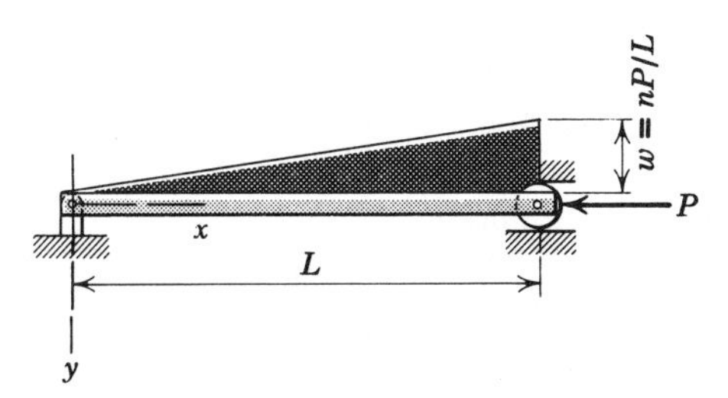

Fig. P9–77

9–78* By solving the differential equation of the elastic curve, verify the effective length given for the column of Fig. 9–7*d*.

9–79* Determine the critical load for the column of Fig. P9–79 in terms of E, I, and L.

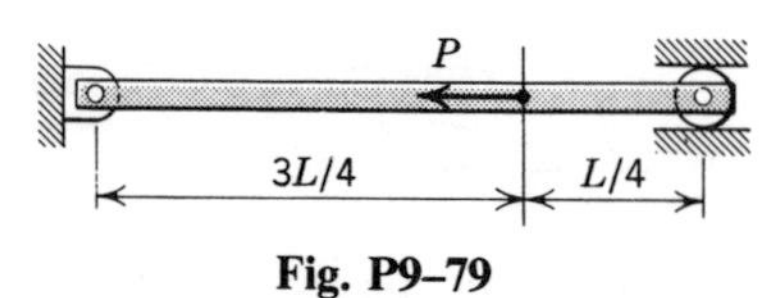

Fig. P9–79

9–80 Develop the equation of the elastic curve for the beam column of Fig. P9–80 in terms of n, P, L, E, I, x, and y.

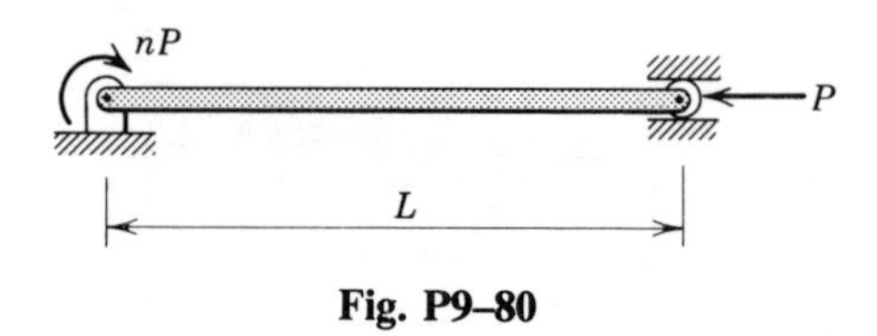

Fig. P9–80

9–9 INELASTIC COLUMN THEORY

The column formulas discussed in Arts. 9–3, 9–5, and 9–6 were based on the assumption that the action was elastic; that is, the stresses were assumed to be below the proportional limit. The Euler formula can thus be used for an *ideal, axially* loaded column for any value of P/A less than the proportional limit of the material.

The first theory for predicting the buckling loads of columns loaded in the inelastic range was proposed by F. R. Engesser about 1890 and is known as the tangent-modulus theory. It is based on the assumption that the Euler formula can be used in the inelastic range, provided that Young's modulus is replaced by the tangent modulus of the material. The tangent modulus is defined as the slope of the stress-strain diagram for a material at *a particular* stress, and it is thus a function of the stress (and strain) for stresses greater than the proportional limit. For stresses less than the proportional limit, the tangent modulus, of course, is the same as Young's modulus. For an ideal elastoplastic material the tangent modulus becomes zero at the proportional limit since the stress-strain diagram becomes horizontal. If E_T represents the tangent modulus, the buckling load, according to the tangent-modulus theory, is

$$\frac{P}{A} = \frac{\pi^2 E_T}{(L/r)^2} \tag{9–9}$$

In order to use Eq. (9–9), values of E_T corresponding to various values of P/A must be obtained from a stress-strain diagram. It is difficult to determine precise values of the slope at specific points on the curve; therefore, it is usually desirable to draw a supplementary curve showing E_T plotted against P/A and use this curve to obtain values for E_T. When corresponding values of E_T and P/A are available, values of L/r can be computed from Eq. (9–9) and used to plot a graph indicating the load-carrying capacity of columns when loaded in the plastic region of the stress-strain diagram. The Euler formula was derived on the assumption that the modulus of elasticity was constant over the cross section of the column, whereas the tangent-modulus formula assumes a variable modulus; therefore, the latter formula cannot be considered to provide a rigorous solution. If the assumption is made that for any given value of the load the *variation* of strain across any section of the column is small, the slope of the stress-strain diagram (the tangent modulus) may be considered a constant for the material at all points in any given section. Since the strains will vary linearly (plane section remains plane), the result will be a linear stress distribution, thus justifying the use of the differential equation for bending. On the basis of this reasoning, it appears that the modified Euler (tangent-modulus) formula should give reasonable results, particularly for materials for which the tangent modulus does not vary too abruptly.

The shape of the tangent-modulus curve is influenced to a considerable degree by the shape of the stress-strain diagram near the proportional limit. Three idealized stress-strain diagrams are shown in Fig. 9–17*a* for three hypothetical materials with identical diagrams to the proportional limit. Material *A* is elastoplastic, *B* has a diagram represented by two

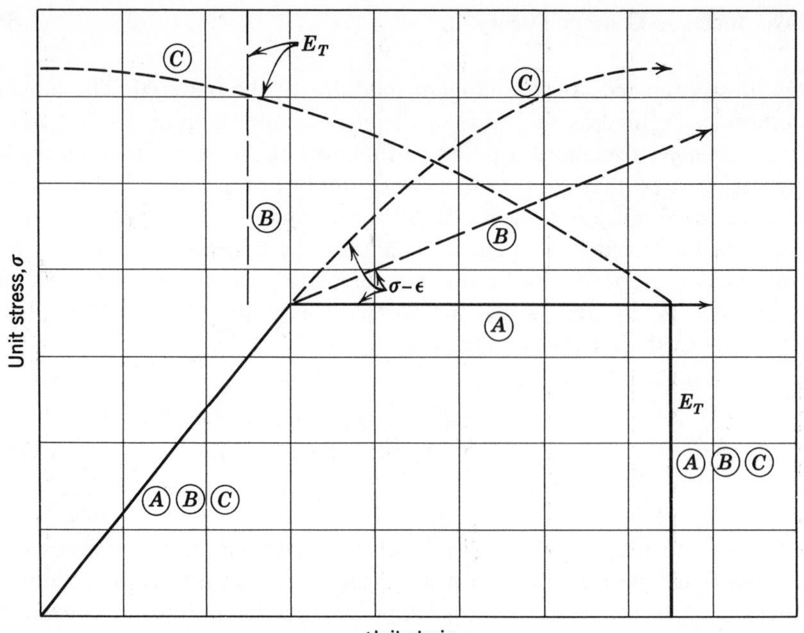

Fig. 9–17*a*

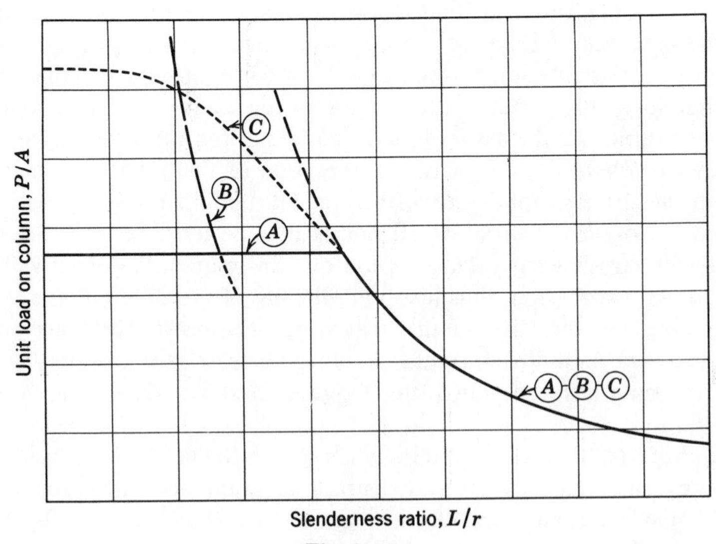

Fig. 9–17*b*

straight lines, and C has a gradual curve at the knee. The variation of the tangent modulus with the stress is shown on the same figure, with the modulus represented by horizontal distances. The buckling loads, as predicted by the tangent modulus, are shown in Fig. 9–17*b*. Note that the buckling load predicted by the tangent modulus formula (Eq. 9–9) is the same as that predicted by the Euler formula (Eq. 9–3) in the elastic range and that this load is usually greater than loads obtained by actual experiment. However, when sufficient care is exercised to assure a centric load, the buckling loads indicated by the tangent-modulus formula are fairly close to experimental results[11] for columns loaded in the elastic and plastic range.

The application of the Euler-Engesser equation (Eq. 9–9) in the inelastic range may require an iterative solution. Such a solution can be obtained with the aid of the digital computer if the time and expense are justified. However, many problems can be solved by sliderule methods with a precision consistent with ordinary engineering applications. The following paragraphs outline three of the possible approaches to such solutions.

1. The actual compression stress-strain diagram can be approximated by a series of straight lines. One then solves a given problem by assuming the stress range is entirely within a particular segment of the diagram, using the value of E_T for this segment in Eq. 9–9, solving for σ_{cr} (or P_{cr}/A), and checking this value against the values of σ for which the curve segment is valid. If the computed value of σ_{cr} does not lie in the range selected, the process is repeated using a different segment of the diagram. Since Eq. 9–9 is identical with Eq. 9–3 in the elastic range, the curve segments to be considered include the elastic range. Note that at the junction of two segments, there are two different values of E_T, leading to an abrupt discontinuity in the P/A vs L/r curve, as indicated for material B in Fig. 9–17*b*. The Euler-Engesser equation is not valid at such points of discontinuity, since the derivation is based on the assumption that the variation of the tangent modulus over the cross section of the member is small.

2. The nonlinear part of the actual stress-strain diagram can be approximated by one or more continuous curves, each of which is represented by a differentiable function of stress and strain. Within certain limits, the tangent modulus ($E_T = d\sigma/d\epsilon$) is a continuous function of σ, which can be substituted in Eq. 9–9, resulting in an expression for σ_{cr} as a function of L/r (where σ_{cr} is P/A). Here again, one should assume a certain segment of the curve is applicable, then check the value of the computed stress against the range of stresses for which the curve segment is applicable. Note that although the stress-strain functions may provide a

[11]"Evaluation of Aeroplane Metals," J. A. Van den Broek, *Journal of the Royal Aeronautical Society,* 1946.

reasonably continuous curve, the derivatives of these functions may not be continuous through the change from one function to another. In other words, at the junction of two curves, there can be two different values for E_T, thus invalidating Eq. 9–9, as discussed in the paragraph above.

3. A solution can be obtained by making use of a stress vs tangent modulus curve, such as the one shown in Fig. P9–87. The solution can be an iterative process, assuming values for E_T, solving for σ_{cr}, and checking with the diagram. Some time might be saved by solving Eq. 9–9 for σ/E_T for the given problem and reading the corresponding value of σ_{cr} from a σ vs σ/E_T curve (such as the one shown in Fig. P9–87) if such a curve is available. This latter curve is usually not provided; hence, it would be necessary to construct the σ vs σ/E_T curve from the given curves or data. However, in actual design practice, it is more practical to construct a P/A vs L/r curve by solving Eq. 9–9 for values of L/r corresponding to values of σ and E_T selected from given curves or data. This curve can then be used as follows: if for a given problem, L/r is known and the critical buckling load is wanted, the value of P/A corresponding to the L/r can be read directly from the curve. If the problem is one of selecting a section to carry a specified load, the trial-solution approach is necessary. In other words, the techniques of Art. 9–6 are followed, with the formulas replaced by graphs. Since there are usually no abrupt discontinuities in the σ vs E_T curves, the ambiguity referred to in the above paragraphs does not, in general, exist.

PROBLEMS

9–81 The data below were obtained from a stress-strain diagram for aluminum alloy 7075-T6. Young's modulus for the alloy is 10,600 ksi, and the proportional limit is 47 ksi. Plot a curve showing the relation of the buckling load P/A to the slenderness ratio L/r, according to the tangent-modulus formula, for columns made of this material. Determine the buckling load for a 7-ft column made of a 3-in. diameter solid rod of 7075-T6:

σ, ksi	E_T, ksi
50	10,300
55	9,000
60	7,000
65	4,800
70	2,000
72	1,400

9–82 Plot the tangent-modulus curve for a 9.25 percent nickel-steel alloy. Use the stress-strain diagram of App. A, and assume the diagram applies to compressive loading as well as tensile loading.

9–83* A compression strut is fabricated from an aluminum alloy 2024-T4 tube having an outside diameter of 1 in., a wall thickness of 1/16 in., and a length between fittings of 10 in. End conditions at the fittings reduce the effective length to $10/\sqrt{3}$ in. The cross-sectional area of the tube is 0.184 sq in. and the principal radius of gyration is 0.332 in. The material has the stress-strain diagram shown in Fig. P9–83. Determine the axial buckling load.

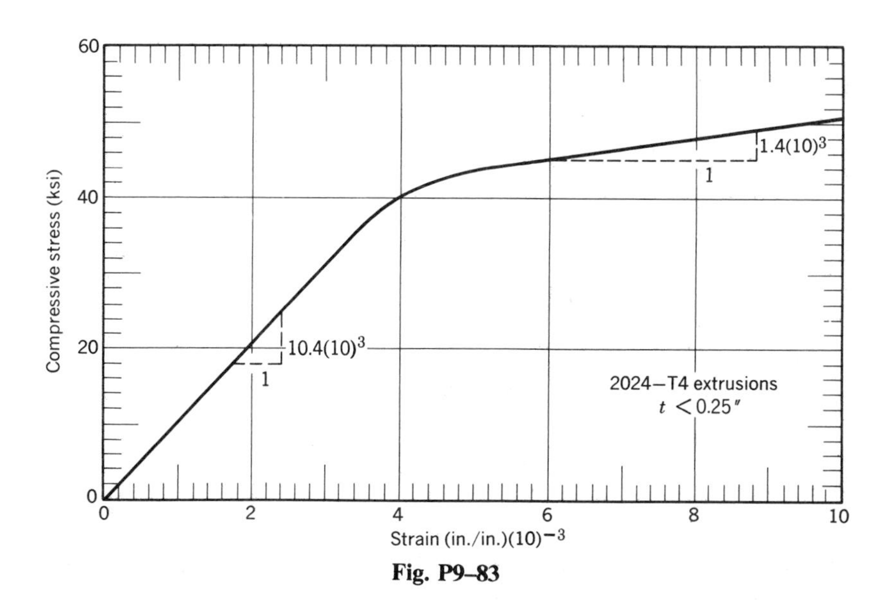

Fig. P9–83

9–84* Determine, for the strut of Prob. 9–83, the length (between fittings) at which buckling would probably occur under an axial load of 7.5 kips C.

9–85 A 3-in. diameter tube with a wall thickness of 0.25 in. is to be used as a compression strut with a length of 50 in. between fittings. The cross-sectional area of the tube is 2.160 sq in. and the principal radius of gyration is 0.976 in. End conditions at the fittings reduce the effective length to 0.68 L. Determine the axial buckling load if the strut is made from an aluminum alloy having the stress-strain diagram shown in Fig. P9–85.

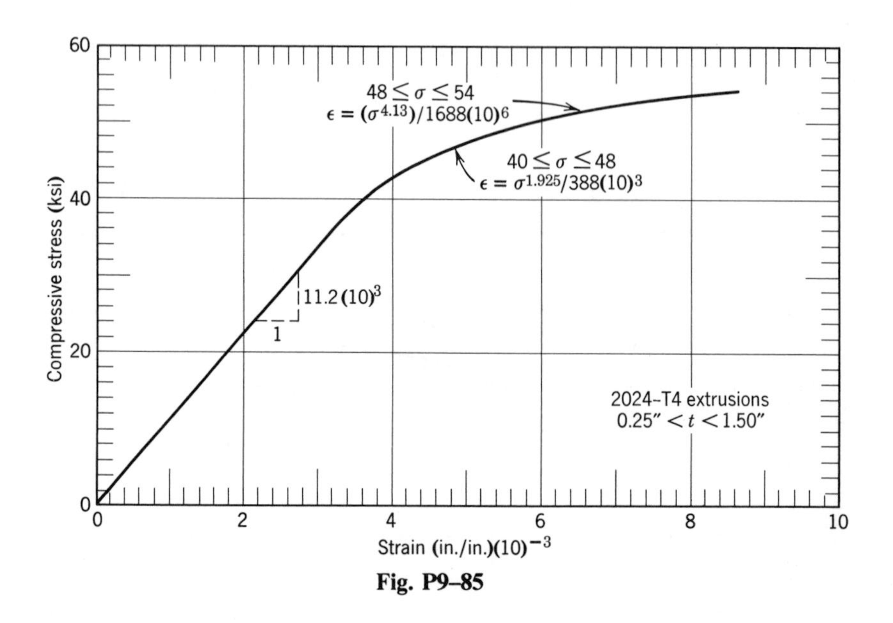

Fig. P9–85

9–86 A compression strut with a length of 16 in. between end fittings is to have a solid rectangular cross section $2.00 \times b$ in. The strut must carry an axial working load of 160 kips with a factor of safety of 1.5. End conditions at the fittings reduce the effective length to $0.7L$. Determine the dimension b if the strut is made from the aluminum alloy of Fig. P9–85.

9–87* A compression strut with a length of 30 in. between end fittings is composed of two $2 \times 2 \times 1/4$-in. angles (see App. C) fastened together to form a T-section. End conditions at the fittings reduce the effective length to $0.7L$. Determine the maximum permissible axial load if the strut is made from the aluminum alloy of Fig. P9–87 and a factor of safety of 2 is specified.

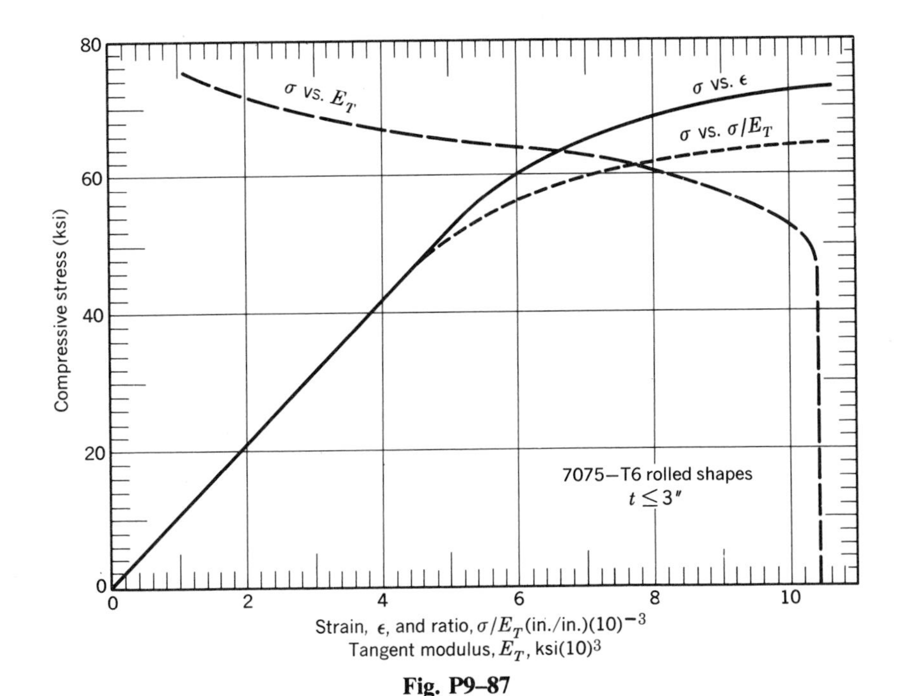

Fig. P9–87

9–88 A single 2×2×1/4-in. angle (see App. C) is to support an axial compressive load of 30 kips with a factor of safety of 2. The pertinent material properties can be obtained from Fig. P9–87. What is the maximum allowable total length if the end fittings reduce the effective length to 3/4 the total length?

10 Repeated Loading

10–1 INTRODUCTION

The discussion in previous chapters of this book has been limited to static loading, for which the behavior of a structure under load can be predicted on the basis of the static properties of the material. However, many machine and structural elements are subjected to a type of loading known as *repeated loading*, under which the load is applied and removed, or the magnitude (and frequently the direction) is changed, many thousands of times during the life of the member. For example, the diesel engine rocker arms of Fig. 10–1, as well as the valves, push rods, connecting rods, and crankshaft, are subjected to repeated loading. In many instances the repeated loading is accidental, as a result of vibrations induced by unpredictable or uncontrollable circumstances.

If the applied load changes from any magnitude in one direction to the same magnitude in the opposite direction, the loading is termed *completely reversed*, whereas if the load changes from one magnitude to another (the direction does not necessarily change), the load is said to be a *fluctuating load*.

The physical effect of a repeated load on a material is different from

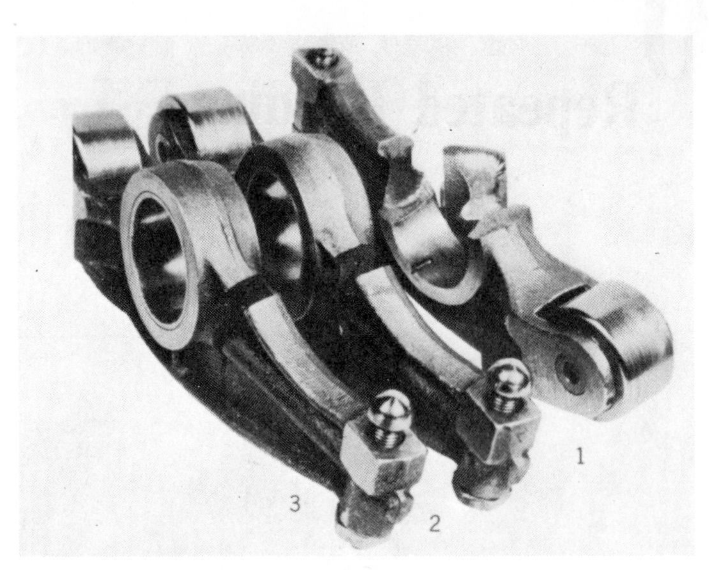

Fig. 10–1 Fatigue failures of exhaust valve rocker levers. Courtesy: Cleveland Diesel Engine Division. General Motors Corp.

that of a static load, the failure always being a fracture (or brittle type of failure) regardless of whether the material is brittle or ductile. Under repeated loading, a small crack forms in a region of high localized stress, and very high stress concentration accompanies the crack. As the load fluctuates, the crack opens and closes and progresses across the section. Frequently this crack propagation continues until there is insufficient cross section left to carry the load and the member ruptures, the failure being given the very descriptive term *fatigue failure*. It is unfortunate that fatigue failures occur at stresses well below the static *elastic* strength of the material.[1]

Figures 10–2, 10–3, and 10–4 show characteristic fatigue failures. Figure 10–2 is a photograph of a torsional fatigue failure at the front end of a crankshaft. This failure started in an area of fretting corrosion (point A) under a hub assembled at this point. From here the fracture progressed to points B and C, before complete separation occurred. Figure 10–3*b* shows the fatigue failure of one of the turbine blades of Fig. 10–3*a*. Here the failure started at the trailing edge (point A) and progressed to BC, the final fracture showing as the dark area.

The failure of one of the compressor blades, such as B, of Fig. 10–4*a* is shown in Fig. 10–4*b*. The fracture appears to have started at A and

[1]For some ductile metals in the fully annealed condition (low elastic strength), the endurance limit may be slightly higher than the elastic strength.

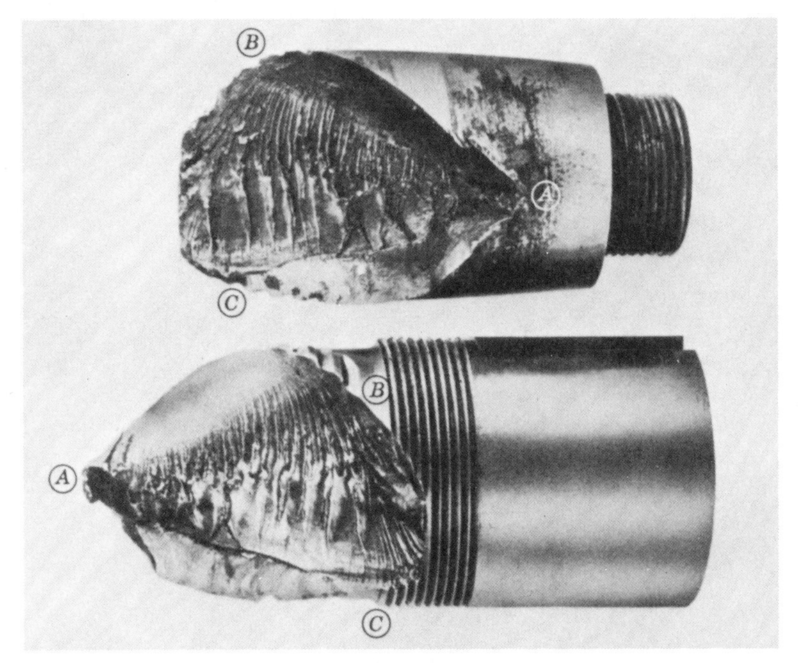

Fig. 10–2 Courtesy: Prof. T. J. Dolan, University of Illinois

continued, as indicated by the light area. The material to fracture last is identified by the dark area.

The importance of repeated loading was recognized in the first half of the nineteenth century. It is thought[2] that J. V. Poncelot (1788–1867), a French professor of engineering mechanics, mentioned the failure of a spring by fatigue in a book published in 1839. Also, W. J. M. Rankine (1820–1872), a Scotch engineer, in a paper published in 1843 discusses the fracture of railroad locomotive axles. Through the years 1843–1870 the railroad wheel and axle assembly occupied much attention and evoked considerable discussion, including the debate regarding whether the metal crystallized in service. The greatest single contribution to the investigation is that of *A*. Wöhler (1819–1914), a German engineer who designed fatigue-testing machines and conducted many fatigue tests that included the effect of stress concentration due to changes in cross section. The argument that the iron changed from a fibrous to a crystalline structure under the repeated loading apparently was successfully refuted during this same period. Although the idea of change in the metal structure under repeated stressing persists to this day with some individuals, metallurgical

[2]*History of Strength of Materials*, S. Timoshenko, McGraw-Hill, New York, 1953.

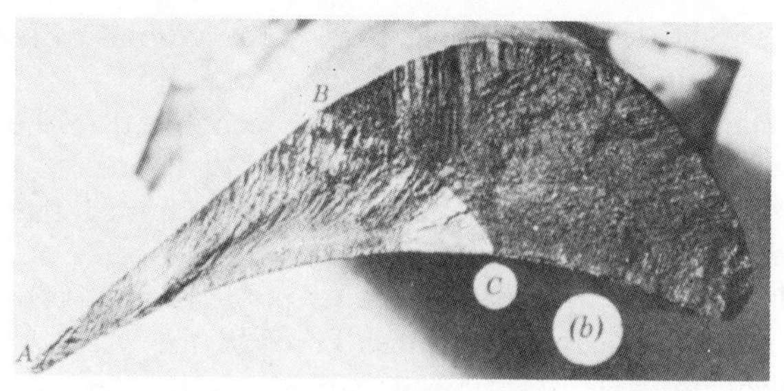

Fig. 10–3

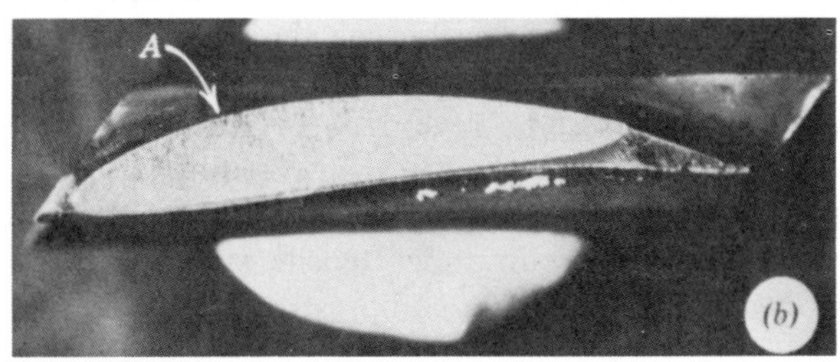

Fig. 10–4

investigation has shown that the crystalline structure of solid metals undergoes no extensive change under such conditions.

Much attention has been devoted to the question of what causes fatigue failure, and, although the question is not completely answered, the electron microscope and X-ray techniques have done much to promote an understanding of the mechanism of fatigue damage. Apparently, in a region of high localized stress (usually at or near the surface), localized slip bands form, and under repeated stressing these bands increase in number and width, rapidly at first but more slowly later. As the damage (slip) progresses, microscopic cracks are formed, and eventually these grow and join to form visible cracks.

Although the mechanism of progressive damage seems to be understood, the fundamental reasons for such action are not known. One explanation advanced is that strain hardening accompanies the local inelastic deformation (slip) and that the restraint of adjacent crystals prevents the localized strain from becoming zero as the applied load becomes zero. This results in the localized stress becoming higher with each cycle and the localized material becoming more brittle. For some combination of stress and material, it would be possible for the magnitude of the localized stress eventually to reach the static ultimate strength of the material, and a local crack would result. Another theory suggests that certain dislocations (defects in atomic structure) are displaced (localized slip) with relative ease and, under cyclic loading, migrate through the material until a number of them coalesce at some region where further movement is inhibited. Such a collection of dislocations in a region at or near the surface may be responsible for the action described in the preceding paragraph.

Although it is possible to have fatigue failure in a region of compressive stress, the usual failures occur under tensile stress and at stress levels considerably lower than those in a compressive stress failure.

10–2 RESISTANCE TO FAILURE UNDER COMPLETELY REVERSED LOADING

The index of resistance to failure under completely reversed repeated loading is known as the *endurance limit*, defined as *the maximum completely reversed stress to which a material may be subjected millions of times without fracture.* The magnitude of the endurance limit is determined from repeated load (or fatigue) tests. One commonly used machine for determining endurance limit is the R. R. Moore rotating beam machine shown schematically in Fig. 10–5. Observe that the loading simulates that imposed on a railroad car axle between the wheels, the specimen being subjected to pure bending and the flexural stresses becoming completely

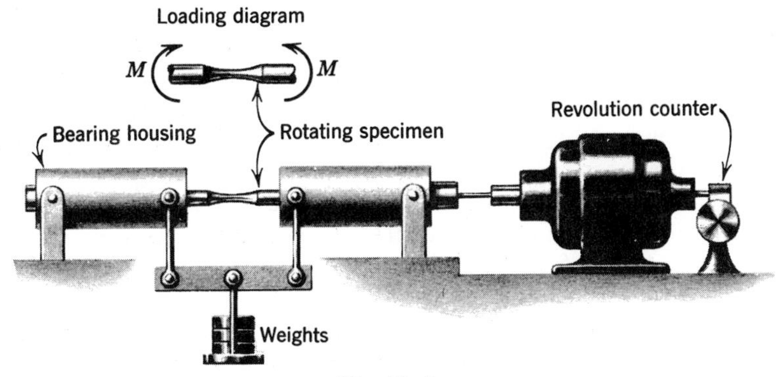

Fig. 10–5

reversed each one-half revolution. There are many other types of machines available, including rotating cantilever beams, nonrotating simple and cantilever beams, and axially loaded members. For the last three, the load may be applied by means of a crank and connecting rod, one machine being indicated diagrammatically in Fig. 10–6. For such experimental setups, the loading may be varied so that fluctuating loads of various ranges may be imposed.

The testing technique is to subject a series of identical specimens to loads of different magnitudes and note the number of cycles of stress (or load) n necessary to fracture the specimen. The data are plotted on a stress-cycle ($\sigma - n$) diagram, and the endurance limit can be determined from the diagram. In order to space the points on the diagram far apart for clarity, it is customary to plot the data on semilogarithmic paper, the stress

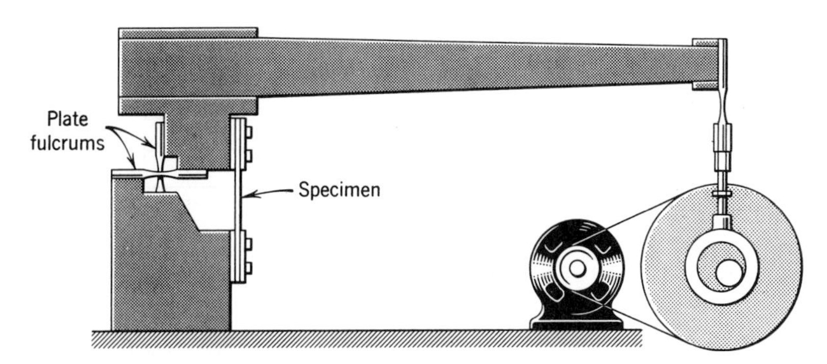

Fig. 10–6

being plotted to a linear scale and the cycles to a logarithmic scale, as indicated in Fig. 10–7. For some materials (particularly the ferrous materials) the curve seems to break sharply, as in curves *A* and *B* of Fig. 10–7, indicating that if the stress has the magnitude of σ_{EA}, material *A* could probably be subjected to an unlimited number of cycles without failure.[3] On the other hand, the alloys of aluminum and magnesium exhibit curves such as *C* and *D* of Fig. 10–7, which are apparently asymptotic to the n axis. For such materials, the stress corresponding to some arbitrary number of cycles, such as $5(10^8)$ for the aluminum alloys, is taken as the endurance limit.

It must be observed that the very nature of the fatigue failure—dependent as it is on local imperfections or peculiar configuration of grain structure as well as on the precision necessary for machining identical specimens—would dictate that the results obtained from a large number of specimens might show considerable scatter. The results could be better represented by a series of curves representing different probabilities of

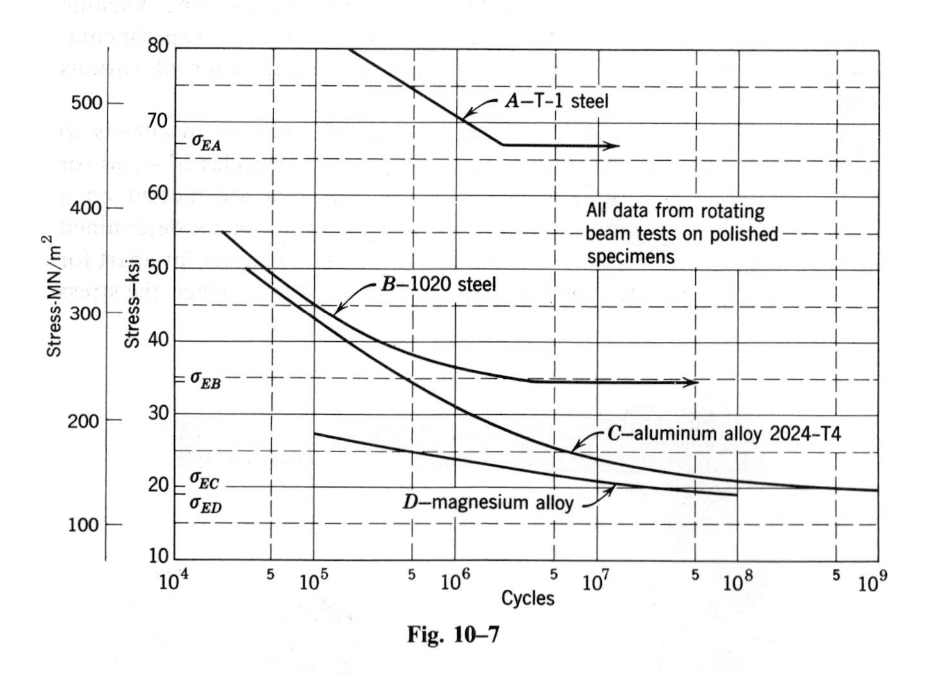

Fig. 10–7

[3]The data for Fig. 10–7 were obtained from polished specimens. Some endurance limits in App. A are for the material in the as-rolled condition and therefore have lower values.

failure or as a band such as that in Fig. 10–8.[4] The customary $\sigma - n$ diagram (such as shown in Fig. 10–7) is quite likely an approximate 50 percent probability of failure curve; that is, a specimen tested at a particular maximum stress will have a 50–50 chance of failure when subjected to the number of cycles indicated by the intersection of the line representing the test stress with the 50 percent curve.

If a machine part is to be designed for a limited life, a strength index higher than the endurance limit can be used. Theoretically, one can estimate the number of cycles of stress to which the member will probably be subjected during its lifetime, enter the $\sigma - n$ diagram with this value, and read the corresponding stress from the stress scale. If the member were stressed to this value, one could expect a fracture failure at the end of the expected lifetime, and the design of the machine part could be based on

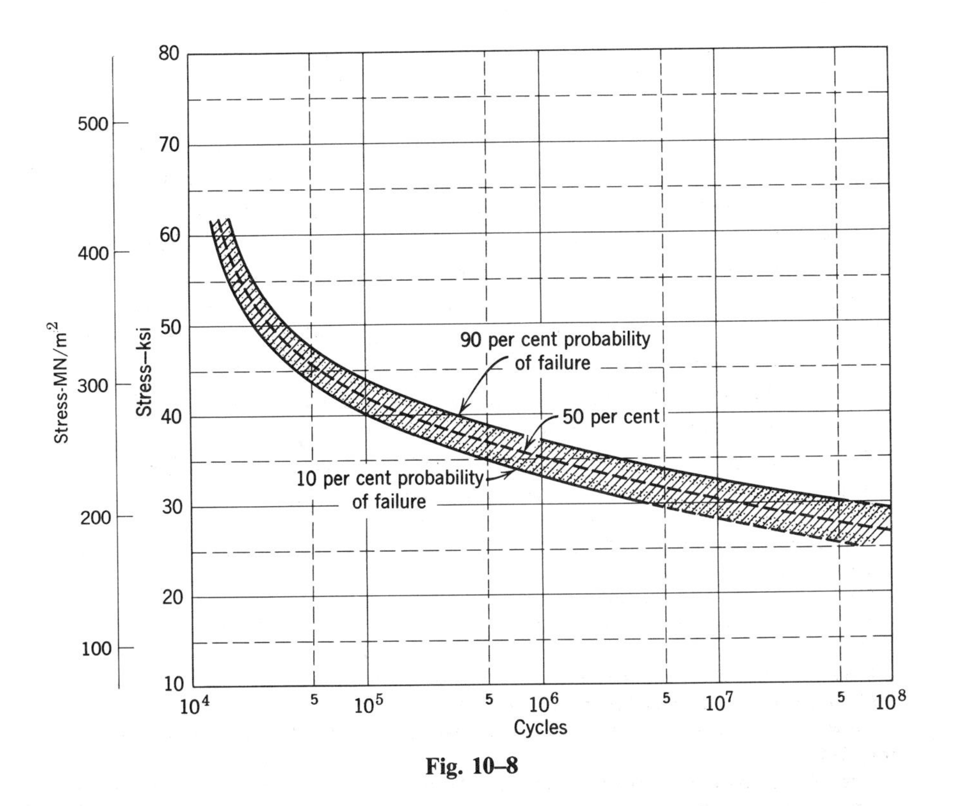

Fig. 10–8

[4] "Effect of Stress Amplitude on Statistical Variability in Fatigue Life of 75S-T6 Aluminum Alloy," G. M. Sinclair and T. J. Dolan, *Trans. A.S.M.E.*, Vol. 75, 1953.

this value reduced by an appropriate factor of safety. In this connection, however, the statistical nature of the $\sigma - n$ data should be emphasized since the life of any particular manufactured part may be much less than that indicated by the ordinary $\sigma - n$ diagram.

10–3 DESIGN FACTORS AFFECTING FATIGUE STRENGTH OF STRUCTURAL ELEMENTS

Since the fatigue failure starts at a point of high localized stress, any discontinuity, accidental or intended, is very apt to initiate such a failure. The accidental discontinuities must be taken care of by the factor of safety; however, the discontinuities due to deliberate design must be included in stress computations by an appropriate stress concentration factor (see Art. 1–10). This factor modifies the nominal equations for stress computations; thus,

$$\sigma = \frac{KP}{A} \qquad \tau = \frac{KT\rho}{J} \qquad \sigma = \frac{KMy}{I}$$

where K is a factor greater than unity and may be obtained from the curves of Fig. 10–9, 10–10, and 10–11.[5] In general, the factors may be based on either the gross or the net section (the minimum cross section at the discontinuity), and it is necessary for the designer to know whether the factors at his disposal apply to the gross or the net section. The factors K_t of Fig. 10–9, 10–10, and 10–11 are all based on the net section except those of curve ④ of Fig. 10–9, which are to be applied to the gross section. Sometimes the solution of a problem is expedited by the use of the factor based on the gross section (K_g), and for this purpose conversion expressions are given with the various curves.

Instead of the theoretical stress concentration factor, a similar experimental factor known as the *endurance limit reduction factor* may be used.

[5]Curves 1, 2, 5, 6, and 8: "Photoelastic Studies in Stress Concentration," M. M. Frocht, *Mechanical Engineering*, August 1936.

Curve 3: "Stress Concentration Factors Around a Central Circular Hole in a Plate Loaded Through Pin in the Hole," M. M. Frocht and H. N. Hill, *Trans. A.S.M.E. Journal*, March 1940.

Curve 4: "Plane Stress and Plane Strain in Bipolar Co-ordinates," G. B. Jeffery, *Philosophical Transactions of the Royal Society of London*, Vol. 221, 1921.

Curve 7: "How Holes and Notches Affect Flat Springs," A. M. Wahl, *Machine Design*, May 1940.

Curve 9: "Stress Concentration Factors and Their Effect on Design," G. H. Neugebauer, *Product Engineering*, Feb. 1943.

Curve 10: "Torsional Stress Concentrations in Shafts of Circular Cross-Section and Variable Diameter," L. S. Jacobsen, *Trans. A.S.M.E.*, Vol. 47, 1925.

The endurance limit reduction factor is determined from repeated load tests and is the ratio of the endurance limit as determined from standard test specimens to the endurance limit as determined from specimens containing a particular discontinuity. The magnitude of this factor depends on the size of the specimen, on the kind of material, and on the care exercised in machining of the test specimen. In general, the endurance limit reduction factor is smaller than the theoretical stress concentration factor; however, the larger the specimen and the stronger and more fine-grained the material, the less will be the difference between the two factors. Naturally, if a designer has at his disposal an appropriate endurance limit reduction factor (that is, the factor was obtained from a specimen of comparable size and material to that of the proposed design), such a factor should be used. In case the size and material limitations are not met or are unknown, the stress concentration factor should be used since it will be larger than the other factor and the resulting design will be more conservative.

The remarks in Art. 1–10 concerning the use of the stress concentration factor in design will bear repeating here. Stress concentration is ignored in a ductile material subjected to static loading. However, if the loading is repeated, stress concentration must not be disregarded, regardless of whether the material is brittle or ductile. Example 10–1 will serve to illustrate this point. For the problems in this chapter, unless otherwise indicated, assume that the details of construction are such that stress concentration is negligible at supports and points of application of loads.

Example 10–1 The simple beam AB of Fig. 10–12a is constructed of aluminum alloy 2024-T4 and has a square cross section 2 in. wide by 2 in. deep with a pair of notches (detail in insert) at the center. Determine the maximum permissible magnitude for the load P:

(a) If applied as a completely reversed repeated load with a factor of safety of 3.

(b) If applied as a steady load with a factor of safety of 2 with respect to failure by slip.

Solution (a) The critical stress is the maximum tensile stress, which occurs at the surface and is given by the flexure formula modified for stress concentration as necessary. The free-body and bending moment diagrams of Fig. 10–12b indicate a maximum moment at C; however, the stress concentration at the notches may produce a greater flexural stress at D than at C. Hence it is necessary to obtain the value for K from curve ⑧ of Fig. 10–10 as follows: $h/(2r)$ is 4, d/r is $1/2$, and r/b is 0.143, from which K_t is about 1.9, which is the factor based on the section 1.75 in. wide by 2 in. deep. In this instance, the work is expedited by converting K_t to

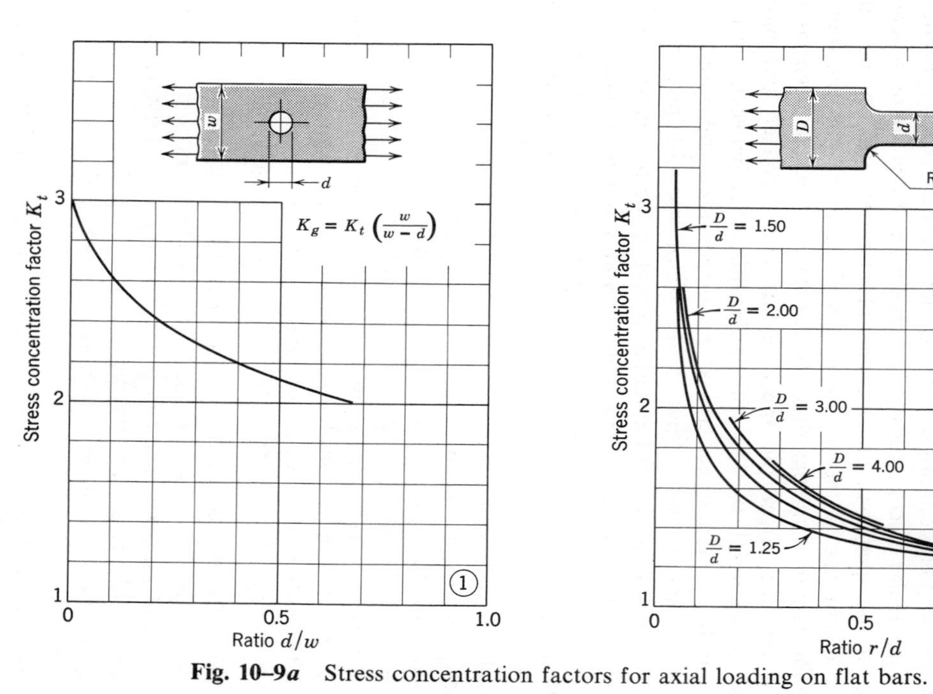

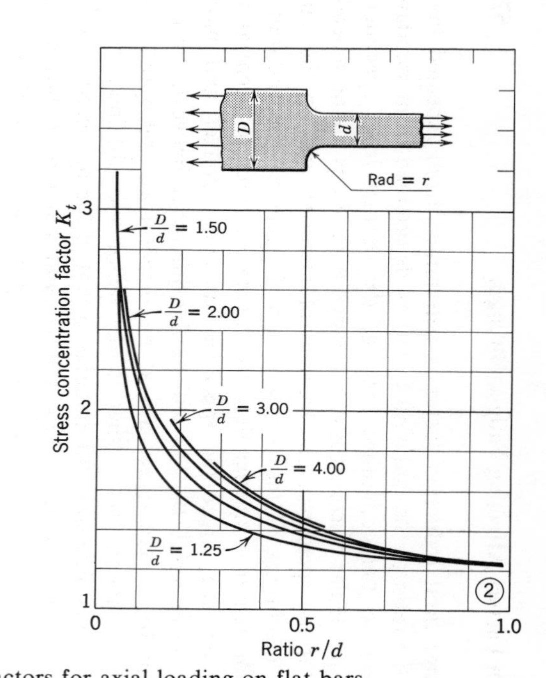

Fig. 10–9*a* Stress concentration factors for axial loading on flat bars.

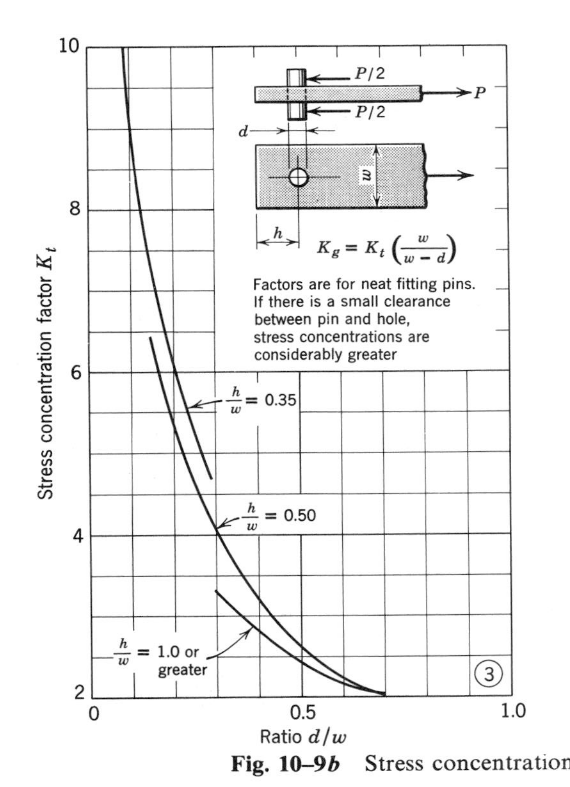

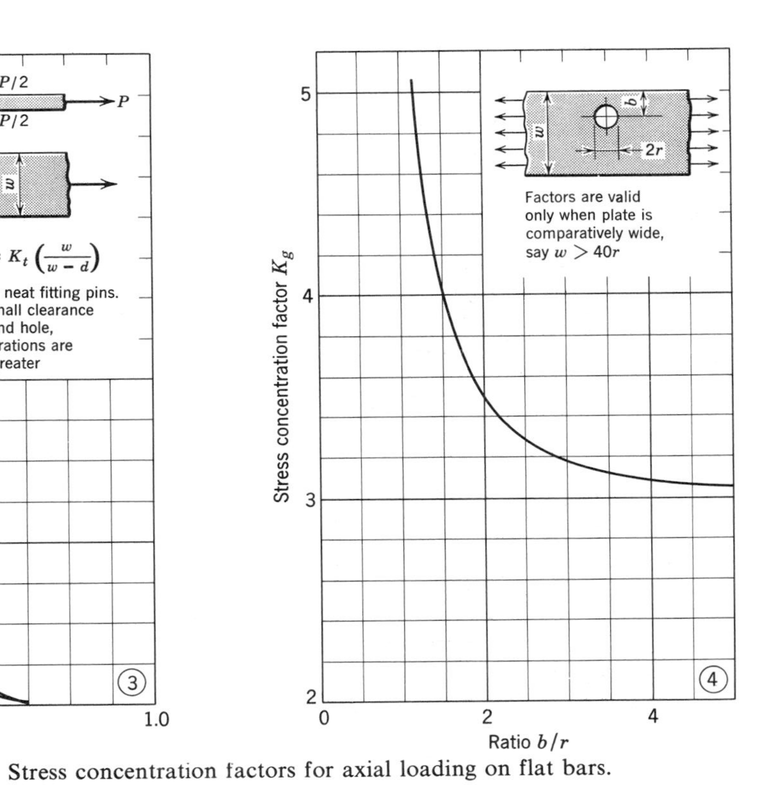

Fig. 10–9b Stress concentration factors for axial loading on flat bars.

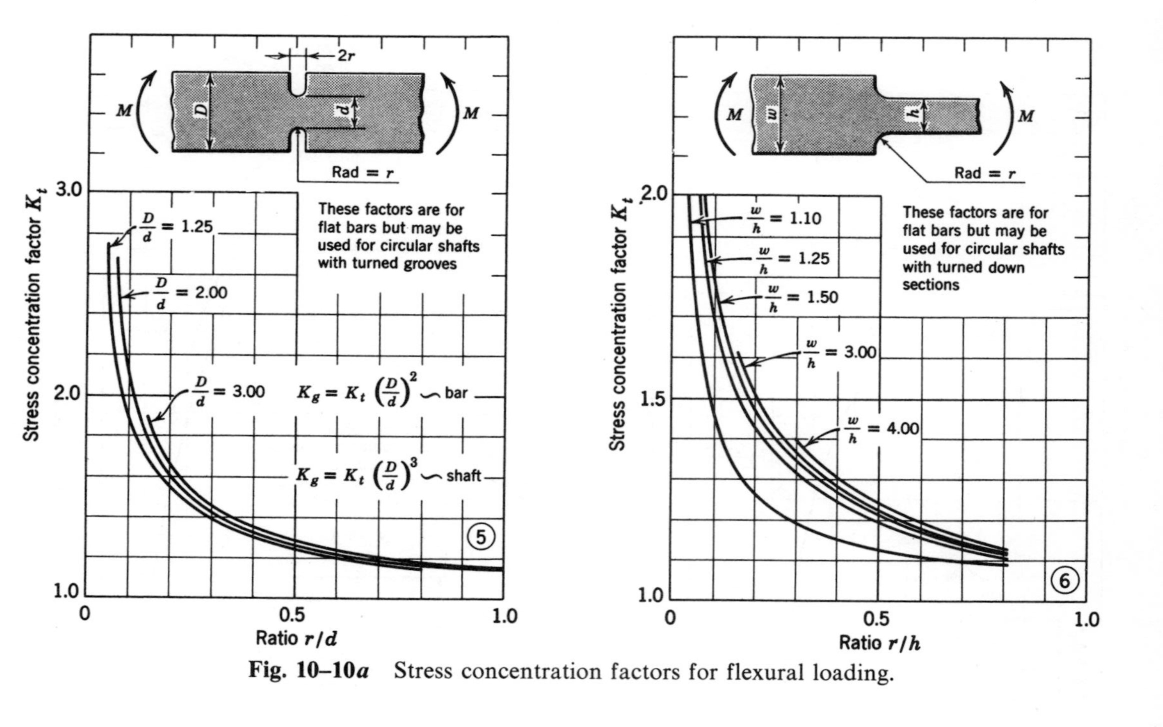

Fig. 10–10*a* Stress concentration factors for flexural loading.

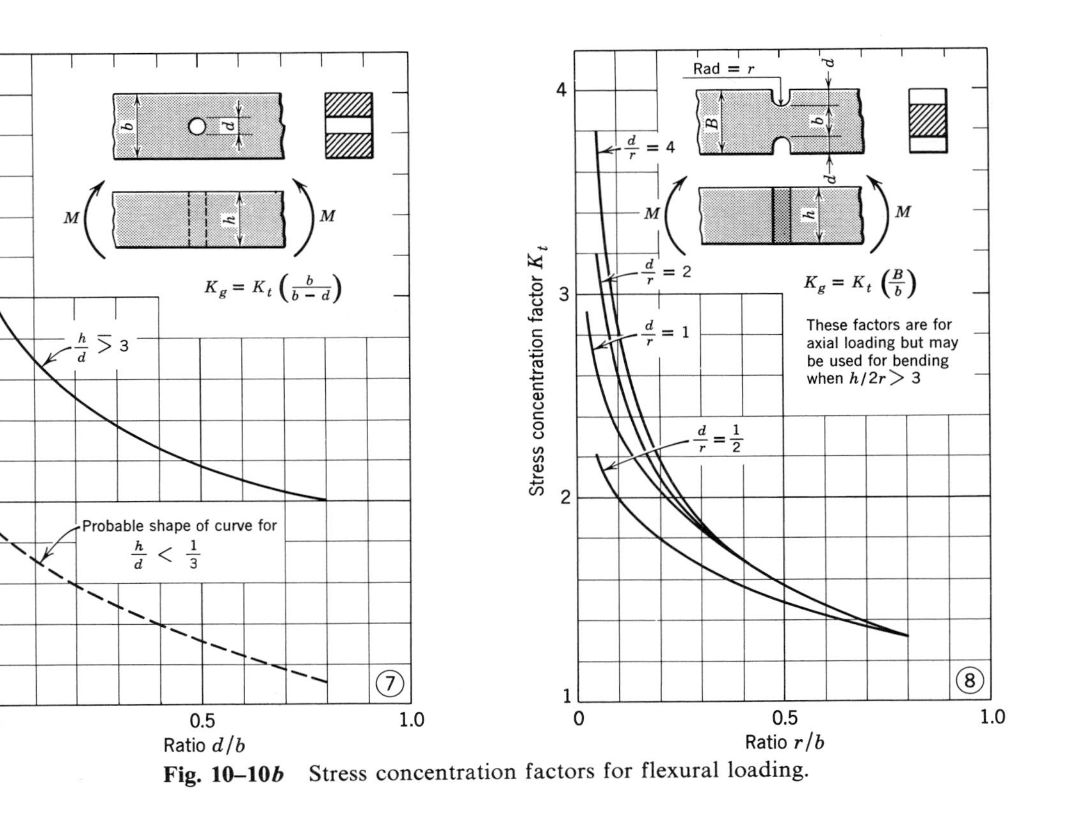

Fig. 10–10*b* Stress concentration factors for flexural loading.

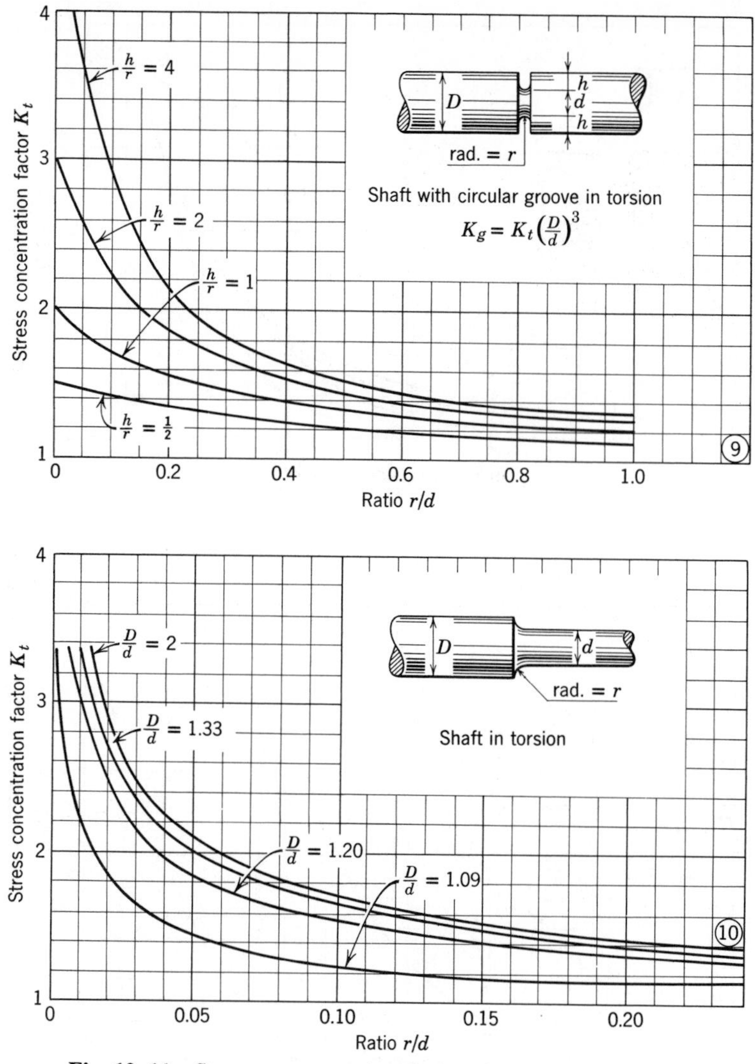

Fig. 10–11 Stress concentration factors for torsional loading.

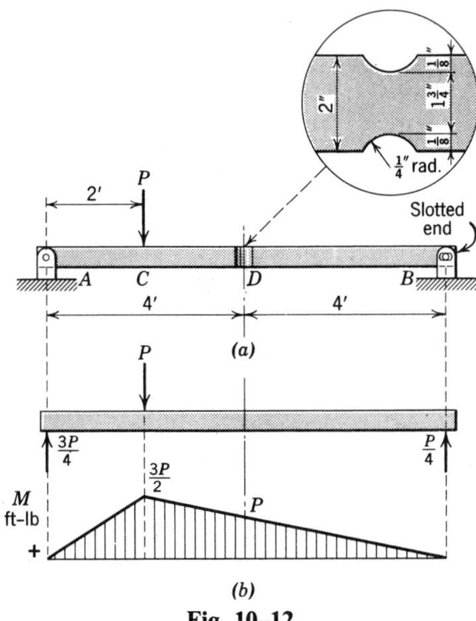

Fig. 10–12

K_g, the factor based on the gross section, since it will then be possible to compare KM at sections C and D. Thus K_g is $1.9(2)/1.75$ or 2.17. The values for KM are $1(3/2)P$ or $1.5\ P$ ft-lb at C and $2.17P$ ft-lb at D. This analysis indicates that D is the critical section, and, when the endurance limit from App. A with a factor of safety k of 3 is used, the expression for the stress is

$$\frac{\sigma}{k} = \frac{KMc}{I} = \frac{18{,}000}{3} = \frac{2.17P\,(12)(1)}{2(2^3)/12}$$

from which

$$P = \underline{307 \text{ lb}} \qquad \text{Ans.}$$

(b) Since the material has an ultimate elongation of 19 percent (see App. A), it is a ductile material, and under static loading the stress concentration is to be neglected. Comparing M/I at C and D, we obtain

$$\frac{1.5P}{2(8)/12} > \frac{P}{1.75(8)/12}$$

therefore, the critical section is now at C. Since the mode of failure is slip, the elastic strength will be used; thus

$$\frac{48{,}000}{2} = \frac{(3P/2)(12)(1)}{2(2^3)/12}$$

from which

$$P = \underline{1780 \text{ lb}} \qquad \text{Ans.}$$

PROBLEMS

10–1* A machine part of 1020 steel is accidentally subjected to a completely reversed repeated stress of 45 ksi at 250 cycles per minute. Estimate the probable service life of the member based on Fig. 10–7.

10–2* If a panel of aluminum alloy 2024-T4 in an airplane were subjected to a completely reversed stress (due to vibration) of 210 MN/m^2, what would be the probable service life of the panel, based on Fig. 10–7, if it vibrated at 200 cycles per minute?

10–3* If the estimated service life of a part made of T-1 steel in a piece of army ordnance is 600,000 cycles, determine a suitable working stress with a factor of safety of 1.25. Use Fig. 10–7.

10–4* The required service life of a machine component made of 1020 steel is 500,000 cycles. If the component is subjected to a completely reversed repeated stress of 100 MN/m^2, determine the factor of safety based on Fig. 10–7.

10–5* The machine part shown in Fig. P10–5 is 1.0 in. thick and is made of 0.4 percent carbon hot-rolled steel (see App. A). Determine the maximum safe axial load if:

(a) The load is a completely reversed repeated load (use a factor of safety of 3).
(b) The load is static (use a factor of safety of 2 with respect to failure by slip).

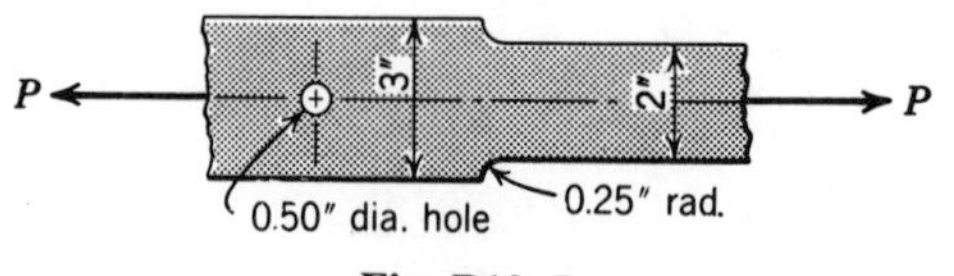

Fig. P10–5

10–6* Solve Prob. 10–5 if the material is aluminum alloy 2024-T4.

10–7* The bar of Fig. P10–7 is 10 mm thick and is made of annealed stainless steel with an endurance limit of 275 MN/m^2, a yield strength of 250 MN/m^2, an ultimate strength of 580 MN/m^2, and an elongation of 55 percent in 50 mm.

Determine the maximum permissible axial load that may be applied to the bar if:

(a) The load is completely reversed and repeated and a factor of safety of 2.5 is specified.

(b) The load is static and a factor of safety of 2 with respect to failure by slip is specified.

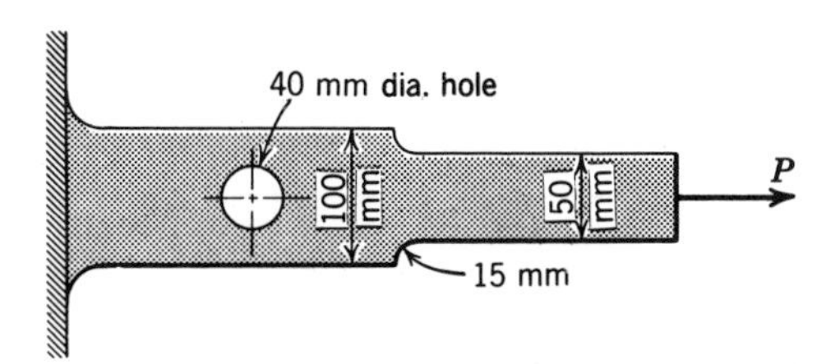

Fig. P10–7

10–8* Solve Prob. 10–7 if the material is heat-treated SAE 4150 steel with an endurance limit of 625 MN/m^2, a yield strength of 900 MN/m^2, an ultimate strength of 1325 MN/m^2, and an elongation of 10 percent in 50 mm.

10–9 The structural steel plate of Fig. P10–9 is 1/2 in. thick and is subjected to a tensile load of 30 kips uniformly distributed across the plate. Using a factor of safety of 2.5 with respect to failure by fracture or slip, determine the minimum allowable width of the plate if:

(a) The load is completely reversed and repeated.

(b) The load is static.

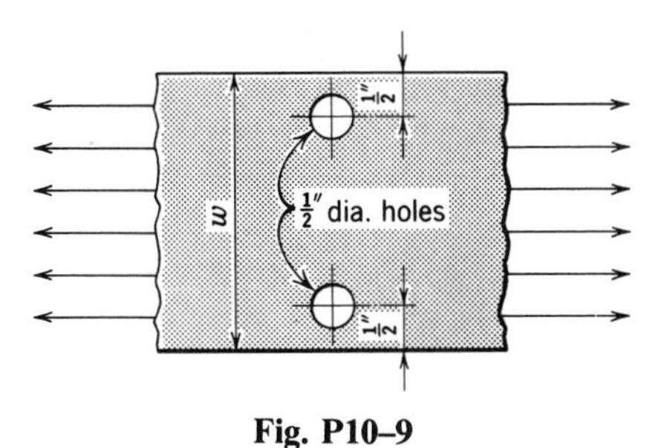

Fig. P10–9

10–10 The bar in Fig. P10–10 is to be made of 0.4 percent carbon hot-rolled steel with an endurance limit of 260 MN/m^2, a yield strength of 365 MN/m^2, an ultimate strength of 580 MN/m^2, and an elongation of 29 percent in 50 mm. If a factor of safety of 2.5 with respect to failure by fracture or slip is specified, determine the minimum permissible thickness h of the bar when:

(a) The load P is a static load of 1500 N.

(b) The load P is a completely reversed and repeated load of 1200 N.

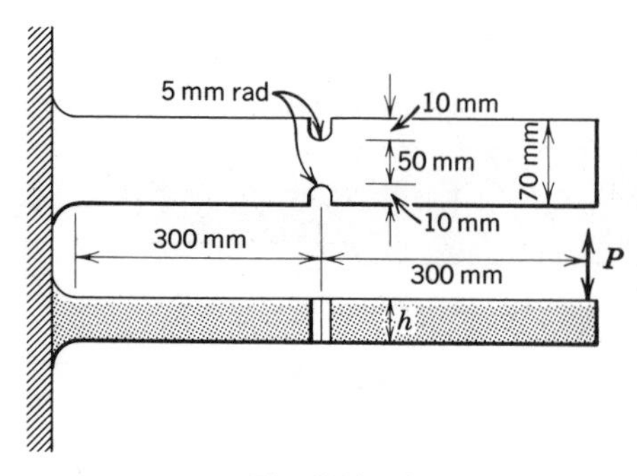

Fig. P10–10

10–11* A 4-in. diameter shaft subjected only to bending has a U-shaped groove 1.0-in. deep (radius of groove is 0.4 in.) turned in it at the section of maximum moment. The shaft is made of structural steel. Determine the maximum permissible bending moment that may be applied if:

(a) The load is static and the factor of safety is 2.5 with respect to failure by slip.

(b) The load is completely reversed and repeated and the factor of safety is 3.

10–12* The cantilever beam shown in Fig. P10–12 is made of structural steel. Determine the maximum safe value of the load P if:

(a) Applied as a steady load. Use a factor of safety of 2 with respect to failure by slip.

(b) Applied as a completely reversed repeated load. Use a factor of safety of 2.5.

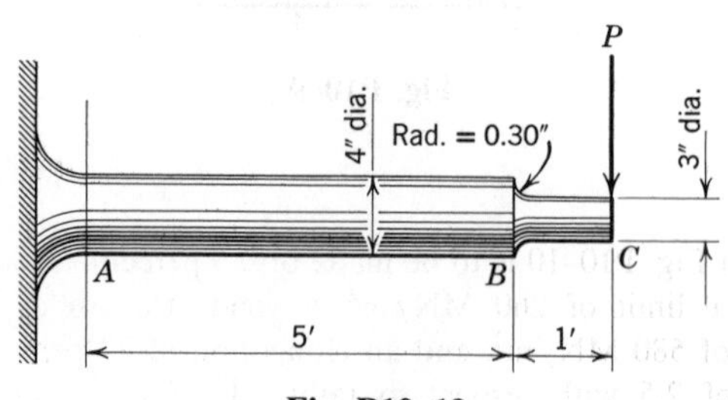

Fig. P10–12

10–13* Figure P10–13 shows a cantilever spring 50 mm wide. If P is a completely reversed repeated load of 200 N and the material is heat-treated SAE 4340 steel with an endurance limit of 525 MN/m^2, a yield strength of 900 MN/m^2, and an ultimate strength of 1025 MN/m^2, determine the maximum safe values for the distances a and b (b in combination with the maximum value for a). Use a factor of safety of 2.5.

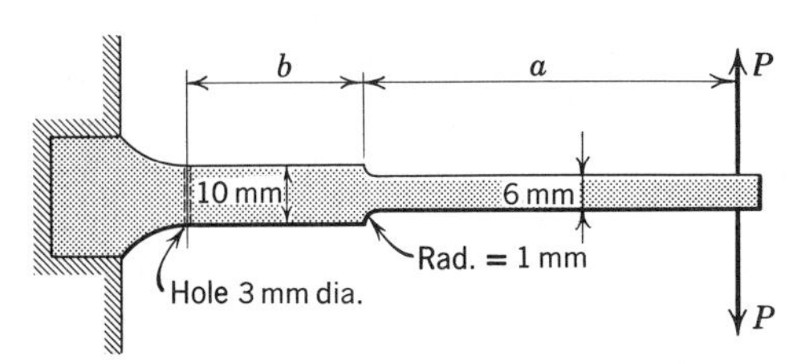

Fig. P10–13

10–14 A simple beam is 150 mm wide by 50 mm deep and has a span of 3 m. It is loaded with a completely reversed repeated concentrated load P at the center of the span and has a vertical hole 15 mm in diameter through the beam 1 m from one end. Determine the maximum safe value for P using a factor of safety of 3. The endurance limit for the material is 300 MN/m^2, the yield strength is 365 MN/m^2, and the utimate strength is 640 MN/m^2.

10–15 The notched cantilever beam of Fig. P10–15 is 2 in. thick and made of aluminum alloy 6061-T6. Determine the maximum completely reversed repeated load P that may be applied if a factor of safety of 3 is specified.

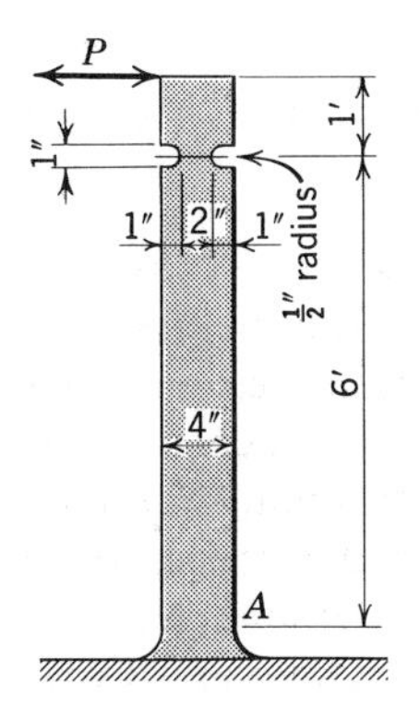

Fig. P10–15

10–16* A simple beam is 4.0 in. wide by 2.0 in. deep and has a span of 10 ft. It is loaded with a completely reversed repeated concentrated load P at the center of the span. The beam has two vertical U-shaped grooves opposite each other on the edges of the beam 1.5 ft from one end. The grooves are 0.5 in. deep and have a radius of 0.3 in. Determine the maximum safe value for P using a factor of safety of 3. The endurance limit for the material is 40 ksi, the yield strength is 36 ksi, and the ultimate strength is 85 ksi.

10–17* A 100-mm diameter shaft is to have one end turned down to a 75-mm diameter. The shaft is to be made of annealed stainless steel with an endurance limit in torsion of 140 MN/m^2. If the shaft is to be subjected to a completely reversed repeated torque of 2000 N-m and a factor of safety of 3 is specified, what should be the minimum radius of the fillet at the junction of the two sections?

10–18 A 4-in. diameter shaft has one end turned down to a 2-in. diameter. The fillet at the junction of the two sections has a 1/4-in. radius. The shaft is made of 3.5 percent nickel steel, hardened and tempered (use 32 ksi for the endurance limit in torsion). A factor of safety of 3 is specified.

(a) What maximum completely reversed repeated torque may be applied?
(b) If a U-shaped groove 0.4 in. wide by 0.6 in. deep (radius of bottom of groove 0.2 in.) is turned in the 4-in. section, what would be the maximum completely reversed repeated torque?

10–19* The bar in Fig. P10–7 is made of heat-treated 4130 steel with an endurance limit of 520 MN/m^2, a yield strength of 950 MN/m^2, an ultimate strength of 1100 MN/m^2, and an elongation of 15 percent in 50 mm. Determine the minimum thickness for the bar if it must initially support a static load of 500 kN with a factor of safety of 2 with respect to failure by rupture and later support a completely reversed and repeated load of 200 kN with a factor of safety of 3.

10–20* The bar in Fig. P10–5 is made of structural steel and must support a static load of 40 kips with a factor of safety of 3, with respect to failure by rupture, during erection of a structure. After the erection is completed, the bar will be subjected to a completely reversed repeated load of 10 kips. Determine the minimum allowable thickness of the bar if it is to have a factor of safety of 2.5.

10–4 RESISTANCE TO FAILURE UNDER FLUCTUATING LOAD

The maximum stress necessary to cause fracture under a fluctuating load is usually larger than the endurance limit. Considerable data have been accumulated from tests performed in machines similar to that shown schematically in Fig. 10–6, in which the stress in the specimen may vary from some minimum to a different maximum value. However, the task of experimentally determining the ranges of stress that will cause fatigue failure for all possible (or reasonable) combinations of maximum and minimum stresses and for the numerous materials in use today would be unending. Hence, some method of predicting the strength based on available data is essential. The approach to the problem is made through the interpretation of data from a limited number of tests in various ranges and is based on the concept that a fluctuating stress can be considered the sum of a steady stress and a completely reversed (or alternating) stress, as indicated for normal stresses in Fig. 10–13, where σ_a, the steady stress, is one-half the sum of $\sigma_{\max}$ and $\sigma_{\min}$, and where σ_r, the alternating stress, is one-half the difference between these two stresses (compressive stresses are negative). The steady and alternating stresses may not be the same kind of stress; for example, one could have a steady torsional shearing stress combined with an alternating flexural stress. However, for the restricted scope of this book, the discussion will be limited to the case where the steady and alternating stresses at a point are both normal (or both shearing) stresses on the same plane.

The data from tests can be plotted on a diagram, and a curve can be drawn to represent the strength of the material for various combinations of steady and alternating stresses. Two curves that have received widespread attention are shown in Fig. 10–14, in which σ_E is the endurance limit, σ_Y is the static elastic[6] tensile strength, and σ_U is the static ultimate tensile strength. Any point with coordinates (σ_a, σ_r), such as A on the diagram, would represent a particular combination of steady and repeated stresses, and, if the point is above the straight line (Soderberg[7] or Goodman,[8] depending on which line is accepted as the failure criterion), failure is presumed to result. Most of the available test data plot above the two lines shown, indicating that both curves are conservative.

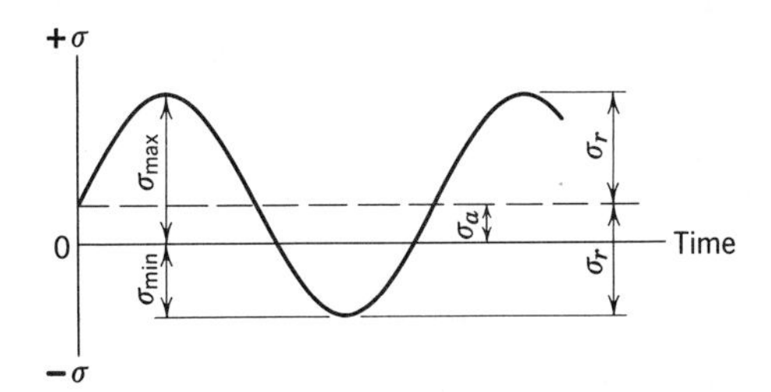

Fig. 10–13

[6]Yield point, yield strength, or proportional limit.

[7]See "Working Stresses," C. R. Soderberg, *Trans. A.S.M.E.*, Vol. 55, APM-55-16, 1933, p. 131.

[8]Goodman's diagram does not resemble the one of Fig. 10–14. However, since the two diagrams yield the same results, it has become customary to identify the upper line of Fig. 10–14 as the Goodman criterion.

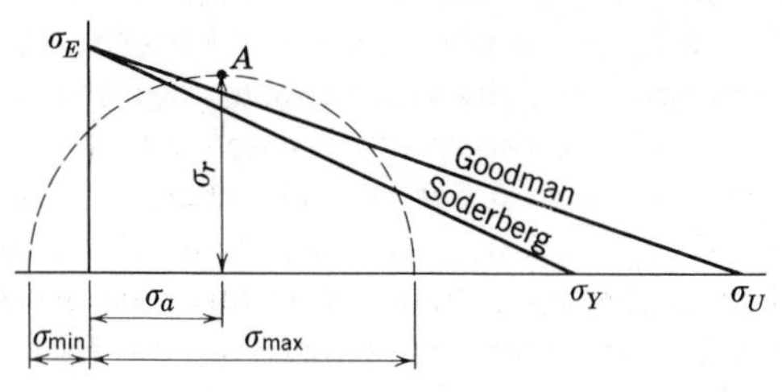

Fig. 10–14

When a design problem involves machine elements that will not function properly if the static elastic strength of the material is exceeded (a slip failure), the Soderberg diagram should be regarded as the criterion of failure. The reason for using the more conservative diagram is that $\sigma_{\max}$ might exceed the static elastic strength on the first application of the maximum load, whereupon the usefulness of the mechanism would be impaired or destroyed. If fracture is the only mode of failure to be considered, slip is of no concern, and the less conservative Goodman diagram should be used. Needless to say, the approximate criteria of strength are to be used only if fluctuating strength data from mechanical tests are unavailable. For designs that warrant the expense, fatigue tests should be conducted on structural components loaded to simulate actual service conditions. When the approximate failure criteria are applied, if the endurance limit is unknown *and the material is ferrous,* one-half of the ultimate tensile strength will give a reasonable value for the endurance limit. The use of the Soderberg and Goodman diagrams is illustrated by Exams. 10–2 and 10–3. The choice of diagrams is usually indicated by a statement concerning the mode of failure. For example, a factor of safety with respect to failure by slip or fracture means that both slip (due to the application of the maximum load) and fracture (due to the alternating load) must be avoided; hence, the Soderberg criterion should be used.

Example 10–2 A machine element of heat-treated SAE 4340 steel is subjected to stresses that fluctuate from 50 ksi T to 10 ksi C. Determine the factor of safety with respect to failure by slip or fracture.

Solution Failure by slip or fracture indicates that the Soderberg diagram should be used. The diagram is shown in Fig. 10–15, for which the properties were obtained from App. A. The first step in the solution is to compute the ratio of σ_r to σ_a as follows:

$$\sigma_{a1} = \frac{50 + (-10)}{2} = 20\,\text{ksi}\,T$$

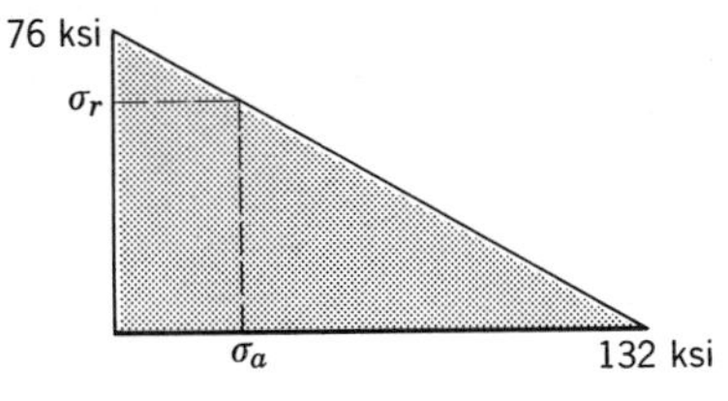

Fig. 10–15

and

$$\sigma_{r1} = \frac{50-(-10)}{2} = 30\,\text{ksi}\ T \text{ and } C$$

from which

$$\sigma_r = \frac{3\sigma_a}{2}$$

The equation of the line in Fig. 10–15 is

$$\sigma_r = 76 - \left(\frac{76}{132}\right)\sigma_a$$

and, since

$$\sigma_r = \frac{3\sigma_a}{2}$$

$$\frac{3\sigma_a}{2} = 76 - \left(\frac{76}{132}\right)\sigma_a$$

from which

$$\sigma_a = 36.6 \text{ ksi}$$

This value represents the limit of the steady stress. Because of the fixed ratio between σ_a and σ_r, 36.6 ksi can be used as a strength index; hence, the factor of safety is

$$k = \frac{36.6}{20} = \underline{1.83} \qquad \text{Ans.}$$

Observe that this value is obtained regardless of whether one uses the steady, the alternating, the maximum, or the minimum stress as the strength index; for example,

$$\sigma_{max} = \sigma_a + \sigma_r = \sigma_a + \frac{3\sigma_a}{2} = \frac{5\sigma_a}{2}$$

and the factor of safety is

$$k = \frac{36.6}{20} = \frac{(3/2)(36.6)}{(3/2)(20)} = \frac{(5/2)(36.6)}{(5/2)(20)} = \underline{1.83}$$

Example 10–3 The load on the beam of Exam. 10–1 fluctuates from $3P$ downward to $2P$ upward. Determine the maximum permissible magnitude for P if a factor of safety of 3 with respect to failure by fracture is specified.

Solution Since the only criterion of failure is fracture,[9] the Goodman diagram will be used with σ_E of 18 ksi and σ_U of 68 ksi. Instead of these strength values being used, they may be divided by the factor of safety, thus converting the failure diagram to a working stress diagram. Whether the factor of safety is introduced in the diagram or later is optional. To get the ratio of stresses,

$$P_a = \frac{3P + (-2P)}{2} = \frac{P}{2}$$

and

$$P_r = \frac{3P - (-2P)}{2} = \frac{5P}{2}$$

from which

$$P_r = 5P_a$$

Therefore, since the stresses will be below the proportional limit,

$$\sigma_r = 5\sigma_a$$

[9]So far as the static strength is concerned, fracture here denotes that the maximum static load is reached when the maximum flexural stress is equal to the ultimate tensile strength. Actually, under a load of this magnitude, the beam would not collapse since the material is ductile (see Art. 5–2).

and, when the Goodman diagram is used with working stresses instead of strengths,

$$\sigma_{r_w} = \frac{18}{3} - \frac{18}{68}\sigma_{a_w} = 5\sigma_{a_w}$$

from which

$$\sigma_{a_w} = 1.14 \text{ ksi}$$

The loading is repeated; therefore the stress concentration will not be ignored, and the analysis of part *a* of Exam. 10–1 is valid. Thus

$$\frac{\sigma_a}{k} = \frac{KMc}{I} = 1140 = \frac{2.17P_a(12)(1)}{2(2^3)/12}$$

from which

$$P_a = 58.4$$

and

$$P = 2P_a = \underline{117 \text{ lb}} \qquad \text{Ans.}$$

Observe that the correct solution is obtained whether $\sigma_a, \sigma_r, \sigma_{max}$, or σ_{min} is used, as long as the corresponding value for the load is used. For example, if σ_r is used, the above equation becomes

$$1140(5) = 2.17(5P_a)(9)$$

giving the same result as above.

When this result is compared with Exam. 10–1, the fluctuating load is seen to vary from -234 lb to $+350$ lb, while a completely reversed load varies from -307 to $+307$ lb.

PROBLEMS

10–21* Determine the factor of safety with respect to failure by fracture or slip for a machine part for which the stress varies from 12,000 psi *T* to 8000 psi *C*. The material is aluminum alloy 2024-T4.

10–22* Determine the factor of safety with respect to failure by fracture or slip for a machine part for which the stress varies from 85 MN/m^2 *T* to 30 MN/m^2 *C*. The material is a magnesium alloy with an endurance limit of 130 MN/m^2, a yield strength of 240 MN/m^2, and an ultimate strength of 335 MN/m^2.

10–23* A machine member is to be subjected to a repeated load varying from 2000 lb in tension to 7000 lb in tension. The material is 3.5 percent nickel steel–hardened and tempered with an endurance limit of 63 ksi, a proportional limit of 86 ksi, and an ultimate strength of 118 ksi. Using Soderberg's method, determine the combination of maximum and minimum stresses to which the member could probably be subjected millions of times without fracture.

10–24* An axial load varies from 20 kN compression to 40 kN tension. Using Soderberg's method, determine the combination of maximum and minimum stresses to which the member could probably be subjected millions of times without fracture if the proposed material is 0.93 percent carbon spring-steel hardened and tempered with an endurance limit of 385 MN/m^2, a proportional limit of 585 MN/m^2, and an ultimate strength of 670 MN/m^2.

10–25 A machine element is subjected to a bending load varying from 6000 lb tension to 3000 lb compression. The material is structural steel (see App. A). Determine the magnitude of a suitable working stress using a factor of safety of 2 with respect to failure by fracture or slip.

10–26 A machine part is subjected to a bending load varying from 25 kN tension to 15kN compression. The material is structural steel with an endurance limit of 190 MN/m^2, a proportional limit of 250 MN/m^2, and an ultimate strength of 450 MN/m^2. Determine the magnitude of a suitable working stress using a factor of safety of 2.5 with respect to failure by fracture or slip.

10–27* Determine the maximum and minimum values for the fluctuating axial load P of Fig. P10–27 if the magnitude of P in tension is twice the magnitude in compression. The material is heat-treated SAE 4340 steel (see App. A), and a factor of safety of 2 with respect to failure by fracture or slip is to be maintained.

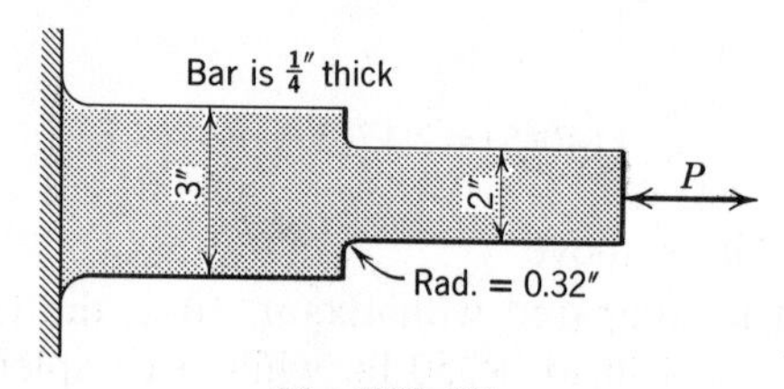

Fig. P10–27

10–28* The connecting rod of Fig. P10–28 is made of normalized 0.57 percent carbon steel with an endurance limit of 225 MN/m^2, a proportional limit of 240 MN/m^2, and an ultimate strength of 495 MN/m^2. The rod is of rectangular cross section and is 50 mm thick. Determine the maximum completely reversed repeated load P that may be applied with a factor of safety of 3 with respect to failure by fracture or slip. Note that the tension on the net section at the pin fluctuates from zero to a maximum.

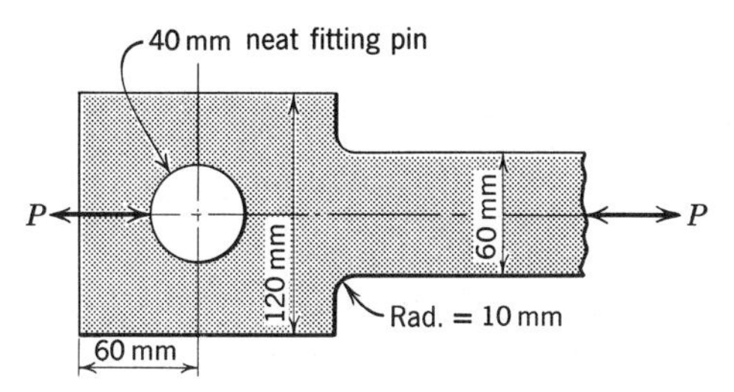

Fig. P10–28

10–29 The bar of Fig. P10–5 is made of 1/2-in. thick structural steel and is subjected to an axial load fluctuating from $4P$ in tension to P in compression. Determine the maximum permissible value for P if a factor of safety of 2 with respect to failure by fracture or slip is specified.

10–30 A simple beam of aluminum alloy 2024-T4 is 100 mm wide by 50 mm deep, has a span of 4 m, and has a 10-mm vertical hole drilled through it 0.5 m from one support. The beam is loaded at the center of the span with a concentrated load varying from P upward to $4P$ downward. Determine the magnitude of P if a factor of safety of 2.5 with respect to failure by fracture or slip is specified. The material has an endurance limit of 125 MN/m^2, a yield strength of 330 MN/m^2, and an ultimate strength of 470 MN/m^2.

10–31* A simple beam of structural steel is 6 in. wide by 2 in. deep and has a span of 10 ft. At the center of the span is a pair of symmetrical notches for which the stress concentration factor is 1.5 based on the gross section. A concentrated fluctuating load is applied 3 ft from one support. Determine the maximum and minimum values of the load if the magnitude in one direction is three-halves that in the other direction. Use a factor of safety of 2 with respect to failure by fracture or slip.

10–32* The cantilever beam of Fig. P10–32 is made of heat-treated spring steel with the properties of Prob. 10–24. The beam is 50 mm wide with depths and stress

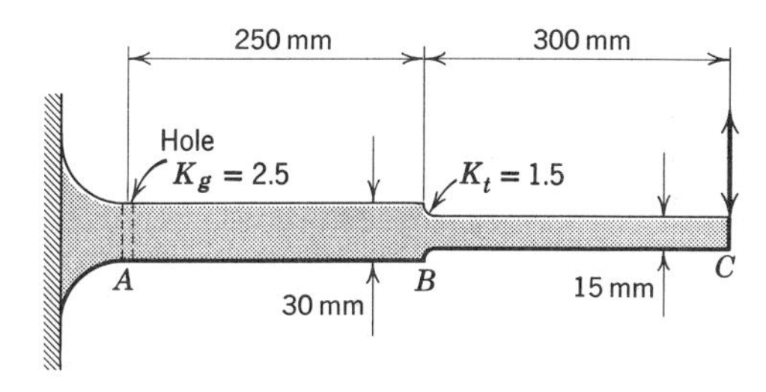

Fig. P10–32

concentration factors as shown. The load at C varies from $3P$ downward to P upward. Determine the maximum permissible magnitude for P using a factor of safety of 3 with respect to failure by fracture or slip.

10–33 If the load of Prob. 10–15 fluctuates from $3P$ to the right to P to the left and a factor of safety of 2.5 with respect to failure by fracture or slip is specified, what is the maximum permissible value of P?

10–34 A spring for a latch mechanism may be approximated by a cantilever beam with a rectangular cross section 10 mm wide by 1.0 mm thick with a span of 60 mm. The free end of the spring is deflected from zero to 7.0 mm as the latch is operated. If the spring is made of heat-treated SAE 4340 steel with the properties of Prob. 10–13, what is the factor of safety with respect to failure by fracture?

10–35 The steel cantilever spring of Fig. P10–35 is deflected repeatedly from zero to 1/4 in. by means of the rotating cam. The spring has a rectangular cross section 0.2 in. thick and 1.2 in. wide. The material is cold-rolled 18–8 stainless steel. Determine the factor of safety with respect to failure by fracture.

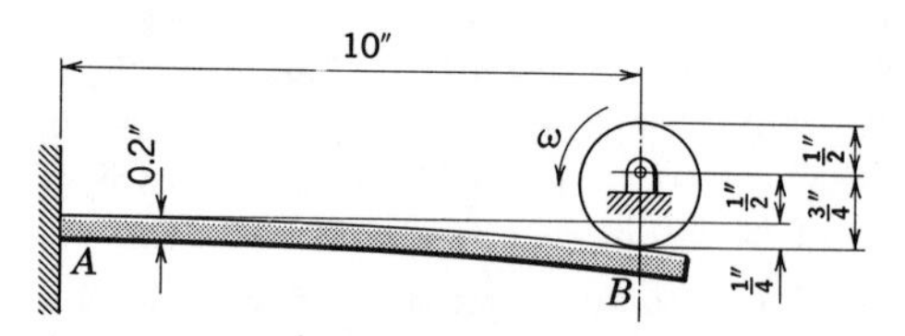

Fig. P10–35

10–36* The unbalanced disk of Fig. P10–36 has a mass of 150 kg and the eccentricity of the mass center induces an effective force of 400 N directed outward from the shaft which is attached to and rotates with the disk. Neglect the mass of the shaft and any torsional moment, assume the end bearings to be simple supports, and consider bending only in a vertical plane. Determine the minimum permissible diameter for the shaft if it is to be made of cold-rolled 18–8 stainless steel with an endurance limit of 620 MN/m^2, a yield strength of 1125 MN/m^2, and an ultimate strength of 1300 MN/m^2. A factor of safety of 2.5 with respect to failure by fracture or slip is specified.

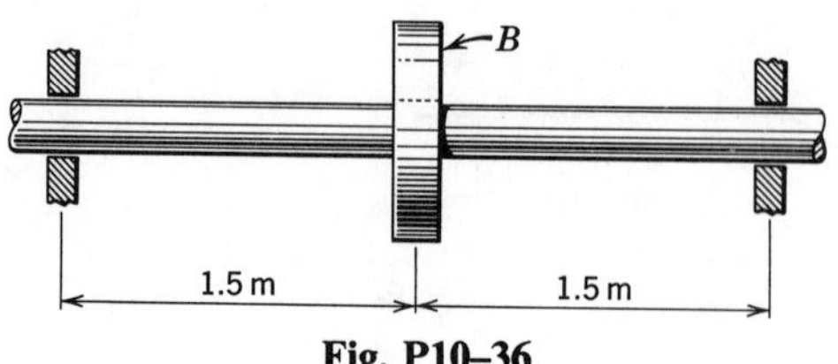

Fig. P10–36

10–37* The large flywheel in Fig. P10–37 weighs 10 kips and is supported by and rotates about the shaft AB, which is fixed at B. A misalignment e of the shaft and the mass center of the wheel (see insert) causes the bottom of the shaft to move in a circular path during rotation of the wheel. Assume that the wheel rotates about its mass center, which remains directly below the center of the shaft at support B. Neglect the weight of the shaft and bearing friction. The shaft is made of 0.4 percent carbon hot-rolled steel and a factor of safety of 2.5 with respect to failure by fracture or slip is specified. Determine the maximum permissible value for e.

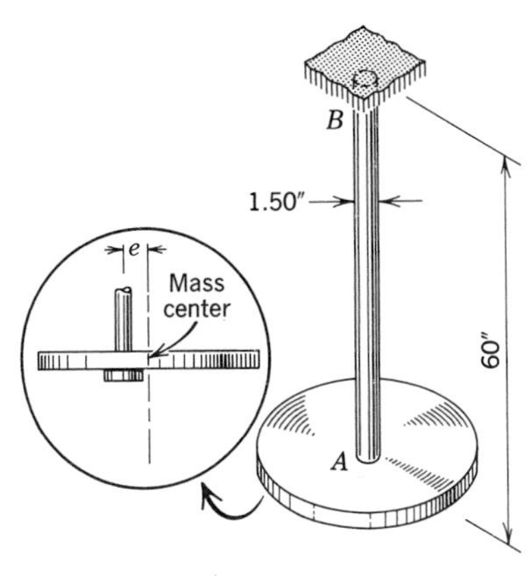

Fig. P10–37

10–38 Solve Prob. 10–36 when the disk rotates on a shaft that does not rotate.

10–39* The machine part shown in Fig. P10–39 is 0.5 in. thick and is made of structural steel. Determine the maximum safe axial load P if the magnitude of the load in tension is twice the magnitude in compression and a factor of safety of 2.5 with respect to failure by fracture or slip is to be maintained.

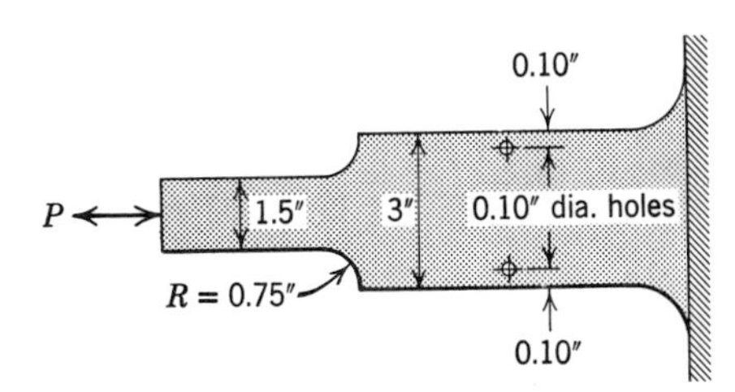

Fig. P10–39

10–40* The strut of Fig. P10–40 is made of 0.4 percent carbon hot-rolled steel with the properties of Prob. 10–10 and is subjected to a completely reversed and repeated load of 7.5 kN. Determine the minimum thickness of the rectangular cross section that may be used if the factor of safety is to be 2 with respect to failure by fracture or slip. Note that the tension on the net cross section at the pin fluctuates from zero to a maximum.

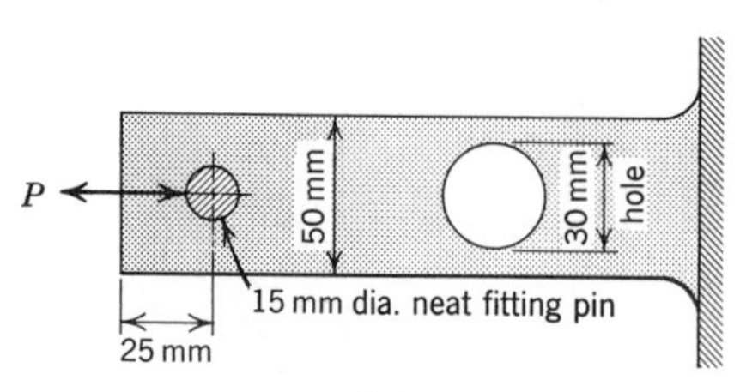

Fig. P10–40

10–41* Solve Prob. 10–35 if a 1/16-in. diameter hole is drilled vertically through the spring at a point 5 in. from A.

10–42 A simple beam is 25 mm wide by 40 mm deep and has a span of 1 m. It is loaded with a uniformly distributed static load of 4000 N/m over its entire length and a completely reversed repeated concentrated load of 1500 N at the center of the span. The beam is made of structural steel with the properties of Prob. 10–26. Determine the factor of safety with respect to failure by fracture or slip for the beam.

10–43 The structural steel beam shown in Fig. P10–43 is 3 in. wide. It is subjected to a completely reversed and repeated uniformly distributed load of 50 lb/in. Determine the factor of safety.

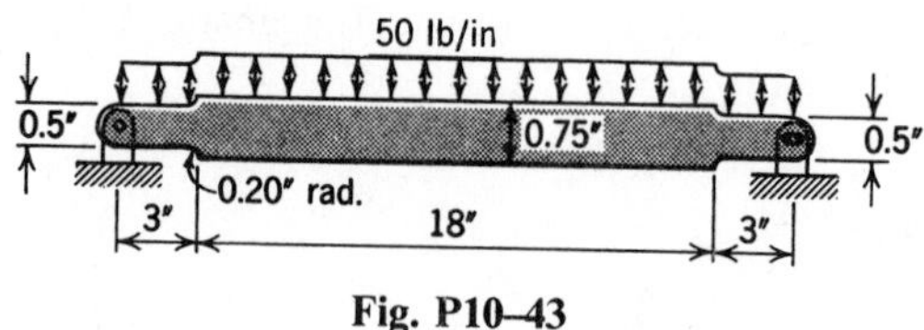

Fig. P10–43

11 Dynamic Loading

11–1 INTRODUCTION

When the motion of a body is changed (accelerated), the force necessary to produce this acceleration is called a *dynamic force* or *load*. For example, the force an elbow in a pipeline exerts on the fluid in the pipe to change its direction of flow, the pressure on the wings of an airplane pulling out of a dive, the collision of two automobiles, and a man jumping on a diving board are all examples of dynamic loading. A suddenly applied load is called an *impact load*. The last two of the preceding examples are considered impact loads. Under impact loading, if there is elastic action, the loaded system will vibrate until equilibrium is established. A dynamic load may be expressed in terms of mass times the acceleration of the mass center, in terms of the rate of change of the momentum, or in terms of the change of the kinetic energy of the body.

For many problems in mechanics of materials, the energy approach, introduced in Art. 2–9, is most effective. The magnitude of the dynamic load is often expressed in terms of the kinetic energy delivered to the loaded system; hence, it is often referred to as an *energy load*. For example, a particle of mass m moving with a speed v possesses a kinetic energy

$mv^2/2$; if this particle is stopped by a body, the energy absorbed by the body (the loaded system) is some fractional part of $mv^2/2$, the balance of the energy being converted into sound, heat, and permanent deformation of the striking particle.

In the loaded system, dynamic loading produces stresses and strains, the magnitude and distribution of which will depend not only on the usual parameters encountered previously but also on the velocity of propagation of the strain waves through the material of which the system is composed. This latter consideration, although very important when loads are applied with high velocities, may often be neglected when the velocity of application of the load is low. The velocities are considered to be low when the loading time permits the material to act in the same manner as it does under static loading; that is, the relations between stress and strain and between load and deflection are essentially the same as those already developed for static loading.

11–2 ENERGY OF CONFIGURATION

Since dynamic loading is conveniently considered to be the transfer of energy from one system to another, the concept of energy of configuration (strain energy) as an index of resistance to failure is important enough to be discussed at some length. In Art. 2–9 the concept of strain energy was introduced by considering the work done by a slowly applied axial load P in elongating a bar of uniform cross section by an amount δ_2. From the load-deformation diagram of Fig. 2–19b it was observed that the work done in elongating an axially loaded bar is

$$W_k = \int_0^{\delta_2} P\,d\delta \tag{a}$$

Since the work done on the bar must equal the strain energy stored in the bar, the expression for strain energy in terms of axial stress and axial strain became

$$W_k = U = \int_0^{\epsilon_2} (\sigma)(A)(L)\,d\epsilon = AL\int_0^{\epsilon_2} \sigma\,d\epsilon \tag{b}$$

When the stress remains below the elastic limit of the material, Hooke's law applies and $d\epsilon$ may be expressed as $d\sigma/E$. Equation (b) then becomes

$$U = \left(\frac{AL}{E}\right)\int_0^{\sigma_1} \sigma\,d\sigma$$

or

$$U = AL\left(\frac{\sigma_1^{\ 2}}{2E}\right) \tag{11–1}$$

When the energy load U applied to a resisting structural member comes from a body falling from rest, the kinetic energy of the falling weight W equals its change in potential energy, which is $W(h+\delta)$, where h is the distance of free fall and δ the deflection of the resisting member due to the impact. For the special case of an axially loaded bar, the maximum axial stress and the maximum elongation of the bar can be determined by equating the kinetic energy of the falling body to the strain energy in the bar at the instant of maximum deflection if it is assumed that all of the kinetic energy of the body is effective in straining the bar. The loss of energy during impact is discussed in the next article.

If, instead of axial loading, one considers a differential element of a material in a region where the stresses vary with respect to the position of the element, there will, in general, be three normal and three shearing stresses acting on the element. The energy intensity expression for this general case was developed in Art. 2–9 (see Eq. 2–15b).

When the state of stress in a body can be represented by one nonzero normal stress, Eq. 11–1 can be written as

$$U = \int_{\text{volume}} \left[\sigma^2/(2E)\right] dx\, dy\, dz \tag{c}$$

Equation (c) is valid for a beam subjected to pure bending, and, if the energy (usually small) resulting from the shearing stresses is neglected, the expression applies to any beam with transverse loading. The stress in Eq. (c) must be expressed as a function of one or more of the other variables before integration.

For the case of pure bending in a beam of constant cross section A and length L, the value of σ_x is

$$\sigma_x = My/I$$

and the strain energy (when x is measured along the axis of the beam) is

$$U = \frac{1}{2E}\int\int\left(\frac{My}{I}\right)^2 dA\, dx = \frac{1}{2E}\int_{\text{area}}\int_0^L\left(\frac{M^2}{I^2}\,dx\right)y^2\,dA$$

$$= \frac{1}{2E}\int_0^L \frac{M^2}{I^2}\,dx \int_{\text{area}} y^2\,dA$$

which reduces to

$$U = \int_0^L \frac{M^2}{2EI}\,dx \tag{11-2}$$

The procedure leading to Eq. (c) for a normal stress can be applied in exactly the same manner to a shearing stress to give

$$U = \int_{\text{volume}} \frac{\tau^2}{2G}\,dx\,dy\,dz \tag{d}$$

where τ must be expressed as a function of x, y, and z before integration. For the case of a circular shaft of length L and cross section A subjected to a torque T, the expression for τ is

$$\tau = \frac{T\rho}{J}$$

and the energy becomes

$$U = \int\int\left(\frac{T\rho}{J}\right)^2 \frac{dA\,dx}{2G} = \int_0^L \frac{T^2\,dz}{2GJ^2}\int_{\text{area}} \rho^2\,dA$$

which gives

$$U = \int_0^L \frac{T^2\,dx}{2GJ} \tag{11-3}$$

One of the important concepts disclosed by Eqs. (c) and (d) is that the energy-absorbing capacity of a member (that is, the resistance to failure) is a function of the *volume* of material available, in contrast to the resistance to failure under static loading, which is a function of cross-sectional area or section modulus. Hence, if the length of the bar of Fig. 2–19 is doubled, the capacity for absorbing energy is doubled (end conditions neglected), whereas the static load capacity remains unchanged (except for the additional load imposed by the added weight of the bar).

The integral $\int \sigma\,d\epsilon$ of Eq. (b) represents the area under the stress-strain curve, and, if evaluated from zero to the elastic limit of the material, it yields the maximum strain energy per unit volume that a material will absorb without inelastic deformation (modulus of resilience). The area under the entire curve from zero to rupture denotes energy per unit volume necessary to produce rupture of the material (modulus of toughness).

PROBLEMS

Note. In the following problems, the mass of the loaded members is to be neglected, and, unless stated otherwise, all of the applied energy is to be assumed to be absorbed elastically by the loaded members. Neglect stress concentration.

11–1* A solid circular steel bar 36 in. long is to be used as a tension member subject to axial energy loads. The allowable axial tensile stress is 20 ksi. Determine the minimum diameter required for an axial energy load of 50 ft-lb.

11–2* A rolled bronze bar with a cross-sectional area of 2000 mm^2 is to be used as a tension member subject to axial energy loads. The allowable axial tensile stress is 70 MN/m^2 and the modulus of elasticity is 100 GN/m^2. Determine the minimum length of bar required for an axial energy load of 100 N-m.

11–3* A bar with cross-sectional area A and length L is made of a homogeneous material with a specific weight γ and a modulus of elasticity E. Determine the total elastic strain energy stored in the bar (as a result of its own weight) if it hangs vertically while suspended from one end. Express the result in terms of γ, L, A, and E.

11–4* Derive an expression for the total elastic strain energy stored in a round cantilever beam of length L when it is loaded by a concentrated load P gradually applied at the free end. Neglect the energy resulting from shearing stresses. Express the result in terms of P, L, E, and I.

11–5* A 6-ft steel beam is simply supported at the ends. A concentrated load P gradually applied at the center of the span produces a maximum fiber stress of 8000 psi. Determine the total elastic strain energy stored in the beam in terms of I and c. Neglect the energy resulting from shearing stresses.

11–6* A rectangular beam of width b, depth $2c$, and span L is simply supported at the ends and is subjected to end couples that produce a constant bending moment for the full length of the beam. Compare the total elastic strain energy stored in this beam with the total elastic strain energy stored in an axially loaded bar of the same size at the same maximum tensile stress levels.

11–7 A rectangular beam of width b, depth $2c$, and span L is simply supported at the ends and carries a concentrated load P at midspan. Compare the total elastic strain energy stored in this beam with the total elastic strain energy stored in an axially load bar of the same size at the same maximum tensile stress levels. Neglect the energy resulting from the shearing stresses.

11–8* A rectangular cantilever beam of width b, depth $2c$, and span L carries a concentrated load P at the free end. Compare the total elastic strain energy stored in this beam with the total elastic strain energy stored in an axially loaded bar of the same size at the same maximum tensile stress levels. Neglect the energy resulting from the shearing stresses.

11–9 A rectangular cantilever beam of width b, depth $2c$, and span L carries a uniformly distributed load w over its entire length. Compare the total elastic strain energy stored in this beam with the total elastic strain energy stored in an axially loaded bar of the same size at the same maximum tensile stress levels. Neglect the energy resulting from shearing stresses.

11–10* A rectangular cantilever beam of width b, depth $2c$, and span L has a couple M applied at the free end. Compare the total elastic strain energy stored in this beam with the total elastic strain energy stored in an axially loaded bar of the same size at the same maximum tensile stress levels.

11–11* A solid circular steel shaft of diameter d and length L is subjected to a constant torque T. Compare the total elastic strain energy stored in this shaft with the total elastic strain energy stored in an axially loaded bar of the same size at the same maximum stress level.

11–12 A 1-in. diameter steel rod 3 ft long is supported at the top end and fitted with a loading flange (stop) at the bottom end, as shown in Fig. P11–12. A collar W slides freely on the rod. Determine the maximum weight that the collar may have without exceeding the yield strength (36,000 psi) of the steel if:

(a) The collar is dropped 2 ft before striking the stop.
(b) The collar is suddenly applied to the stop (dropped from a very small height).
(c) The collar is slowly applied to the stop (static load).

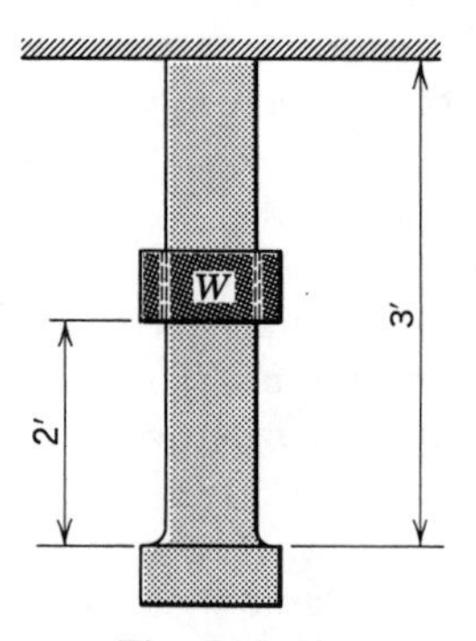

Fig. P11–12

11–13 A timber cantilever beam 150 mm wide, 50 mm deep, and 2 m long is subjected to an impact load as shown in Fig. P11–13. The block W has a mass of 5 kg and is dropped 50 mm onto the end of the beam. The modulus of elasticity of the timber is 8.4 GN/m^2. Determine the maximum deflection of the beam due to the falling mass.

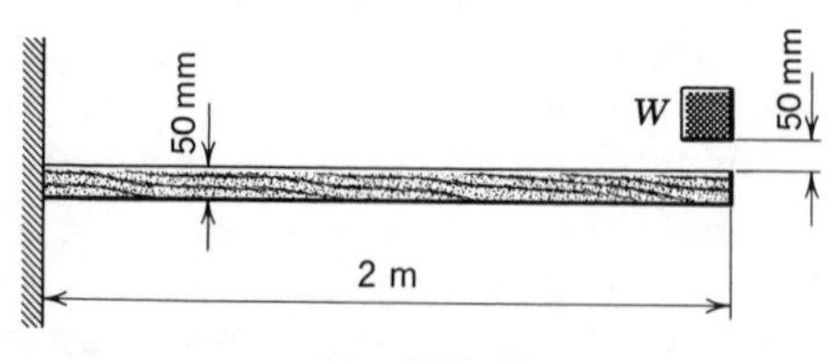

Fig. P11–13

11–14 A weight of 10 lb is dropped from a height of 8 in. onto the free end of a steel cantilever beam. The beam is 2 in. wide, 3 in. deep, and 30 in. long. Determine the maximum fiber stress caused by the falling weight.

11–15* A 100-kg block is dropped from a height of 250 mm onto the center of a simply supported structural aluminum beam. The beam is 150 mm wide by 50 mm deep, and the span is 3 m. The modulus of elasticity of the aluminum is 70 GN/m^2. Determine the maximum deflection of the beam caused by the falling block.

11–16* A timber beam 6 in. wide, 4 in. deep, and 10 ft long is simply supported at the ends. The modulus of elasticity for the timber is $1.2(10^6)$ psi. From what maximum height can a 30-lb weight be dropped onto the center of the beam without causing the maximum fiber stress to exceed 1500 psi?

11–17 A steel shaft 75 mm in diameter and 1 m long is subjected to a torsional energy load of 100 N-m. Determine the maximum shearing stress in the shaft.

11–18 A hollow steel shaft is 15 ft long and has an outside diameter of 8 in. and an inside diameter of 6 in. Determine the torsional energy load required to produce a maximum shearing stress of 12,000 psi in the shaft.

11–19* An elevator cage weighing 3000 lb is being lowered at the rate of 5 ft/sec. The elevator cable has an effective cross-sectional area of 1 sq in. and an effective modulus of elasticity of $25(10^6)$ psi. If the cage is suddenly stopped by the cable drum (not by brakes on the side of the cage) after descending 200 ft, determine the maximum tensile stress developed in the cable. Neglect the weight of the cable. Recall that the mass of the cage is W/g.

11–20* A flat piece of steel 10 mm thick and 2 m long has a width of 25 mm for 0.5 m of its length and a width of 50 mm for the remaining 1.5 m. A 2-kg collar W is dropped on the end of the bar, as shown in Fig. P11–20. Determine the maximum height from which the collar may be dropped on the end of the bar without causing an axial tensile stress greater than 100 MN/m^2.

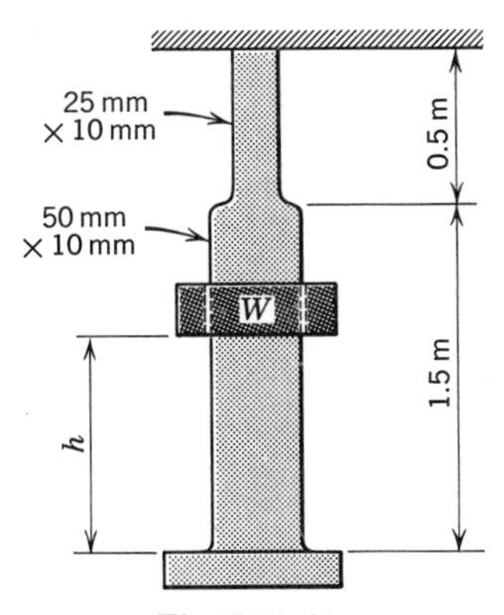

Fig. P11–20

11–3 THE EQUIVALENT STATIC LOAD

As noted in Art. 11–1, stresses and strains produced by dynamic loading depend on the velocity of propagation of the strain wave through the material. However, for many engineering applications (when velocities of the applied load are not too high), one may neglect such complications and assume that the material behaves as it does under slowly applied loads. This will be the approach taken in this book, and with such a simplifying assumption there is no need to evaluate Eqs. (c) or (d) of Art. 11–2. Instead, one may apply an *equivalent static load* (that is, a slowly applied load that will produce the same maximum deflection as that due to the dynamic load) and *assume that the material behaves identically under the two loadings*. With this assumption, the stress-strain diagram (Fig. 2–19*c*) for any point in the loaded system, as well as the strain distribution, will be identical under both loads.

As an illustration, consider the steel cantilever beam of Fig. 11–1*a* on which is dropped the weight W, as indicated. The variation of the end deflection of the beam with respect to time is shown in Fig. 11–1*b* where Δ, equal to δ_{max}, is the maximum deflection resulting from the falling weight transferring part of its kinetic energy ($mv^2/2$) to the beam and where δ_{stat} is the static deflection resulting when the beam and weight finally reach equilibrium. The static load P, applied as in Fig. 11–1*c*, produces the same maximum deflection Δ; the deflection curve and stress-strain diagrams are assumed to be the same in both cases. Hence, the strain energy in the beam is the same in both cases, and, since the strain energy is equal to the effective applied energy (work done on the beam), the following relation may be stated:

$$\begin{Bmatrix} \text{effective} \\ \text{energy} \\ \text{applied} \end{Bmatrix} = \begin{Bmatrix} \text{work done by} \\ \text{equivalent} \\ \text{static load} \end{Bmatrix} \qquad \text{(a)}$$

The effective energy is that portion of the applied energy that produces the maximum deflection. In addition to energy lost in the form of sound, heat, and local distortion, some is absorbed by the supports.

From Art. 11–2, the strain energy in terms of the static load equals $\int P\, d\delta$ or $P_{avg}(\delta_{max})$. Equation (a) may then be written as

$$U = P_{avg}(\Delta)$$

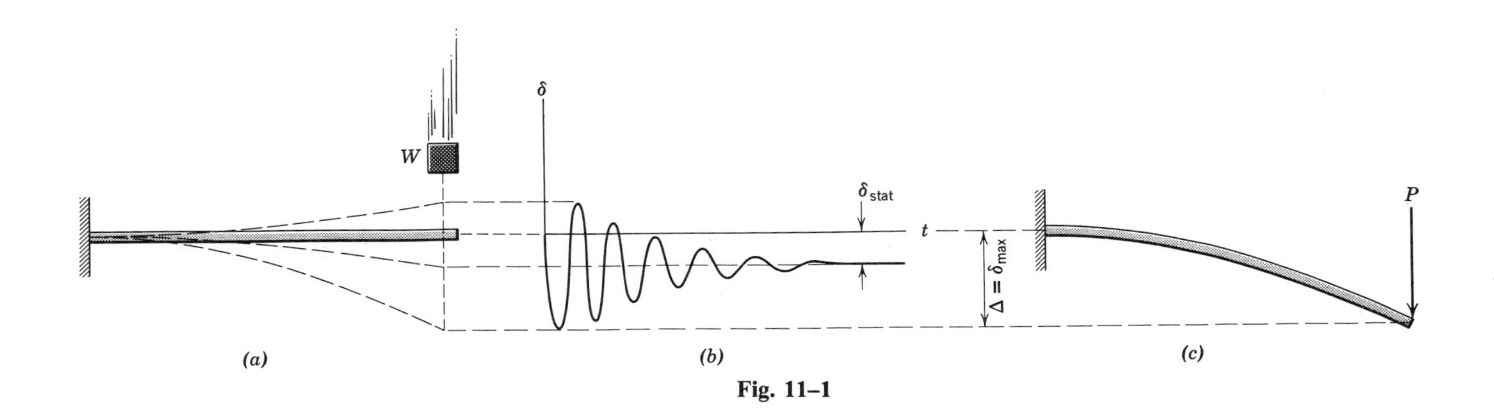

Fig. 11–1

and, if stress and strain are proportional (load-deflection diagram is a straight line), P_{avg} is equal to one-half P_{max}. Hence, for a material possessing a linear stress-strain relationship,

$$U = P\Delta/2 \tag{11-4}$$

The right-hand side of Eq. 11–4 involves the relations between load, stress, and strain previously encountered under static loading and applies to any type of loading. If the effective energy applied, U, is due to a change of velocity of a translating body, the left side of Eq. 11–4 becomes

$$\eta m v^2/2$$

where η is the effective portion of the applied energy, m is the mass of the body, and v is the velocity of the moving body just before impact. If the effective energy applied comes from a body falling from rest and being stopped by the resisting structure, the left-hand side of Eq. 11–4 comes from the change in potential energy of the body and is

$$\eta W h$$

where W is the weight of the body and h is the total vertical distance through which it falls.

For torsional energy loading, Eq. (a) can be written as

$$U = T\theta/2 \tag{11-5}$$

where T is the slowly applied torque that will produce the same maximum angle of twist θ, as the suddenly applied torque if the maximum shearing stress is less than the proportional limit.

A term sometimes encountered in working with dynamic loading is load factor (or impact factor), defined as the *ratio of the equivalent static load to the magnitude of the force the loading mass would exert on the loaded system if slowly applied*. For example, the equivalent static loading on the wings of an airplane pulling out of a dive divided by the weight of the plane (loading on horizontal tail surfaces neglected) is the load factor.

Examples 11–1 and 11–2 illustrate the application of Eq. 11–4.

Example 11–1 The upper 3 ft of the steel bar AB in Fig. 11–2a is 2 in. wide by $\frac{1}{2}$ in. thick, and the lower 2 ft is 1 in. wide by $\frac{1}{2}$ in. thick. When the weight W is released from rest from a height of 20 in. onto the end A of the bar, the axial stress in the bar (neglecting stress concentration) is not to exceed 16 ksi. Assume that 80 percent of the energy of the falling block is

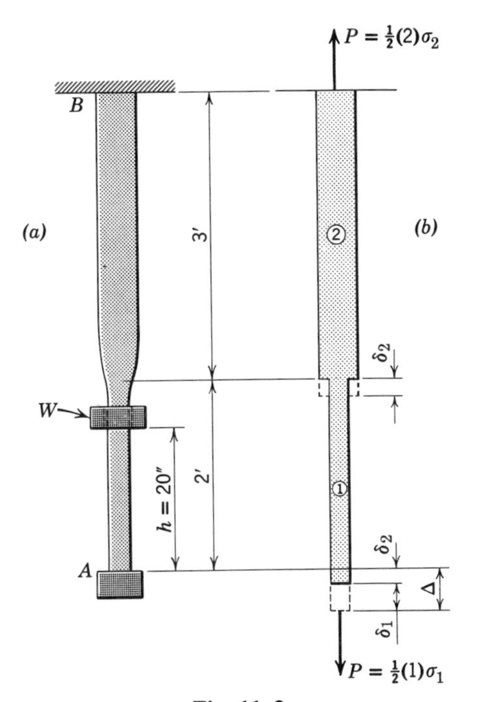

Fig. 11–2

absorbed elastically by member AB. Determine:

(a) The maximum permissible weight for block W.
(b) The maximum permissible weight for W if the bar AB is 1 in. wide through its entire length.

Solution (a) Figure 11–2*b* indicates the static load applied to the bar and the resulting maximum deflections. When Eq. 11–4 is used, the following expression is obtained:

$$\eta Wh = (0.8)W(20+\Delta) = P\Delta/2 \tag{b}$$

The quantities P and Δ can be calculated directly from the given data. From Fig. 11–2*b*

$$1(\sigma_2) = 0.5(\sigma_1) \quad \text{or} \quad \sigma_1 = 2\sigma_2$$

This expression indicates that σ_1 (equal to 16 ksi) is the limiting stress and that σ_2 must be 8 ksi. The load P is

$$P = \sigma_1 A_1 = 16(1)(0.5) = 8\,\text{kips}$$

and the deflection Δ is

$$\Delta=\delta_2+\delta_1=36\epsilon_2+24\epsilon_1=(36\sigma_2+24\sigma_1)/30{,}000$$

$$=\frac{36(8)+24(16)}{30{,}000}=0.0224 \text{ in.}$$

Substituting these values of P and Δ into Eq. (b) gives the expression

$$0.8(W)(20.0224)=(8)(0.0224)/2$$

from which

$$W=\underline{5.59 \text{ lb}} \qquad \text{Ans.}$$

(b) If the bar is 1 in. wide for its entire length,

$$\Delta=\frac{(36+24)(16)}{3(10^4)}=0.032 \text{ in.}$$

and

$$0.8(W)(20.032)=(1/2)(8)(0.032)$$

from which

$$W=\underline{7.99 \text{ lb}} \qquad \text{Ans.}$$

In view of the development of Art. 11–2, which indicates that the energy-absorbing capacity of a member is a function of the amount of material present, the results in parts (a) and (b) seem paradoxical since the strength of the member is increased by removing some material. The apparent discrepancy is probably best explained by referring to the stress-strain diagram of Fig. 11–3. For part (a) of the problem, the energy absorbed per unit volume by the lower 2 ft of the bar is represented by the area *ocd*, which gives

$$u=\frac{1}{2}\left(\frac{16}{E}\right)16=\frac{8(16)}{3(10^4)}=0.00427 \text{ in-kip/in.}^3$$

The energy absorbed per unit volume by the upper 3 ft of the bar is represented by area *oab* of Fig. 11–3 and is seen to be only one-fourth as much because both the stress and the strain are one-half as much as in the

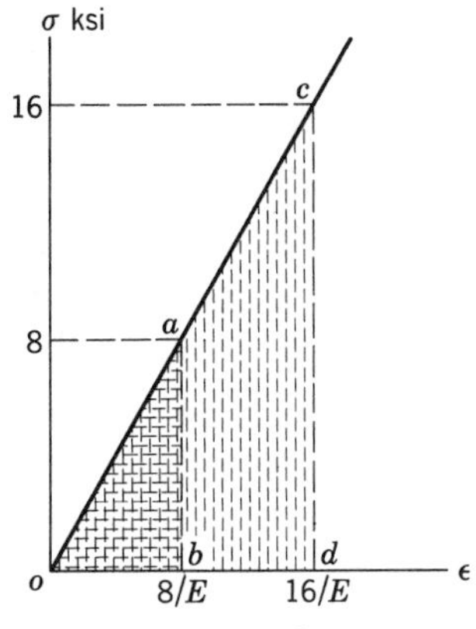

Fig. 11–3

lower 2 ft. When one-half of the volume of the upper 3 ft is removed, the energy absorbed by each unit volume of the remaining material is quadrupled (area *ocd* as compared to area *oab*), resulting in a net gain in energy-absorbing capacity. For this reason, bolts designed to carry dynamic loads are sometimes turned down so that the diameter of the shank of the bolt is the same as the diameter at the root of the threads. In this way the stress distribution in the bolt is approximately uniform, and the capacity of the bolt to absorb energy is a maximum.

Example 11–2 The simply supported 6061-T6 aluminum alloy beam A of Fig. 11–4*a* is 3 in. wide by 1 in. deep. The center support is a helical spring with a modulus of 100 lb/in. The spring is initially unstressed and in contact with the beam. When the 50-lb block drops 2 in. onto the top of the beam, 90 percent of the energy is absorbed elastically by the system consisting of beam A and spring B. Determine:

(a) The maximum flexural stress in the beam for this loading.
(b) The load factor.
(c) The proportion of the effective energy absorbed by each of the components A and B.

Solution In Fig. 11–4*b*, *c*, and *d* are shown a deflection diagram for the system and free-body diagrams for the component parts. Since the action is assumed to be elastic and the materials obey Hooke's law, the load-deflection relation will be linear, and Eq. 11–4 is valid; hence,

$$50(2+\Delta)(0.9) = P\Delta/2 \tag{c}$$

The fiber stress is wanted; therefore the forces P and Q will be evaluated,

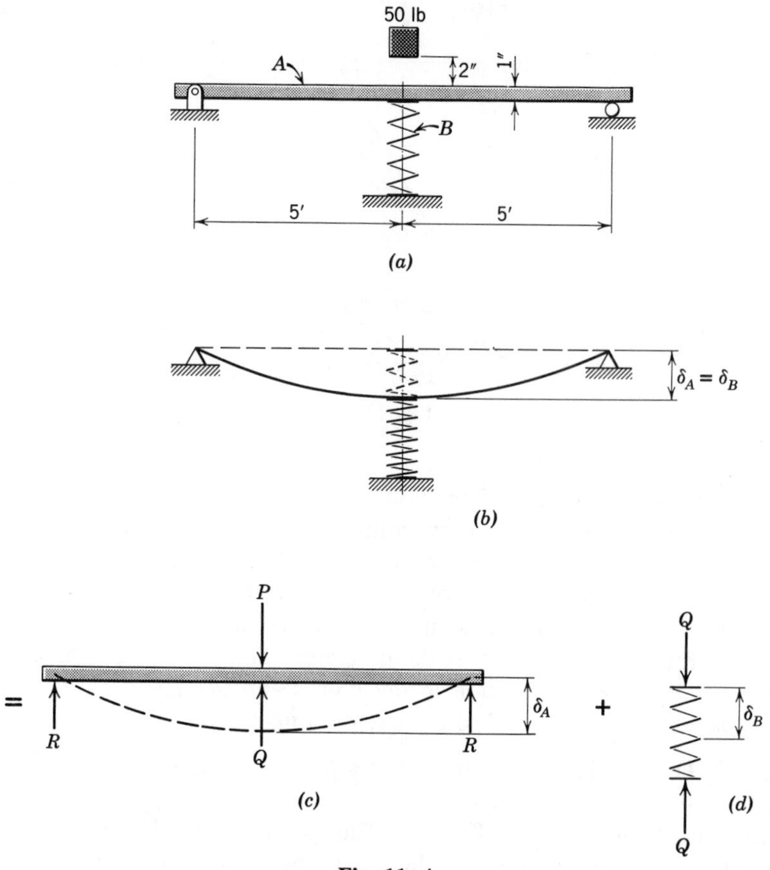

Fig. 11–4

and from these forces and Fig. 11–4*c* the fiber stress can be determined. From Fig. 11–4*c* and *d* and App. D,

$$\delta_A = \frac{(P-Q)L^3}{48EI} = \frac{(P-Q)(10^3)(12^3)}{48(10^7)(3)(1^3)/12} = \frac{(P-Q)(144)}{(10^4)}$$

and

$$\delta_B = \frac{Q}{100}$$

From Fig. 11–4*b*,

$$\delta_A = \delta_B = \Delta$$

or

$$\frac{(P-Q)(144)}{10{,}000} = \frac{Q}{100}$$

from which

$$Q = \frac{36P}{61}$$

then

$$\Delta = \frac{Q}{100} = \frac{36P}{6100}$$

and Eq. (c) becomes

$$50\left(2+\frac{36P}{6100}\right)0.9 = \left(\frac{1}{2}\right)\frac{36P^2}{6100}$$

or

$$P^2 - 90P = 30{,}500$$

from which

$$P = 225.3 \quad \text{and} \quad -135.3$$

The positive root is the magnitude of P necessary to produce the maximum downward deflection of the beam, whereas the negative root is the force necessary to produce the maximum upward deflection (see Fig. 11–1*b*) of the vibrating system.

When the value of 225.3 lb for P is used,

$$Q = \frac{36(225.3)}{61} = 133.0 \text{ lb}$$

and

$$P - Q = 92.3 \text{ lb}$$

(a) From Fig. 11–4*c*, R is 46.15 lb, and the maximum bending moment occurs at the center of the beam and is 46.15(5)(12) in-lb; therefore, the maximum flexural stress is

$$\sigma = \frac{Mc}{I} = \frac{46.15(60)/2}{3(1^3)/12}$$

$$= \underline{5540\,\text{psi}\ T \text{ and } C} \qquad \text{Ans.}$$

which is well below the proportional limit, and therefore the action is elastic.

(b) The load factor is

$$\frac{225.3}{50} = \underline{4.51} \qquad \text{Ans.}$$

(c) The resultant equivalent static load at the center of beam A is $(P - Q)$ and is 92.3 lb, and the energy absorbed elastically by the beam is

$$U_A = \frac{(92.3)(\Delta)}{2}$$

The elastic energy absorbed by the helical spring is

$$U_B = \frac{(133.3)(\Delta)}{2}$$

The total energy absorbed elastically by the system is

$$U = \frac{(225.3)(\Delta)}{2}$$

therefore the proportions absorbed by each component are

$$\frac{92.3}{225.3} = \underline{41 \text{ percent for } A} \qquad \text{Ans.}$$

and

$$\frac{133.0}{225.3} = \underline{59 \text{ percent for } B} \qquad \text{Ans.}$$

To recapitulate, Eq. 11–4 is applicable only to systems having a linear relationship between load and deformation, and only when the velocity of application of the dynamic load is not too high.

PROBLEMS

Note. In the following problems, the mass of the loaded members is to be neglected, and, unless stated otherwise, all of the applied energy is to be assumed to be absorbed elastically by the loaded members. Neglect stress concentration.

11–21* A load of 50 lb is dropped from a height of 4 in. onto a helical spring. The modulus of the spring is 100 lb/in. Determine:

(a) The maximum deflection of the spring.
(b) The static load that would produce the same deflection.

11–22* A force of 5000 N compresses a given car spring 2.5 mm. What static load would be required to produce the same deflection as would be produced by a mass of 150 kg dropped from a height of 300 mm?

11–23* A weight of 40 lb is dropped from a height of 3 ft onto the center of a small rigid platform, as shown in Fig. P11–23. The two steel rods supporting the platform are 1 in.×2 in.×8 ft long. If one-half of the energy is effective in producing stress in the rods, determine:

(a) The equivalent static load.
(b) The deflection of the platform.

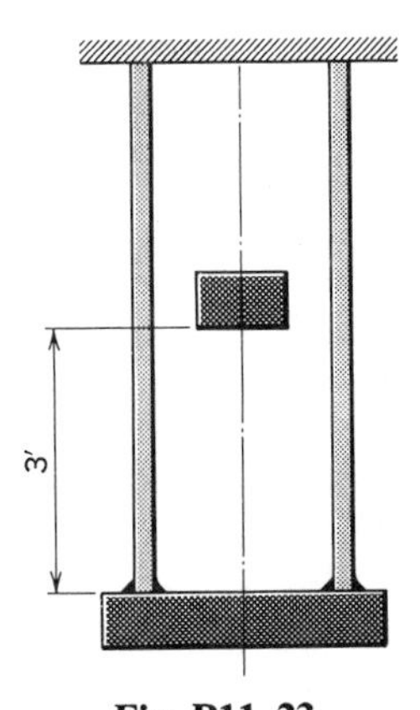

Fig. P11–23

11–24* A hot-rolled Monel bar with a cross-sectional area of 2000 mm^2 is to be used as a tension member subject to axial energy loads. The allowable axial tensile stress is 70 MN/m^2. Determine the minimum length of bar required for an axial energy load of 100 N-m when eight-tenths of the energy is effective in producing stress in the bar. The modulus of elasticity of Monel is 180 GN/m^2.

11–25 A flat bar of steel 2 in. thick and 6 ft long has a width of 6 in. for 2 ft of its length and a width of 4 in. for the remaining 4 ft. Determine the maximum axial stress developed in the bar by an axial energy load of 120 ft-lb applied at one end of the bar. Assume that four-fifths of the energy produces stress in the bar.

11–26 A mass W is dropped 250 mm onto the end of a steel bar, as shown in Fig. P11–26. If 80 percent of the energy is assumed to be effective in producing stress in the bar and the maximum stress is not to exceed 120 MN/m^2 tension, determine:

(a) The mass W if the bar is as shown in Fig. P11–26.
(b) The mass W if the bar has a uniform cross section of 1200 mm^2 throughout its entire length.

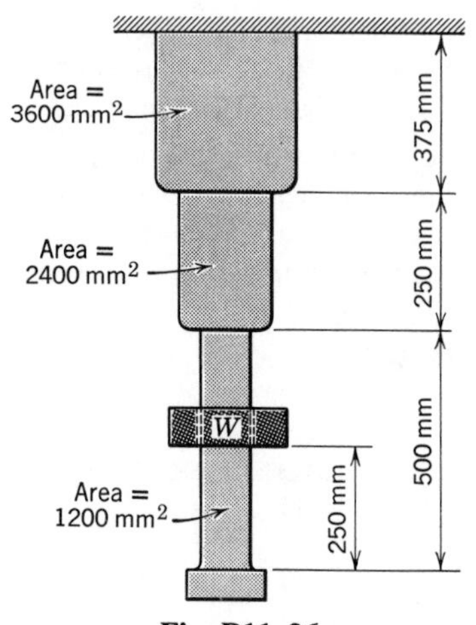

Fig. P11–26

11–27* An S12×40.8 steel section is used as a simply supported beam 10 ft long. What weight falling on the center of the span from a height of 4 in. will cause a maximum fiber stress of 15,000 psi in this beam? Consider three-fourths of the energy to be effective in producing the given stress. The web of the beam is vertical.

11–28* A block having a mass of 1 kg falls 100 mm and strikes the middle of a simply supported wooden beam 50 mm wide, 25 mm deep, on a 1-m span. Assuming that the beam absorbs eight-tenths of the energy elastically and using E for this timber as 10 GN/m^2, determine:

(a) The equivalent static load.
(b) The maximum deflection due to the falling mass.

11–29 A weight of 40 lb is dropped from a height of 4 in. onto the free end of a steel cantilever beam. The beam is 2 in. wide, 3 in. deep, and 30 in. long. Assume that one-half of the energy of the falling weight is effective in producing deflection in the beam. Determine:

(a) The maximum deflection of the beam due to the falling weight.
(b) The static load necessary to produce the same deflection.
(c) The maximum fiber stress caused by the falling weight.

11-30 An aluminum cantilever beam is 100 mm wide, 50 mm deep, and 2 m long. Young's modulus is 70 GN/m^2. Determine the height from which a 10-kg mass should be dropped upon the free end of the beam in order to produce a maximum deflection three times as great as that produced by the same mass gradually applied.

11-31 A 10-ft steel shaft 4 in. in diameter is subjected to a 400-ft-lb energy load. Determine:

(a) The maximum normal stress in the shaft when the load is an axial tensile energy load.

(b) The maximum shearing stress in the shaft when the load is a torsional energy load.

11-32* A steel shaft 5 m long with a diameter of 250 mm is solid for 2 m of its length and hollow with an inside diameter of 200 mm for the remaining 3 m. Determine:

(a) The torsional energy load required to produce a maximum shearing stress of 70 MN/m^2 in the shaft.

(b) The magnitude of the angle of twist caused by the above energy load.

11-33* An effective torsional energy load of 25 ft-lb is applied to the steel shaft shown in Fig. P11-33. Determine the maximum shearing stress in the shaft. Neglect stress concentration.

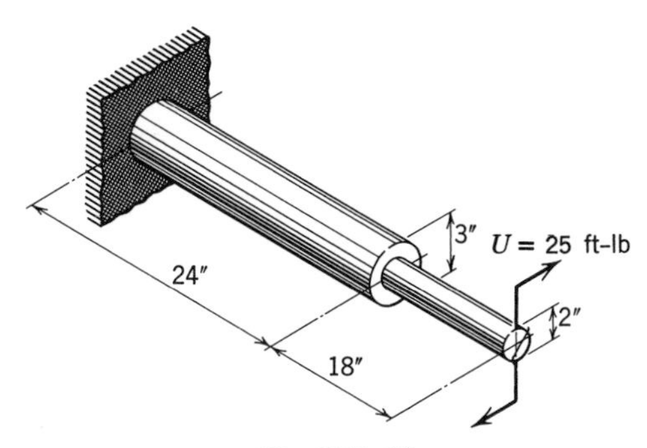

Fig. P11-33

11-34 The beam of Fig. P11-34 is 75 mm wide and 25 mm deep and the supporting coil springs each have a modulus of 10 kN/m. When the block W is slowly applied to the center of the beam, the ends of the beam are observed to move downward 2.5 mm and the center of the beam moves an additional 3.0 mm downward. Determine:

(a) The mass of the block W.

(b) The height h from which W must be dropped to produce a fiber stress in the beam of 20 MN/m^2.

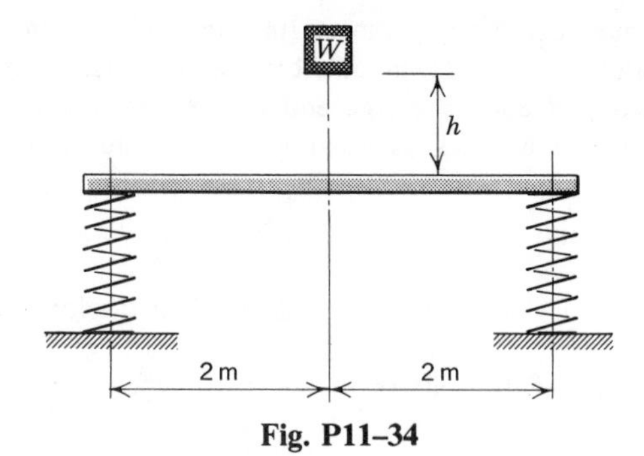

Fig. P11–34

11–35 The bronze beam of Fig. P11–35 is 3 in. wide by 1 in. deep, and the supporting coil springs each have a modulus of 50 lb/in. From what height should the 10-lb weight W be dropped in order to produce a total deflection at the center four times that which would result from the same weight slowly applied. The modulus of elasticity of the bronze is $12(10^6)$ psi.

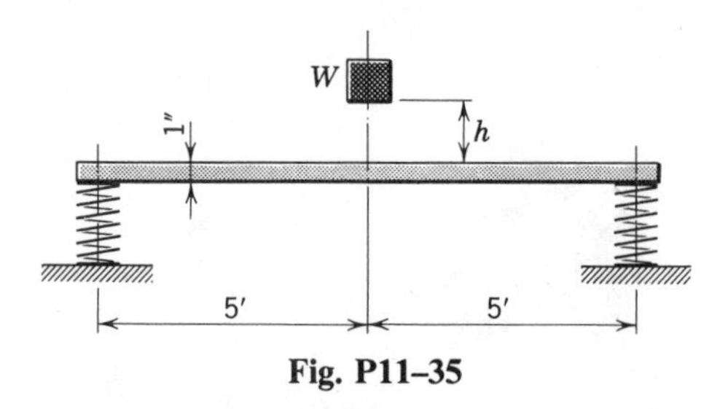

Fig. P11–35

11–36* A vertical energy load is applied downward to the coil spring of Fig. P11–36. If the spring itself (modulus is 20 kN/m) absorbs 8 N-m of energy, how much energy does the wooden beam (E is 8.4 GN/m²) absorb?

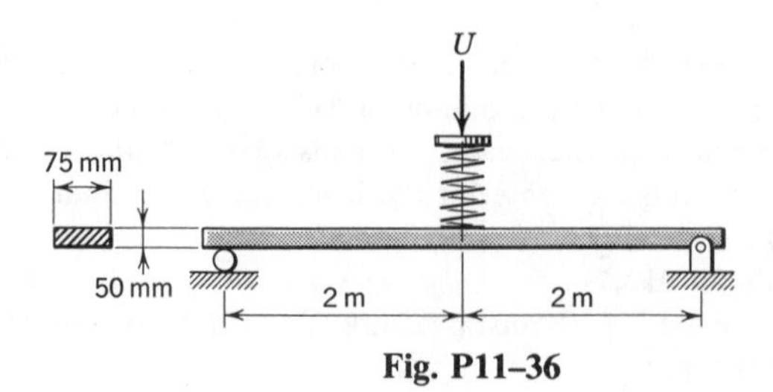

Fig. P11–36

11–37* The 60-lb block B of Fig. P11–37 is dropped from an elevation 1/2 in. above the top of the coil spring C (modulus is 200 lb/in.), and the resulting load factor is 2. Determine the percentage of the energy in the block B that is absorbed elastically by: (1) the coil spring C, and (2) the beam A. The beam is of aluminum alloy for which E is 10^7 psi.

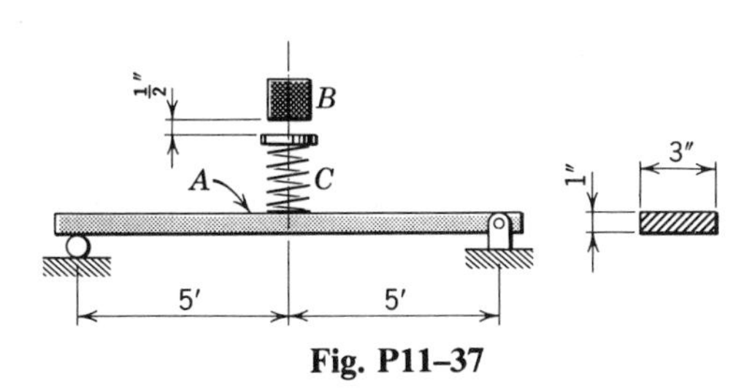

Fig. P11–37

11–38 When the 75-lb block B of Fig. P11–38 was dropped from a point 2 in. above the top of the coil spring C (modulus is 150 lb/in.), the point D on the steel cantilever beam A was observed to deflect 2.00 in. downward. Determine:

(a) The percentage of the energy in block B that was absorbed by the coil spring and beam combination (A and C).

(b) The load factor.

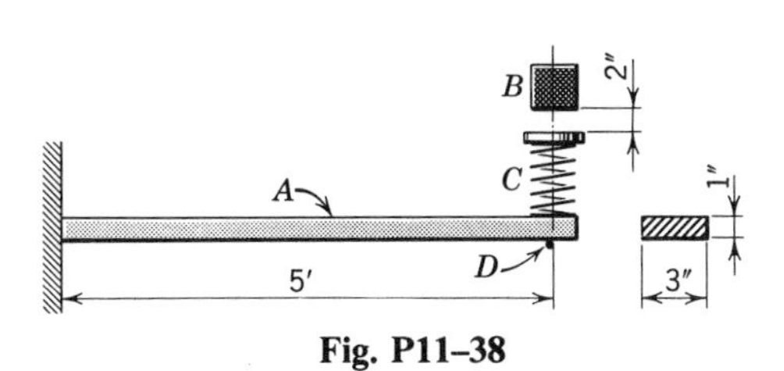

Fig. P11–38

11–39 Figure P11–39 shows a steel beam with a box cross section 150 mm deep and 200 mm wide with a coil spring on the end of it. The spring modulus is 500 kN/m and the cross-sectional moment of inertia of the beam is $37.5(10^6)$ mm^4. Determine the maximum height from which the 100-kg mass W may be dropped on the spring without exceeding a fiber stress of 100 MN/m^2 in the beam. Assume that 80 percent of the energy of the falling mass is absorbed elastically by the beam-spring combination.

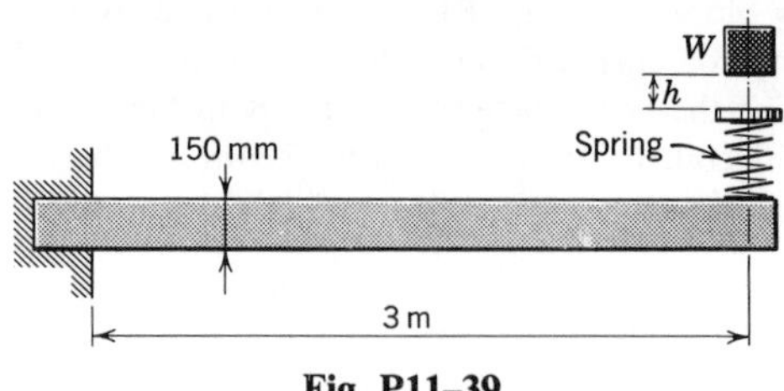

Fig. P11–39

11–40* The timber beam A in Fig. P11–40 is 6 in. wide and 2 in. deep and is connected to the rigid arm B, which rotates about the horizontal pin at 0. The modulus of elasticity of the timber is 1200 ksi and moduli of the helical springs are given on the figure. Body W weighs 20 lb. Determine:

(a) The maximum height h from which W can be dropped on the beam without producing a flexural stress in the beam of more than 1.2 ksi. Assume that 80 percent of the energy of W is effective in producing elastic deflections in the beam-spring system.

(b) The load factor corresponding to the h of part (a).

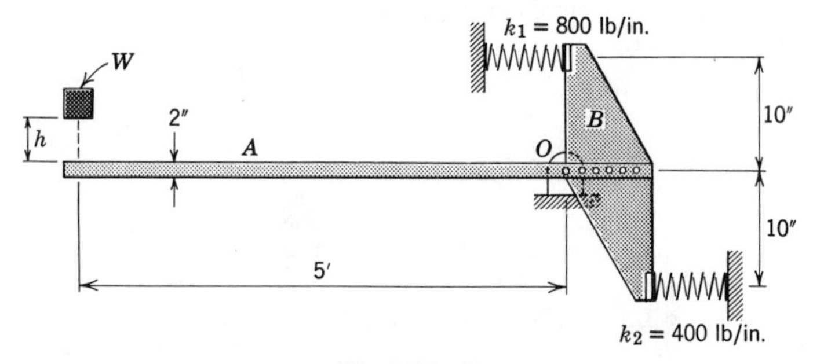

Fig. P11–40

11–41* Beam AB of Fig. P11–41 has a flexural rigidity of EI and is securely fastened to the rigid arm C. The modulus of the helical spring is $6EI/L^3$. The structure rotates about a shaft at B. Determine the height h, in terms of W, L, E, and I, from which W must be dropped to have a load factor of 3. Assume 80 percent of the energy of W is effective in producing elastic deformation in the beam-spring system.

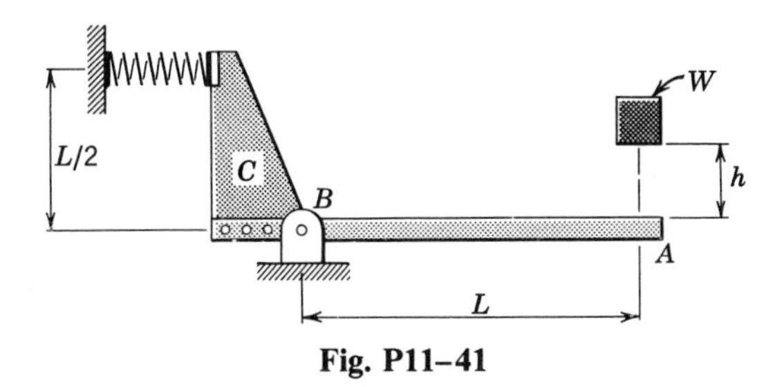

Fig. P11–41

11–42* The structure in Fig. P11–42 consists of a solid steel rod AB rigidly connected to the steel beam BC. The rod is fixed at A and supported by a smooth bearing D at B. The 50-kg mass W is dropped 25 mm onto the bar and 80 percent of the energy of the falling mass is absorbed elastically in AB and BC. Determine the maximum fiber stress in BC and the maximum normal stress on the top surface of AB at the wall.

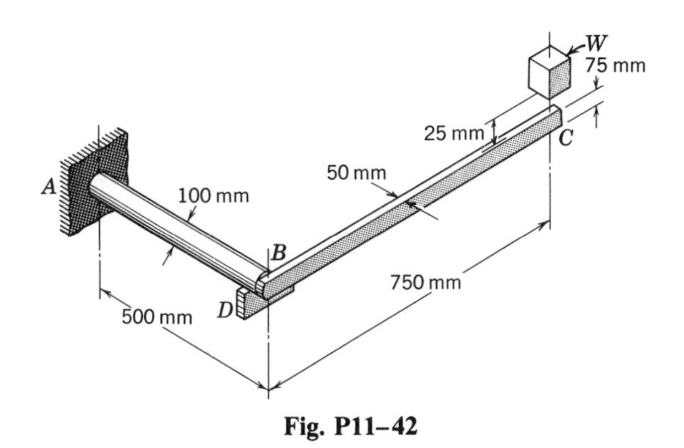

Fig. P11–42

11–43* Beams A and B of Fig. P11–43 are of wood (E is 1200 ksi) and are 2 in. deep by 6 in. wide. When the 25-lb weight W is dropped on the beam A, 90 percent of the energy is absorbed elastically by beams A and B, and the maximum deflection is 1.5 in. Determine the distance h, the load factor, and the amount of energy absorbed by each beam.

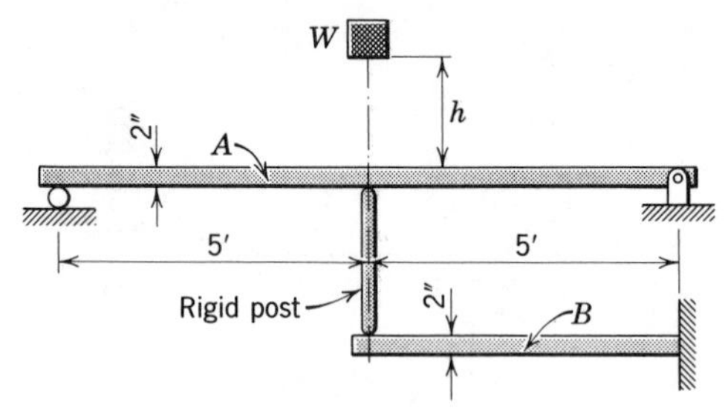

Fig. P11–43

11–44* The wooden beams A and B of Fig. P11–44 are 50 mm deep and 150 mm wide (E is 8.4 GN/m^2). Assume nine-tenths of the energy of the 10-kg mass W to be absorbed elastically by beams A and B. If under the impact the maximum fiber stress in beam B is 10 MN/m^2 T and C, determine:

(a) The height h from which W is dropped.
(b) The load factor.
(c) The energy absorbed by each beam.

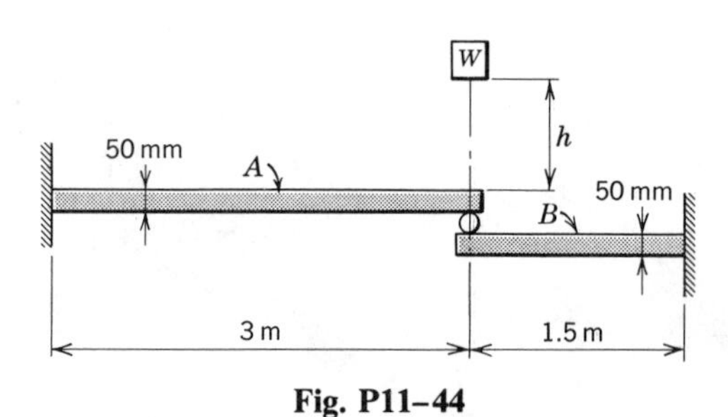

Fig. P11–44

11–45* The aluminum alloy 6061-T6 (E is 70 GN/m^2) beams A and B of Fig. P11–45 are 25 mm deep by 100 mm wide, and the helical spring C has a modulus

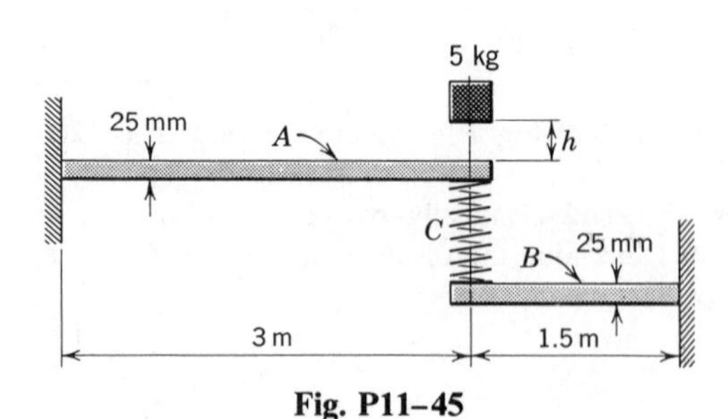

Fig. P11–45

of 20 kN/m. The 5-kg block is dropped onto beam A with a load factor of 4. If 80 percent of the energy is absorbed elastically by A, B, and C:

(a) What is the distance, h?

(b) How much energy is absorbed elastically by each of the components?

11–4 SOME ELEMENTARY DESIGN CONCEPTS

Some design considerations for beams subjected to energy loads are readily obtained from the equivalent static load procedure of Art. 11–3. For example, consider the energy absorbed by a cantilever beam of length L with a concentrated load P at the free end. The strain energy (see App. D) is

$$U = P\Delta/2 = \frac{P(PL^3)}{2(3EI)}$$

With the load P replaced by its value in terms of the maximum fiber stress in the beam, σ equals PLc/I, the expression for the energy becomes

$$U = \frac{\sigma^2 IL}{6c^2E} = \frac{\sigma^2 L}{6E}\left(\frac{I}{c^2}\right)$$

For a rectangular beam I/c^2 is equal to one-third the cross-sectional area, regardless of whether the long or short dimension of the rectangle is parallel to the load. When the parenthetical quantity is replaced by its value in terms of the area of the rectangle, the expression for the energy becomes

$$U = (AL)\frac{\sigma^2}{18E} \tag{a}$$

which shows that for a rectangular cross section the capacity to absorb energy is independent of the moment of inertia of the cross section. Comparison of Eq. (a) with Eq. 11–1 of Art. 11–2 reveals that the rectangular cantilever beam considered has only one-ninth as much capacity to absorb or store elastic energy as an axially loaded member of the same volume. Too much of the volume of the cantilever beam has low stresses for efficient energy load capacity. The situation can be improved by using a tapered beam whose width varies directly as the distance from the free end.

For this purpose consider the cantilever beam of constant depth h of Fig. 11–5. The maximum fiber stress at any section is

$$\sigma = Mc/I = \frac{Pxh/2}{2zh^3/12} = \frac{6PL}{bh^2}$$

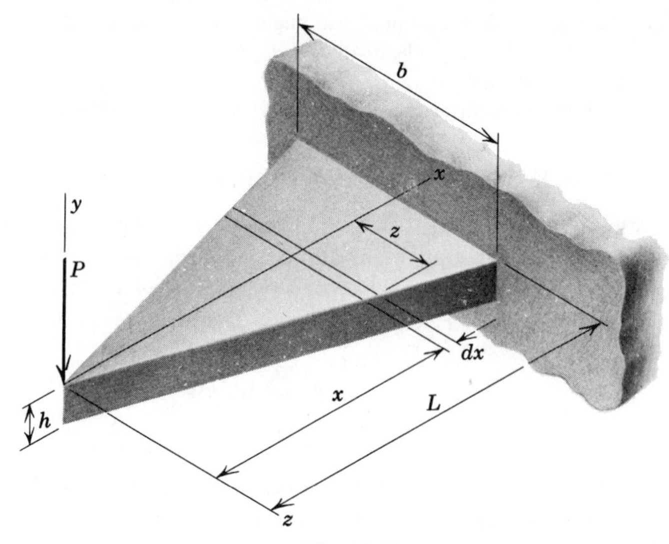

Fig. 11–5

which shows that the maximum fiber stress is independent of the distance x along the beam. The elastic curve differential equation for the beam is

$$E\frac{d^2y}{dx^2} = -\frac{Px}{I} = -\frac{12PL}{bh^3}$$

After integration twice and evaluation of the constants of integration, the magnitude of the maximum deflection at the free end of the beam is found to be

$$\Delta = \frac{6PL^3}{bh^3E}$$

The energy U then becomes

$$U = \frac{P\Delta}{2} = \frac{P}{2}\frac{6PL^3}{bh^3E}$$

or in terms of the maximum fiber stress the energy is

$$U = (\text{volume})\left[\sigma^2/(6E)\right] \qquad \text{(b)}$$

Comparison of Eqs. (a) and (b) shows that the tapered cantilever beam of Fig. 11–5 can absorb three times as much elastic energy as a cantilever beam of the same volume with a constant rectangular cross section. However, comparison with Eq. 11–1 of Art. 11–2 shows that the beam of Fig. 11–5 is still only one-third as effective as an axially loaded member of the same volume.

For torsion members subjected to energy loads, a hollow shaft will provide higher energy-absorbing capacity per unit volume than a solid circular shaft.

12 Connections

12–1 GENERAL DISCUSSION OF RIVETED CONNECTIONS

One of the most vexing problems confronting the structural designer is the proportioning and detailing of connections to transmit forces from one structural element to another. Among the common types of fastenings are rivets, bolts, welds, pins, and threaded couplings. This article is concerned with riveted (or bolted) connections such as the joints of Fig. 12–1.

A riveted joint can fail as a result of the cross shearing stresses in the rivets, the bearing stresses between the rivets and the plates, the tensile stresses in the plates, or perhaps a combination of these stresses.

Before discussing the analysis of these connections, it is best to define, with the aid of Fig. 12–2, some of the terms used. Figure 12–2*a* represents a *double-riveted butt joint* with one *cover* (*or splice*) *plate*. The term *double-riveted* (or *single-riveted*, etc.) refers to the number of rows of rivets that transfer the total load from one element to another (in this case from main plate to cover plate, or from cover plate to main plate). A butt joint may also be fabricated with two cover plates, one on each side of the main plates. In some butt joints, both cover plates engage all rows of rivets; in

(a) Double-riveted lap joint. Shear failure.

(b) Single-riveted lap joint. Bearing failure.

(c) Triple-riveted lap joint. Tension failure in outer row.

(d) Triple-riveted lap joint. Tension failure in inner row.

(e) Triple-riveted butt joint. Tension failure in inner rows of cover plates (top plate on left side of joint and bottom plate on right side).

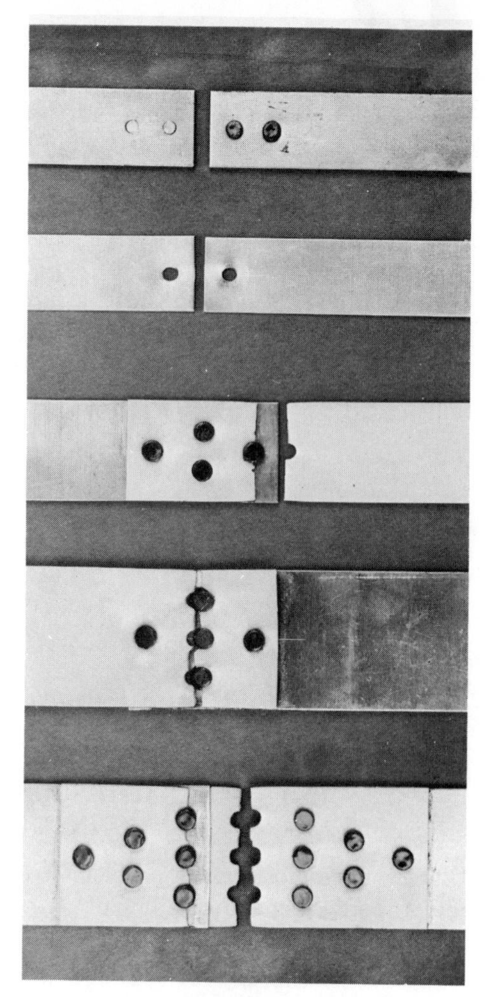

Fig. 12–1

others, one plate may engage only the inner rows and the other plate all rows (see Fig. 12–2*b*). Figure 12–2*c* represents a *triple-riveted lap joint*. The various terms used are indicated in the diagrams. Note carefully the designation of inner and outer rows in the two different types of joints. The term *pitch* alone always denotes spacing between rivets in a row.

When analyzing long joints such as those in boilers, water tanks, and air compressor tanks, a representative length of the joint, called the *repeating group*, is investigated rather than the entire joint. The repeating group of

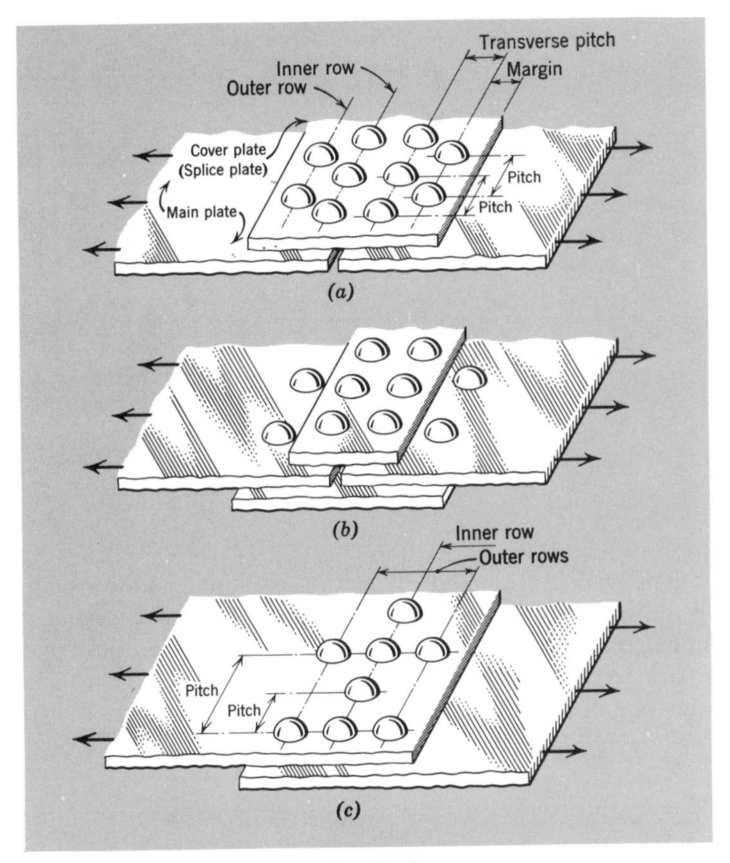

Fig. 12–2

rivets is the shortest pattern that repeats itself along the joint, and the length of the group is the least common multiple of the rivet pitches in the joint. It is recommended that the rivet pattern be sketched as an aid to obtaining the correct number of rivets (or shear areas) in the repeating group. Such a sketch is given in Fig. 12–3, which is an example of a triple-riveted joint with pitches of 2, 3, and 4 in. It will be observed that there are thirteen rivets in the length of 12 in. This sketch is extended to show that the patterns for two consecutive 12-in. lengths are identical.

The analysis of a riveted or bolted connection based on elastic action[1] is unsatisfactory because the solution is so laborious, and the results of the

[1]For a discussion of elastic analysis, see "Work of Rivets in Riveted Joints," A. Hrennikoff, *Trans. Am. Soc. of Civil Engrs.*, Vol. 99, 1934, p. 437. The accompanying discussions include an extensive bibliography on the subject.

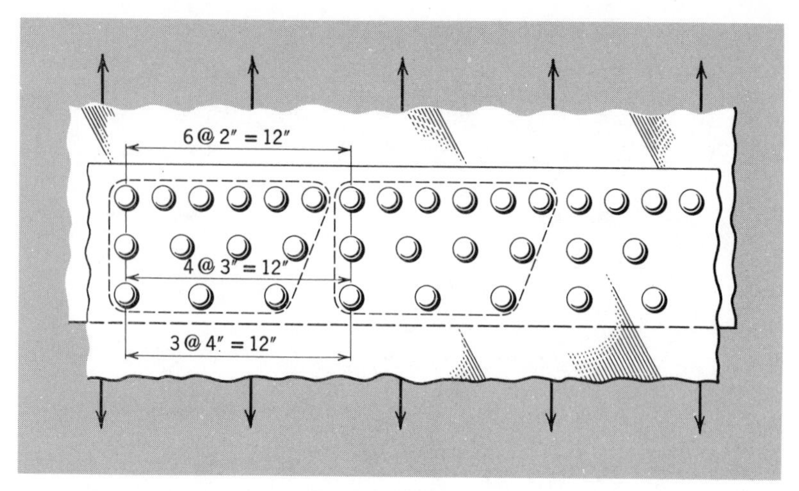

Fig. 12–3

computations (which are based on certain assumptions) are usually considerably different from experimental results, which of necessity involve indeterminate variables due to fabrication methods. For a brief exploration into the field of elastic analysis, consider section a–a of the quadruple-riveted lap joint of Fig. 12–4 in which the rivet distortions are exaggerated.

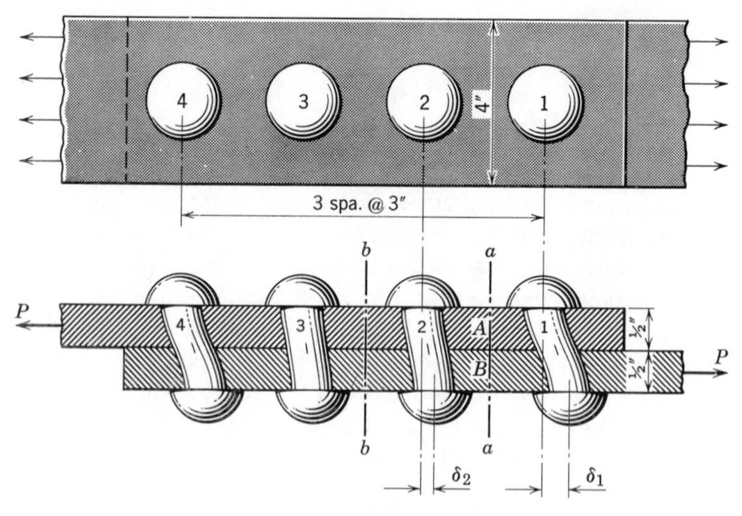

Fig. 12–4

If it is assumed that the distortions in rivets 1 and 2 are equal (equivalent to assuming that the forces on the two rivets are equal) and if there is no slip in the joint (the rivets fill the holes completely and elastic action prevails), the axial deformation between rivets 1 and 2 in plates A and B is the same. Therefore, the average axial strain in A is equal to that in B; if Hooke's law applies and the plates are of the same size and material, the tensile force in A equals that in B, equals $P/2$. When this result is applied to the free-body diagrams of Fig. 12–5, the condition of equilibrium

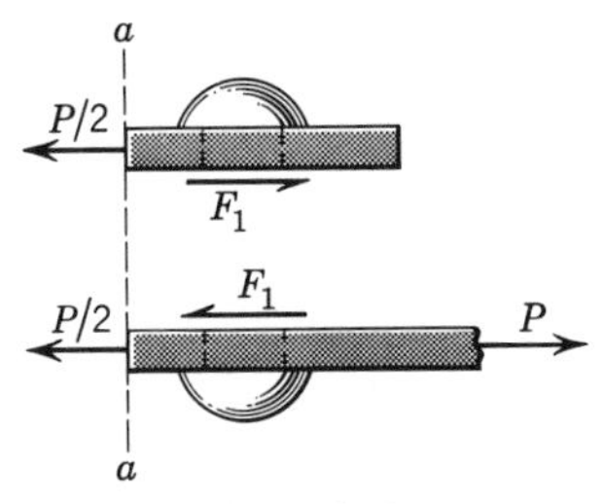

Fig. 12–5

indicates that F_1 is equal to $P/2$. Now, when the same reasoning is applied to section b–b and the free-body diagrams of Fig. 12–6 are drawn, it becomes apparent that if F_1 is equal to $P/2$, F_2 must be zero. This statement is not consistent with the original assumption, indicating that, if elastic action prevails throughout the joint and the rivets completely fill the holes, a uniform distribution of load to rivets is impossible (except for the special case of the double-riveted joint with balanced plate areas).

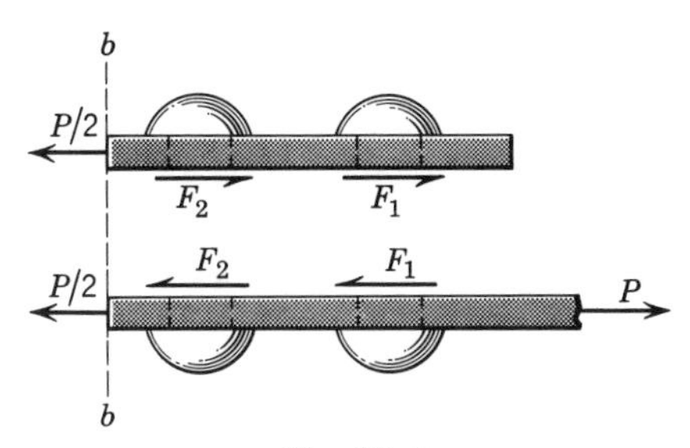

Fig. 12–6

One possible elastic analysis of the joint of Fig. 12–4, based on the assumptions of negligible friction, all rivets bearing on the plates immediately on application of load, and the force on any rivet being a constant[2] times the rivet distortion δ, would indicate that for a steel joint having the dimensions shown, at working load rivets 1 and 4 would each carry about 35 percent of the total force P and rivets 2 and 3 would each carry about 15 percent. These results, although closer to the truth than those obtained under the first assumption, are still in error because experiments[3,4,5] show that slip may begin upon application of even a small load, indicating that although friction may be negligible, the rivets do not completely fill the holes (which is to be expected if the cooling of the rivet after driving is considered); therefore, the assumption of no slip is invalidated. Furthermore, bending of the plates was neglected, and, as can be seen from Fig. 12–4, the applied forces P cannot become collinear until the plates are bent.

The foregoing discussion will serve to indicate the need for some simplifying assumption or concept regarding the action of a riveted or bolted connection leading to a reliable easily applied method of analysis. Two widely used methods are outlined in the two succeeding articles. For both methods, the following approximations are made:

1. *The nominal size of the rivet or bolt* is used for *shear* and *bearing* stresses in the design of *steel bridges* and *buildings*. In the design of *pressure vessels* and *aircraft structures*, the actual diameter of the hole in the plate is used (the rivet when driven is assumed to fill the hole).

2. The net section for computing the tensile stress in the plates is calculated using the nominal size of the rivet or bolt plus 1/8 in. (3 mm) for *steel bridge* and *building design*. The actual (specified) size of the hole is used to calculate the net section in the design of *aircraft structures and pressure vessels*.

3. *Bending* is neglected in all joints. *Friction* is neglected in *riveted* or *ordinary bolted* joints. In *friction type* bolted joints, the high-strength bolts are torqued to produce sufficient friction to prevent the joint from slipping into bearing; hence, bearing stress need not be considered in such joints.

A term frequently encountered in discussions of riveted connections is the *efficiency* of a joint, here defined as *the maximum permissible load on a joint divided by the maximum permissible axial load on the solid plate.*

[2]This constant is experimentally determined and varies considerably with materials, joints, loads, and investigators. For this analysis the constant $12.5(10^6)$ lb/in. was taken from reference 3.

[3]"Riveted and Pin-connected Joints of Steel and Aluminum Alloys," L. S. Moisseiff, E. C. Hartmann, and R. L. Moore, *Trans. Am. Soc. of Civil Engrs.*, Vol. 109, 1944, p. 1359.

[4]"Tension Tests of Large Riveted Joints," R. E. Davis, G. B. Woodruff, and H. E. Davis, *Trans. Am. Soc. of Civil Engrs.*, Vol. 105, 1940, p. 1193.

[5]"Tests of Joints in Wide Plates," W. M. Wilson, J. Mather, and C. O. Harris, *Bulletin No.* 239, Engr. Exp. Sta., Univ. of Ill., 1931.

Efficiency may also be defined as *the strength of the joint divided by the strength of the solid plate*. One may speak of the efficiency in shear, bearing, or tension as the permissible load (or strength) in shear, bearing, or tension divided by the safe load capacity (or strength) of the solid plate; however, the *efficiency of the joint* is as defined above.

12–2 ANALYSIS OF RIVETED OR BOLTED JOINTS—UNIFORM SHEAR METHOD

One of the common assumptions made regarding load distribution in riveted and bolted connections is that the cross shearing stress in the rivets or bolts is uniformly distributed over all the shear areas (the cross-sectional areas of the rivets or bolts). After the discussion in Art. 12–1, this may appear to be a somewhat indefensible assumption; however, tests have been conducted which seem to indicate that such a condition is possibly attained with only a small amount of slip in the joint.[6] The tensile stress in the plate is assumed to be uniformly distributed over the net area of the plate at each row of rivets. The bearing stress between the rivets and the plates is assumed to be uniformly distributed over the projection of the contact area; that is, the plate thickness multiplied by the rivet diameter.

In order to design or check the safety of riveted joints, working stresses must be specified. Table 12–1 lists allowable stresses from some widely used specifications.

To illustrate the analysis of riveted joints by assuming a uniform shearing stress, the joint of Fig. 12–1*d* and a boiler joint will be discussed in Exams. 12–1 and 12–2.

[6]"Riveted and Pin-connected Joints of Steel and Aluminum Alloys," L. S. Moisseiff, E. C. Hartmann, and R. L. Moore, *Trans. Am. Soc. of Civil Engrs.*, Vol. 109, 1944, p. 1359. Included in these tests were symmetrical triple-riveted butt joints having $\frac{3}{4}$-in. main plates and two $\frac{3}{8}$-in. cover plates with twelve $\frac{7}{8}$-in. rivets in each half. At design loads (average shearing stress = 13,500 psi), the rivets in the inner row were found to carry 33–42 percent of the load, those in the intermediate row 16–26 percent, and those in the outer row 36–43 percent. However, the ultimate loads for the joints designed to fail in shear were within 2 percent of the predicted loads based on the uniform shearing stress assumption and on shearing strengths obtained from auxiliary control specimens having one and two rivets in each half. These results indicate that although at working loads there is definitely a nonuniform distribution of load to the rivets, the inequality disappears by the time the ultimate load is reached. Joints designed to fail in tension averaged within 4 percent of the loads predicted on the basis of the tensile properties of the plate material (as obtained from control specimens) on the net section. The writers of this paper concluded that, "The amount of plastic yielding necessary to level off high stress concentrations and to effect a uniform distribution of load among the rivets must be relatively small." Of course, this means that a reasonably ductile material is necessary (see also references 4 and 5 of Art. 12–1).

TABLE 12–1 Allowable Stresses for Riveted Joints

Specification	Tensile Stress, Net Section of Plate (psi)	Shearing Stress in Rivet (psi)	Bearing Stress on Plate (psi)
Am. Soc. of Mech. Engrs. Boiler Construction Code for Unfired Pressure Vessels, 1971; for temps to 650 F SA-515 plate, SA-31A rivets	13,700	9,000	18,000
Am. Inst. of Steel Const. for Struct. Steel Bldgs., 1970 A36 steel, A502-1 rivets, A325 bolts[a]	21,600	15,000	48,600
Am. Assoc. of State Highway Officials for Steel Bridges, 1973 A36 steel, A502-1 rivets, A325 bolts[a]	20,000	13,500	40,000
Am. Ry. Engrs. Assoc., 1973 A 36 steel, A502-1 rivets, A325 bolts[a]	20,000	13,500[b] 20,000	27,000[c] 36,000
Am. Soc. of Civil Engrs. Spec. for Alum. Alloy 2014-T6 (14S-T6), 1952	22,000	10,000[d] 8,000	36,000[e]

[a]The shearing stress applies and the bearing stress is not restricted for friction-type joints with A325 bolts.

[b]The smaller value for rivets, the other for A325 bolts in nonfriction joints.

[c]The smaller value for rivets in single shear, the other for rivets in double shear.

[d]The larger value for alloy 2117-T3 driven cold, the smaller value for alloy 6061-T43 driven hot. If the rivet is used in relatively thin plates, the allowable shear stress shall be reduced (see spec.).

[e]If the margin is less than two rivet diameters, the stress shall be reduced (see spec.).

Example 12–1 The triple-riveted lap joint of Fig. 12–1*d* is composed of $1\frac{3}{8}$-in. wide by 0.025-in. thick aluminum alloy 2024-T4 plates connected with $\frac{5}{32}$-in. aluminum alloy 2117-T4 rivets. The holes and rivets are essentially the same size for this joint. The joint ruptured under an axial load of 1640 lb. Determine for this loading:

(a) The average shearing stress in the rivets.

(b) The average bearing stress between rivet and plate.

(c) The maximum average tensile stress, and state where it occurs.

Solution (a and b) From the free-body diagram of Fig. 12–7*a* and the equilibrium equation $\Sigma F_x = 0$, the force on each rivet is

$$F_s = 1640/5 = 328 \text{ lb}$$

and, from Fig. 12–7*c*, the average shearing stress is

$$\tau = \frac{328}{(\pi/4)(5/32)^2} = \underline{17{,}100 \text{ psi}} \qquad \text{Ans.}$$

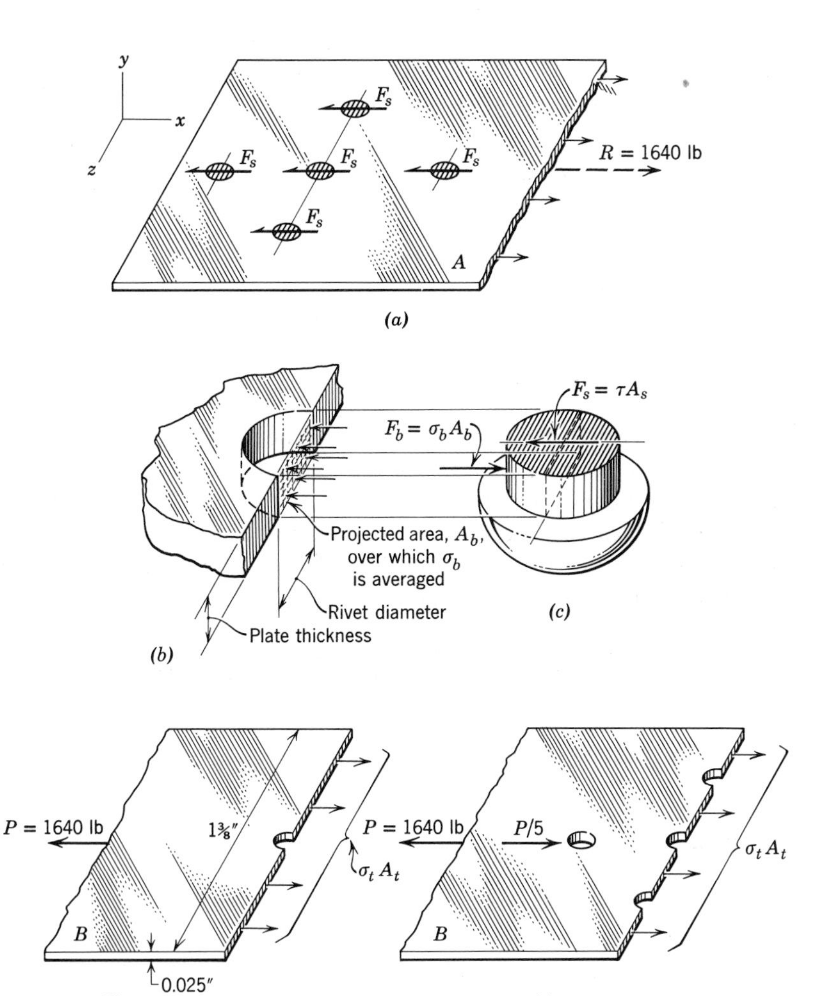

Fig. 12–7

From Fig. 12–7b and c, the average bearing stress is

$$\sigma_b = \frac{328}{0.025(5/32)} = \underline{84{,}000 \text{ psi}} \qquad \text{Ans.}$$

The specified strengths of these materials in shear and bearing are 27,000 psi and 90,000 psi, respectively;[7] therefore, this joint should not fail in shear or bearing under the given load.

(c) The average tensile stress (axial stress) may be maximum at either the outer row where there is only one hole or the inner row where there are three holes and a reduction in the load. The section at the outer row will be checked first with the aid of the free-body diagram of Fig. 12–7d. Summing forces in the horizontal direction gives

$$P = \sigma_t A_t = \sigma_t\left(\frac{11}{8} - \frac{5}{32}\right)(0.025)$$

$$1640 = \sigma_t\left(\frac{39}{32}\right)(0.025)$$

from which

$$\sigma_t = 53{,}900 \text{ psi } T \text{ in the outer row}$$

The section through the inner row will be checked with the aid of Fig. 12–7e. Note that in order to separate the joint at the inner row, not only must the plate be cut at this section, but the rivet ahead of the section must also be removed (or cut); according to the assumption of uniform shearing stress, the load on this rivet is one-fifth of the total load. Again, summing forces in the horizontal direction gives

$$P = \frac{P}{5} + \sigma_t A_t \quad \text{or} \quad \frac{4P}{5} = \sigma_t\left[\frac{11}{8} - 3\left(\frac{5}{32}\right)\right](0.025)$$

$$\frac{4(1640)}{5} = \sigma_t\left(\frac{29}{32}\right)(0.025)$$

from which

$$\sigma_t = \underline{57{,}900 \text{ psi } T \text{ at the inner row}} \qquad \text{Ans.}$$

There is no need to check the next row because the situation at the outer row of the other plate is identical with that already checked. The results of

[7]ANC-5, Strength of Aircraft Elements, December, 1942. Note that specified strengths are guaranteed minimum values and are usually less than typical values.

these computations indicate that, in tension, the joint is weaker at the inner row than at the outer row; this fact is verified by Fig. 12–1*d*, which shows the plate ruptured at the inner row. The specified tensile strength of this material is 62,000 psi (see reference 7); therefore, the average stress obtained was about 7 percent below the specified strength.[8]

Example 12–2 A 50-in. diameter boiler is made of $\frac{3}{4}$-in. plate. The longitudinal seam is a double-riveted butt joint with two $\frac{9}{16}$-in. cover plates and with $\frac{7}{8}$-in. diameter rivets in $\frac{15}{16}$-in. holes. The outer rows of rivets have a pitch of 5 in. and the inner rows a pitch of 2 in. Using the ASME boiler code, determine:

(a) The maximum allowable internal pressure.
(b) The efficiency of the joint.

Solution (a) From the sketch of the rivet pattern shown in Fig. 12–8*a*, it is evident that the repeating group has two rivets in each outer row and five rivets in each inner row. The free-body diagram of Fig. 12–8*b* shows a section of the main plate with a repeating group of rivets and indicates the fourteen shearing forces F_s, two forces for each rivet. The free-body diagram in Fig. 12–8*c* indicates that $F_b = 2F_s$ since each rivet is in double shear. The maximum allowable values of F_b and F_s are

$$(F_b)_{\max} = \sigma_b A_b = 18{,}000\left(\frac{3}{4}\right)\left(\frac{15}{16}\right) = 12{,}660 \text{ lb}$$

and

$$(F_s)_{\max} = \tau A_s = 9{,}000(\pi)\left(\frac{15}{32}\right)^2 = 6{,}210 \text{ lb}$$

It is evident that the bearing capacity of the rivets and cover plate is greater than that of the main plate because the thickness of the two cover plates exceeds that of the main plate. Since twice the allowable shearing load is less than the maximum allowable bearing load, the shearing strength of the rivets is the critical factor as far as shear and bearing are concerned. The maximum load that can be carried by the rivets is, from equilibrium (see Fig. 12–8*b*),

$$P_s = 7(2F_s) = 7(2)(6{,}210) = 86{,}900 \text{ lb}$$

[8]This tends to verify the conclusions in references 4 and 5 that protecting the minimum net section by increasing the pitch in the outer rows is not as effective as the analysis would indicate. This joint would very likely have carried a larger load if it had been fabricated with two $\frac{1}{8}$-in. rivets in each of the three rows.

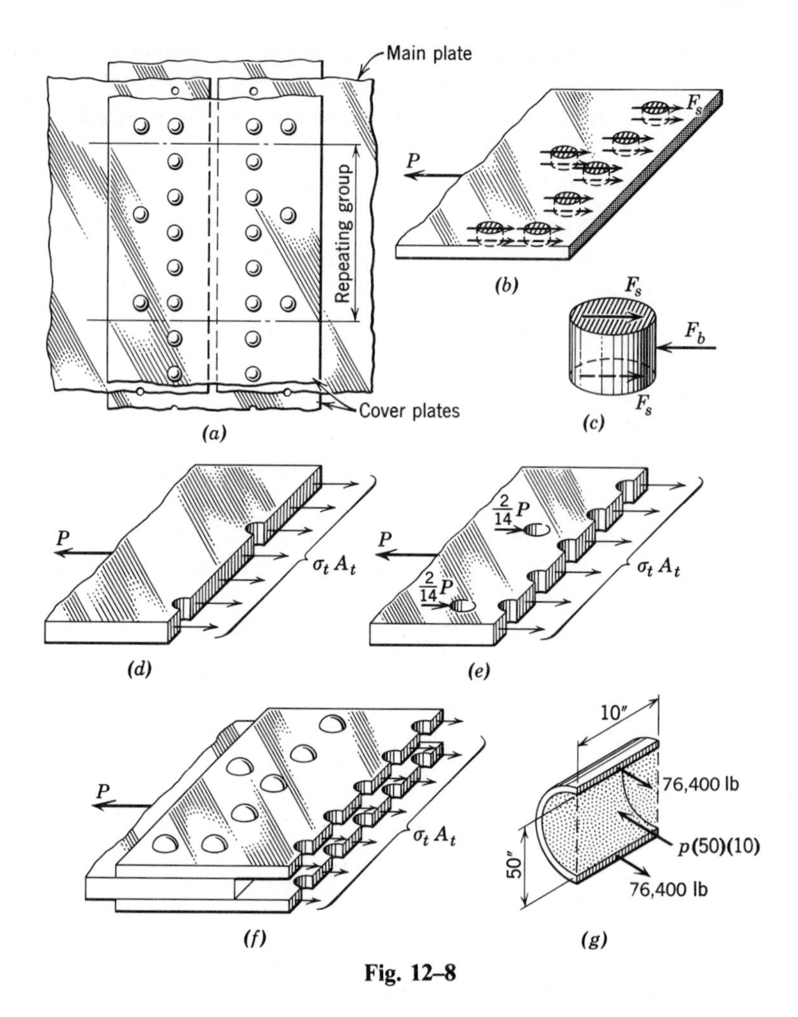

Fig. 12–8

Three different sections must be checked for tensile capacity: (1) the main plate at the outer row of rivets, (2) the main plate at the inner row of rivets with some of the load already transferred to the cover plates, and (3) the cover plates at the inner row of rivets. The summation of forces in the direction of the load for the free-body diagram of Fig. 12–8*d* gives

$$P_t\,(\text{outer row}) = \left[10 - 2\left(\frac{15}{16}\right)\right]\left(\frac{3}{4}\right)13{,}700 = 83{,}500 \text{ lb}$$

The same equilibrium equation for the free-body diagram of Fig. 12–8*e* for tension in the inner row with $(2/7)P$ already transferred to the cover plates gives

$$P_t\,(\text{inner row}) = \left[10 - 5\left(\frac{15}{16}\right)\right]\left(\frac{3}{4}\right)13{,}700 + \left(\frac{2}{7}\right)P_t$$

from which

$$P_t = 76{,}400 \text{ lb}$$

as the maximum allowable load for tension at the inner row of the main plate. The summation of forces in the direction of the load for the free-body diagram of Fig 12–8*f*, which shows the tension in the cover plates at the inner row, gives

$$P_t\,(\text{cover}) = (2)\left[10 - 5\left(\frac{15}{16}\right)\right]\left(\frac{9}{16}\right)13{,}700 = 81{,}900 \text{ lb}$$

The 76,400-lb load that can be carried by the the main plate at the inner row of rivets is the smallest of the four loads computed and is therefore the maximum allowable load on the 10-in. portion of the joint investigated.

The free-body diagram of Fig. 12–8*g* shows a 10-in. length of one-half of the boiler where p is the internal pressure in the boiler. The applicable equation of equilibrium gives

$$p500 = 2(76{,}400)$$

or

$$p = \underline{306 \text{ psi}} \qquad \text{Ans.}$$

(b) The efficiency of the joint is the allowable load on the joint divided by the allowable load on the main plate or

$$\eta = \frac{76{,}400}{(3/4)(10)13{,}700} = 0.744 = \underline{74.4 \text{ percent}} \qquad \text{Ans.}$$

A type of failure not yet considered here is a *marginal failure*, frequently called *tearout*. If a joint is designed to satisfy the usual specifications, the margins will be wide enough ($1\frac{1}{2}$ to 2 rivet diameters is usually specified) to prevent tearout failures. Should a narrow margin be needed, the shear tearout investigation should be made as indicated in Fig. 12–9. For the problems in this book, the margins will be assumed to be wide enough to preclude marginal failures, and the transverse pitch great enough to prevent failure along a diagonal (or zigzag) line between rivets in different rows.

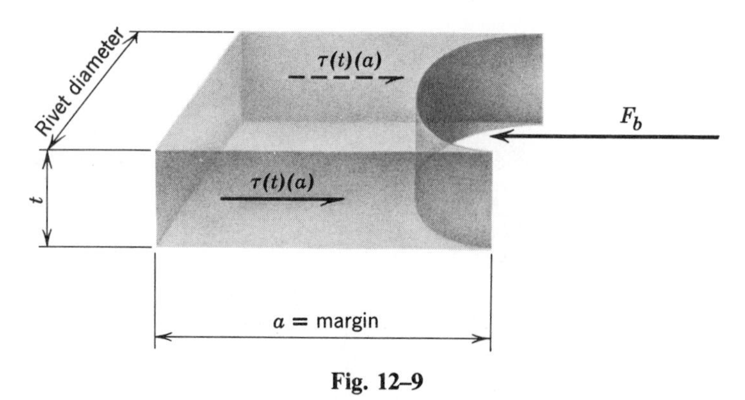

Fig. 12–9

PROBLEMS

Note. In the following problems, when the size of the hole is given, use the hole diameter for shear, bearing, and tension investigations. When the size of the rivet or bolt is given and no hole diameter is specified, use the nominal diameter of the rivet (or bolt) for shear and bearing investigations, and the nominal diameter plus $\frac{1}{8}$ in. (3 mm) for the size of the hole in tension investigations.

12–1* A bolted joint from a truss is shown in Fig. P12–1. Member A consists of two $3 \times \frac{5}{16}$-in. bars; B and C each consist of two $L3 \times 3 \times \frac{5}{16}$ angles. The diameter of

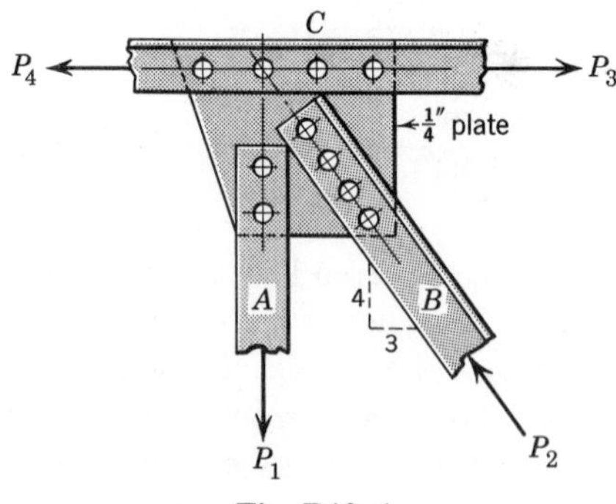

Fig. P12–1

the bolts is $\frac{3}{4}$ in. The total tensile force $P_1 = 20{,}000$ lb. Determine:

(a) The highest average tensile stress in A, and state where it occurs.
(b) The maximum average bearing stress on the connection of B.
(c) The average shearing stress in the bolts at C.

12–2* A C254×35.74 steel channel is riveted to a gusset plate as shown in Fig. P12–2. The rivets are 20 mm in diameter. Determine:

(a) The average shearing stress in the rivets.
(b) The average bearing stress between rivets and channel web.
(c) The highest average tensile stress in the channel, and state where it occurs.

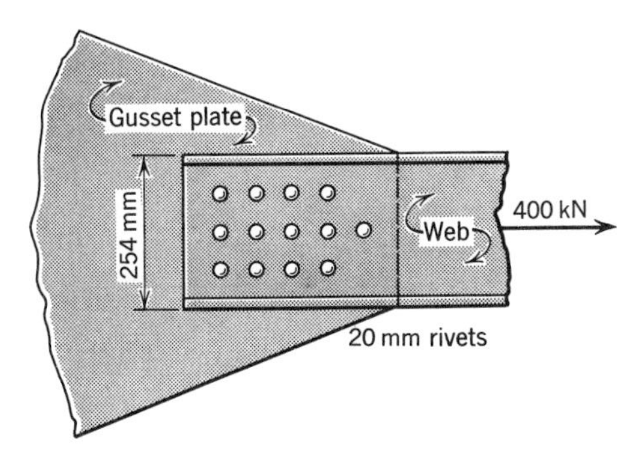

Fig. P12–2

12–3 Figure P12–3 represents a compression member in a pin-connected truss. The member is composed of two C10×25 steel channels latticed together. To provide enough bearing area on the pin, two $\frac{3}{8}$-in. pin plates (a and b) are riveted to each channel web with $\frac{3}{4}$-in. diameter rivets as shown. The total axial load on the member (two channels) is 220 kips. Assume the bearing stress on the pin to be uniformly distributed, and compute the average shearing stress in the rivets:

(a) On the plane of contact between the pin plates and channel web.
(b) On the plane of contact between the two pin plates.

12–4* A long steel conduit 2 m in diameter is subjected to an internal pressure of 620 kPa. It is made of 10-mm steel plates joined with a longitudinal double-riveted lap joint. The rivets are 20 mm in diameter in 22-mm holes, and the pitch in each row is 75 mm. Determine:

(a) The average shearing stress in the rivets.
(b) The average bearing stress on the plates.
(c) The highest average tensile stress in the plates.

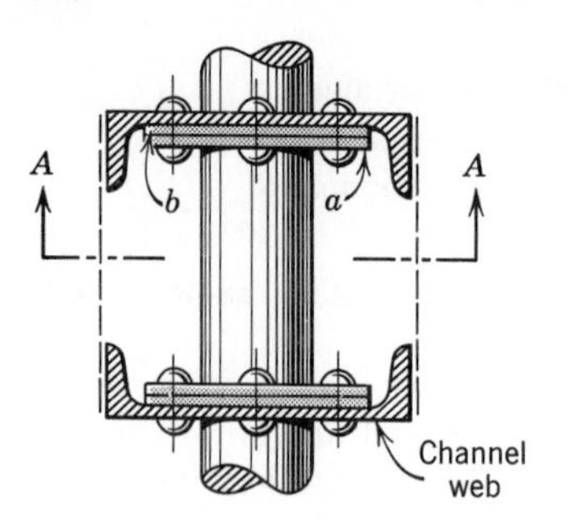

Transverse section

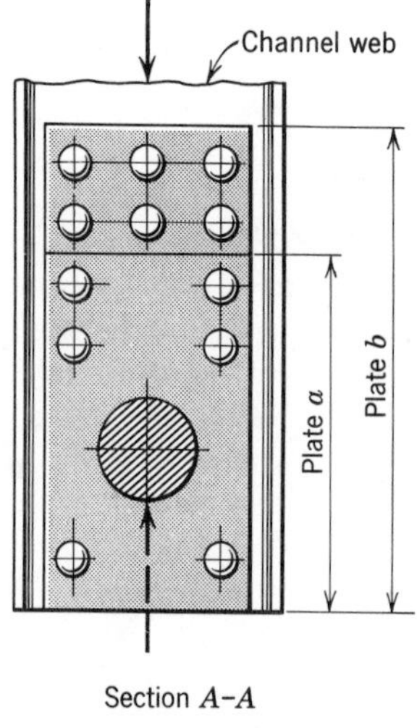

Section A–A

Fig. P12–3

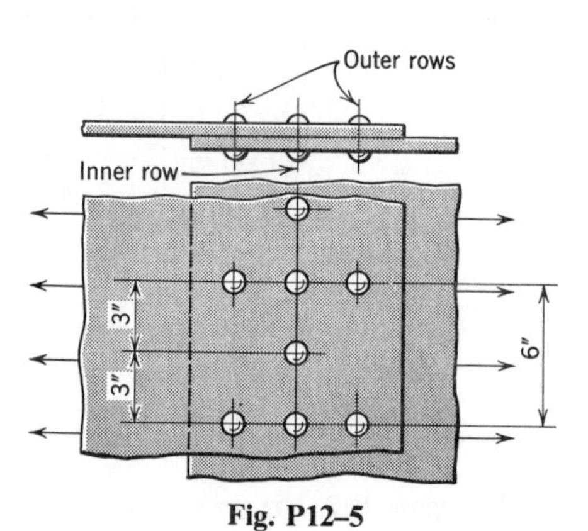

Fig. P12–5

12–5* A triple-riveted lap joint (Fig. P12–5) unites two $\frac{1}{2}$-in. plates. The rivets are 1 in. in diameter and are spaced 6 in. in the outer rows and 3 in. in the inner row. When the tensile stress is 15 ksi in the gross section of the plate, determine:

(a) The average intensity of shearing stress in the rivets.
(b) The average bearing stress on the plates.
(c) The highest average tensile stress, and state where it occurs.

12–6 Figure P12–6 shows a riveted lap splice in the stressed skin of an airplane wing. The material is aluminum alloy. The rivet holes are 4 mm in diameter. Determine:

(a) The average shearing stress in the rivets.
(b) The maximum average bearing stress between rivets and sheet.
(c) The maximum average tensile stress, and state where it occurs.

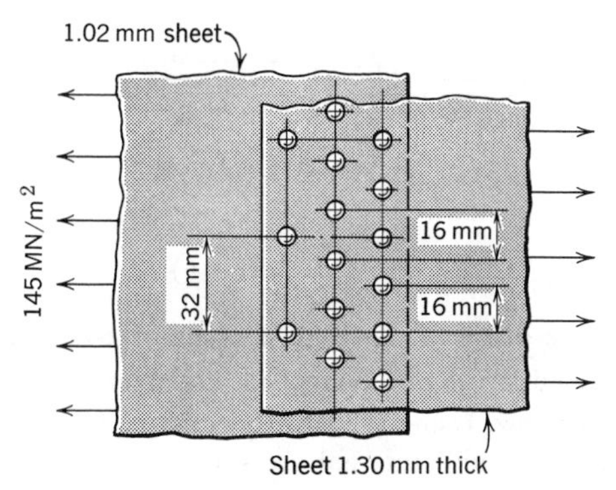

Fig. P12–6

12–7* A cylindrical compressed air tank with an inside diameter of 36 in. is made of $\frac{3}{8}$-in. plate and is subjected to an internal pressure of 300 psi. The longitudinal seam is a double-riveted butt joint with two $\frac{1}{4}$-in. cover plates. The $\frac{3}{4}$-in. diameter rivets in $\frac{13}{16}$-in. holes are spaced at $2\frac{1}{2}$ in. in the inner rows and 5 in. in the outer rows. Determine:

(a) The average shearing stress in the rivets.
(b) The highest average bearing stress between rivets and any plate.
(c) The highest average tensile stress in the plates, and state where it occurs.

12–8 Two $\frac{7}{16}$-in. steel plates are connected by a triple-riveted butt joint with two $\frac{1}{4}$-in. cover plates. The $\frac{5}{8}$-in. diameter rivets are spaced at 4 in. in the outer rows, 2 in. in the intermediate rows, and 3 in. in the inner rows. The joint is loaded in tension, and each shear area carries a total force of 3000 lb. Determine the highest average tensile stress in the joint, and state where it occurs.

12–9* Two 200×20-mm plates are connected by a triple-riveted butt joint with unequal length cover plates, as shown in Fig. P12–9. The rivets are 22 mm in diameter. When $P = 300$ kN, determine:

(a) The average shearing stress in the rivets.
(b) The average bearing stress on the rivets in row 3.
(c) The average tensile stress in the main plate at each row of rivets.

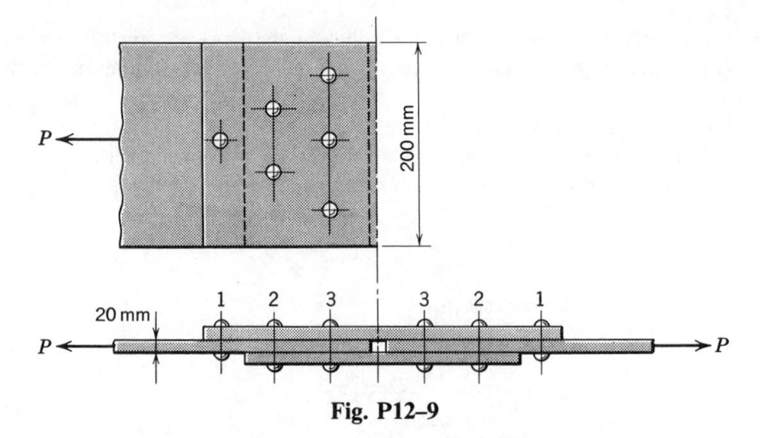

Fig. P12–9

12–10 A boiler with a diameter of 40 in. is made of $\frac{1}{2}$-in. plate.The longitudinal seam is a double-riveted lap joint having 1-in. diameter rivets in $\frac{17}{16}$-in. holes, at a pitch of 3.5 in. in each row. Using the ASME boiler code, determine:

(a) The maximum allowable internal pressure.
(b) The efficiency of the riveted joint.

12–11* A triple-riveted lap joint is used to connect two 10-mm plates with 20-mm rivets in 22-mm holes. The pitch in the outer rows is 180 mm and the pitch in the inner row is 60 mm. Determine the efficiency of the joint, using the following allowable stresses in MN/m^2: 65 in shear, 130 in bearing, and 90 in tension.

12–12 A triple-riveted lap joint is to be made of steel plates and $\frac{3}{4}$-in. rivets at pitches of $4\frac{1}{2}$ in. in the outer rows and $2\frac{1}{4}$ in. in the inner row. Determine the minimum permissible thickness of the plate for the maximum allowable load on the joint, assuming the AISC building code can be applied to this joint.

12–13 A 1.5-m diameter boiler with an internal steam pressure of 1200 kPa is made of 15-mm plate. The double-riveted longitudinal butt joint is made with two splice plates. The pitch in the inner row of rivets is 60 mm and in the outer row the pitch is 100 mm. The rivet diameter is 20 mm (22-mm holes), and the splice plates are each 9 mm thick. Determine the efficiency of the joint based on the stresses of Prob. 12–11.

12–14* A cylindrical pressure tank with a diameter of 40-in. is made of $\frac{5}{16}$-in. steel plate. The longitudinal seam is a triple-riveted butt joint with one $\frac{3}{8}$-in. cover plate and $\frac{3}{4}$-in. diameter rivets in $\frac{13}{16}$-in. holes, at pitches of $2\frac{1}{2}$ in. in the inner rows, $3\frac{3}{4}$ in. in the intermediate rows, and $7\frac{1}{2}$ in. in the outer rows. Using the AISC specification of 1970, determine:

(a) The maximum allowable pressure in the tank.
(b) The efficiency of the joint.

12–15* The triple-riveted butt joint of Fig. 12–1*e* is composed of 0.040-in. main plates spliced with two 0.025-in. cover plates and 5/32-in. rivets in 0.159-in. holes. The plates are 1.375 in. wide and are of aluminum alloy 2024-T3, and the rivets are 2117-T3. Use the strength values given in Table 12–2 and reduce the shearing strength value to 20,600 psi because the D/t ratio is greater than 1.5. Determine the probable ultimate (failure) load for the specimen and the efficiency of the joint.

12–16 The joint of Prob. 12–9 results in stresses varying from 37–61 percent of the stresses allowed by the AISC specifications (103 in shear; 335 in bearing; 149 in tension, all in MN/m^2).

(a) What would be the effect on each of the stresses (increase, decrease, or no change) due to a decrease in (1) the rivet diameter, (2) the plate thickness, (3) the plate width?

(b) Determine the maximum permissible load on the joint of Prob. 12–9 (AISC spec.) with the following changes: 200 × 12-mm main plates, two 200 × 8-mm splice plates, and 20-mm rivets.

12–17 Two $6 \times \frac{3}{4}$-in. steel plates are spliced with $\frac{7}{16}$-in. plates and $\frac{7}{8}$-in. diameter A325 bolts to form a friction-type joint as shown in Fig. P12–17. Using the AASHO specification of 1973, determine the maximum allowable load P.

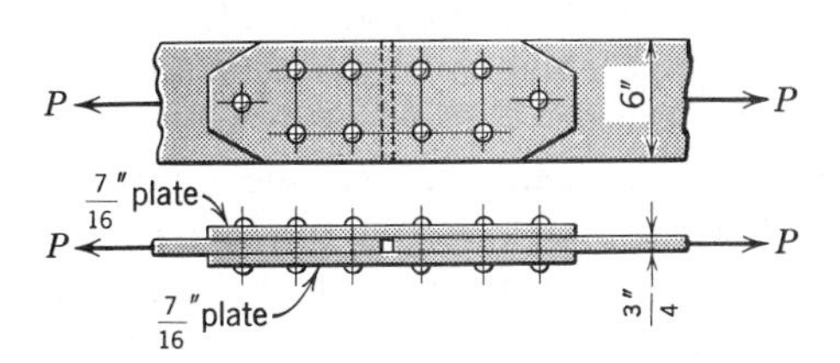

Fig. P12–17

12–18* Determine the efficiency of the double-riveted butt joint of Fig. P12–18. The rivets are 20 mm in diameter with 22-mm holes, and the allowable stresses are given in Prob. 12–11.

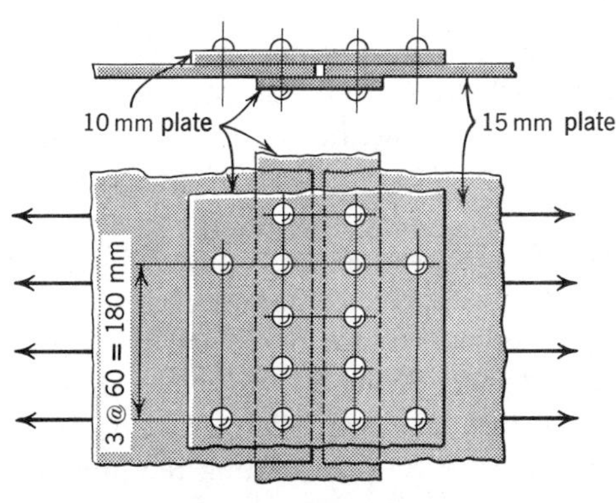

Fig. P12–18

12–19 A structural joint is composed of two $9 \times \frac{3}{4}$-in. plates connected by a friction-type bolted butt joint with two cover plates and 1-in. diameter A325 bolts. The joint is to carry an axial tensile load of 95 kips. Using the 1973 AASHO specifications, determine:

(a) The minimum allowable number of bolts.
(b) The minimum allowable number of rows of bolts.
(c) The best distribution of bolts in the rows.
(d) The minimum allowable thickness for each cover plate, using the distribution of part (c).

12–3 ANALYSIS OF RIVETED AND BOLTED JOINTS—ULTIMATE STRENGTH METHOD

A widely used method for the design and investigation of riveted connections is based on the concept that if the load necessary to produce excessive inelastic action in a joint (ultimate load) can be predicted, a safe working load can then be determined by introducing an appropriate factor of safety. The determination of stresses at design loads (working loads) is of no consequence; hence, the question of distribution of load to the various elements of a connection at design loads merits no consideration.

The ultimate strength of the joint is determined by considering the various methods and combinations of failure that could occur. Any rivet can fail in either shear or bearing, and these possibilities are usually investigated first. When a joint contains some rivets in single shear and some in double shear, it is possible that the shearing strength of the material could be the controlling factor for the rivets in single shear and that the bearing strength could limit the load carried by the rivets in double shear.

The joint can fail in tension at the outer row if rivets in a lap joint, at the outer row of rivets in the main plate of a butt joint, or at the inner row of rivets in the cover plate of a butt joint. Failure can also result from a tensile failure of the main plate, at a row other than the outer row, combined with rivet failure (shear or bearing), which would have to occur if the joint were to separate. All possible combinations of tension failure with shear or bearing failure must be visualized and checked to determine the weakest combination. The smallest of the computed loads is the strength of the joint.

Table 12–2 lists a few representative values of ultimate strengths that may be used in the analysis of riveted connections. Although the preceding discussion concerned ultimate loads on joints, the same analysis can be made using working stresses, instead of ultimate strengths, to obtain the maximum working load on the joint. Such a procedure amounts to dividing by the factor of safety before calculating the joint strength instead of afterward.

The ultimate strength method of analyzing riveted joints is illustrated in Exams. 12–3 and 12–4.

TABLE 12–2 Ultimate Strengths for Some Selected Materials

Specification	Tensile Strength of Plate (psi)	Bearing Strength of Plate (psi)	Shearing Strength of Rivet (psi)
Am. Soc. of Mech. Engrs. Boiler Construction Code for Power Boilers, 1971			
SA 515-grade 55 plate	55,000	95,000	
SA 31, Grade A rivets			44,000
MIL-HDBK-5 Met. Matls. and Elements for Flight Vehicle Struct., Feb. 1966			
2024-T3 sheet	64,000[a]	124,000[b]	
2117-T3 rivets			30,000[a,c]
2024-T31 rivets			41,000[a,c]
Structural steel (1025)	55,000	90,000[b]	35,000
Stainless steel, AISI 301			
Annealed	75,000	150,000[b]	
Sheet or strip, $\frac{1}{4}$ hard	125,000	250,000[b]	
Monel rivets			49,000

[a]Probability (not guaranteed) values.

[b]For ratio of margin to rivet diameter = 2, and ratio of rivet diameter to sheet thickness < 5.5.

[c]For ratio of rivet diameter to sheet thickness $\leqslant 3$ for single shear and 1.5 for double shear. See specification for reduction factors.

Example 12–3 The triple-riveted butt joint of Fig. 12–1*e* is composed of $1\frac{3}{8} \times 0.040$-in. main plates spliced with two 0.025-in. cover plates and $\frac{5}{32}$-in. diameter rivets. The diameter of the hole is essentially the same as that of the rivet. The plate material is aluminum alloy 2024-T4, and the rivets are 2117-T6. If the guaranteed[9] ultimate strengths of these materials are 27,000

[9]ANC-5, Strength of Aircraft Elements, December 1942. Note that guaranteed values are minimum acceptable values and are usually less than typical values.

psi shear, 90,000 psi bearing, and 62,000 psi tension, determine the ultimate load for the specimen.

Solution The modes of failure to be investigated are: shearing of all rivets, bearing on all rivets, and tension at any net section in combination with shear or bearing failure of any rivets assisting the net section.

Ultimate shearing and bearing loads will be computed first with the aid of Fig. 12–10*a* and *b* (note that in Fig. 12–10*a* there are six additional forces F_s, not shown, on the opposite side of the plate) as follows:

$$2F_s = 2(27{,}000)\left(\frac{\pi}{4}\right)\left(\frac{5}{32}\right)^2 = 1035 \text{ lb shear per rivet}$$

$$F_b = 90{,}000\left(\frac{5}{32}\right)(0.040) = 562 \text{ lb bearing per rivet}$$

The cover plates will carry more load in bearing than the main plate since

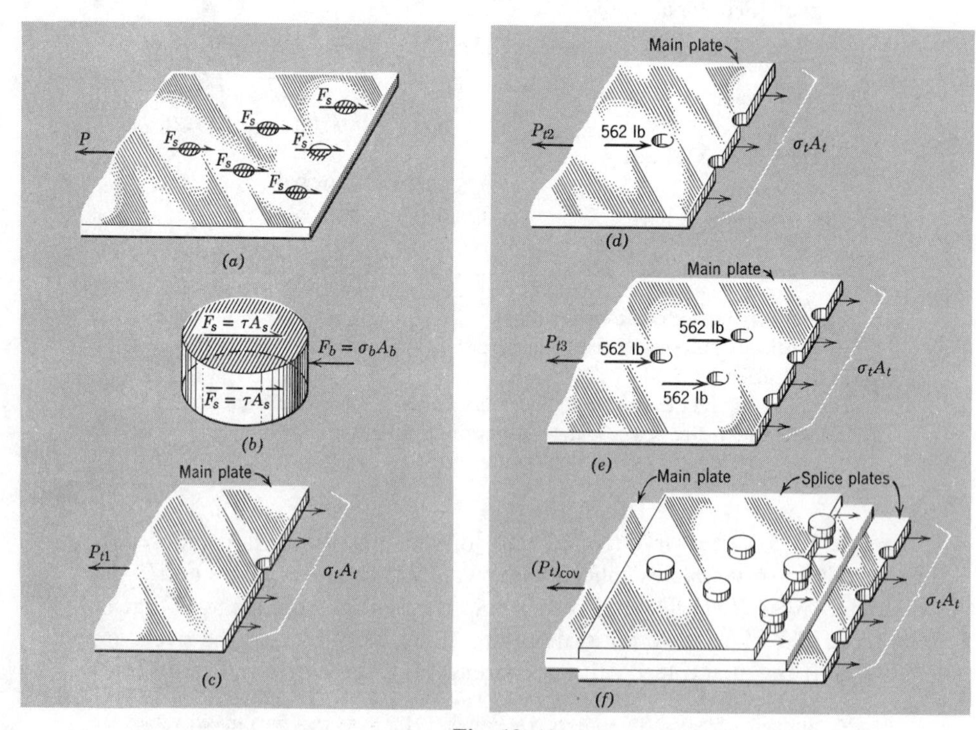

Fig. 12–10

the bearing area is greater. The results above indicate that the shearing load is not critical; therefore, as far as shear and bearing are concerned, the ultimate load on the joint is

$$P_b = 6F_b = 6(562) = 3370 \text{ lb}$$

Next, the ultimate tensile load will be determined with the aid of the free-body diagrams of Fig. 12–10*c*, *d*, *e*, and *f*. The load for each rivet assisting the net section in the main plate (Fig. 12–10*d* and *e*) is 562 lb, since the previous shear and bearing analysis indicates that the rivet will fail in bearing before the ultimate shear load is attained. Summing forces in the horizontal direction on each of the free-body diagrams in order yields the following:

$$(c)\quad P_{t1} = \sigma_t A_t = 62{,}000\left(\frac{11}{8} - \frac{5}{32}\right)(0.040) = 3020 \text{ lb}$$

$$(d)\quad P_{t2} = 562 + 62{,}000\left[\frac{11}{8} - 2\left(\frac{5}{32}\right)\right](0.040) = 3200 \text{ lb}$$

$$(e)\quad P_{t3} = 562(3) + 62{,}000\left[\frac{11}{8} - 3\left(\frac{5}{32}\right)\right](0.040) = 3930 \text{ lb}$$

$$(f)\quad (P_t)_{\text{cov}} = 62{,}000\left(\frac{11}{8} - \frac{15}{32}\right)(0.025)(2) = \underline{2810 \text{ lb}} \qquad \textit{Ans.}$$

As the load on the cover plates is the least of the five loads computed, the joint might be expected to fracture at a load of 2810 lb. The average of two such joints, one of which is shown in Fig. 12–1*e*, loaded to fracture was 2790 lb.

Example 12–4 Figure 12–11*a* represents a riveted lap splice in the stressed skin of an airplane wing. The sheet material is 2024-T3, and the $\frac{5}{32}$-in. diameter rivets in 0.159-in. holes are 2024-T31. Use the MIL-HDBK-5 specifications with the shearing strength reduced to 0.964 times the value given (because the rivet diameter to plate thickness ratio is greater than 3), and determine:

(a) The maximum permissible load per linear inch of joint if a factor of safety of 1.5 is specified.

(b) The efficiency of the joint.

Solution (a) From Fig. 12–11*a* the length of a repeating group of rivets is seen to be $\frac{5}{4}$ in. with one rivet in the left row and two rivets in each of the other rows. Each rivet is in single shear, and the free-body diagram in Fig. 12–11*b* indicates that $F_b = F_s$. The bearing load is limited by the lower plate since the thickness of this plate is smaller. The failure loads per rivet in shear and bearing are

$$(F_s)_{\max} = (\pi/4)(0.159^2)(41{,}000)(0.964) = 785 \text{ lb}$$

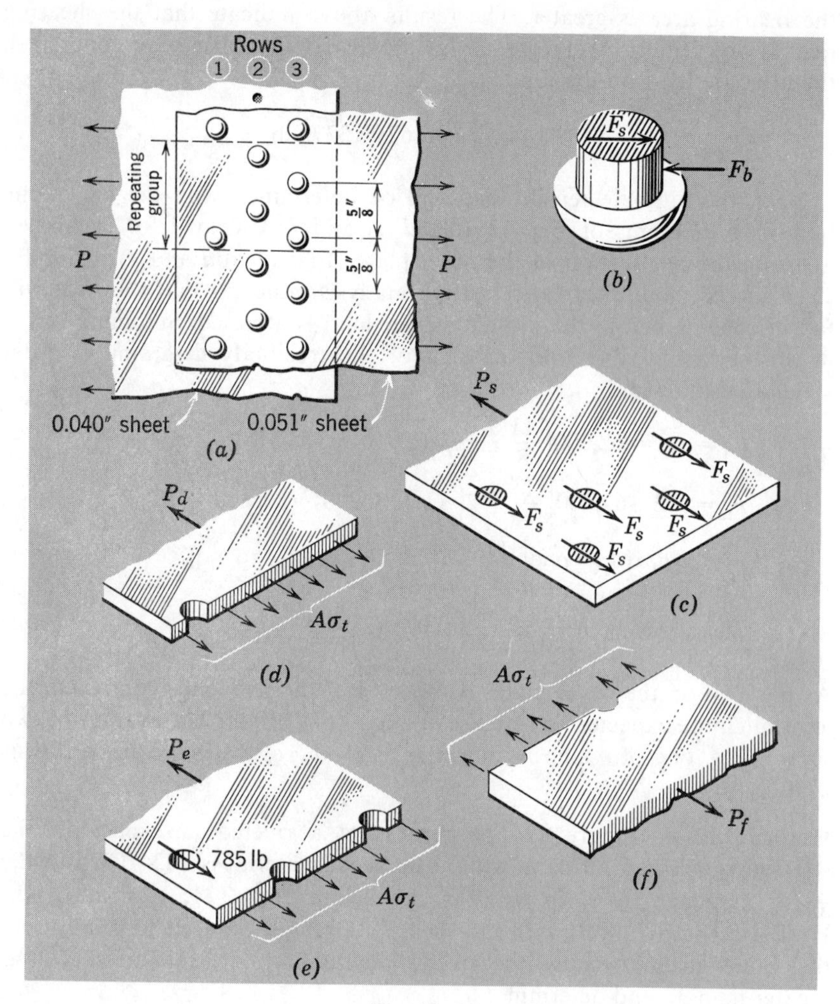

Fig. 12–11

and

$$(F_b)_{\max} = (0.040)(0.159)(124{,}000) = 789 \text{ lb}$$

There are five rivets in each repeating group, and the free-body diagram in Fig. 12–11*c* indicates that the failure load in the joint, as far as shear and bearing are concerned, is

$$P_s = 5(785) = 3920 \text{ lb}$$

There are three possibilities of joint failure involving tensile failure of the plates to be investigated, The joint might fail by tearing the lower plate at row 1; the joint might fail by tearing the lower plate at row 2 and shearing the rivet in row 1; or the joint might fail by tearing the top plate at row 3. These three possibilities are shown in the free-body diagrams in Fig. 12–11*d*, *e*, and *f*. The equations of equilibrium for these free-body diagrams give

From (*d*) $P_d = 0.040(1.25 - 0.159)64{,}000 = 2790$ lb
From (*e*) $P_e = 785 + 0.040[1.25 - 2(0.159)]64{,}000 = 3170$ lb
From (*f*) $P_f = 0.051[1.25 - 2(0.159)]64{,}000 = 3040$ lb

It is not necessary to check other failure combinations because the plates have as much or more areas at all other rows and more rivets would have to shear for other combinations. The controlling load on a repeating group is the 2790-lb load from Fig. 12–11*d*. With a factor of safety of 1.5 the load per inch becomes

$$P = \left(\frac{4}{5}\right)\frac{2790}{1.5} = \underline{1488 \text{ lb/in.}} \qquad \text{Ans.}$$

(b) The efficiency of the joint is

$$\eta = \frac{2790}{(5/4)(0.040)(64{,}000)} = 0.872 = \underline{87.2 \text{ percent}} \qquad \text{Ans.}$$

PROBLEMS

Note. In the following problems, when the size of the hole is given, use the hole diameter for shear, bearing, and tension investigations. When the size of the rivet or bolt is given and no hole diameter is specified, use the nominal diameter of the rivet (or bolt) for shear and bearing investigations, and the nominal diameter plus $\frac{1}{8}$ in. (3 mm) for the size of the hole in tension investigations.

12–20* A boiler with a diameter of 40 in. is made of $\frac{1}{2}$-in. plate. The longitudinal seam is a double-riveted lap joint having $\frac{3}{4}$-in. diameter rivets in $\frac{13}{16}$-in. holes, at a pitch of $2\frac{1}{2}$ in. in each row. Using ultimate strengths from the ASME boiler code and a factor of safety of 5, determine:

(a) The maximum allowable internal pressure.
(b) The efficiency of the riveted joint.

12–21* The triple-riveted lap joint in Fig. P12–21 connects two 10-mm structural steel plates with 20-mm diameter rivets. Determine the efficiency of the joint and the allowable load if a factor of safety of 3 is specified. Use the following ultimate strengths (in MN/m^2): 380 in tension, 620 in bearing, and 240 in shear.

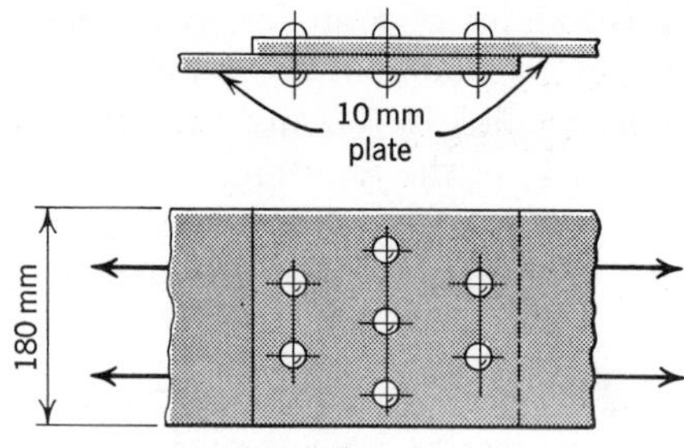

Fig. P12–21

12–22* A triple-riveted lap joint is used to join two $\frac{1}{4}$-in. plates with $\frac{5}{8}$-in. diameter rivets in $\frac{11}{16}$-in. holes. The pitch in the outer rows is 5 in. and that in the inner row is 2 in. Using the ultimate strengths from the ASME boiler code, determine the efficiency of the joint.

12–23* A triple-riveted lap joint is to be made of structural steel plates and 22-mm diameter rivets at pitches of 130 mm in the outer rows and 65 mm in the inner row. Using the values for strength given in Prob. 12–21 and a factor of safety of 3, determine the minimum permissible thickness of the plate for the maximum allowable load on the rivets.

12–24 A 4-ft diameter boiler with an internal steam pressure of 180 psi is made of $\frac{3}{8}$-in. plate. The double-riveted longitudinal butt joint is made with two $\frac{1}{4}$-in. splice plates and $\frac{3}{4}$-in. diameter rivets in $\frac{13}{16}$-in. holes, at pitches of $2\frac{1}{4}$ in. in the inner and 6 in. in the outer rows. Determine the factor of safety of the joint using the ASME boiler code.

12–25* A 1.2-m diameter boiler is made of 20-mm plate. The longitudinal seam is a double-riveted butt joint with two 15-mm cover plates and with 22-mm diameter rivets in 24-mm holes. The outer rows of rivets have a pitch of 150 mm and the inner rows a pitch of 60 mm. Using the ASME boiler code (converted to MN/m^2: 380 in tension; 655 in bearing; 305 in shear) and a factor of safety of 5, determine the maximum allowable internal pressure.

12–26 Two 130×18-mm structural steel plates are spliced with 12-mm cover plates and 20-mm high-strength bolts to form the friction-type joint shown in Fig.

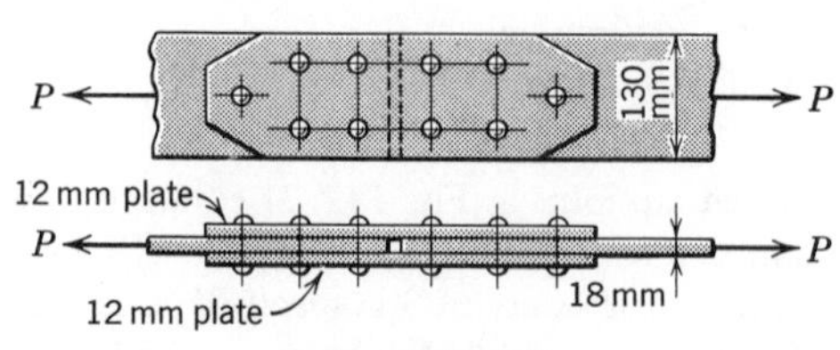

Fig. P12–26

P12–26. Using the strength values of Prob. 12–21, determine the maximum allowable load P if a factor of safety of 3 is specified.

12–27* A double-riveted butt joint with two equal cover plates has a 3.0-in. pitch in the inner rows and a 4.5 in. pitch in the outer rows. The ultimate strengths are: 40,000 psi shear, 64,000 psi tension, and 96,000 psi bearing.

(a) Determine the rivet diameter and main plate thickness that will result in equal strength in shear, bearing, and tension in the outer row of rivets in the main plate. Assume rivets and holes are the same size.

(b) Using the results of part (a), determine the ratio of the joint strength computed on the basis of a tension failure at the inner row of rivets coupled with rivet failure to the joint strength in bearing. Assume the cover plates are thick enough to prevent tearing of the cover plate.

12–28 The cover plates in Fig. P12–28 are of equal thickness, the rivets are 1 in. in diameter, and the material is structural steel. Using MIL-HDBK-5 specifications and a factor of safety of 3, determine:

(a) The maximum permissible load on the joint.

(b) The minimum permissible thickness for each cover plate.

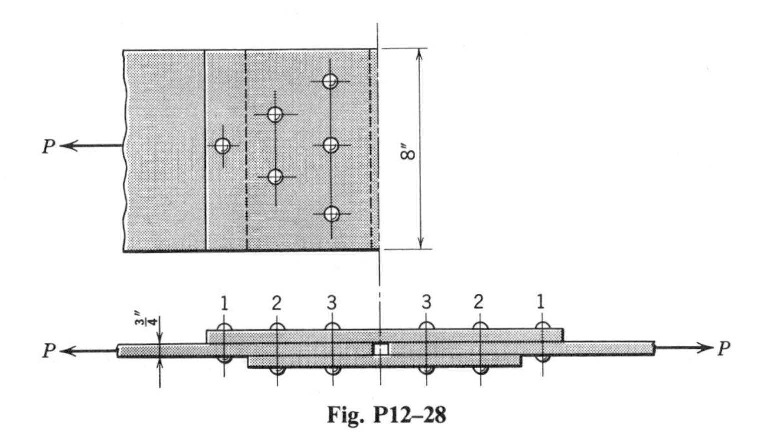

Fig. P12–28

12–29* A butt joint is fabricated like that of Exam. 12–3 except for the following changes: the cover plates are each 0.040 in. thick and only one plate engages all three rows of rivets while the other engages only the inner and intermediate rows. The rivets are of monel metal $\frac{5}{32}$ in. in diameter in 0.159-in. holes and the plates are of annealed AISI 301 stainless steel. Determine the efficiency of the joint based on the MIL-HDBK-5 strength values.

12–30 Two 240×20-mm steel plates are to be connected by a friction-type bolted butt joint with two cover plates and 22-mm bolts. The joint is to carry an axial load of 275 kN. Using the ASME boiler code (see Prob. 12–25) and a factor of safety of 5, determine:

(a) The minimum allowable number of bolts.

(b) The minimum allowable number of rows of bolts.
(c) The best distribution of bolts in the rows.
(d) The minimum allowable thickness for each cover plate, for the distribution of part (c).

12–4 WELDED CONNECTIONS

Connections between metal plates, angles, pipes, and other structural elements are frequently made by welding. Fusion welds are made by melting portions of the materials to be joined with an electric arc, a gas flame, or with thermit. In fusion welds, additional welding material is usually added to the melted metal to fill the space between the two parts to be joined or to form a fillet. A gas shield is provided when welding certain metals to prevent rapid oxidation of the molten metal. A good welded joint will usually develop the full strength of the material being joined unless the high temperature necessary for the process changes the properties of the materials.

Metals may also be joined by resistance welding in which a small area or spot is heated under high localized pressure. The material is not melted with this type of welding. Other joining methods for metals include brazing and soldering, in which the joining metal is melted but the parts to be connected are not melted. Such connections are usually much weaker than the materials being connected. Fusion welding is the most effective method when high strength is an important factor, and it will be discussed more in detail.

Figure 12–12 illustrates the two general types of fusion-welded joints. The joints illustrated in Fig. 12–12*a*, *b*, and *c* are called *butt* joints and are made with groove welds. The weld may be square, for fairly thin materials, or there may be one or two vees for thicker metals. When approved welding techniques and materials are used and complete penetration of the material is obtained, the strength of the joint is assumed to be the same as that of the material being joined.

A fillet weld is illustrated by Fig. 12–12*d* and is used to join a plate, angle, or other strucural shape to another structural member. The end weld may be omitted in some applications. Ideally, the fillet should be slightly concave, rather than convex or straight as shown, to reduce possible stress concentrations at reentrant corners.

From an analysis similar to that for the riveted connection (see Art. 12–1), it can be shown that the shearing stress in the side weld cannot be uniformly distributed along the weld as long as the action is elastic, and tests[10] indicate the distribution is somewhat as indicated in Fig. 12–13.

[10]"Distribution of Shear in Welded Connections," H. W. Troelsch, *Trans. Am. Soc. of Civil Engrs.*, Vol. 99, 1934, p. 409.

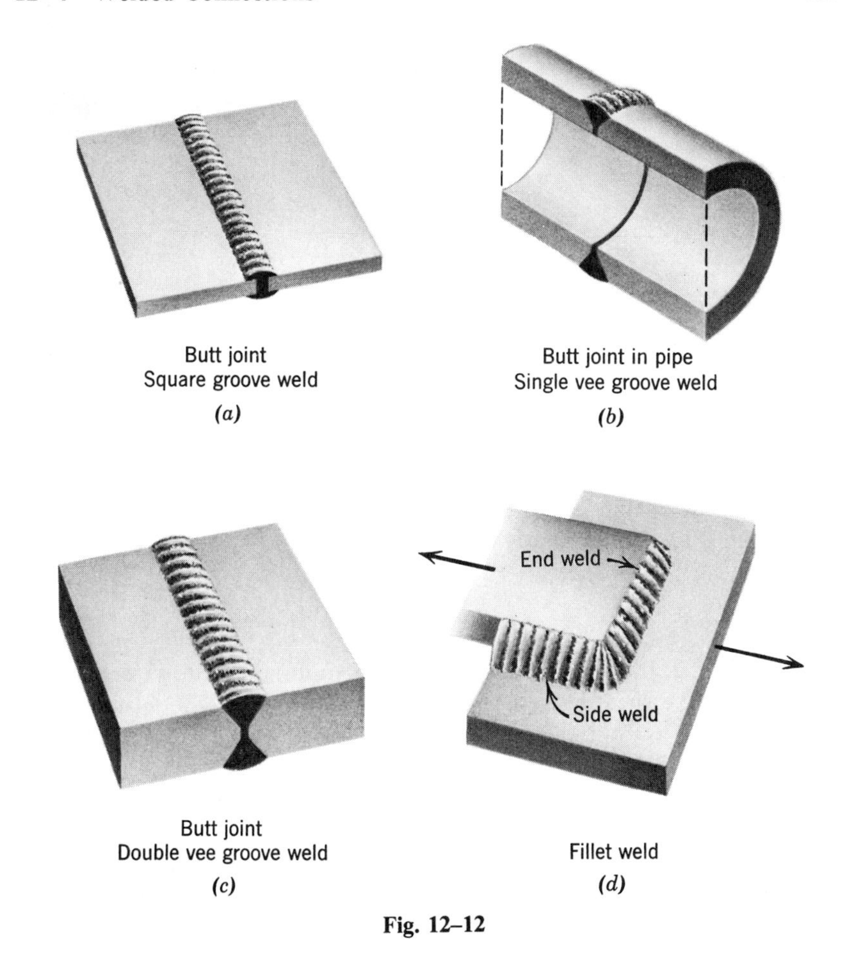

Fig. 12–12

However, if the material is ductile (the usual condition for welded connections), a slight amount of inelastic action at the ends of the welds will result in essentially a uniform shearing stress distribution; therefore, in the design of welded connections, a uniform distribution is assumed. Fillet welds are designed on the assumption that failure will occur by shearing the minimum section of the weld, which is called the throat of the weld (section AB in Fig. 12–14a) whether the weld is a side (longitudinal) weld or an end (transverse) weld. The strength of side fillet welds can be determined from Fig. 12–14a and b. Although it is desirable to make fillet welds slightly concave, calculations are usually based on the assumption that the weld surface is straight and makes an angle of 45° with the two legs. The maximum allowable load P that can be carried by a side fillet

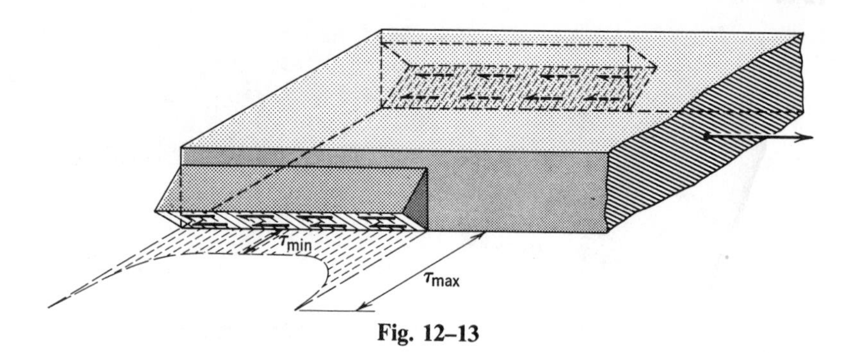

Fig. 12–13

weld with a length L and legs t (it is common practice to designate the size of fillet welds by the length of leg t) can be obtained from the free-body diagram in Fig. 12–14*b* and is

$$P = 0.707\,tL\tau$$

where τ is the allowable shearing stress permitted in the weld material. Some allowable shearing stresses for steel welds using A233 class E60XX electrodes are as follows: AISC, 1970, 18 ksi; AASHO, 1973, 12.4 ksi; AREA, 1973, 12.4 ksi.

Tests indicate that the strength of a transverse fillet weld is about 30 percent greater than that of a side fillet weld with the same dimensions. It is common practice, however, to assume that the strength of a transverse weld is the same as that of a side weld with the same dimensions. This practice will be followed in this book.

The following example illustrates the analysis of a centrically loaded fillet weld.

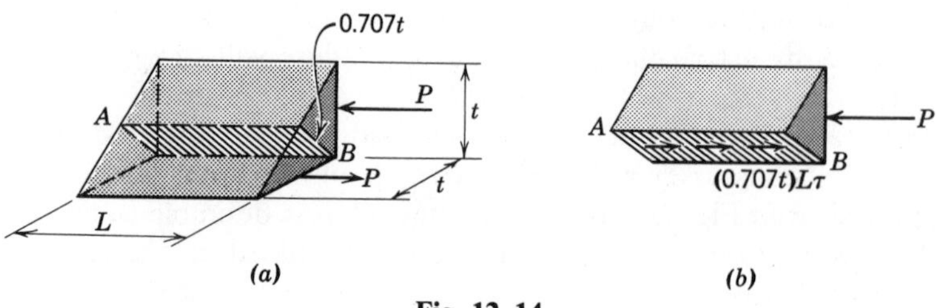

Fig. 12–14

Example 12–5 An L8×6×$\frac{1}{2}$ steel angle is to be welded to a flat plate with the long side of the angle against the plate as shown in Fig. 12–15*a* and *b*.

(a) Determine the minimum lengths L_1 and L_2 that will cause the angle to carry the maximum allowable axial load. The allowable tensile stress for the material in the angle is 18,000 psi, and the allowable shearing stress in the weld material is 13,600 psi. Each leg of the weld is 0.50 in.

(b) Solve part (a) if a transverse weld is placed across the end of the angle.

Solution (a) The maximum longitudinal force in the angle will be developed when the tensile stress is uniformly distributed and will act through the centroid of the cross section of the angle. The area of the cross section and the location of its centroid can be obtained from the table of properties of sections of structural shapes in App. C. The area is 6.75 sq

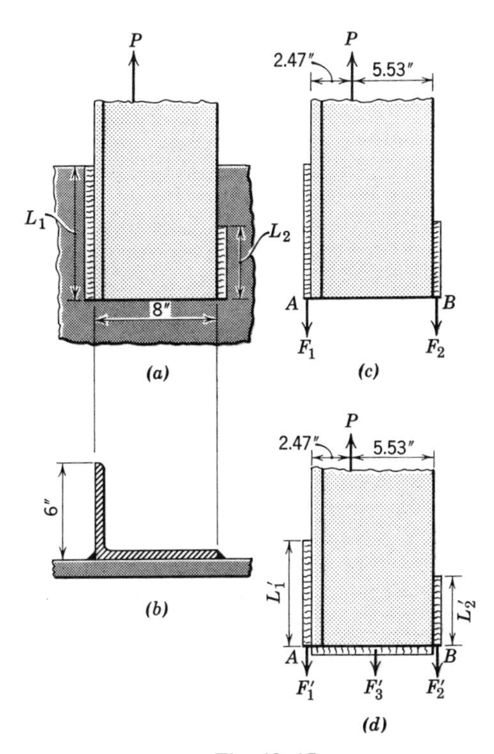

Fig. 12–15

in., and the location of the centroid is shown on Fig. 12–15*c*, which is a free-body diagram of a part of the angle. The forces F_1 and F_2 are assumed to act at the edges of the angle. The force P is

$$P=\sigma A=18{,}000(6.75)=121{,}500 \text{ lb}$$

The equation $\Sigma M_A=0$, applied to the free-body diagram, gives

$$\circlearrowleft + \qquad \Sigma M_A=8F_2-2.47P=0$$

from which

$$F_2=37{,}500 \text{ lb as shown}$$

Similarly,

$$F_1=84{,}000 \text{ lb as shown}$$

The shear area of the throat that must carry the load F_2 is $L_2(0.707)(0.50)$, and the shearing stress multiplied by this area must equal the applied load. Thus,

$$F_2=37{,}500=\tau A=13{,}600L_2(0.707)(0.50)$$

from which

$$L_2=\underline{7.80 \text{ in.}} \qquad \text{Ans.}$$

In the same manner

$$L_1=\underline{17.45 \text{ in.}} \qquad \text{Ans.}$$

(b) When a weld is placed across the end of the angle, an additional force must be shown on the free-body diagram, as in Fig. 12–15*d*. The force F_3' is

$$F_3'=\tau A=13{,}600(8)(0.707)(0.50)=38{,}500 \text{ lb}$$

Note that the transverse weld is assumed to carry the same load per inch of weld as the side weld. The equation $\Sigma M_A=0$ applied to Fig. 12–15*d* gives

$$\circlearrowleft + \qquad \Sigma M_A=8F_2'+4(38{,}500)-2.47P=0$$

from which

$$F_2' = 18{,}250 \text{ lb} = 13{,}600 L_2'(0.707)(0.50)$$

as in part a of the solution. This equation gives

$$L_2' = \underline{3.80 \text{ in.}} \qquad \text{Ans.}$$

Similarly

$$L_1' = \underline{13.45 \text{ in.}} \qquad \text{Ans.}$$

PROBLEMS

12–31* Design a fillet-welded joint, for attaching an $L5 \times 5 \times \frac{3}{4}$ steel angle to a gusset plate, that will develop the full working tensile strength of the angle. Use a tensile working stress of 15,000 psi in the angle and a maximum allowable force per linear inch of weld of 6000 lb. Weld along the end of the angle in order to shorten the gusset plate (see App. C for additional data).

12–32* Design a fillet-welded joint, for attaching an $L100 \times 100 \times 12$ steel angle to a gusset plate, that will develop the full working tensile strength of the angle. Use a tensile working stress of 120 MN/m^2 in the angle and a maximum allowable force per linear millimetre of weld of 600 N. Weld along the end of the angle in order to shorten the gusset plate (see App. C for additional data).

12–33* Design a fillet-welded joint, for attaching an $L5 \times 3 \times \frac{1}{2}$ steel angle to a gusset plate with the 5-in. side against the plate, that will develop the full working tensile strength of the angle. Use a tensile working stress of 24 ksi in the angle and an allowable shearing stress in the $\frac{7}{16}$-in. weld of 12.4 ksi. Weld along the end of the angle in order to shorten the gusset plate (see App. C for additional data).

12–34* Design a fillet-welded joint, for attaching an $L150 \times 90 \times 15$ steel angle to a gusset plate with the 150-mm side against the plate, that will develop the full working tensile strength of the angle. Use a tensile working stress of 160 MN/m^2 in the angle and an allowable shearing stress in the 10-mm weld of 85 MN/m^2. Weld along the end of the angle in order to shorten the gusset plate (see App. C for additional data).

12–35 A cylindrical pressure tank is to be constructed by welding a $\frac{1}{2}$-in. thick plate on the ends of a 6-ft piece of steel pipe. The pipe has an outside diameter of 18 in. and a thickness of 0.25 in. Determine the size of a fillet weld to develop the full tensile working strength of the pipe. The working stress in the pipe is 20,000 psi, and the allowable force per inch of weld is 10,600t where t is the size of the weld.

12–36* An improvised tension member is to be made by splicing two lengths of $L6 \times 4 \times \frac{9}{16}$ angle. Since the angles are too short to permit groove welding, a joint composed of plate A and bar B fillet welded to the angles as shown in Fig. P12–36 is proposed. The working stress for the angles is 18 ksi, and the maximum allowable load on each inch of fillet weld is 4 kips. Determine the length of each weld a, b, and c necessary to develop the full axial tensile capacity of the angles with no bending about any axis in the joint.

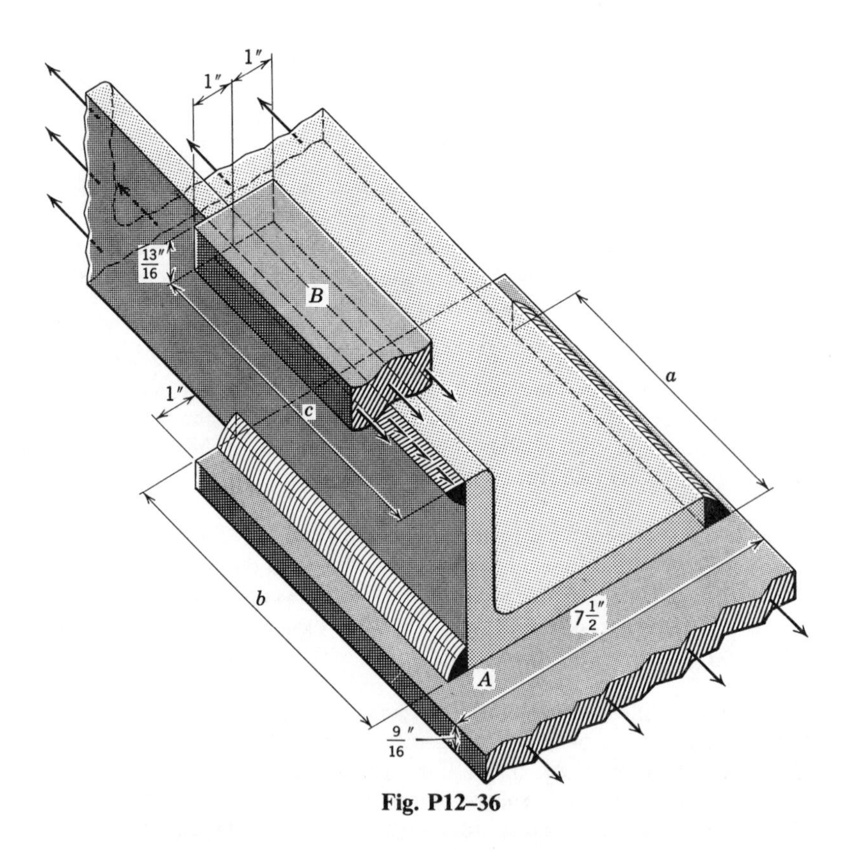

Fig. P12–36

12–37* A cylindrical pressure tank is fabricated by butt welding 12-mm plate with a spiral seam, as shown in Fig. P12–37. The pressure in the tank is 1400 kPa, and an axial load of 21.5π kN is applied to the end of the tank through a rigid bearing plate. Determine the normal and shearing stresses in the weld on the plane of the weld.

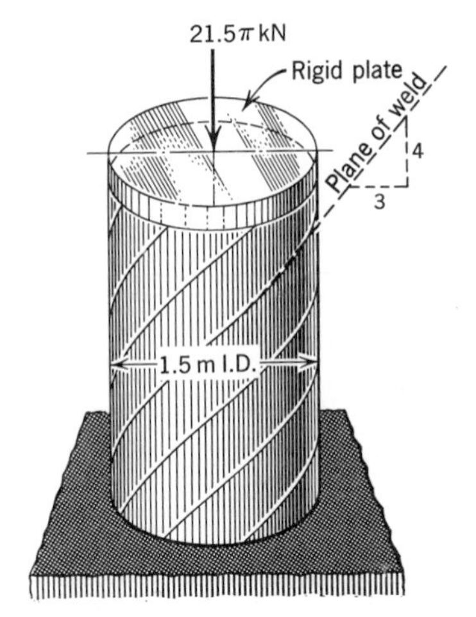

Fig. P12–37

12–38 A steel pipe 40 in. in diameter has a welded spiral seam, as shown in Fig. P12–38. The pressure inside the pipe is 100 psi. Determine the normal and shearing forces that must be carried by each linear inch of the weld. Assume that there is no longitudinal stress in the pipe.

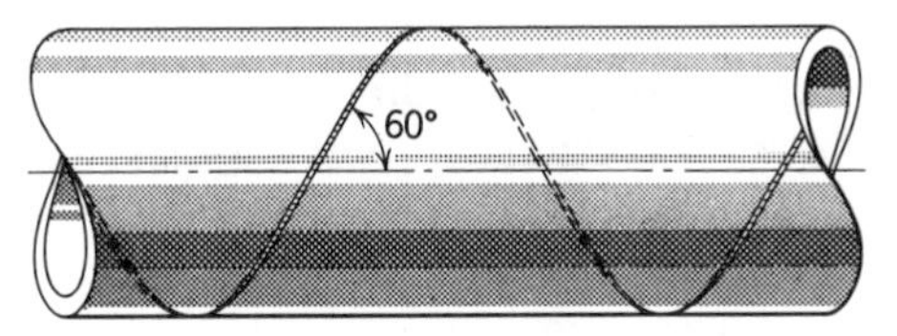

Fig. P12–38

12–39* The flanges of the steel cantilever beam AB of Fig. P12–39 are to be connected to the column with 1×14-in. butt welds, and the web connection is to be $\frac{3}{8} \times 2$-in. intermittent fillet welds on both sides of the web. Recall that the 75-kip load at B may be replaced with a vertical force at A and a couple. Assume that the couple is resisted only by the flange welds and that the vertical force is resisted only by the web welds, and determine:

(a) The average tensile stress in the flange weld (on the vertical plane).

(b) The number of $\frac{3}{8}\times 2$-in. welds to be used on each side of the web (use AISC specifications).

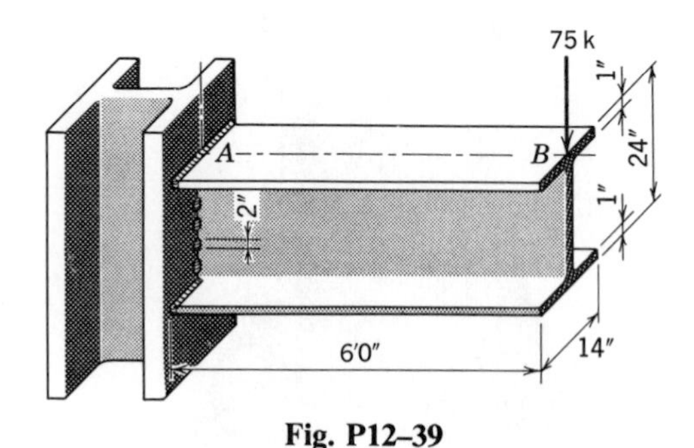

Fig. P12–39

12–40 Prove that when a structural angle is to be welded to a plate in such a manner that the maximum tensile strength of the angle is developed, the use of a transverse end weld has the effect of shortening each of the side welds by equal amounts.

12–5 CONNECTIONS UNDER TORSIONAL LOADING

Power-transmitting shafts are frequently connected by flange couplings (see, for example, Fig. 4–2), and the design of the coupling is, of course, important. The flange coupling is not the only device called on to transmit a torque. Many structural connections are subjected to torsional loading, often combined with centric loading. The latter case is taken up in the next article. This article is limited to pure torsion.

When a torque is applied to the plate C of Fig. 12–16a that is riveted to another structural element, the plate will rotate (a slight amount for elastic action) about some point such as O in the figure. If the assumption is made that any straight line in the plate (such as OB) before the torque is applied remains straight after application of the torque, the deformations of the rivets, and hence the average cross shearing strain in each rivet, will be proportional to the distance from O. If, in addition, stress and strain are proportional, the rivet cross shearing stress will be proportional to the distance from O. This means that for rivets (or bolts) of the same diameter the force on each rivet is proportional to the distance from O and is in the direction of the strain, that is, perpendicular to the radius from O.

The equations of equilibrium can be used to locate the point of rotation O. The resultant of the resisting forces $F_1, F_2, \cdots$ in Fig. 12–16b is a

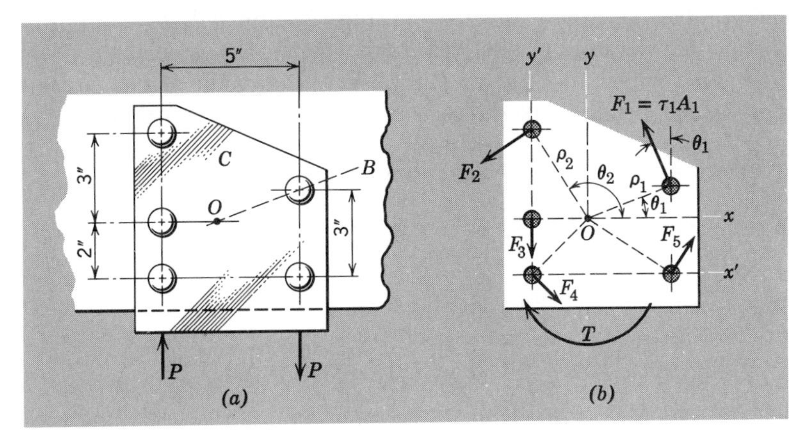

Fig. 12–16

couple; therefore $\Sigma_{i=1}^{n} F_i$ is equal to zero in any direction, where F_i is one of the n resisting forces. The summation of forces in the y direction gives

$$\Sigma F_y = F_1 \cos\theta_1 + F_2 \cos\theta_2 + \cdots = \sum_{i=1}^{n} F_i \cos\theta_i = 0$$

Also,

$$F_i = \tau_i A_i = (\tau_i/\rho_i)(\rho_i A_i) = k\rho_i A_i$$

because the cross shearing stress is proportional to the radius ρ. The summation thus becomes

$$\Sigma F_y = k \sum_{i=1}^{n} \rho_i \cos\theta_i A_i = k \sum_{i=1}^{n} x_i A_i = k\bar{x}A_t = 0$$

where $\bar{x}$ is the x coordinate of the centroid of the rivet areas and A_t is the sum of the rivet areas.

This expression proves that $\bar{x}$ is zero because neither k nor A_t can be zero. The summation of forces in the x direction will indicate that $\bar{y}$ must also be zero, which means that the center of rotation of the plate and the centroid of the rivet areas coincide. The following is an example of a riveted joint loaded in pure torsion.

Example 12–6 The plates and rivets of Fig. 12–16*a* are of structural steel, the rivets are all $\frac{1}{2}$ in. in diameter, and the applied torque is 2000 ft-lb. Determine the magnitude of the maximum cross shearing stress in the rivets.

Solution The free-body diagram of Fig. 12–16*b* will be used for this problem. The centroid of the rivet areas, which is the same as the center of rotation, is located by applying the principle of moments of areas with respect to the x' and y' axes. The coordinates of the centroid are

$$\bar{x}=2 \text{ in.} \quad \text{and} \quad \bar{y}=2 \text{ in.}$$

The summation of moments of forces with respect to the O axis gives

$$\circlearrowleft + \quad \Sigma M_0=\sqrt{10}\,F_1+\sqrt{13}\,F_2+2F_3 + \sqrt{8}\,F_4+\sqrt{13}\,F_5-2000(12)=0$$

Since the rivet areas are all equal, the forces on the rivets are proportional to the radial distances from O; thus,

$$\frac{F_1}{\sqrt{10}}=\frac{F_2}{\sqrt{13}}=\cdots$$

The maximum forces are F_2 and F_5; if the forces are written in terms of F_2, they become

$$F_1=\frac{\sqrt{10}\,F_2}{\sqrt{13}}, \quad F_3=\frac{2F_2}{\sqrt{13}}, \quad F_4=\frac{\sqrt{8}\,F_2}{\sqrt{13}}, \quad F_5=F_2$$

When these values are substituted in the moment equation, it becomes

$$\sqrt{10}\left(\frac{\sqrt{10}\,F_2}{\sqrt{13}}\right)+\sqrt{13}\,F_2+2\left(\frac{2F_2}{\sqrt{13}}\right)+\sqrt{8}\left(\frac{\sqrt{8}\,F_2}{\sqrt{13}}\right)+\sqrt{13}\,(F_2)$$

$$=24{,}000$$

from which $F_2=1803$ lb. The maximum cross shearing stress is

$$\tau_{\max}=\frac{F_2}{A}=\frac{1803}{(\pi/4)(1/2)^2}=\underline{9180 \text{ psi}} \qquad \text{Ans.}$$

PROBLEMS

12–41* A hollow circular shaft is subjected to a torque of 7500 ft-lb. The inside diameter is 3 in. and the outside diameter is 4 in. Determine:

(a) The maximum shearing stress in the shaft.
(b) The number of $\frac{3}{4}$-in. bolts needed for a coupling on a bolt circle of 6-in. diameter for a cross shearing stress in the bolts of 10,000 psi.

12–42* Four equal-sized bolts, placed as shown in Fig. P12–42, are used to transmit a 15 kN-m torque from one machine part to another. Determine:
(a) The magnitude of the force on bolt A.
(b) The minimum required diameter of the bolts if the allowable cross shearing stress is 100 MN/m^2.

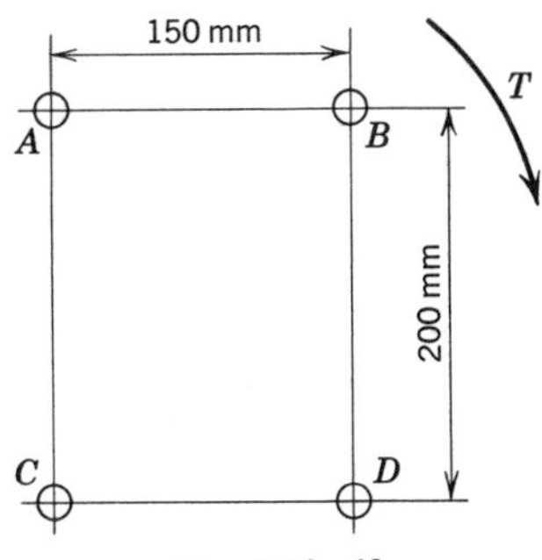

Fig. P12–42

12–43* Six bolts, placed as shown in Fig. P12–43, are used to transmit a torque of 40,000 ft-lb from one plate to another. Determine:
(a) The magnitude of the force on bolt A.
(b) The magnitude of the force on bolt B.

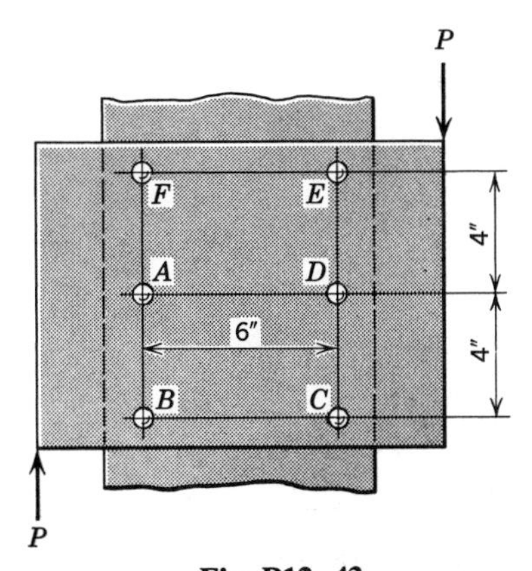

Fig. P12–43

12–44* A shaft coupling has six 12-mm diameter bolts on a 100-mm diameter bolt circle. Determine the maximum power the coupling can transmit at 300 rpm if the cross shearing stress in the bolts is limited to 50 MN/m^2.

12–45* The riveted structural connection in Fig. P12–45 is subjected to a torque as shown. All rivets are the same size.

(a) Which rivet transmits the largest force?
(b) Determine the magnitude of the force of part (a).

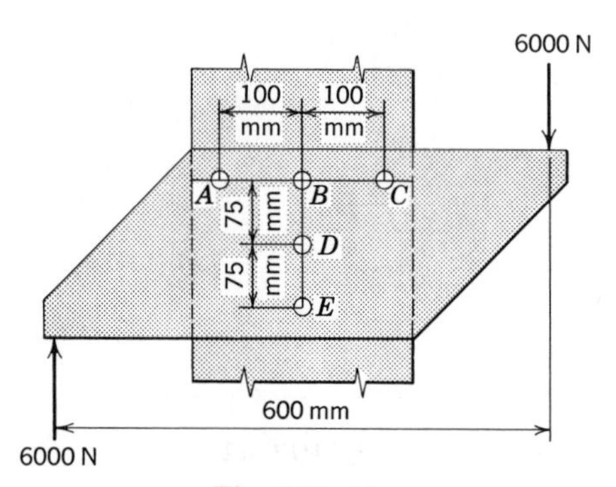

Fig. P12–45

12–46 A riveted structural connection is subjected to the torque shown in Fig. P12–46. All rivets are the same size.

(a) Which rivet transmits the largest force?
(b) Determine the magnitude of the force of part (a).

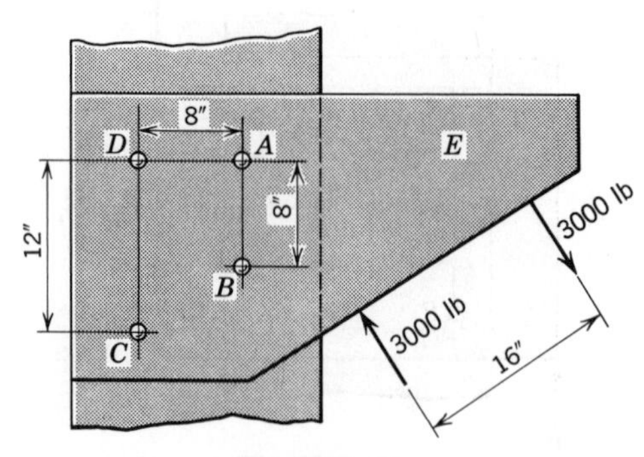

Fig. P12–46

12–47 A bolted structural connection has the dimensions shown in Fig. P12–45. Bolts A, B, and C are 25 mm in diameter, and bolts D and E are each 20 mm in diameter. The allowable shearing stress in the bolts is 100 MN/m^2. Determine the allowable moment that can be applied to the joint as shown.

12–48 Two plates are riveted together as shown in Fig. P12–48. The couple produced by the forces P ($P = 4000$ lb) on one plate is transmitted to the other plate by the nine $\frac{1}{2}$-in. diameter rivets. Determine the cross shearing stress in each rivet.

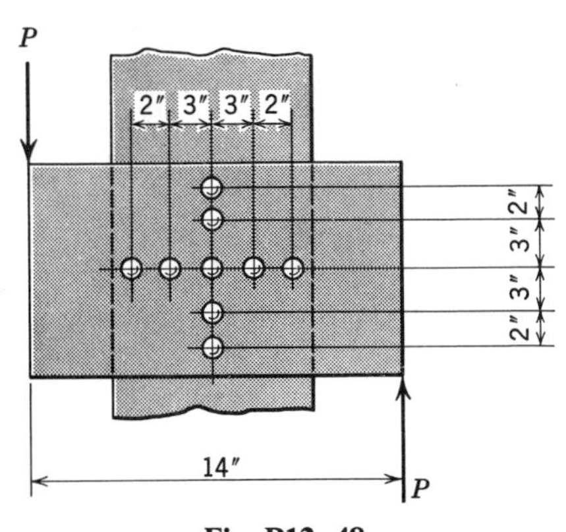

Fig. P12–48

12–49* A motor supplies the torque Q to the driving pulley B, which is mounted on the bronze shaft of Fig. P12–49. Pulleys A and C are for power takeoff with 200 hp being taken off at C and the remainder at A. The allowable shearing stress in the hollow steel segment AB is 15 ksi. The angle of twist at A with respect to C is not to exceed 0.01 rad counterclockwise looking from A toward C. The bronze ($G = 6000$ ksi) and steel segments are rigidly connected by a rigid flange coupling with a bolt circle of 8-in. diameter. When the shaft is rotating at 200 rpm, determine:

(a) The maximum permissible power that can be supplied to pulley B.

(b) The number of $\frac{3}{4}$-in. diameter bolts required for the flange coupling with an allowable cross shearing stress of 10 ksi in the bolts.

12–50* A hollow steel shaft has diameters of 150 mm and 100 mm and an allowable shearing stress of 65 MN/m^2. Determine:

(a) The power transmitted at 25 rad per s.

(b) The magnitude of the angle of twist in a length of 3 m.

(c) The required number of 20-mm diameter bolts for a flange coupling in which two-thirds of the bolts are on a 300-mm diameter bolt circle and one-third are on a 200-mm diameter circle. The allowable cross shearing stress in the bolts is 85 MN/m^2.

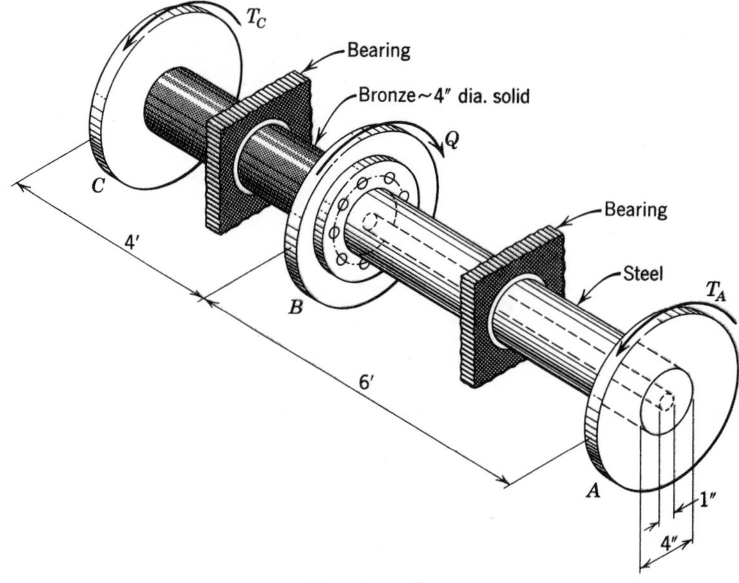

Fig. P12–49

12–6 ECCENTRIC LOAD ON A RIVETED OR BOLTED JOINT

A common and important combination of torsional and centric loads occurs when a riveted or bolted joint is subjected to an eccentric load. The presentation in Arts. 12–1 through 12–3 was limited to centric loads on riveted or bolted joints. The present discussion will be limited to eccentric loads in the plane of the shear areas. In Art. 12–5 bolted or riveted connections subjected to torsional loads were analyzed. The proof that a riveted joint rotates or tends to rotate about the centroid of the rivet areas when the joint is subjected to a couple was given in Art. 12–5.

Any load on a riveted or bolted joint that does not act through the centroid of the rivet areas is defined as an eccentric load. Fortunately, any eccentric load can be replaced by a centric load and a couple in the plane of the shear areas. Furthermore, the force acting on any given shear area can be readily obtained by superposition of the component forces due to the torsional and centric loads. The following example illustrates the procedure.

Example 12–7 Determine the cross shearing stress in the rivet of the joint

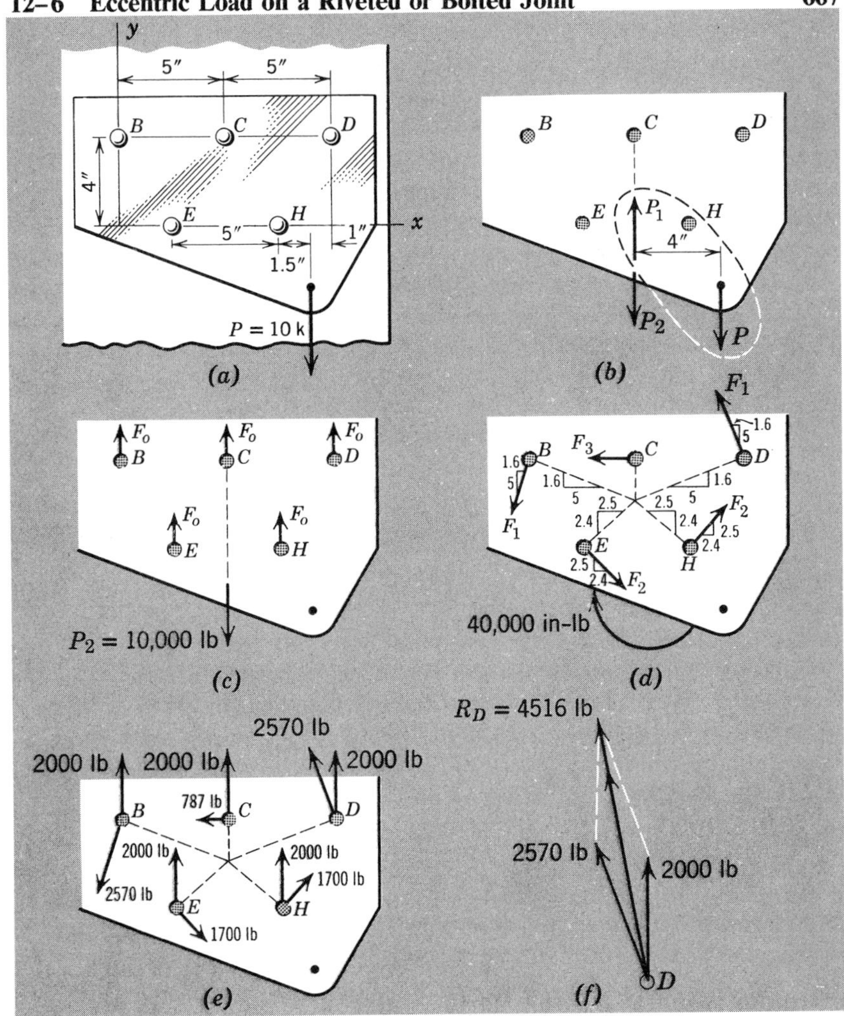

Fig. 12–17

of Fig. 12–17a carrying the greatest load. The diameter of each rivet is $\frac{3}{4}$ in.

Solution By symmetry the x coordinate of the centroid of the rivet areas is $\bar{x} = 5.00$ in. By the principal of moments the y coordinate of the centroid of the rivet areas is

$$\bar{y} = \frac{\Sigma M_x}{\Sigma A} = \frac{(3A)4}{5A} = 2.40 \text{ in.}$$

where A is the cross-sectional area of one rivet.

Two equal opposite centric loads P (labeled P_1 and P_2) are introduced as shown in Fig. 12–17*b*. By this means the eccentric load P is replaced by a centric load P_2 of 10,000 lb and a clockwise couple P_1P of 40,000 in-lb, as shown in Fig. 12–17*b*. An upward resisting force of F_O, equal to 2000 lb, is required on each shear area, as shown in Fig. 12–17*c*, to hold the centric load P_2 in equilibrium. The five resisting forces shown in Fig. 12–17*d* are required to hold the 40,000-in-lb couple PP_1 in equilibrium. As shown in Art. 12–5,

$$\frac{F_1}{\sqrt{27.56}} = \frac{F_2}{\sqrt{12.01}} = \frac{F_3}{1.6}$$

and

$$2F_1\sqrt{27.56} + 2F_2\sqrt{12.01} + 1.6F_3 = 40{,}000$$

The simultaneous solution of these equations gives

$$F_1 = 2570 \text{ lb} \quad F_2 = 1700 \text{ lb} \quad \text{and} \quad F_3 = 787 \text{ lb, all as shown}$$

The component forces acting on each rivet as shown in Fig. 12–17*e* reveal that rivet D is carrying the greatest load. The two resisting forces on rivet D are added vectorally to give the resultant resisting force R_D of 4516 lb, as shown in Fig. 12–17*f*. The cross shearing stress on rivet D is

$$\tau = \frac{4516}{\pi(3/8)^2} = \underline{10{,}240 \text{ psi}} \qquad \text{Ans.}$$

PROBLEMS

12–51* A gusset plate carrying a load of 3000 lb is held by three $\frac{1}{2}$-in. diameter bolts, as shown in Fig. P12–51.

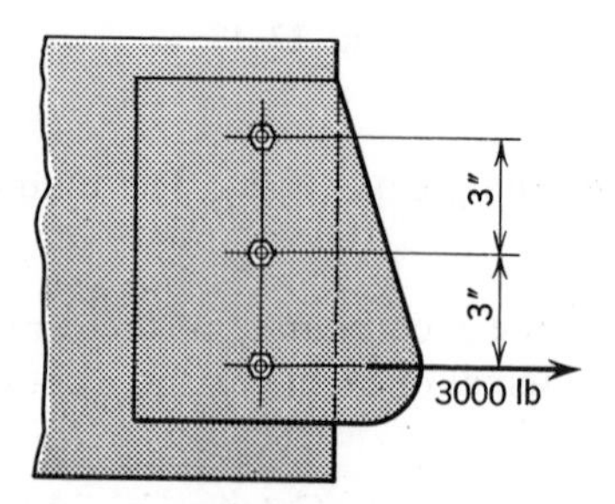

Fig. P12–51

(a) Which bolt carries the greatest load?
(b) Determine the cross shearing stress in the bolt carrying the greatest load.

12–52* Given the bolted connection in Fig. P12–52, determine the cross shearing stress in bolt A. All bolts are 25 mm in diameter.

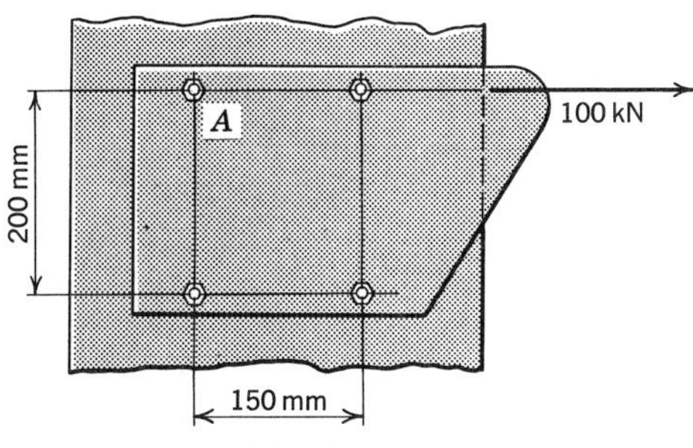

Fig. P12–52

12–53* The diameter of each rivet in the joint of Fig. P12–53 is 1 in. Determine the cross shearing stresses in rivets A and B.

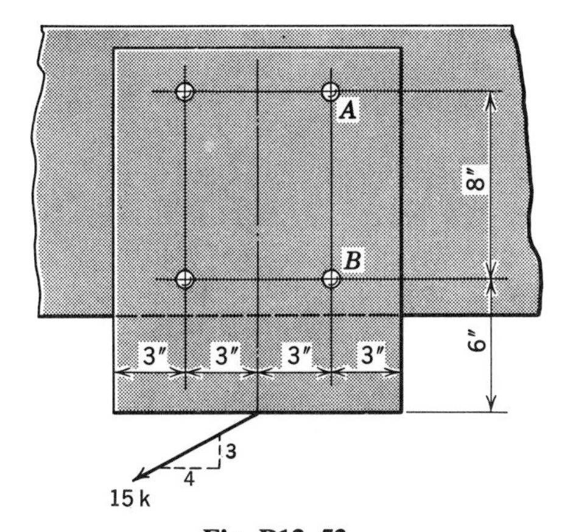

Fig. P12–53

12–54* The bolts in the joint in Fig. P12–54 each have a diameter of 15 mm. Determine the maximum cross shearing stress in any of the bolts.

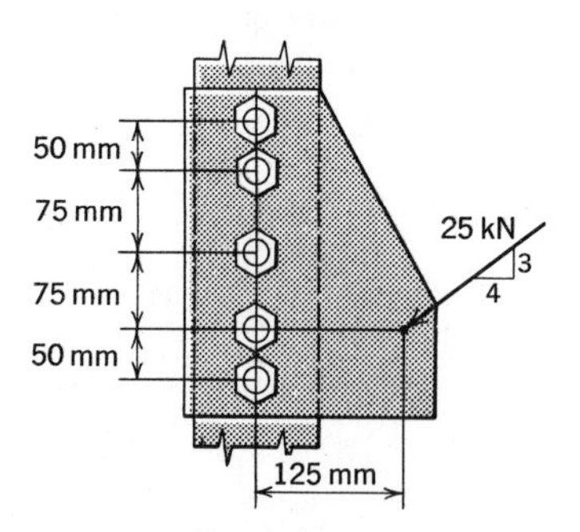

Fig. P12–54

12–55* In the joint of Fig. P12–55 the diameter of each bolt is $\frac{1}{2}$ in. Determine the cross shearing stresses in bolts A and B.

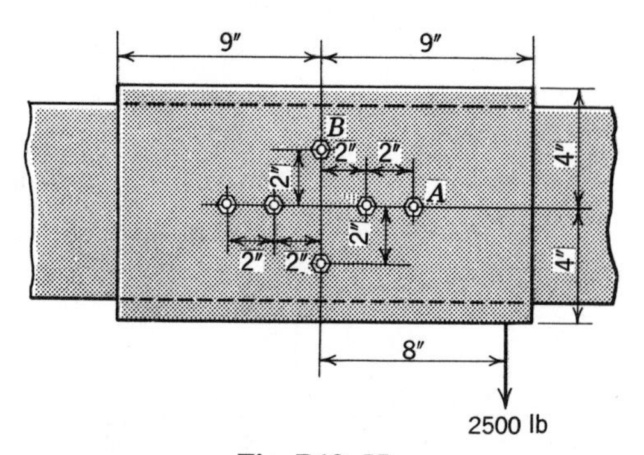

Fig. P12–55

12–56* The diameter of each rivet in the joint in Fig. P12–56 is 20 mm. Determine the cross shearing stresses in the top and bottom rivets.

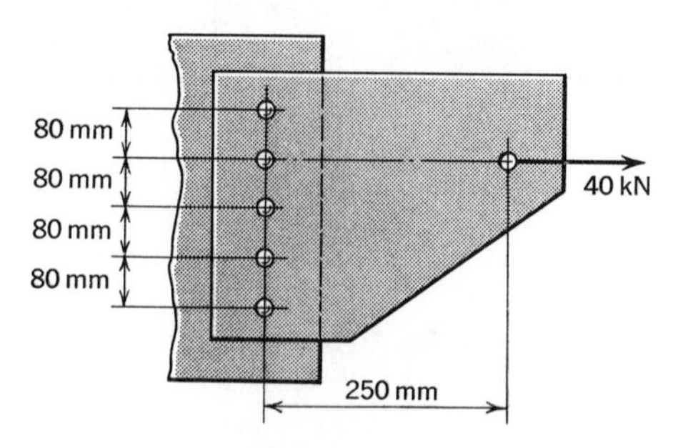

Fig. P12–56

12–57 Each of the rivets in the joint shown in Fig. P12–57 has a diameter of 1 in. Determine the cross shearing stress in the rivet carrying the greatest load.

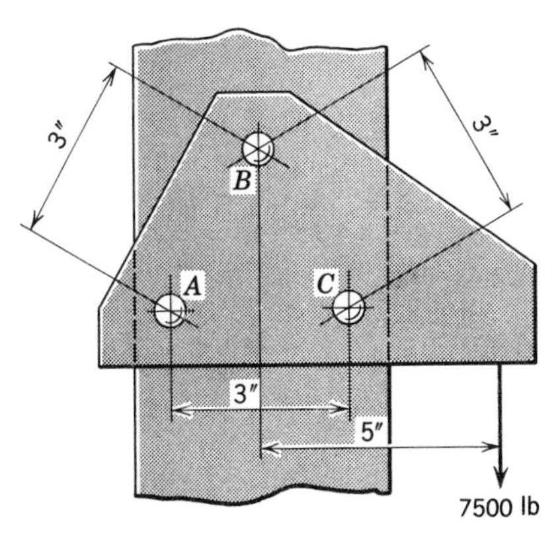

Fig. P12–57

12–58 A load P is applied to the plate riveted to a support, as shown in Fig. P12–58. Determine the maximum value of P if the maximum allowable cross shearing stress in the rivets is 85 MN/m^2. All rivets are 25 mm in diameter.

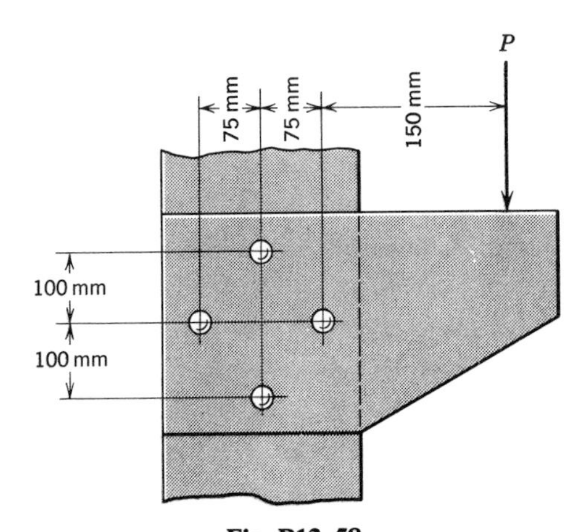

Fig. P12–58

12–59* The bolts in Fig. P12–59 are $\frac{1}{2}$ in. in diameter. If the allowable cross shearing stress in the bolts is 10,000 psi, determine the maximum permissible value for the load P.

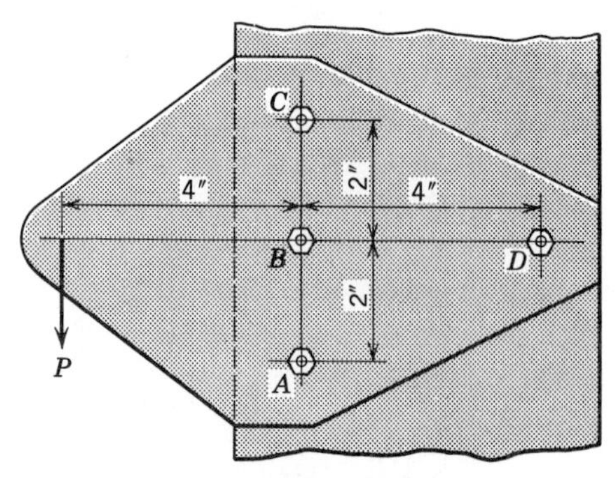

Fig. P12–59

12–60* The bolted joint in Fig. P12–60 is to be designed to support the 15-kN eccentric load as shown. The diameter of the bolts is 15 mm and the allowable cross shearing stress in the bolts is 60 MN/m^2. Determine the minimum permissible distance a.

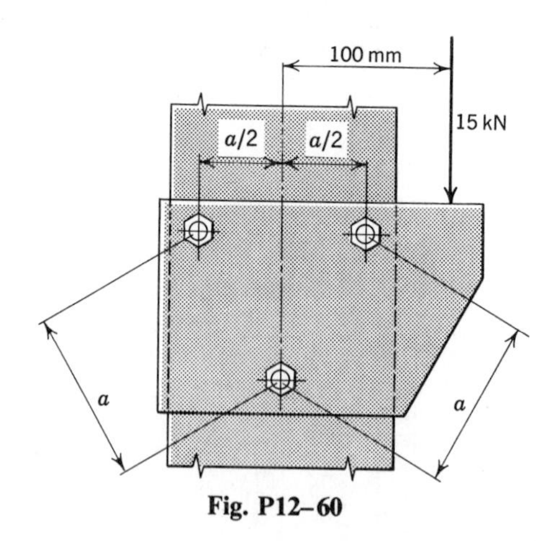

Fig. P12–60

APPENDIX A

1. Average Properties of Selected Materials

2. Stress-Strain Diagrams for Typical Metals

TABLE A–1 Average Properties of Selected Engineering Materials. Exact values may vary widely with changes in composition, heat treatment, and mechanical working. More precise information can be obtained from manufacturers.

Materials	Specific Weight (lb/in.^{3})	Elastic Strength[a]			Ultimate Strength			Endurance Limit[c] (ksi)	Modulus of Elasticity (1000 ksi)	Modulus of Rigidity (1000 ksi)	Percent Elongation in 2 in.	Coefficient of Thermal Expansion (10^{-6}in./in./F)
		Tension (ksi)	Comp. (ksi)	Shear (ksi)	Tension (ksi)	Comp. (ksi)	Shear (ksi)					
Ferrous metals												
Wrought iron	0.278	30	[b]		48	[b]	25	23	28		30[d]	6.7
Structural steel	0.284	36	[b]		66	[b]		28	29	11.0	28[d]	6.6
Steel, 0.2% C hardened	0.284	62	[b]		90	[b]			30	11.6	22	6.6
Steel, 0.4% C hot-rolled	0.284	53	[b]		84	[b]		38	30	11.6	29	
Steel, 0.8% C hot-rolled	0.284	76	[b]		122	[b]			30	11.6	8	
Cast iron—gray	0.260				25	100		12	15		0.5	6.7
Cast iron—malleable	0.266	32	[b]		50	[b]			25		20	6.6
Cast iron—nodular	0.266	70			100				25		4	6.6
Stainless steel (18–8) annealed	0.286	36	[b]		85	[b]		40	28	12.5	55	9.6
Stainless steel (18–8) cold-rolled	0.286	165	[b]		190	[b]		90	28	12.5	8	9.6
Steel, SAE 4340, heat-treated	0.283	132	145		150	[b]	95	76	29	11.0	19	
Nonferrous metal alloys												
Aluminum, cast, 195-T6	0.100	24	25		36		30	7	10.3	3.8	5	
Aluminum, wrought, 2014-T4	0.101	41	41	24	62	[b]	38	18	10.6	4.0	20	12.5
Aluminum, wrought, 2024-T4	0.100	48	48	28	68	[b]	41	18	10.6	4.0	19	12.5
Aluminum, wrought, 6061-T6	0.098	40	40	26	45	[b]	30	13.5	10.0	3.8	17	12.5

Magnesium, extrusion, AZ80X	0.066	35	26		49	[b]	21	19	6.5	2.4	12	14.4
Magnesium, sand cast, AZ63-HT	0.066	14	14		40	[b]	19	14	6.5	2.4	12	14.4
Monel, wrought, hot-rolled	0.319	50	[b]		90	[b]		40	26	9.5	35	7.8
Red brass, cold-rolled	0.316	60			75				15	5.6	4	9.8
Red brass, annealed	0.316	15	[b]		40	[b]			15	5.6	50	9.8
Bronze, cold-rolled	0.320	75			100				15	6.5	3	9.4
Bronze, annealed	0.320	20	[b]		50	[b]			15	6.5	50	9.4
Titanium alloy, annealed	0.167	135	[b]		155	[b]			14	5.3	13	
Invar, annealed	0.292	42	[b]		70	[b]			21	8.1	41	0.6
Nonmetallic materials												
Douglas fir, green[e]	0.022	4.8	3.4			3.9	0.9		1.6			
Douglas fir, air dry[e]	0.020	8.1	6.4			7.4	1.1		1.9			
Red oak, green[e]	0.037	4.4	2.6			3.5	1.2		1.4			1.9
Red oak, air dry[e]	0.025	8.4	4.6			6.9	1.8		1.8			
Concrete, medium strength	0.087		1.2			3.0			3.0		5.0	6.0
Concrete, fairly high str.	0.087		2.0			5.0			4.5			6.0

[a]Elastic strength may be represented by proportional limit, yield point, or yield strength at a specified offset (usually 0.2 percent for ductile metals).

[b]For ductile metals (those with an appreciable ultimate elongation) it is customary to assume the properties in compression have the same values as those in tension.

[c]Rotating beam.

[d]Elongation in 8 in.

[e]All timber properties are parallel to the grain.

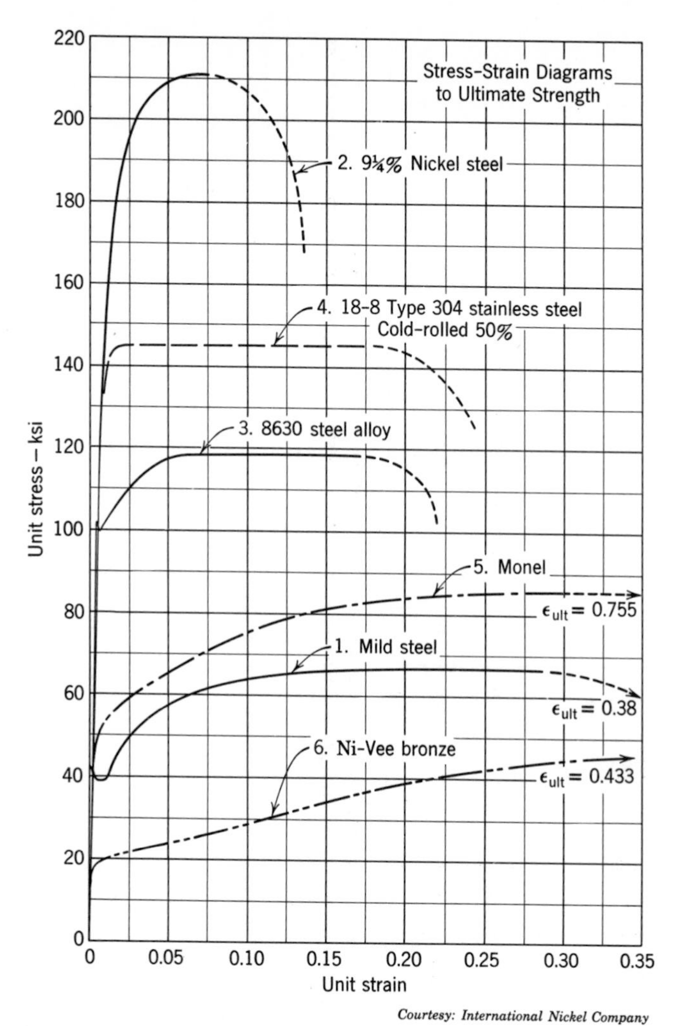

Stress-strain diagrams to ultimate strength. Courtesy: International Nickel Company.

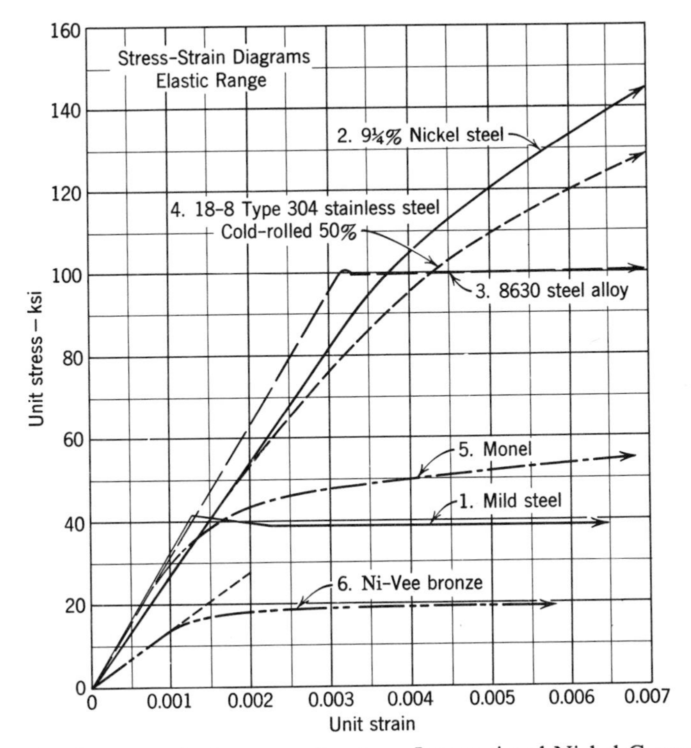

Stress-strain diagrams, elastic range. Courtesy: International Nickel Company.

APPENDIX B

1. Properties of Selected Areas

2. Second Moment of Plane Areas

TABLE B–1 Properties of Selected Areas

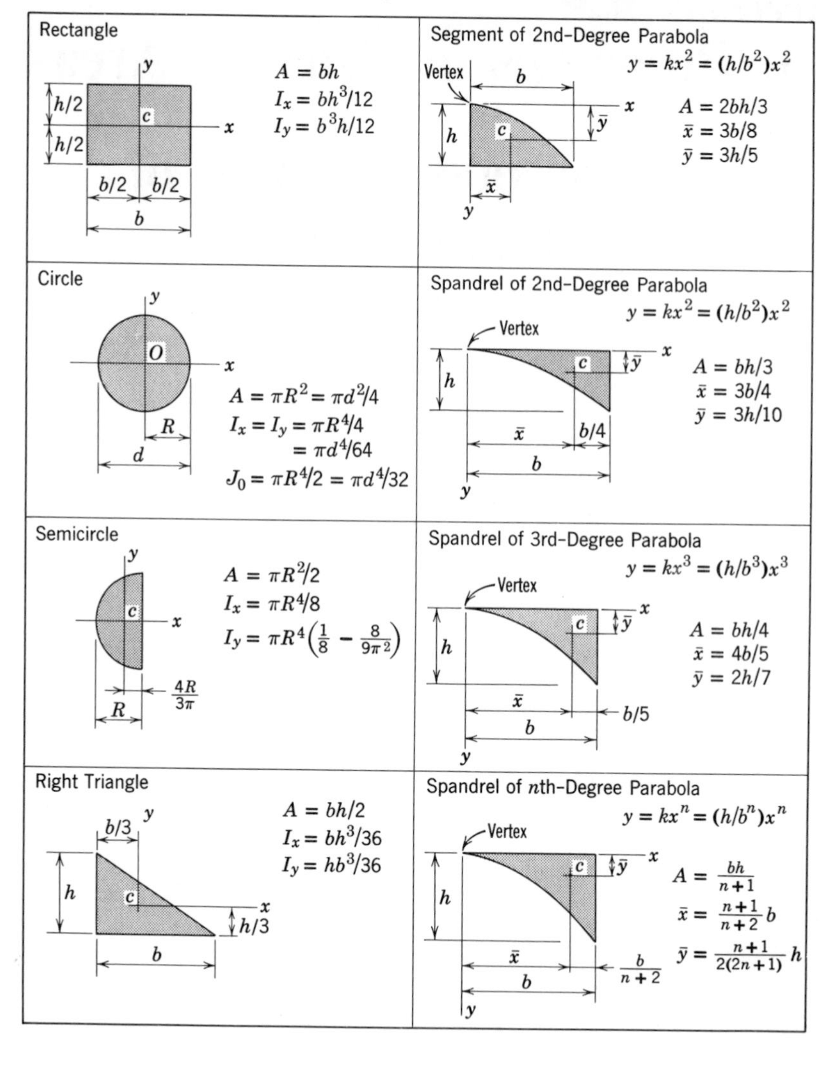

SECOND MOMENT OF PLANE AREAS

B–1 DEFINITIONS. The second moment of the area in Fig. B–1 with respect to the x axis is defined as the integral $\int_A y^2\,dA$ where the integration is carried out over the area A. The quantity is called the second moment of the area to distinguish it from the first moment about the axis, which is defined as $\int_A y\,dA$. The second moment is frequently referred to as *moment of inertia* of the area because of the similarity of the defining equation to that used to express the moment of inertia of the mass of a body and the terms *second moment* and *moment of inertia* are used interchangeably.

From the definition in the preceding paragraph, it may be noted that a similar equation can be written for the second moment of the area with respect to any axis in the plane of the area. The symbol I is used for the second moment with respect to any axis in the plane of the area and is sometimes called the rectangular moment of inertia, thus

$$I_x = \int_A y^2\,dA \qquad I_y = \int_A x^2\,dA \tag{B–1}$$

When the reference axis is normal to the plane of the area, for example through O in Fig. B–1, the integral is called the polar second moment or polar moment of inertia J_O and can be written as

$$J_O = \int_A r^2\,dA = \int_A (x^2+y^2)\,dA = \int_A x^2\,dA + \int_A y^2\,dA$$

Thus,

$$J_O = I_y + I_x \tag{B–2}$$

Note that the x and y axes can be any two mutually perpendicular axes intersecting at O.

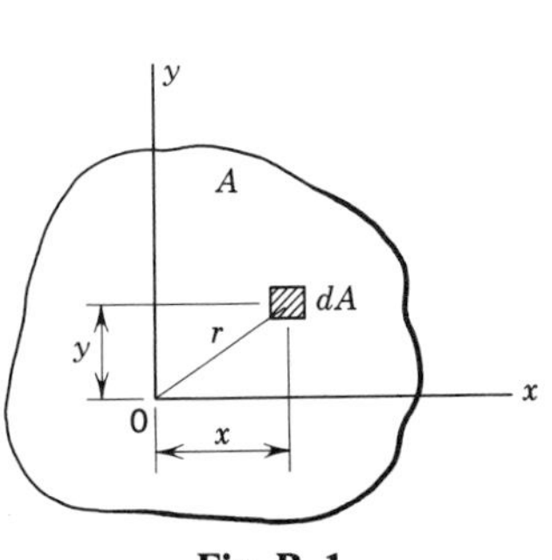

Fig. B–1

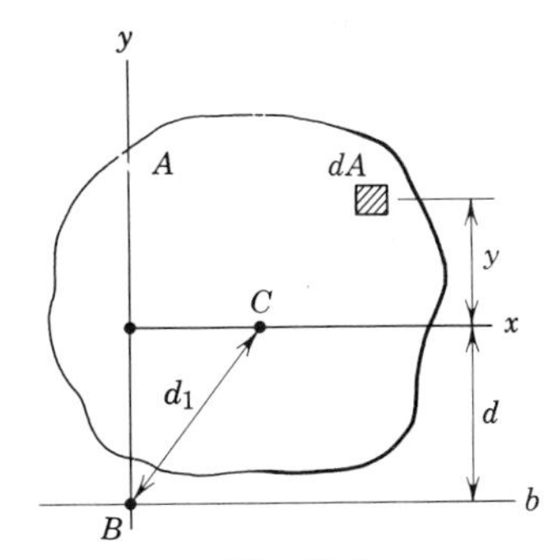

Fig. B–2

From the definition, the second moment of an area is the product of a distance squared multiplied by an area and thus is always positive and has the dimension of a length raised to the fourth power (L^4). Common units are in.^{4} and cm^4.

B–2 PARALLEL AXIS THEOREM. When the second moment of an area has been determined with respect to a given axis, the second moment with respect to a parallel axis can be obtained by means of the *parallel axis theorem* (sometimes called the transfer formula), provided one of the axes passes through the centroid of the area.

The second moment of the area in Fig. B–2 about the b axis is

$$I_b = \int_A (y+d)^2 dA = \int_A y^2 dA + 2d\int_A y\,dA + d^2\int_A dA$$

$$= I_x + 2d\int_A y\,dA + Ad^2 \tag{a}$$

The integral $\int_A y\,dA$ is the first moment of the area with respect to the x axis. If the x axis passes through the centroid of the area, the first moment is zero and Eq. (a) becomes

$$I_b = I_c + Ad^2 \tag{B–3}$$

where I_c is the second moment of the area with respect to an axis parallel to b and passing through the centroid C, and d is the distance between the two axes. In a similar manner it can be shown that

$$J_B = J_c + Ad_1^2 \tag{B–4}$$

The parallel-axis theorem states that the *second moment of an area with respect to any axis is equal to the second moment of the area with respect to a parallel axis through the centroid of the area added to the product of the area and the square of the distance between the two axes*. The theorem demonstrates that the second moment of an area with respect to an axis through the centroid of the area is less than that for any parallel axis. Thus, solving Eq. B–3 for I_c gives

$$I_c = I_b - Ad^2$$

B–3 SECOND MOMENTS OF AREAS BY INTEGRATION. When determining the second moment of an area by integration, it is first necessary to select an element of area. If all parts of the element are the same distance from the axis, the second moment can be determined directly from the definition (see Art. B–1). When any other element is selected, the second moment of the *element* with respect to the axis must either be known or be obtainable from a known result by the parallel-axis theorem.

Either single or double integration may be involved, depending on the element selected. When double integration is used, all parts of the element will be the same distance from the moment axis, and the second moment of the element can be written directly. Special care must be taken in establishing the limits for the two integrations to see that the correct area is included. If a strip element is selected, the second moment can usually be obtained by a single integration, but the element must be properly selected in order for its second moment about the reference axis to be either known or readily calculated. Either cartesian or polar coordinates can be used for some problems. The choice of a coordinate system and the selection of an element is a matter of either personal preference or previous experience. The following examples illustrate the procedure for determining the second moments of areas by integration.

Example B–1 A rectangle has a base b and a height h. Determine the second moment of the area with respect to:

(a) The base of the rectangle.
(b) An axis through the centroid parallel to the base.
(c) An axis through the centroid normal to the area (polar second moment).

Solution (a) The rectangular area is shown in Fig. B–3, together with an element of area parallel to the reference axis. The second moment of the area of the element about the x axis is

$$dI_x = y^2\,dA = y^2 b\,dy$$

and the second moment of the entire area is

$$I_x = b\int_0^h y^2\,dy = \underline{\frac{bh^3}{3}} \qquad \text{Ans.}$$

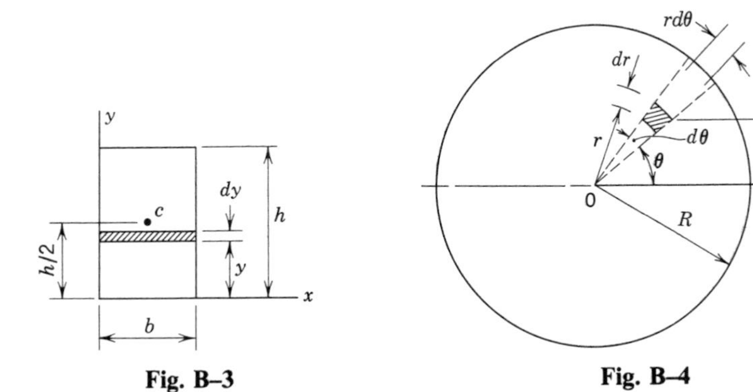

Fig. B–3

Fig. B–4

(b) The parallel-axis theorem can be used to determine the second moment about an axis through C parallel to the x axis. Thus

$$I_{xc} = I_x - Ad^2 = \frac{bh^3}{3} - (bh)\left(\frac{h}{2}\right)^2 = \underline{\frac{bh^3}{12}} \qquad \text{Ans.}$$

(c) The second moment with respect to the vertical axis through C can be obtained from the preceding solution by interchanging b and h; that is,

$$I_{yc} = \frac{hb^3}{12}$$

and the polar second moment of the area about an axis at C is

$$J_c = I_{xc} + I_{yc} = \frac{bh^3}{12} + \frac{hb^3}{12} = \underline{\frac{bh}{12}(h^2 + b^2)} \qquad \text{Ans.}$$

Example B–2 Determine the second moment of a circular area with a radius R with respect to a diameter of the circle.

Solution Figure B–4 shows a circular area with a radius R. The x axis is the moment axis. Polar coordinates are convenient for this problem. The second moment of the element about the x axis is

$$dI_x = (r \sin\theta)^2 dr (r\,d\theta) = r^3 \sin^2\theta\, dr\, d\theta$$

The second moment of the entire area is

$$I_x = \int_0^{2\pi}\int_0^R r^3 \sin^2\theta\, dr\, d\theta = \int_0^{2\pi} \frac{R^4}{4} \sin^2\theta\, d\theta$$

$$= \frac{R^4}{4}\int_0^{2\pi} \frac{1 - \cos 2\theta}{2} d\theta = \frac{R^4}{4}\left[\frac{\theta}{2} - \frac{\sin 2\theta}{4}\right]_0^{2\pi} = \underline{\frac{\pi R^4}{4}} \qquad \text{Ans.}$$

Example B–3 The equation of the parabolic curve in Fig. B–5 is $cy = x^2$ where c is a constant. Determine the second moment of the shaded area with respect to the x and y axes.

Solution The constant in the equation can be obtained by substituting $x = b$ and $y = h$ in the equation, which gives $c = b^2/h$ and the equation becomes

$$b^2 y = hx^2$$

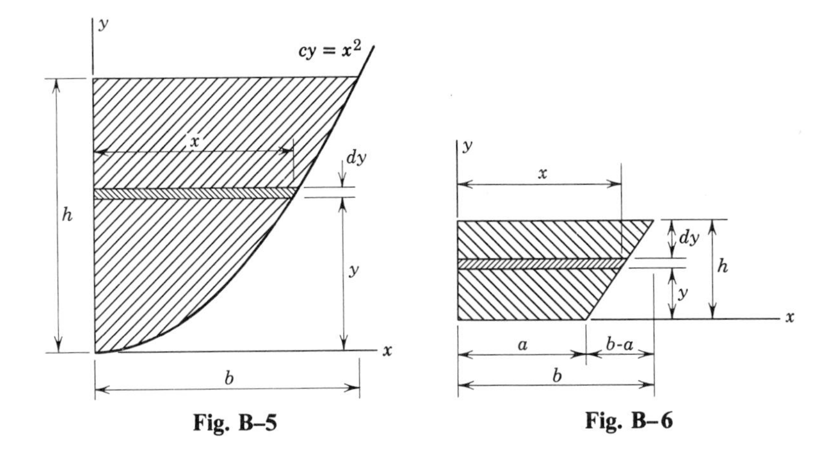

Fig. B–5 **Fig. B–6**

A horizontal element (Fig. B–5) is selected for this example. The area of the element is

$$dA = x\,dy = bh^{-1/2}y^{1/2}\,dy$$

and the second moment of the element about the x axis is

$$dI_x = y^2\,dA = bh^{-1/2}y^{5/2}\,dy$$

The second moment of the area is

$$I_x = bh^{-1/2}\int_0^h y^{5/2}dy = bh^{-1/2}\left[\frac{2}{7}y^{7/2}\right]_0^h = \underline{\frac{2}{7}bh^3} \qquad \text{Ans.}$$

Since all parts of the element are *not* the same distance from the y axis, the horizontal element can be used only if its second moment with respect to the y axis is known. The element is a rectangle of height x and width dy and the reference axis is the base of the rectangle. From Exam. B–1, the second moment of the element about the y axis is

$$dI_y = \frac{1}{3}(dy)x^3 = \frac{1}{3}\left(\sqrt{\frac{b^2y}{h}}\right)^3 dy = \frac{b^3y^{3/2}}{3h^{3/2}}\,dy$$

and the second moment of the area is

$$I_y = \frac{b^3}{3h^{3/2}} \int_0^h y^{3/2} dy = \frac{2}{15} b^3 h \qquad \text{Ans.}$$

Example B–4 Determine the second moment of the area in Fig. B–6 with respect to the x axis.

Solution An element of area parallel to the x axis is selected as shown in Fig. B–6. The area of the element is $dA = x\,dy$ where

$$x = a + \frac{b-a}{h} y$$

and the second moment of the element with respect to the x axis is

$$dI_x = y^2\,dA = \left(ay^2 + \frac{b-a}{h} y^3 \right) dy$$

The second moment of the area about the x axis is

$$I_x = \int_0^h \left(ay^2 + \frac{b-a}{h} y^3 \right) dy = \frac{(a+3b)h^3}{12} \qquad \text{Ans.}$$

PROBLEMS

B–1* Determine the polar second moment of the circular area in Exam. B–2 with respect to an axis at the origin O.

B–2* Solve Exam B–3 using a vertical element of area.

B–3* Determine the second moment of the triangular area in Fig. PB–3 with respect to:

(a) The x axis.

(b) An axis through the centroid of the triangle parallel to the x axis.

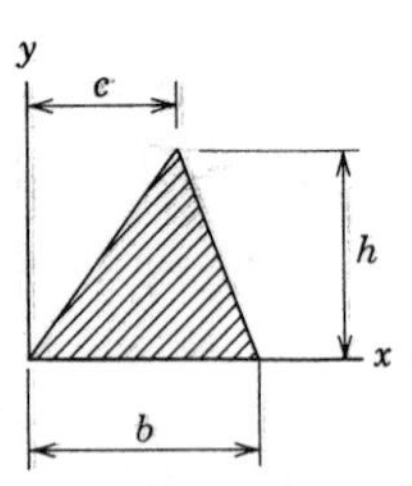

Fig. PB–3

B–4*

(a) Determine the polar second moment of the shaded area in Fig. PB–4 with respect to an axis at O.

(b) Determine the second moment of the area in part (a) with respect to the y axis.

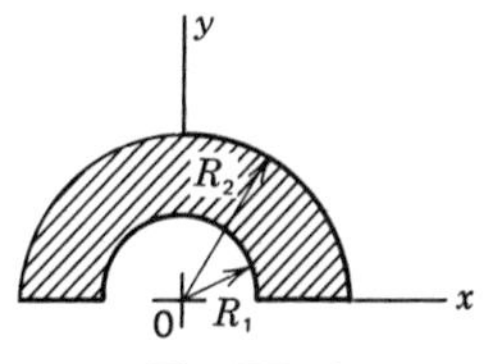

Fig. PB–4

B–5* Determine the second moments of the shaded area in Fig. PB–5 with respect to the x and y axes.

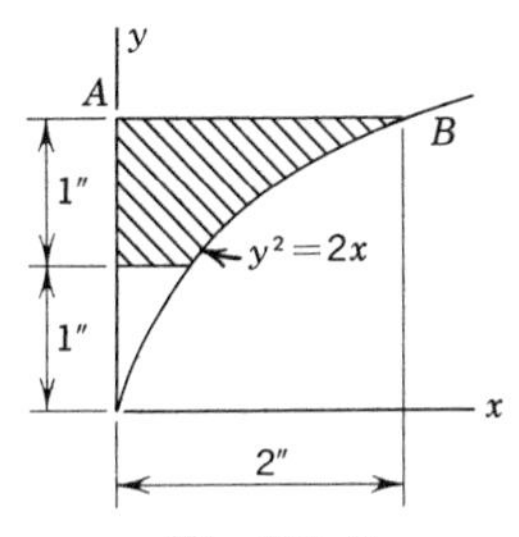

Fig. PB–5

B–6* Determine the second moment of the shaded area in Prob. B–5 with respect to the line AB.

B–7* Determine the second moment of the shaded area in Fig. PB–7 with respect to the x axis. The equation of the curve is $y^2 = 2bx$.

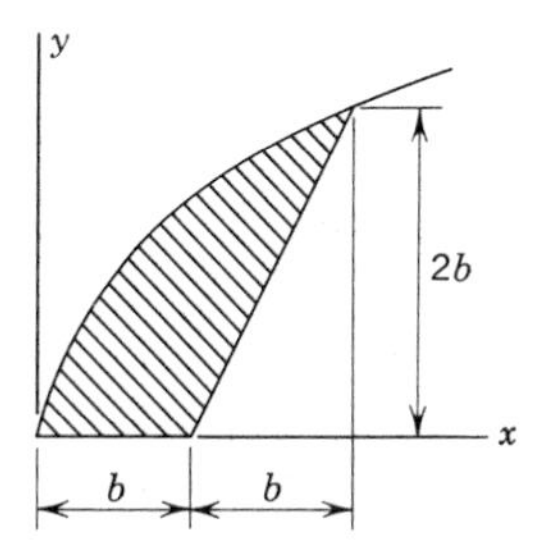

Fig. PB–7

B–8* Determine the polar second moment of the shaded area in Fig. PB–8 about an axis at the origin. The equation of the curve is $x^{1/2}+y^{1/2}=a^{1/2}$.

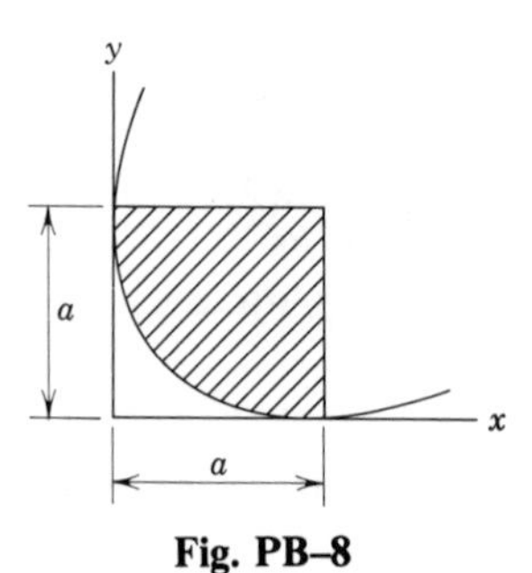

Fig. PB–8

B–9* The equation of the curve in Fig. PB–9 is $ax^2=y^3$. Determine the second moment of the shaded area with respect to (1) the x axis and (2) the y axis.

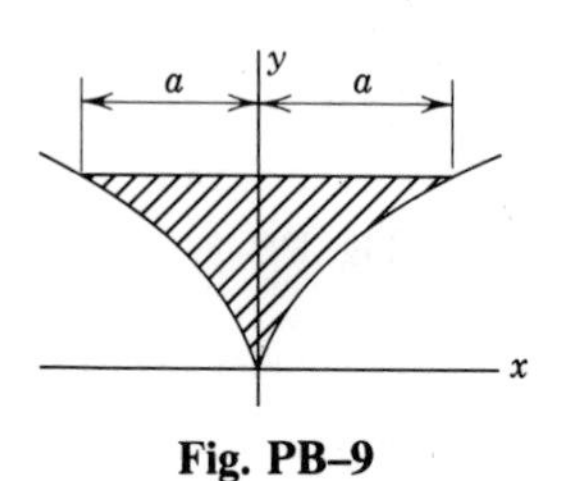

Fig. PB–9

B–10* The equation of the parabola in Fig. PB–10 is $b^2y=h(b^2-x^2)$. Determine the second moment of the area with respect to (1) the x axis and (2) the y axis.

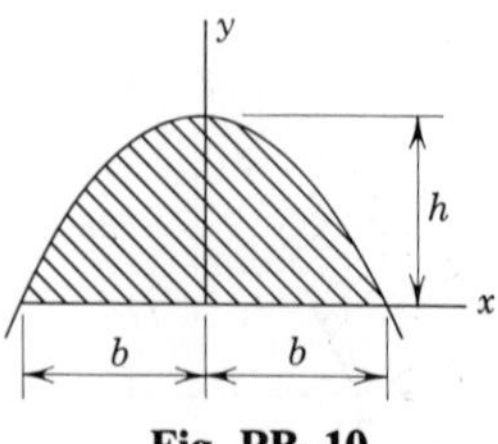

Fig. PB–10

B–4 RADIUS OF GYRATION OF AREAS. It is sometimes convenient to express the second moment of area (moment of inertia) as a function of the area and a length. Since the second moment of an area has the dimensions of length to the fourth power, it can be expressed as the product of the area multiplied by a length squared. *The radius of gyration of an area with respect to a specified axis is defined as the length that when squared and multiplied by the area, will give the second moment of the area with respect to the specified axis.* This definition can be expressed by the equations

$$Ak_b^2 = I_b \quad \text{and} \quad Ak_O^2 = J_O \tag{B–5}$$

where k_b and k_O are the radii of gyration with respect to the b axis and the polar axis through O respectively. The radius of gyration is *not* the distance from the reference axis to a fixed point in the area (such as the centroid), but it is a useful property of the area and the specified axis.

The radius of gyration of an area with respect to any axis is always greater than the distance from the axis to the centroid of the area. From Fig. B–7, the moment of inertia of the area with respect to the b axis is

$$I_b = I_c + Ad^2 = Ak_b^2 = Ak_c^2 + Ad^2$$

or

$$k_b^2 = k_c^2 + d^2$$

from which

$$k_b = \left(k_c^2 + d^2\right)^{1/2}$$

This equation indicates that k_b will always be greater than the distance d from the b axis to the centroid. If the distance d is increased, the value of k_b will also be increased, but the increase in k_b will be less than the increase in d indicating that k_b is *not* the distance from the b axis to a fixed point in the area.

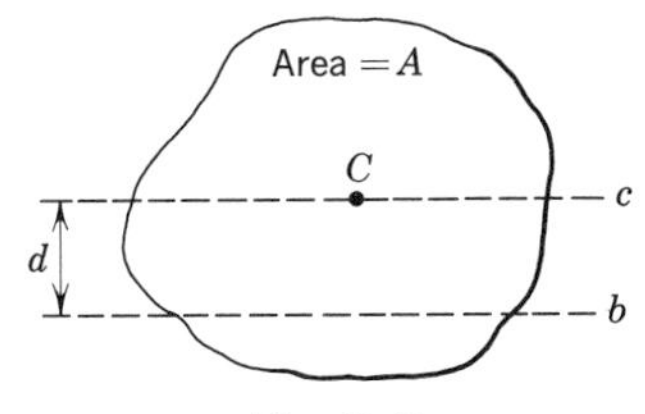

Fig. B–7

The radius of gyration of an area is a convenient means of expressing a property of an area and is often useful in formulas developed in mechanics—for example, formulas giving the strength of columns.

B–5 SECOND MOMENTS OF COMPOSITE AREAS. It is often necessary to calculate the second moment of an irregularly shaped area. When such an area is made up of a series of simple shapes such as rectangles, triangles, and circles, the second moment can be determined by use of the parallel-axis theorem rather than by integration. The second moment of the irregular area, the *composite area*, with respect to any axis is equal to the sum of the second moments of the separate parts of the area with respect to the specified axis. When an area such as a hole is removed from a larger area, its second moment must be subtracted from the second moment of the larger area to obtain the resulting second moment.

It is useful to have formulas available giving second moments of areas frequently encountered in routine work. Table B–1 lists properties of several geometric shapes frequently encountered. Tables listing second moments of the cross-sectional areas of various structural shapes are found in engineering handbooks and in data books prepared by industrial organizations such as the American Institute of Steel Construction and the Aluminum Company of America. Properties of a few selected shapes are included in App. C for use in solving problems.

Example B–5 Determine the second moment of the shaded area in Fig. B–8 with respect to the x axis.

Solution The shaded area can be divided into a 10×6-in. rectangle A and a 10×6-in. triangle B with a 4-in. radius semicircle C removed. The location of the centroids of the three areas, as determined from Table B–1, is shown to the right of the figure. The second moments of the three areas with respect to the x axis can be obtained, by using information from Table B–1, as follows:
For area A

$$I_x = \frac{bh^3}{3} = \frac{10(6)^3}{3} = 720 \text{ in.}^4$$

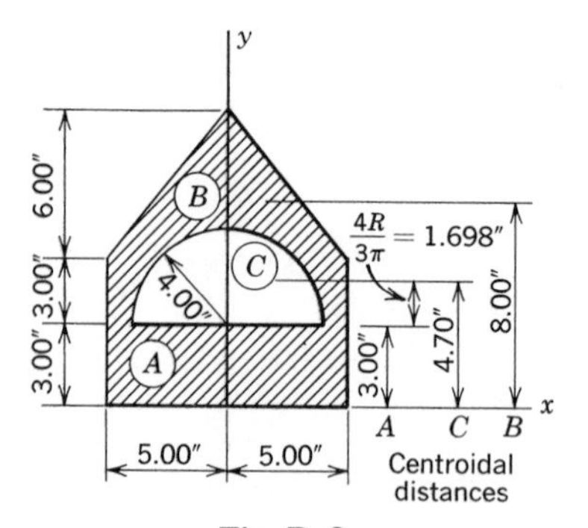

Fig. B–8

For area B

$$I_x = I_c + Ad^2 = \frac{bh^3}{36} + Ad^2$$

$$= \frac{10(6)^3}{36} + \frac{10(6)}{2}(8)^2 = 1980 \text{ in.}^4$$

For area C—although I_c for the semicircle is given in Table B–1, the development of the expression here will expand the coverage of the subject matter. Since the area is half of a circle, its second moment with respect to a diameter is half that of a full circle, and from the parallel-axis theorem

$$I_c = I_{\text{dia}} - Ad^2 = \frac{1}{2}\frac{\pi R^4}{4} - \frac{\pi R^2}{2}\left(\frac{4R}{3\pi}\right)^2$$

$$= \frac{\pi(4)^4}{8} - \frac{8(4)^4}{9\pi} = 28.10 \text{ in.}^4$$

and

$$I_x = I_c + Ad^2 = 28.10 + \frac{\pi(4)^2}{2}(4.698)^2 = 582.8 \text{ in.}^4$$

The second moment of the composite area is

$$I_x = 720 + 1980 - 583 = \underline{2117 \text{ in.}^4} \qquad \text{Ans.}$$

Example B–6 A column (cross section in Fig. B–9) is constructed of a W10×77 wide flange section and a C15×33.9 standard channel. Determine the second moments and radii of gyration of the cross-sectional area with respect to horizontal and vertical axes through the centroid of the section.

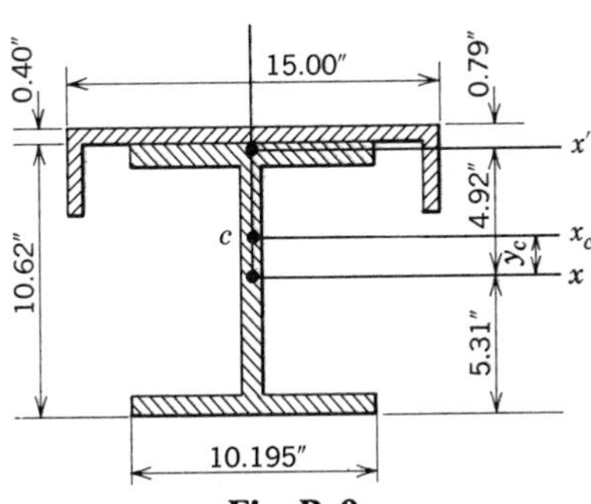

Fig. B–9

Solution The significant dimensions of the section are shown on the figure. Centroidal distances are shown to the right. Properties of the structural shapes can be obtained from App. C. In Fig. B–9, the x axis passes through the centroid of the wide-flange section and the x' axis passes through the centroid of the channel. The x_c axis is located using the principle of moments as applied to areas. The total area is

$$A = A_{WF} + A_C = 22.67 + 9.90 = 32.57 \text{ in.}^2$$

and the moment of the area about the x axis is

$$M_x = \Sigma A y_c = 22.67(0) + 9.90(4.92) = 32.57 y_c$$

from which the distance from the x axis to the centroid, y_c, is 1.50 in. The second moment of each area with respect to the x_c axis is

for the wide-flange section

$$I_{x_c} = I_c + Ad^2 = 457.2 + 22.67(1.50)^2 = 508.2 \text{ in.}^4$$

for the channel

$$I_{x_c} = I_c + Ad^2 = 8.2 + 9.90(4.92 - 1.50)^2 = 124.0 \text{ in.}^4$$

for the composite area

$$I_{x_c} = 508.2 + 124.0 = \underline{632.2 \text{ in.}^4} \qquad \text{Ans.}$$

The radius of gyration about the x_c axis is

$$k_{x_c} = \left(\frac{I}{A}\right)^{1/2} = \left(\frac{632.2}{32.57}\right)^{1/2} = \underline{4.41 \text{ in.}} \qquad \text{Ans.}$$

The y axis passes through the centroid of each of the areas; thus, the parallel axis theorem is not needed to obtain I_y; that is,

$$I_y = 153.4 + 312.6 = \underline{466.0 \text{ in.}^4} \qquad \text{Ans.}$$

and

$$k_y = \left(\frac{I}{A}\right)^{1/2} = \left(\frac{466.0}{32.57}\right)^{1/2} = \underline{3.78 \text{ in.}} \qquad \text{Ans.}$$

PROBLEMS

B–11* Determine the second moment of the shaded area in Fig. PB–11 with respect to the x and y axes through the centroid.

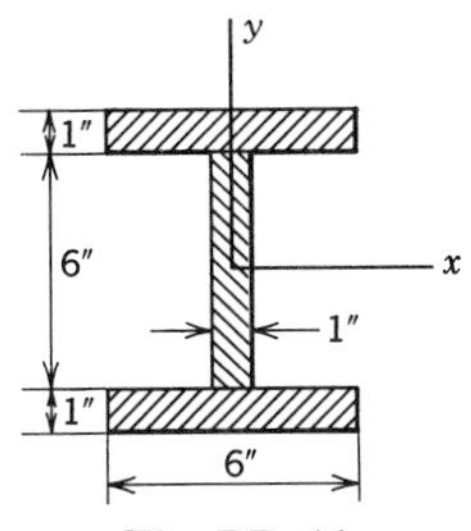

Fig. PB–11

B–12* Determine the polar moment of inertia of the shaded area in Fig. PB–12 with respect to the centroid.

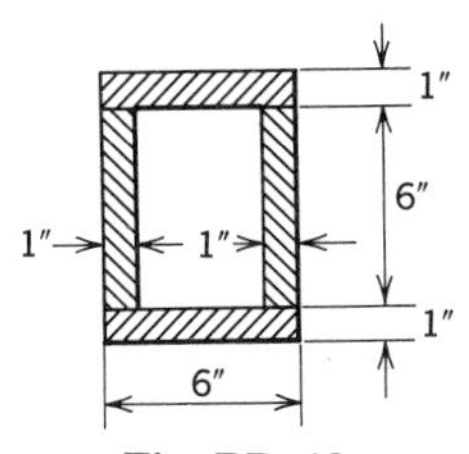

Fig. PB–12

B–13* Determine the moment of inertia of the shaded area in Fig. PB–13 with respect to the horizontal and vertical centroidal axes.

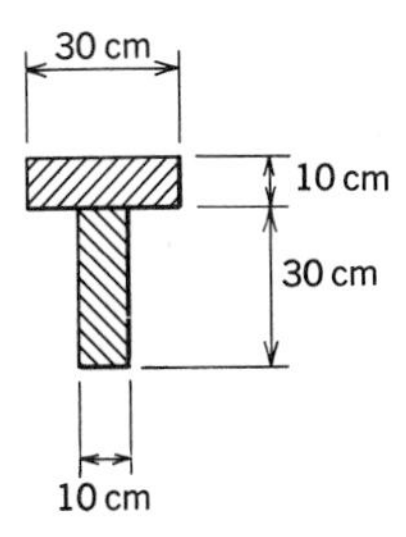

Fig. PB–13

B–14* Determine the moment of inertia of the shaded area in Fig. PB–14 with respect to the horizontal and vertical centroidal axes.

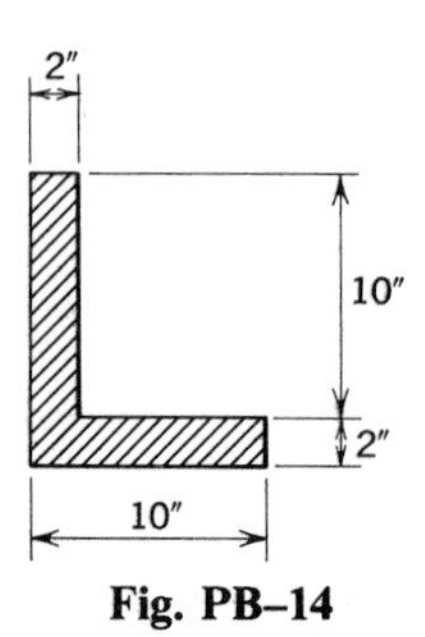

Fig. PB–14

B–15* Determine the polar moment of inertia of the shaded area in Fig. PB–15 with respect to the centroid.

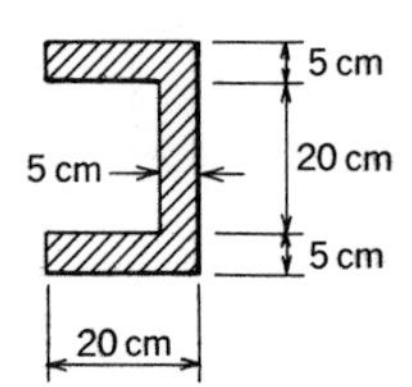

Fig. PB–15

B–16* Determine the radii of gyration of the shaded area in Fig. PB–16 with respect to the x and y axes through the centroid.

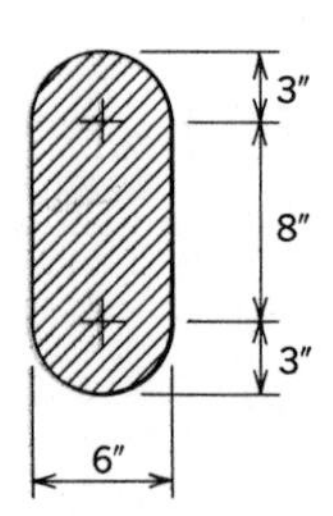

Fig. PB–16

B–17* Determine the polar radius of gyration of the area in Fig. PB–17 with respect to an axis at the centroid of the area.

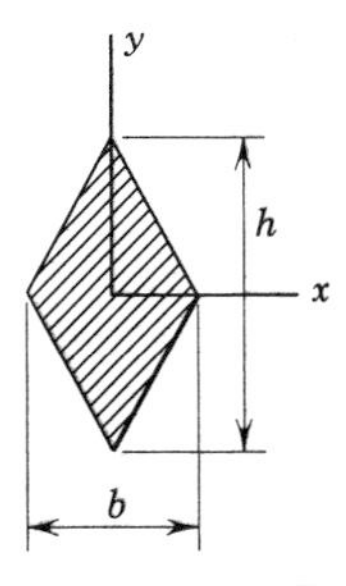

Fig. PB–17

B–18* Determine the radius of gyration of the shaded area in Fig. PB–18 with respect to a horizontal axis through the centroid.

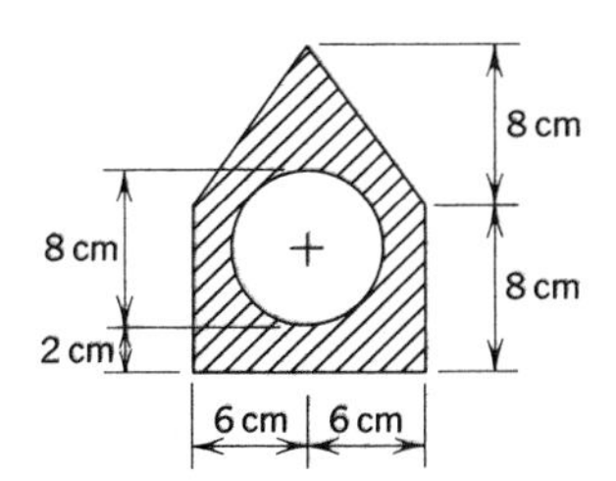

Fig. PB–18

B–19* Determine the radii of gyration of the area in Fig. PB–19 with respect to the horizontal and vertical centroidal axes.

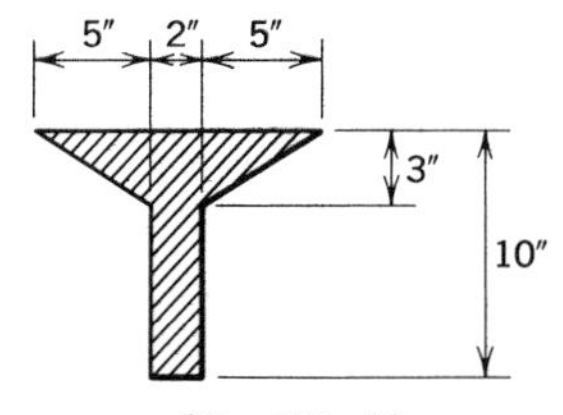

Fig. PB–19

B–20* The structural shape in Fig. PB–20 is made up of a 24×1-in. plate and four $6 \times 6 \times \frac{1}{2}$-in. angles. Determine the centroidal radii of gyration with respect to horizontal and vertical axes.

Fig. PB–20

B–21* Determine the centroidal moments of inertia of the structural section in Fig. PB–21, with respect to horizontal and vertical axes.

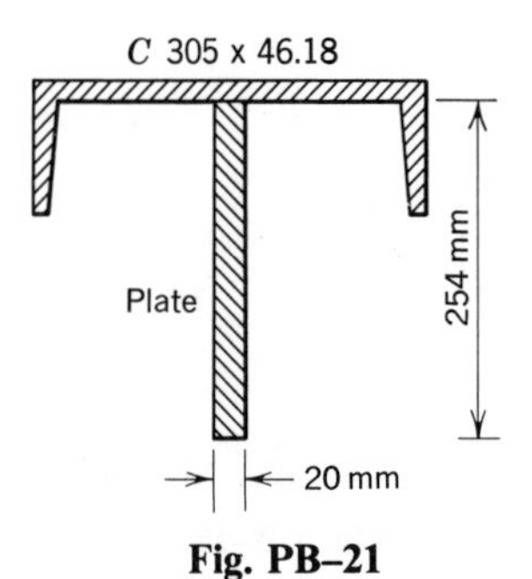

Fig. PB–21

B–22* The section in Fig. PB–22 is made up of a $C12 \times 30$ channel and two $L6 \times 3\frac{1}{2} \times \frac{3}{8}$-in. angles as shown. Determine the second moment of the combined area with respect to the horizontal centroidal axis.

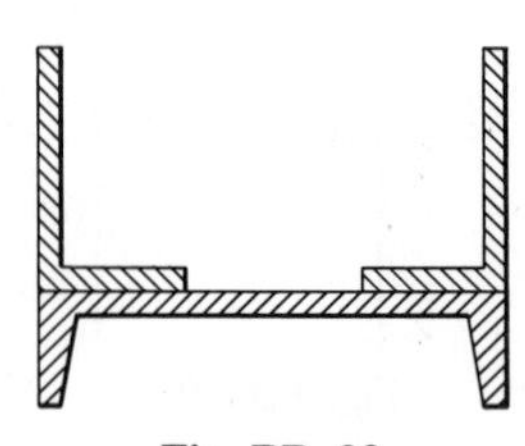

Fig. PB–22

B–23* The section in Fig. PB–23 is made up of a UB610 × 140 universal beam and two L150 × 90 × 15-mm angles. Determine the second moment of the combined area with respect to the horizontal centroidal axis.

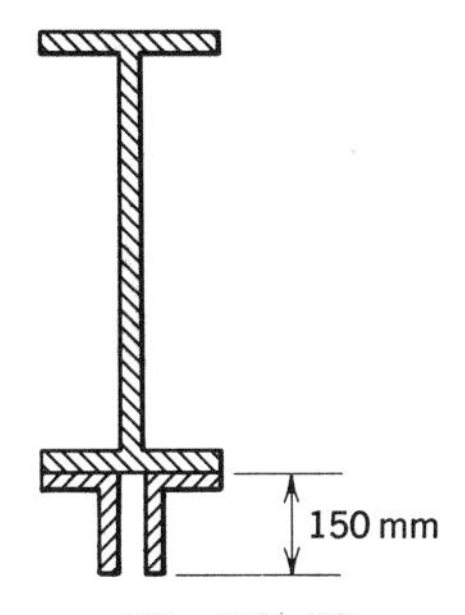

Fig. PB–23

B–24* The eave strut in Fig. PB–24 is made of a C8 × 11.5 channel, an L3 × 3 × $\frac{1}{4}$-in. angle, and an L3 × $2\frac{1}{2}$ × $\frac{5}{16}$-in. angle. Determine the radius of gyration of the area with respect to the horizontal centroidal axis.

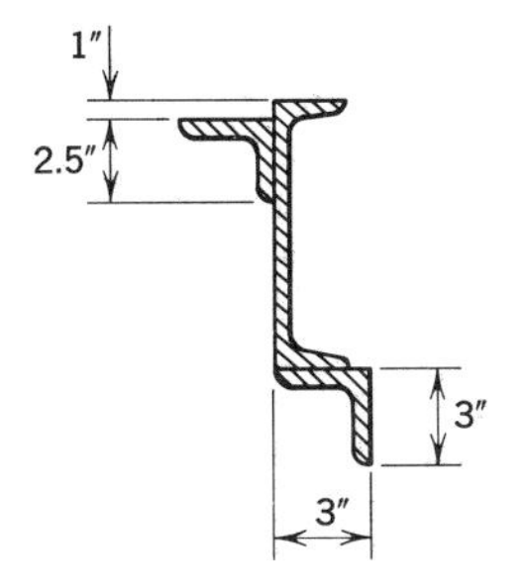

Fig. PB–24

B–6 PRODUCTS OF INERTIA OF AREAS. The product of inertia dI_{xy} of the element of area dA in Fig. B–10 with respect to the x and y axes is defined as the product of the two coordinates of the element multiplied by the area of the element; the product of inertia of the total area A about the x and y axes is the sum of the products of inertia of the elements of the area; thus,

$$I_{xy} = \int_A xy\,dA \qquad \text{(B–6)}$$

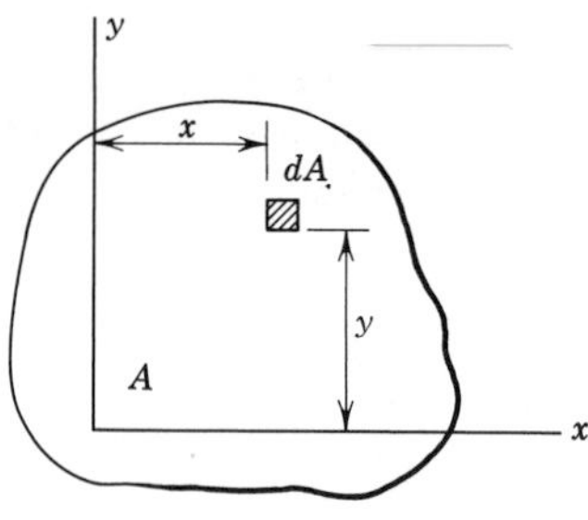

Fig. B–10

The dimensions of the product of inertia are the same as for the moment of inertia (second moment), but since the product xy can be either positive or negative, the product of inertia can be positive, negative, or zero as contrasted with the second moment, which is always positive.

The product of inertia of an area with respect to any two orthogonal axes is zero when either of the axes is an axis of symmetry. This statement can be demonstrated by means of Fig. B–11, which is symmetrical with respect to the x axis. The products of inertia of the elements dA and dA' on opposite sides of the axis of symmetry will be equal in magnitude and opposite in sign and will thus cancel each other in the summation. Therefore, the resulting product of inertia will be zero.

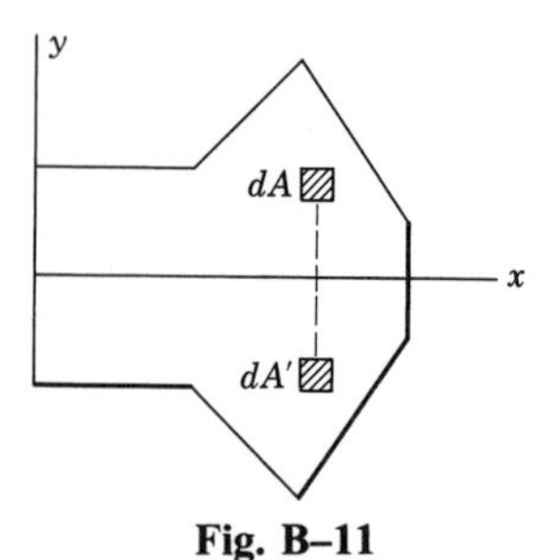

Fig. B–11

The parallel-axis theorem for products of inertia can be derived from Fig. B–12 in which the x' and y' axes pass through the centroid C and are parallel to the x and y axes. The product of inertia with respect to the x and y axes is

$$I_{xy} = \int_A xy\,dA = \int_A (x_c + x')(y_c + y')\,dA$$

$$= x_c y_c \int_A dA + x_c \int_A y'\,dA + y_c \int_A x'\,dA + \int_A x'y'\,dA$$

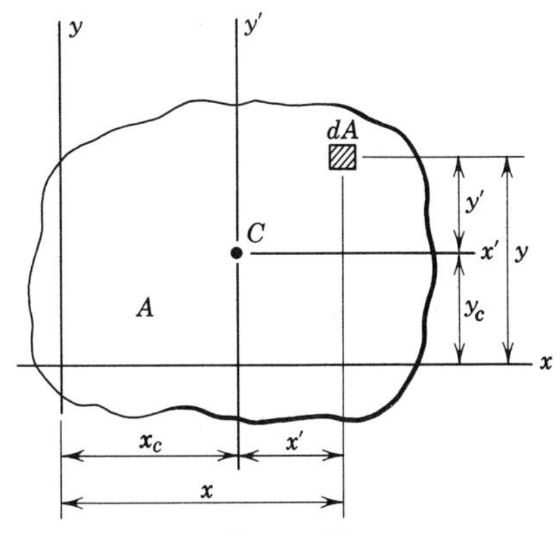

Fig. B–12

The second and third integrals in the preceding equation are zero since x' and y' are centroidal axes and the last integral is the product of inertia with respect to the centroidal axes. Consequently, the product of inertia is

$$I_{xy} = I_{x'y'} + x_c y_c A \tag{B–7}$$

The parallel-axis theorem for products of inertia can be stated as follows: *the product of inertia of an area with respect to any two perpendicular axes x and y is equal to the product of inertia of the area with respect to a pair of centroidal axes parallel to the **x** and **y** axes added to the product of the area and the two centroidal distances from the **x** and **y** axes.*

The product of inertia is useful in determining principal axes of inertia, as discussed in the following article. The determination of the product of inertia is illustrated in the next two examples.

Example B–7 Determine the product of inertia of the triangular area shown in Fig. B–13 with respect to:

(a) The x and y axes.
(b) A pair of centroidal axes parallel to the x and y axes.

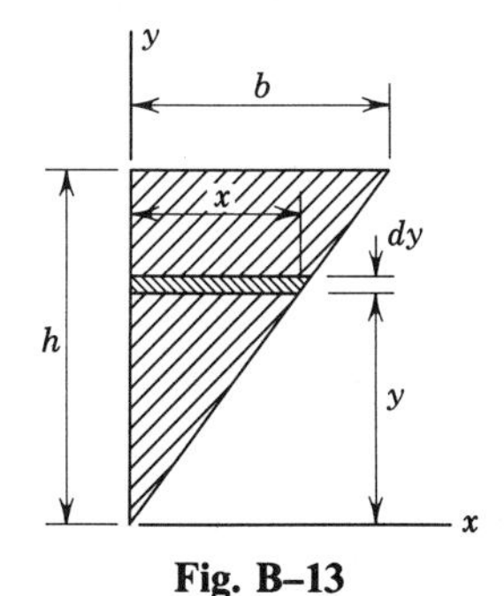

Fig. B–13

Solution (a) The horizontal element in Fig. B–13 has a length x and a height dy. From symmetry, the product of inertia of the element with respect to axes through the centroid of the element parallel to the x and y axes is zero. Thus the product of inertia of the element with respect to the x and y axes is

$$dI_{xy} = \frac{x}{2}\, y\, dA = \frac{x}{2}\, y(x\, dy)$$

From similar triangles, x is equal to $(b/h)y$ and the product of inertia of the element becomes

$$dI_{xy} = \frac{b^2}{2h^2}\, y^3\, dy$$

The product of inertia of the triangular area is

$$I_{xy} = \frac{b^2}{2h^2}\int_0^h y^3\, dy = \underline{\frac{b^2h^2}{8}} \qquad \text{Ans.}$$

(b) The product of inertia with respect to the centroidal axes can be obtained from the parallel-axis theorem; thus,

$$I_{x'y'} = I_{xy} - Ax_c y_c = \frac{b^2h^2}{8} - \frac{bh}{2}\left(\frac{b}{3}\right)\left(\frac{2h}{3}\right) = \underline{\frac{b^2h^2}{72}} \qquad \text{Ans.}$$

Example B–8 Determine the product of inertia of the quadrant of the circle in Fig. B–14 with respect to the x and y axes.

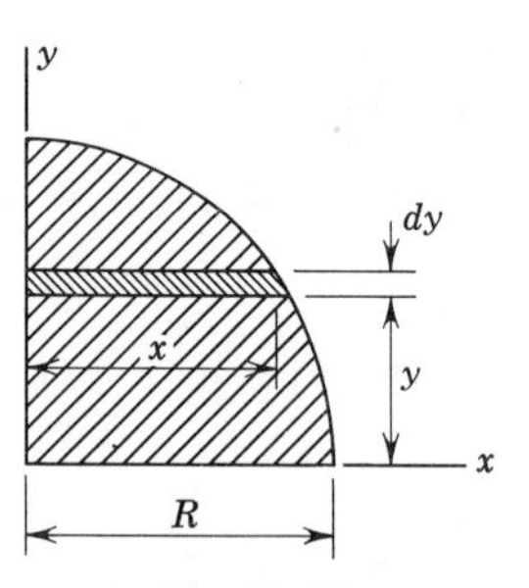

Fig. B–14

Solution The horizontal element selected is shown in Fig. B–14. As in Exam. B–7, the product of inertia of the element with respect to its centroidal axes is zero and its product of inertia with respect to the x and y axes is

$$dI_{xy} = \frac{x}{2}(y)(x\, dy) = \frac{y}{2}x^2\, dy$$

Since $R^2 = x^2 + y^2$, the preceding expression becomes

$$dI_{xy} = \frac{y}{2}(R^2 - y^2)dy$$

and the product of inertia of the area is

$$I_{xy} = \frac{1}{2}\int_0^R (R^2y - y^3)dy = \frac{R^4}{8} \qquad \text{Ans.}$$

PROBLEMS

B–25* Determine by integration the product of inertia of the shaded area in Fig. PB–25 with respect to the x and y axes.

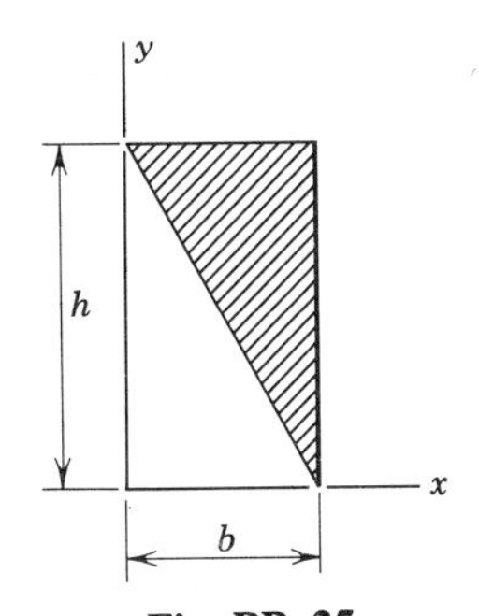

Fig. PB–25

B–26* Determine the product of inertia of the shaded area in Fig. PB–26 with respect to the x and y axes.

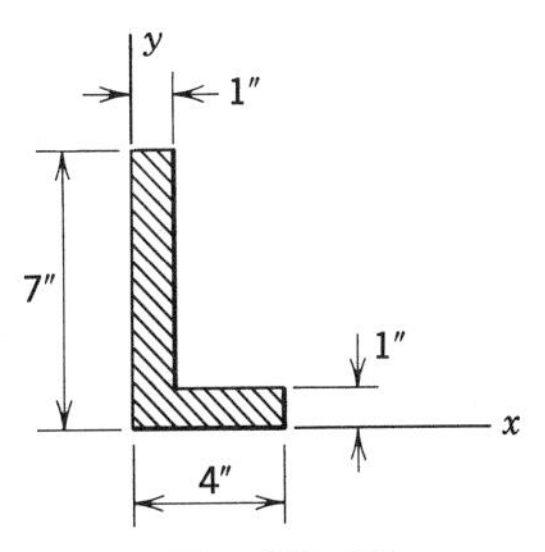

Fig. PB–26

B–27* Determine the product of inertia of the area in Fig. PB–27 with respect to the x and y axes.

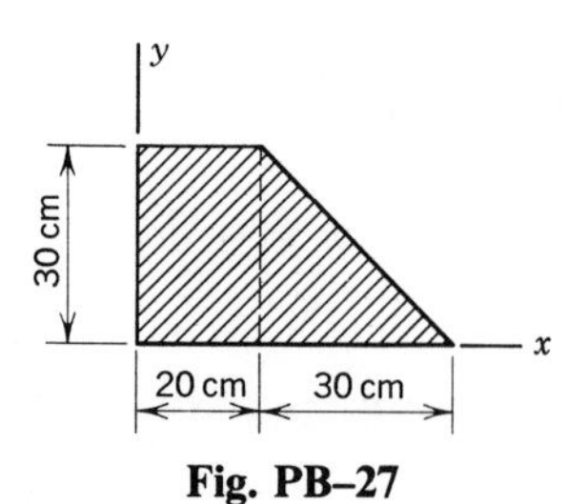

Fig. PB–27

B–28* Determine the product of inertia of the Z section in Fig. PB–28 with respect to horizontal and vertical axes through the centroid.

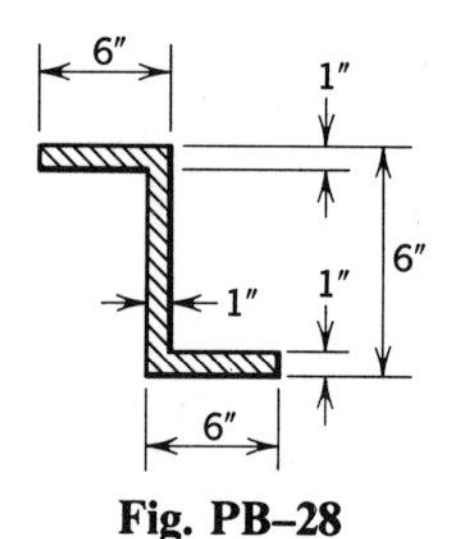

Fig. PB–28

B–29* Determine the product of inertia of the parallelogram in Fig. PB–29 with respect to horizontal and vertical axes through the centroid of the area.

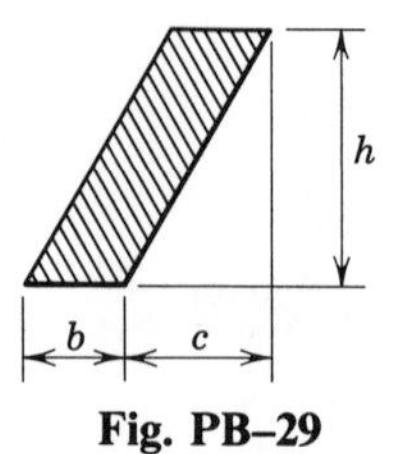

Fig. PB–29

B–30* Determine the product of inertia of the shaded area in Fig. PB–30 with respect to the x and y axes. The curve is a parabola.

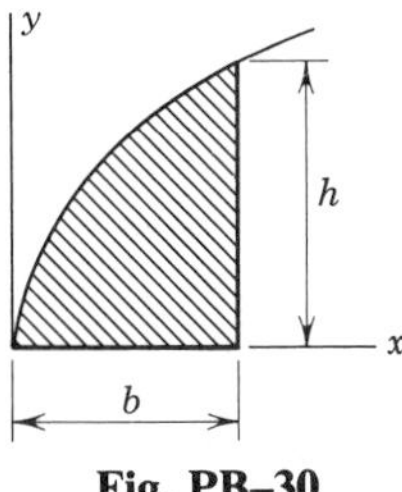

Fig. PB–30

B–7 PRINCIPAL MOMENTS OF INERTIA. The moment of inertia of the area A in Fig. B–15 with respect to the x' axis through O will, in general, vary with the angle θ. The x and y axes used to obtain Eq. B–2 were any pair of orthogonal axes in the plane of the area passing through O; therefore,

$$J_O = I_x + I_y = I_{x'} + I_{y'}$$

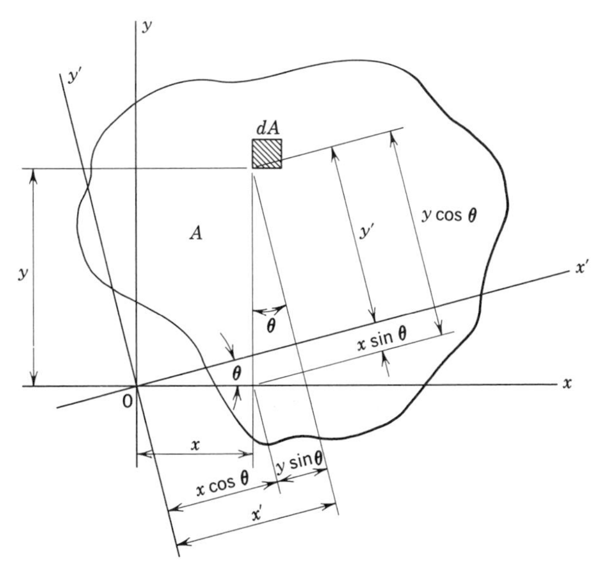

Fig. B–15

where x' and y' are any pair of orthogonal axes through O. Since the sum of $I_{x'}$ and $I_{y'}$ is a constant, $I_{x'}$ will be maximum and the corresponding $I_{y'}$ will be minimum for one particular value of θ.

The set of axes for which the second moments are maximum and minimum are called the *principal axes* of the area through point O and are designated as the u and v axes. The second moments of the area with respect to these axes are called the *principal moments of inertia* of the area and are designated I_u and I_v. There is only one set of principal axes for any point in an area unless all axes have the same second moment, such as the diameters of a circle.

A convenient way to determine the principal moments of inertia of an area is to express $I_{x'}$ as a function of I_x, I_y, I_{xy}, and θ and then set the derivative of $I_{x'}$ with respect to θ equal to zero to obtain the value of θ giving the maximum and minimum second moments. From Fig. B–15,

$$dI_{x'} = y'^2\,dA = (y\cos\theta - x\sin\theta)^2\,dA$$

and

$$I_{x'} = \cos^2\theta\int_A y^2 dA - 2\sin\theta\cos\theta\int_A xy\,dA + \sin^2\theta\int_A x^2 dA$$

$$\left.\begin{aligned} &= I_x\cos^2\theta - 2I_{xy}\sin\theta\cos\theta + I_y\sin^2\theta \\ &= \frac{1}{2}(I_x+I_y) + \frac{1}{2}(I_x-I_y)\cos 2\theta - I_{xy}\sin 2\theta \end{aligned}\right\} \tag{B–8}$$

The angle 2θ for which $I_{x'}$ is maximum can be obtained by setting the derivative of $I_{x'}$ with respect to θ equal to zero; thus,

$$\frac{dI_{x'}}{d\theta} = -2(1/2)(I_x-I_y)\sin 2\theta - 2I_{xy}\cos 2\theta = 0$$

from which

$$\tan 2\beta = -\frac{I_{xy}}{(I_x-I_y)/2} \tag{B–9}$$

where β represents the two values of θ that locate the principal axes u and v. Equation B–9 gives two values of 2β 180° apart, and thus two values of β 90° apart. The principal moments of inertia can be obtained by substituting these values of β in Eq. B–8. From Eq. B–9

$$\cos 2\beta = \mp\frac{(I_x-I_y)/2}{\left\{\left[(I_x-I_y)/2\right]^2 + I_{xy}{}^2\right\}^{1/2}}$$

and

$$\sin 2\beta = \pm \frac{I_{xy}}{\{[(I_x - I_y)/2]^2 + I_{xy}{}^2\}^{1/2}}$$

When these expressions are substituted in Eq. B–8, the principal moments of inertia reduce to

$$I_{u,v} = \frac{I_x + I_y}{2} \mp \sqrt{\left(\frac{I_x - I_y}{2}\right)^2 + I_{xy}{}^2} \qquad \text{(B–10)}$$

The product of inertia of the element of area in Fig. B–15 with respect to the x' and y' axes is

$$dI_{x'y'} = x'y'\,dA = (x\cos\theta + y\sin\theta)(y\cos\theta - x\sin\theta)dA$$

and the product of inertia of the area is

$$\begin{aligned} I_{x'y'} &= (\cos^2\theta - \sin^2\theta)\int_A xy\,dA + \sin\theta\cos\theta\int_A (y^2 - x^2)dA \\ &= I_{xy}\cos 2\theta + \frac{1}{2}(I_x - I_y)\sin 2\theta \qquad \text{(B–11)} \end{aligned}$$

The product of inertia $I_{x'y'}$ will be zero when

$$\tan 2\theta = -\frac{I_{xy}}{(I_x - I_y)/2}$$

which is the same as Eq. B–9, which locates the principal axes. Consequently, *the product of inertia is zero with respect to the principal axes.* Since the product of inertia is zero with respect to any axis of symmetry, it follows that any axis of symmetry must be a principal axis for any point on the axis of symmetry.

The following example illustrates the procedure for determining second moments of areas with respect to the principal axes.

Example B–9 Determine the maximum and minimum second moments of the area of the unequal-leg angle in Fig. B–16 with respect to axes through the centroid of the area.

Solution The area is divided into two rectangles A and B, and the location of the centroid of each area is indicated by the dimensions. The centroid of the composite area is at C. The values of I_x and I_y for the two areas are obtained by applying the

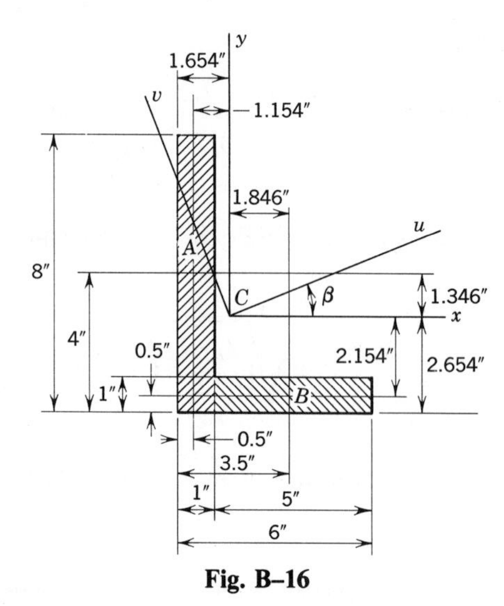

Fig. B–16

parallel-axis theorem to the expression for I_c from Table B–1; thus,

$$I_x = \frac{bh^3}{12} + Ay_c^2$$

$$I_x = I_{Ax} + I_{Bx} = \frac{1(8)^3}{12} + 1(8)(1.346)^2 + \frac{5(1)^3}{12} + 5(1)(2.154)^2$$

$$= 80.8 \text{ in.}^4$$

$$I_y = I_{Ay} + I_{By} = \frac{8(1)^3}{12} + 8(1)(1.154)^2 + \frac{1(5)^3}{12} + 1(5)(1.846)^2$$

$$= 38.8 \text{ in.}^4$$

The products of inertia are determined as indicated in Art. B–6; thus,

$$I_{xy} = I_{x'y'} + Ax_c y_c = 0 + Ax_c y_c$$

$$I_{xy} = I_{Axy} + I_{Bxy} = 1(8)(-1.154)(1.346) + 5(1)(1.846)(-2.154)$$

$$= -32.3 \text{ in.}^4$$

From Eq. B–9

$$\tan 2\beta = -\frac{-32.3}{(80.8-38.8)/2} = 1.538$$

$$2\beta = 57.0° \quad \text{or} \quad 237° \quad \text{therefore} \quad \beta = 28.5° \quad \text{or} \quad 118.5°$$

From Eq. B–8 with $\beta = 28.5°$, the maximum second moment is

$$I_u = 80.8(0.879)^2 - 2(-32.3)(0.477)(0.879) + 38.8(0.477)^2$$

$$= \underline{98.3 \text{ in.}^4} \qquad \text{Ans.}$$

With $\beta = 118.5°$, the minimum second moment is

$$I_v = 80.8(-0.477)^2 - 2(-32.3)(0.879)(-0.477) + 38.8(0.879)^2$$

$$= \underline{21.3 \text{ in.}^4} \qquad \text{Ans.}$$

The principal moments of inertia can also be determined by means of Eq. B–10.

$$I_{u,v} = \frac{80.8+38.8}{2} \pm \sqrt{\left(\frac{80.8-38.8}{2}\right)^2 + (-32.3)^2}$$

$$= \underline{98.3 \text{ in.}^4} \quad \text{and} \quad \underline{21.3 \text{ in.}^4} \qquad \text{Ans.}$$

PROBLEMS

B–31* Determine the principal moments of inertia of the Z section in Prob. B–28 with respect to axes through the centroid of the section.

B–32* Determine the minimum radius of gyration of a $5 \times 3\frac{1}{2} \times \frac{1}{2}$-in. angle section with respect to axes through the centroid of the section.

B–33* Locate the principal axes through the centroid of the triangular area in Fig. PB–33 and determine the minimum radius of gyration.

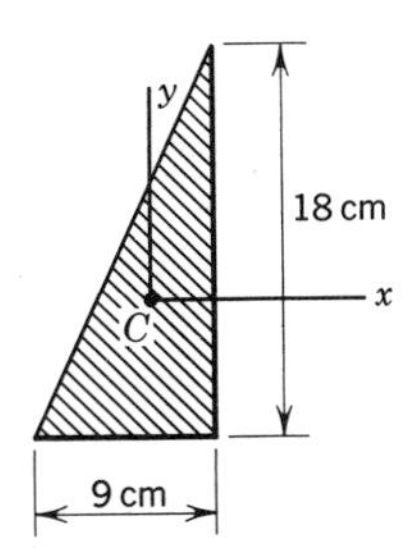

Fig. PB–33

B–34* Determine the second moments of the C254×35.74 channel in Fig. PB–34 with respect to the x' and y' axes.

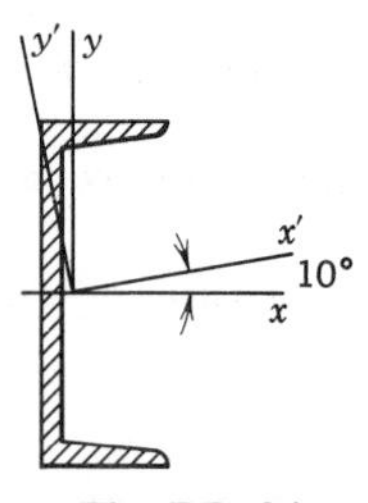

Fig. PB–34

B–35* Locate the axis through point O of Fig. PB–35 for which the second moment of the area will be maximum and calculate the corresponding second moment.

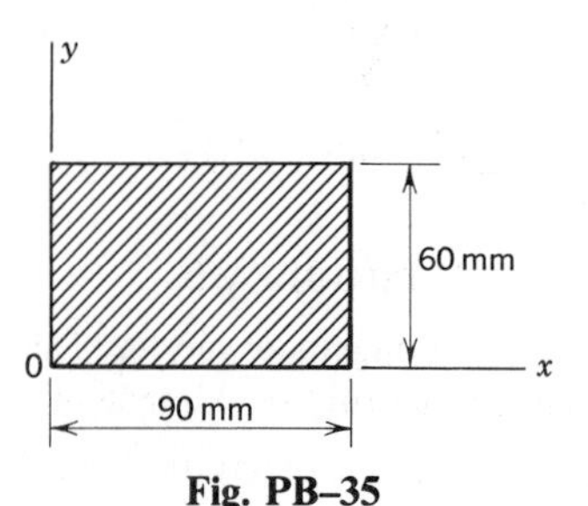

Fig. PB–35

B–8 MOHR'S CIRCLE FOR SECOND MOMENTS OF AREAS. The use of Mohr's circle for determining principal stresses was discussed in Art. 1–8. A comparison of Eq. 1–4a and 1–4b with Eq. B–8 indicates that a similar procedure can be used to obtain the principal moments of inertia of an area.

Figure B–17 illustrates the use of Mohr's circle for second moments. Assume that I_x is greater than I_y and that I_{xy} is positive. Moments of inertia are plotted along the horizontal axis, and products of inertia are plotted along the vertical axis. Second moments are always positive and are plotted to the right of the origin. Products of inertia can be either positive or negative; positive values are plotted above the horizontal axis. The distance OA' along the I axis is equal to I_x and $A'A$, parallel to the I_{xy} axis, is equal to I_{xy}. Similarly, OB' is laid off equal to I_y and $B'B$ is equal to $-I_{xy}$. The line AB intersects the I axis at C and AB is the diameter

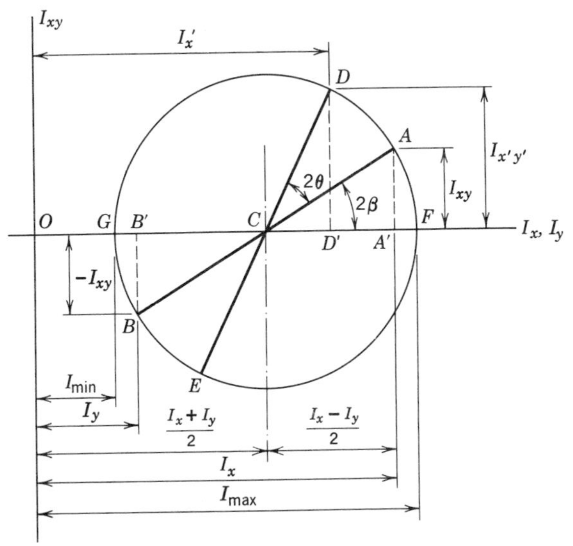

Fig. B–17

of Mohr's circle for second moments. Each point on the circle represents $I_{x'}$ and $I_{x'y'}$ for one particular orientation of the x' and y' axes. To demonstrate this statement, the diameter DE is drawn at an angle 2θ counterclockwise from line AB. From the figure

$$OD' = OC + CD\cos(2\beta + 2\theta)$$

which, since CD is equal to CA, reduces to

$$OD' = OC + CA\cos 2\beta\cos 2\theta - CA\sin 2\beta\sin 2\theta$$

Since $CA\cos 2\beta$ and $CA\sin 2\beta$ are equal to CA' and $A'A$, respectively, the distance OD' becomes

$$OD' = \frac{1}{2}(I_x + I_y) + \frac{1}{2}(I_x - I_y)\cos 2\theta - I_{xy}\sin 2\theta = I_{x'}$$

from Eq. B–8. In a similar manner

$$D'D = CD\sin(2\beta + 2\theta)$$

$$= \frac{1}{2}(I_x - I_y)\sin 2\theta + I_{xy}\cos 2\theta = I_{x'y'}$$

from Eq. B–11.

Since the horizontal coordinate of each point on the circle represents a particular value of $I_{x'}$, the maximum second moment is represented by OF and its value is

$$I_{max} = OF = OC + CF = OC + CA$$

$$= \frac{I_x + I_y}{2} + \sqrt{\left(\frac{I_x - I_y}{2}\right)^2 + I_{xy}^2}$$

which agrees with Eq. B–10. Figure B–18 represents an area and a set of axes for which the data used to construct Fig. B–17 are valid. In deriving Eq. B–8, the angle between the x and x' axes was θ. In obtaining the same equation from Mohr's circle, the angle between the radii to points (I_x, I_{xy}), line CA, and $(I_{x'}, I_{x'y'})$, line CD, is 2θ. Thus all angles on Mohr's circle are twice the corresponding angles for the given area.

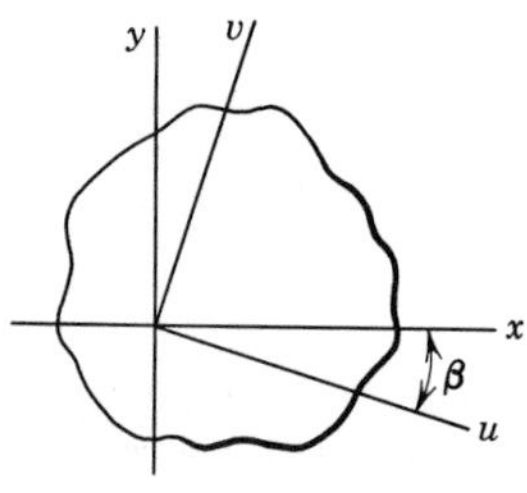

Fig. B–18

Example B–10 Solve Exam. B–9 by means of Mohr's circle.

Solution Values of I_x, I_y, and I_{xy} must be computed in the same manner as in Exam. B–9. These values are

$$I_x = 80.8 \text{ in.}^4, \quad I_y = 38.8 \text{ in.}^4, \quad I_{xy} = -32.3 \text{ in.}^4$$

The circle in Fig. B–19 is constructed as indicated from these values. Point A is located at I_x and I_{xy} (which is negative) and B is at I_y and $-I_{xy}$. Point C is the center of the circle and is 59.8 units from O. The radius of the circle is

$$CA = \sqrt{(21.0)^2 + (32.3)^2} = 38.5 \text{ in.}^4$$

and the principal moments of inertia are

$$I_{max} = OC + CF = 59.8 + 38.5 = \underline{98.3 \text{ in.}^4} \qquad \text{Ans.}$$

$$I_{min} = OC - CG = 59.8 - 38.5 = \underline{21.3 \text{ in.}^4} \qquad \text{Ans.}$$

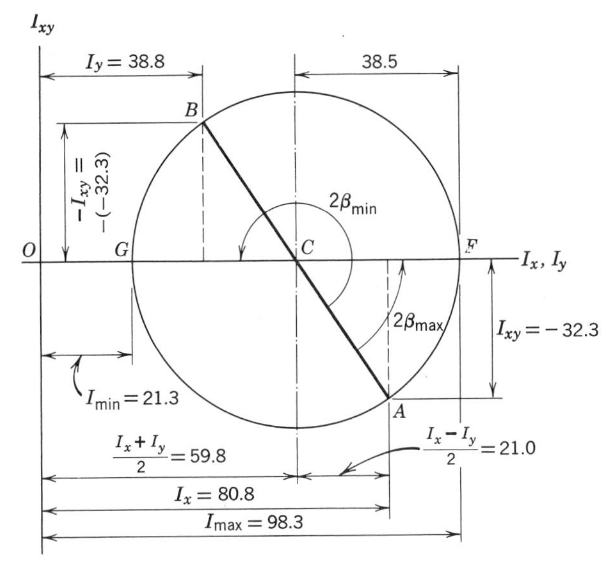

Fig. B–19

Twice the angle from the x axis to the principal axis with the maximum moment of inertia is shown on the sketch as $2\beta_{max}$ and

$$\beta_{max} = \frac{1}{2}\tan^{-1}\frac{32.3}{21.0} = 28.5° \text{ counterclockwise from the } x \text{ axis.}$$

The angle from the x axis to the principal axis with the minimum moment of inertia is β_{min} and is

$$\beta_{min} = \beta_{max} + 90° = 118.5° \text{ counterclockwise from the } x \text{ axis.}$$

These values were obtained analytically from the geometry of Mohr's circle. They could also have been measured directly from the figure; however, the accuracy of the results obtained by scaling the distances from the figure would depend on the scale used and the care employed in constructing the figure.

PROBLEMS

B–36* Solve Prob. B–32 using Mohr's circle.

B–37* Solve Prob. B–33 using Mohrs circle.

B–38* Solve Prob. B–35 using Mohr's circle.

B–39* A lintel is made by welding an L7×4×$\frac{3}{8}$-in. angle to a C12×30 channel as shown in Fig. PB–39. Determine the minimum radius of gyration with respect to an axis through the centroid of the section.

Fig. PB–39

B–40* Locate the principal axes of inertia through the origin and determine the corresponding second moments of the area in Prob. B–30 when $b=h=10$ cm.

APPENDIX C

Properties of Rolled Steel Shapes

WIDE FLANGE SHAPES

PROPERTIES FOR DESIGNING

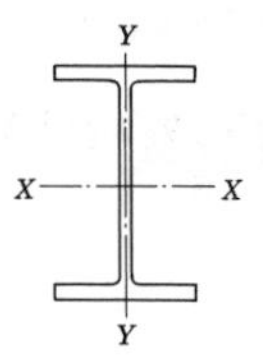

Designation	Weight	Area	Depth	Flange		Web	Axis $X-X$			Axis $Y-Y$		
	(lb/ft)	(in^2)	(in.)	Width (in.)	Thickness (in.)	Thickness (in.)	I ($in.^4$)	Z ($in.^3$)	r (in.)	I ($in.^4$)	Z ($in.^3$)	r (in.)
W36×230	230	67.73	35.88	16.475	1.260	0.765	14988.4	835.5	14.88	870.9	105.7	3.59
×160	160	47.09	36.00	12.000	1.020	0.653	9738.8	541.0	14.38	275.4	45.9	2.42
W33×200	200	58.79	33.00	15.750	1.150	0.715	11048.2	669.6	13.71	691.7	87.8	3.43
×130	130	38.26	33.10	11.510	0.855	0.580	6699.0	404.8	13.23	201.4	35.0	2.29
W30×172	172	50.65	29.88	14.985	1.065	0.655	7891.5	528.2	12.48	550.1	73.4	3.30
×108	108	31.77	29.82	10.484	0.760	0.548	4461.0	299.2	11.85	135.1	25.8	2.06
W27×145	145	42.68	26.88	13.965	0.975	0.600	5414.3	402.9	11.26	406.9	58.3	3.09
×94	94	27.65	26.91	9.990	0.747	0.490	3266.7	242.8	10.87	115.1	23.0	2.04
W24×100	100	29.43	24.00	12.000	0.775	0.468	2987.3	248.9	10.08	203.5	33.9	2.63
×84	84	24.71	24.09	9.015	0.772	0.470	2364.3	196.3	9.78	88.3	19.6	1.89
W21×112	112	32.93	21.00	13.000	0.865	0.527	2620.6	249.6	8.92	289.7	44.6	2.96
×82	82	24.10	20.86	8.962	0.795	0.499	1752.4	168.0	8.53	89.6	20.0	1.93
×62	62	18.23	20.99	8.240	0.615	0.400	1326.8	126.4	8.53	53.1	12.9	1.71
W18×96	96	28.22	18.16	11.750	0.831	0.512	1674.7	184.4	7.70	206.8	35.2	2.71
×77	77	22.63	18.16	8.787	0.831	0.475	1286.8	141.7	7.54	88.6	20.2	1.98
×60	60	17.64	18.25	7.558	0.695	0.416	984.0	107.8	7.47	47.1	12.5	1.63
W16×64	64	18.80	16.00	8.500	0.715	0.443	833.8	104.2	6.66	68.4	16.1	1.91
×50	50	14.70	16.25	7.073	0.628	0.380	655.4	80.7	6.68	34.8	9.8	1.54
W14×78	78	22.94	14.06	12.000	0.718	0.428	851.2	121.1	6.09	206.9	34.5	3.00
×43	43	12.65	13.68	8.000	0.528	0.308	429.0	62.7	5.82	45.1	11.3	1.89
×30	30	8.81	13.86	6.733	0.383	0.270	289.6	41.8	5.73	17.5	5.2	1.41
W12×53	53	15.59	12.06	10.000	0.576	0.345	426.2	70.7	5.23	96.1	19.2	2.48
×40	40	11.77	11.94	8.000	0.516	0.294	310.1	51.9	5.13	44.1	11.0	1.94
×27	27	7.97	11.95	6.500	0.400	0.240	204.1	34.1	5.06	16.6	5.1	1.44
W10×77	77	22.67	10.62	10.195	0.868	0.535	457.2	86.1	4.49	153.4	30.1	2.60
×60	60	17.66	10.25	10.075	0.683	0.415	343.7	67.1	4.41	116.5	23.1	2,57
×45	45	13.24	10.12	8.022	0.618	0.350	248.6	49.1	4.33	53.2	13.3	2.00
×25	25	7.35	10.08	5.762	0.430	0.252	133.2	26.4	4.26	12.7	4.4	1.31
×21	21	6.19	9.90	5.750	0.340	0.240	106.3	21.5	4.14	9.7	3.4	1.25
W8×40	40	11.76	8.25	8.077	0.558	0.365	146.3	35.5	3.53	49.0	12.1	2.04
×31	31	9.12	8.00	8.000	0.433	0.288	109.7	27.4	3.47	37.0	9.2	2.01
×28	28	8.23	8.06	6.540	0.463	0.285	97.8	24.3	3.45	21.6	6.6	1.62
×17	17	5.00	8.00	5.250	0.308	0.230	56.4	14.1	3.36	6.7	2.6	1.16

UNIVERSAL BEAMS

PROPERTIES FOR DESIGNING

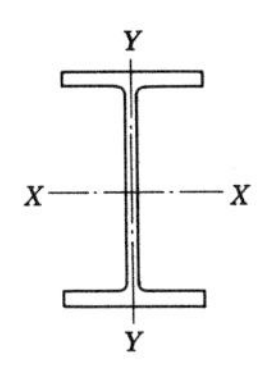

Designation	Mass (kg/m)	Area (cm^2)	Depth (mm)	Flange		Web	Axis $X-X$			Axis $Y-Y$		
				Width (mm)	Thickness (mm)	Thickness (mm)	I (cm^4)	Z (cm^3)	r (cm)	I (cm^4)	Z (cm^3)	r (cm)
UB914×343	343	436.9	911.4	418.5	32.0	19.4	623866	13691	37.8	36251	1733	9.11
×201	201	256.1	903.0	303.4	20.2	15.2	324715	7192	35.6	8632	569.1	5.81
UB838×226	226	288.4	850.9	293.8	26.8	16.1	339130	7971	34.3	10661	725.9	6.08
×176	176	223.8	834.9	291.6	18.8	14.0	245412	5879	33.1	7111	487.6	5.64
UB762×197	197	250.5	769.6	268.0	25.4	15.6	239464	6223	30.9	7699	574.6	5.54
×147	147	187.8	753.9	265.3	17.5	12.9	168535	4471	30.0	5002	377.1	5.16
UB686×170	170	216.3	692.9	255.8	23.7	14.5	169843	4902	28.0	6225	486.8	5.36
×125	125	159.4	677.9	253.0	16.2	11.7	117700	3472	27.2	3992	315.5	5.00
UB610×140	140	178.2	617.0	230.1	22.1	13.1	111673	3620	25.0	4253	369.6	4.88
×113	113	144.3	607.3	228.2	17.3	11.2	87260	2874	24.6	3184	279.1	4.70
×82	82	104.4	598.2	177.8	12.8	10.1	55779	1865	23.1	1203	135.3	3.39
UB533×189	189	241.2	539.5	331.7	25.0	14.9	125618	4657	22.8	14093	849.6	7.64
×122	122	155.6	544.6	211.9	21.3	12.8	76078	2794	22.1	3208	302.8	4.54
×73	73	93.0	528.8	165.6	13.5	9.3	40414	1528	20.8	1027	124.1	3.32
UB457×98	98	125.2	467.4	192.8	19.6	11.4	45653	1954	19.1	2216	229.9	4.21
×82	82	104.4	460.2	191.3	16.0	9.9	37039	1610	18.8	1746	182.6	4.09
×60	60	75.9	454.7	152.9	13.3	8.0	25464	1120	18.3	794	104.0	3.23
UB406×74	74	94.9	412.8	179.7	16.0	9.7	27279	1322	17.0	1448	161.2	3.91
×46	46	58.9	402.3	142.4	11.2	6.9	15603	775.6	16.3	500	70.26	2.92
UB356×67	67	85.3	364.0	173.2	15.7	9.1	19483	1071	15.1	1278	147.6	3.87
×45	45	56.9	352.0	171.0	9.7	6.9	12052	684.7	14.6	730	85.39	3.58
×33	33	41.7	348.5	125.4	8.5	5.9	8167	468.7	14.0	257	40.99	2.48
UB305×54	54	68.3	310.9	166.8	13.7	7.7	11686	751.8	13.1	988	118.5	3.80
×48	48	60.8	310.4	125.2	14.0	8.9	9485	611.1	12.5	438	69.94	2.68
×33	33	41.8	312.7	102.4	10.8	6.6	6482	414.6	12.5	189	37.00	2.13
UB254×43	43	55.0	259.6	147.3	12.7	7.3	6546	504.3	10.9	633	85.97	3.39
×31	31	39.9	251.5	146.1	8.6	6.1	4427	352.1	10.5	406	55.53	3.19
×28	28	36.2	260.4	102.1	10.0	6.4	4004	307.6	10.5	174	34.13	2.19
×22	22	28.4	254.0	101.6	6.8	5.8	2863	225.4	10.0	116	22.84	2.02
UB203×30	30	38.0	206.8	133.8	9.6	6.3	2880	278.5	8.71	354	52.85	3.05
×25	25	32.3	203.2	133.4	7.8	5.8	2348	231.1	8.53	280	41.92	2.94

AMERICAN STANDARD BEAMS

PROPERTIES FOR DESIGNING

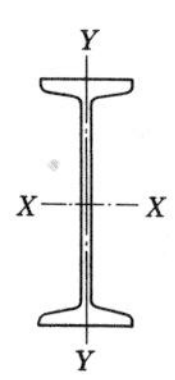

Designation	Weight (lb/ft)	Area (sq in.)	Depth (in.)	Flange		Web Thickness (in.)	Axis $X-X$			Axis $Y-Y$		
				Width (in.)	Thickness (in.)		I (in.^{4})	Z (in.^{3})	r (in.)	I (in.^{4})	Z (in.^{3})	r (in.)
S24×120	120.0	35.13	24.00	8.048	1.102	0.798	3010.8	250.9	9.26	84.9	21.1	1.56
×100	100.0	29.25	24.00	7.247	0.871	0.747	2371.8	197.6	9.05	48.4	13.4	1.29
×90	90.0	26.30	24.00	7.124	0.871	0.624	2230.1	185.8	9.21	45.5	12.8	1.32
×79.9	79.9	23.33	24.00	7.000	0.871	0.500	2087.2	173.9	9.46	42.9	12.2	1.36
S20×95	95.0	27.74	20.00	7.200	0.916	0.800	1599.7	160.0	7.59	50.5	14.0	1.35
×85	85.0	24.80	20.00	7.053	0.916	0.653	1501.7	150.2	7.78	47.0	13.3	1.38
×75	75.0	21.90	20.00	6.391	0.789	0.641	1263.5	126.3	7.60	30.1	9.4	1.17
×65.4	65.4	19.08	20.00	6.250	0.789	0.500	1169.5	116.9	7.83	27.9	8.9	1.21
S18×70	70.0	20.46	18.00	6.251	0.691	0.711	917.5	101.9	6.70	24.5	7.8	1.09
×54.7	54.7	15.94	18.00	6.000	0.691	0.460	795.5	88.4	7.07	21.2	7.1	1.15
S15×50	50.0	14.59	15.00	5.640	0.622	0.550	481.1	64.2	5.74	16.0	5.7	1.05
×42.9	42.9	12.49	15.00	5.500	0.622	0.410	441.8	58.9	5.95	14.6	5.3	1.08
S12×50	50.0	14.57	12.00	5.477	0.659	0.687	301.6	50.3	4.55	16.0	5.8	1.05
×40.8	40.8	11.84	12.00	5.250	0.659	0.460	268.9	44.8	4.77	13.8	5.3	1.08
×31.8	31.8	9.26	12.00	5.000	0.544	0.350	215.8	36.0	4.83	9.5	3.8	1.01
S10×35	35.0	10.22	10.00	4.944	0.491	0.594	145.8	29.2	3.78	8.5	3.4	0.91
×25.4	25.4	7.38	10.00	4.660	0.491	0.310	122.1	24.4	4.07	6.9	3.0	0.97
S8×23	23.0	6.71	8.00	4.171	0.425	0.441	64.2	16.0	3.09	4.4	2.1	0.81
×18.4	18.4	5.34	8.00	4.000	0.425	0.270	56.9	14.2	3.26	3.8	1.9	0.84
S7×20	20.0	5.83	7.00	3.860	0.392	0.450	41.9	12.0	2.68	3.1	1.6	0.74
×15.3	15.3	4.43	7.00	3.660	0.392	0.250	36.2	10.4	2.86	2.7	1.5	0.78
S6×17.25	17.25	5.02	6.00	3.565	0.359	0.465	26.0	8.7	2.28	2.3	1.3	0.68
×12.5	12.5	3.61	6.00	3.330	0.359	0.230	21.8	7.3	2.46	1.8	1.1	0.72
S5×14.75	14.75	4.29	5.00	3.284	0.326	0.494	15.0	6.0	1.87	1.7	1.0	0.63
×10	10.0	2.87	5.00	3.000	0.326	0.210	12.1	4.8	2.05	1.2	0.82	0.65
S4×9.5	9.5	2.76	4.00	2.796	0.293	0.326	6.7	3.3	1.56	0.91	0.65	0.58
S3×7.5	7.5	2.17	3.00	2.509	0.260	0.349	2.9	1.9	1.15	.59	0.47	0.52

STANDARD CHANNELS

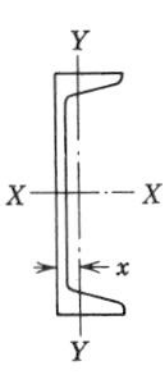

Designation	Weight (Mass) (lb/ft)	Area (in.²)	Depth (in.)	Flange Width (in.)	Flange Average Thickness (in.)	Web Thickness (in.)	Axis X – X I (in.⁴)	Axis X – X Z (in.³)	Axis X – X r (in.)	Axis Y – Y I (in.⁴)	Axis Y – Y Z (in.³)	Axis Y – Y r (in.)	Axis Y – Y x (in.)
C15×33.9	33.9	9.90	15.00	3.400	0.650	0.400	312.6	41.7	5.62	8.2	3.2	0.91	0.79
C12×30	30.0	8.79	12.00	3.170	0.501	0.510	161.2	26.9	4.28	5.2	2.1	0.77	0.68
×20.7	20.7	6.03	12.00	2.940	0.501	0.280	128.1	21.4	4.61	3.9	1.7	0.81	0.70
C10×30	30.0	8.80	10.00	3.033	0.436	0.673	103.0	20.6	3.42	4.0	1.7	0.67	0.65
×25	25.0	7.33	10.00	2.886	0.436	0.526	90.7	18.1	3.52	3.4	1.5	0.68	0.62
×15.3	15.3	4.47	10.00	2.600	0.436	0.240	66.9	13.4	3.87	2.3	1.2	0.72	0.64
C9×20	20.0	5.86	9.00	2.648	0.413	0.448	60.6	13.5	3.22	2.4	1.2	0.65	0.59
×15	15.0	4.39	9.00	2.485	0.413	0.285	50.7	11.3	3.40	1.9	1.0	0.67	0.59
C8×18.75	18.75	5.49	8.00	2.527	0.390	0.487	43.7	10.9	2.82	2.0	1.0	0.60	0.57
×11.5	11.5	3.36	8.00	2.260	0.390	0.220	32.3	8.1	3.10	1.3	0.79	0.63	0.58
C6×10.5	10.5	3.07	6.00	2.034	0.343	0.314	15.1	5.0	2.22	0.87	0.57	0.53	0.50
×8.2	8.2	2.39	6.00	1.920	0.343	0.200	13.0	4.3	2.34	0.70	0.50	0.54	0.52
C4×7.25	7.25	2.12	4.00	1.720	0.296	0.320	4.5	2.3	1.47	0.44	0.35	0.46	0.46
C3×4.1	4.1	1.19	3.00	1.410	0.273	0.170	1.6	1.1	1.17	0.20	0.21	0.41	0.44
	(kg/m)	(cm²)	(mm)	(mm)	(mm)	(mm)	(cm⁴)	(cm³)	(cm)	(cm⁴)	(cm³)	(cm)	(cm)
C381×55.10	55.10	70.19	381.0	101.6	16.3	10.4	14894	781.8	14.6	579.8	75.87	2.87	2.52
C305×46.18	46.18	58.83	304.8	101.6	14.8	10.2	8214	539.0	11.8	499.5	66.60	2.91	2.66
×41.69	41.69	53.11	304.8	88.9	13.7	10.2	7061	463.3	11.5	325.4	48.49	2.48	2.18
C254×35.74	35.74	45.52	254.0	88.9	13.6	9.1	4448	350.2	9.88	302.4	46.71	2.58	2.42
×28.29	28.29	36.03	254.0	76.2	10.9	8.1	3367	265.1	9.67	162.6	28.22	2.12	1.86
C229×32.76	32.76	41.73	228.6	88.9	13.3	8.6	3387	296.4	9.01	285.0	44.82	2.61	2.53
×26.06	26.06	33.20	228.6	76.2	11.2	7.6	2610	228.3	8.87	158.7	28.22	2.19	2.00
C203×29.78	29.78	37.94	203.2	88.9	12.9	8.1	2491	245.2	8.10	264.4	42.34	2.64	2.65
×23.82	23.82	30.34	203.2	76.2	11.2	7.1	1950	192.0	8.02	151.4	27.59	2.23	2.13
C152×23.84	23.84	30.36	152.4	88.9	11.6	7.1	1166	153.0	6.20	215.1	35.70	2.66	2.86
×17.88	17.88	22.77	152.4	76.2	9.0	6.4	851.6	111.8	6.12	113.8	21.05	2.24	2.21
C102×10.42	10.42	13.28	101.6	50.8	7.6	6.1	207.7	40.89	3.96	29.10	8.16	1.48	1.51
C76×6.70	6.70	8.53	76.2	38.1	6.8	5.1	74.14	19.46	2.95	10.66	4.07	1.12	1.19

ANGLES
EQUAL LEGS

PROPERTIES FOR DESIGNING

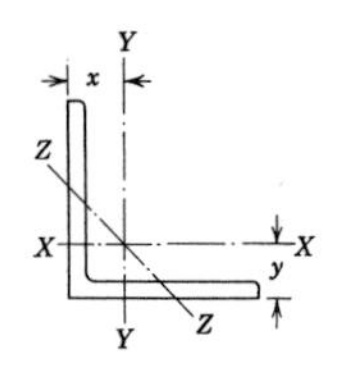

Size and Thickness (in.)	Weight (lb/ft)	Area (in.^{2})	Axis $X-X$ and Axis $Y-Y$: I (in.^{4})	Z (in.^{3})	r (in.)	x or y (in.)	Axis $Z-Z$: r (in.)
L8×8×$\frac{3}{4}$	38.9	11.44	69.7	12.2	2.47	2.28	1.57
×$\frac{1}{2}$	26.4	7.75	48.6	8.4	2.50	2.19	1.59
L6×6×$\frac{3}{4}$	28.7	8.44	28.2	6.7	1.83	1.78	1.17
×$\frac{1}{2}$	19.6	5.75	19.9	4.6	1.86	1.68	1.18
L5×5×$\frac{3}{4}$	23.6	6.94	15.7	4.5	1.51	1.52	0.97
×$\frac{7}{16}$	14.3	4.18	10.0	2.8	1.55	1.41	0.98
L4×4×$\frac{1}{2}$	12.8	3.75	5.6	2.0	1.22	1.18	0.78
×$\frac{3}{8}$	9.8	2.86	4.4	1.5	1.23	1.14	0.79
L3$\frac{1}{2}$×3$\frac{1}{2}$×$\frac{3}{8}$	8.5	2.48	2.9	1.2	1.07	1.01	0.69
×$\frac{1}{4}$	5.8	1.69	2.0	0.79	1.09	0.97	0.69
L3×3×$\frac{1}{2}$	9.4	2.75	2.2	1.1	0.90	0.93	0.58
×$\frac{3}{8}$	7.2	2.11	1.8	0.83	0.91	0.89	0.58
×$\frac{1}{4}$	4.9	1.44	1.2	0.58	0.93	0.84	0.59
L2$\frac{1}{2}$×2$\frac{1}{2}$×$\frac{1}{4}$	4.1	1.19	0.70	0.39	0.77	0.72	0.49
×$\frac{3}{16}$	3.07	0.90	0.55	0.30	0.78	0.69	0.49
L2×2×$\frac{1}{4}$	3.19	0.94	0.35	0.25	0.61	0.59	0.39
×$\frac{1}{8}$	1.65	0.48	0.19	0.13	0.63	0.55	0.40
L1$\frac{1}{2}$×1$\frac{1}{2}$×$\frac{1}{4}$	2.34	0.69	0.14	0.13	0.45	0.47	0.29
(mm)	(kg/m)	(cm^2)	(cm^4)	(cm^3)	(cm)	(cm)	(cm)
L200×200×20	59.9	76.3	2850	199	6.11	5.68	3.92
×16	48.5	61.8	2340	162	6.16	5.52	3.94
L150×150×18	40.1	51.0	1050	98.7	4.54	4.37	2.92
×12	27.3	34.8	737	67.7	4.60	4.12	2.95
L120×120×15	26.6	33.9	445	52.4	3.62	3.51	2.33
×10	18.2	23.2	313	36.0	3.67	3.31	2.36
L100×100×12	17.8	22.7	207	29.1	3.02	2.90	1.94
×8	12.2	15.5	145	19.9	3.06	2.74	1.96
L90×90×12	15.9	20.3	148	23.3	2.70	2.66	1.75
×10	13.4	17.1	127	19.8	2.72	2.58	1.76
×6	8.30	10.6	80.3	12.2	2.76	2.41	1.78
L80×80×10	11.9	15.1	87.5	15.4	2.41	2.34	1.55
×6	7.34	9.35	55.8	9.57	2.44	2.17	1.57
L60×60×10	8.69	11.1	34.9	8.41	1.78	1.85	1.16
×6	5.42	6.91	22.8	5.29	1.82	1.69	1.17
L50×50×8	5.82	7.41	16.3	4.68	1.48	1.52	0.96
×5	3.77	4.80	11.0	3.05	1.51	1.40	0.97

L

ANGLES
UNEQUAL LEGS

PROPERTIES FOR DESIGNING

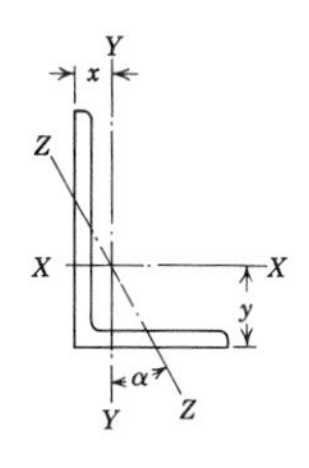

Size and Thickness (in.)	Weight (lb/ft)	Area (in.^{2})	Axis $X-X$				Axis $Y-Y$				Axis $Z-Z$	
			I (in.^{4})	Z (in.^{3})	r (in.)	y (in.)	I (in.^{4})	Z (in.^{3})	r (in.)	x (in.)	r (in.)	$\tan\alpha$
L8×6×1	44.2	13.00	80.8	15.1	2.49	2.65	38.8	8.9	1.73	1.65	1.28	0.543
×$\frac{1}{2}$	23.0	6.75	44.3	8.0	2.56	2.47	21.7	4.8	1.79	1.47	1.30	0.558
L8×4×$\frac{3}{4}$	28.7	8.44	54.9	10.9	2.55	2.95	9.4	3.1	1.05	0.95	0.85	0.258
×$\frac{1}{2}$	19.6	5.75	38.5	7.5	2.59	2.86	6.7	2.2	1.08	0.86	0.86	0.267
L7×4×$\frac{3}{8}$	13.6	3.98	20.6	4.4	2.27	2.37	5.1	1.6	1.13	0.87	0.88	0.339
L6×4×$\frac{9}{16}$	18.1	5.31	19.3	4.8	1.90	2.01	6.9	2.3	1.14	1.01	0.87	0.438
×$\frac{3}{8}$	12.3	3.61	13.5	3.3	1.93	1.94	4.9	1.6	1.17	0.94	0.88	0.446
L6×3$\frac{1}{2}$×$\frac{3}{8}$	11.7	3.42	12.9	3.2	1.94	2.04	3.3	1.2	0.99	0.79	0.77	0.350
L5×3$\frac{1}{2}$×$\frac{5}{8}$	16.8	4.92	12.0	3.7	1.56	1.70	4.8	1.9	0.99	0.95	0.75	0.472
×$\frac{1}{2}$	13.6	4.00	10.0	3.0	1.58	1.66	4.1	1.6	1.01	0.91	0.75	0.479
L5×3×$\frac{1}{2}$	12.8	3.75	9.5	2.9	1.59	1.75	2.6	1.1	0.83	0.75	0.65	0.357
×$\frac{5}{16}$	8.2	2.40	6.3	1.9	1.61	1.68	1.8	0.75	0.85	0.68	0.66	0.368
L4×3$\frac{1}{2}$×$\frac{1}{2}$	11.9	3.50	5.3	1.9	1.23	1.25	3.8	1.5	1.04	1.00	0.72	0.750
×$\frac{5}{16}$	7.7	2.25	3.6	1.3	1.26	1.18	2.6	1.0	1.07	0.93	0.73	0.757
L4×3×$\frac{3}{8}$	8.5	2.48	4.0	1.5	1.26	1.28	1.9	0.87	0.88	0.78	0.64	0.551
×$\frac{1}{4}$	5.8	1.69	2.8	1.0	1.28	1.24	1.4	0.60	0.90	0.74	0.65	0.558
L3×2$\frac{1}{2}$×$\frac{5}{16}$	5.6	1.62	1.4	0.69	0.94	0.93	0.90	0.49	0.74	0.68	0.53	0.680
L3×2×$\frac{5}{16}$	5.0	1.47	1.3	0.66	0.95	1.02	0.47	0.32	0.57	0.52	0.43	0.435
(mm)	(kg/m)	(cm^2)	(cm^4)	(cm^3)	(cm)	(cm)	(cm^4)	(cm^3)	(cm)	(cm)	(cm)	
L200×150×18	47.1	60.0	2376	174.0	6.29	6.33	1146	103.0	4.37	3.85	3.[illegible]1	0.548
×15	39.6	50.5	2022	147.0	6.33	6.21	979	86.9	4.40	3.73	3.23	0.550
L150×90×15	26.6	33.9	761	77.7	4.74	5.21	205	30.4	2.46	2.23	1.93	0.354
×10	18.2	23.2	533	53.3	4.80	5.00	146	21.0	2.51	2.04	1.95	0.360
L150×75×15	24.8	31.6	713	75.3	4.75	5.53	120	21.0	1.94	1.81	1.58	0.254
×10	17.0	21.6	501	51.8	4.81	5.32	85.8	14.6	1.99	1.61	1.60	0.261
L125×75×12	17.8	22.7	354	43.2	3.95	4.31	95.5	16.9	2.05	1.84	1.61	0.353
×10	15.0	19.1	302	36.5	3.97	4.23	82.1	14.3	2.07	1.76	1.62	0.356
8	12.2	15.5	247	29.6	4.00	4.14	67.6	11.6	2.09	1.68	1.63	0.359
L100×75×10	13.0	16.6	162	23.8	3.12	3.19	77.6	14.0	2.16	1.95	1.59	0.544
×8	10.6	13.5	133	19.3	3.14	3.10	64.1	11.4	2.18	1.87	1.60	0.547
L100×65×10	12.3	15.6	154	23.2	3.14	3.36	51.0	10.5	1.81	1.63	1.39	0.410
×7	8.77	11.2	113	16.6	3.17	3.23	37.6	7.53	1.83	1.51	1.40	0.415
L75×50×8	7.39	9.41	52.0	10.4	2.35	2.52	18.4	4.95	1.40	1.29	1.07	0.430
×6	5.65	7.19	40.5	8.01	2.37	2.44	14.4	3.81	1.42	1.21	1.08	0.435
L65×50×50×8	6.75	8.60	34.8	7.93	2.01	2.11	17.7	4.89	1.44	1.37	1.05	0.569
×6	5.16	6.58	27.2	6.10	2.03	2.04	14.0	3.77	1.46	1.29	1.06	0.575

APPENDIX D
Table of Beam Deflections and Slopes

TABLE OF BEAM DEFLECTIONS AND SLOPES

Case	Load and Support (Length L)	Slope at End (+ ↺)	Maximum Deflection (+ upward)
1	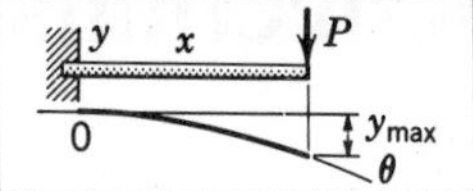	$\theta = -\frac{PL^2}{2EI}$ at $x = L$	$y_{max} = -\frac{PL^3}{3EI}$ at $x = L$
2	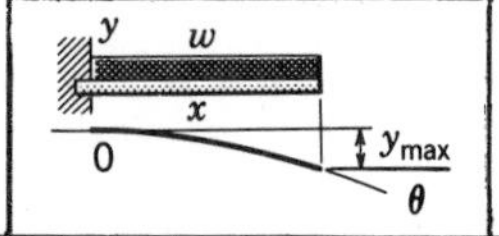	$\theta = -\frac{wL^3}{6EI}$ at $x = L$	$y_{max} = -\frac{wL^4}{8EI}$ at $x = L$
3	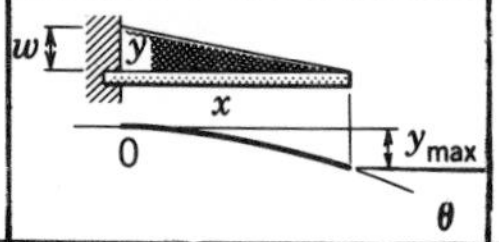	$\theta = -\frac{wL^3}{24EI}$ at $x = L$	$y_{max} = -\frac{wL^4}{30EI}$ at $x = L$
4	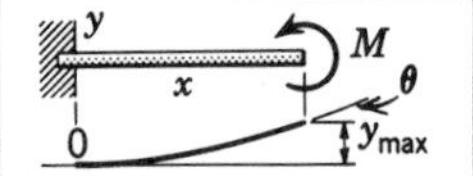	$\theta = +\frac{ML}{EI}$ at $x = L$	$y_{max} = +\frac{ML^2}{2EI}$ at $x = L$
5	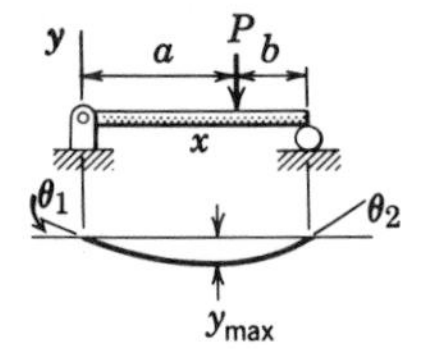	$\theta_1 = -\frac{Pb(L^2 - b^2)}{6LEI}$ at $x = 0$ $\theta_2 = +\frac{Pa(L^2 - a^2)}{6LEI}$ at $x = L$	$y_{max} = -\frac{Pb(L^2 - b^2)^{3/2}}{9\sqrt{3}\,LEI}$ at $x = \sqrt{(L^2 - b^2)/3}$ $y_{center} = -\frac{Pb(3L^2 - 4b^2)}{48EI}$ not max
6	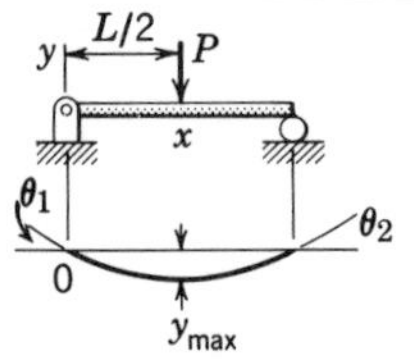	$\theta_1 = -\frac{PL^2}{16EI}$ at $x = 0$ $\theta_2 = +\frac{PL^2}{16EI}$ at $x = L$	$y_{max} = -\frac{PL^3}{48EI}$ at $x = L/2$
7	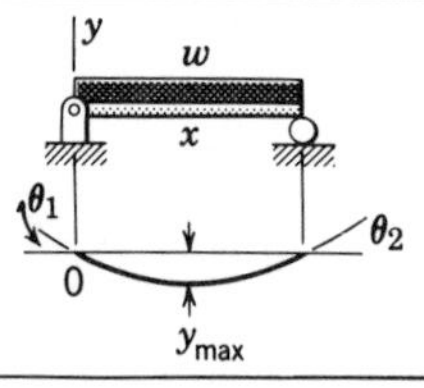	$\theta_1 = -\frac{wL^3}{24EI}$ at $x = 0$ $\theta_2 = +\frac{wL^3}{24EI}$ at $x = L$	$y_{max} = -\frac{5wL^4}{384EI}$ at $x = L/2$
8	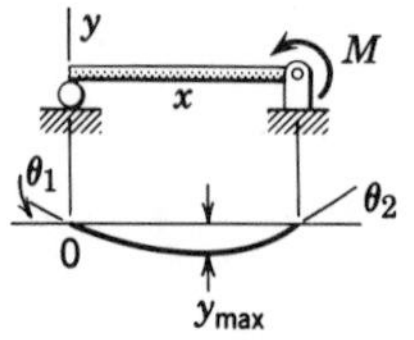	$\theta_1 = -\frac{ML}{6EI}$ at $x = 0$ $\theta_2 = +\frac{ML}{3EI}$ at $x = L$	$y_{max} = -\frac{ML^2}{9\sqrt{3}\,EI}$ at $x = L/\sqrt{3}$ $y_{center} = -\frac{ML^2}{16EI}$ not max

APPENDIX E
The International System of Units (SI)

THE INTERNATIONAL SYSTEM OF UNITS

E–1 INTRODUCTION The importance of the regulation of weights and measures was recognized as early as 1787 when Article 1, Section 8, of the United States Constitution was written. The metric system was legalized in the United States in 1866; in 1893, the international metre and kilogram became the fundamental standards of length and mass both for metric and customary weights and measures. International standardization began with an International Metric Convention in 1875, which established a permanent International Bureau of Weights and Measures. The National Bureau of Standards represents the United States in this international body.

The original metric system provided a set of units for the measurement of length, area, volume, capacity, and mass based on two fundamental units: the metre and the kilogram. With the addition of a unit of time, practical measurements began to be based on the metre-kilogram-second (MKS) system. In 1960, the Eleventh General Conference on Weights and Measures formally adopted the *International System of Units*, for which the abbreviation is *SI* in all languages, as the international standard. Thirty-six countries, including the United States, participated in this conference.

E–2 SI UNITS The International System of Units adopted by the conference includes three classes of units: (1) base units, (2) supplementary units, and (3) derived units. The system is founded on the seven base units listed in Table E–1.

Table E–1

Quantity	Name of Base SI Unit	Symbol
Length	metre	m
Mass	kilogram	kg
Time	second	s
Electric current	ampere	A
Thermodynamic temperature	kelvin	K
Amount of substance	mole	mol
Luminous intensity	candela	cd

Certain units of the international system have not been classified under either base units or derived units. These units, listed in Table E–2, are called supplementary units and may be regarded either as base units or as derived units.

Table E–2

Quantity	Name of Supplementary SI Unit	Symbol
Plane angle	radian	rad
Solid angle	steradian	sr

Derived units are expressed algebraically in terms of base units and/or supplementary units. Their symbols are obtained by means of the mathematical signs of multiplication and division. For example, the SI unit for velocity is metre per second (m/s) and the SI unit for angular velocity is rad per second (rad/s). For some of the derived units, special names and symbols exist; those of interest in mechanics are listed in Table E–3.

E–3 MULTIPLES OF SI UNITS Prefixes are used to form names and symbols of multiples (decimal multiples and sub-multiples) of SI units. The choice of the appropriate multiple is governed by convenience and should usually be chosen so that the numerical values will be between 0.1 and 1000. Only one prefix should be used in forming a multiple of a compound SI unit, and prefixes in the denominator should be avoided. Approved prefixes with their names and symbols are listed in Table E–4.

Table E–3

Quantity	Derived SI Unit	Special Name	Symbol
Area	square metre	—	m^2
Volume	cubic metre	—	m^3
Linear velocity	metre per second	—	m/s
Angular velocity	radian per second	—	rad/s
Linear acceleration	metre per second squared	—	m/s^2
Frequency	(cycle) per second	hertz	Hz
Density	kilogram per cubic metre	—	kg/m^3
Force	kilogram-metre per second squared	newton	N
Moment of force	newton-metre	—	N–m
Pressure	newton per metre squared	pascal	Pa
Stress	newton per metre squared	pascal	Pa or N/m^2
Work, energy	newton-metre	joule	J
Power	joule per second	watt	W

Table E–4

Factor by which Unit is Multiplied	Prefix	
	Name	Symbol
10^{12}	tera	T
10^{9}	giga	G
10^{6}	mega	M
10^{3}	kilo	k
10^{2}	hecto*	h
10	deca*	da
10^{-1}	deci*	d
10^{-2}	centi*	c
10^{-3}	milli	m
10^{-6}	micro	μ
10^{-9}	nano	n
10^{-12}	pico	p
10^{-15}	femto	f
10^{-18}	atto	a

*To be avoided where possible.

E–4 CONVERSION BETWEEN THE SI AND GRAVITATIONAL ENGLISH SYSTEMS. As the use of SI becomes more commonplace in the United States, engineers will be required to be familiar with both SI and the English system in common use today. As an aid to interpreting the physical significance of answers in SI units for those more accustomed to the English system, the following conversion factors are provided:

CONVERSION FACTORS

	English to SI	**SI to English**
Length:	1 in. = 25.40 mm	1 m = 39.37 in.
	1 ft = 0.3048 m	1 m = 3.281 ft
Area:	1 in.^{2} = 6.452 cm^2	1cm^2 = 0.1550 in.^{2}
	1 ft^2 = 0.09290 m^2	1 m^2 = 10.76 ft^2
Volume:	1 in.^{3} = 16.39 cm^3	1 cm^3 = 0.06102 in.^{3}
	1 ft^3 = 0.02832 m^3	1 m^3 = 35.31 ft^3
Moment of Inertia:	1 in.^{4} = 41.62 cm^4	1 cm^4 = 0.02403 in.^{4}
Force:	1 lb = 4.448 N	1 N = 0.2248 lb
Distributed Load:	1 lb/ft = 14.59 N/m	1 kN/m = 68.53 lb/ft
Pressure or Stress:	1 psi = 6.895 kPa	1 MPa = 145.0 psi
	1 ksi = 6.895 MN/m^2	1 MN/m^2 = 145.0 psi
Bending Moment or Torque:	1 ft-lb = 1.356 N-m	1 N-m = 0.7376 ft-lb
Work or Energy:	1 ft-lb = 1.356 J	1 J = 0.7376 ft-lb
Power:	1 hp = 745.7 W	1 kW = 1.341 hp

Numerical solutions to engineering problems often require use of unit conversion factors, since for convenience the various quantities involved in the calculations are expressed in different size units. For example, the dimensions of a cross section may be expressed in millimetres (mm) while a bending moment or a torque will likely be expressed in newton-metres (N-m). A procedure for handling this situation, when SI units are used, is illustrated in the following examples.

Example E–1 A hollow steel shaft 2 m long with outside and inside diameters of 100 mm and 80 mm is fixed at one end and subjected to a torque of 15 kN-m at the other (free) end. Determine:

(a) The maximum shearing stress in the shaft.

(b) The magnitude of the angle of twist of the free end of the shaft.

Solution (a) The elastic torsion formula $\tau = T\rho/J$ will give the maximum shearing stress (see Art. 4–2). All terms will be expressed in terms of the base units of metres and newtons; thus

$$T = 15 \text{ kN-m} = 15(10^3) \text{ N-m}$$

$$\rho = 50 \text{ mm} = 0.050 \text{ m}$$

$$J = \frac{\pi}{32}(0.100^4 - 0.080^4) = 5.796(10^{-6})\text{m}^4$$

from which

$$\tau = \frac{15(10^3)(0.050)}{5.796(10^{-6})} = 1.294(10^8)\text{N/m}^2 = \underline{129.4\ \text{MN/m}^2} \qquad \text{Ans.}$$

(b) The angle of twist is given by Eq. 4–3*b*, $\theta = TL/JG$, where $G = 80\ \text{GN/m}^2 = 80(10^9)\text{N/m}^2$; thus,

$$\theta = \frac{15(10^3)(2)}{5.796(10^{-6})80(10^9)} = \underline{0.0647\ \text{rad}} \qquad \text{Ans.}$$

Example E–2 A simple beam has a hollow rectangular cross section with outside dimensions of 120 mm wide by 200 mm deep and inside dimensions of 60 mm wide by 140 mm deep. The beam has a span of 4 m and carries a uniformly distributed load of w N/m over the entire span. The material is douglas fir having a modulus of elasticity of 13 GN/m^2. The maximum allowable flexural stress is 10 MN/m^2 and the maximum allowable deflection is 10 mm. Determine the maximum permissible value for w.

Solution The maximum bending moment (at the center of the span) obtained by methods of Arts. 5–6, 5–7, and 5–8 is $wL^2/8$. All quantities will be expressed in newtons and metres. The elastic flexure formula $\sigma = My/I$ relates the bending moment and stress (see Art. 5–4). The second moment of the area is

$$I = \frac{1}{12}\left[0.120(0.200)^3 - 0.060(0.140)^3\right] = 66.28(10^{-6})\text{m}^4$$

which can be substituted into the flexure formula to give

$$\sigma = 10(10^6) = \frac{\left[w(4^2)/8\right](0.100)}{66.28(10^{-6})}$$

from which

$$w = 3314\ \text{N/m}$$

The maximum deflection (see App. D) is

$$y_{max} = \frac{5wL^4}{384EI}$$

$$0.010 = \frac{5w(4^4)}{384\left[13(10^9)\right]\left[66.28(10^{-6})\right]}$$

from which $w = 2585$ N/m, which is less than 3314 N/m; therefore,

$$w_{max} = \underline{2585\ \text{N/m}} \qquad \text{Ans.}$$

Answers

CHAPTER 1

1–1 (a) 15.0 ksi T.
(b) 12.0 ksi C.

1–2 (a) 25.0 MN/m^2 C.
(b) 15.0 MN/m^2 C.
(c) 22.5 MN/m^2 C.

1–3 14,250 lb. ↓

1–4 653 kN. ↓

1–6 $A_A = 500$ mm^2.
$A_B = 667$ mm^2.

1–7 10.0 in.

1–10 100.0 mm.

1–12 12.60 ksi T.

1–14 15.25 MN/m^2 C.

1–17 10,000 psi.

1–18 314 kN.

1–20 (a) 1200 mm^2.
(b) 20.9 mm.

1–21 (a) $\sigma = 5.00$ ksi T.
$\tau = 8.67$ ksi.
(b) $\sigma_{max} = 20.0$ ksi T.
$\tau_{max} = 10.0$ ksi.

1–22 (a) $\sigma = 30.0$ MN/m^2 T.
$\tau = 52.0$ MN/m^2.
(b) $\sigma_{max} = 120.0$ MN/m^2 T.
$\tau_{max} = 60.0$ MN/m^2.

1–23 (a) 107,500 lb.
(b) 1287 psi.
(c) 1680 psi.

1–24 (a) 467 kN.
(b) 3.64 MN/m^2 C.
(c) 15.56 MN/m^2.

1–26 9840 N.

1–27 4700 lb.

1–29 9.61 ksi T.

1–32 $\sigma = 30.0$ MN/m^2 C.
$\tau = 86.6$ MN/m^2.

1–34 14.40 MN/m^2 T.

1–35 4400 psi T.

1–37 28.9 MN/m^2 T.

1–39 $\sigma_x = 21.6$ ksi C.
$\tau_{xy} = -31.2$ ksi.

1–41 Partial Results:
$\sigma_{p1} = 14.0$ ksi T.
$\sigma_{p2} = 6.00$ ksi C.
$\sigma_{p3} = 0$.
$\tau_{max} = 10.0$ ksi.
$\theta_p = 18.43°$. ↻

1–42 Partial Results:
$\sigma_{p1} = 170.0$ MN/m^2 T.
$\sigma_{p2} = 90.0$ MN/m^2 C.
$\sigma_{p3} = 0$.
$\tau_{max} = 130.0$ MN/m^2.
$\theta_p = 11.31°$. ↻

1–43 Partial Results:
$\sigma_{p1} = 31.0$ ksi T.
$\sigma_{p2} = 19.0$ ksi C.
$\sigma_{p3} = 0$.
$\tau_{max} = 25.0$ ksi.
$\theta_p = 8.13°$. ↻

1–44 $\sigma_{p1} = 70.0$ MN/m^2 T.
$\sigma_{p2} = 100.0$ MN/m^2 C.
$\sigma_{p3} = 0$.
$\tau_{max} = 85.0$ MN/m^2.
$\theta_p = 14.04°$. ↻

1–47 (a) 800 psi.
(b) 75° ↻, 15° ↺.
(c) −693 psi.

1–48 Partial Results:
$\sigma_y = 30.0$ MN/m^2 C.
$\tau_{xy} = -60.0$ MN/m^2.
$\sigma_{p2} = 50.0$ MN/m^2 C.
$\theta_p = 18.43°$. ↻

1–50 Partial Results:
$\tau_{xy} = \pm 12.0$ ksi.
$\sigma_{p2} = 16.0$ ksi C.
$\tau_{max} = 13.0$ ksi.
$\theta_p = \pm 33.7°$.

1–52 Partial Results:
$\sigma_x = 28.7$ ksi T.
$\sigma_y = 11.34$ ksi T.
$\sigma_{p2} = 10.0$ ksi T.
$\tau_{max} = 15.0$ ksi.

1–54 Partial Results:
$\sigma_y = 6.00$ MN/m^2 T.
$\tau_{xy} = \pm 33.5$ MN/m^2.
$\sigma_{p2} = 14.00$ MN/m^2 C.
$\theta_p = \pm 30.9°$.

1–56 See 1–41.

1–57 See 1–42.

1–58 See 1–43.

1–59 See 1–44.

1–62 See 1–47.

1–63 See 1–48.

1–64 See 1–50.

1–67 See 1–54.

1–69 Partial Results:
(a) $\sigma = 9.62$ MN/m^2 T.
$\tau = 23.3$ MN/m^2.
(b) $\sigma_{p1} = 17.70$ MN/m^2 T.
$\sigma_{p2} = 57.7$ MN/m^2 C.
$\sigma_{p3} = 0$.
$\tau_{max} = 37.7$ MN/m^2.
$\theta_p = 10.90°$. ↻

1–71 Partial Results:
$\sigma_x = 134.0$ MN/m^2 C.
$\tau_{xy} = 84.0$ MN/m^2.
$\sigma_{p2} = 190.0$ MN/m^2 C.
$\theta_p = 33.7°$. ↻

1–73 Partial Results:
$\sigma_x = 95.0$ MN/m^2 T.
$\sigma_y = 25.0$ MN/m^2 T.
$\tau_{max} = 65.0$ MN/m^2.
$\theta_p = 11.31°$. ↻

1–76 Partial Results:
$\sigma_x = 15.50$ ksi T.
$\tau_{xy} = -7.50$ ksi.
$\sigma_{p2} = 3.00$ ksi T.

1–79 $\sigma = 3.78$ ksi T.
$\tau = 2.74$ ksi.

1–80 $\sigma = 71.5$ MN/m^2 T.
$\tau = 20.7$ MN/m^2.

1–83 $\sigma_{p1} = 6.39$ ksi T.
$\sigma_{p2} = 2.00$ ksi T.
$\sigma_{p3} = 4.39$ ksi C.
$\tau_{max} = 5.39$ ksi.

1–84 $\sigma_{p1} = 8.15$ ksi T.
$\sigma_{p2} = 1.000$ ksi C.
$\sigma_{p3} = 5.15$ ksi C.
$\tau_{max} = 6.65$ ksi.

1–85 $\sigma_{p1} = 106.2$ MN/m^2 T.
$\sigma_{p2} = 10.00$ MN/m^2 T.
$\sigma_{p3} = 3.77$ MN/m^2 T.
$\tau_{max} = 51.2$ MN/m^2.

1–86 $\sigma_{p1} = 6.00$ ksi T.
$\sigma_{p2} = 2.00$ ksi T.
$\sigma_{p3} = 4.00$ ksi C.
$\tau_{max} = 5.00$ ksi.

1–87 $\sigma_{p1} = 90.0$ MN/m^2 T.
$\sigma_{p2} = 60.0$ MN/m^2 T.
$\sigma_{p3} = 30.0$ MN/m^2 T.
$\tau_{max} = 30.0$ MN/m^2.

1–90 (a) $\sigma = 22.4$ ksi T.
$\tau = 15.15$ ksi.
(b) $\sigma_{p1} = 32.4$ ksi T.
$\sigma_{p2} = 10.00$ ksi T.
$\sigma_{p3} = 12.36$ ksi C.
$\tau_{max} = 22.4$ ksi.
(c) $\theta_x = 45.0°$.
$\theta_y = 50.8°$.
$\theta_z = 71.6°$.

1–91 (a) $\sigma_{p1} = 146.9$ MN/m^2 T.
$\sigma_{p2} = 41.9$ MN/m^2 T.
$\sigma_{p3} = 8.77$ MN/m^2 C.
$\tau_{max} = 77.8$ MN/m^2.
(b) $\theta_x = 41.2°$.
$\theta_y = 64.4°$.
$\theta_z = 60.2°$.

CHAPTER 2

2–1 $\epsilon_8 = 0.3125$.
$\epsilon_2 = 0.565$.
$\epsilon_2/\epsilon_8 = 1.808$.

2–2 0.320.

2–3 $\epsilon_C = 1.0256\epsilon_B - 0.0256$.

2–4 0.00480.

2–6 0.00300.

2–11 1550 μrad.

2–12 2820 μrad.

2–13 3990 μrad.

2–14 2830 μrad.

2–16 Partial Results:
$\epsilon_{p1} = 1400\mu$.
$\epsilon_{p2} = -600\mu$.
$\epsilon_{p3} = 0$.
$\gamma_{max} = 2000\mu$.
$\theta_p = 18.43°$. ↻

2–17 Partial Results:
$\epsilon_{p1} = 1700\mu$.
$\epsilon_{p2} = -900\mu$.
$\epsilon_{p3} = 0$.
$\gamma_{max} = 2600\mu$.
$\theta_p = 11.31°$. ↻

2–19 Partial Results.
$\epsilon_{p1} = 1410\mu$.
$\epsilon_{p2} = 390\mu$.
$\epsilon_{p3} = 0$.
$\gamma_{max} = 1410\mu$.
$\theta_p = 39.3°$. ↺

2–22 Partial Results:
$\gamma_{xy} = \pm 2400\mu$.
$\epsilon_{p2} = -1600\mu$.
$\gamma_{max} = 2600\mu$.
$\theta_p = \pm 33.7°$.

2–23 Partial Results:
$\epsilon_y = 250\mu$.
$\gamma_{xy} = \pm 1520\mu$.
$\epsilon_{p2} = -300\mu$.
$\theta_p = \pm 35.9°$.

2–24 Partial Results:
$\epsilon_x = 2866\mu$.
$\epsilon_y = 1134\mu$.
$\epsilon_{p2} = 1000\mu$.
$\gamma_{max} = 3000\mu$.

2–28 Partial Results:
$\epsilon_{p1} = 240\mu$.
$\epsilon_{p2} = -34.3\mu$.
$\epsilon_{p3} = -160\mu$.
$\gamma_{max} = 400\mu$.
$\theta_p = 18.43°$. ↺

2–30 Partial Results:
$\epsilon_{p1} = 955\mu$.
$\epsilon_{p2} = -648\mu$.
$\epsilon_{p3} = -131.6\mu$.
$\gamma_{max} = 1604\mu$.
$\theta_p = 1.92°$. ↺

2–32 Partial Results:
$\epsilon_{p1} = 531\mu$.
$\epsilon_{p2} = 69\mu$.
$\epsilon_{p3} = -257\mu$.
$\gamma_{max} = 788\mu$.
$\theta_p = 15.0°$. ↻

2–34 Partial Results:
$\epsilon_{p1} = 420\mu$.
$\epsilon_{p2} = -100\mu$.
$\epsilon_{p3} = -137.1\mu$.
$\gamma_{max} = 557\mu$.
$\theta_p = 11.31°$. ↺

2–36 Partial Results:
$\epsilon_{p1} = 750\mu$.
$\epsilon_{p2} = -550\mu$.
$\epsilon_{p3} = -85.7\mu$.
$\gamma_{max} = 1300\mu$.
$\theta_p = 11.31°$. ↺

2–37 150μ.

2–41 (a) $10.7(10^6)$ psi.
(b) 0.326.
(c) 48 ksi.

2–42 $\nu = 0.35$.
$E = 172$ GN/m^2.

2–44 (a) 0.35.
(b) 95.5 GN/m^2.
(c) 35.4 GN/m^2.

2–45 (a) $26.7(10^6)$ psi.
(b) 38 ksi.
(c) 73 ksi.

2–49 Partial Results:
$\sigma_{p1} = 2230$ psi T.
$\sigma_{p2} = 972$ psi C.
$\sigma_{p3} = 0$.
$\tau_{max} = 1600$ psi.
$\theta_p = 18.43°$. ↺

2–50 Partial Results:
$\sigma_{p1} = 15.33$ MN/m^2 T.
$\sigma_{p2} = 6.57$ MN/m^2 C.
$\sigma_{p3} = 0$.
$\tau_{max} = 10.95$ MN/m^2.
$\theta_p = 18.43°$. ↺

2–52 Partial Results:
$\sigma_{p1} = 18.55$ ksi T.
$\sigma_{p2} = 10.05$ ksi C.
$\sigma_{p3} = 0$.
$\tau_{max} = 14.30$ ksi.
$\theta_p = 11.31°$. ↺

2–53 Partial Results:
(a) $\sigma_x = 200$ MN/m^2 T.
$\sigma_y = 0$.
$\tau_{xy} = -100.0$ MN/m^2.
(b) $\sigma_{p1} = 241$ MN/m^2 T.
$\sigma_{p2} = 41.4$ MN/m^2 C.
$\sigma_{p3} = 0$.
$\tau_{max} = 141.4$ MN/m^2.
$\theta_p = 22.5°$. ↻

2–55 Partial Results:
(a) $\sigma_x = 6320$ psi T.
$\sigma_y = 942$ psi T.
$\tau_{xy} = 1120$ psi.
(b) $\sigma_{p1} = 6540$ psi T.
$\sigma_{p2} = 718$ psi T.
$\sigma_{p3} = 0$.
$\tau_{max} = 3270$ psi.
$\theta_p = 11.31°$. ↺

2–58 Partial Results:
(a) $\epsilon_{p1}=190\mu$.
$\epsilon_{p2}=90\mu$.
$\epsilon_{p3}=-150.8\mu$.
$\gamma_{max}=341\mu$.
$\theta_p=18.43°$. ↻
(b) $\sigma_{p1}=11.36$ MN/m^2 T.
$\sigma_{p2}=8.03$ MN/m^2 T.
$\sigma_{p3}=0$.
$\tau_{max}=5.68$ MN/m^2.

2–61 (a) 30.1 in.-lb/in.^{3}.
(b) 30.1 in.-lb/in.^{3}.

2–62 (a) 1441 N-m/m^3.
(b) 1441 N-m/m^3.

2–64 (a) 0.345 in.-lb/in.^{3}.
(b) 0.345 in.-lb/in.^{3}.

2–66 (a) 26.6 kN-m/m^3.
(b) 26.6 kN-m/m^3.

CHAPTER 3

3–1 37.5 kips.

3–2 (a) 240 kN.
(b) 2.08.

3–4 (a) 2.25.
(b) 3.89 mm.

3–5 1.886 kips.

3–7 (a) 2.56.
(b) 2.44.

3–9 $\frac{WL}{2AE}$.

3–12 $\frac{5\gamma L^2}{6E}$.

3–13 0.0503 in. to the right.

3–14 1.375 mm upward.

3–15 (a) 0.0479 in. longer.
(b) 1.80.

3–17 (a) 10.83 ksi T.
(b) 5.42 ksi.
(c) 0.00318 in. decrease.

3–19 2.83 ksi T.

3–21 (a) $\frac{100\alpha E}{3}$.
(b) 4170 psi C.

3–23 0.0012.

3–24 0.00090.

3–26 0.00048.

3–28 $k_A=2.50$.
$k_B=2.67$.

3–29 $k_A=4.17$.
$k_B=2.22$.

3–31 (a) $\sigma_s=45$ ksi C.
$\sigma_o=2.7$ ksi C.
(b) 367 kips.
(c) 1.426 in. to the right.

3–32 (a) $\sigma_A=60$ MN/m^2 T.
$\sigma_B=20$ MN/m^2 T.
(b) 0.06 mm downward.
(c) 66.7 mm to the right of A.

3–33 (a) 37,500 lb.
(b) 8.00 in. to the right of A.
(c) 51,000 lb.
5.88 in. to the right of A.

3–34 88,800 lb.

3–35 $\sigma_A=40$ ksi T.
$\sigma_B=46$ ksi C.
$\delta=0.12$ in. downward.

3–37 $\sigma_A=25$ ksi T.
$\sigma_B=30$ ksi T.

3–39 46 kN.

3–41 40 kips.

3–43 (a) 3680 psi T.
(b) 5680 lb.

3–45 (a) 25.53 ksi T.
(b) $d=0.141$ in.
$\delta_A=0.072$ in. T.
$\delta_B=0.096$ in. T.
(c) $\sigma_A=36$ ksi T.
$\sigma_B=24$ ksi T.

3–47 (a) 5.96 kips upward.
(b) 10.80 kips downward.

3–50 66 kN.

3–52 153 kN.

3–54 $\sigma_A = 52$ ksi T.
$\sigma_B = 36$ ksi T.
$y_C = 0.216$ in. downward.

3–56 301 kN.

3–59 44.4 kN.

3–61 (a) 15,590 psi T.
(b) 4.49.

3–62 $t = 15$ mm.
$k = 5.00$.

3–64 150 kN T.

3–67 $\sigma = 6340$ psi T.
$\tau = 2450$ psi.

3–69 (a) 370μ T.
(b) 86 $\mathrm{MN/m^2}$ T.
(c) 1.72 MPa.
(d) 5.23.

3–71 $\sigma_a = 252$ psi T.
$\sigma_h = 907$ psi T.

3–73 $\sigma_{aA} = 1000$ psi T.
$\sigma_{hA} = 2110$ psi T.
$\sigma_{aB} = 1000$ psi T.
$\sigma_{hB} = 1909$ psi T.

3–75 (a) 26,000 psi T.
(b) $\sigma_t = 19{,}520$ psi T.
$\sigma_r = 3520$ psi C.

3–77 $\sigma_t = 258\ \mathrm{MN/m^2}$ C.
$\sigma_r = 126.3\ \mathrm{MN/m^2}$ C.

3–79 (a) 0.0102 in.
(b) $\sigma_T = 37.5$ ksi T.
$\sigma_J = 46.6$ ksi T.

3–80 20.9 mm.

3–83 2.90 in.

3–84 (a) 179.7 MPa.
(b) 30.1 MPa.

CHAPTER 4

4–1 2010 in.-lb.

4–2 262 kN-m.

4–3 $\dfrac{65\pi c^3 \tau_{\max}}{512}$.

4–4 25.1 ft-kips.

4–5 (a) 0.004 rad.
(b) 1.292.

4–7 (a) 0.003 rad.
(b) 292 in.-kips.

4–9 $T = 6.80$ kN-m.
$\theta = 0.5$ rad.

4–12 $T = 458$ in.-kips.
$\theta = 0.0885$ rad.

4–14 42.4 in.-kips.

4–17 (a) $T_{ABa} = 20$ N-m.
$T_{BCa} = 12$ N-m.
$T_{ABb} = 12$ N-m.
$T_{ACb} = 8$ N-m.
(b) System b.

4–19 (a) 91.7 $\mathrm{MN/m^2}$.
(b) 3.67 mm.

4–21 3360 N-m.

4–23 37,300 in.-lb.

4–25 (a) $\dfrac{8.15}{d^4}$ rad.
(b) 3.71 in.

4–26 (a) 127.3 $\mathrm{MN/m^2}$.
(b) 0.0223 rad.

4–28 $T_1 = 10{,}470$ N-m.
$T_2 = 5240$ N-m.

4–29 (a) 320 in.-kips.
(b) 0.024 rad.

4–32 (a) 12 ksi.
(b) 109.6 in.-kips.
(c) 0.040 rad.

4–33 (a) 18 ksi.
(b) 0.014 rad.

4–35 $\dfrac{3TL}{16Gt\pi r^3}$.

4–39 $\dfrac{qL^2}{3\pi Gc^4}$.

4–40 4310 in.-lb.

4–41 1611 N-m.

4–42 Partial Results:
(a) $\sigma = 8860$ psi C.
$\tau = 1563$ psi.
(b) $\sigma_{p1} = 9000$ psi T.
$\sigma_{p2} = 9000$ psi C.
$\theta_p = 45°$.

4–43 (a) 50 mm.
(b) 57.6 MN/m^2 T.

4–48 Partial Results:
(a) $\sigma_{p1} = 16{,}000$ psi T.
$\sigma_{p2} = 4000$ psi C.
$\sigma_{p3} = 0$.
$\tau_{max} = 10{,}000$ psi.
$\theta_p = 26.6°$. ↺
(b) $\sigma = 14{,}000$ psi T.
$\tau = 6000$ psi.

4–49 Partial Results:
$\sigma_{p1} = 97.9$ MN/m^2 T.
$\sigma_{p2} = 33.9$ MN/m^2 C.
$\sigma_{p3} = 0$.
$\tau_{max} = 65.9$ MN/m^2.
$\theta_p = 30.5°$. ↺

4–50 Partial Results:
(a) $\sigma_{p1} = 7390$ psi T.
$\sigma_{p2} = 3390$ psi C.
$\sigma_{p3} = 0$.
$\tau_{max} = 5390$ psi.
$\theta_p = 34.1°$. ↺
(b) 0.001 rad.

4–51 (a) Anywhere on the surface of the left segment.
(b) $\sigma_{p1} = 29.2$ MN/m^2 T.
$\sigma_{p2} = 7.87$ MN/m^2 C.
$\sigma_{p3} = 0$.
$\tau_{max} = 18.54$ MN/m^2.
$\theta_p = 27.4°$. ↺

4–54 27.3 kN-m.

4–55 199 kips.

4–56 (a) 449 hp.
(b) 0.048 rad.

4–57 (a) 165.8 kW.
(b) 132.6 kW.

4–58 (a) 2.89 in.
(b) 500 rpm.

4–61 101.1 mm.

4–62 3.58 in.

4–63 $\tau_{max} = 12{,}220$ psi.
$\theta = 0.01630$ rad.

4–64 128 MN/m^2.

4–65 0.0400 rad.

4–66 748 in.-kips.

4–67 16,990 N-m.

4–69 69,300 N-m.

4–71 6140 N-m.

4–72 9.17 ksi.

4–74 534 in.-kips.

4–76 224 N-m.

4–78 (a) 6220 psi.
(b) 0.00228 rad.

4–81 5500 N-m.

4–83 48,900 N-m.

4–85 $\tau_s = 18$ ksi.
$\tau_b = 23.8$ ksi.

4–86 (a) $\tau_s = 120.0$ MN/m^2.
$\tau_b = 151.1$ MN/m^2.
(b) 0.1259 rad.

4–87 (a) $T_1 = 63.7$ in.-kips.
$T_2 = 50.0$ in.-kips.
$T_3 = 48.0$ in.-kips.
(b) $\theta_1 = 0.0262$ rad.
$\theta_2 = 0.0333$ rad.
$\theta_3 = 0.0347$ rad.

4–88 (a) $T_1 = 14.61$ kN-m.
$T_2 = 11.48$ kN-m.
$T_3 = 8.83$ kN-m.
(b) $\theta_1 = 0.0334$ rad.
$\theta_2 = 0.0425$ rad.
$\theta_3 = 0.0552$ rad.

4–90 (a) 12.24 kN-m.
(b) 0.0278 rad.

4–92 (a) 124.9 kN-m.
(b) 0.00575 rad.

4–93 (a) 448 in.-kips.
(b) $\tau_{AB}=12.50$ ksi.
$\tau_{BC}=7.81$ ksi.
(c) 0.00387 rad.

CHAPTER 5

5–1 (a) 64,000 in.-lb.
(b) 21.1%

5–2 (a) 54.4 σ_{max}.
(b) 1088 in.-kips.

5–3 13.38 kN-m.

5–8 9910 in.-kips.

5–9 4270 kN-m.

5–11 (a) 6.67 in. above the bottom of the stem.
(b) 1920 in.-kips.
(c) 1.714.

5–13 (b) 190.8 kN-m.

5–19 490 psi T.

5–20 232 psi C.

5–21 20.3 MN/m^2 T.

5–22 21.2 MN/m^2 C.

5–24 22.9 MN/m^2 T and C.

5–28 (a) 936 psi.
(b) 75.3° from the Z axis. ↶

5–29 (a) 9550 psi.
(b) 67.4° from the Z axis. ↷

5–30 (a) 8.40 MN/m^2 T and C.
(b) 66.6° from the Z axis. ↶

5–31 (a) 42.3 MN/m^2 T.
(b) 55.8 MN/m^2 C.
(c) 75.4° from the Z axis. ↷

5–33 (a) 2.22 ksi T.
(b) 61.8° from the Z axis. ↶

5–35 (a) 9.37 ksi C.
(b) 53.1° from the Z axis. ↷

5–37 $V=5600-1000x$.
$M=-500x^2+5600x-8000$.

5–38 $V=35-15x$.
$M=-7.5x^2+35x-17.5$.

5–41 $V=2100-300x$.
$M=-150x^2+2100x-2000$.

5–42 $V=9-7.5x$.
$M=-3.75x^2+9x+1.5$.

5–43 (a) 66.0 ft.
(b) $V=970x-24{,}000$.
$M=485x^2-24{,}000x+192{,}000$.

5–49 (a) With origin at the right end:
$V=8x-6$.
$M=-4x^2+6x-30$.
(b) Partial Results:
$M_{10\text{ ft}}=-370$ and -350 ft-kips.
$M_{15\text{ ft}}=-720$ ft-kips.

5–50 (a) $V=2x^2-6$.
$M=-\frac{2}{3}x^3+6x$.
(b) Partial Results:
$M=6.93$ kN-m at $x=\sqrt{3}$.

5–51 (a) $V=1100$.
$M=1100x-4000$.
(b) Partial Results:
$M_0=-4000$ ft-lb.
Inflection point at $x=3.64$ ft.

5–52 (a) $V=2400-400x$.
$M=-200x^2+2400x-6000$.
(c) -6000 ft-lb.

5–53 (a) $V=3x-7$.
$M=-1.5x^2+7x-6$.
(c) -12 kN-m.

5–60 Partial Results:
$M_{5\text{ ft}}=-4000$ ft-lb.
$M_{9\text{ ft}}=2400$ and -600 ft-lb.
$M_{15\text{ ft}}=3000$ ft-lb.
Inflection Points:
2.25 ft to the right of B.
0.523 ft to the right of C.

With the origin at B and x positive to the right:
$M = -100x^2 + 2000x - 4000$.

5–61 Partial Results:
$M_{2\text{ m}} = -2$ kN-m.
$M_{4.5\text{ m}} = 23$ and 15 kN-m.
$M_{7\text{ m}} = 27.5$ kN-m.
Inflection Points:
0.367 m to the left of B.
0.200 m to the right of B.
With the origin at B and x positive to the right:
$M = -2x^2 + 20x - 22.5$.

5–70 93.8 MN/m^2 C at the bottom at A.

5–71 Partial Results:
(a) 136.2 MN/m^2 T.
(b) 97.3 MN/m^2 C.

5–72 Partial Results:
(a) 5440 psi T.
(b) 8170 psi C.

5–77 (a) $M_2 = 3600$ N-m.
$F_2 = 22.7$ kN T.
$F_1 = 15.12$ kN T.
(b) 7.56 kN.
(c) 0.315 MN/m^2.

5–78 (a) 2160 lb.
(b) 90 psi.

5–79 (a) 720 lb.
(b) 30 psi.

5–81 118.4 psi (top joint).
101.5 psi (bottom joint).

5–82 (a) 114.1 psi at the neutral axis.
(b) 95.8 psi.

5–83 (a) 0.751 MN/m^2.
(b) 0.927 MN/m^2 at the neutral surface at the supports.

5–84 (a) 161.5 psi T.
(b) 111.1 psi.
(c) 111.1 psi.

5–85 $\sigma = 3.50$ MN/m^2 C.
$\tau = 1.084$ MN/m^2.

5–87 85.3 kN.

5–90 (a) 1200 psi T and C.
(b) 140 psi.

5–92 (a) 11.73 MN/m^2 C.
(b) 0.694 MN/m^2.
(c) 0.1481 MN/m^2.
(d) 6.17 MN/m^2 T.

5–94 (a) 1522 psi C.
(b) 108.8 psi.
(c) 947 psi T.

5–96 14.58 kN.

5–98 8700 N/m.

5–101 215 lb/ft.

5–103 (a) 138.1 lb/ft.
(b) 117.6 psi.

5–104 Partial Results:
$\sigma_{p1} = 12{,}240$ psi T.
$\sigma_{p2} = 1312$ psi C.
$\sigma_{p3} = 0$.
$\tau_{max} = 6780$ psi.
$\theta_p = 18.13°$. ↺

5–105 Partial Results:
$\sigma_{p1} = 0.0537$ MN/m^2 T.
$\sigma_{p2} = 2.14$ MN/m^2 C.
$\sigma_{p3} = 0$.
$\tau_{max} = 1.097$ MN/m^2.
$\theta_p = 9.00°$. ↺

5–111 1.192 in. outside of the channel.

5–112 $F = 65.6$ kN.
$e = 42.6$ mm to the left.

5–114 9.124 in.

5–115 15 mm.

5–118 $d = 2R$.

5–119 $\dfrac{b(2h+3b)}{2(h+3b)}$ to the left.

5–121 $\sigma_w = 1286$ psi T.
$\sigma_a = 12{,}860$ psi T.

5–122 $\sigma_w = 3.79$ MN/m^2 T and C.
$\sigma_s = 94.7$ MN/m^2 T and C.

5–125 $\sigma_{s\,max} = 15{,}700$ psi T and C.

$\sigma_{w\,max} = 837$ psi T and C.

5–126 $\sigma_{s\,max} = 153.7\ MN/m^2$ T and C.
$\sigma_{w\,max} = 6.99\ MN/m^2$ T and C.

5–129 $\sigma_c = 7.52\ MN/m^2$ C.
$\sigma_s = 146.8\ MN/m^2$ T.

5–131 $\sigma_i = 18.29$ ksi C.
$\sigma_o = 15.35$ ksi T.

5–132 156 mm.

5–133 $\sigma_i = 3.73$ ksi C.
$\sigma_o = 3.01$ ksi T.

5–134 $\sigma_i = 58.3\ MN/m^2$ T.
$\sigma_o = 61.0\ MN/m^2$ C.

5–135 $\sigma_i = 6.10$ ksi C.
$\sigma_o = 10.14$ ksi T.

CHAPTER 6

6–1 (a) $\rho = 7620$ in.
$\sigma = 7880$ psi T and C.
(b) $\rho = 12{,}980$ in.
$\sigma = 7510$ psi T.

6–2 (a) $\rho = 583$ m.
(b) $\sigma = 25.7\ MN/m^2$ T and C.
(c) Circle.

6–5 $\rho = 3110$ in.
$\delta = 0.833$ in. upward.

6–6 $\sigma = 36.3\ MN/m^2$ T.
$\delta = 0.660$ mm upward.

6–9 (a) $y = \dfrac{M}{2EI}(x^2 - Lx)$.
(b) Parabola
(c) $\dfrac{1}{\rho} \approx \dfrac{d^2y}{dx^2}$.

6–10 $y = \dfrac{w}{48EI}(3Lx^3 - 2x^4 - L^3x)$.

6–11 With the origin at the left end:
$y = \dfrac{w}{24EI}(Lx^3 - x^4)$.

6–12 $y = \dfrac{w}{48EI}(3Lx^3 - 2x^4 - L^3x)$.

6–13 $y = \dfrac{w}{16EI}(Lx^3 - 4L^2x^2 + 3L^3x)$.

6–14 $y = \dfrac{w}{48EI}(3Lx^3 - 3L^2x^2 - 2x^4 + 2L^3x)$.

6–17 2.27 in. to the right of C.

6–18 (a) 16.67 mm. ↑
(b) 16.67 mm. ↓

6–21 $\dfrac{wL^4}{24EI}$. ↓

6–22 $y = \dfrac{w}{72EI}(5Lx^3 - 3L^2x^2 - 3x^4 + L^3x)$.

6–23 0.517L ft.

6–24 $\dfrac{2PL^3}{27EI}$. ↓

6–27 (a) $0 \leqslant x \leqslant L/5$
$y = \dfrac{P}{375EI}(50x^3 - 18L^2x)$.
$L/5 \leqslant x \leqslant L$
$y = \dfrac{P}{750EI}\left[100x^3 - 125\left(x - \dfrac{L}{5}\right)^3 - 36L^2x\right]$.
(b) $x = 0.434L$.
(c) $y_{max} = 0.01207\dfrac{PL^3}{EI}$. ↓

6–28 (a) $x = L$.
(b) $\dfrac{PL^3}{24EI}$. ↓

6–31 (a) $y = \dfrac{w}{24EI}[2ax^3 - (x-a)^4 - 13a^3x]$.
(b) 0.0769 in. ↓

6–32 (a) $y = \dfrac{w}{120EIL}[10L^2x^3 - 40L^3x^2 - 5L(x-L)^4 + (x-L)^5]$.
(b) 3.65 mm. ↓

6–33 $\dfrac{67PL^3}{162EI}$. ↓

6–34 $\dfrac{29wL^4}{48EI}$. ↓

6–35 (a) $y = \frac{P}{750EI}\left[100x^3 - 125\left\langle x - \frac{L}{5}\right\rangle^3 - 36L^2x\right].$
(b) $0.434L$.
(c) $y_{max} = 0.01207\frac{PL^3}{EI}.\downarrow$

6–36 (a) $x = L$.
(b) $\frac{PL^3}{24EI}.\downarrow$

6–39 (a) $y = \frac{w}{24EI}[2ax^3 - \langle x-a\rangle^4 - 13a^3x].$
(b) 0.0769 in. $\downarrow$

6–40 (a) $y = \frac{w}{120EIL}[10L^2x^3 - 40L^3x^2 - 5L\langle x-L\rangle^4 + \langle x-L\rangle^5].$
(b) 3.65 mm. $\downarrow$

6–43 4.20 mm. $\downarrow$

6–44 0.334 in. $\uparrow$

6–45 11.31 mm. $\downarrow$

6–56 $\frac{PL^3}{2EI}.\uparrow$

6–57 $\frac{17PL^3}{81EI}.\downarrow$

6–58 $\frac{PL^3}{2EI}.\downarrow$

6–59 (a) $\frac{31PL^3}{6EI}.\downarrow$
(b) $\frac{1489PL^3}{256EI}.\downarrow$

6–61 $\frac{11wL^4}{120EI}.\downarrow$

6–64 $\frac{35wL^4}{6144EI}.\downarrow$

6–65 (a) 1.515 kN.
(b) 179.4 MN/m^2 T and C.

6–66 $W10\times25$.

6–67 57.6 kN.

6–69 $\frac{5wL^4}{144EI}.\downarrow$

6–71 $\frac{101wL^4}{128EI}.\downarrow$

6–72 $\frac{2wL^4}{3EI}.\downarrow$

6–75 0.0769 in. $\downarrow$

6–76 $\frac{4PL^3}{9\sqrt{3}\,EI}.\uparrow$

6–77 $\frac{5PL^3}{48EI}.\downarrow$

6–78 (a) 0.733 in. $\downarrow$
(b) 0.098 in. $\uparrow$

6–79 (a) 26.3 mm. $\downarrow$
(b) 22.7 mm. $\downarrow$

6–80 $x = L$.

6–81 $0.566L$ to the left of the right support.

6–83 $0.01107\frac{wL^4}{EI}.\downarrow$

6–84 $0.0394\frac{wL^4}{EI}.\uparrow$

6–85 2.4 in. $\downarrow$

6–86 27.6 mm. $\downarrow$

6–87 1.274 in. $\downarrow$

6–89 $\frac{2PL^3}{3EI}.\downarrow$

6–91 $\frac{29wL^4}{48EI}.\downarrow$

6–94 $\frac{9wL^4}{128EI}.\downarrow$

6–95 $\frac{5wL^4}{128EI}.\uparrow$

6–96 $\frac{5wL^4}{96EI}.\downarrow$

6–97 $\frac{wL^4}{96EI}.\uparrow$

6–100 $\frac{3PL^3}{256EI}.\downarrow$

6–101 0.

6–104 $\frac{ML}{9EI}$. ↶

6–105 $\frac{PL^2}{4EI}$. ↶

6–107 23.1 mm. ↓

6–111 $\frac{\pi PR^3}{4EI}$. ↓

6–112 $\frac{\pi PR^3}{2EI}$. ↓

6–113 $\frac{2PR^3}{EI}$. →

6–114 $\frac{2PR^2}{EI}$. ↶

CHAPTER 7

7–1 (a) $\frac{wL^2}{6}$ ft-lb.
(b) $\frac{wL^4}{24EI}$ ft. ↓

7–2 $\frac{wL^2}{3}$ N-m. ↷

7–3 (a) $\frac{7wL^2}{24}$ N-m. ↷
(b) $\frac{wL^4}{48EI}$ m. ↑

7–5 (a) $\frac{3wL}{4}$ N. ↓
(b) $0.75L$ m right of the left support.

7–6 (a) 2 kN. ↓
(b) 1.333 m right of the left support.

7–7 $\frac{3wL}{16}$ lb. ↑

7–8 $V_A = \frac{3wL}{8}$ N. ↑
$M_A = \frac{wL^2}{24}$ N-m. ↶

7–11 $\frac{17wL}{16}$ N. ↑

7–13 $\frac{wL^2}{60}$ ft-lb.

7–14 $V_A = \frac{wL}{4}$ N. ↑
$M_A = 0$.
$R_B = \frac{3wL}{2}$ N. ↑

7–16 $\frac{351wL}{512}$ N. ↑

7–18 $\frac{2wL}{3}$ N. ↓

7–20 $V = \frac{9wL}{10}$ N. ↑
$M = -\frac{wL^2}{4}$ N-m.

7–22 $\frac{3wL}{20}$ N. ↓

7–23 $+\frac{wL^2}{4}$.

7–24 wL N. ↓

7–25 $\frac{3P}{8}$. ↑

7–28 $2P$. ↓

7–30 (a) 12 kN. ↓
(b) 3 m left of the right support.

7–32 $V_A = 10.67$ kN. ↑
$M_A = 16.00$ kN-m. ↶
$V_B = 25.33$ kN. ↑
$M_B = 24.0$ kN-m. ↷

7–33 (a) $\frac{4M}{9L}$. ↑
(b) $\frac{3L}{2}$ right of the left support.
(c) $\frac{ML^2}{8EI}$. ↑

7–35 $R_A = 13$ kN. ↑
$R_B = 33$ kN. ↑
$R_C = 2$ kN. ↑

7–37 6000 lb. ↑

7–38 20 kN.

7–41 $\frac{wL}{2}$ lb. ↑

7–42 $R_A = 10.5$ kN. ↑
$R_B = 27.0$ kN. ↑

7–44 1796 lb. ↑

7–47 $R = 696$ lb. $\uparrow$
$\delta = 0.488$ in. $\downarrow$

7–48 (a) $\frac{wL}{4}$ N. $\uparrow$
(b) 15.00 mm. $\downarrow$

7–49 464 lb/in.

7–51 1436 lb C.

7–53 0.1388 in. $\downarrow$

7–54 (a) $\frac{wL^2}{6}$.
(b) $\frac{wL^4}{24EI}$. $\downarrow$

7–55 $\frac{wL^2}{3}$ N-m. $\circlearrowright$

7–56 $\frac{3wL}{16}$ lb. $\uparrow$

7–57 $\frac{3wL}{4}$ N. $\downarrow$

7–58 $\frac{wL^2}{60}$ ft-lb. $\circlearrowleft$

7–59 $V_A = \frac{wL}{4}$ N. $\uparrow$
$M_A = 0$.
$R_B = \frac{3wL}{2}$ N. $\uparrow$

7–60 $V_A = \frac{3wL}{8}$ N. $\uparrow$
$M_A = \frac{wL^2}{24}$ N-m. $\circlearrowleft$

7–61 $\frac{3wL}{20}$ N. $\downarrow$

7–64 $\frac{351wL}{512}$ N. $\uparrow$

7–65 $\frac{4M}{9L}$. $\uparrow$

7–67 6000 lb. $\uparrow$

7–68 1796 lb. $\uparrow$

7–69 1.404P. $\leftarrow$

7–70 (a) $\frac{15wL}{32}$ lb. $\uparrow$
(b) $\frac{wL^4}{64EI}$ ft. $\leftarrow$

7–71 $R_y = \frac{3wL}{7}$ lb. $\uparrow$
$R_x = \frac{3wL}{28}$ lb. $\rightarrow$

7–72 $W21 \times 62$.

7–73 48.8 kN/m.

7–76 32.1 kN/m.

7–77 50.1 kN/m.

7–78 1.520 kips/ft.

CHAPTER 8

8–1 $\sigma_A = 875$ psi T.
$\sigma_B = 125$ psi T.
$\sigma_C = 1375$ psi C.
$\sigma_D = 625$ psi C.

8–2 $\sigma_A = 8.00$ MN/m^2 T.
$\sigma_B = 11.20$ MN/m^2 C.

8–5 3430 psi C at the bottom on the right.

8–6 67.3 MN/m^2 C.

8–9 (a) 12,080 psi C.
(b) 12,180 psi C.

8–12 Partial Results:
$\sigma_A = 8.83$ MN/m^2 T.
$\sigma_B = 3.83$ MN/m^2 T.
$\sigma_C = 12.17$ MN/m^2 C.
$\sigma_D = 7.17$ MN/m^2 C.

8–13 2750 psi C.

8–14 Partial Results:
$\sigma_{p1} = 242$ MN/m^2 T.
$\sigma_{p2} = 2.09$ MN/m^2 C.
$\sigma_{p3} = 0$.
$\tau_{max} = 122.1$ MN/m^2.
$\theta_p = 5.31°$. $\circlearrowleft$

8–15 1.404 ksi.

8–16 Partial Results:
$\sigma_{p1} = 1.637$ MN/m^2 T.
$\sigma_{p2} = 31.2$ MN/m^2 C.
$\sigma_{p3} = 0$.
$\tau_{max} = 16.40$ MN/m^2.
$\theta_p = 12.91°$. $\circlearrowright$

8–18 Partial Results:
$\sigma_{p1} = 1231$ psi T.
$\sigma_{p2} = 20{,}600$ psi C.
$\sigma_{p3} = 0$.
$\tau_{max} = 10{,}930$ psi.
$\theta_p = 13.73°$. ↺

8–19 Partial Results:
$\sigma_{p1} = 2.23$ psi T.
$\sigma_{p2} = 649$ psi C.
$\sigma_{p3} = 0$.
$\tau_{max} = 326$ psi.
$\theta_p = 3.35°$. ↺

8–20 Partial Results:
at G: $\sigma_{p1} = 11.04\ MN/m^2$ C.
$\sigma_{p2} = 0$.
$\sigma_{p3} = 0$.
$\tau_{max} = 5.52\ MN/m^2$.
$\theta_p = 0$.
at H: $\sigma_{p1} = 5.17\ MN/m^2$ T.
$\sigma_{p2} = 0.008\ MN/m^2$ C.
$\sigma_{p3} = 0$.
$\tau_{max} = 2.59\ MN/m^2$.
$\theta_p = 2.24°$. ↻

8–21 12,040 lb.

8–22 350 kN.

8–23 (a) $\frac{d}{3}$.
(b) $0.5\sigma_y$.
(c) 0.1708.

8–26 Partial Results:
$\sigma_{p1} = 11{,}460$ psi T.
$\sigma_{p2} = 1273$ psi C.
$\sigma_{p3} = 0$.
$\tau_{max} = 6370$ psi.
$\theta_p = 18.43°$. ↺

8–27 Partial Results:
$\sigma_{p1} = 53.3\ MN/m^2$ T.
$\sigma_{p2} = 4.38\ MN/m^2$ C.
$\sigma_{p3} = 0$.
$\tau_{max} = 28.8\ MN/m^2$.
$\theta_p = 16.00°$. ↺

8–28 Partial Results:
$\sigma_{p1} = 10{,}270$ psi T.
$\sigma_{p2} = 4090$ psi C.
$\sigma_{p3} = 0$.
$\tau_{max} = 7180$ psi.
$\theta_p = 32.2°$. ↺

8–29 Partial Results:
$\sigma_{p1} = 299\ MN/m^2$ T.
$\sigma_{p2} = 3.55\ MN/m^2$ C.
$\sigma_{p3} = 0$.
$\tau_{max} = 151.2\ MN/m^2$.
$\theta_p = 6.22°$. ↺

8–32 Partial Results:
$\sigma_{p1} = 624$ psi T.
$\sigma_{p2} = 4060$ psi C.
$\sigma_{p3} = 0$.
$\tau_{max} = 2340$ psi.
$\theta_p = 21.4°$. ↻

8–33 Partial Results:
$\sigma_{p1} = 0.228\ MN/m^2$ T.
$\sigma_{p2} = 41.0\ MN/m^2$ C.
$\sigma_{p3} = 0$.
$\tau_{max} = 20.6\ MN/m^2$.
$\theta_p = 4.27°$. ↻

8–35 Partial Results:
$\sigma_{p1} = 109.4\ MN/m^2$ T.
$\sigma_{p2} = 37.4\ MN/m^2$ C.
$\sigma_{p3} = 0$.
$\tau_{max} = 73.4\ MN/m^2$.
$\theta_p = 30.3°$.

8–36 Partial Results:
$\sigma_{p1} = 3740$ psi T.
$\sigma_{p2} = 10{,}940$ psi C.
$\sigma_{p3} = 0$.
$\tau_{max} = 7340$ psi.
$\theta_p = 30.3°$.

8–37 Partial Results:
$\sigma_{p1} = 15{,}920$ psi T.
$\sigma_{p2} = 4{,}200$ psi T.
$\sigma_{p3} = 125$ psi C.
$\tau_{max} = 8020$ psi.
$\theta_p = 16.30°$. ↻

8–39 Partial Results:
$\sigma_{p1} = 23.2\ MN/m^2$ T.
$\sigma_{p2} = 44.0\ MN/m^2$ C.
$\sigma_{p3} = 0$.
$\tau_{max} = 33.6\ MN/m^2$.
$\theta_p = 36.0°$. ↻

8–41 Maximum-shear-stress.
Maximum-distortion-energy.

8–42 Maximum-shear-stress.

8–43 (a) Normal-stress $k=1.60$
Shear-stress $k=1.60$
Normal-strain $k=2.00$
Distortion-energy $k=1.83$
(b) Normal-stress $k=1.60$
Shear-stress $k=1.00$
Normal-strain $k=1.33$
Distortion-energy $k=1.14$

8–44 Normal-stress $\sigma_f=350\ \text{MN/m}^2$
Shear-stress $\sigma_f=550\ \text{MN/m}^2$
Normal-strain $\sigma_f=410\ \text{MN/m}^2$
Distortion-energy $\sigma_f=482\ \text{MN/m}^2$

8–51 (a) $\sigma=96.6\ \text{MN/m}^2$ T and C.
$\tau=72.4\ \text{MN/m}^2$.
(b) $\sigma_{p1}=135.3\ \text{MN/m}^2$ T.
$\sigma_{p2}=38.8\ \text{MN/m}^2$ C.
$\sigma_{p3}=0$.
$\tau_{max}=87.1\ \text{MN/m}^2$.
(c) 2.41.

8–52 4.93 in.

8–53 Normal-stress $P=4440$ lb.
Shear-stress $P=4270$ lb.
Distortion-energy $P=4350$ lb.

8–56 11,820 psi.

8–57 281 mm.

8–58 5.88 ksi.

CHAPTER 9

9–1 200 lb.

9–2 225 N.

9–3 1440 lb.

9–4 13.33 kN.

9–5 $r=0.901$ in.
$L/r=106.5$.

9–6 $r=9.60$ mm.
$L/r=156.2$.

9–7 $r=1.683$ in.
$L/r=85.6$.

9–8 132.3 mm.

9–11 $L/r=798$.
$P_{cr}=4.89$ lb.

9–12 (a) 384.
(b) 394 lb.
(c) 2010 psi.

9–13 (a) 160.
(b) 14.84 kN.
(c) 605 kg.

9–14 (a) 185.2.
(b) 71,000 lb.
(c) 39,500 lb.

9–17 (a) 26.6 kN.
(b) 84.1 kN.

9–18 (a) 123.5.
(b) 283 kN.
(c) 12,820 kg.

9–23 $L'=L$.

9–24 (a) 409.
(b) 318 lb.

9–25 (a) 110.8.
(b) 9600 lb.

9–26 (a) 149.2.
(b) 17,170 kg.

9–27 $P_w=18{,}400$ lb.
$e=0.0167$ in.

9–28 112 kN.

9–29 (a) $d/32$.
(b) $d/16$
(c) 0.128 in.

9–32 0.16 in.

9–33 2.8.

9–36 (a) 16.20 kips.
(b) 155.3 kips.

9–37 (a) 72.9 kips.
(b) 17.24 kips.

9–38 (a) 820 kN.
(b) 94.9 kN.

9–39 (a) 189.4 kN.
(b) 64.8 kN.

9–41 188.8 kips.

9–43 (a) 134.6 kips.
(b) 285 kips.

9–45 (a) 148.3 kips.
(b) 44.1 kips.

9–46 (a) 181.0 mm.
(b) 576 kN.

9–52 $a = 1.237$ in.

9–55 65.2 kips.

9–56 390 kN.

9–57 286 kips.

9–58 218 kN.

9–59 42.0 kips.

9–60 263 kN.

9–62 124.6 kN.

9–63 48.7 kips.

9–65 (a) 31.8 kips.
(b) 38.9 in.-kips.

9–67 123.9 kips.

9–68 114.6 kN.

9–70 97.9 mm.

9–72 111.1 kN.

9–75 $\delta = \dfrac{n}{\sqrt{P/EI}} \tan\sqrt{\dfrac{P}{EI}}\,(L) - nL.$

9–76 $\delta = \dfrac{n}{\sqrt{P/EI}}\left[\dfrac{1}{2} - \dfrac{EI}{PL^2}\right] \tan\sqrt{\dfrac{P}{EI}}\,(L) - \dfrac{nL}{6} + \dfrac{nEI}{PL}.$

9–78 $L' \approx 0.7L.$

9–79 $1.553\dfrac{\pi^2 EI}{L^2}.$

9–83 8.41 kips.

9–84 19 in.

9–87 58 kips.

CHAPTER 10

10–1 6.67 hr.

10–2 108 hr.

10–3 59 ksi.

10–4 2.65.

10–5 (a) 12.7 kips.
(b) 53 kips.

10–6 (a) 6.0 kips.
(b) 48 kips.

10–7 (a) 30 kN.
(b) 62.5 kN.

10–8 (a) 68.2 kN.
(b) 225 kN.

10–11 (a) 11.31 in.-kips.
(b) 4.55 in.-kips.

10–12 (a) 1.571 kips.
(b) 0.977 kips.

10–13 $a = 203$ mm.
$b = 90.6$ mm.

10–16 1185 lb.

10–17 4.5 mm.

10–19 42.3 mm.

10–20 0.909 in.

10–21 1.674.

10–22 1.796.

10–23 $\sigma_{max} = 76.1$ ksi T.
$\sigma_{min} = 21.7$ ksi T.

10–24 $\sigma_{max} = 421\ \text{MN/m}^2$ T.
$\sigma_{min} = 210\ \text{MN/m}^2$ C.

10–27 $P_{max} = 11.8$ kips T.
$P_{min} = 5.9$ kips C.

10–28 82.6 kN.

10–31 $P_{max} = 2.15$ kips.
$P_{min} = 1.436$ kips.

10–32 201 N.

10–36 37.6 mm.

10–37 0.594 in.

10–39 2.54 kips.

10–40 5.91 mm.

10–41 3.94.

CHAPTER 11

11–1 1.784 in.

11–2 2.04 m.

11–3 $\frac{\gamma^2 A L^3}{6E}$.

11–4 $\frac{P^2 L^3}{6EI}$.

11–5 $25.6I/c^2$.

11–6 $U_{bar} = 3U_{beam}$.

11–8 $U_{bar} = 9U_{beam}$.

11–10 $U_{bar} = 3U_{beam}$.

11–11 $U_{shaft} = (1+\nu)U_{bar}$.

11–15 55.5 mm. ↓

11–16 9.25 in.

11–19 17,060 psi T.

11–20 398 mm.

11–21 (a) 2.56 in.
(b) 256 lb.

11–22 43.5 kN.

11–23 (a) 42,400 lb.
(b) 0.0339 in. ↓

11–24 2.94 m.

11–27 364 lb.

11–28 (a) 229 N.
(b) 7.34 mm. ↓

11–32 (a) 1855 N-m.
(b) 0.0293 rad.

11–33 14,200 psi.

11–36 32.5 N-m.

11–37 (a) 21.2%.
(b) 61.1%.

11–40 (a) 5.4 in.
(b) 4.

11–41 $\frac{21wL^3}{8EI}$.

11–42 $\sigma_{beam} = 87.9$ MN/m^2 T and C.
$\sigma_{rod} = 21.0$ MN/m^2 T and C.

11–43 $h = 8.5$ in.
$L.F. = 12$.
$U_A = 150$ in.-lb.
$U_B = 75$ in.-lb.

11–44 (a) 59.1 mm.
(b) 4.78.
(c) $U_A = 0.93$ N-m.
$U_B = 7.44$ N-m.

11–45 (a) 43.4 mm.
(b) $U_A = 0.424$ N-m.
$U_S = 0.696$ N-m.
$U_B = 1.719$ N-m.

CHAPTER 12

12–1 (a) 15.06 ksi T.
(b) 33.3 ksi C.
(c) 4.24 ksi.

12–2 (a) 97.9 MN/m^2.
(b) 169.1 MN/m^2 C.
(c) 94.1 MN/m^2 T.

12–4 (a) 61.1 MN/m^2.
(b) 105.7 MN/m^2 C.
(c) 87.7 MN/m^2 T.

12–5 (a) 14.32 ksi.
(b) 22.5 ksi C.
(c) 18.46 ksi T.

12–7 (a) 8.68 ksi.
(b) 29.5 ksi C.
(c) 17.19 ksi T.

12–9 (a) 71.7 MN/m^2.
(b) 124.0 MN/m^2 C.
(c) $\sigma_1 = 85.7$ MN/m^2 T.
$\sigma_2 = 90.9$ MN/m^2 T.
$\sigma_3 = 65.5$ MN/m^2 T.

12–11 76.3%.

12–14 (a) 273 psi.
(b) 81.0%.

12–15 $P_{ult} = 2.87$ kips.
$\eta = 81.6\%$.

12–18 61.8%.

12–20 (a) 182.4 psi.
(b) 66.4%.

12–21 $\eta = 74.4\%$.
$P = 169.7$ kN.

12–22 86.3%.

12–23 9.15 mm.

12–25 2130 kPa.

12–27 (a) $d = 0.947$ in.
$t = 0.620$ in.
(b) 1.266.

12–29 88.4%.

12–31 $L_1 = 2.77$ in.
$L_3 = 9.58$ in.

12–32 $L_1 = 81.7$ mm.
$L_3 = 272$ mm.

12–33 $L_1 = 5.71$ in.
$L_3 = 12.75$ in.

12–34 $L_1 = 238$ mm.
$L_3 = 514$ mm.

12–36 $a = 8.00$ in.
$b = 9.86$ in.
$c = 6.03$ in.

12–37 $\sigma = 71.3\ \text{MN/m}^2$ T.
$\tau = 21.6\ \text{MN/m}^2$.

12–39 (a) 16.77 ksi.
(b) 4 on each side.

12–41 (a) 10,480 psi.
(b) 7 bolts.

12–42 (a) 30 kN.
(b) 19.54 mm.

12–43 (a) 12,200 lb.
(b) 20,300 lb.

12–44 53.3 kW.

12–45 (a) $F_A = F_C = F_{max}$.
(b) 10.39 kN.

12–49 (a) 598 hp.
(b) 8 bolts.

12–50 (a) 864 kW.
(b) 0.0325 rad.
(c) 12.

12–51 (a) Bolt C.
(b) 12.73 ksi.

12–52 $87.0\ \text{MN/m}^2$.

12–53 $\tau_A = 7.79$ ksi.
$\tau_B = 12.41$ ksi.

12–54 $80.6\ \text{MN/m}^2$.

12–55 $\tau_A = 10.61$ ksi.
$\tau_B = 4.75$ ksi.

12–56 $50.9\ \text{MN/m}^2$ and 0.

12–59 2.78 kips.

12–60 145.0 mm.

APPENDIX B

B–1 $\pi R^4/2$.

B–2 $I_x = 2bh^3/7$.
$I_y = 2hb^3/15$.

B–3 (a) $bh^3/12$.
(b) $bh^3/36$.

B–4 (a) $\pi(R_2^4 - R_1^4)/4$.
(b) $\pi(R_2^4 - R_1^4)/8$.

B–5 $I_x = 3.10\ \text{in}^4$.
$I_y = 0.756\ \text{in}^4$.

B–6 $0.267\ \text{in}^4$.

B–7 $1.467b^4$.

B–8 $9a^4/14$.

B–9 (a) $4a^4/9$.
(b) $4a^4/33$.

B–10 (a) $32bh^3/105$.
(b) $4hb^3/15$.

B–11 $I_x = 166\ \text{in}^4$.
$I_y = 36.5\ \text{in}^4$.

B–12 $296\ \text{in}^4$.

B–13 $I_x = 85{,}000\ \text{cm}^4$.
$I_y = 25{,}000\ \text{cm}^4$.

B–14 $I_x = 533\ \text{in}^4$.
$I_y = 333\ \text{in}^4$.

B–15 45,625 cm^4.

B–16 $k_x = 3.73$ in.
$k_y = 1.650$ in.

B–17 $\sqrt{(b^2 + h^2)/24}$.

B–18 4.50 cm.

B–19 $k_x = 2.98$ in.
$k_y = 1.958$ in.

B–20 $k_x = 8.85$ in.
$k_y = 2.02$ in.

B–21 $I_x = 6570$ cm^4.
$I_y = 8230$ cm^4.

B–22 59.5 in^4.

B–23 177,100 cm^4.

B–24 3.42 in.

B–25 $5b^2h^2/24$.

B–26 16 in^4.

B–27 213,750 cm^4.

B–28 -75.0 in^4.

B–29 $bch^2/12$.

B–30 $b^2h^2/6$.

B–31 $I_u = 172.8$ in^4.
$I_v = 19.85$ in^4.

B–32 0.755 in.

B–33 $k = 1.771$ cm.
$\theta = -16.85°$.

B–34 $I_{x'} = 4323$ cm^4.
$I_{y'} = 427.4$ cm^4.

B–35 $\theta = -59.3°$.
$I_{max} = 1887$ cm^4.

B–36 0.755 in.

B–37 $k = 1.771$ cm.
$\theta = -16.85°$.

B–38 $\theta = 59.5°$ ↓ from the x axis.
$I_{max} = 1887$ cm^4.

B–39 1.835 in.

B–40 $\theta_1 = 32.7°$. ↺
$I_1 = 263$ cm^4.
$\theta_2 = 122.7°$. ↺
$I_2 = 3928$ cm^4.

Index